COLLEGE ALGEBRA

COLLEGE ALGEBRA

A VIEW OF THE WORLD AROUND US

DAVID WELLS
Penn State University–New Kensington

LYNN SCHMITT TILSON
Penn State University–University Park

Prentice Hall
Upper Saddle River, New Jersey 07458

Library of Congress Cataloging-in-Publication Data

Wells. Dave.
College algebra : a view of the world around us / David Wells,
Lynn Schmitt Tilson.
p. cm.
Includes index.
ISBN 0-13-571019-7 (hc : alk. paper)
1. Algebra. I. Schmitt Tilson, Lynn. II. Title.
QA152.2.W436 1997
512.9—dc20 96-42042
CIP

Acquisition Editor: Sally Denlow
Editorial Assistant: Joanne Wendelken
Supplement Editor: Audra Walsh
Development Editor: Tony Palermino
Marketing Manager: Jolene Howard
Marketing Assistant: Jennifer Pan
AVP, Production and Manufacturing: David W. Riccardi
Production Editor: Elaine W. Wetterau
Senior Managing Editor: Linda Mihatov Behrens
Executive Managing Editor: Kathleen Schiaparelli
Buyer: Alan Fischer
Manufacturing Manager: Trudy Pisciotti
Creative Director: Paula Maylahn
Art Director, Interior: Amy Rosen
Assistant to the Art Director: Rod Hernandez
Art Manager: Gus Vibal
Art Director, Cover: Jayne Conte

Published by Prentice-Hall, Inc.
Simon & Schuster/A Viacom Company
Upper Saddle River, New Jersey 07458

Printed in the United States of America
10 9 8 7 6 5 4 3 2 1

ISBN 0-13-571019-7

Prentice-Hall International (UK) Limited, *London*
Prentice-Hall of Australia Pty, Limited, *Sydney*
Prentice-Hall Canada, Inc., *Toronto*
Prentice-Hall Hispanoamericana, S.A., *Mexico*
Prentice-Hall of India Private Limited, *New Delhi*
Prentice-Hall of Japan, Inc., *Tokyo*
Simon & Schuster Asia Ptc. Ltd., *Singapore*
Editora Prentice-Hall do Brasil, Ltda., *Rio de Janeiro*

Dave's Dedication

To Alice and Sarah with love—
who kept my life alive while I was writing.

Lynn's Dedications

To Papa with love—
for my father, Dr. Ernest Watts Harvey, Jr., who died February 20, 1994 but who still lives within me and within the pages of this book.

To Mom with love—
for my mother Harriet Virginia Peaseley Harvey, who died November 7, 1995—This book would not be, had you not been. It is your legacy to me, through me.

CONTENTS

CHAPTER 3 LINEAR FUNCTIONS 73

CHAPTER 4 LINEAR SYSTEMS 115

CHAPTER 5 QUADRATIC FUNCTIONS 149

APPENDIX A BASIC ALGEBRA REFERENCE A1

APPENDIX B TIPS FOR GRAPHING FUNCTIONS WITH A CALCULATOR B1

PREFACE

"The answer is in the attempt,"

a quote from the character Celine in the movie, *Before Sunrise.*

This idea permeates our view of mathematics and is found throughout the book.

Students who take college algebra as a terminal course will find that this book shows them some ways in which algebra can be used to "view the world around us." It brings out an appreciation for mathematical ideas generated from that view by not only showing how algebra can be applied to real physical problems but how algebra can provide insight into these problems through the process of mathematical modeling. They will also see that algebra is of interest in its own right, not *just* as a course needed for the study of trigonometry or calculus.

Students who take college algebra as a prelude to calculus will see algebraic functions presented dynamically. Specifically, we discuss where functions are increasing and decreasing and introduce the idea of average rate of change to foreshadow the idea of instantaneous rate of change in calculus. We also introduce the idea of the limit both numerically and graphically without giving its calculus definition.

One principal goal in this book is to present algebra using the "rule-of-three," by introducing concepts numerically, analytically, and graphically. The rule-of-three not only helps students gain a solid grasp of the indispensable tools of algebra but creates an ability to "view the world around us" in several ways. We have also designed appropriate *writing to learn* exercises expecting students to communicate mathematical ideas. The inclusion of communication skills into the rule-of-three is now known in some circles as the "rule-of-four." Using these various mathematical methods to investigate and solve both routine and nonroutine problems will help the students gain mathematical power.

Another principal goal is to "engage" students, both in action and in interest, instead of allowing them to acquire a passive role. The book not only teaches skills but creates a curiosity and understanding of "the world around us" through what we call Mathematical Looking Glasses. The Mathematical Looking Glasses present a rich variety of situations in which the students can see the mathematics as interesting, timely, and useful. These situations use mathematics both as a problem-solving technique and as a tool for gaining additional insight. We also present algebra as part of our historical heritage and as a pattern-seeking device that can discover underlying principles. This leads students to a deeper understanding of algebra and a broader view of its applicability.

We also wanted to write something that we could both use in our own teaching. Since our teaching styles are very different, the book needed the flexibility to be the supporting material for Dave's lectures as well as for Lynn's group work. It needed to have an appeal to Dave, who always wanted to be a mathematician and a rock musician—and is—and to Lynn, who always wanted to be an actress or veterinarian—and isn't. We each benefit from the book's depth and breadth.

This broad scope is necessary partly because there is no universal agreement on what any text should contain, algebra texts being no exception, and partly because we needed to strike a balance between tradition and innovation.

These ideas, which we formulated more than a decade ago, just happened to align closely with both the AMATYC and NCTM standards that came out later. We realize that our text will be associated with the "reform" movement in calculus and will precede such a course nicely, but it was not written with that primarily in mind. It therefore has the tone of a text that will also move smoothly into a "traditional" calculus class.

Structurally, this book is flexible enough for instructors to adapt it to their own focus. A class that is terminal or precedes a reform calculus course might place more emphasis on the Mathematical Looking Glasses. A class that precedes a more traditional calculus course might emphasize algebraic skill.

While trying to keep the balance just described, we also kept in mind the potential use of this text as a reference book. Its visual appearance and topical sequence lends itself to easy accessibility. It includes an algebra reference that is meant for a quick review but not a contiguous course in elementary algebra. It also has a graphing calculator guide meant to focus on the elementary techniques common to most of these calculators but not meant to replace any individual manual. These references appear in a form designed for individual reading, leaving classroom discussion of these ideas to a minimum.

This book includes the use of graphing calculators, but as a means to an end, not an end in itself. We guide the students, at first quite often, as to whether the calculator is the most efficient, or most comfortable, path to a solution. Later, we usually leave that decision up to the individual.

Pedagogically, the book lays the foundations at a slower pace near the beginning so that the students can concentrate on the fundamental ideas of the rule-of-three, functions, and Pólya strategies. Exercises follow directly from examples, giving the book a natural flow from concept to skill. These exercises allow the students to determine whether their understanding is sufficient to attempt the ones at the end of the section.

Many textbooks have a *letter to the student.* Our first section takes the place of this letter. It details what the students need coming into the book and what they will have going out. It gives them instructions and hints on how to use our book in the most effective way and some personal notes on what we think success in this course means. Most important, it will let them see what this book is all about.

ACKNOWLEDGMENTS

LYNN'S ACKNOWLEDGMENTS

My new husband James—for his patience and love as he discovers what the world of book writing and publishing are all about.

My daughter Carrie—for her conjured-up enthusiasm about the book and for now being old enough to read it and drink at the same time.

My mother "Boo" Peaseley Harvey—for her faith in me, her role model as teacher, and her financial assistance.

My father Ernest Watts Harvey—for saying, "I didn't know you were so smart," when he first read some of the manuscript, and for keeping me on my intellectual toes with philosophical conversation.

Steve Albert and company (Brian, Mike, & Gene)—for keeping my old Mac running.

Tim Schmitt—for supporting me through graduate school and for his many suggestions that helped form the focus of this book.

The staff (especially Mary Beth Kleist) and patrons at the Hawthorne Club in Hawthorne, Nevada—for believing in me and giving me a second home.

Kathi Arnold—for her warmth and model of strength.

Steve Merrill—for seeing this theatre major through graduate school.

My students—for being themselves and helping me grow.

My co-author Dave—for being there, always.

Barbara Schmitt—for her laughter and love.

Lee Koenig—for her continuous affirmation of friendship.

Ann Crowe—for being my best friend during the last 25 years.

Mary Hembree—for her kindness when I needed it most.

DAVE'S ACKNOWLEDGMENTS

Dave Brown—who showed me what good teaching is.

Roy Myers—for the benefit of his wisdom during the many projects we have worked on together.

My wife Alice—for being a one-person support staff.

My daughter Sarah—for making her life an inspiration to her father.

My mother Jacqueline Gaye Wells—for encouraging me to be a mathematician.

My father Edgar Franklin Wells—for his excitement about the originality of our approach in this project.

My co-author Lynn—for helping me rediscover my sense of wonder.

My long-suffering students—for class testing the manuscript during 1994–1996.

All my friends at the Unitarian Universalist Church of the North Hills, and at the Pittsburgh Songwriters Guild—for keeping me from going over the edge.

OUR ACKNOWLEDGMENTS

Sally Denlow—for being our lovely and bright mathematics editor who in the face of these two inexperienced authors has "put it all together." Thanks Sally—we did it!

Bob Pirtle—for his unwavering support and constant direction for this project.

Elaine Wetterau—for being a unique blend of production editor, stern advisor, and friend.

Tony Palermino—for being our third author, who is disguised by Prentice Hall as a development editor.

All those people that we met in our travels over the last eight years who seemed excited about our project—for regenerating our enthusiasm.

Our reviewers—for their honest comments. (We listened, by the way.) In particular, John Fink and Nancy Prosenjak for exceptionally thorough and valuable reviews.

Billy Kratsa, Sr.—for keeping us in lunch and cookies.

All our friends and colleagues—for listening to book woes and sagas.

Lee Row, III—for staying in constant touch with us for almost eight years and providing background for several of our Mathematical Looking Glasses.

Fred Rickey—for patiently answering untimely questions about mathematical history.

Steve Monk—for making some valuable suggestions in the early stages of development of this book.

The many people whose interests and activities provided material for the Mathematical Looking Glasses—for having interesting lives and letting us in on them. (We generally have not changed names to protect the guilty—but sometimes *did* change their lives.)

David Wells
Lynn Schmitt Tilson

COLLEGE ALGEBRA

CHAPTER 1

MODELING AND PROBLEM SOLVING

1-1 A CASE FOR ALGEBRA

PREREQUISITES MAKE SURE YOU ARE FAMILIAR WITH:

The Coordinate Plane (Section A-3)
Basic Graphing Techniques (Section A-5)
Graphing Linear Equations (Section A-6)

(Prerequisites are indicated at the beginning of many sections in this book. They indicate skills from previous courses that you will need in these sections. In the present section you will need to know the standard methods for graphing linear equations and finding equations of linear graphs. Refer to the sections cited and review them as needed before continuing.)

Mathematics lives in the world around us. For example, mathematics is needed to build a laser, bake a cake, and calculate the Dow Jones average. Because this is so, this book contains descriptions of the world viewed through a "mathematical looking glass," along with mathematical questions that real people might ask. The descriptions are detailed, so that you can see the mathematics embedded in the situation and view the questions in a meaningful context. The accompanying tables, equations, and graphs will help you to answer the questions.

Let's use the following situation to illustrate the process of formulating mathematical questions in a physical context, answering the questions, and interpreting the results. Read through it once to understand the context, without worrying about the mathematics.

A MATHEMATICAL LOOKING GLASS
A Case for Algebra

One evening in the summer of 1988, a young man whom we will call Bill was socializing at the Greentree Holiday Inn in Pittsburgh. He drank seven beers between 10:00 P.M. and 2:00 A.M. before leaving with two friends and a young woman they had met. En route to Primanti Brothers' Sandwich Shop at 2:10 A.M., Bill's Volkswagen GTI struck a steel pole. The woman was thrown forward and sustained a fatal head injury.

Technicians at Mercy Hospital took a reading of Bill's blood alcohol concentration (BAC) at 2:52 A.M. This reading showed a level of 0.136, indicating that his bloodstream contained 0.136% alcohol. A second reading taken by the police at 4:40 A.M. showed a level of 0.100. Both were at or above Pennsylvania's legal threshold of intoxication, which is 0.100. Bill was therefore charged with vehicular homicide while DUI.

Dr. Charles Winek, chief toxicologist of the Allegheny County Crime Lab, testified at the trial that for a person of Bill's weight, elimination of alcohol would have lowered his BAC at a rate of about 0.020 per hour. Based on this assumption, his BAC at the time of the accident would have been higher than 0.136, as indicated in Figure 1-1.

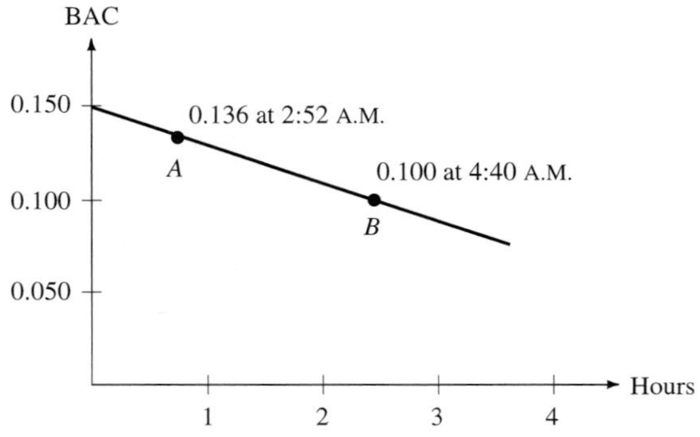

Figure 1–1

During cross-examination Bill's lawyer, Thomas Schuchert, reminded Dr. Winek that Bill had consumed at least two beers between 1:30 and 2:00. He then asked how Dr. Winek's conclusions would be affected if neither of those beers had been absorbed into Bill's bloodstream prior to the accident. Dr. Winek replied that in this case, Bill's BAC would have risen by about 0.064 per hour between 2:10 and 2:52, so that Figure 1-2 would present a more appropriate picture.

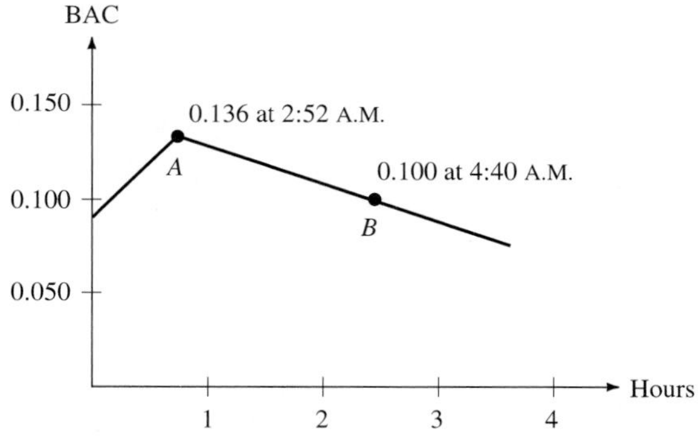

Figure 1–2

The critical problem for both the prosecution and the defense was to determine Bill's BAC at the time of the accident as reliably as possible. ■

The problem described here certainly has mathematical aspects. Let's look at the situation again and try to formulate it as a mathematical question. Read on, but more slowly now, with pencil in hand.

The graphs in Figures 1-1 and 1-2 each provide a rough estimate of Bill's BAC at 2:10 A.M. How can we improve the accuracy of those estimates? Let's focus on Figure 1-1 first. Since the graph in that figure is a straight line, we should be able to find its equation if we know the coordinates of the points A and B. (See pages A14, A15 to review this idea.) Let's define the variables

t = time in hours after the accident

y = Bill's BAC

We have been given some information about these variables. In particular, we were told that Bill's BAC was 0.136 at 2:52 A.M. This was 42 minutes, or 0.7 hours, after the accident. (Don't just take our word for this. Check your understanding of the situation by performing the calculations yourself.) Therefore, point A in Figure 1-1 has coordinates (0.7, 0.136). Similarly, 4:40 A.M. was 2.5 hours after the accident, so point B has coordinates (2.5, 0.100).

The physical question,

> "What was Bill's BAC at the time of the accident?"

has now been formulated as a mathematical question,

> "What is the equation of the line through the points (0.7, 0.136) and (2.5, 0.100)?"

You can answer this mathematical question in Exercise 1.

EXERCISE

1. Find the equation of the line through (0.7, 0.136) and (2.5, 0.100).

Let's interpret our results. In slope-intercept form, your equation in Exercise 1 should be $y = -0.020t + 0.150$. The slope and the y-intercept of this line each have a physical meaning. The slope of -0.020 represents the rate at which alcohol was removed from his bloodstream, and the y-intercept of 0.150 represents Bill's BAC at the time of the accident. Thus, according to the graph in Figure 1-1, Bill was legally intoxicated when the accident occurred.

Now let's focus on the line in the left part of Figure 1-2. It has a slope of 0.064, representing the rate at which Bill's BAC would have increased according to Dr. Winek's testimony. Thus in Figure 1-2, you can calculate Bill's BAC at the time of the accident if you know the equation of the line through (0.7, 0.136) with a slope of 0.064.

EXERCISES

2. Find the equation of the line passing through (0.7, 0.136) with a slope of 0.064.

3. *(Interpreting Mathematics)* According to the graph in Figure 1-2, what was Bill's BAC at the time of the accident?

4. *(Writing to Learn)*

a. Which of the two graphs would Mr. Schuchert be more likely to use in his arguments? Explain your answer.

b. As a juror, would you accept either graph as evidence? If so, which one?

c. As a juror, would you need additional information before reaching a decision? If so, what kind of information?

5. *(Creating Models)* Dr. Winek's testimony was based on an assumption that Bill had not eaten. However, one of Bill's friends testified that Bill had some food at the shrimp bar after his last drink.

a. The presence of food slows the absorption of alcohol into the bloodstream. How would this affect the appearance of the graph in Figure 1-2?

b. Does Bill's friend's testimony favor the prosecution or the defense?

Many students see no value in a college algebra course except that it fulfills a degree requirement. However, ***A Case for Algebra*** indicates that algebra can be useful to you even if you do not use it in your career. To help you understand our reasons for asking you to commit your time to this book, let's discuss a few reasons for studying algebra.

ALGEBRA IS AN ANALYTICAL TOOL

A Case for Algebra illustrates dramatically that no matter what your interests or your chosen occupation may be, you sometimes need to process quantitative information to make decisions. Even if you never serve on a jury, you will certainly make personal decisions about mortgages, credit cards, and investments. You will also evaluate the credibility of quantitative information provided by politicians and others, concerning subjects such as the economy, the environment, and national security.

If you are reading as critically as we hope you are, the preceding paragraph has probably not convinced you of the value of algebra. After all, your great aunt Bertha never went beyond the third grade and was still able to make shrewd financial investments and intelligent political decisions. We are not claiming that algebra is the only road to quantitative enlightenment. However, you can analyze the dynamics of quantitative relationships systematically only through algebra and other mathematics based on algebra.

ALGEBRA IS RICH IN CONCEPTS AND CONTEXTS

Many students believe that algebra consists only of computation and symbol pushing. By contrast, this book will discuss algebra in the context of situations which have personally touched your authors, Lynn and Dave. This approach will help you to appreciate the importance of algebra, and to integrate the subject into your life.

Many students also believe that algebra is a rigid, inflexible discipline. Studying algebra in context will dissolve this image and reveal the richness of the subject. Behind the mathematical operations, there are assumptions which are subject to opinions and judgments, and beyond the operations are results which are open to varying interpretations. For example, in completing Exercises 4 and 5, you formed opinions and judgments based partly on the results of mathematical operations. Several other aspects of Bill's trial also integrate algebra into a larger context. For example, the actual BAC readings were taken from Bill's blood serum rather than whole blood. Since Pennsylvania's legal standard for intoxication is based on whole blood samples, Dr. Winek needed to base his testimony on estimates of readings that would have resulted from whole blood samples.

Even within the narrower context of mathematical operations, where questions have only one correct answer, there are usually many paths to that answer. For example, in ***A Case for Algebra***, questions about Bill's BAC can be answered by using either the graphs in Figures 1-1 and 1-2 or the equations in Exercises 1 and 2.

ALGEBRA IS PART OF OUR INTELLECTUAL HERITAGE

Like literature or music, algebra has been created over the course of centuries by fascinating people who were passionate about their activities as ends in themselves, regardless of the practical applicability of their ideas. You will become acquainted with some of these people as you study their contributions to our mathematical heritage. You will meet, among others, an astronomer who cast astrological charts for kings to avoid starvation while calculating planetary orbits and a political activi˘t who developed an important new body of mathematical theory before he was killed in a duel at the age of 20.

Like literature or music, algebra has also had a profound effect on the cultures which created it. For example, the computers which affect our everyday lives operate on the principles of Boolean algebra, an area of abstract mathematics developed about 150 years ago by George Boole (1815–1864). The study of algebra, like that of literature or music, enriches your understanding of your cultural environment.

YOU MAY ACTUALLY USE ALGEBRA!

We live in an increasingly global economy with changing needs. It is impossible to predict whether the career you choose today will be viable 10 or 20 years from now. Furthermore, our increasing reliance on technology ensures that many career paths will require you to think quantitatively. If you know algebra, you will have more career options.

EXERCISE

6. *(Writing to Learn)* Choose one of our statements that describes a benefit to be gained by studying algebra. Write a paragraph citing your reasons for either agreeing or disagreeing with it. (We won't be offended if you disagree with us. What is important is that you read critically and support your opinions.)

SOME SUGGESTIONS FOR FEELING AT HOME WITH THIS BOOK

By beginning this book, you have given yourself an opportunity to become familiar with the many faces of algebra. Here are some suggestions for making the most of that opportunity.

- Regard the book as a continuous journey, and not a collection of isolated facts. It is best read as you would read a novel, that is, without skipping immediately to the exercises. (Your author Lynn confesses that she reads the last page of a novel first, but swears that she reads every page carefully afterward.)

- Our suggestion to read this book like a novel is not meant to imply that you can read it as quickly as you would read a contemporary romance. A more appropriate comparison would be with Shakespeare. Both reading and doing mathematics take time, and temporary failures are normal. Be patient!
- Exercises are scattered through each section, rather than being grouped all at the end. Read with pencil and paper in hand, and participate by completing exercises as you come to them. Feel free to scribble in the wide margins.
- Because of the cumulative nature of the subject, you will sometimes need to refer to your work from an earlier exercise. It will be important to keep all your prior work organized for ready reference.
- The Mathematical Looking Glasses you will encounter fall broadly into two categories, each having its own purpose.

  Some, such as ***Ultrasound*** (page 225) and ***Standing Room Only*** (page 323), describe uses of algebra by mathematicians, scientists, and engineers to establish scientific principles and obtain information. All descriptions and data are accurate to our best knowledge, except where we admit to meddling with facts to enhance an illustration.

  Some, such as ***Split-Rail Fence*** (page 19) and ***Satellite Dish*** (page 221), describe quantitative problems encountered in everyday life. Although we have sometimes altered details in order to make it more natural to use algebra, the problems are presented in essence as they actually occurred. In some cases, the algebraic solution is the one used by the person in the situation. Even when this is not the case, applying algebra to the problem often provides insight into the situation, beyond the answer to the specific question being asked.
- Exercises are labeled according to their purpose.

  Exercises marked *(Review)* both reinforce skills presented in previous sections and prepare you to use them in other exercises.

  Exercises marked *(Extension)* explore concepts beyond those discussed in the text, or at a depth beyond that already considered. Their purpose is to indicate some possible continuations of the lines of thought developed in the text and to arouse your interest in the further study of mathematics.

  Exercises marked *(Creating Models)* ask you to represent a physical situation by means of an equation, a table, or a graph. Exercises marked *(Interpreting Mathematics)* ask you to interpret an equation (or its solution), a table, or a graph in terms of a physical situation. Exercises marked *(Problem Solving)* ask you to do both. All serve to place algebraic ideas in a physical context.

  Exercises marked *(Writing to Learn)* require responses written in complete sentences. Their purpose is to help you synthesize and express mathematical ideas.

  Exercises marked *(Making Observations)* ask you to make observations about material you have just read, or exercises you have just completed. Their purpose is to help to identify patterns, and to use logic to connect mathematical ideas.

  Exercises not otherwise marked are designed to reinforce technical skills developed in the section where the exercise appears.

By the way, Bill was convicted of vehicular homicide while DUI in the fall of 1989. In 1991, he was granted a new trial on the grounds that the court had failed to admit evidence that the steel pole which Bill's car struck had been the location of numerous accidents and had subsequently been removed. His conviction was reversed.

In working with ***A Case for Algebra***, you have begun a journey of quantitative problem solving. We hope you remembered to pack your toothbrush and lots of pencils. Bon voyage!

1–2 MODELS OF QUANTITATIVE RELATIONSHIPS

PREREQUISITE MAKE SURE YOU ARE FAMILIAR WITH:
Accuracy and Precision (Section A-1)

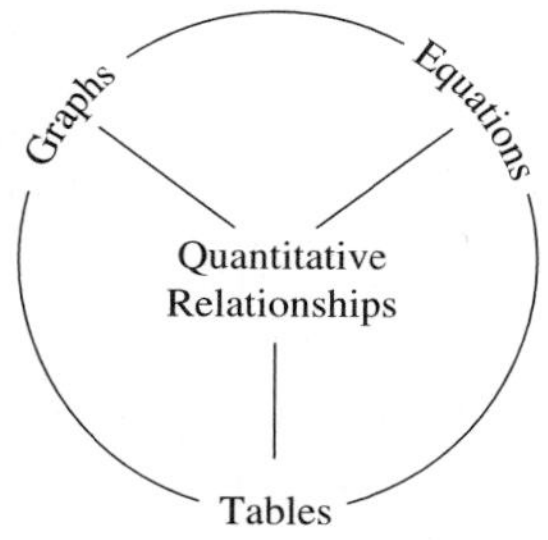

Many questions that affect people's lives involve **quantitative relationships**, that is, relationships among things which can be counted or measured. You have already seen a quantitative relationship in ***A Case for Algebra*** (page 1). The quantities being related were the time in hours after the accident and Bill's blood alcohol count (BAC). Quantitative relationships are frequently studied by creating mathematical models, such as **tables**, **graphs**, and **equations**, also referred to as **numerical**, **graphical**, and **analytical models**, respectively. Although you will study tables, graphs, and equations for their own sakes, we will use them primarily to study quantitative relationships.

In this section we will look at numerical, graphical, and analytical models of the relationship in ***A Case for Algebra***, so that you will have a specific example of each type of model. Afterward we will discuss three other quantitative relationships, focusing on the process of creating one model from another. The wheel at the left will serve as a visual guide through the section.

THREE TYPES OF MODEL

A Numerical Model The data from the BAC readings available to Dr. Winek can be summarized in the form of a table, such as Table 1.

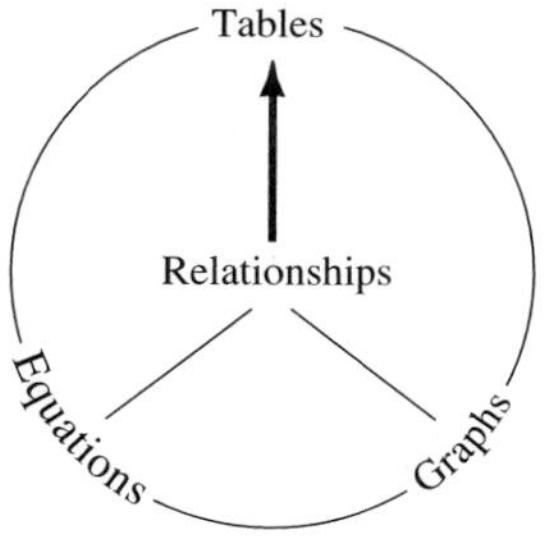

TABLE 1

Time since accident, in hours (t)	BAC (y)
0.7	0.136
2.5	0.100

In most relationships you will encounter in this book, it is natural to think of one variable as the **input**, or **independent variable**, and the other as the **output**, or **dependent variable**. In this case, since Bill's BAC *depends on* the elapsed time since the accident, it is natural to think of the variable t as

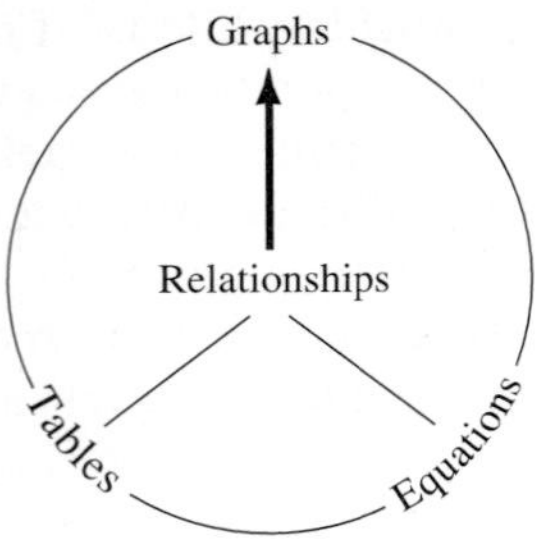

independent and y as dependent. When a relationship is described by a table, it is traditional to list values of the independent variable in the left column, as has been done in Table 1.

A Graphical Model The graph from Figure 1-1 is reproduced in Figure 1-3.

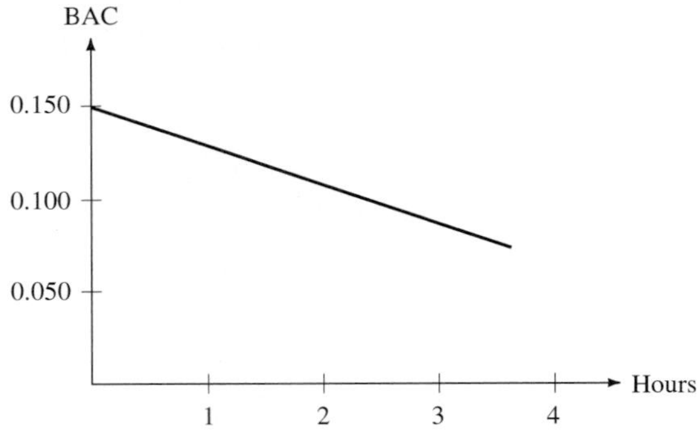

Figure 1–3

When a relationship is described by a graph, it is traditional to measure values of the independent variable along the horizontal axis, as has been done in Figure 1-3.

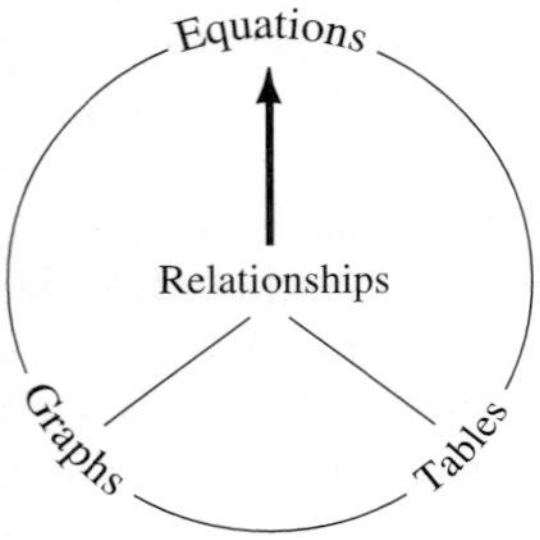

An Analytical Model In Section 1-1 the relationship in Figure 1-3 was described by the equation

$$y = -0.020t + 0.150$$

When a relationship is described by an equation, it is traditional to express the dependent variable in terms of the independent variable, as has been done here.

EXERCISE

1. *(Interpreting Mathematics)* From newspapers, magazines, or your own experience, find numerical, analytical, and graphical models of quantitative relationships. Identify the independent and dependent variables for each relationship.

INFORMATION PROVIDED BY THE THREE TYPES OF MODEL

Now that you have seen examples of tables, equations, and graphs, let's focus on the information each model provides.

Complete Versus Incomplete Information An equation provides complete quantitative information about the relationship it describes. The equation $y = -0.020t + 0.150$ in ***A Case for Algebra*** allows you to calculate Bill's BAC at any time. By contrast, information provided by tables is often incomplete. Table 1 provides BAC readings only at two specific times.

Graphs provide complete information about values of the input variable within some interval. Figure 1-3 shows Bill's BAC over a time interval of about 4 hours, but does not provide any information outside that interval.

Exact Versus Approximate Information Some equations provide exact information. For example, the equation $A = s^2$ describes the relationship between the side length s and the area A of a square. It allows you to calculate that a square 8 inches on a side has an area of exactly 64 square inches.

Other equations provide only approximate information. For example, in ***A Case for Algebra***, the BAC readings of 0.136 and 0.100 are almost certainly approximations. Since the equation $y = -0.020t + 0.150$ is based on those readings, any BAC calculated from the equation is also an approximation.

Tables can also provide either exact or approximate information. For example, the information in Table 1 is approximate. However, a table always provides exact information to the extent that its entries are exact.

Graphs provide only approximate information. In Figure 1-3 you can estimate that Bill's BAC was about 0.090 three hours after the accident, but you cannot calculate it exactly from the graph alone.

Explicit Versus Hidden Information The entries in a table display explicit information. For example, Table 1 explicitly displays a BAC of 0.100 at a time of 2.5 hours after the accident. By contrast, information provided by equations and graphs is hidden. The BAC after 2.5 hours is not displayed explicitly in the equation $y = -0.020t + 0.150$, but you can calculate it. Similarly, it is not displayed explicitly in the graph in Figure 1-3, but you can estimate it.

CREATING ONE MODEL FROM ANOTHER

Which type of model is the most appropriate way to represent a relationship? This depends on the questions you are trying to answer, or the type of information you wish to convey. Newspapers and magazines abound with quantitative relationships modeled as tables or graphs, because they must present information which readers can absorb quickly and easily. Scientific journals are usually more concerned with precision, and therefore model many relationships as equations, rather than graphs or tables.

To answer a specific question or gain insight into a relationship, you may need to create a model different from the one presented. Let's look at three different quantitative relationships, and ask some questions that are more easily answered by using a different model. The first of our three relationships is initially described by a table.

A MATHEMATICAL LOOKING GLASS
Pulleys

John Corriere, a friend of your author Dave, owns a woodworking shop. His equipment includes a centrifugal fan, which is part of a dust collection system. The fan is driven by a 0.25-horsepower motor, which turns at 1700 revolutions per minute (rpm) and is connected to the fan by a fan belt of fixed length. The pulleys M and F come in standard diameters ranging from 2 to 6 inches in 1-inch increments, but John can make pulleys of other sizes.

Changing the pulley sizes will change the fan speed. In particular, if pulley F (the fan pulley) in Figure 1-4 has a fixed diameter of 5 inches, then increasing the diameter of pulley M (the motor pulley) will cause the fan to turn more rapidly. There is a quantitative relationship between the diameter of the motor pulley and the speed of the fan. Since the pulley diameter controls the fan speed, it is natural to think of the diameter as the input, and the speed as the output.

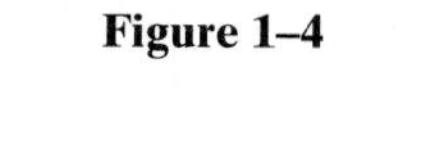

Figure 1–4

Table 2 contains data from the owner's manual for John's motor and is a numerical model of the quantitative relationship.

TABLE 2

Motor pulley diameter, in inches (input)	Fan speed, in rpm (output)
1.5	510
3	1020
5	1700
8	2720
10	3400

Higher fan speeds collect more dust, but speeds above 3600 rpm will damage the motor. John has a pulley 11 inches in diameter, and he would like to know whether the corresponding fan speed will exceed the motor's limit. Since the numerical model in Table 2 provides incomplete information about the relationship, John cannot answer his question from the table alone. Let's see how he might create an analytical or graphical model of the relationship, and how each model will allow him to determine the speed by "filling in the gaps" in the table. ■

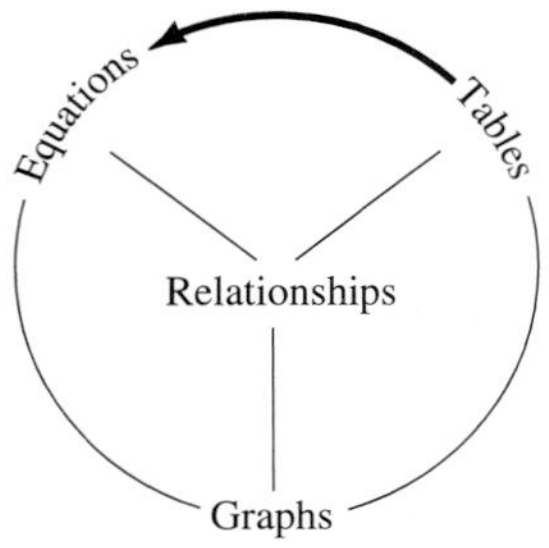

Tables to Equations John would like to know what fan speed is produced by a pulley diameter of 11 inches. One strategy for answering this question is to create an equation by studying the data and searching for patterns. By studying his table, John observed that the fan speed divided by the motor pulley diameter is always 340. If he uses the variable names M and S to represent the motor pulley diameter and fan speed, he can write this observation as an equation

$$S = 340M$$

EXERCISES

2. *(Interpreting Mathematics)* Use John's equation to find the fan speed for a motor pulley diameter of 11 inches. Will this speed exceed the motor's limit of 3600 rpm?

3. *(Interpreting Mathematics)* What motor pulley diameter will produce a fan speed of exactly 3600 rpm?

4. *(Writing to Learn)* Explain why Exercises 2 and 3 can be completed more efficiently by using an equation rather than a table.

Creating an equation from a table is often more difficult than it was in this example. Throughout the book, you will learn techniques for dealing with more complicated cases.

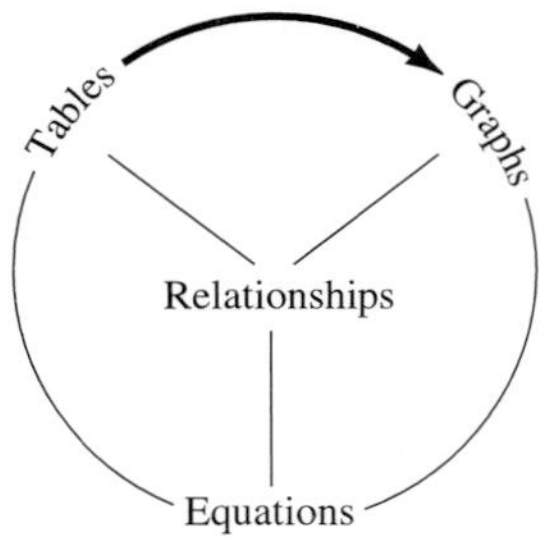

Tables to Graphs A different strategy John can use for answering his question in ***Pulleys*** is to draw a graph of the data, as in Figure 1-5a. The values of the independent variable, pulley diameter, are on the horizontal axis.

So far, the graph allows John to see the correspondence of the five motor pulley diameters in Table 2 with the five fan speeds. To answer his ques-

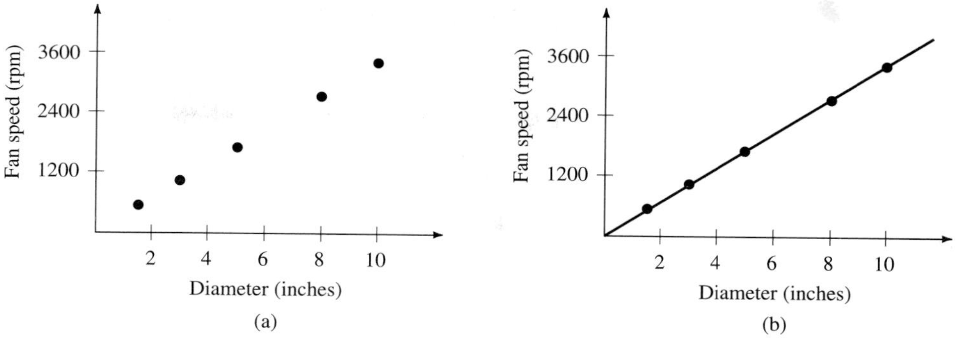

Figure 1–5

tion using his graph, John has sketched a smooth curve through the data points in Figure 1-5b. (Actually, John's graph is a line, but mathematicians often use the word *curve* to include lines.) Drawing a smooth curve "fills in the gaps" in Table 2.

EXERCISES

5. *(Interpreting Mathematics)* Use John's graph to estimate the fan speed for a motor pulley diameter of 11 inches. How does your result compare with that from Exercise 2?
6. *(Interpreting Mathematics)* Use John's graph to estimate the motor pulley diameter which will produce a fan speed of exactly 3600 rpm. How does your result compare with that from Exercise 3?
7. *(Writing to Learn)* Compare John's table, equation, and graph in terms of accuracy and efficiency in answering the questions in Exercises 5 and 6.

The second of our three relationships is initially described by an equation.

A MATHEMATICAL LOOKING GLASS
Swimming Pool

Skyview Pools, Inc., in Pittsburgh, PA, sells outdoor swimming pool in several standard sizes, with diameters of 15, 18, 21, 24, and 27 feet. The pools are 4 feet deep and are designed to be filled to a depth of 46 inches. See Figure 1-6.

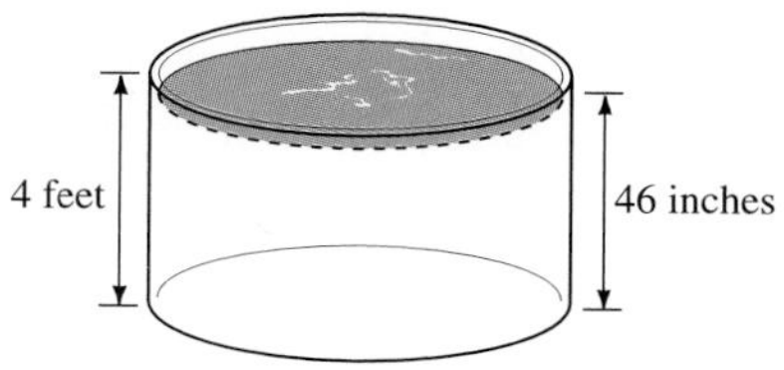

Figure 1–6

To provide information to customers, the salespeople need to know the number of gallons of water required to fill a pool of each standard diameter. ■

EXERCISE **8.** *(Problem Solving)*

a. If a pool with a diameter of d feet is filled to a depth of 46 inches, what is the volume of water in cubic feet? Express your answer as an equation $V = (\text{coefficient})d^2$, and round the coefficient to two decimal places. (The volume V of a cylinder with radius r and height h is $V = \pi r^2 h$.)

b. A cubic foot of water contains about 7.80 gallons. How many gallons of water does the pool contain? Express your answer as an equation $G = (\text{coefficient})d^2$, and round the coefficient to two decimal places.

c. How many gallons are required to fill a pool of each of the five standard diameters listed in ***Swimming Pool***?

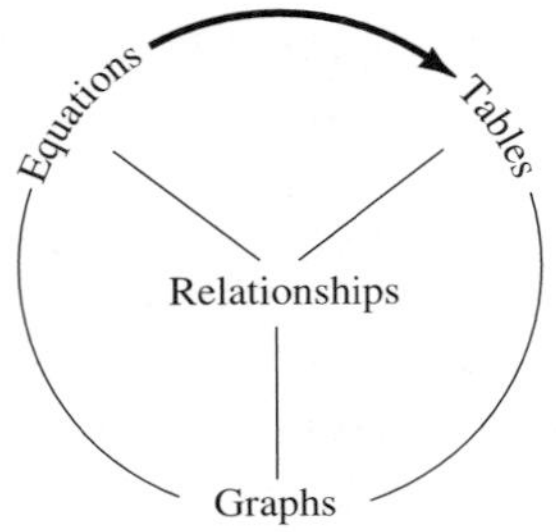

Equations to Tables The information provided by the analytical model in Exercise 8 is hidden. That is, to find the capacity of a pool with a given diameter, you must perform some calculations. It is more convenient for Skyview's salespeople to have immediate access to explicit information, such as a table can provide. Your results from Exercise 8c are reproduced in Table 3.

TABLE 3

Diameter, in feet	Capacity, in gallons
15	5,283
18	7,608
21	10,355
24	13,524
27	17,117

Table 3 will not give the capacity of pools with every possible diameter. However, a salesperson armed with Table 3 can immediately answer some commonly asked questions about pool capacities without performing any calculations. For example, the table tells immediately that a pool 21 feet in diameter has a capacity of 10,355 gallons.

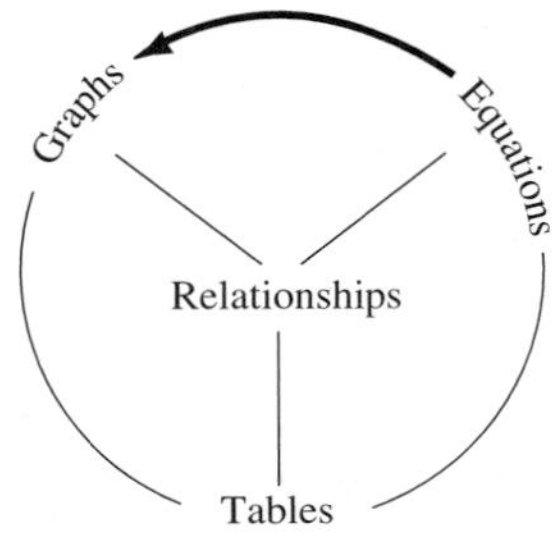

Equations to Graphs Would a graph of the relationship in ***Swimming Pool*** be helpful to Skyview's salespeople? One way to obtain a graph from the equation is to make a table of values, plot points, and draw a smooth curve through the points. We have used Table 3, obtained from the equation in Exercise 8, to draw the graph in Figure 1-7a. Figure 1-7b shows a graph of the same equation, using a table which extends over diameters from 0 to 30 feet.

The graphs provide visual estimates without computations, but also without the equation's accuracy. For example, the equation allows us to calculate the capacity of a pool 20 feet in diameter as 9392 gallons. Using the graph in Figure 1-7a, we can only estimate the capacity as a little more than 9000 gallons. An estimate based on the graph in Figure 1-7b would be even less accurate, because its scale is larger.

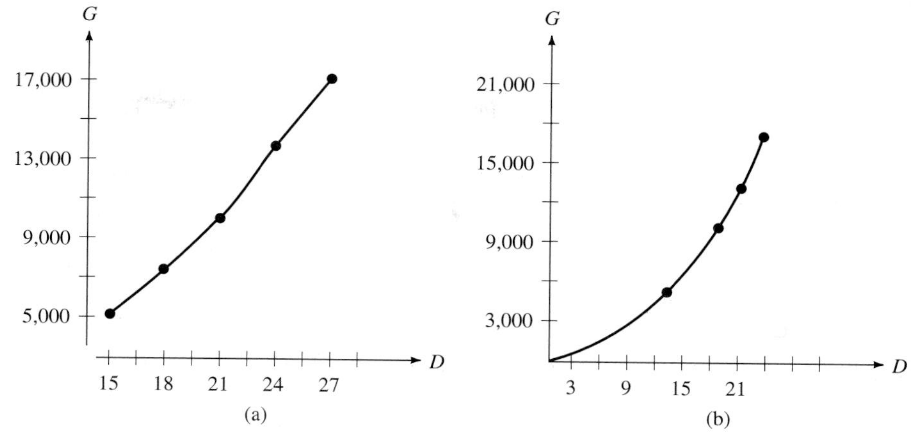

Figure 1–7

EXERCISE **9.** *(Writing to Learn)* As a salesperson for Skyview, would you rather have access to Table 3 or the graphs in Figure 1-7? Explain your answer.

It is worth noting that by their visual nature, graphs allow us to observe the dynamic behavior of relationships. For example, in Figure 1-7a we see that larger diameters correspond to larger capacities. In Figure 1-7b we can observe that the capacity grows more rapidly as the diameter increases. Such observations can be used to predict results beyond the scope of the graph.

Several techniques for graphing equations without extensive point plotting will be presented later in this book. In particular, you will learn a more efficient way to graph the equation from ***Swimming Pool*** in Section 5-1.

The last of our three relationships is initially described by a graph.

A MATHEMATICAL LOOKING GLASS
Greenhouse Effect

Our planet's atmosphere contains nitrogen (75% by weight) and oxygen (23%), along with much smaller concentrations of other gases, such as carbon dioxide. A small change in this balance of elements can have drastic consequences for everything that lives on earth. For example, let's look at some reasons for the recent concern over the buildup of carbon dioxide (CO_2), caused largely by the burning of fossil fuels in automobiles and other vehicles.

Part of the heat produced on earth by sunlight is reflected through the skies and back out into space. Much of this reflected heat could not pass through an atmosphere containing a high concentration of CO_2, and would thus be trapped near ground level. The resulting increase in temperature is often called the **greenhouse effect**, because it would produce the warm, humid conditions found in greenhouses. Increased temperatures would melt the polar icecaps, causing water levels to rise in the oceans and flooding all coastal cities. The greenhouse effect could eventually make the planet too hot to sustain life at all.

The graph in Figure 1-8a shows the level of CO_2 in the atmosphere since 1960. The waviness occurs because the concentration of CO_2 varies with the season. The graph in Figure 1-8b shows the average trend in CO_2 concentration over the same period.

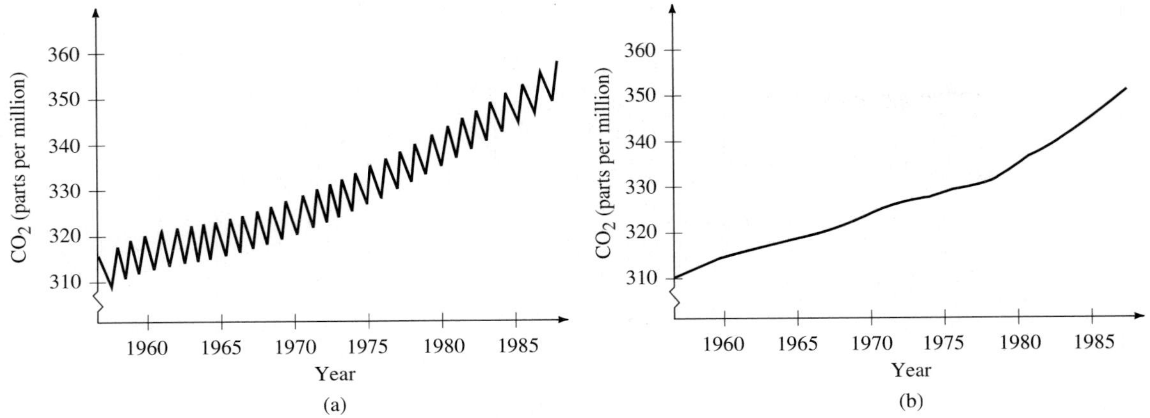

Figure 1–8

Adapted from Michael Seeds, *Foundations of Astronomy*, Wadsworth, 1990.

Although the graphs in Figure 1-8 provide immediate visual information about the greenhouse effect, they are of limited value in predicting the concentration of CO_2 at any time in the future. ■

EXERCISES **10.** *(Writing to Learn)*

a. What fact about CO_2 concentration can be observed in Figure 1-8a, but not Figure 1-8b?

b. What fact about CO_2 concentration can be observed more easily in Figure 1-8b?

c. From which graph can future concentration levels be predicted more easily?

11. *(Interpreting Mathematics)* Use Figure 1-8 to estimate CO_2 concentration in the year 2000, and in 2050. Which estimate do you trust more? Why?

12. *(Writing to Learn)* Scientists who want to estimate future CO_2 levels often try to fit equations to the graphs in Figure 1-8, and use the equations to arrive at their estimates. Why do you think they prefer to work with equations rather than graphs?

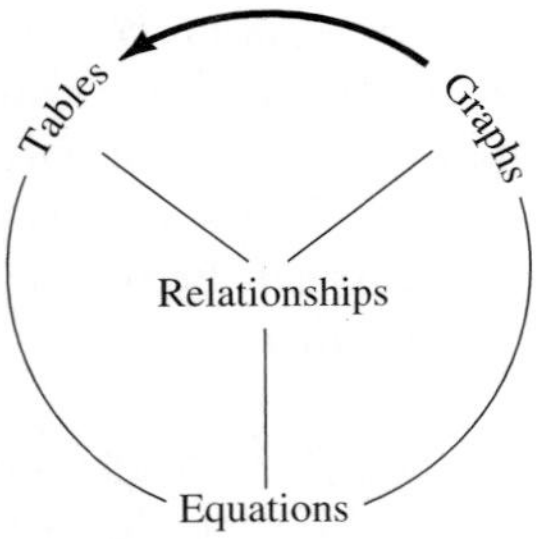

Graphs to Tables You can obtain a table from any graph by estimating the coordinates of several points.

EXERCISE **13.** *(Problem Solving)*

a. Use the graph in Figure 1-8b to complete the following table.

Year	CO_2 concentration, in parts per million
1960	
1970	
1980	
1985	

b. Use the table to estimate the concentration in the year 2000, and in 2050.

c. Do you trust these estimates more than the ones you obtained in Exercise 11? Why or why not?

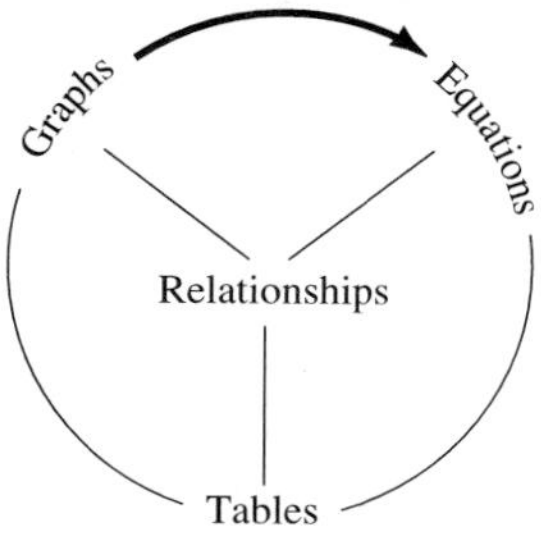

Graphs to Equations There is no all-purpose procedure for constructing equations from *every* graph. However, it is possible to construct equations from *some* graphs. For example, in ***A Case for Algebra***, we were able to construct an equation from the graph in Figure 1-3. The process required us to assume that the graph was a straight line, as it appeared to be in the figure. We will return periodically to the problem of constructing equations from graphs, beginning in Chapter 3.

Additional Exercises

For each table in Exercises 14–17,

a. Write an equation that might describe the relationship between x and y. (*Hint*: Look for a pattern in each table, as John did to construct his equation in ***Pulleys***.)

b. Sketch a possible graph of the relationship.

14.

x	y
1	3
2	4
3	5
4	6
5	7
6	8

15.

x	y
1	9
2	8
3	7
4	6
5	5
6	4

16.

x	y
5	10
10	20
15	30
20	40
25	50
30	60

17.

x	y
1	12
2	6
3	4
4	3
6	2
12	1

For each equation in Exercises 18–21,

a. Complete the following table.

x	y
−4	
−2	
0	
2	
4	

b. Use your table to sketch a possible graph of the relationship between x and y.

18. $y = 2x + 3$

19. $y = 10 - x$

20. $y = \frac{1}{2}x^2$

21. $y = 10 - x^2$

For each graph in Exercises 22–25,

a. Complete a table showing the values of y when $x = -2, -1, 0, 1,$ and 2.

b. Make reasonable assumptions about the behavior of the graph outside the region shown, and estimate the value of y when $x = 5$. (There is no single correct answer.)

22.

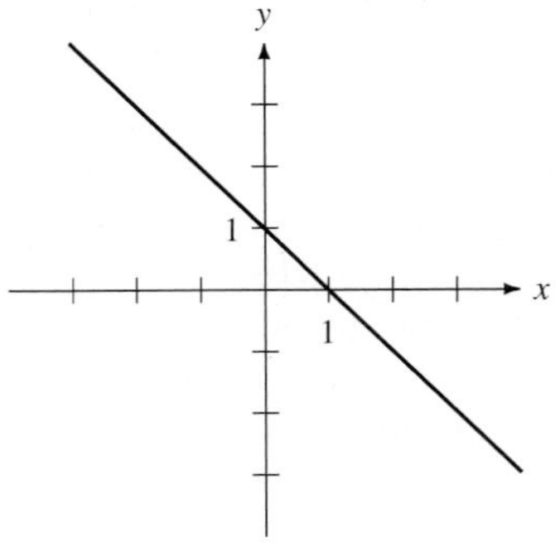

23.

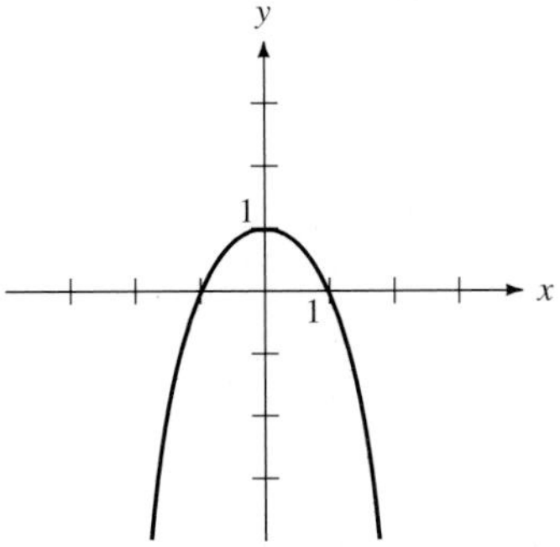

24.

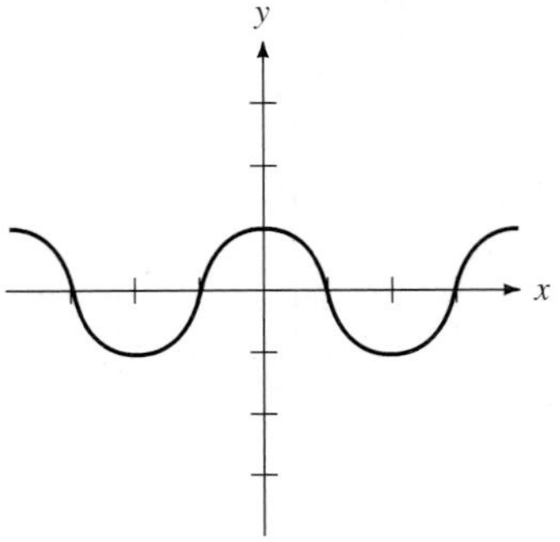

25.

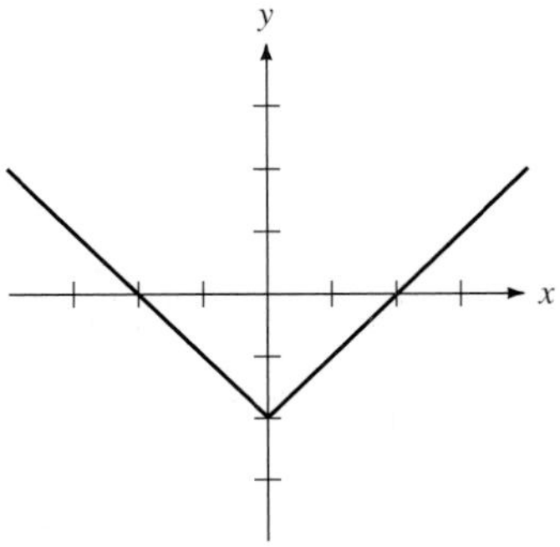

In Exercises 26–29 (Writing to Learn), tell whether the information presented by the model of the relationship is

a. complete or incomplete

b. exact or approximate

c. explicit or hidden

Explain your answers. (Arguments in favor of both answers can often be given, so your reasons are important.)

26. The following table shows how much mortgage can be carried by a family with a given income.

G = gross annual family income, in dollars

M = maximum monthly PITI (*p*rincipal, *i*nterest, *t*axes, and *i*nsurance), in dollars

G	*M*
40,000	1200
50,000	1500
60,000	1800
70,000	2100
80,000	2400

Adapted from Joel Lerner, *Financial Planning for the Utterly Confused*, McGraw-Hill, 1994.

27. The following table shows the life expectancy of older females in the United States.

A = present age

E = expected years of remaining life

A	**E**
65	18.4
70	14.8
75	11.6
80	8.7
85	6.4
90	4.7

28. If you want to invest a sum of money for a relatively short time, you might buy a Treasury bill, also called a T-bill, from the U.S. Treasury Department. Because the interest on such an investment is paid "up front," the actual rate of interest you earn is higher than the stated rate. For example, suppose you buy a 1-year T-bill for \$10,000 (the minimum amount allowed) at 5% interest. After you pay for it, the Treasury Department will send you a check for the interest (5% of \$10,000 = \$500). In effect, you have paid \$9500 and will receive \$10,000 at the end of 1 year. As a percentage of your investment, the interest is

$$(100)\left(\frac{500}{9500}\right) \cong 5.3$$

In general, let

D = the dollar amount of interest on a \$10,000 1-year T-bill

r = the actual interest rate

Then the relationship between D and r is given by the equation

$$r = \frac{100D}{10{,}000 - D}$$

29. Most archaeologists believe that the first natives of North America migrated on foot from Siberia to Alaska. Since the Bering Sea presently separates Siberia from Alaska, the migration could have occurred only during a time when the world's sea levels were much lower than they are today. Therefore, scientists attempting to estimate the date of the earliest migration have an interest in past sea levels. In the following graph,

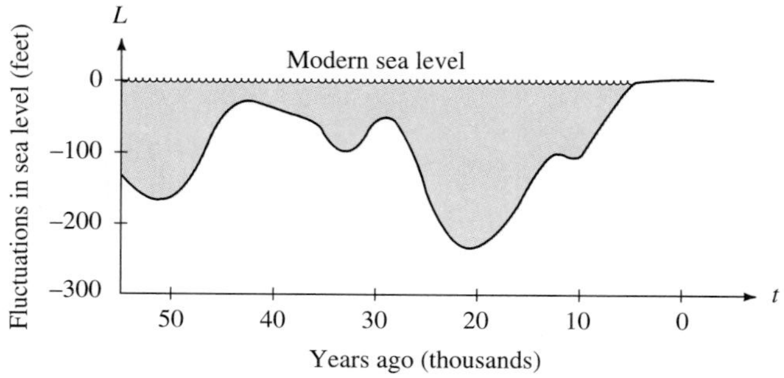

(Adapted from Brian Fagan, *The Great Journey*, Thames and Hudson, 1987.)

t = time in years before the present

L = sea level, in feet above or below present level

30. *(Problem Solving)* Refer to the table in Exercise 26.

a. Sketch a possible graph of the relationship between G and M.

b. Write an equation that might describe the relationship.

c. Using either your equation or your graph, decide how much mortgage could be carried by a family with a gross annual income of \$120,000.

d. Using either your equation or your graph, decide what gross annual income is needed to carry a mortgage of \$1000.

31. *(Creating Models)*

a. From the table in Exercise 27, sketch a possible graph of the relationship between A and E.

b. Estimate the expected years of remaining life for a 100-year-old female.

32. Refer to the table in Exercise 27.

a. What is the expected age at death for a 65-year-old woman?

b. What is the expected age at death for a 70-year-old woman?

c. *(Writing to Learn)* Why do you think your answers in parts (a) and (b) are different?

33. *(Creating Models)* Refer to the equation in Exercise 28.

a. Complete the following table.

D	r
200	
300	
400	
500	
600	

b. Sketch a possible graph of the relationship between D and r.

34. *(Interpreting Mathematics)* Refer to the graph in Exercise 29. A migration from Siberia to Alaska would have been possible only during periods when the sea level was at least 100 feet below its present level. During what times was this the case?

35. *(Writing to Learn)* Both your purpose in communicating information and your audience determine the most appropriate model to use. Describe actual situations in which quantitative information should be communicated by

a. using a graph **b.** using a table

c. using an equation

36. *(Writing to Learn)* Compare tables, equations, and graphs as models of quantitative relationships. Include the advantages and limitations of each.

37. *(Writing to Learn)* Refer to one of the models you found in Exercise 1, and make up a question which would be more easily answered using a model different from the one given. Tell what type of model would be appropriate, and why. You do not need to create the model or answer your question.

1–3 PROBLEM-SOLVING STRATEGIES

PREREQUISITES MAKE SURE YOU ARE FAMILIAR WITH:

Linear Equations (Section A-2)

The Pythagorean Theorem (Section A-4)

Throughout this book you will construct mathematical models and interpret the information they provide to answer questions about quantitative relationships. You can do so more effectively by arming yourself with a systematic approach to problem solving. In this section we will develop a general problem-solving method developed by George Pólya (*How to Solve It*, Princeton University Press, 1945). We will also consider several specific strategies. These strategies are not rigid rules for solving a narrowly defined type of problem, such as motion or mixture problems. You can think of them as a general and flexible process you can use to solve the wide variety of quantitative problems you encounter in this book, as well as many everyday, nonquantitative problems. The method and the strategies will be developed in the context of the following Mathematical Looking Glass.

A MATHEMATICAL LOOKING GLASS
Split-Rail Fence

The property around your author Lynn's house includes a rectangular pasture enclosed by a split-rail fence, as in Figure 1-9.

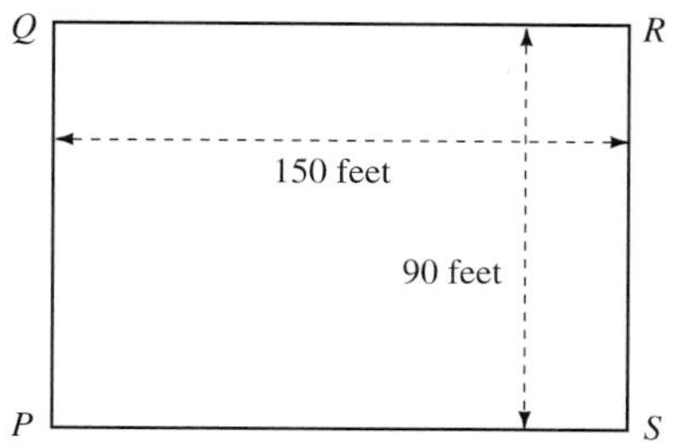

Figure 1–9

A few years ago, Lynn's daughter Carrie wanted to expand the pasture to give her horse, Illiad, more room. She considered moving one side of the fence from *QR* to *TU* and installing additional fence along *QT* and *RU*, as in Figure 1-10a. She also considered moving one side from *SR* to *VW*, installing additional fence along *RV* and *SW*, as in Figure 1-10b.

Carrie called contractor Steve Ford of Ford Enterprises, Inc. Steve agreed to do the job at a cost of \$1.50 per foot to move existing fence, and \$3.00 per foot to install new fence. Carrie had several hundred dollars set aside for the project, but was not sure how much she wanted to spend. To

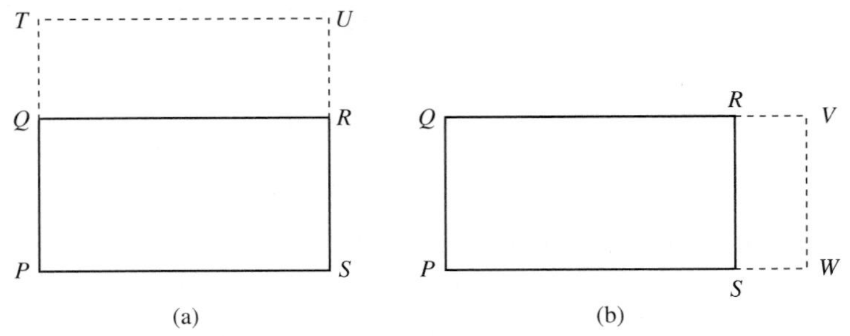

Figure 1–10

help her decide, she wanted to know which of the alternatives in Figure 1-10 would give her the greater additional area for a given cost. ■

Summarized briefly, Pólya's approach to problem solving consists of four steps.

- Understand the problem.
- Devise a plan.
- Carry out the plan.
- Look back.

Let's look at each step in detail as it relates to ***Split-Rail Fence***.

FIRST STEP: UNDERSTAND THE PROBLEM

Here are several questions you can ask yourself to make sure you understand a quantitative problem.

a. Can you state the problem in your own words?

In the last sentence of ***Split-Rail Fence***, we have stated Carrie's problem in our own words. You can check your understanding of the problem by doing the same in Exercise 1.

EXERCISE

1. *(Writing to Learn)* Reread ***Split-Rail Fence***. Then close your book and state Carrie's problem in a sentence or two. (Don't just memorize our statement. Use *your own* words.)

b. What information is given?

The information that is likely to be relevant to Carrie's problem can be summarized in three statements.

- The present dimensions of the pasture are 90 by 150 feet.
- It will cost \$1.50 per foot to move existing fence.
- It will cost \$3.00 per foot to install new fence.

c. What information, if any, is given but not needed?

Nonquantitative information in physical situations can often be used to decide what quantitative questions are worth asking. Once a quantitative question has been asked, most nonquantitative information is not needed for the solution. For example, the fact that Carrie owns a horse indicates the reason for her concern about the size of the pasture, but does not help her solve the problem.

Some quantitative information may also be unneeded. For example, in ***Pulleys*** (page 9), the fact that John's motor has 0.25 horsepower has no bearing on the solution to his problem.

d. What information, if any, is needed but not given?

In ***Split-Rail Fence***, it appears that we have all the information we need to solve Carrie's problem. However, we often encounter situations in which we need to gather more information before solving a problem. For example, suppose that the land around the pasture were heavily wooded. (Fortunately, that is not the case.) Then Carrie could not analyze her problem without measuring the distance from each edge of the pasture to the nearest tree. She would also need to decide whether to spend any of her money on tree removal.

e. What are the constants and the variables in the problem? That is, what quantities are fixed, and what quantities can vary?

This book deals with relationships between two varying quantities, under the condition that all other quantities remain fixed. Often, you have the freedom to choose which two quantities are variables. For example, since Carrie wants to know how much additional area she can obtain for a given cost, she might regard additional area and cost as her variables. Alternatively, she might recognize that the area of a rectangle is the product of its length and width. Thus in Figure 1-10a, where the length is fixed, she could regard width and cost as the variables. Similarly, in Figure 1-10b, she could regard length and cost as the variables.

SECOND STEP: DEVISE A PLAN

According to Pólya in *How to Solve It*, "the main achievement in the solution of a problem is to conceive the idea of a plan." After we list some frequently used strategies for doing this, we will see how you might use some of them to solve Carrie's problem. Here is the list.

- *Drawing a Picture or Diagram.* This strategy often takes the form of *Drawing a Graph.*
- *Making a Table.*
- *Looking for a Pattern.* This strategy often takes the form of *Looking for Symmetry.*
- *Guessing and Checking.*
- *Working Backwards.*
- *Examining a Special Case.* This strategy often takes the form of *Examining Extreme Cases.*
- *Examining a Related Problem.* This strategy often takes the form of *Examining a Simpler Problem.*
- *Breaking the Problem into Parts.*
- *Looking for Relationships Among Variables.* This strategy often takes the form of *Writing an Equation* or *Looking for a Formula.*

Of course, you will not use all these strategies on every problem you solve, and you will sometimes use others not listed here. Most problems can be solved in more than one way, and there is no foolproof method for deciding which strategies work best for a given problem. The only way to become a better problem solver is to get plenty of practice.

Now let's return to ***Split-Rail Fence***. In Figure 1-9, Carrie has already approached the problem by *Drawing a Picture or Diagram*. Figure 1-10 suggests that she can continue by *Breaking the Problem into Parts*, considering first the plan in Figure 1-10a, then the one in Figure 1-10b.

Next let's recall that in Figure 1-10a, the variables were the width of the pasture and the cost of enlarging it. Carrie needs to know how much additional area she can get for a given cost, so she should try *Looking for Relationships Among the Variables*, probably in the form of *Writing an Equation*.

These strategies provide a starting point for the third step of Pólya's process. Other strategies may come into play as we continue.

THIRD STEP: CARRY OUT THE PLAN

Most of the mathematics occurs here. You must often simplify algebraic expressions, solve equations, and draw and interpret graphs. You will need to give some thought to the order of these if the problem is complex. Let's carry out the plan we just devised for the solution of Carrie's problem.

First, let's define the variables

$$A = \text{additional area obtained}$$

$$C = \text{cost of enlarging the pasture}$$

To gain some insight into the relationship between A and C, Carrie might begin by *Examining A Special Case*. Specifically, suppose she expands the pasture as in Figure 1-10a, and increases the width by 5 feet. The additional area she obtains is then $(150)(5) = 750$ square feet. What is the cost of the additional area? The length of new fence to be installed is $(2)(5) = 10$ feet, at a cost of $(3)(10) = 30$ dollars. The length of existing fence to be moved is 150 feet, at a cost of $(1.50)(150) = 225$ dollars. The total cost is \$255.

The insight gained from the special case suggests a line of thought for constructing an equation relating A and C. Suppose Carrie expands the pasture as in Figure 1-10a and increases the width by x feet. The additional area she obtains is $(150)(x) = 150x$ square feet. The length of new fence to be installed is $(2)(x) = 2x$ feet, at a cost of $(3)(2x) = 6x$ dollars. The length of existing fence to be moved is 150 feet, at a cost of $(1.50)(150) = 225$ dollars. The total cost is $6x + 225$ dollars. Carrie now knows that if she expands the width by x feet, then

$$A = 150x$$

and

$$C = 6x + 225$$

To calculate the additional area obtainable for a given cost, she would like to express A in terms of C. She can do so by means of the following calculation.

$$C = 6x + 225 \text{ (from our previous work)}$$

$$6x = C - 225$$

$$x = \frac{C - 225}{6}$$

$$A = 150x \text{ (from our previous work)}$$

$$A = 150\left(\frac{C - 225}{6}\right)$$

$$A = 25C - 5625$$

Now Carrie can calculate the additional area she will obtain for any given cost if she increases the width of the pasture. What if she increases the length instead, as in Figure 1-10b? In Exercise 2 you can find out by *Examining a Related Problem.*

EXERCISE **2.** *(Creating Models)* Suppose Carrie increases the length of the pasture, as in Figure 1-10b. Construct an equation expressing A in terms of C by using a line of reasoning similar to the one just given.

At this point Carrie has two equations relating cost to additional area:

$A = 25C - 5625$, if she increases the width

$A = 15C - 2025$, if she increases the length

To solve her problem, she needs to decide which plan gives her more area for her money. She might continue by *Making a Table,* such as Table 4.

TABLE 4

C (cost)	$A = 25C - 5625$ (width is increased)	$A = 15C - 2025$ (length is increased)
250	625	1725
300	1875	2475
350	3125	3225
400	4375	3975
450	5625	4725
500	6875	5475

She can now apply the strategy of *Looking for a Pattern.* Table 4 indicates that for a small expansion, it is better to increase the length, while for a larger expansion it is better to increase the width.

Although Carrie can probably use Table 4 to decide how much she wants to spend, she might also want to know where the "crossover" occurs. That is, at what cost will the two plans give her the same amount of area? In Exercise 3 you can answer this question by using the strategy of *Guessing and Checking.*

EXERCISE **3.** By trial and error, find a cost for which the two plans give equal areas. Use Table 4 to make efficient guesses.

FOURTH STEP: LOOK BACK

One purpose of looking back at the problem-solving process is to verify that your solution is correct. Another is to sharpen your problem-solving skills by studying both the problem and the process you used to solve it. Understanding why your plan worked may help you to solve other problems in the future. It is helpful to ask yourself the following questions.

a. Is your solution mathematically accurate?

We hope you have checked the accuracy of all computations in ***Split-Rail Fence***, both yours and ours. Contrary to popular belief, authors are human, and we sometimes make errors.

In general, your checking should include making sure each step leads to an equivalent model. For example, if you are solving an equation, remember that you may lose solutions if you divide both sides by a variable quantity. You should also check the validity of your solution in the original model. In particular, if you use several equations in the process of solving, substitute each solution into the original equation.

b. Is your solution reasonable in the context of the problem?

The pattern Carrie observed in Table 4 can be supported by a "common sense" line of reasoning. To increase the width of the pasture by *any* amount, Carrie must pay to have 150 feet of existing fence moved. To increase the length requires moving only 90 feet of existing fence. If the area is to be increased by only a small amount, the cost of new fence will be comparatively small with either plan, so it will be cheaper to increase the length. However, each time the width is increased by a foot, Carrie gains 150 square feet of additional area. Increasing the length by a foot gives her only 90 square feet. Thus, if the additional area is to be sufficiently large, it will be cheaper to increase the width.

Sometimes mathematically correct solutions fail to be reasonable in the context of a problem. In ***Swimming Pool***, the equation relating diameter to capacity was $G = 23.48d^2$. If a pool has a capacity of 10,000 gallons, we can find its diameter by solving the equation $10{,}000 = 23.48d^2$ to obtain $d \cong \pm 20.64$ feet. Both are valid solutions to the equation, but the negative one is not reasonable in the context of the problem.

c. Could you have solved the problem in a different way?

This is an invitation to look at the same problem from several different perspectives. This often gives you greater insight into the problem. Solving the problem in a different way also supports the accuracy of your result more reliably than reworking it in the same way. This is because you are extremely unlikely to arrive at the same wrong answer in several different ways.

After obtaining her equations, Carrie could have continued by solving the system of linear equations

$$A = 25C - 5625$$

$$A = 15C - 2025$$

Methods of solving linear systems are reviewed in Section A-10 and discussed in greater detail in Chapter 4.

d. Is the problem or solution similar to others that you have solved previously? If so, can you generalize the method of solution?

By contrast with (c), this is an invitation to look at several problems from the same perspective. This will help you to understand the common mathematical structure of seemingly diverse problems. As we build a foundation of common experience in the coming chapters, we will often ask you to reflect on the common features of the problems you solve. In particular, you will see that ***Health Insurance*** (page 101) and ***Breakeven Analysis*** (page 66) have mathematical structures similar to ***Split-Rail Fence*** (page 19).

ADDITIONAL EXERCISES

In Exercises 4–5, use the given information to answer as many of the questions as possible, and tell what additional information you would need to answer the others.

4. *(Problem Solving)* Three sprinters started a 100-meter dash at the same time. The fastest one averaged 9 meters per second, and the slowest averaged 8 meters per second.

a. What was the winning time?

b. What was the winner's margin of victory, in seconds?

c. Where was the slowest runner when the winner crossed the finish line?

d. Which runner was ahead at the 50-meter mark?

5. *(Problem Solving)* The following table shows percentage increases from 1990 to 1991 in three categories of health care expenditures in Idaho, Oklahoma, and Wyoming. For example, total expenditures for hospital care in Idaho rose 14.6% between 1990 and 1991.

Category	ID	OK	WY
Hospital care	14.6	12.1	10.3
Physician care	7.0	1.7	3.6
Prescription drugs	12.8	11.8	10.1

Adapted from *Health Care State Rankings*, Morgan Quitno Corp., 1994.

a. In which state was the least money spent on prescription drugs in 1991?

b. Which category of health care experienced the smallest percentage increase in cost between 1990 and 1991?

c. Which category of health care experienced the smallest dollar increase in cost between 1990 and 1991?

d. Which state had the lowest overall percentage increase (in all three categories combined)? (*Hint*: A state's overall percentage increase is *not* the sum of the percentage increases in each category.)

In Exercises 6–20, identify each step in the Pólya process as you solve the problem and each strategy being used from the second step. Focus on your thought processes as you solve the problems, and not just on the final solution. Even if you can solve the problem with a "flash of insight," identify the strategies that triggered the flash.

6. *(Problem Solving)* A Swimman (like a Walkman, but watertight with submersible headphones) is on sale for $180 after having been marked down 20%. What was its original price?

7. *(Problem Solving)* Fifteen International Master chess players took part in a recent tournament with a round-robin format, so that every participant played every other participant. How many games were played altogether?

8. *(Problem Solving)* If the sponsor of a round-robin chess tournament (see Exercise 7) wants to limit the number of games played to no more than 80, how many players can participate?

9. *(Problem Solving)* A well-known mathematical brain-teaser challenges you to measure out exactly four liters of water using only a 3-liter container and a 5-liter container. Describe the procedure for doing so.

10. *(Problem Solving)* Can you solve the brain-teaser in Exercise 9 using only a 3-liter and a 6-liter container? If so, how? If not, why not?

11. *(Problem Solving)* A number of elephants and ostriches are in an enclosure at a zoo. Slats looks over the fence and counts 60 heads. Shorty looks under the fence and counts 194 legs. How many elephants are there?

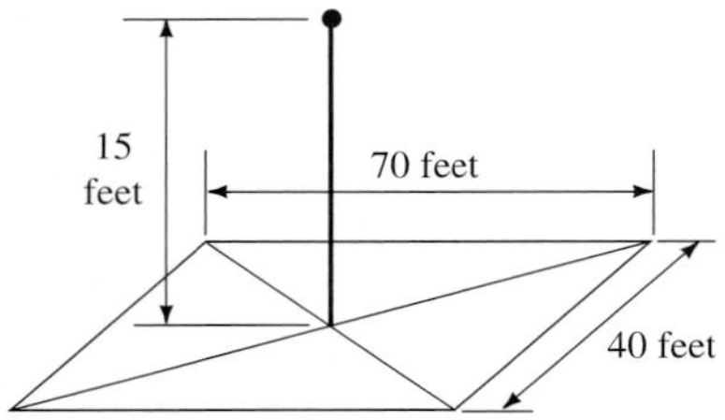

Figure 1–11

12. *(Problem Solving)* A few years ago your author Lynn helped Robert Kelly install a ham radio antenna on a trailer owned by Lynn's father. They installed it at the center of the rectangular roof, which was 70 feet long and 40 feet wide. They had no tape measure, but had 165 feet of string with them. Is this enough to locate the center by stretching across the two diagonals, as in Figure 1-11?

13. *(Problem Solving)* The antenna in Exercise 12 is 15 feet high and is secured by wires running from the top to each corner of the roof. How long should each wire be?

14. *(Problem Solving)* Your absent-minded author Dave recently had to drive to Atlantic City, NJ, to attend a conference. Thirty minutes after he left, his wife, Alice, noticed that his suitcase was still in the living room. Their daughter, Sarah, grabbed the suitcase and took off in pursuit. She knew that her father would drive at a steady 50 mph, since he had just received a speeding ticket. Risking her license by doing 70, she tried to estimate how far she would need to drive before starting to watch for him. How far did she drive before overtaking Dave?

15. *(Problem Solving)* Lynn's friend Robert Kelly is in the habit of feeding the ducks at North Park Lake. He has discovered through trial and error that they adore pumpernickel bread from Periwinkle's Bakery. However, at a cost of \$1.75 a loaf, a month's supply of 14 loaves costs more than he is willing to spend. A precise budgeter, Robert has set aside \$17.00 per month as his allowance for duck food. He will therefore buy as much pumpernickel bread as he can, subject to the 14-loaf total, with the remainder being day-old bread at \$0.50 a loaf. How much of each will he buy?

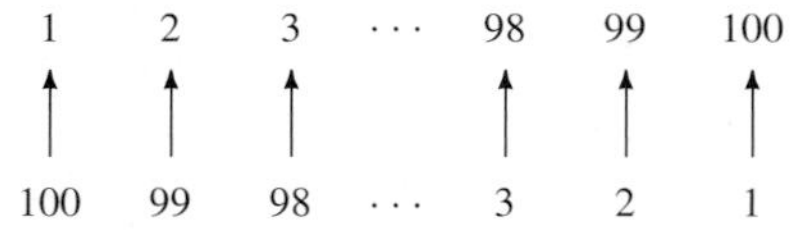

Figure 1–12

16. *(Problem Solving)* When the German mathematician Carl Friedrich Gauss was a child, his class was given the task of finding the sum of the first 100 positive integers as a punishment for some misbehavior. While his classmates labored over the addition, Carl shouted out the answer in a few seconds. Use Figure 1-12 to duplicate his feat.

17. *(Problem Solving)* The graphs in Figure 1-13 show the relationship between time and the height of two elevators. Which one would you rather ride, and why?

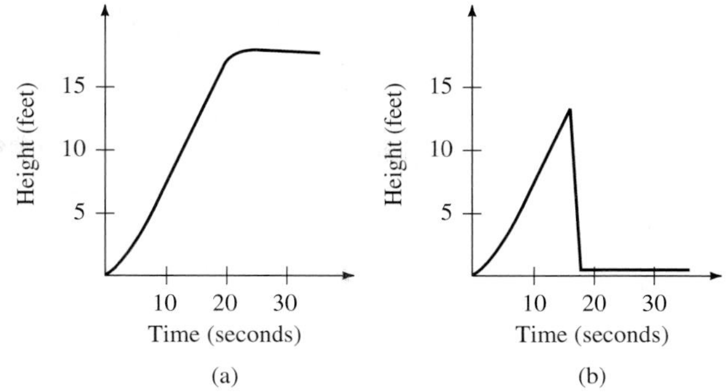

Figure 1–13

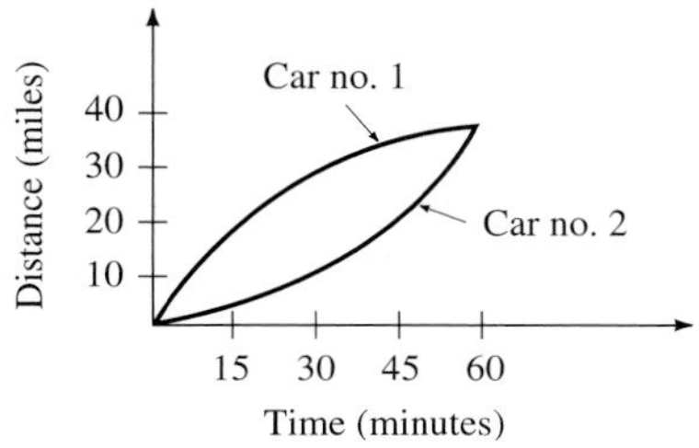

Figure 1–14

Figure 1–15

18. (This exercise was adapted from a group of problems developed by Steve Monk, of the University of Washington.)

(Problem Solving) The graph in Figure 1-14 shows the relationship between time and the distance traveled for each of two cars. Which car has gone farther after an hour? Which is going faster after an hour?

19. *(Problem Solving)* Suppose that the graph in Figure 1-14 shows the relationship between time and the *speed* of the two cars. Which car has gone farther after an hour? Which is going faster after an hour?

20. *(Problem Solving)* Imagine two glasses as in Figure 1-15, one containing water and the other containing wine. Transfer a teaspoon of wine to the water glass; then transfer a teaspoon of the mixture back to the wine glass. Is there now more water in the wine or more wine in the water?

21. *(Extension)* Find the book *But the Crackling Is Superb*, Nicholas and Giana Kurti, eds. (Adam Hilger, Bristol, 1989), and read the article, "Wine and Water," by H. B. G. Casamir. The conclusion of the article differs from the one you obtained in Exercise 20. Ask a chemist or physicist to explain why.

22. *(Extension)* Think of three problems, mathematical or otherwise, you have encountered in your "real life" (that is, outside of mathematics classrooms). Identify at least three Pólya strategies you used in solving these problems, and describe how they were used.

CHAPTER REVIEW

Complete Exercises 1–15 *(Writing to Learn) before referring to the indicated pages.*

WORDS AND PHRASES

In Exercises 1–3, *explain the meaning of the words or phrases in your own words.*

1. (page 7) **quantitative relationship**

2. (pages 7, 8) **numerical**, **analytical**, and **graphical** models of relationships

3. (page 7) **independent variable**, **dependent variable**

IDEAS

4. (pages 7, 8) When you look at a table, graph, or equation, how can you tell which variable is the independent variable?

5. (pages 7, 8) Describe the process for obtaining output values from input values in a table, in a graph, and in an equation.

6. (pages 8, 9) Discuss the nature of the information provided by a table, a graph, and an equation. In particular discuss whether the information is complete or incomplete, exact or approximate, and explicit or hidden.

7. (pages 10, 11) Describe the processes of creating a graph or equation from a table. Make up a table and use it to illustrate the processes.

8. (page 12) Describe the processes of creating a table or graph from an equation. Make up an equation and use it to illustrate the processes.

9. (pages 14, 15) Describe the process of creating a table from a graph. Make up a graph and use it to illustrate the process. Why do you think it is usually difficult to create an equation from a graph?

10. (page 9) If a relationship is represented by a table, an equation, or a graph, why might we want to represent it by a different model?

11. (page 22) Why is it useful to develop a systematic approach to problem solving?

12. (pages 23–25) What are the four steps in Pólya's approach to problem solving? Explain what is involved in each step.

13. (page 20) What questions can you ask yourself to make sure you understand a problem? How does each question check your understanding?

14. (pages 21–26) Name several frequently used strategies for solving problems, and describe how you have used some of them in solving problems in this chapter.

15. (page 24) What questions can you ask yourself to look back at your solution of a problem? How is each question helpful?

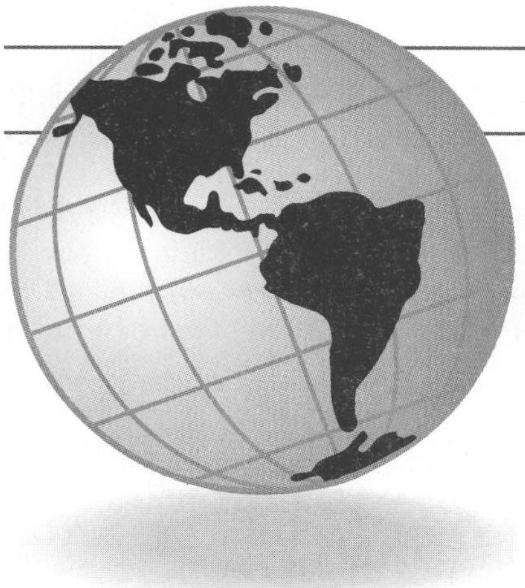

CHAPTER 2

FUNCTIONS

2-1 THREE VIEWS OF FUNCTIONS

PREREQUISITES MAKE SURE YOU ARE FAMILIAR WITH:
The Laws of Exponents (Section A-11)
Factoring (Section A-12)
Quadratic Equations (Section A-13)

THE DEFINITION OF FUNCTION

In Chapter 1 you saw many examples of relationships in which it is natural to think of one variable as the independent variable (input) and the other as the dependent variable (output). For example, in ***Swimming Pool*** (page 11), the independent variable was the pool's diameter, and the dependent variable was its capacity in gallons.

In all the relationships in Chapter 1, each input value generates a unique output value. In this chapter you will learn how to tell whether a table, a graph, or an equation represents a relationship with uniqueness of output. You will discover why such relationships are both convenient and useful, and you will explore them in greater depth. Let's begin by noticing how uniqueness of output is exhibited numerically, analytically, and graphically.

- In the numerical model for ***Pulleys*** (page 9), each entry in the first column of John's table (input) generates a unique entry in the second column (output).
- In the analytical model for ***Swimming Pool***, each value of d in the equation $G = 23.48d^2$ (input) generates a unique value of G (output).
- In the graphical model for ***Greenhouse Effect*** (page 13), each horizontal coordinate on the graph (input) generates a unique vertical coordinate (output).

We admit that we chose these situations carefully, so that they would clearly illustrate input–output behavior. However, even in the messier quantitative relationships we encounter in our lives, one variable often determines the other. This behavior occurs so frequently that mathematicians have chosen a word to describe it. The word is **function**. Informally, a function is a relationship with uniqueness of output.

Function is an appropriate word, because one of its everyday definitions is, "something closely related to another thing, and dependent upon it for its existence, value, or significance: *growth is a function of nutrition*." (*The American Heritage Dictionary*). One way to state its mathematical definition is as follows.

> If x and y are so related that each value assumed by x determines a unique value for y, then y is a function of x.

We will spend the next several pages answering two questions.

- If a relationship between x and y is modeled by a table, equation, or graph, how can you tell whether y is a function of x?
- Why is it useful to know whether y is a function of x?

Let's deal with the first question first, in the context of a physical situation.

A MATHEMATICAL LOOKING GLASS
Temperature Scales

On a July afternoon, radio station WCSX in Detroit reports that the local temperature is 86°. Ten miles away in Windsor, Ontario, station CKLW reports that the temperature is 20°. Both temperatures are correct. The apparent difference occurs because meteorologists in the United States use the **Fahrenheit scale** in reporting temperatures, while those in Canada use the **Celsius scale**.

The Fahrenheit scale was invented by Gabriel Fahrenheit (1685–1735) to provide a standard for measuring and recording the temperatures he encountered in his scientific work. Several conflicting and unconfirmed stories tell how he arrived at his method of assigning numbers to temperatures. One is that on an extremely cold day, when his wife had a fever, he assigned a value of 0° to the outside temperature and 100° to her body temperature.

The Celsius scale was developed somewhat later by Anders Celsius (1701–1744), who wanted to have convenient numbers to refer to the two most important temperatures in most scientific work. These are the freezing and boiling points of water under certain standard atmospheric conditions, approximately 32° and 212° Fahrenheit. They are assigned values of 0° and 100° on the Celsius scale. (Celsius actually assigned them values of 100° and 0°, in that order, but other scientists of his day insisted that higher numbers should describe hotter temperatures.) ■

EXERCISE

1. *(Writing to Learn)* Since each Fahrenheit temperature F corresponds to a unique Celsius temperature C, we know that C is a function of F.

a. For this function, which variable is the independent variable?

b. Can we also regard F as a function of C? Explain your answer.

Let's see how a table, graph, or equation can tell us that C is a function of F. You will need to know that the relationship can be described by the equation

$$C = \frac{5}{9}(F - 32)$$

A NUMERICAL VIEW OF FUNCTIONS

How can we decide whether a table represents a function? Your experience with ***Pulleys*** suggests that a table describes a function as long as no two rows have the same first entry. The suggestion is accurate, with two slight modifications. Examples 1 and 2, related to ***Temperature Scales***, illustrate the need for the modifications.

EXAMPLE 1 *A table that describes a function*

TABLE 5

F	C
68	20
77	25
77	25
86	30
86	30
77	25
68	20

Table 5 shows Fahrenheit and Celsius high temperatures recorded at the same place each day for a week. Although several rows in the table have the same first entry, they also have the same second entry, so that C is a function of F. ■

Table 5 is **complete**, that is, it contains *all* data points for the relationship. Example 1 suggests the following way to decide whether a complete table represents a function.

> A complete table represents a function if and only if, whenever the first entries in two or more rows are the same, so are the second.

In general, we cannot decide whether an incomplete table represents a function unless we know the underlying relationship. If there is a function that could generate it, the table is said to **fit** that function.

EXAMPLE 2 *A table that fits a function*

TABLE 6

x	y
32	58
35	62
41	64
45	62

In Table 6 no two rows have the same first entry. However, we cannot see *all possible* tables that might have been generated by the relationship between x and y. Without more information, we cannot decide whether Table 6 represents y as a function of x. To see why, let's look at two situations which might have generated the data.

If x represents Dave's age and y his resting pulse during his annual checkup, then y is a function of x, because no matter how the table continues, he will never be the same age twice.

If x and y represent daily low and high temperatures at Logan International Airport in Boston, then y is almost certainly not a function of x. This is because some day not listed could have a low temperature of 32 and a high temperature other than 58. ■

EXERCISES *In Exercises 2–3,*

a. Assume that the table is complete, and then draw one of the following conclusions.
- y is a function of x.
- y is not a function of x.
- More information is needed to decide whether y is a function of x.

b. Do your conclusions change if the tables are incomplete? What are your new conclusions?

2.

x	y
5	8
9	12
−1	4
2	8
7	0

3.

x	y
8	5
12	9
4	−1
8	2
0	7

4. a. *(Creating Models)* Use the equation $C = \frac{5}{9}(F - 32)$ from ***Temperature Scales*** to complete the following table.

F	*C*
−4	
14	
32	
50	
68	

b. *(Writing to Learn)* Explain why the table fits a function.

c. *(Writing to Learn)* Without knowing what the variables F and C represent, could you conclude from the table alone that F is a function of C? Why or why not?

AN ANALYTICAL VIEW OF FUNCTIONS

How can we decide whether an equation represents a function? Let's begin by observing that some equations provide instructions for calculating unique output values from input values. In ***Temperature Scales***, the equation $C = \frac{5}{9}(F - 32)$ instructs you to subtract 32 from F and multiply the result by $\frac{5}{9}$. Because this process yields a unique value of C for each value of F, C is a function of F.

No procedure always tells you whether an equation in x and y represents y as a function of x. However, one strategy that often works is to solve the equation for y.

EXAMPLE 3 *Solving for y to decide whether y is a function of x*

To decide whether the equation $2y - x^2 = 1$ represents y as a function of x, solve it for y.

$$2y - x^2 = 1$$
$$2y = x^2 + 1$$
$$y = \frac{x^2 + 1}{2}$$

In this form, the equation describes a process for assigning a unique value of y to every value assumed by x. The process is to square x, add one to the result, and divide that result by 2. Since this computation provides only one output for each input, y is a function of x. ■

EXAMPLE 4 *Solving for y to decide whether y is a function of x*

To decide whether the equation $2x - y^2 = 1$ represents y as a function of x, solve for y.

$$\begin{aligned} 2x - y^2 &= 1 \\ -y^2 &= -2x + 1 \\ y^2 &= 2x - 1 \\ y &= \pm\sqrt{2x - 1} \end{aligned}$$

The $\pm$ indicates that the same input may lead to two different outputs. For example, the input $x = 5$ leads to the outputs $y = 3$ and $y = -3$. Thus y is not a function of x. ■

EXERCISES *In Exercises 5–8, solve for y to decide whether y is a function of x.*

5. $4x + 2y = 9$ **6.** $0.2x = 3 - 0.1y$

7. $x^4 + y = 1$ **8.** $x^3 + y^2 = 1$

Many equations, such as $x = y^5 - y + 1$, are impossible to solve for y. Other strategies must therefore be used to decide whether y is a function of x. In Exercise 17, you will apply a graphical strategy to this equation.

By asking whether y is a function of x in the last several examples and exercises, we have assigned the role of independent variable to x. It is equally reasonable, although not traditional, to regard y as the independent variable and ask whether x is a function of y.

EXERCISES

9. As parts (a)–(d) of this exercise, tell whether x is a function of y in each of Exercises 5–8.

10. a. *(Creating Models)* Solve the equation $C = \frac{5}{9}(F - 32)$ in ***Temperature Scales*** for F. Does the equation represent F as a function of C?

b. *(Writing to Learn)* Give a prose description of the process of obtaining output (F) from input (C), similar to that given in Example 3.

A GRAPHICAL VIEW OF FUNCTIONS

How can we decide whether a graph represents a function? Your experience with ***Greenhouse Effect*** suggests a test for deciding. If the variables are x and y and the x-axis is horizontal, the test can be stated as follows.

The Vertical Line Test

A graph represents y as a function of x if and only if no vertical line intersects it more than once.

The vertical line test is illustrated by the following two examples.

EXAMPLE 5 *A graph for which the vertical variable is a function of the horizontal variable*

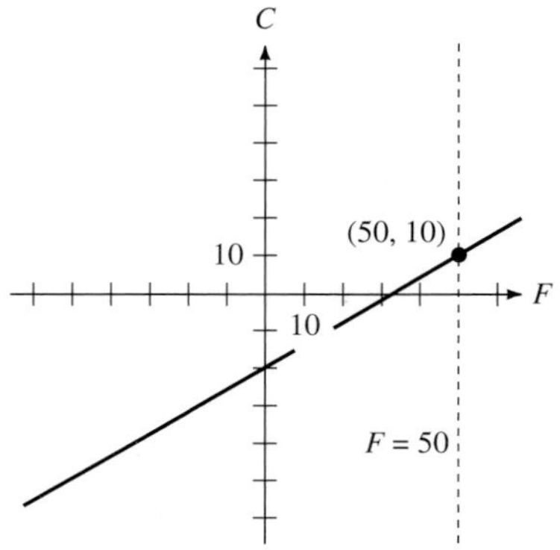

Figure 2–1

Figure 2-1 shows a graph of the equation $C = \frac{5}{9}(F - 32)$ from ***Temperature Scales***. Vertical lines represent values of F. We can determine that $C = 10$ when $F = 50$ because the vertical line $F = 50$ contains only one point on the graph. Similarly, every vertical line contains only one point on the graph, so C is a function of F. ■

EXAMPLE 6 *A graph for which the vertical variable is not a function of the horizontal variable*

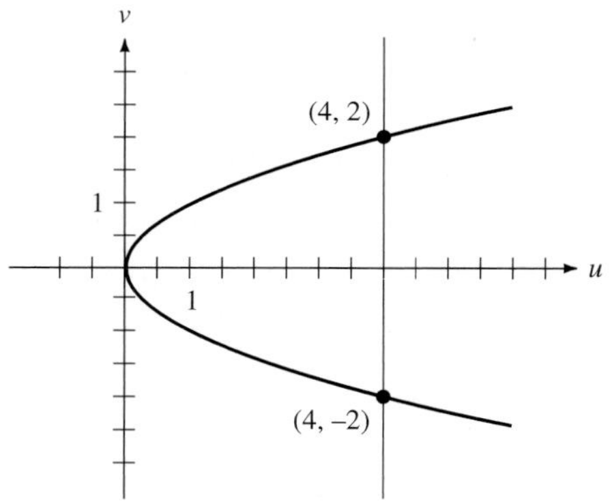

Figure 2–2

Figure 2-2 shows the graph of the equation $u = v^2$. The highlighted vertical line $u = 4$ intersects the graph at both (4, 2) and (4, −2). This tells us that in the equation $u = v^2$, the input (u-value) of 4 leads to both of the outputs (v-values) 2 and −2. Because the graph does not pass the vertical line test, v is not a function of u. ■

Exercises *In Exercises* 11–14, *decide whether y is a function of x.*

11.

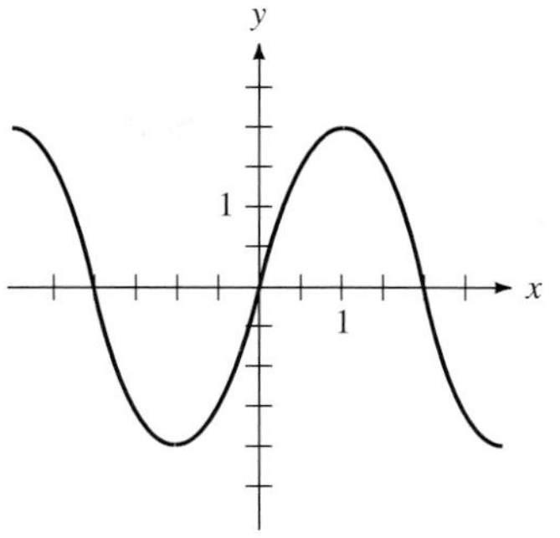

12.

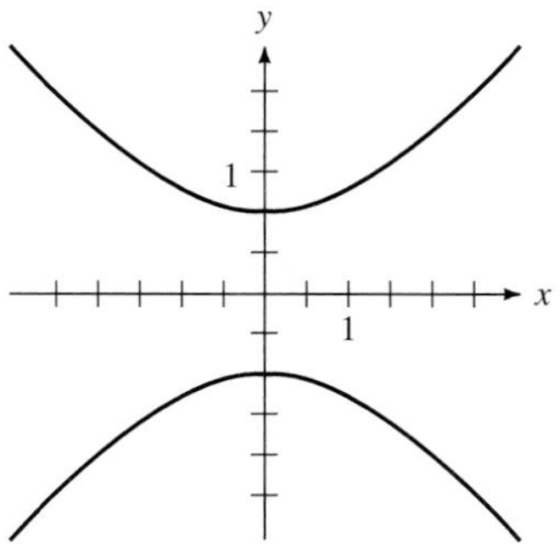

13.

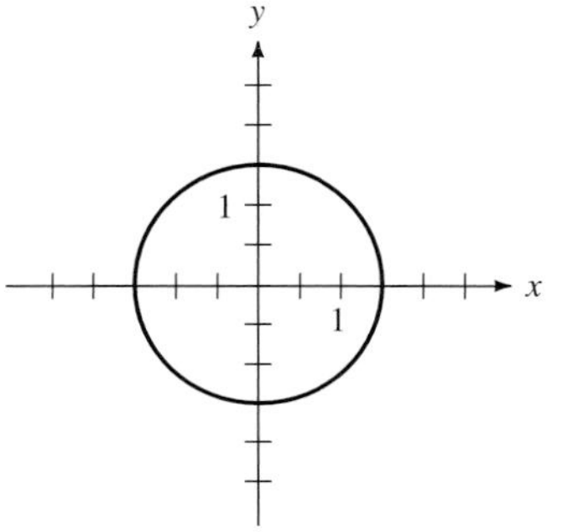

14.

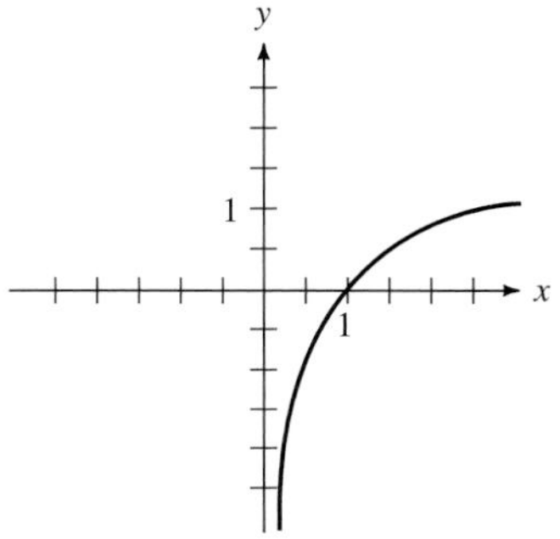

15. *(Creating Models)* Your equation from Exercise 10a should be equivalent to $F = \frac{9}{5}C + 32$. Sketch the graph of this equation, making the C-axis horizontal. Does the graph show that F is a function of C? Does your answer support your result from Exercise 10?

16. *(Making Observations)* For each of the tables in Exercises 2 and 3, plot the data points on graph paper. In each case, if possible, connect them with a smooth curve which passes the vertical line test. If that is impossible, what can you conclude about the data?

17. *(Writing to Learn)* Figure 2-3(a) shows the graphs of $x = y^5 - y + 1$ (discussed earlier in this section) and Figure 2-3b shows the graph of $x = y^5 + y + 1$. Which equation represents y as a function of x? How do you know?

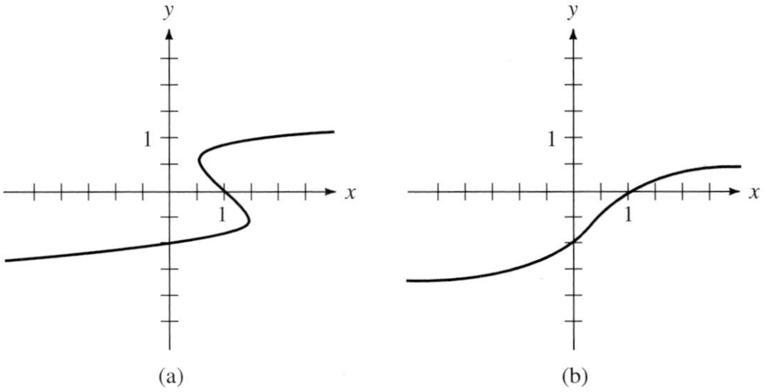

Figure 2–3

REASONS TO STUDY FUNCTIONS AND THEIR MODELS

Now that you have discovered how to tell whether y is a function of x, let's turn to the question of what good it does to know. Functions are useful for two primary reasons.

- Functions reflect a great deal of the physical world. This fact has been recognized at least since the seventeenth century, when mathematicians such as Sir Isaac Newton (1642–1727) were attempting to discover the laws governing gravity and optics. The mathematics of calculus, based heavily on the concept of function, played a fundamental role in developing these laws. For example, the effect of gravity on a falling object is shown by the function $s = 16t^2$, where s is the number of feet the object has fallen after t seconds.
- Functions have uniqueness of output. In fact, you have seen that an equation of a function provides step-by-step instructions for operating on a given input to produce a unique output. Since calculators and computers are designed precisely to execute such step-by-step instructions, they are efficient function-evaluating machines. Thus the more we use technology to perform our calculations, the more we need both an awareness of functions in the physical world and an ability to construct mathematical models for them.

There are also several important reasons to study the connections among analytical, numerical, and graphical models of functions. In particular:

- The connections among the three types of models allow you to change from one to another. For example, as you will begin to see in Chapter 3, the equation of a function can provide information about its graph. This information can be used to check the accuracy of calculator-generated graphs and to solve problems graphically.
- The model of a function presented to you may not be the most convenient one for solving your problem. For example, in ***Swimming Pool*** (page 11), the relationship between the diameter and capacity of a swimming pool was presented as an equation, but it was more convenient for the salespeople to develop a table.

ADDITIONAL EXERCISES *In Exercises* 18–21,

a. Assume that the table is complete, and draw one of the following conclusions.

- y is a function of x.
- y is not a function of x.
- More information is needed to decide whether y is a function of x.

b. Do your conclusions change if the table is incomplete? What are your new conclusions?

18.

x	y
5.3	2.15
0.009	1.645
1.645	0.009
−2.1	7.0

19.

x	y
0	0
−17	−17
2π	2π
$\sqrt{3}$	$\sqrt{3}$

20.

x	y
42	2
42	4
42	6
42	8
42	10

21.

x	y
2	42
4	42
6	42
8	42
10	42

In Exercises 22–27 solve for y, if possible, to decide whether y is a function of x. Identify the equations for which you cannot solve for y.

22. $2x - 3y = 11$

23. $y - 2x^5 + 3x = 0$

24. $x^2 + y^2 = 25$

25. $x = y^7 + 4y + 17$

26. $y = 6$ (*Hint*: The equation is true when y is 6, regardless of the value of x.)

27. $x = 6$

28. As parts (a)–(f) of this exercise, solve each of the equations in Exercises 22–27 for x, if possible, to decide whether x is a function of y.

In Exercises 29–32, decide whether y is a function of x.

29.

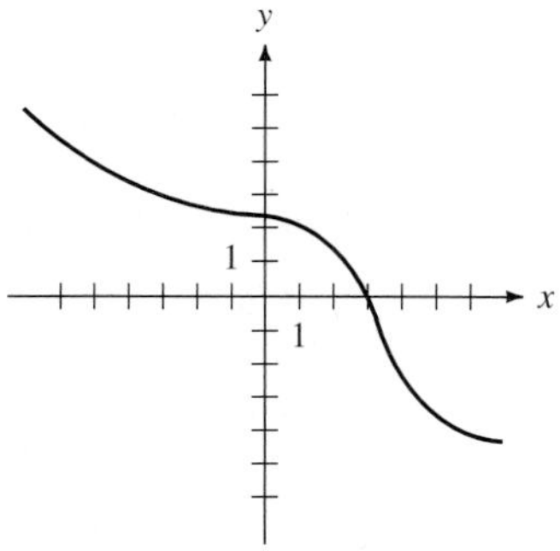

30.

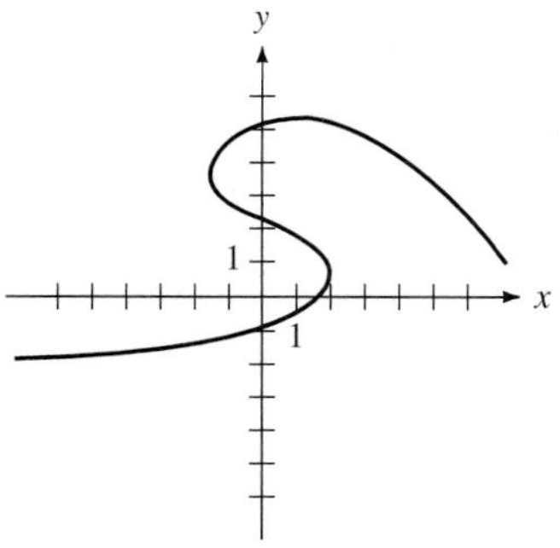

31.

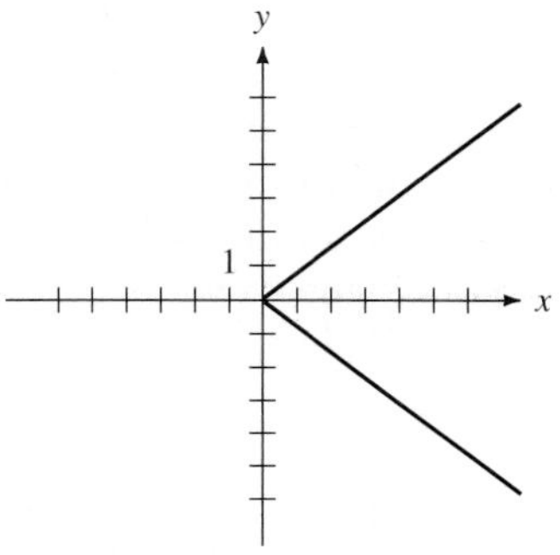

32.

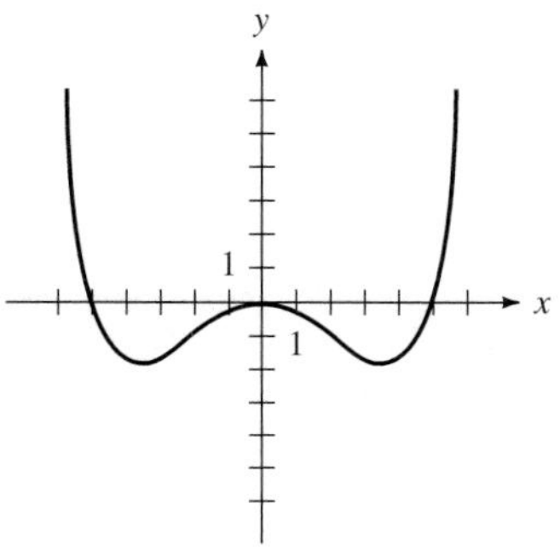

33. The following table lists the height and weight of each person in one of Lynn's college algebra classes.

h = height, in inches	w = weight, in pounds
68	150
70	155
67	130
71	155
64	105
70	145

a. Is w a function of h? Is h a function of w?

b. *(Interpreting Mathematics)* During the semester, one student withdrew from the class. Afterward, w was a function of h and vice versa. Which student withdrew?

34. *(Writing to Learn)* For each relationship, decide whether the second variable is a function of the first, and write a few sentences to justify your answer.

a. G = number of gallons of paint required to paint a house
C = cost of painting the house

b. C = cost of painting a house
G = number of gallons of paint required to paint the house

c. n = the serial number on a Texas Instruments TI-85 graphing calculator
p = the selling price of the calculator

d. p = the selling price of a Texas Instruments TI-85 graphing calculator
n = the serial number of the calculator

e. W = number of gallons of water used by members of your household since 6:00 this morning
t = time today, between 6:00 A.M. and 10:00 P.M.

f. t = time today, between 6:00 A.M. and 10:00 P.M.
W = number of gallons of water used by members of your household since 6:00 this morning

g. a = your age in years
w = your weight in pounds

35. The Kelvin temperature scale was invented by William Thomson (Lord Kelvin, 1824–1907) in 1848, following his discovery that there is a "lowest possible" temperature at which all molecular motion stops. This temperature, which is about $-459.7°$ F and about $-273.2°$ C, is called **absolute zero** and is assigned a value of 0 on the Kelvin scale. Kelvin temperatures are related to Fahrenheit and Celsius temperatures by the equations

$$F = \frac{9}{5}K - 459.7, \qquad C = K - 273.2$$

a. *(Writing to Learn)* Do these equations indicate that F and C are functions of K? Explain your answer.

b. *(Creating Models)* Sketch the graph of each equation, making the K-axis horizontal. Explain how your graphs support your answer in part (a).

c. *(Creating Models)* Solve each equation for K. Is K a function of both F and C?

36. In the following graph,

$t =$ time in years since 1970

$M =$ millions of metric tons of fish caught in the Pacific Ocean

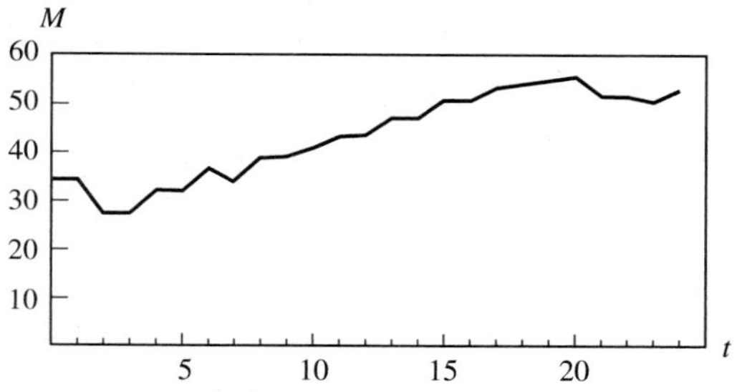

(Adapted from *National Geographic*, November 1995.)

a. *(Writing to Learn)* The graph shows that M is a function of t. What does this statement say in physical terms (that is, in the language of time and fish)?

b. *(Extension)* How does the graph show that t is not a function of M?

37. *(Interpreting Mathematics)* Find a table that models a physical relationship, different from those discussed in this section, for which y is a function of x. Explain how you know that y is a function of x.

38. *(Interpreting Mathematics)* Find a table that models a physical relationship, different from those discussed in this section, for which y is not a function of x. Explain how you know that y is not a function of x.

39. *(Writing to Learn)* Explain why many tables that are generated by physical relationships do not allow you to decide whether one variable is a function of the other.

40. *(Writing to Learn)* Another student has just said to you, "I don't understand why we just have a *vertical* line test to tell whether y is a function of x. Why doesn't it matter if a *horizontal* line crosses a graph more than once?" Write a short paragraph to answer your classmate's question.

(Actually, there *is* a horizontal line test for graphs. In Section 9-3 we will see what information it provides.)

41. *(Writing to Learn)* In this section, you have learned numerical, analytical, and graphical techniques for distinguishing functions from other relationships. Explain how each allows you to decide whether a table, equation, or graph fits the mathematical definition of function.

SUPPLEMENTARY TOPIC
Nonnumerical Functions

The word *function* has a more general mathematical definition than the one given at the beginning of this section. A **function** from a set A of objects to a set B is a correspondence with the property that no object in A corresponds to more than one object in B.

EXAMPLE 7 *Functional and nonfunctional relationships among sets of objects*

a. Define a correspondence between the two sets

$$S = \{\text{Illinois, Indiana, Michigan, Minnesota, New York, Ohio, Pennsylvania, Wisconsin}\}$$

and

$$L = \{\text{Erie, Huron, Michigan, Ontario, Superior}\}$$

by the rule that a state in S corresponds to a lake in L if and only if the state borders the lake. The correspondence is not a function from S to L, since Wisconsin borders two lakes and Michigan borders four. (Extra credit if you can tell which lakes.) It is also not a function from L to S, since Lakes Erie, Michigan, and Superior are all bordered by more than one state.

b. Ludwig von Köchel (1800–1877) devoted much of his life to numbering in chronological order the 626 largely undated musical compositions of Wolfgang Amadeus Mozart (1756–1791). Let M denote the set of all musical pieces composed by Mozart, and let K denote the set of integers from 1 to 626, inclusive. The correspondence between M and K that associates each composition with its **Köchel number** is a function both from M to K and from K to M. This is because no composition has more than one Köchel number, and no number refers to more than one composition. ■

In Exercises 42–43, *decide whether the correspondence is a function from A to B and whether it is a function from B to A.*

42. *(Extension)*

$$A = \{\text{lazy, malevolent}\}$$

$$B = \{\text{sluggish, shiftless, evil, idle, wicked}\}$$

An element of A corresponds to an element of B if the two are synonyms.

43. *(Extension)* A question on an art history quiz asks students to match artists in set A with their works in set B.

$$A = \{\text{Michelangelo, Picasso, Van Gogh, Jefferson}\}$$

$$B = \{\text{Monticello, Guernica, David, Self-portrait}\}$$

44. *(Writing to Learn)* Explain why the question $y = x^2$ defines y as a function of x according to *both* of the definitions given in this section.

2-2 THE CONCEPT OF FUNCTION AS PROCESS

A function can be viewed as a type of relationship between two variables. It can also be viewed as a process that acts on the variables and is distinct from them. For example, we can think of the function in ***Temperature Scales*** (page 30) as the equation

$$C = \frac{5}{9}(F - 32)$$

relating the variables F and C. We can also think of it as the process of subtracting 32 from a value of F and multiplying the result by $\frac{5}{9}$.

Since we assign symbolic names to the variables, it makes sense to give a symbolic name to the process as well. The symbolic naming of the process is referred to as **functional notation**. This section will introduce you to this notation and the rules governing its use.

FUNCTIONAL NOTATION

A MATHEMATICAL LOOKING GLASS
Cricket Chirps

At some time during your life you have undoubtedly found yourself outside on a warm evening, listening to crickets. Have you noticed that they chirp faster on warmer nights? In fact, if a cricket is chirping at the rate of r times per minute, the Fahrenheit temperature F is closely approximated by the equation

$$F = \frac{1}{4}r + 40$$

Functions are typically given names such as f, g, and h. Let's call this function h. Then h represents the process of multiplying the independent variable r by $\frac{1}{4}$ and adding 40 to the result. The value of the dependent variable is written $h(r)$ and is an instruction to apply the process h to the input r. For example,

$$h(\mathbf{100}) = \frac{1}{4}(\mathbf{100}) + 40 = 65$$

More generally,

$$h(\boldsymbol{r}) = \frac{1}{4}\boldsymbol{r} + 40$$

■

To see how functional notation compares with other ways of making statements about relationships, let's make an English statement about cricket chirps and temperatures.

> If a cricket is chirping 100 times per minute, the Fahrenheit temperature is 65°.

Now let's make the same statement using variables r and F.

> If $r = 100$, then $F = 65$.

Finally, let's make the same statement in functional notation.

> $h(100) = 65$

The functional statement is the shortest of the three. In general, functional notation allows us to make quantitative statements concisely. The notation has other advantages that you will discover later.

Here are a few important points concerning functional notation and its use.

- The function represents a process, in contrast to the variables, which represent numbers. Thus $h(r)$ does *not* mean a number h multiplied by a number r, but the process h applied to the number r.

- The symbols h and $h(r)$ have different meanings. The symbol h refers to the function, and $h(r)$ refers to the output obtained by applying the function h to the input r. While $h(r)$ has the same meaning as F, h by itself does not.
- Although it is common practice to use the letters f, g, and h to represent functions, we are free to choose any symbol at all to represent either a function or a variable. As authors, it is our responsibility to make clear the meaning of any symbol we use in this book. You have a similar responsibility whenever you communicate a mathematical thought to someone.
- The quantity in parentheses denotes an input to the function. It can be a number or a variable expression representing a number.

EXAMPLE 1 *Using functional notation*

a. If $f(x) = x^2 + 2x$, then

$$f(\mathbf{-5}) = (\mathbf{-5})^2 + 2(\mathbf{-5}) = 15$$

b. If $f(x) = x^2 + 2x$, then

$$\begin{aligned} f(\mathbf{2z - 5}) &= (\mathbf{2z - 5})^2 + 2(\mathbf{2z - 5}) \\ &= 4z^2 - 16z + 15 \end{aligned}$$

The expression $4z^2 - 16z + 15$ represents the result of the process f applied to the number $2z - 5$. ■

EXERCISES *In Exercises* 1–4, *evaluate* $H(7)$, $H(0)$, $H(A)$, *and* $H(A - 1)$.

1. $H(x) = 5x - 4$ **2.** $H(z) = z^2$

3. $H(T) = \dfrac{T}{T + 1}$ **4.** $H(w) = \sqrt{w + 9}$

5. *(Interpreting Mathematics)* If we give the name g to the process of converting Fahrenheit temperatures to Celsius, then

$$g(F) = \frac{5}{9}(F - 32)$$

a. Evaluate $g(41)$. What physical quantity is represented by this output?

b. Verify that $g(-40) = -40$. What physical information is provided by this statement?

c. When the Fahrenheit temperature is 86°, the Celsius temperature is 30°. Make the same statement in functional notation.

COMBINING FUNCTIONS

Adding, Subtracting, Multiplying, and Dividing Functions The outputs of functions can be combined by adding, subtracting, multiplying, or dividing them. If we begin with functions f and g, then we write the combined functions as $f + g$, $f - g$, fg, and $\dfrac{f}{g}$, respectively.

EXAMPLE 2 *Combining functions using addition, subtraction, multiplication, and division*

If $f(x) = 2x - 5$ and $g(x) = x^2 + 4x$, then

$$\begin{aligned}
(f + g)(x) &= (2x - 5) + (x^2 + 4x) \\
&= x^2 + 6x - 5 \\
(f - g)(x) &= (2x - 5) - (x^2 + 4x) \\
&= -x^2 - 2x - 5 \\
(fg)(x) &= (2x - 5)(x^2 + 4x) \\
&= 2x^3 + 3x^2 - 20x \\
\left(\frac{f}{g}\right)(x) &= \frac{2x - 5}{x^2 + 4x}
\end{aligned}$$

■

EXERCISES *In Exercises 6–9, write expressions to represent* $(f + g)(x)$, $(f - g)(x)$, $(fg)(x)$, *and* $\left(\frac{f}{g}\right)(x)$.

6. $f(x) = x + 3$ and $g(x) = x - 3$

7. $f(x) = x^2 - 2$ and $g(x) = 4 - x$

8. $f(x) = \frac{3}{x}$ and $g(x) = \frac{2}{x}$

9. $f(t) = \sqrt{t - 1}$ and $g(t) = \sqrt{t^2 - 1}$

Sums, differences, products, and quotients of functions often occur in physical contexts, such as the study of economics. Here is a situation in which combinations of several functions provide useful information.

A MATHEMATICAL LOOKING GLASS

Fat Cats on Campus

Your economics professor wants your class to gain some experience in how to run a business and has agreed to provide up to \$1000 of his own money to help you get started. He expects that you and your classmates will reimburse him out of your profits and split any remaining profits. (A similar situation actually occurs at Penn State–New Kensington, where Professor Mark DeHainaut gives his classes \$1000 to invest in the stock market.) After doing some marketing research, the class decides to start a T-shirt company. (An account of a real student-run T-shirt company appeared in "Fat Cats on Campus," *USA Weekend*, March 26–28, 1993.)

To run your business profitably, you need to know what it will cost to produce the shirts and how many you can expect to sell for a given price. You have identified the following costs.

Purchase of heat press	\$612.50
White T-shirts in bulk	2.50 each
Transfers to press onto each shirt	1.00 each

You can now express the cost as a function of the number of shirts you produce. The cost of producing x shirts is your **cost function** $C(x)$. In Exercise 10 you can verify that your cost function is $C(x) = 3.50x + 612.50$.

You have also surveyed students on your campus and operators of other T-shirt companies to help you decide on a reasonable level of production and an appropriate price. For simplicity, let's assume that there are no competing companies nearby. However, to sell more shirts you will need to charge a lower price, as indicated in Table 7.

TABLE 7

Target shirt sales for the semester (x)	Price per shirt (p)
500 shirts	\$13.50
900	11.50
1300	9.50
1700	7.50
2100	5.50
2500	3.50

The price for which you can sell x shirts during the semester is called your **demand function** $p(x)$. In Exercise 11 you can verify that the function $p(x) = 16 - 0.005x$ fits each data point in Table 7, so that it is a plausible demand function.

The functions C and p can help you identify the conditions under which you can make a profit. First, you can construct a **revenue function** $R(x)$ to express the number of dollars produced from the sale of x shirts. You can gain insight into this function by *Examining a Special Case*, that is, by looking at a specific value of x. Suppose you aim to sell 1000 shirts during the semester. Then you will need to charge a dollar price of

$$p(1000) = 16 - 0.005(1000) = 11$$

The revenue produced by the sale of 1000 shirts at \$11 each is \$11,000. That is,

$$R(1000) = (1000)(11) = 11{,}000$$

In general, if you aim to sell x shirts, you will need to charge a dollar price of

$$p(x) = 16 - 0.005x$$

The revenue produced by the sale of x shirts at $p(x)$ dollars each is

$$\begin{aligned} R(x) &= xp(x) \\ &= x(16 - 0.005x) \\ &= 16x - 0.005x^2 \end{aligned}$$

In Exercise 12 you can construct a **profit function** $P(x)$ to express your profit, in dollars, from the sale of x T-shirts. ■

EXERCISES

10. *(Writing to Learn)* Explain why your cost function is

$$C(x) = 3.50x + 612.50$$

11. *(Creating Models)* Verify that the demand function $p(x) = 16 - 0.005x$ fits each data point in Table 7.

12. *(Creating Models)*

a. How much does it cost to produce 1000 T-shirts?

b. How much profit will you make if you sell 1000 T-shirts?

c. Explain why the profit function P is equal to $R - C$.

d. Find an expression for $P(x)$.

You can continue to plan your company's operations in the chapters ahead.

Composing Functions Sometimes you need to use the output from one function as the input to another, so that two functional processes are applied in succession.

A MATHEMATICAL LOOKING GLASS

Canadian Crickets

Thad Rude, a friend of your author Dave, does a lot of camping. He sometimes estimates the temperature by listening to crickets, as in ***Cricket Chirps***. However, Thad lived in Canada for many years, so it is more natural for him to express the *Celsius* temperature as a function of the number of chirps per minute. The required process uses the functions

$$h(r) = \frac{1}{4}r + 40$$

to calculate the Fahrenheit temperature for a rate of r chirps per minute and

$$g(F) = \frac{5}{9}(F - 32)$$

to convert the Fahrenheit temperature F to Celsius. For example, if the cricket chirps 112 times in a minute, Thad evaluates

$$h(\mathbf{112}) = \frac{1}{4}(\mathbf{112}) + 40 = 68$$

to obtain the Fahrenheit temperature of 68°. Then he evaluates

$$g(\mathbf{68}) = \frac{5}{9}(\mathbf{68} - 32) = 20$$

to obtain the Celsius temperature of 20°.

More generally, if the cricket chirps r times in a minute, Thad evaluates $h(r) = \frac{1}{4}r + 40$ and uses that number as an input to g. The resulting output is

$$g\left(\frac{\mathbf{1}}{\mathbf{4}}\boldsymbol{r} + \mathbf{40}\right) = \frac{5}{9}\left[\left(\frac{\mathbf{1}}{\mathbf{4}}\boldsymbol{r} + \mathbf{40}\right) - 32\right]$$

$$= \frac{5}{36}r + \frac{40}{9}$$

Thus the function that describes the Celsius temperature C as a function of r is described by the equation

$$C = \frac{5}{36}r + \frac{40}{9}$$

The function in ***Canadian Crickets*** is called the **composition** of g with h and is written $g \circ h$. It represents the process g applied to the input $h(r)$, so

$$(g \circ h)(r) = g[h(r)]$$

Figure 2-4 illustrates the relationship of $g \circ h$ to g and h for an input of 112 chirps per minute and for an arbitrary input.

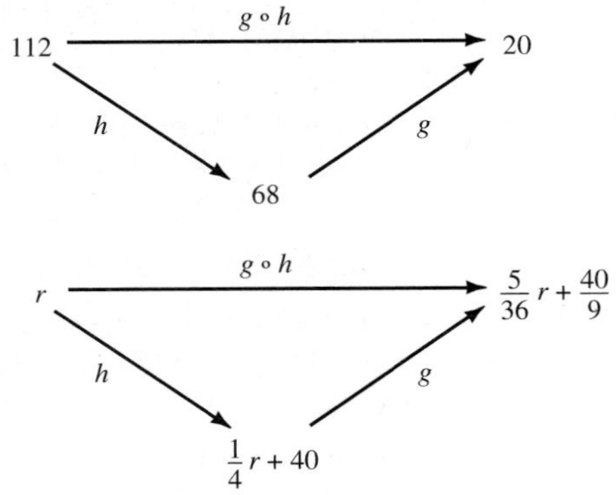

Figure 2–4

FYI When you are composing functions, order matters! For example, in ***Cricket Chirps***, the output from the function $h \circ g$ is

$$\begin{aligned}(h \circ g)(F) &= h[g(F)] \\ &= h\left[\frac{5}{9}(F - 32)\right] \\ &= \frac{1}{4}\left[\frac{5}{9}(F - 32)\right] + 40 \\ &= \frac{5}{36}F + \frac{320}{9}\end{aligned}$$

which is not the same as the output from $g \circ h$. Also, $h \circ g$ makes no sense in the context of the problem, because the output from g is a Fahrenheit temperature and the input to h must be a frequency of chirps.

Example 3 shows how to find an analytical expression to represent the composition of two functions.

EXAMPLE 3 *Finding an expression for a composed function*

Suppose $U(x) = 3x + 100$ and $f(h) = \dfrac{2h}{h - 100}$. To obtain an expression for $(f \circ U)(x)$, begin by rewriting

$$(f \circ U)(x) = f[U(x)] = f(3x + 100)$$

To evaluate $f(3x + 100)$, substitute $3x + 100$ for every h in the expression for $f(h)$.

$$f(3x + 100) = \frac{2(3x + 100)}{(3x + 100) - 100}$$

$$= \frac{6x + 200}{3x}$$

■

EXERCISES **13.** Find an expression for $(U \circ f)(h)$ in Example 2.

In Exercises 14–17, *find expressions for* $(u \circ v)(x)$ *and* $(v \circ u)(x)$.

14. $u(x) = 2x + 1$, $v(x) = 5x$

15. $u(x) = 8 - 4x$, $v(x) = x^2$

16. $u(x) = 4x - 5$, $v(x) = \frac{x + 5}{4}$

17. $u(x) = x^2 - 3x$, $v(x) = \frac{1}{x}$

EXAMPLE 4 *Representing a single function as a composition*

Every function is a sequence of operations performed on a variable, and each operation is itself a function. For example, to evaluate $f(x) = 5x^2 + 3$, we apply first the squaring function, then the multiplying-by-five function, and finally the adding-three function. The sequence of operations can be represented schematically as follows.

$$x \to x^2 \to 5x^2 \to 5x^2 + 3$$

■

EXERCISES *In Exercises* 18–21, *draw a schematic diagram as in Example* 4 *to represent the sequence of functions applied to x in the given order. Then write an equation in functional notation to represent the composed function f.*

18. the multiplying-by-five function, the adding-three function, and the squaring function

19. the squaring function, the adding-three function, and the multiplying-by-five function

20. the adding-three function, the squaring function, and the multiplying-by-five function

21. the adding-three function, the multiplying-by-five function, and the squaring function

The functions in Exercises 22–25 *are all compositions of the square-root function, the subtracting-two function, and the multiplying-by-six function. In each case, write down the order in which these functions are applied.*

22. $g(x) = 6\sqrt{x - 2}$

23. $h(x) = \sqrt{6x - 2}$

24. $j(x) = \sqrt{6x} - 2$

25. $k(x) = 6\sqrt{x} - 2$

ADDITIONAL EXERCISES **26.** *(Interpreting Mathematics)* If the sides of a square are s centimeters long, its area in square centimeters is

$$A(s) = s^2$$

A square whose sides are 9 cm long has an area of 81 cm^2. Make the same statement using functional notation.

27. *(Interpreting Mathematics)* The number of hours required to drive 100 miles at r miles per hour is

$$T(r) = \frac{100}{r}$$

It takes 2 hours to drive 100 miles at 50 miles per hour. Make the same statement using functional notation.

Shutter speed, in seconds	f-stop
1/125	2
1/60	2.8
1/30	4
1/15	5.6

Adapted from Lefty Kreh, *The L.L. Bean Guide to Outdoor Photography*, Random House, 1988.

28. *(Interpreting Mathematics)* In photography, the amount of light striking the film depends on two numbers. The first, called **shutter speed**, refers to the amount of time the camera's shutter is opened to admit light. The second, called the **f-stop**, refers to the size of the shutter opening. The higher the f-stop number, the smaller the opening. Although modern cameras are programmed to adjust the shutter speed and f-stop for proper exposure, experienced photographers often prefer to adjust them manually. Under certain conditions, each combination in the table at the left will result in proper exposure.

a. Let $F(s)$ be the f-stop that goes with a shutter speed s. Then $f(1/30) = 4$. What does this say in physical terms?

b. The table does not show the value of $f(1/8)$. Is $f(1/8) > 5.6$ or is $f(1/8) < 5.6$? Explain your answer.

For each function in Exercises 29–38,

a. *(Writing to Learn)* Describe in words the process represented by the function h.

b. Evaluate $h(4)$, $h(0)$, and $h(-0.5)$.

c. Write an expression for $h(3Z)$.

d. If $g(x) = x$, write expressions for $(h + g)(x)$, $(h - g)(x)$, $(hg)(x)$, and $\left(\frac{h}{g}\right)(x)$.

e. If $f(x) = x + 4$, write expressions to represent $(h \circ f)(x)$ and $(f \circ h)(x)$.

29. $h(x) = 20x + 13$

30. $h(y) = y - 0.1$

31. $h(t) = 100t - 4t^2$

32. $h(\&) = \&^3 - 5\&^2 - 6$

33. $h(W) = \dfrac{W}{W + 1}$

34. $h(a) = a + \dfrac{1}{a}$

35. $h(\partial) = \sqrt{2\partial + 10}$

36. $h(p) = \sqrt{4 - p}$

37. $h(r) = r$

38. $h(x) = 6$ (*Hint*: Look at Exercise 26 of Section 2-1.)

39. If $u(P) = (2P - 6)^3 - 5(2P - 6)^2 + 4(2P - 6) - 7$ and $v(Q) = 0.5Q + 3$, write expressions to represent $(v \circ u)(P)$ and $(u \circ v)(Q)$.

In Exercises 40–43, *write an equation for the function that results when the given sequence of functions is applied to x in order.*

40. the dividing-by-two function, the subtracting-five function, and the cubing function

41. the cubing function, the subtracting-five function, and the dividing-by-two function

42. the subtracting-five function, the cubing function, and the dividing-by-two function

43. the subtracting-five function, the dividing-by-two function, and the cubing function

The functions in Exercises 44–47 are all compositions of the multiplying-by-three function, the adding-one function, and the taking-the-reciprocal function. In each case, write down the order in which these functions are applied.

44. $g(x) = 3\left(\frac{1}{x}\right) + 1$ **45.** $h(x) = \frac{1}{3x} + 1$

46. $j(x) = 3\left(\frac{1}{x + 1}\right)$ **47.** $k(x) = \frac{1}{3x + 1}$

In Exercises 48–49, decide which of the given compositions make sense in a physical context. For those that do, describe the physical information provided by the output.

48. *(Interpreting Mathematics)*

$N(t)$ = number of sales by a telemarketer in t days

$I(s)$ = income to the telemarketer from s sales

a. $(I \circ N)(t)$ **b.** $(N \circ I)(s)$

49. *(Interpreting Mathematics)*

$p(A)$ = air pressure in pounds per square inch at an altitude of A feet above sea level

$q(D)$ = altitude in feet above sea level at a distance of D miles west of St. Louis

a. $(q \circ p)(A)$ **b.** $(p \circ q)(D)$

Use the table at the left to complete Exercises 50–51.

x	$f(x)$	$g(x)$
1	3	0
2	5	5
3	1	4
4	2	8
5	4	1

50. Evaluate

a. $(f + g)(2)$ **b.** $(fg)(4)$

c. $(f \circ g)(3)$ **d.** $(g \circ f)(3)$

51. *(Writing to Learn)* Explain why each of the following is undefined:

a. $\left(\frac{f}{g}\right)(1)$ **b.** $(f \circ g)(4)$

52. Use the following graph to estimate the value of each expression in parts (a)–(d).

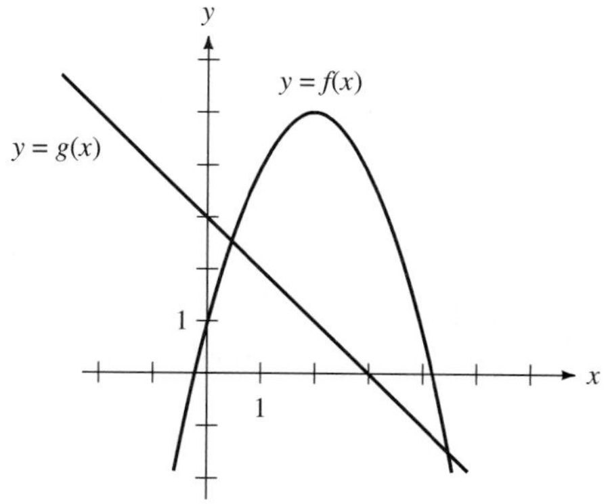

a. $(g - f)(3)$

b. $\left(\frac{f}{g}\right)(2)$

c. $(f \circ g)(1)$

d. $(g \circ f)(1)$

In Exercises 53–56, the functions $u(C) = \frac{9C}{5} + 32$ *and* $v(C) = C + 273.2$ *express Fahrenheit and Kelvin temperatures, respectively, as functions of Celsius temperature C.*

53. *(Problem Solving)* The thermometer on Dave's back porch registers Celsius temperatures, but always reads 5° too low. Suppose his thermometer reading is T, and he wants to determine the actual Fahrenheit temperature. Since the actual Celsius temperature is $T + 5$, the actual Fahrenheit temperature is $u(T + 5)$.

a. Write an expression in T to represent $u(T + 5)$.

b. Use this expression to find the Fahrenheit temperature when his thermometer reads 23°.

c. Express the Kelvin temperature as a function of Dave's thermometer reading T.

d. Use this expression to find the Kelvin temperature when his thermometer reads 23°.

54. *(Making Observations)* When a kettle of water at room temperature is put onto a hot stove, its Celsius temperature C after t seconds depends on several factors, including how much water is in the kettle and how high the burner is turned up. The Celsius temperature might be approximated for the first few minutes by the equation

$$C = c(t) = 22 + 0.5t$$

a. Write an expression in T to represent $(u \circ c)(t)$, and evaluate $(u \circ c)(60)$.

b. Now obtain the answer to part (a) in a different way. Use the function $c(t) = 22 + 0.5t$ to calculate the Celsius temperature of the water after 1 minute, and use the result as an input for the function u.

c. Write an expression in t to represent the Kelvin temperature of the water after t seconds.

d. Evaluate this expression when $t = 60$.

55. *(Writing to Learn)* Explain what physical information is given by the calculations in Exercise 54a.

56. *(Making Observations)* Obtain the results from Exercise 54d in a different way. Use the equation $C = 22 + 0.5t$ to calculate the Celsius temperature of the water after 1 minute, and use the result as an input for the function v defined by $v(C) = C + 273.2$.

57. *(Problem Solving)* A car dealer offers a 15% discount on all new cars on his lot. At the same time, the auto manufacturer offers a $1000 rebate.

a. Let P represent the sticker price of a car. Write a function f to represent its price if only the 15% discount is applied.

b. Write a function g to represent its price if only the $1000 rebate is applied.

c. When both the discount and the rebate are applied, the purchase price of the car is either $(f \circ g)(P)$ or $(g \circ f)(P)$, depending on the order in which they are applied. Which would you ask the dealer to apply first? Which composition represents your choice?

58. A well-known mathematical question states,

> Suppose a string is tied around the earth at the equator, so that it fits snugly at all points. If the length of the string is increased by 10 feet, how far off the surface of the earth could it be held at each point? (See the figure at the left.)

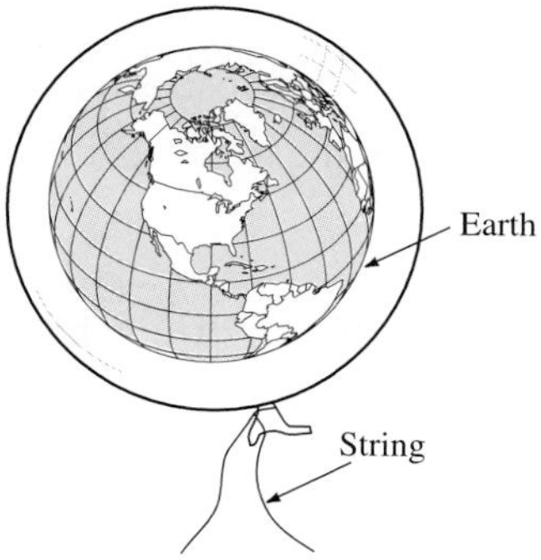

Before reading on, make a guess at the answer.

a. *(Writing to Learn)* The function R defined by $R(c) = \dfrac{c}{2\pi}$ gives the radius of a circle whose circumference is c. If c_0 denotes the circumference of the earth, in feet, explain why the answer to the question is given by $R(c_0 + 10) - R(c_0)$.

b. *(Interpreting Mathematics)* Answer the question by evaluating $R(c_0 + 10) - R(c_0)$. How does the answer compare with what you expected?

59. *(Problem Solving)* Periwinkle's, Lynn's favorite bakery, bakes cakes at a cost of \$1.50 each for ingredients. Baker Chris Kozubal devotes half her time to cakes, so that the cost of producing cakes includes half her weekly salary of \$425. Because there are several bakeries in the area, Periwinkle's cannot set a price for cakes independently, but must sell them for \$10.00 each. Find Periwinkle's profit function for cakes.

2–3 DOMAIN AND RANGE

PREREQUISITES MAKE SURE YOU ARE FAMILIAR WITH:
Interval Notation (Section A-7)
Linear Inequalities (Section A-8)

You have already observed that certain inputs to functions sometimes fail to yield reasonable results. This can happen for one of three reasons.

- The *input* is not *physically reasonable*. In ***Swimming Pool*** (page 11), the equation $G = 23.48d^2$ expresses the pool's capacity G as a function of its diameter d. Since diameters must be positive, an input of -10 is not physically reasonable.
- The *output* is not *physically reasonable*. The equation $S = 340M$ in ***Pulleys*** (page 9) expresses the speed of John's fan as a function of the pulley diameter. Since the fan has a maximum speed of 3600 rpm, an output of 3700 is not physically reasonable.
- The *output* is not *mathematically reasonable*. If $f(x) = \dfrac{1}{x - 1}$ and we attempt to calculate $f(1)$, we must divide by zero. There is no reasonable output, regardless of whether the function occurs in a physical context.

We will use the word *reasonable* to mean "mathematically reasonable" for an abstract function and "physically reasonable" in a physical context. The collection of all reasonable values for the independent variable is called the **domain** of a function. The collection of all reasonable values of the dependent variable is called its **range**.

You can avoid much mathematical and physical nonsense by identifying the domain and range of any function you work with. The domain and range depend on whether you consider the function in the abstract, or as a model of a physical situation.

DOMAIN AND RANGE OF ABSTRACT FUNCTIONS

The domain and range of a function may be found by numerical, graphical, or analytical methods.

TABLE 8

x	y
1	5
2	4
3	3
4	5
5	6
6	8

Numerical Methods When a function is represented by a table, we cannot find its domain and range without additional information. However, the entries of a *complete* table provide the domain and range. For example, if Table 8 is complete, it represents a function with domain $\{1, 2, 3, 4, 5, 6\}$ and range $\{3, 4, 5, 6, 8\}$.

Graphical Methods Example 1 shows how to estimate the domain and range of a function from its graph.

EXAMPLE 1 *Finding the domain and range of a function graphically*

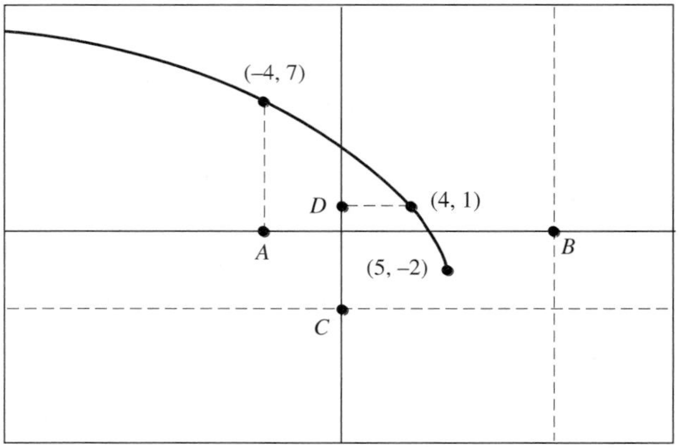

Figure 2–5

Figure 2-5 shows the graph of a function $y = f(x)$. To find the domain of f, let's take an imaginary walk in the positive direction along the x-axis. As we walk, we look straight up and down (directly above and below our location on the x-axis) for points on the graph. From A, at $x = -4$, we see the point $(-4, 7)$ on the graph, so -4 is in the domain of f. On the other hand, no point on the graph is visible from B, at $x = 10$, so 10 is not in the domain of f. Since the graph is visible from any point in $(-\infty, 5]$, the domain of f is $(-\infty, 5]$.

Similarly, we can find the range of f by climbing up the y-axis. If we look directly to our left and right as we climb, the graph is not visible from C, at

$y = -4$, so -4 is not in the range of f. On the other hand, the point (4, 1) is visible from D, at $y = 1$, so 1 is in the range of f. Since the graph is visible from any point in $[-2, \infty)$, the range of f is $[-2, \infty)$. ■

FYI We cannot read graphs of functions with perfect accuracy, so to "find" the domain and range graphically means to *estimate* them. Furthermore, we can't estimate without knowing what the entire graph looks like. We often need to make reasonable guesses based on what we can see. For example, it is reasonable to guess that the graph in Figure 2-5 extends infinitely far up and to the left.

EXERCISES *In Exercises 1–4, find the domain and range of the function. Make reasonable guesses about the portion of the graph not shown.*

1.

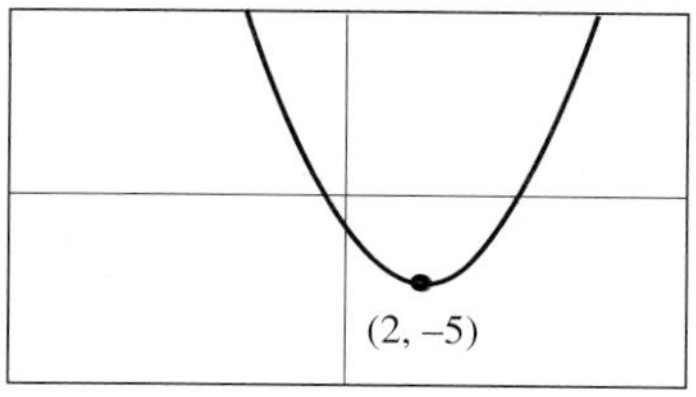

2.

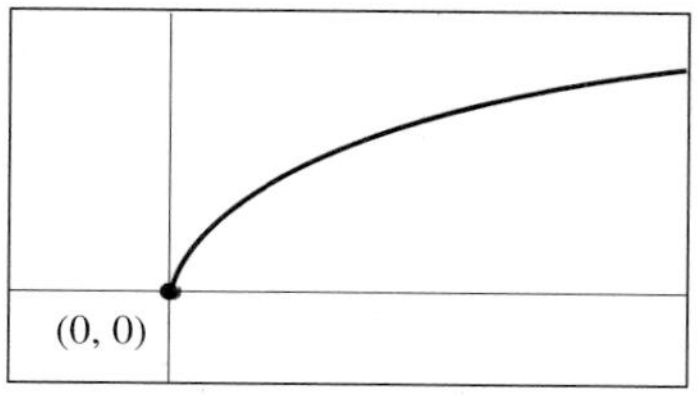

3.

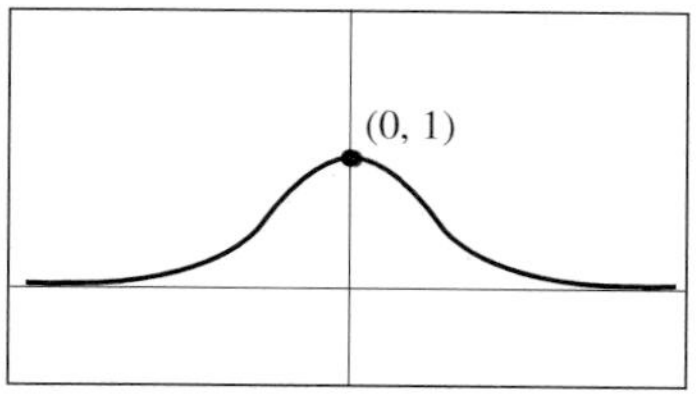

4.

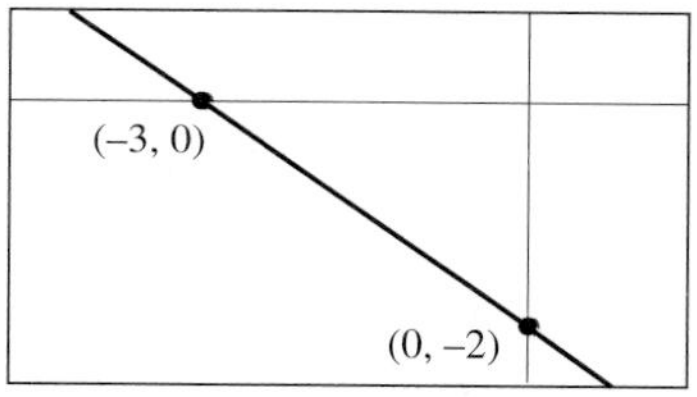

Analytical Methods The functions in Chapters 1–8 of this book use only the operations of addition, subtraction, multiplication, division, and taking powers and roots. Such functions are called **algebraic functions**. These operations produce undefined results in two instances:

- Dividing by zero
- Taking an even-indexed root of a negative number

The only mathematical restrictions on the domain and range of algebraic functions result from those two operations.

EXAMPLE 2 *Finding the domain and range of a function analytically*

Suppose that $g(x) = \dfrac{x}{2x + 12{,}800}$.

To find the domain, notice that $g(x)$ is undefined when the denominator is zero. This will happen if and only if

$$\begin{aligned} 2x + 12{,}800 &= 0 \\ 2x &= -12{,}800 \\ x &= -6400 \end{aligned}$$

so the domain of g is $(-\infty, -6400) \cup (-6400, \infty)$. To find the range, begin by writing the equation as

$$y = \frac{x}{2x + 12{,}800}$$

The range of g consists of all possible values for y. We can more easily see what these values are if we solve the equation for x.

$$\begin{aligned}(2x + 12{,}800)y &= x \\ 2xy + 12{,}800y &= x \\ 2xy - x &= -12{,}800y \\ x(2y - 1) &= -12{,}800y \\ x &= \frac{-12{,}800y}{2y - 1}\end{aligned}$$

The expression in y is undefined when the denominator is zero. This will happen if and only if

$$\begin{aligned}2y - 1 &= 0 \\ 2y &= 1 \\ y &= \frac{1}{2}\end{aligned}$$

so the range of g is $\left(-\infty, \frac{1}{2}\right) \cup \left(\frac{1}{2}, \infty\right)$. ■

It is usually easy to find the domain of an algebraic function analytically. You can find the range analytically if you can solve for the independent variable, as in Example 2. Otherwise you will usually need to use another method, such as graphing the function.

EXERCISES *In Exercises 5–8, find the domain and range of each function analytically.*

5. $y = 3x + 24$

6. $y = \dfrac{1}{3x + 24}$

7. $Q(x) = \sqrt{3x + 24}$

8. $y = \dfrac{1}{\sqrt{3x + 24}}$

In Exercises 9–10,

a. Graph the function.

b. Make a table of values for $f(x)$, using x-values of 0, 1, 2, 3, 4, and 5. If f is undefined at a particular x-value in the table, say so.

c. *(Writing to Learn)* Find the domain of the function, and tell where the restrictions on the domain appear in the equation, in the graph, and in the table.

9. $f(x) = \dfrac{x}{x - 2}$

10. $f(x) = \sqrt{3 - x}$

So far, we have looked at restrictions on domains resulting from division by zero or taking even roots of negative numbers. The domain of a function can also be restricted by an explicit statement, as in Example 3.

EXAMPLE 3 *A function whose domain is explicitly restricted*

Suppose $h(x) = x^2 - 1$ for x in $[-2, 3)$. Although the expression $x^2 - 1$ is defined for all values of x, the statement indicates that the domain of h is $[-2, 3)$.

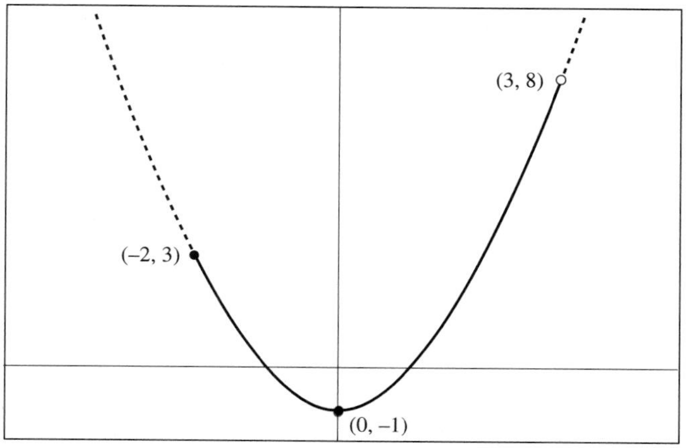

Figure 2–6

Figure 2-6 shows the graph of $y = x^2 - 1$. The solid portion is the graph of $y = h(x)$. A climb up the y-axis shows that the range is $[-1, 8)$. ■

EXERCISES *The domain of each function in Exercises* 11–14 *is explicitly restricted. Find its range graphically.*

11. $V(x) = x + 4$ for x in $(-3, 5)$

12. $\Omega(x) = x^2 - 4$ for x in $[-2, 2]$

13. $y = 3x - x^3$ for x in $(0, \infty)$

14. $y = \dfrac{1}{1 + x^2}$ for x in $[1, \infty)$

DOMAIN AND RANGE OF FUNCTIONS IN A PHYSICAL CONTEXT

Now let's explore how the domain and range of a function can be restricted in a physical context.

A MATHEMATICAL LOOKING GLASS
Big Blue Marble

How much of the earth's surface can be seen from an airplane in flight? From a satellite? From the moon? With the use of trigonometry, a formula can be developed to provide the answers. (The formula is not quite accurate because it assumes that the earth is a perfect sphere with no mountains or valleys.) At a distance of x kilometers from the surface, the fraction that can be seen is

$$g(x) = \frac{x}{2x + 12{,}800}$$

For example,

$$g(1600) = \frac{1600}{2(1600) + 12{,}800} = 0.1$$

so a satellite camera at an altitude of 1600 km can see one-tenth of the surface.

In Example 2 you saw that the domain of g in the abstract is $(-\infty, -6400) \cup (-6400, \infty)$. However, in a physical context the independent variable represents distance from the earth's surface, which cannot be negative. Thus the domain of g is $[0, \infty)$. We do not place an upper limit on the domain because, although no one has observed the earth from as far as 1,000,000,000 km, it is theoretically possible to do so.

In Exercises 15–17 you can find the range of g in context. ■

EXERCISES

15. *(Interpreting Mathematics)* What fraction of the earth's surface can be seen from each distance?

a. 0.001 km (eye level of a child)

b. 10 km (cruising altitude of an airplane)

c. 386,000 km (distance to the moon)

d. 4×10^{13} km (distance to the nearest star)

16. *(Writing to Learn)*

a. According to the formula, $g(0) = 0$. Does this make sense in context? Explain.

b. Is there any height from which you can see half the surface? How do your results from Exercise 15 support your conclusion?

c. What is the range of g in context? How do your results in parts (a) and (b) support your conclusion?

17. *(Making Observations)* Use the graph of $y = g(x)$, shown here, to estimate the range of g in context. Does your result agree with the one you obtained in Exercise 16c?

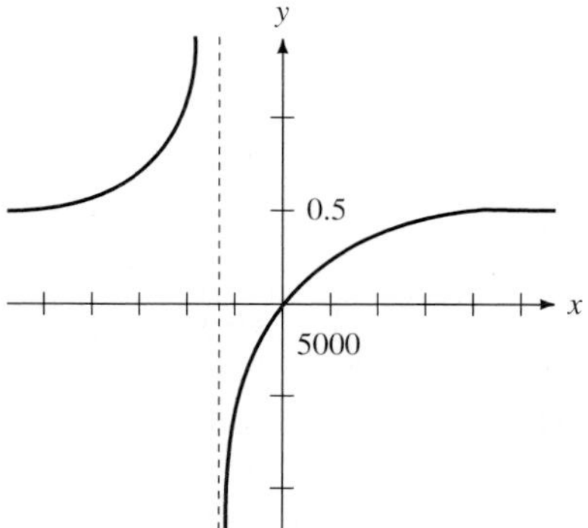

SEQUENCES

Sometimes the independent variable in a function represents a quantity that is counted rather than measured. Such a function is called a **sequence**. Specifically, a sequence is a function whose domain is a set $\{K, K+1, K+2, \ldots\}$ for some integer K. In this section you may assume that $K = 1$, so that the domain is the set of all positive integers. The following Mathematical Looking Glass describes an especially famous sequence.

A MATHEMATICAL LOOKING GLASS
Fibonacci's Rabbits

Leonardo Fibonacci (c. 1170–1250), also known as Leonardo of Pisa, had a profound influence on the course of mathematical history. At the time of his birth, the world's mathematical expertise was concentrated in northern Africa, the Middle East, and Asia. After being educated in Africa, Fibonacci returned to Italy, where in 1202 he published a free translation of Greek and Arabic mathematics into Latin. His treatise, called *Liber Abaci*, introduced Europeans to Hindu methods of calculating with integers and fractions and extracting square and cube roots. It also helped to popularize the use of modern Arabic numbers in place of Roman numerals. Today, however, Fibonacci is remmbered primarily for the function he encountered while studying the following problem.

A particular breed of rabbit grows and reproduces according to a precise schedule. Each female rabbit becomes an adult and produces a pair of babies (one male and one female) at the age of two months and produces exactly one pair every month thereafter. Suppose you begin with one pair of adult rabbits who produce a pair of babies at the end of the first month. If no rabbits die, what function describes the number of adult female rabbits you will have after a given number of months?

It is likely that Fibonacci first approached this problem by using the strategy of *Making a Table* such as Table 9.

TABLE 9

Month	Newborns	1-Month-olds	Adults
1	1	0	1
2	1	1	1
3	2	1	2
4	3	2	3
5	5	3	5
6	8	5	8

If Table 9 is continued indefinitely, it defines a function called the **Fibonacci sequence**. Its domain is the set of integers $\{1, 2, 3, 4, 5, 6, \ldots\}$ in the left column, and its range is the set of integers $\{1, 2, 3, 5, 8, \ldots\}$ in the right column.

The Fibonacci sequence has many fascinating mathematical properties, some of which you will learn in Exercises 83–84. Its numbers also show up unexpectedly often in the physical world. For example, the leaves of many common plants are arranged in patterns involving Fibonacci numbers.

Mathematician Rudy Rucker cites several examples in his book, *Mind Tools* (Houghton Mifflin, 1987). In a completely different direction, baseball statistician Bill James's book, *The Politics of Glory* (Macmillan, 1994), gives a fascinating account of his formula, "Fibonacci win points," for evaluating career performances of major league pitchers.

By the way, more than 700 years after Fibonacci's death a group of mathematicians formed the Fibonacci Association and began publication of a quarterly journal dedicated to research on the Fibonacci sequence. ■

EXERCISE **18.** *(Making Observations)*

a. Continue Table 9 for months 7–10. What pattern do you observe in the last column?

b. Use your observation in part (a) to list the number of adult female rabbits each month for the first two years.

Sequences are usually given functional names using letters at the beginning of the alphabet. Their independent variables are usually denoted by letters in the middle of the alphabet. Furthermore the values of the dependent variable, called the **terms** of the sequence, are usually written using subscripts such as a_n rather than $a(n)$. This variant of standard functional notation is illustrated in Example 4.

EXAMPLE 4 *Functional notation for sequences*

a. The equation

$$a_k = k^2$$

describes a sequence. For this sequence,

$$a_1 = 1^2 = 1$$
$$a_2 = 2^2 = 4$$
$$a_3 = 3^2 = 9$$
$$a_4 = 4^2 = 16$$

etc.

The same sequence can be described by listing its terms

$$\{1, 4, 9, 16, \ldots\}$$

b. In the Fibonacci sequence, each term beginning with the third is the sum of the preceding two. If the nth term is denoted by c_n, the sequence can be described by the equations

$$c_1 = 1$$
$$c_2 = 1$$
$$c_n = c_{n-1} + c_{n-2}, \quad \text{for } n \geq 3$$

The Fibonacci sequence can also be described by listing its terms

$$\{1, 1, 2, 3, 5, 8, \ldots\}$$

■

Example 4a illustrates that some sequences can be defined by displaying an explicit formula for a given term. The formula $a_k = k^2$ allows us to calculate a_k for any value of k. For example, $a_{1000} = 1{,}000{,}000$. By contrast, the formula $c_n = c_{n-1} + c_{n-2}$ given for the Fibonacci sequence in Example 4b is **recursive.** That is, each term beyond the first few is expressed as a function of the preceding terms. This formula does not allow us to calculate c_{1000} without first calculating c_1 through c_{999}.

EXERCISES *In Exercises 19–26, write out the first ten terms of the sequence.*

19. $a_n = 2n$ **20.** $b_k = \dfrac{1}{k}$

21. $c_j = 2 + j - j^2$ **22.** $A_m = \dfrac{m}{m+1}$

23. $a_1 = 1$

$a_k = ka_{k-1}$, for $k \geq 2$

24. $C_1 = 1$

$C_2 = -1$

$C_n = C_{n-1}C_{n-2}$, for $n \geq 3$

25. $b_1 = 1024$

$b_2 = 0$

$b_n = \dfrac{b_{n-1} + b_{n-2}}{2}$, for $n \geq 3$

26. $B_1 = 1$

$B_j = B_{j-1} + j$, for $j \geq 2$

27. *(Making Observations)*

a. Write out the first ten terms of the sequence $a_1 = 3$, $a_n = a_{n-1} + 4$ for $n \geq 2$.

b. Write an explicit formula for a_n, and use it to calculate a_{1000}.

28. *(Making Observations)*

a. Write out the first four terms of the sequence $b_1 = 2$, $b_k = 2b_{k-1}$ for $k \geq 2$.

b. Write an explicit formula for b_k, and use it to calculate b_{20}.

The sequence in Exercise 27 is an **arithmetic sequence**, in which the difference between consecutive terms is constant. The sequence in Exercise 28 is a **geometric sequence**, in which the ratio of consecutive terms is constant. You will study these types of sequences in greater detail in Sections 3-2 and 9-1, respectively.

DYNAMIC BEHAVIOR: INCREASING AND DECREASING FUNCTIONS

Imagine yourself walking along the x-axis through the domain of a function, as you did in Example 1. As you walk, the values of the independent variable increase. The values of the dependent variable may increase over some intervals and decrease over others, as in Figure 2-7.

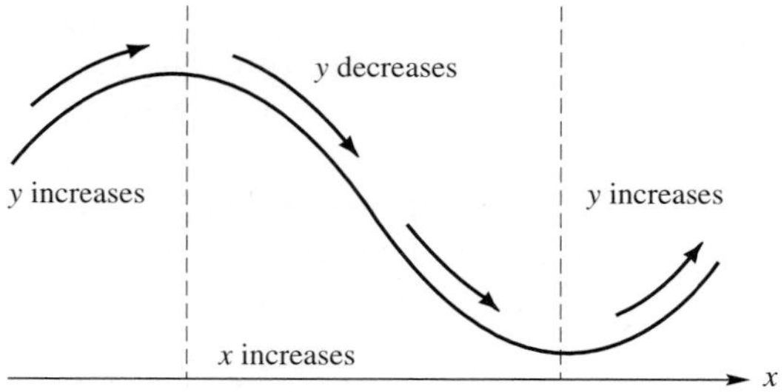

Figure 2–7

If the values of y increase as x increases over an interval, we say that the function is **increasing** over the interval. If the values of y decrease as x increases, the function is **decreasing** over the interval.

EXAMPLE 5 *Identifying intervals where a function is increasing or decreasing*

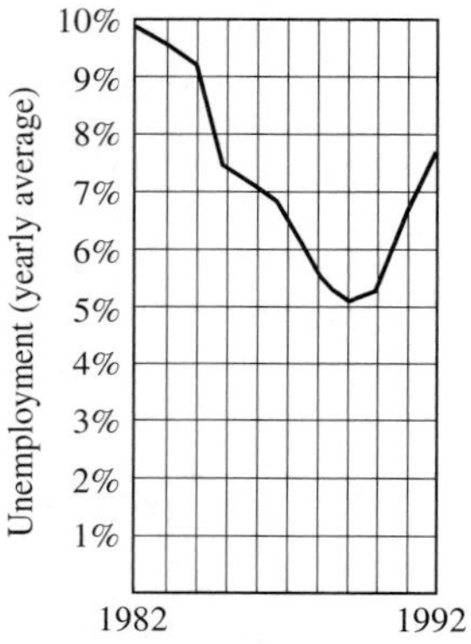

Figure 2–8
Adapted from *Time*, July 13, 1992.

Figure 2-8 shows the rate of unemployment in the United States as a function of time between 1982 and 1992. The graph indicates that the unemployment rate fell from near 10% to just above 5% between 1982 and 1989, then rose to almost 8% by 1992. Therefore the function is decreasing on the interval (1982, 1989), and increasing on (1989, 1992). ■

EXERCISES *In Exercises 29–32, identify the intervals on which the function is increasing and those on which it is decreasing.*

29.

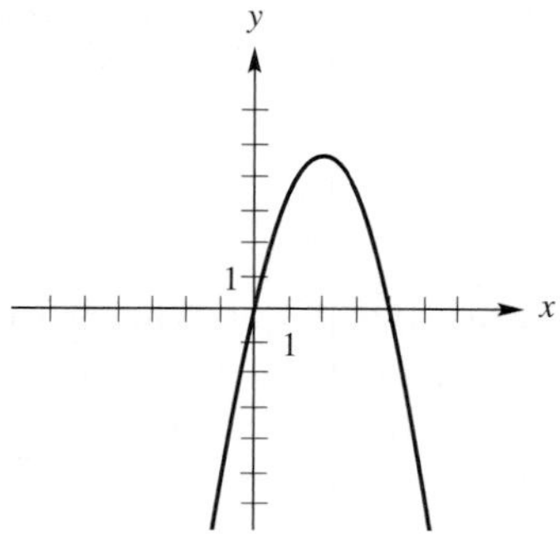

30.

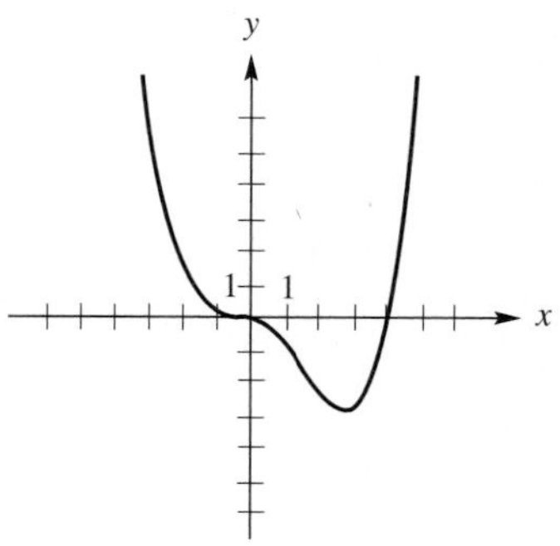

31.

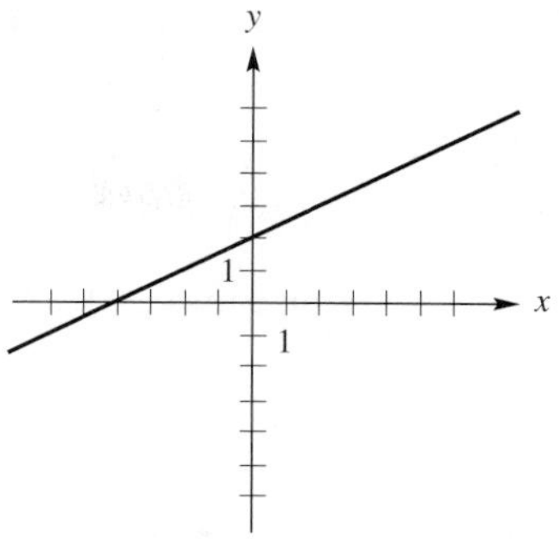

32.

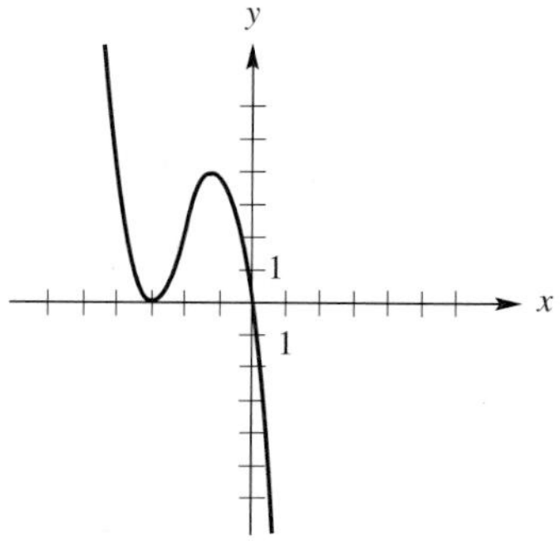

ADDITIONAL EXERCISES

In Exercises 33–36, find the domain and range of the function. You may assume that each table is complete.

33.

x	y
0	3
1	2
2	1
3	0

34.

x	y
−3	9
0	0
3	9
6	36

35.

x	y
−2	11
0	8
2	5
4	2
6	−1

36.

x	y
1	3
10	3
100	3
1,000	3
10,000	3

In Exercises 37–40, find the domain and range of the function. Make reasonable guesses about the graph outside the viewing window.

37.

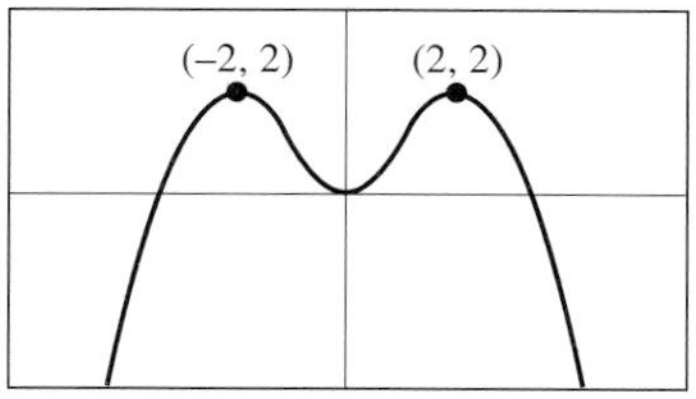

38.

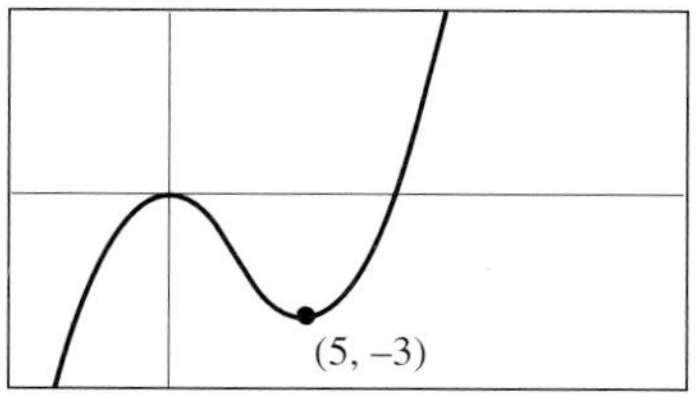

39.

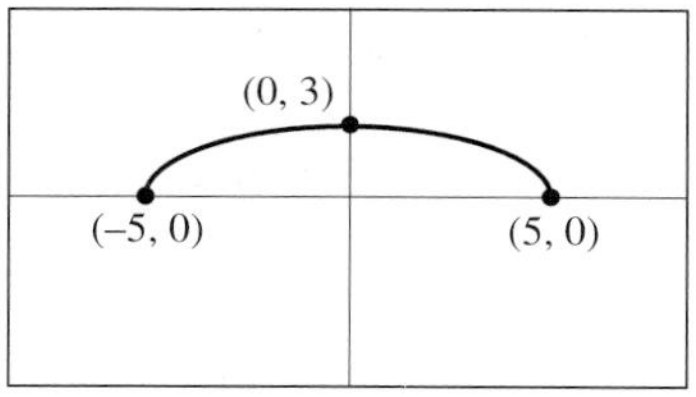

40.

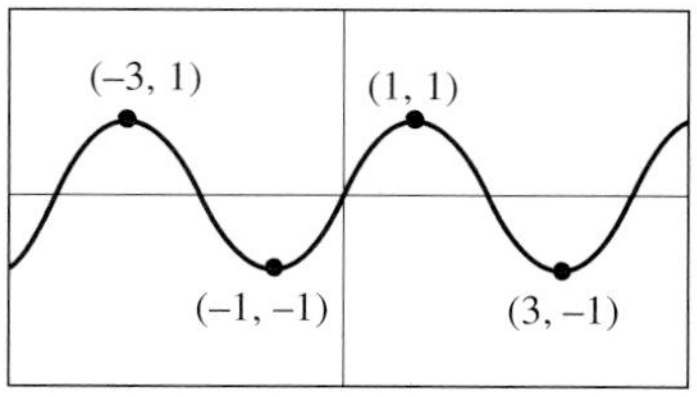

In Exercises 41–44, find the domain of each function analytically, and find the range by graphing each in the window $[-10, 10]$ *by* $[-10, 10]$. *You can obtain a clearer graph for Exercise 43 by using your calculator's "dot mode."*

41. $f(x) = 8 - 2x$

42. $g(x) = x^2 + x - 6$

43. $c(x) = \dfrac{3x}{x - 2}$

44. $k(x) = \dfrac{3x}{\sqrt{x - 2}}$

In Exercises 45–48, find the domain and range of each function analytically.

45. $z = 3x + 600$

46. $y = 4 - x^2$

47. $W = 2 + \sqrt{Z - 2}$

48. $Q = \dfrac{P - 3}{P + 3}$

The domain of each function in Exercises 49–52 is explicitly given. Find its range graphically.

49. $w = x^2 + 2x$, for x in $[-2, 1]$

50. $H(x) = \sqrt{x + 1}$, for x in $[0, 8]$

51. $L(x) = \dfrac{x^2 + 1}{x}$, for x in $(0, \infty)$

52. $y = \dfrac{x}{x^2 + 1}$, for x in $(-3, 3)$

In Exercises 53–58, find the domain and range of each function both analytically and graphically. Be sure to compare your two answers for compatibility.

53. $y = x^2$

54. $y = x^3$

55. $y = \dfrac{1}{x^4}$

56. $y = 20$

57. $y = \sqrt{1 - x^2}$

58. $y = \dfrac{1}{1 + x^2}$

In Exercises 59–64, find the domain and range of each function by any method.

59. $f(x) = x^4$

60. $y = \dfrac{1}{x^3}$

61. $y = \dfrac{1}{\sqrt{1 - x^2}}$

62. $\theta(x) = (1 - x^2)^2$

63. $h(x) = x^3 + 1$, for x in $[-1, 4]$

64. $Q = \dfrac{1}{x}$, for x in $(-\infty, -1]$

In Exercises 65–70, find the domain and range of the function in the given context.

65. $S = 340M$
M and S are as in ***Pulleys*** (page 9).

66. $C = 23.48d^2$
d and C are as in ***Swimming Pool*** (page 11).

67. $F = \dfrac{9C}{5} + 32$
C = Celsius temperature
F = Fahrenheit temperature
Temperatures must be above absolute zero (see Exercise 35 in Section 2-1).

68. $C = \dfrac{5(F - 32)}{9}$
F = Fahrenheit temperature
C = Celsius temperature
Temperatures must be above absolute zero.

69. $A = x(2000 - x)$
x = length, in feet, of a rectangular pasture you are planning to fence off
A = area of the pasture, in square feet
There are 4000 feet of fencing available.

70. $P = 2x + \dfrac{2{,}000{,}000}{x}$

x = length, in feet, of a rectangular pasture you are planning to fence off
P = perimeter of the pasture, in feet
The pasture must have an area of 1,000,000 square feet, and there is an unlimited supply of fencing.

In Exercises 71–78, write out the first ten terms of the sequence.

71. $A_j = 5 - j$ **72.** $c_k = |k - 5|$

73. $b_m = m^3$ **74.** $f_n = 3^n$

75. $C_1 = 1$
$C_i = 0.1C_{i-1}$, for $i \geq 2$

76. $d_1 = 1$
$d_2 = 2$
$d_n = d_{n-1} - d_{n-2}$, for $n \geq 3$

77. $F_1 = 1$
$F_2 = -1$
$F_k = F_{k-2}$, for $k \geq 3$

78. $a_1 = 1$
$a_2 = 1$
$a_n = a_1 + a_2 + \cdots + a_{n-1}$, for $n \geq 3$

In Exercises 79–82, identify the intervals on which the function is increasing and those on which it is decreasing.

79.

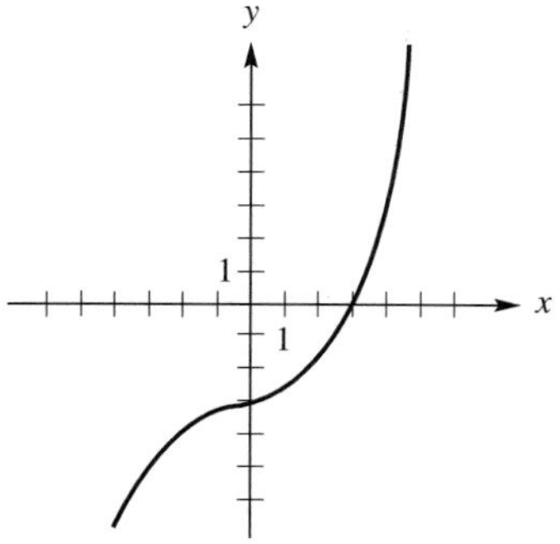

80.

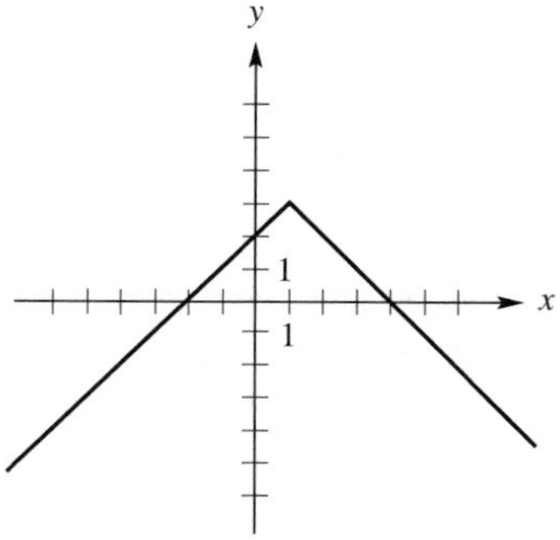

81.

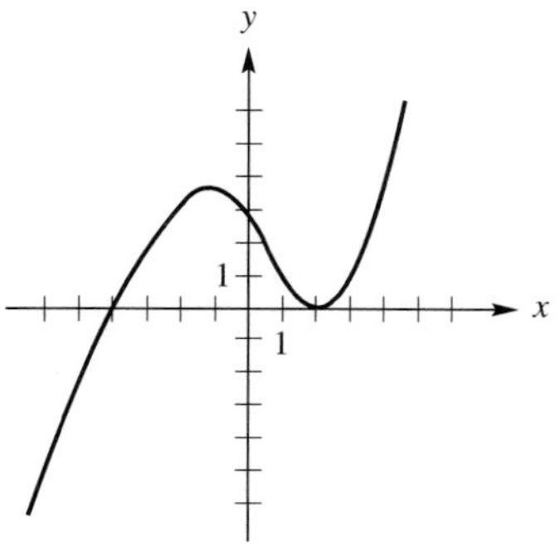

82.

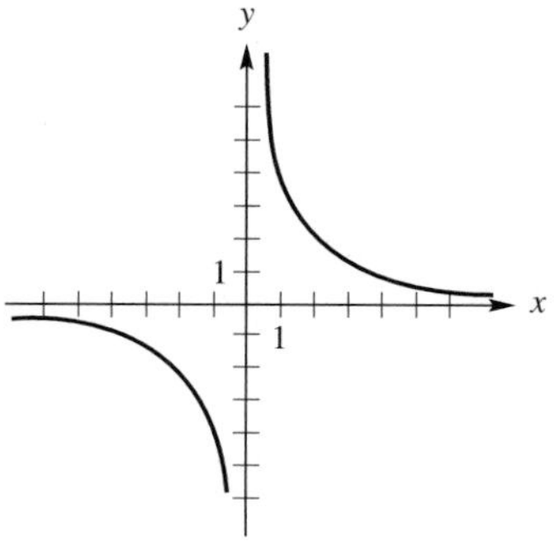

83. If c_n is the nth term of the Fibonacci sequence, it can be shown that for even values of n,

$$c_n = \frac{(\sqrt{5}+1)^n - (\sqrt{5}-1)^n}{2^n\sqrt{5}}$$

Verify that this formula is correct for $n = 2, 4,$ and 6.

84. *(Extension)*

a. If c_n is the nth term of the Fibonacci sequence, show that

$$c_1 + c_2 + \cdots + c_k = c_{k+2} - 1$$

Hint: First explain why you can write

$$\begin{aligned} c_1 &= c_3 - c_2 \\ c_2 &= c_4 - c_3 \\ &\vdots \\ c_k &= c_{k+2} - c_{k+1} \end{aligned}$$

b. Use the formula in part (a) to find the sum of the first 20 terms in the Fibonacci sequence. (Use your calculator and the formula in Exercise 83 to evaluate c_{22}.)

85. *(Extension)* Every finite sequence can be continued in more than one way. For example, the fourth term of the sequence $\{1, 2, 4, \ldots\}$ might be 8 (using the rule $a_n = 2a^{n-1}$) or 7 (using the rule $a_n = a_{n-1} + n - 1$). Other continuations are also possible. Describe at least two rules that lead to different continuations of each of the following sequences.

a. $\{1, 1, 2, \ldots\}$ **b.** $\{1, 2, 3, \ldots\}$

c. $\{1, 2, 6, \ldots\}$ **d.** $\{1, -1, 2, \ldots\}$

86. *(Extension)* The following sequence was created by John Conway of Princeton University. Can you guess the rule for generating the numbers in each row?

1
11
21
1211
111221
312211
13112221
⋮

87. *(Interpreting Mathematics)* In a newspaper or magazine, find a complete table of a function, and write down its domain and range.

88. *(Interpreting Mathematics)* Find a graph in a newspaper or magazine. Identify the intervals where it is increasing and those where it is decreasing. Explain what physical information your answers provide.

89. *(Creating Models)* As parts (a)–(g) of this exercise, sketch a possible graph for each relationship in Exercise 34 of Section 2-1. Then identify the intervals where each function is increasing, and those where it is decreasing.

SUPPLEMENTARY TOPIC
Domain and Range of Combinations of Functions

In Exercises 90–95, determine the domain of each of the following functions.

a. f **b.** g
c. $f + g$ **d.** $f - g$
e. fg **f.** $\frac{f}{g}$
g. $f \circ g$ **h.** $g \circ f$

90. $f(x) = 3x,\ g(x) = x + 2$

91. $f(x) = 3x - 2,\ g(x) = x^2 + x - 6$

92. $f(x) = \frac{1}{x},\ g(x) = \frac{x + 2}{x - 5}$

93. $f(x) = \sqrt{x},\ g(x) = \sqrt{4 - x}$

94. $f(x) = \sqrt{x},\ g(x) = \frac{1}{x - 2}$

95. $f(x) = \sqrt[4]{x},\ g(x) = \frac{1}{x - 2}$

96. *(Making Observations)* Suppose that the domain of a function S is $[-4, \infty)$ and the domain of a second function T is $(-\infty, 1) \cup (1, \infty)$.

a. What is the domain of each of the functions $S + T$, $S - T$, and ST? (*Hint*: Look at your results from Exercises 90–95.)

b. What other information would you need in order to find the domain of $\frac{S}{T}$?

97. *(Writing to Learn)* Make a general statement describing the domains of the functions $f + g$, $f - g$, fg, and $\frac{f}{g}$ in terms of the domains of f and g.

98. *(Making Observations)* If you know the domain of f and g, is that enough information to find the domain of $g \circ f$? (*Hint*: Compare your results in Exercises 94 and 95.)

99. *(Making Observations)* In general, it is true that the domain of $g \circ f$ is contained in the domain of f, and the range of $g \circ f$ is contained in the range of g. Verify these facts for the functions in Exercise 93.

2–4 SOLVING EQUATIONS AND INEQUALITIES GRAPHICALLY

Although quantitative relationships involve two variables, questions about such relationships can often be answered by solving equations and inequalities in one variable. In the past you have learned analytical methods for solving certain equations and inequalities. (For example, see Sections A-2, A-8, and A-13.) In the present section you will learn a graphical method. Like all graphical methods, it yields only approximate solutions. However, it is powerful in that it can be applied when analytical methods fail. Let's apply the graphical method to answer some questions about your T-shirt company in ***Fat Cats on Campus*** (page 43).

SOLVING EQUATIONS GRAPHICALLY

A MATHEMATICAL LOOKING GLASS

Breakeven Analysis (Fat Cats 2)

One of the most fundamental questions that can be asked about any business is, "At what sales levels will the company make a profit?" The first step toward an answer is usually to answer the question, "At what sales levels will the company break even, with neither a profit nor a loss?" These sales levels are referred to as **breakeven points**, and the process of finding them is called **breakeven analysis**.

Under the simple conditions described in ***Fat Cats on Campus***, your company's profit is a function $P(x)$ depending only on the number x of shirts you aim to sell during the semester. In Exercise 12 of Section 2-2, you found that

$$P(x) = -0.005x^2 + 12.50x - 612.50$$

For example,

$$P(100) = -0.005(100)^2 + 12.50(100) - 612.50 = 587.50$$

indicating that your profit from the sale of 100 shirts will be \$587.50.

To find your breakeven points, you must solve the equation

$$-0.005x^2 + 12.50x - 612.50 = 0$$

Let's compare analytical and graphical methods that can be used to solve this equation.

a. To solve

$$-0.005x^2 + 12.50x - 612.50 = 0$$

analytically, you might factor the left side, but it is probably easier to use the quadratic formula.

$$x = \frac{-12.50 + \sqrt{12.50^2 - 4(-0.005)(-621.50)}}{2(-0.005)}$$

$$x = 50 \quad \text{or} \quad x = 2450$$

b. To solve

$$-0.005x^2 + 12.50x - 612.50 = 0$$

graphically, begin by graphing

$$P(x) = -0.005x^2 + 12.50x - 612.50$$

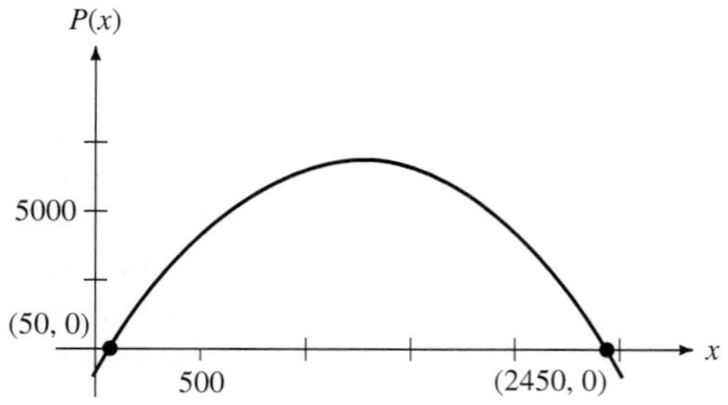

Figure 2–9

The graph is shown in Figure 2-9. On the graph the expressions y, $P(x)$, and $-0.005x^2 + 12.50x - 612.50$ are identical. Therefore, solving the equation means finding all points on the graph at which $y = 0$. This, in turn, means finding all x-intercepts on the graph. The graph has an x-intercept at about (50, 0), so that the equation has a solution at about $x = 50$. To see whether this is the exact solution, substitute 50 for x in the equation.

$$-0.005(50)^2 + 12.50(50) - 612.50 = 0$$
$$-12.50 + 625 - 612.50 = 0$$
$$0 = 0$$

Similarly, you can verify that the other x-intercept is at (2450, 0). Therefore, the equation has solutions $x = 50$ and $x = 2450$. ■

For the equation in ***Breakeven Analysis***, the analytical solution is much simpler than the graphical one. Example 1 illustrates that sometimes the graphical solution is simpler.

EXAMPLE 1 *A comparison of analytical and graphical solutions*

a. To solve $x^3 + 8 = 2x^2 + 10x$ analytically, begin by writing the equation with zero on one side.

$$x^3 - 2x^2 - 10x + 8 = 0$$

Since the left-hand side does not factor easily, the analytical solution involves methods we will not discuss until Section 7-2. For now, we need only observe that it has already become complicated.

b. To solve $x^3 + 8 = 2x^2 + 10x$ graphically, begin by writing the equation with zero on one side.

$$x^3 - 2x^2 - 10x + 8 = 0$$

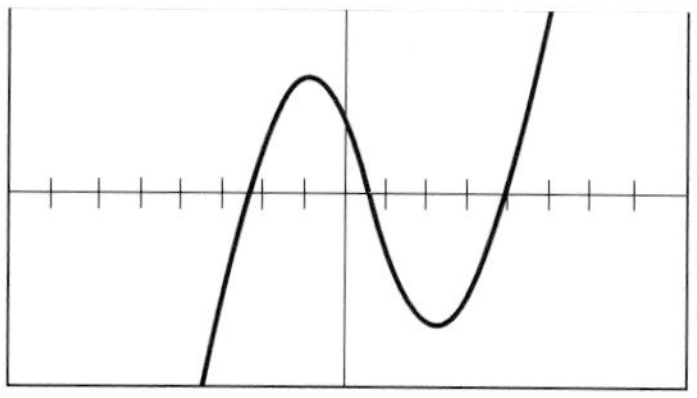

[–8, 8] by [–20, 20]

Figure 2–10

The graph of $g(x) = x^3 - 2x^2 - 10x + 8$ is shown in Figure 2-10. It has three x-intercepts, one of which appears to occur at about (4, 0). Substitution in the equation will verify that $x = 4$ is a solution. One of the other x-intercepts occurs in the interval $[-3, -2]$. To locate it more precisely, you can zoom in on the graph, as in Figure 2-11a.

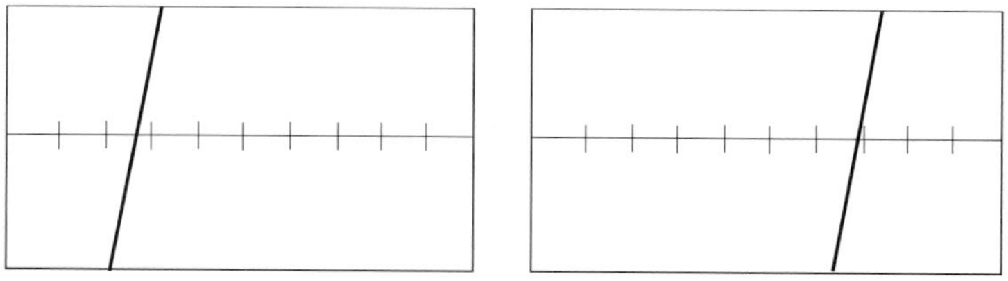

[–3, –2] by [–1, 1]

(a)

[–2.8, –2.7] by [–0.1, 0.1]

(b)

Figure 2–11

Figure 2-11a shows that the intercept is in $[-2.8, -2.7]$. Zooming in again, Figure 2-11b shows that it is in $[-2.74, -2.73]$. You can now say that the intercept occurs at $(-2.73, 0)$, and the solution is $x = -2.73$, with an error of no more than 0.01. Although we can locate the intercept with increasing precision by zooming in repeatedly, we may never locate it exactly. It is therefore practical to stop at some point, and settle for an estimate.

For this equation, the graphical solution is simpler. ■

FYI Unless otherwise indicated, an approximate solution to an equation should have at least two significant digits and should have an error of no more than 0.01.

Example 1 illustrates that as equations become more complicated, graphical methods become more useful. However, it is still worthwhile to learn some analytical methods for solving complicated equations. As we will see in Chapter 7, a combination of analytical and graphical methods is often more effective than either one alone. Furthermore, you can identify only real solutions graphically. Nonreal solutions must be found analytically. (Nonreal solutions are discussed in Section A-13.)

EXERCISES

1. In Example 1, find the third solution of $x^3 + 8 = 2x^2 + 10x$.

Solve the equations in Exercises 2–5 by graphical methods. Graph in the indicated viewing window.

2. $10 - 5x = 2x$ ([−10, 10] by [−10, 10])
3. $x^2 + x = 10$ ([−10, 10] by [−10, 10])
4. $x^2 + 400 = 100x - 2000$
   ([0, 100] by [−200, 200])
5. $x^3 + 408x - 1728 = 34x^2$
   ([−20, 20] by [−1000, 1000])

The process of solving equations graphically depends on a hidden assumption. To see what it is, trace along the graphs of the functions in Exercises 2–5 near their x-intercepts. You will notice that in most cases, your cursor does not show the exact intercept. Instead, it shows two pixels in adjacent columns whose y-coordinates are near 0, but opposite in sign. The hidden assumption is that the graph between those points is an unbroken curve, and thus contains an x-intercept. Informally, a function whose graph is an unbroken curve over an interval is said to be **continuous** over that interval. Otherwise, it is **discontinuous**. (You will learn formal definitions of these terms in a calculus course.) The equations you will solve between here and the end of Chapter 7 involve continuous functions, so you may solve them using the methods of this section. We will return to the idea of continuity in Sections 3-4 and 8-1.

SOLVING INEQUALITIES GRAPHICALLY

To find your profitable levels of production in ***Breakeven Analysis***, you must solve the inequality

$$-0.005x^2 + 12.50x - 612.50 > 0$$

An analytical method for solving this particular inequality is discussed in Section 5-2. For now, let's solve it graphically. As Example 2 shows, the first step is to solve the equation

$$-0.005x^2 + 12.50x - 612.50 = 0$$

as we did in ***Breakeven Analysis***. (This is an instance of *Solving a Simpler Problem First*.)

EXAMPLE 2 *Solving an inequality graphically*

To solve

$$-0.005x^2 + 12.50x - 612.50 > 0$$

graphically, graph $P(x) = -0.005x^2 + 12.50x - 612.50$, as in Figure 2-9. The graph is reproduced in Figure 2-12.

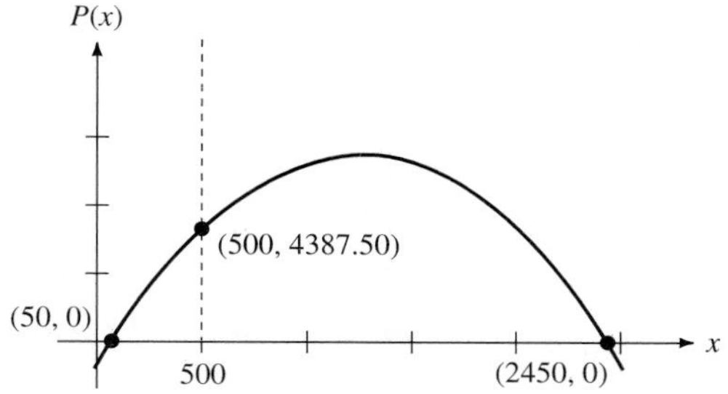

Figure 2–12

The expressions y, $P(x)$, and $-0.005x^2 + 12.50x - 612.50$ are identical on the graph. Thus solving the inequality means finding all x-values for which $y > 0$ on the graph. To find these values, take an imaginary walk in the positive direction along the x-axis. Look up because $y > 0$ above you. The graph becomes visible above the x-axis when $x > 50$. For example, at $x = 500$ you can see the point (500, 4387.50). Since the graph is visible above any point in (50, 2450), the solution of the inequality is (50, 2450). Thus you will make a profit if you sell between 50 and 2450 T-shirts. ■

EXAMPLE 3 *Solving an inequality graphically*

To solve

$$x^3 - 2x^2 - 10x + 8 \leq 0$$

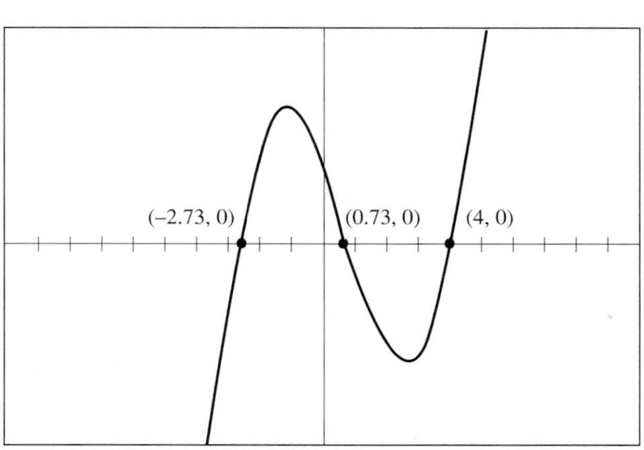

Figure 2–13

graphically, graph $g(x) = x^3 - 2x^2 - 10x + 8$, as in Figure 2-10. The graph is reproduced in Figure 2-13.

This time, as you walk along the x-axis, look directly on and below your path, since $y = 0$ on the x-axis and $y < 0$ below it. The graph intersects your path when $x \cong -2.73, 0.73$, and 4. It is below you in $(-\infty, -2.73) \cup (0.73, 4)$. The solution to the inequality is $(-\infty, -2.73] \cup [0.73, 4]$. ■

Example 3 shows that to solve inequalities graphically it is necessary to find x-intercepts and to find x-intercepts is to solve equations graphically. Thus, *to solve an inequality, you must first solve an equation.*

EXERCISES *Solve the inequalities in Exercises* 6–9 *graphically. Use the viewing windows and the solutions from Exercises* 2–5.

6. $10 - 5x > 2x$ **7.** $x^2 + x \le 10$

8. $x^2 + 400 < 100x - 2000$

9. $x^3 + 408x - 1728 \ge 34x^2$

10. *(Writing to Learn)* You have just read that to solve an inequality graphically, you must first solve the corresponding equation by locating the x-intercepts. Explain how locating the x-intercepts helps you solve the inequality.

ADDITIONAL EXERCISES *In each of Exercises* 11–20,

a. Solve the equation graphically. Use a viewing window of $[-10, 10]$ by $[-10, 10]$ except where another is indicated.

b. Solve the inequality graphically.

11. $x^2 - 4x = 5$
$x^2 - 4x < 5$

12. $3x - 13 = 2x + 3$
$3x - 13 > 2x + 3$
$[-20, 20]$ by $[-20, 20]$

13. $x^3 = 4$
$x^3 \ge 4$

14. $x^3 = 4x$
$x^3 \le 4x$

15. $x^2 + 2 = 2x$
$x^2 + 2 > 2x$

16. $4 - 3x = 3(x + 7)$
$4 - 3x \le 3(x + 7)$

17. $25x^2 = x^4$
$25x^2 \ge x^4$
$[-10, 10]$ by $[-200, 200]$

18. $x^2 + 98 = 20x$
$x^2 + 98 < 20x$
$[0, 20]$ by $[-10, 10]$

19. $|2x - 5| = 3$
$|2x - 5| \le 3$

20. $\dfrac{1}{x^2 + 1} = 2$
$\dfrac{1}{x^2 + 1} > 2$

21. Solve the equation in Exercise 11 by the following alternate method.

a. Graph both $y = x^2 - 4x$ and $y = 5$ in the window $[-10, 10]$ by $[-10, 10]$.

b. *(Writing to Learn)* Explain why the solutions to the equation are the x-coordinates at the points where the two graphs intersect.

c. *(Making Observations)* Find the points of intersection, and compare your results with those you obtained in Exercise 11.

22. Solve the inequality in Exercise 11 by the following alternate method.

a. Graph both $y = x^2 - 4x$ and $y = 5$ in the window $[-10, 10]$ by $[-10, 10]$.

b. *(Writing to Learn)* Explain why the solutions to the inequality are the x-coordinates at the points where the first graph is below the second.

c. *(Making Observations)* Find the solutions to the inequality, and compare your results with those you obtained in Exercise 11.

In Exercises 23–28, solve each equation and each inequality by the methods of Exercises 21–22. Use a viewing window of $[-10, 10]$ *by* $[-10, 10]$ *except where another is indicated.*

23. $3x - 13 = 2x + 3$
$3x - 13 > 2x + 3$
$[0, 30]$ by $[0, 60]$

24. $x^3 = 4$
$x^3 \geq 4$

25. $x^3 = 4x$
$x^3 \leq 4x$

26. $\dfrac{4x}{x^2 + 4} = 1$
$\dfrac{4x}{x^2 + 4} \leq 1$
$[-10, 10]$ by $[-2, 2]$

27. $x^2 = x + 2$
$x^2 < x + 2$

28. $x^2 = x - 2$
$x^2 < x - 2$

29. *(Problem Solving)* In Exercise 59 of Section 2-2 you found the profit function

$$P(x) = 8.50x - 212.50$$

related to cake sales at Periwinkle's Bakery. Use graphical methods to find

a. Periwinkle's breakeven point

b. Periwinkle's profitable levels of cake sales

CHAPTER REVIEW

Complete Exercises 1–24 (Writing to Learn) before referring to the indicated pages.

WORDS AND PHRASES

In Exercises 1–11, explain the meaning of the words or phrases in your own words.

1. (pages 29, 30) **function**

2. (page 31) **complete table**

3. (page 41) **functional notation**

4. (page 46) **composition** of functions

5. (page 51–55) **domain, range**

6. (page 57) **sequence**

7. (page 58) **terms** of a sequence

8. (page 59) **recursive** definition of a sequence

9. (page 59) **arithmetic sequence, geometric sequence**

10. (page 60) **increasing, decreasing**

11. (page 68) **continuous, discontinuous**

IDEAS

12. (page 31) Under what conditions do we say that a table fits a function? How does that statement differ from a statement that a table represents a function?

13. (pages 31–34) Describe some methods for deciding whether a table, equation, or graph represents a function. Include statements about the conditions under which each method can be applied. Illustrate each method with two examples, one that represents a function and one that does not.

14. (page 36) Why are functions important both in mathematics and as models of physical relationships?

15. (page 36) What do we gain by studying the connections among analytical, graphical, and numerical models of functions?

16. (page 40, 41) Using a specific function of your choice, illustrate the meaning of the statement that a function is a process that acts on variables and is distinct from them.

17. (page 41, 42) If $y = f(x)$, explain the meaning of the symbol $f(x)$ and how it differs from f alone.

18. (pages 42–47) Describe the processes of adding, subtracting, multiplying, dividing, and composing two functions. Make up two functions and use them to illustrate each process.

19. (page 51, 52) What is meant by the phrases *physically reasonable* and *mathematically reasonable* in connection with inputs and outputs of functions? Give examples of a situation in which an output is not physically reasonable, and in which an output is not mathematically reasonable.

20. (page 52, 53) Describe how to find the domain and range of a function from a table, graph, or equation. Illustrate each process with a specific example.

21. (pages 53) What operations can limit the domain of an algebraic function? Illustrate each type of limitation with a specific example.

22. (pages 57, 58) Describe a specific physical situation in which a relationship between two variables is represented by a sequence. Use your example to explain how sequences differ from other functions.

23. (page 59) Compare the processes of calculating a given term of a sequence when it is defined recursively and when it is defined by an explicit formula.

24. (pages 66, 69) Describe the processes for solving an equation $f(x) = 0$ and an inequality $f(x) > 0$ graphically. Explain the reasoning behind each process.

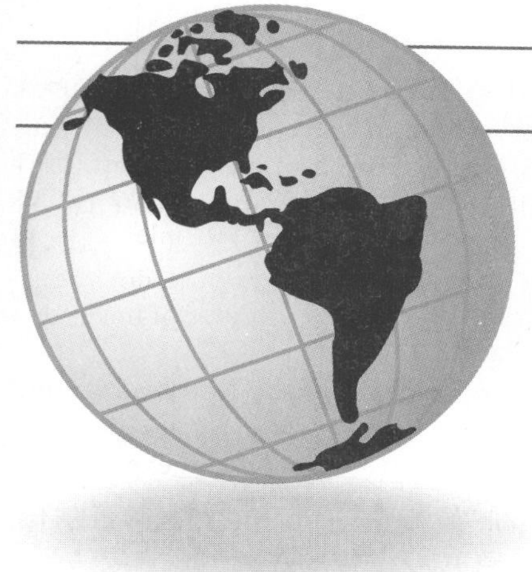

CHAPTER 3

LINEAR FUNCTIONS

3-1 THREE VIEWS OF LINEAR FUNCTIONS

Functions vary so widely as to have few global properties in common. However, several useful classes of functions have graphs, equations, and tables with easily identifiable properties. The simplest such class, which you will study in this chapter, consists of functions whose equations can be written in the form $y = mx + b$. Because the graphs of these functions are straight lines, they are called **linear functions**. (Some basic facts about linear functions and their graphs are discussed in Section A-6.)

In the present section we will look at linear functions from numerical, graphical, and analytical perspectives. In the process we will discover that every linear function exhibits a constant **rate of change** between its variables. That is, as the values of the variables change, the ratio

$$\frac{\text{change in the dependent variable}}{\text{change in the independent variable}}$$

remains the same. Conversely, any function with a constant rate of change must be linear. We will see how this ratio is reflected in tables, graphs, and equations of linear functions, and what information it can provide in a physical context.

A NUMERICAL VIEW OF LINEAR FUNCTIONS

The following Mathematical Looking Glass contains a table taken from a linear function. By exploring it in Exercise 1 you can guess some general properties of tables that fit linear functions.

A MATHEMATICAL LOOKING GLASS
Sound Waves

When you sing, sneeze, whisper, whistle, clap, or stomp, you create vibrations that travel through the air around you as sound waves. Their speeds vary depending on conditions such as your elevation above sea level and the temperature and humidity of the air. In particular, it has been discovered that the speed S of sound at sea level (in meters per second) is related to the Celsius temperature T by the linear equation $S = 0.61T + 331.4$.

Table 10 shows the speed of sound at several Celsius temperatures. Looking for patterns in the data, we observe that an increase of 10 in T is always accompanied by an increase of 6.1 in S. The symbol Δ commonly denotes a change in a variable, so we will denote changes in S and T by ΔS and ΔT, respectively. ■

TABLE 10

T	S
−10	325.3
0	331.4
10	337.5
20	343.6
30	349.7

EXERCISE **1.** *(Making Observations)*

a. Calculate the ratio $\frac{\Delta S}{\Delta T}$ for the values of S and T in the first and second rows of Table 10.

b. Calculate the ratio $\frac{\Delta S}{\Delta T}$ for the values of S and T in the third and fifth rows of Table 10.

c. Use the equation $S = 0.61T + 331.4$ to generate two points not in Table 10.

d. Predict the value of $\frac{\Delta S}{\Delta T}$ for your pair of points in part (c).

e. Verify your answer with a calculation.

Exercise 1 suggests that you have found a property common to all tables from linear functions. It can be stated as follows.

For any table from a linear function $y = f(x)$, the ratio $\frac{\Delta y}{\Delta x}$ is the same between any two rows.

FYI We cannot say that a function is linear on the basis of a table without additional information. This is because even when $\frac{\Delta y}{\Delta x}$ is the same for all *given* pairs of points, it may not be the same for all *possible* pairs of points. However, if $\frac{\Delta y}{\Delta x}$ is not constant, then no linear function could generate the table.

If $\frac{\Delta y}{\Delta x}$ is constant for a given table, we say that the table fits a linear function.

EXERCISES *In Exercises 2–5, decide which tables fit linear functions.*

2.

x	y
13	29
15	26
17	23
19	20
21	17

3.

x	y
−2	−5
−1	−1
0	3
1	5
2	7

4.

x	y
5	7
4	5
2	1
−3	−9
−4	−11

5.

x	y
0.3	17
0.6	11
0.7	9
0.9	5
1.0	3

The quantities Δx and Δy are called **first differences** in x and y, respectively. If the values of x are equally spaced (that is, if the first differences in x are constant), then it is easy to tell whether the table fits a linear function.

> A table with equally spaced x-values fits a linear function if and only if it has y-values with constant first differences.

EXERCISE **6.** *(Making Observations)* The tables in Exercises 2 and 3 have equally spaced x-values. For which of these are the first differences in y constant? Is this the same table that fits a linear function?

Can a table fit more than one linear function? If it contains more than one point, the answer is no. This is because the graph of every linear function that fits the data is a straight line through all of the data points, and two distinct lines cannot have more than one point in common.

FYI If a table contains only two points, the line through those two points fits the data. This alone is not evidence that the data came from a linear function. On the contrary, it says that *every* two-point table "looks" linear, *no matter what function produced it*! Therefore, be wary about concluding that a two-point table was generated by a linear function. You can explore this idea further in Exercise 63.

$\frac{\Delta y}{\Delta x}$ as Rate of Change The physical world is filled with changing quantities, and functions are excellent tools for analyzing the rate at which one quantity changes with respect to another. For a linear function, the constant ratio $\frac{\Delta y}{\Delta x}$ is called the **rate of change in y with respect to x**. When x and y represent physical quantities, the ratio $\frac{\Delta y}{\Delta x}$ also has physical significance.

EXAMPLE 1 *Interpreting $\frac{\Delta y}{\Delta x}$ as a rate of change in a physical context*

Table 11 is reproduced from ***Fat Cats on Campus*** (page 43).

TABLE 11

Target shirt sales for the semester (x)	Price per shirt (p)
500 shirts	\$13.50
900	11.50
1300	9.50
1700	7.50
2100	5.50
2500	3.50

From Table 11, you can calculate that the rate of change in p with respect to x is $\frac{\Delta p}{\Delta x} = -\frac{2}{400} = -0.005$. This says that every change of 1 unit in x produces a change of -0.005 units in p. In more concrete terms, it says that to increase your target sales level by one shirt, you must lower the price per shirt by half a cent ($0.005). ■

EXERCISES

7. *(Writing to Learn)* In ***Sound Waves***, what is the rate of change in S with respect to T? What does the rate of change tell you about the relationship between temperature and the speed of sound?

In Exercises 8–11, *calculate the rate of change in the dependent variable with respect to the independent variable, and explain its physical significance.*

Valentines	Cost
1	$2.50
2	5.00
3	7.50
4	10.00
5	12.50

8. *(Interpreting Mathematics)* Raenell Jones is selling chocolate valentines to anyone at Penn State–Beaver who has $2.50 to spare. To avoid tedious calculations, she has made the table at the left.

9. *(Interpreting Mathematics)* The following table is from ***Pulleys*** (page 9).

Motor pulley diameter, in inches (*M*)	Fan speed, in rpm (*S*)
1.5	510
3	1020
5	1700
8	2720
10	3400

10. *(Interpreting Mathematics)* Tony Zarillo, a student in one of Lynn's classes, devised the following problem to illustrate rates of change.

"I am taking a math test about which I know nothing. The following table describes my state of mind during the test."

Minutes since the test began (*m*)	Percentage of my sanity remaining (*s*)
8	80
16	60
24	40
32	20

11. *(Interpreting Mathematics)* In one of his failed attempts to budget his money, Dave recently analyzed the monthly cost of driving his 1991 Cavalier. The following table includes both mileage-dependent costs, such as gasoline and tires, and time-dependent costs, such as insurance.

Miles driven each month (*m*)	Total cost, in dollars (*d*)
1500	325
2000	400
2500	475
3000	550

12. *(Interpreting Mathematics)* Make up your own problem involving a linear table. Calculate the rate of change in the dependent variable with respect to the independent variable, and explain its physical significance.

A GRAPHICAL VIEW OF LINEAR FUNCTIONS

For a linear function, the ratio $\frac{\Delta y}{\Delta x}$ represents the slope of its graph. If you know the coordinates of two points (x_1, y_1) and (x_2, y_2) on the line, you can calculate its slope as $\frac{y_2 - y_1}{x_2 - x_1}$, as illustrated in Figure 3-1. (See Section A-6 to review the idea of slope.)

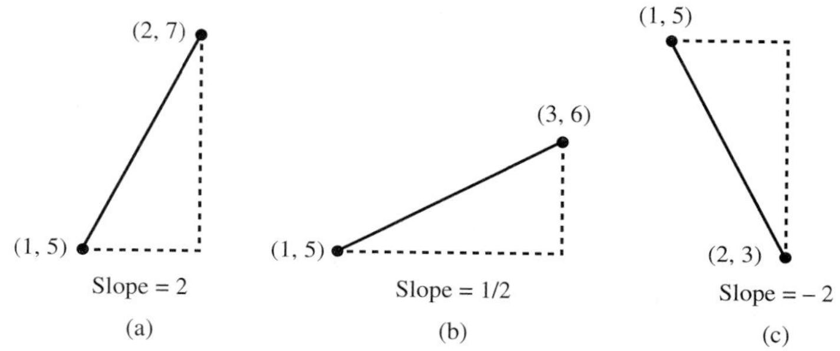

Figure 3–1

EXERCISE **13.** *(Making Observations)*

a. Plot the points from Table 10 in ***Sound Waves***. How do these points support the conclusion that Table 10 fits a linear function?

b. What is the slope of the graph of the function $S = 0.61T + 331.4$ from ***Sound Waves***?

Slope as Rate of Change The slope of a linear graph allows you to see the rate of change in y with respect to x. If the slope $\frac{\Delta y}{\Delta x}$ is positive as in Figure 3-1a and 3-1b, then increases in x lead to increases in y. (This is another way of saying that the function is increasing.) Similarly, if the slope is negative as in Figure 3-1c, increases in x lead to decreases in y. Furthermore, the steeper the graph, the greater the magnitude of the change in y compared to the change in x.

The slope of a horizontal line is zero, reflecting the fact that the value of y does not change. A horizontal line is the graph of a function that is neither increasing nor decreasing. The slope of a vertical line is undefined, reflecting the fact that x does not change. Since a vertical line is not the graph of a function, the terms increasing and decreasing do not apply.

EXERCISES **14.** *For each of the tables in Exercises* 8–11,

a. Plot the points on graph paper, and sketch the line through them.

b. Decide whether the slope is positive or negative.

c. Explain the physical information that is provided by your answer to part (b).

In Exercises 15–18,

a. Decide whether the function is increasing, decreasing or neither.

b. Find the rate of change in y with respect to x.

15.

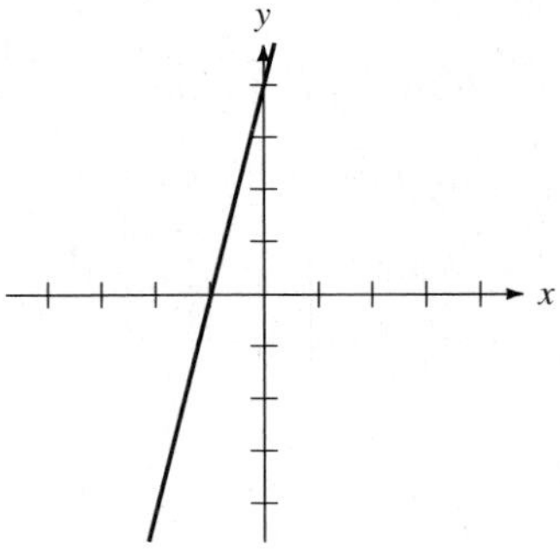

16.

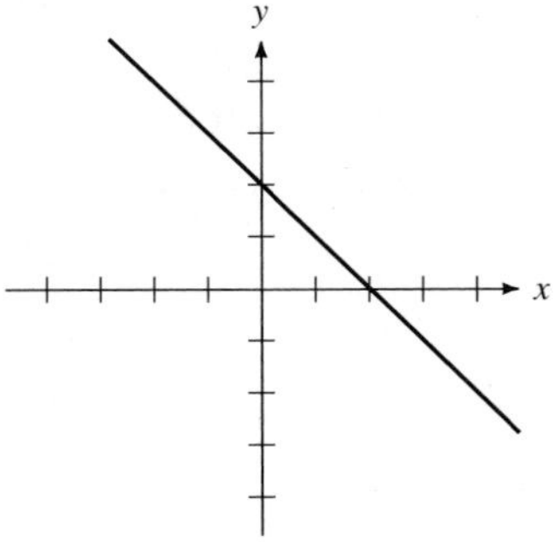

17.

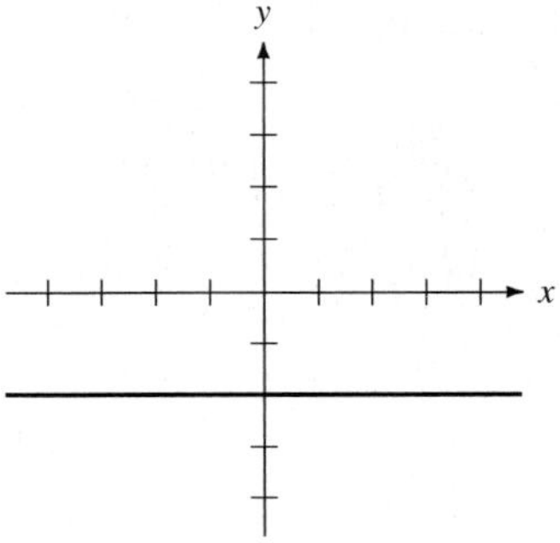

18.

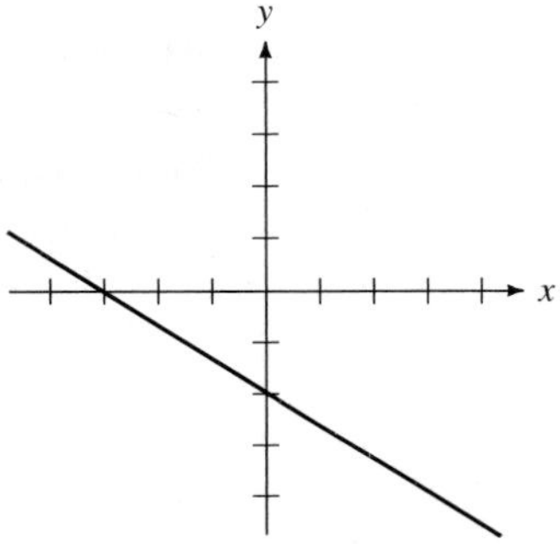

AN ANALYTICAL VIEW OF LINEAR FUNCTIONS

According to our definition a function is linear if and only if its equation can be written in the form $f(x) = mx + b$.

EXAMPLE 2 *Deciding whether a function is linear*

a. To verify that the function

$$y = \frac{7 - 3(x + 4)}{2}$$

is linear, rewrite the equation in the form $y = mx + b$.

$$y = \frac{7 - 3x - 12}{2}$$

$$y = \frac{-3x - 5}{2}$$

$$y = -\frac{3}{2}x - \frac{5}{2}$$

b. The function $Q(x) = x^2 + 4$ is not linear. We cannot eliminate x^2 from the equation, so it cannot be written in the form $Q(x) = mx + b$. ■

EXERCISES *In Exercises* 19–24, *decide whether the function is linear, and write the equation in the form* y = mx + b, *if possible.*

19. $f(x) = 3x - 9$ **20.** $g(x) = 3$

21. $h(x) = 3\sqrt{x} - 9$ **22.** $K(x) = \sqrt{3}\,x - 9$

23. $R(x) = \dfrac{6}{3x - 9}$ **24.** $r(x) = \dfrac{3x - 9}{6}$

25. *(Making Observations)* Support your answers in Exercises 19–24 by graphing each equation on your calculator to see which graphs appear to be lines.

If a table fits a linear function, you can find the function by choosing any two data points and writing the equation of the line which passes through them.

EXERCISES **26.** *(Review)* Find the equation of a linear function to fit

a. the table in Exercise 2

b. the table in Exercise 5

27. *(Review)* As parts (a)–(d) of this exercise, find the equations of the linear functions in Exercises 15–18.

m* as Rate of Change** The coefficient m in the equation $f(x) = mx + b$ represents both the slope of the graph and the ratio $\dfrac{\Delta y}{\Delta x}$. Thus it is easy to find the rate of change in y with respect to x from a linear equation and to tell whether y is an increasing or decreasing function of x. For example, in Exercise 7 you calculated the rate of change in S with respect to T in ***Sound Waves as $\dfrac{\Delta S}{\Delta T} = 0.61$. In Exercise 13 you calculated it as the slope of the graph. The equation $S = 0.61T + 331.4$ allows you to conclude without doing any calculations that the rate of change is 0.61 and that S is an increasing function of T.

EXERCISES *In Exercises* 28–31,

a. Find the rate of change in y with respect to x.

b. Decide whether the function is increasing or decreasing.

28. $f(x) = 5 - 2.7x$ **29.** $h(x) = 3x + 4(10 - x)$

30. $y = \dfrac{x}{2} - 7$ **31.** $y = 3$

In Exercises 32–33,

a. Find the rate of change in the dependent variable with respect to the independent variable.

b. Decide whether the function is increasing or decreasing.

c. Give a physical interpretation of the rate of change.

32. *(Interpreting Mathematics)* See ***Temperature Scales*** (page 30).

F = Fahrenheit temperature
C = Celsius temperature

$$C = \frac{5}{9}(F - 32)$$

33. *(Interpreting Mathematics)* See ***Cricket Chirps*** (page 41).

$$r = \text{number of chirps per minute}$$
$$h(r) = \text{Fahrenheit temperature}$$
$$h(r) = \frac{1}{4}r + 40$$

SUMMARY

We now have three ways to say that a function f is linear.

- *Numerically*, the ratio $\frac{\Delta y}{\Delta x}$ between any two points in any table for f is the same.
- *Graphically*, the graph of f is a nonvertical straight line.
- *Analytically*, we can describe f by an equation $f(x) = mx + b$.

We also have three ways to describe the rate of change in the dependent variable with respect to the independent variable.

- *Numerically*, as the ratio $\frac{\Delta y}{\Delta x}$
- *Graphically*, as the slope of the graph
- *Analytically*, as the coefficient m in the equation $f(x) = mx + b$

ADDITIONAL EXERCISES

In Exercises 34–43,

a. Use an equation, table, or graph to decide whether the function is linear.

If it is, then

b. Write the equation in the form $y = mx + b$.

c. Decide whether the function is increasing or decreasing.

d. Find the rate of change in y with respect to x.

34. $y = 2(x - 5.1) - 3(x + 4.7)$

35. $y = 2x^3 - 7x^2 + 3x + 1$

36. $y = 7 - 13x$

37. $f(x) = 0.5x - \sqrt{2}$

38. $P(x) = x(x - 2)$

39. $\Sigma(x) = \pi x$

40. $y = \frac{3}{x}$

41. $a(x) = \frac{x}{3}$

42. $\beta(x) = \sqrt{x^2 + 9}$

43. $z(x) = (x + 3)^2 - x^2$

44. *(Making Observations)* Support your answers in Exercises 34–43 by graphing each equation on your calculator to see which graphs appear to be lines.

In Exercises 45–50,

a. Tell which tables fit linear functions.

If the table fits a linear function, then

b. State the rate of change in y with respect to x.

c. Decide whether the function is increasing or decreasing.

d. Write an equation for the function.

45.

x	y
1	13
2	17
3	21
4	25
5	29

46.

x	y
0	1
0.1	0
0.2	−1
0.3	0
0.4	1

47.

x	y
32	68
37	63
39	61
42	58
43	57

48.

x	y
1	3π
2	6π
5	15π
7	21π
8	24π

49.

x	y
3	1
3	2
3	3
3	4
3	5

50.

x	y
20	2.3
21	2.3
24.5	2.3
$20\sqrt{3}$	2.3
97	2.3

In Exercises 51–54,

a. Write an equation of a linear function to fit the table in the given exercise.

b. *(Making Observations)* Use the equation to find the rate of change in the dependent variable with respect to the independent variable. Do your results agree with the ones you obtained in the given exercise?

51. Exercise 8 **52.** Exercise 9

53. Exercise 10 **54.** Exercise 11

55. *(Problem Solving)* When a large airliner approaches an airport, it begins its descent from about 100 miles away and takes about 20 minutes to descend from an altitude of 36,000 feet.

a. Assume that the airliner's altitude A in feet is a linear function of the time t minutes after it begins its descent. Express A as a function of t.

b. Find and interpret the rate of change in A with respect to t.

c. Assume that the airliner's altitude A in feet is a linear function of its distance d from the airport in miles. Express A as a function of d.

d. Find and interpret the rate of change in A with respect to d.

56. *(Making Observations)* (Continuation of Exercise 55) If the cabin of a small aircraft is not pressurized (that is, if the air pressure in the cabin varies with the altitude), it must fly at a lower altitude and descend more gradually to avoid discomfort to the passengers. A typical small aircraft might take 15 minutes to descend from an altitude of 10,500 feet.

a. Assume that the aircraft's altitude A in feet is a linear function of the time t minutes after it begins its descent. Express A as a function of t.

b. Compare your equation in part (a) with your equation from Exercise 55a. How do the equations reflect the fact that the smaller aircraft must descend more slowly?

c. Graph your equations from Exercises 55a and 56a. How do the graphs reflect the fact that the smaller aircraft must descend more slowly?

Exercises 57–60 show tables from linear functions with some entries missing. Calculate the rate of change for each table, and fill in the missing numbers.

57.

x	y
1	100
2	92
3	
5	

58.

x	y
5.2	7.6
5.4	
5.6	8.8
5.7	

59.

x	y
14	2
	7
24	22
30	

60.

x	y
0	0
4	
	5
16	8

61. *(Writing to Learn)* If you had not been told that the tables in Exercises 57–60 were linear, could you have filled in the missing entries? If so, how? If not, why not?

62. *(Writing to Learn)* Let $f(x) = x - 1$, and $g(x) = 10x - 10$.

a. Which function has the steeper graph?

b. Graph $y = f(x)$ in the window $[-10, 10]$ by $[-2, 2]$, and graph $y = g(x)$ in the window $[-10, 10]$ by $[-100, 100]$. Which graph appears steeper?

c. Explain the apparent contradiction in parts (a) and (b).

Exercise 63 shows that a nonlinear function can generate a table that fits a linear function.

63. *(Extension)* The following table was generated by the function $f(x) = x^3$.

x	y
-1	-1
0	0
1	1

a. Verify that the value of $\frac{\Delta y}{\Delta x}$ is the same for each pair of data points.

b. By generating more points from $f(x) = x^3$, show that $\frac{\Delta y}{\Delta x}$ is not the same for all pairs of points from this function.

Exercise 64 shows that the graph of a nonlinear function can appear linear in some viewing windows.

64. *(Extension)* Suppose $g(x) = \sqrt{x^2 - 1}$.

a. Graph $y = g(x)$ in the window [10, 25] by [10, 20]. Does the graph look linear?

b. Find a different viewing window to show that g is not linear.

Exercise 65 illustrates that a table fits a linear function precisely when its points are collinear.

65. *(Making Observations)* Plot the points from Exercise 2 in the coordinate plane. Repeat for Exercises 3–5. Verify that the tables with collinear points are those which fit linear functions.

66. *(Extension)* Is the following statement true or false?

A graph through $(-1, -1)$, $(0, 0)$, and $(1, 1)$ must be a straight line with a slope of 1 and a y-intercept of 0.

Explain how you arrived at your conclusion. (*Hint*: Plot the three points, and visualize several graphs that could pass through all three.) (*Second Hint*: See Exercise 63.)

3–2 MODELING AND PROBLEM SOLVING WITH LINEAR FUNCTIONS

Linear functions can model many quantitative relationships. In this section you will learn how to decide whether a relationship is linear. You will also study two special types of linear functions that occur often in physical contexts.

MODELING OF LINEAR RELATIONSHIPS

How can we tell whether a relationship can be modeled by a linear function? The key to answering this question is that linear functions always exhibit a constant rate of change. Therefore, a relationship can be modeled by a linear function as long as one variable changes at a constant rate with respect to the other. The following Mathematical Looking Glass indicates how to decide whether this is the case. The modeling process uses several of the Pólya problem-solving strategies you saw in Section 1-3.

A MATHEMATICAL LOOKING GLASS
Phone Bill

Lynn's daughter Carrie makes frequent phone calls to her friends. Her habit recently became much more expensive when her friend Naomi moved to Nice, France. The wages from her part-time job at Periwinkle's Bakery won't pay for all the calls she would like to make, especially since she has to pay vet bills and meet other expenses. Carrie has therefore put herself on a budget. She now wants a way to calculate, in advance, how long she can talk for a given amount of money. Let's think along with Carrie as she solves her problem using Pólya's four-step process.

To understand the problem, Carrie asked the telephone company for their rates. She was told that economy rates to Nice begin after 6:00 P.M., which is when she should phone Naomi. Economy rates are \$1.15 for the first minute and \$0.67 for each additional minute.

To devise a plan Carrie adopts the strategy of *Looking for Relationships Among Variables.* She observes that after the first minute, when the time M increases by 1 minute, the cost C of the call increases by \$0.67. Thus the rate of change in C with respect to M is constant. Specifically, $\dfrac{\Delta C}{\Delta M} = 0.67$. Carrie's observation tells her that M and C are linearly related and allows her to continue by *Writing an Equation.*

To carry out her plan, Carrie remembers that the equation can be written in the point-slope form

$$C - C_0 = m(M - M_0)$$

She knows that the value of m is the rate of change 0.67. She also knows that she can choose any point (M_0, C_0) on the graph of the function. Since the cost of the first minute is \$1.15, she chooses (1, 1.15) and obtains

$$C - 1.15 = 0.67(M - 1)$$
$$C = 0.67M + 0.48$$

Since she wants to use C as input and M as output, she solves the equation to make C the independent variable.

$$M = \frac{C - 0.48}{0.67}$$

Looking back, Carrie recognizes the 0.48 as the "extra" cost of the first minute (\$1.15, as opposed to \$0.67 for each additional minute). Therefore, subtracting 0.48 from the cost and dividing the result by 0.67 yields the number of minutes. She is now confident that her equation is correct. ■

Carrie could also have solved her problem either by *Making a Table* or *Drawing a Graph* to construct a numerical or graphical model of the situation. You will do this in Section 3-4.

EXERCISES

1. *(Interpreting Mathematics)* Use the equation to calculate how long Carrie can talk for \$10.00. She must pay for a full minute even if she uses only part of it.
2. *(Interpreting Mathematics)* Write and solve an inequality to calculate how long Carrie can talk without spending more than \$10.00.
3. *(Problem Solving)* Dave's daughter Sarah recently spent several months studying in Tanzania, resulting in a large increase in Dave's phone bill. Knowing that his situation is similar to Carrie's, he uses the strategy of *Examining a Related Problem* to write an equation expressing the length of time he and Alice can talk to Sarah for a given cost. Economy rates to Tanzania, from 2:00 A.M. to 7:00 A.M. are \$1.64 for the first minute, and \$1.01 for each additional minute.
   a. Write Dave's equation, and use it to calculate how long he can talk for \$10.00.
   b. Write an inequality to calculate how long Dave can talk for no more than \$10.00.

Relationships that can be modeled by a linear function $y = mx + b$ may also have one or both of the following properties.

- The variable y is a constant multiple of x. In this case the equation has the simpler form $y = mx$.
- The variable x represents a quantity that is counted rather than measured. As you learned in Section 2-3, the function is then an arithmetic sequence.

Let's look at each of these properties in turn to see how they are indicated in the relationship, and how they affect the modeling process.

LINEAR VARIATION

In the physical world, we find that one quantity in a relationship is often a constant multiple of another.

EXAMPLE 1 *Relationships in which one quantity is a constant multiple of another*

a. If an object's mass in kilograms is M and its weight in pounds is W, then M and W are related approximately by the equation $W = 2.20M$.

b. If s is the number of feet an object falls t seconds after being dropped, then t and s are related approximately by the equation $s = 16t^2$.

c. If V is the volume of a sphere of radius r, then V and r are related by the equation $V = \frac{4}{3}\pi r^3$. ■

The relationship in part (a) is described by saying that W **varies directly** with M. Similarly s varies directly with t^2 in part (b) and V varies directly with r^3 in part (c). Alternately, it is sometimes said that W is **directly proportional** to M, that s is directly proportional to t^2, and that V is directly proportional to r^3. The coefficients 2.20, 16, and $\frac{4}{3}\pi$ are called **constants of proportionality**.

Because the equation in part (a) is linear, we can also say that W **varies linearly** with M. The equations in parts (b) and (c) illustrate nonlinear variation, which we will consider in Section 7-1.

Every linear variation can be described by an equation $y = mx$. The following test allows you to decide whether a given relationship is a linear variation.

> If $f(x) = mx$ for some real m, then whenever the input x is doubled, the output $f(x)$ is also doubled.

For the functions in this book, the converse is also true. That is, if doubling the input to f always results in doubling the output, then $f(x) = mx$ for some real m.

EXAMPLE 2 *Constructing a linear variation equation*

Knowing that the people of Allegheny County, PA (population 1,300,000), create 28,000,000 pounds of garbage each week, can we estimate the weekly garbage production of New York City (population 7,320,000)? We might approach this question by constructing an equation to express the number of pounds y of garbage produced by x people. It is reasonable to assume that doubling the population of a city will double the amount of garbage, so that x and y are related by a linear variation. We can now ask our question in abstract language.

> If y varies linearly with x, and $y = 28{,}000{,}000$ when $x = 1{,}300{,}000$, what is the value of y when $x = 7{,}320{,}000$?

We know that x and y are related by an equation $y = mx$. We can find m by using our data to write

$$28{,}000{,}000 = m(1{,}300{,}000)$$

so that $m \cong 21.5$, and the equation is

$$y = 21.5x$$

Then, when $x = 7{,}320{,}000$, the value of y is $21.5(7{,}320{,}000) \cong 157{,}000{,}000$ pounds. ■

Example 2 shows that you can model a relationship from only one data point if you know it to be a linear variation.

EXERCISES

4. *(Interpreting Mathematics)* In Example 2, what does the value $m = 21.5$ represent in physical terms?
5. If y varies linearly with x, and $y = 35$ when $x = 21$, find the value of y when $x = 39$.
6. If W varies directly with Z, and $W = 0.004$ when $Z = 0.001$, find the value of Z when $W = 30{,}000$.
7. If p varies directly with q, and $p = 15$ when $q = 2$, find the value of p when $q = 40$.
8. If δ is directly proportional to Σ, and $\delta = -3$ when $\Sigma = 16$, find the value of δ when $\Sigma = -102.4$.

It is easy to tell whether a table fits a linear variation, and to find an equation for one that does.

> A table fits a function $y = mx$ if and only if the ratio $\frac{y}{x} = m$ for every row.
>
> (*Exception*: A row with $x = 0$ fits if and only if $y = 0$ also.)

EXERCISES

In Exercises 9–12, decide whether the table fits a linear variation. If it does, write an equation relating x and y.

9.

x	y
3	12
6	24
10	40
19	76

10.

x	y
3	5
6	8
10	12
19	21

11.

x	y
−0.3	−0.06
3.1	0.62
5	1
7.3	1.46

12.

x	y
$\sqrt{2}$	2
3	$3\sqrt{2}$
$5\sqrt{2}$	10
$\sqrt{71}$	$\sqrt{142}$

ARITHMETIC SEQUENCES

In the following Mathematical Looking Glass, the independent variable represents a quantity that is counted rather than measured.

A MATHEMATICAL LOOKING GLASS
Great Race

Tim Schmitt has been teased by his parents, Gini and George, for being out of shape, so he has resolved to run Pittsburgh's Great Race in September. Since he must be prepared to run 10 kilometers (6.2 miles), he begins a vigorous exercise program, including a daily jog at Fox Chapel High School's track. During week 1 of his program he runs 1 mile each day. In week 2 he increases to 1.25 miles per day. Each week he continues to increase his daily distance by 0.25 miles. To find out how many weeks it will take him to reach his goal of 6.25 miles per day, he has decided to write an equation to model the relationship between the week number and his daily distance. ■

EXERCISES **13.** *(Writing to Learn)*

Let

$$n = \text{the number of the week since Tim began exercising}$$

$$a_n = \text{the number of miles per day he runs}$$

a. How does the situation indicate that n and a_n are linearly related?

b. Describe the steps you would use in constructing the equation $a_n = 0.25n + 0.75$.

14. *(Interpreting Mathematics)* How many weeks will it take Tim to reach his goal of 6.25 miles per day?

The equation in Exercise 13 defines a_n as a function of n. Since Tim counts his weeks using positive integers, the domain of his function is the set $\{1, 2, 3, \ldots\}$. In Section 2-3 you learned that such a function is called a **sequence**. Tim's sequence is defined by a linear function and is called an **arithmetic sequence**. An arithmetic sequence differs from other linear functions only in its domain. Because the domain of Tim's sequence consists only of positive integers, its graph is the one in Figure 3-2a, not Figure 3-2b.

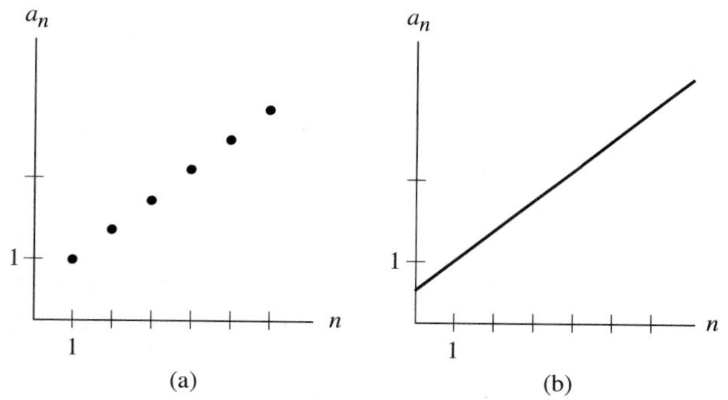

Figure 3–2

Arithmetic sequences, like other linear functions, have the property that $\dfrac{\Delta a_n}{\Delta n}$ is constant. Since $\Delta n = 1$ between any two consecutive terms, the difference Δa_n between consecutive terms is constant. That is,

> A sequence a is arithmetic if and only if $a_{n+1} - a_n$ has the same value for each positive integer n.

For any sequence a, the quantities $a_n{+}1 - a_n$ are called its **first differences**. Thus, a sequence is arithmetic if and only if its first differences are constant.

An arithmetic sequence can be described by the point-slope equation $a_n - a_1 = d(n - 1)$ or, equivalently.

$$a_n = a_1 + d(n - 1)$$

where d is the first difference and n is any positive integer. If you know a_1 and d, you can use this equation to find a_n for any n.

EXAMPLE 3 *Finding a specified term of an arithmetic sequence*

Suppose that we need to find the 247th term in the sequence

$$a = \{3.4, 4.1, 4.8, 5.5, 6.2, \ldots\}$$

This sequence is a linear function whose graph consists of the points (1, 3.4), (2, 4.1) (3, 4.8), . . . We can calculate the slope of the graph as 0.7 and write its equation in point-slope form as

$$a_n = 3.4 + 0.7\,(n - 1)$$

Therefore,

$$a_{247} = 3.4 + 0.7(247 - 1) = 175.6$$ ■

Exercises *In Exercises* 15–18, *decide whether the given sequence is arithmetic. If it is, find a formula for the nth term of the sequence, and find the value of the* 425*th term.*

15. $\{4, 7, 10, 13, 16, \ldots\}$

16. $\{90, 100, 110, 120, 130, \ldots\}$

17. $\{7.5, 6, 4.5, 3, 1.5, \ldots\}$

18. $\{1, 3, 6, 10, 15, \ldots\}$

19. *(Problem Solving)* Alan Medvick and Justina Prenatt, friends of Dave's daughter Sarah, are opening a savings account to finance the college education of their newborn son Lennon. They plan to deposit \$1200 during the first year, \$1320 during the second year, and so on, increasing the amount by \$120 each year. If they follow this plan, how much will they deposit into the account during the 18th year?

In solving problems you sometimes need to add a number of consecutive terms in an arithmetic sequence. An expression $a_1 + a_2 + \cdots + a_N$ is called an **arithmetic series**. In Exercises 20–27 you will develop and apply a formula to find sums of arithmetic series.

Exercises **20.** *(Problem Solving)* In Exercise 19, how much will Lennon's parents have deposited into their savings account during the first 18 years of his life? (*Hint*: The sequence of deposits has been copied down twice in Figure 3-3. To avoid a lot of tedious addition, refer to Figure 1-12 in Section 1-3 and use the strategy of *Examining a Related Problem*.)

1200	1320	1440	. . .	. . .	3000	3120	3240
↑	↑	↑			↑	↑	↑
3240	3120	3000	. . .	. . .	1440	1320	1200

Figure 3–3

21. Use the idea in Exercise 20 to find the sum

$$3 + 5 + 7 + \cdots + 2001$$

To apply the idea, you must first find the number of terms in the series. [*Hint*: The equation of the sequence is $a_n = 3 + 2(n - 1)$.]

22. Find the sum $7 + 4 + 1 - 2 - 5 - \cdots - 2987$.

23. Find the sum of the first 73 terms of the sequence

$$\{6, 10, 14, 18, 22, \ldots\}$$

24. Find the sum of the first 824 terms of the sequence

$$\{3.94, 4.03, 4.12, 4.21, 4.30, \ldots\}$$

In general, the sum of the series $a_1 + a_2 + \cdots + a_N$ is

$$S_N = \frac{N(a_1 + a_N)}{2}$$

25. *(Writing to Learn)* Using Exercises 20–24 as examples, explain why this formula is valid.

26. Find the sum $57 + 59 + 61 + \cdots + 123$.

27. Find the sum of the first 51 terms of the sequence

$$\{-15, -10, -5, 0, 5, \ldots\}$$

ADDITIONAL EXERCISES

28. *(Problem Solving)* A 5-gallon bucket in a well is full of water, and is attached to a rope weighing 2 pounds per foot. The bucket weighs 6 pounds, and a gallon of water weighs 8 pounds. Write an equation expressing the total weight of the bucket, water, and rope as a function of the length of the rope.

29. *(Problem Solving)* You are supervising the construction of an interstate highway along the Atlantic coast in Massachusetts. A portion of the roadbed will descend gradually in a straight line, covering a horizontal distance of 4000 feet. The elevations at the high and low ends will be 170 and 10 feet, respectively.

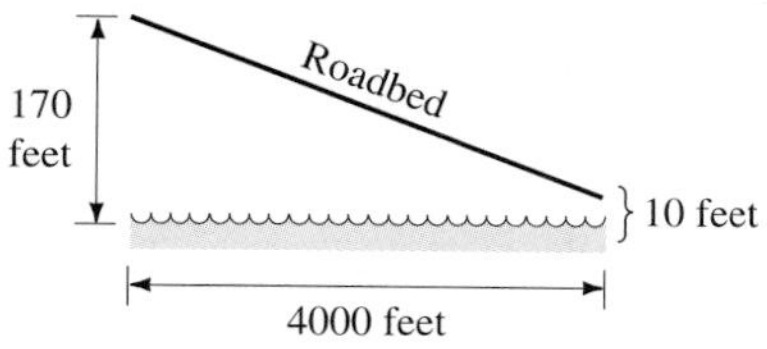

a. Write an equation expressing the elevation as a function of horizontal distance from the low end. We have helped you by using the strategy of ***Drawing a Picture or Diagram***.

b. To prevent erosion, a seawall must be constructed along any portion of the road whose elevation is no more than 20 feet. Use your equation to decide how far the seawall must extend from the low end of the roadway.

c. Graph the function in part (a), and use the graph to answer the question in part (b).

d. List any of Pólya's strategies which you used to complete parts (a)–(c), and describe where and how they were used.

30. *(Problem Solving)* When you apply the brakes on a car, it loses speed at an approximately constant rate. It takes a 1995 Nissan 300ZX 3 seconds to come to a stop from an initial speed of 87 feet per second.

a. Write an equation expressing the initial speed of the car as a function of the time required to stop after the brakes are applied.

b. At what initial speed is a 1995 Nissan 300ZX traveling if it takes 1.5 seconds to come to a complete stop? 4 seconds?

31. *(Problem Solving)* A charter bus service has offered a deluxe 7-day tour of Florida's Disney World at a special rate of \$800 per person for groups of 50 people. The company will also give a discount of \$15 per person for every additional two people in the group. How many people must participate for the group to get a rate of \$635 per person?

32. *(Problem Solving)* In planning a backpacking trip, you have decided that on the first day you will begin hiking at 7:00 A.M. At some point you plan to stop and take 30 minutes to set up a campsite. You will continue hiking without your pack and return to the campsite at 6:30 P.M. Based on prior experience, you expect to walk about 1.6 miles per hour with your pack and 2.8 miles per hour without it.

a. By how much will your total mileage be reduced each time you decide to walk another 15 minutes before setting up camp?

b. By how much will your total mileage be reduced each time you decide to walk another half-mile before setting up camp?

33. If t is directly proportional to s, and $t = 57$ when $s = 126$, find t when $s = 168$.

34. If N varies directly with M, and $N = \frac{1}{3}$ when $M = \frac{4}{5}$, find M when $N = \frac{3}{4}$.

35. If A varies linearly with B, and the constant of proportionality is 0.7, find the value of A when $B = 62$.

36. If q varies directly with p, and the constant of proportionality is 250, find the value of q when $p = 0.004$.

37. *(Writing to Learn)* Decide whether each of the following is a relationship of direct variation, and explain your answer.

a. $H =$ the number of hours a 100-watt light bulb is burned
$C =$ the cost, in cents, of burning the light bulb

b. $N =$ the number of people using a reservoir
$G =$ the number of gallons of water taken from the reservoir each day

c. $P =$ the number of people living in a house
$C =$ the amount of closet space per person

d. $N =$ the number of pages in a telephone book
$W =$ the weight of the book in pounds

38. *(Creating Models)* If the relationship referred to is a direct variation, use the given information to write its equation. If the relationship is not a direct variation, do not attempt to write an equation.

a. (Exercise 37a) It costs \$0.06375 to burn a 100-watt bulb for 5 hours.

b. (Exercise 37b) 30,000 people use a combined total of 2,400,000 gallons of water from the reservoir each day.

c. (Exercise 37c) When 5 people share the house, each one has 23 cubic feet of closet space.

d. (Exercise 37d) A telephone book with 1568 pages weighs 4.8 pounds.

In Exercises 39–46,

a. Decide whether the sequence is arithmetic.

If it is, then

b. Write an equation to describe it.

c. Find the value of the 100th term, and the value of the 425th term.

d. Find the sum of the first 100 terms and the sum of the first 425 terms.

39. $\{7, 10, 13, 16, 19, \ldots\}$

40. $\{-8, -3, 2, 7, 12, \ldots\}$

41. $\{3487.6, 3487.5, 3487.4, 3487.3, 3487.2, \ldots\}$

42. $\{\sqrt{6}, \sqrt{6}, \sqrt{6}, \sqrt{6}, \sqrt{6}, \ldots\}$

43. $\{3, 6, 12, 24, 48, \ldots\}$

44. $\{-1, 0, 1, 0, -1, \ldots\}$

45. $\{\pi - 2, 2\pi - 3, 3\pi - 4, 4\pi - 5, 5\pi - 6, \ldots\}$

46. $\{2, 4, 6, 10, 16, \ldots\}$

47. *(Interpreting Mathematics)* The following table shows the number of feet an object in a vacuum falls each second after it is dropped.

Time, in seconds	Distance, in feet
1	16
2	48
3	80
4	112

a. Verify that the table defines an arithmetic sequence.

b. How far does the object fall during the 60th second?

c. How far does the object fall altogether during the first 60 seconds?

48. *(Writing to Learn)* If mathematicians had named baseball's World Series, it would be called the World Sequence. Explain why.

3–3 LINEAR MODELING OF NONLINEAR RELATIONSHIPS

LINEARIZATION

Try the following experiment. First, graph the function $y = x^2$ in the window $[-10, 10]$ by $[-10, 10]$. Next, zoom in on a small interval around the point (1, 1). Finally, pick any other point on the graph and zoom in on a small interval around it. In each case, what do you see on the screen after zooming in?

As your experiment suggests, any smooth curve can be closely approximated by a line over a small interval. This simple observation has many powerful consequences. In fact, much of calculus is based on it. Its importance to us is suggested by Figure 3-4.

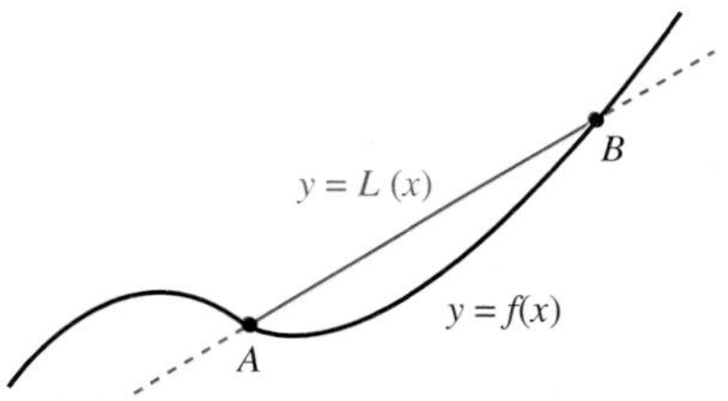

Figure 3–4

A line such as L, joining two points A and B on the graph of a nonlinear function f, is called a **secant line**. If the secant line is close to the nonlinear graph between A and B, we can use values of the linear function $L(x)$ to estimate values of $f(x)$. This process is referred to as **linearization** and is carried out for two primary reasons.

- Linear relationships are simple. Many nonlinear relationships are so complicated that it is impossible to write useful equations to describe them. Even when equations can be written, they are often difficult to analyze.
- In practice, many quantitative questions require only approximate answers. Errors due to linearizing may be acceptable, depending on the question. In addition, it is possible to analyze the size of errors due to linearization. (The analysis requires calculus, so we will not discuss it here.)

In the following situation a nonlinear function f can be closely approximated by a linear function L over an interval.

A MATHEMATICAL LOOKING GLASS
Cold Coffee

As we wrote the preceding paragraphs, Lynn took a sip of her coffee and said, "Yuk! It's cold already." Wondering how long it actually takes a cup of coffee to cool off, we poured another cup and recorded the data in Table 12.

TABLE 12

Time since coffee was poured, in minutes	Temperature, in degrees Fahrenheit
0	175
5	147
10	126
15	111
20	100
25	92
30	86

Table 12 does not fit a linear function. In Chapter 9 you will discover a type of function that fits the data. In the meantime, you can provide an approximate fit by linearizing.

Suppose that you want to know how long it took the coffee to cool to a temperature of 150°. In Exercises 1–6, you will use a graphical approach to find out. ■

EXERCISES

1. *(Review)* Verify that Table 12 does not fit a linear function.
2. *(Making Observations)* Plot the points from Table 12 on a large piece of graph paper. Sketch a smooth curve through the points, and use your graph to complete the following table. Estimate the temperatures to the nearest degree.

x	y
0	175
1	
2	
3	
4	
5	147

3. On your graph paper from Exercise 2, sketch the line through (0, 175) and (5, 147).

Exercise 3 illustrates that the line segment through (0, 175) and (5, 147) is "close" to the segment of the curve between the same two points.

4. Find the equation of the line through (0, 175) and (5, 147), and use it to complete the table in Exercise 2.

5. *(Making Observations)* Complete the following table.

x	y (from Exercise 2)	y (from Exercise 4)	Difference
0	175	175	0
1			
2			
3			
4			
5	147	147	0

Exercise 5 illustrates numerically that the line segment is "close" to the curve. That is, the segment of the curve between (0, 175) and (5, 147) is "approximately linear" and can be modeled by a linear function without great loss of accuracy. You can now use this fact to answer the question posed in ***Cold Coffee***.

EXERCISE

6. *(Interpreting Mathematics)* Use your linear equation from Exercise 4 to estimate how long it takes the coffee to reach a temperature of 150°.

AVERAGE RATE OF CHANGE

Your work in Exercises 1–6 can be viewed from the perspective of rates of change. During the first 5 minutes, the change in the temperature was

$$147 - 175 = -28 \text{ degrees}$$

The change in time was

$$5 - 0 = 5 \text{ minutes}$$

Thus the temperature changed at an average rate of

$$\frac{147 - 175}{5 - 0} = -5.6 \text{ degrees per minute}$$

Example 1 shows how you can use the idea of average rate of change to estimate changes in x and y.

EXAMPLE 1 *Using rates of change to estimate data*

What was the temperature of the coffee in ***Cold Coffee*** after 2 minutes? We now know that it cooled at an average rate of 5.6 degrees per minute during the first 5 minutes. Since its initial temperature was 175 degrees, its temperature after 2 minutes was about $175 + 2(-5.6) = 163.8$ degrees. ■

EXERCISES

7. *(Interpreting Mathematics)* If the coffee cools at the rate of 5.6 degrees per minute, how long does it take to cool from 175 degrees to 150 degrees? Make sure your answer agrees with the one you obtained in Exercise 6.

8. *(Interpreting Mathematics)* Refer to the data from ***Cold Coffee***.

   a. During the time interval between 15 and 20 minutes, the coffee cooled at an average rate of how many degrees per minute?

   b. Use your answer from part (a) to estimate the temperature of the coffee after 18 minutes.

   c. Use your answer from part (a) to estimate the time when the temperature was 109°.

Estimate the answers to the questions in Exercises 9–11 *by assuming that the relationship is approximately linear. Either construct an equation, as in Exercise* 4, *or use the idea of rate of change, as in Exercise* 7.

9. *(Problem Solving)* A house whose appraised value in 1980 was \$52,000 had a value of \$97,000 in 1995. What was it worth in each year during the 1980s?

10. *(Problem Solving)* The city of Yellowknife in Canada's Northwest Territories receives 20.8 hours of daylight on June 21 and 12.0 hours of daylight on September 21 (92 days later). How many hours of daylight does Yellowknife receive on August 12 (52 days after June 21)?

11. *(Problem Solving)* $\sqrt{16} = 4$ and $\sqrt{36} = 6$. Without using your calculator, estimate the square root of each whole number between 16 and 36.

Let's look more carefully at the idea of average rate of change. Graphs of nonlinear functions, such as the one in ***Cold Coffee***, do not have a constant slope. Thus the ratio $\frac{\Delta y}{\Delta x}$ is not constant, but varies according to the two points chosen on the graph. In Figure 3-5 the value of $\frac{\Delta y}{\Delta x}$ is larger between Q and R than between P and Q.

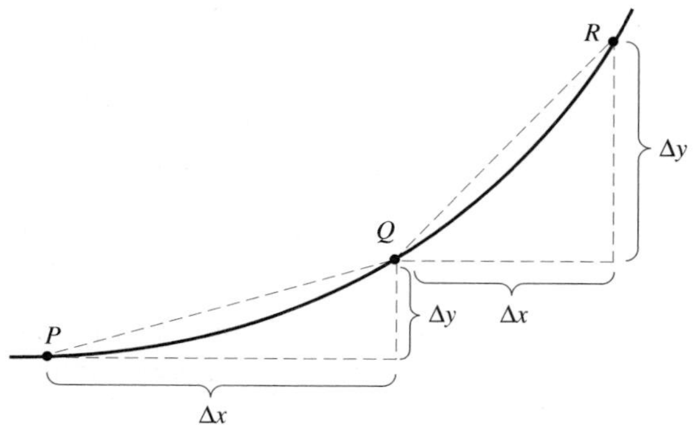

Figure 3–5

The value of $\frac{\Delta y}{\Delta x}$ on the line PQ (its slope) represents the rate of change in y with respect to x along that line. Since P and Q are also on the nonlinear graph, that value of $\frac{\Delta y}{\Delta x}$ also represents the **average rate of change** in $f(x)$ with respect to x between these points. Similarly, the slope of the line QR represents the average rate of change in $f(x)$ with respect to x between Q and R. In general, the average rate of change in $f(x)$ with respect to x over an interval $[a, b]$ is

$$\frac{f(b) - f(a)}{b - a}$$

Example 2 shows how to calculate an average rate of change and illustrates why the word "average" is appropriate in the definition.

EXAMPLE 2 *Calculating and interpreting an average rate of change*

Suppose you are traveling a long stretch of open highway by car. Let

$t =$ time in hours past noon

$y =$ distance driven since noon in miles

The average rate of change in y (distance) with respect to t (time) is your average speed. Let's look at a situation in which the rate of change in y with respect to t is constant, and one in which it is not.

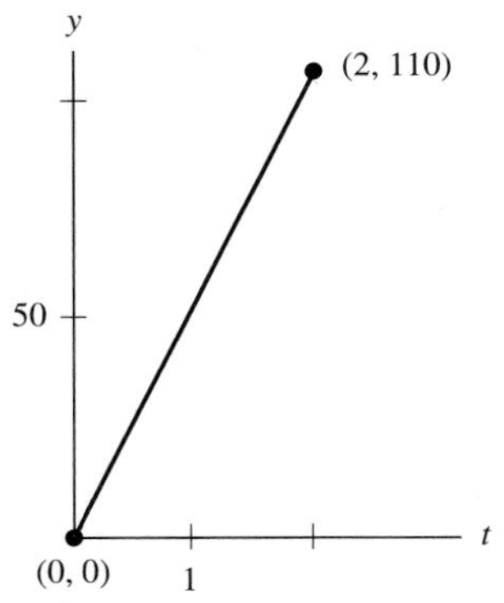

Figure 3–6

a. If you set your cruise control for 55 mph, then t and y are related by the linear equation

$$y = f(t) = 55t$$

Between noon and 2:00 P.M. your average speed is

$$\frac{f(2) - f(0)}{2 - 0}$$

$$= \frac{110 - 0}{2 - 0}$$

$$= 55$$

The result makes sense in the physical context, since your speed is constant at 55 mph. Graphically, the result represents the slope of the graph in Figure 3-6.

b. If you have no cruise control, your speed will probably not be constant. The equation

$$y = g(t) = 5t^2 + 45t$$

gives one possible relationship between t and y. Between noon and 2:00 P.M. your average speed is

$$\frac{g(2) - g(0)}{2 - 0}$$

$$= \frac{110 - 0}{2 - 0}$$

$$= 55$$

This result also makes sense in the physical context, since you have gone 110 miles in 2 hours. Graphically, the result represents the slope of the secant line in Figure 3-7.

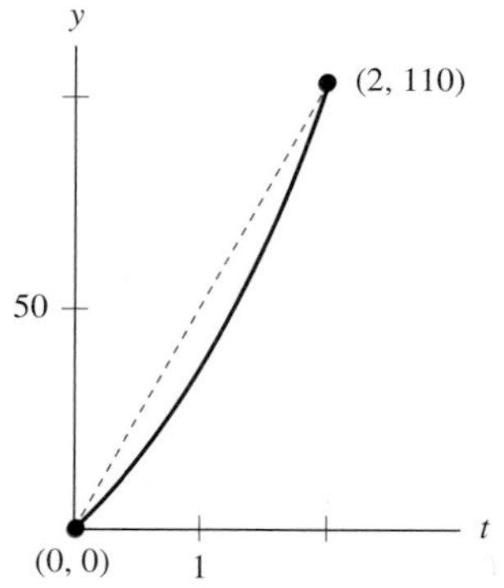

Figure 3–7

Since the graph of $y = g(t)$ is not linear, the average rate of change in y with respect to t will vary from one interval to another. You can confirm this in Exercise 12. ■

EXERCISE **12.** *(Making Observations)*

a. In part (a) of Example 2, calculate the average rate of change in y with respect to t over each of the intervals $[0, 1]$ and $[1, 2]$.

b. In part (b) of Example 2, calculate the average rate of change in y with respect to t over each of the intervals $[0, 1]$ and $[1, 2]$.

c. What do your results tell you about the movement of the car in part (b) compared to that of the car in part (a)?

Example 2 and Exercise 12 illustrate that

- An object traveling at a constant velocity has a linear time-distance graph, such as the line in Figure 3-6. The slope of the line is equal to the velocity.
- An object traveling at a varying velocity has a nonlinear time-distance graph, such as the curve in Figure 3-7. The slope of a secant line between any two points on the graph is equal to the average velocity between those two points.

To generalize the preceding statements:

- If the rate of change in y with respect to x is constant, the graph of $y = f(x)$ is linear. The slope of the line is equal to the rate of change.
- If the rate of change in y with respect to x varies, the graph of $y = f(x)$ is nonlinear. The slope of a secant line between any two points on the graph is equal to the average rate of change in y with respect to x between those two points.

The process of finding and interpreting an average rate of change is illustrated again in Example 3.

EXAMPLE 3 *Finding and interpreting an average rate of change*

a. To find the average rate of change in the function

$$f(x) = \frac{1000}{x}$$

over the interval [40, 50], first calculate $f(40) = 25$ and $f(50) = 20$. Then find the slope of the line through (40, 25) and (50, 20), as shown in Figure 3-8. The average rate of change is

$$\frac{\Delta y}{\Delta x} = \frac{20 - 25}{50 - 40} = -\frac{1}{2}$$

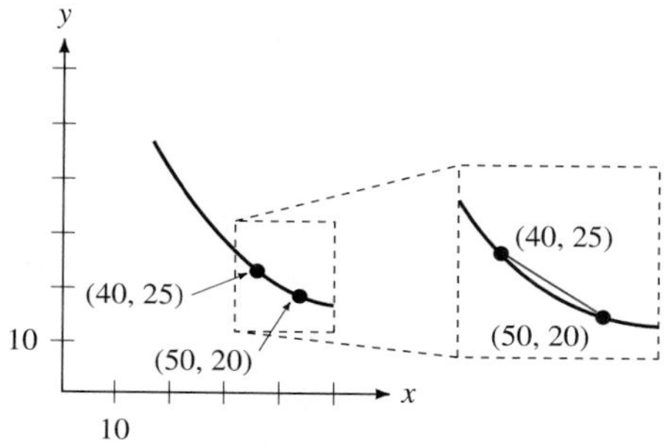

Figure 3–8

b. If $f(x)$ represents the number of hours needed to drive 1000 miles at x miles per hour, then part (a) says that for average speeds between 40 and 50 mph, each increase of 1 mph in your speed reduces your total driving time by about one half hour.

c. If $f(x)$ represents the number of months needed to save \$1000 by saving x dollars per month, then part (a) says that for amounts between \$40 and \$50 per month, each increase of \$1 per month reduces the length of time by half a month. ■

EXERCISES

13. *(Interpreting Mathematics)* In ***Cold Coffee*** (page 92), find and interpret the average rate of change between the points (20, 100) and (30, 86).

14. *(Interpreting Mathematics)* In Exercise 9, find and interpret an appropriate average rate of change.

15. *(Interpreting Mathematics)* In Exercise 10, find and interpret an appropriate average rate of change.

16. *(Interpreting Mathematics)* In ***Big Blue Marble*** (page 55), find and interpret the average rate of change in $g(x)$ with respect to x over the interval [0, 1600].

Economists often linearize nonlinear functions when analyzing cost, revenue, or profit, as in the following Mathematical Looking Glass.

A MATHEMATICAL LOOKING GLASS

Marginal Profit (Fat Cats 3)

Based on the information in ***Fat Cats on Campus*** (page 43), one of your classmates has constructed a table containing your projected profit for the semester at a variety of sales levels. Table 13 shows some of his projections.

TABLE 13

Number of shirts sold (x)	Profit [$P(x)$]
500	$4387.50
510	4462.00
520	4535.50
530	4608.00

Your economics professor has told you that a company's **marginal profit** is the additional profit resulting from the sale of one additional item. She has also said that a company's marginal profit varies from one sales level to another. To gain some insight into her statement, you have decided to find your company's marginal profit at sales levels of 505 and 525 shirts.

Although Table 13 does not list the projected profits for every number of shirts between 500 and 530, you can use the idea of average rate of change to estimate your marginal profits. ■

EXERCISES

17. *(Interpreting Mathematics)*

a. Calculate the average rate of change in $P(x)$ with respect to x over the interval $[500, 510]$.

b. What is your projected marginal profit at a sales level of 505 shirts?

c. What do your calculations in parts (a) and (b) tell you in physical terms?

18. *(Interpreting Mathematics)*

a. Calculate the average rate of change in $P(x)$ with respect to x over the interval $[520, 530]$.

b. What is your projected marginal profit at a sales level of 525 shirts?

19. *(Making Observations)* Does your profit increase more if you increase your target sales level from 505 to 506 shirts, or if you increase it from 525 to 526 shirts?

ADDITIONAL EXERCISES

In Exercises 20–23,

a. Find a linear function L to approximate the nonlinear function f over the interval $[2, 5]$.

b. Verify that $L(2) = f(2)$ and $L(5) = f(5)$.

c. Calculate the differences $L(3) - f(3)$ and $L(4) - f(4)$.

20. $f(x) = \dfrac{1}{x}$ **21.** $f(x) = \sqrt{x - 1}$

22. $f(x) = x^2 + 6x$ **23.** $f(x) = x^2 - 6x$

In Exercises 24–27, find the average rate of change in y with respect to x over the interval [0, 10].

24. $y = x^3$

25. $y = x^3 - 500$

26. $y = \dfrac{1}{x^2 + 100}$

27. $y = 7 + 10x - x^2$

t	*m*
10	1321.51
15	1074.61
20	965.02
25	908.70
30	877.57

28. *(Problem Solving)* In the table at the left *m* is the dollar amount of each monthly payment on a $100,000 mortgage at an interest rate of 10%, to be paid back over a period of *t* years.

a. Find and interpret the average rate of change in *m* with respect to *t* over [25, 30].

b. In how many years could you pay off the mortgage at a rate of $900 per month?

29. *(Interpreting Mathematics)*

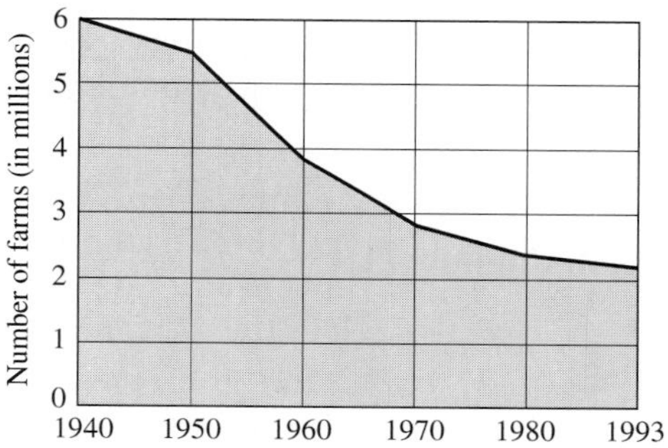

(Adapted from *The World Almanac*, Funk and Wagnalls, 1995.)

a. Estimate and interpret the rate of change in the number *N* of farms in the United States with respect to the year *y* over [1940, 1980].

b. Use your result to estimate the number of farms in the United States in 1993. (The actual number was 2.1 million.)

30. *(Interpreting Mathematics)*

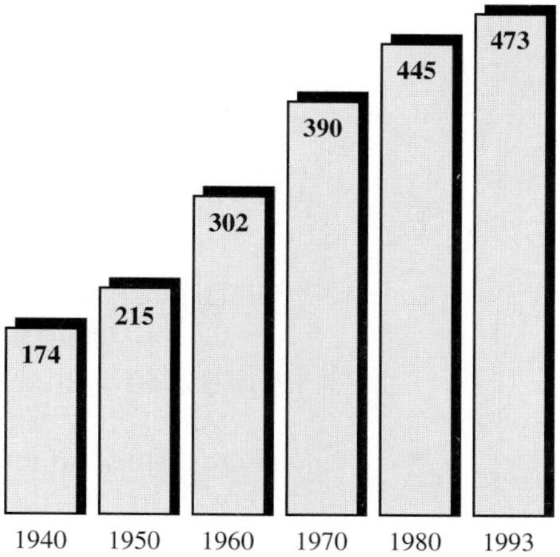

(Adapted from *The World Almanac*, Funk and Wagnalls, 1995.)

a. Estimate and interpret the rate of change in the size S of the average farm in the United States with respect to the year y over [1940, 1980].

b. Use your result to estimate the size of the average farm in the United States in 1993. (The actual average size was 473 acres.)

31. *(Problem Solving)* In 1983, 26.2% of all persons in the United States between the ages of 18 and 24 were enrolled in college. By 1992 the percentage had risen to 34.4%. (*Source*: *The Almanac of Higher Education*, University of Chicago Press, 1995.) Let p represent the percentage in the year y.

a. Find and interpret the average rate of change in p with respect to y over [1983, 1992].

Use your result from part (a) to estimate:

b. the value of p in 1987. (The actual value was 29.6%.)

c. the value of p in 1993. (The actual value was 34.0%.)

d. the value of p in 2063.

e. How accurate do you believe your estimate in part (d) is? Explain your answer.

SUPPLEMENTARY TOPICS

Linearizing a Situation

To model a problem with a linear function, we often need to make some simplifying assumptions. In ***Cold Coffee*** it was assumed that the coffee cooled at a constant rate during the first five minutes. While this was almost certainly not the case, the assumption still yielded reasonable results.

32. *(Writing to Learn)* List at least one simplifying assumption you needed to make to construct a linear model in each exercise.

a. in Exercise 9 **b.** in Exercise 10

c. in Exercise 11 **d.** in Exercise 28

33. *(Writing to Learn)* List at least two assumptions you needed to make to construct the linear functions in Exercise 55 of Section 3-1. Do you think these assumptions closely match the actual behavior of an airliner approaching an airport? Why or why not?

In Exercises 34–36,

a. *(Problem Solving)* Complete the exercise, assuming that the described relationship is linear.

b. *(Writing to Learn)* Explain what the assumption of linearity means in physical terms.

34. If the temperature outside was 45° F at 7:00 A.M., and is 83° F at 2:00 P.M., construct a table giving the approximate temperature at 30-minute intervals between 8:00 A.M. and noon.

35. Driving westward across Kansas on I-70, I passed milepost 421 at 9:45 A.M. and milepost 195 at 1:31 P.M. Where was I at noon?

36. The air pollution level is 136 parts per million when measured at a distance of 50 feet from a major highway and 84 parts per million when measured at a distance of 180 feet. How far must you go from the highway before the level of pollution drops to 130 parts per million? 120? 110? 100? 90?

Local Linearization: A "Propinquity Principle" The word *propinquity* means "nearness in time or place." Our "Propinquity Principle" states that linearizing a function over an interval usually yields better estimates close to the interval than farther away from it.

37. *(Making Observations)* Recopy your graph from Exercise 2. Obtain two estimates of the time when the coffee reaches a temperature of 90°, as follows.

a. Sketch a straight line through the points (0, 175) and (5, 147), and use the x-coordinate when $y = 90$ on this line.

b. Sketch a straight line through the points (20, 100) and (25, 92), and use the x-coordinate when $y = 90$ on this line.

c. Decide which estimate is more accurate, and explain your choice.

3–4 PIECEWISE LINEAR FUNCTIONS

PREREQUISITE MAKE SURE YOU ARE FAMILIAR WITH:

Absolute Value Equations and Inequalities (Section A-9)

A MATHEMATICAL LOOKING GLASS

Health Insurance

After Dave's daughter Sarah graduated from college in 1995, she was employed as a counselor at Bethesda Children's Home. For the first several months her position was part time and included no benefits. In particular, Sarah had to purchase her own health care insurance. She called several providers, including Blue Cross and Blue Shield, who offered her a choice of the three plans summarized in Table 14.

TABLE 14

	Monthly premium	Annual deductible
Plan 1	\$73.30	\$1000
Plan 2	81.60	500
Plan 3	88.80	250

Under Plan 1 Sarah must pay the first \$1000 of any medical expenses each year (the deductible) out of her own pocket. The plans with higher premiums carry smaller deductibles. If medical expenses exceed the deductible for a given plan, Blue Cross and Blue Shield pays 80% of the additional expenses until the total for the year reaches \$5000.

Sarah realized that different annual levels of medical expenses might make different plans more desirable. To decide which plan would be best for her, she began by *Making a Table* showing the total cost of each plan at several levels of medical expenses. *Breaking the Problem into Parts*, she began by considering the plan with the lowest premium. Table 15 shows some of her calculations.

TABLE 15

Total annual medical expenses	Total annual cost of Plan 1
\$ 0	\$ 879.60
250	1129.60
500	1379.60
750	1629.60
1000	1879.60

■

EXERCISES

1. *(Writing to Learn)* Describe how each entry in Table 15 was calculated.
2. *(Writing to Learn)* Explain why Table 15 can be continued as follows.

Total annual medical expenses	Total annual cost of Plan 1
\$1000	\$1879.60
1250	1929.60
1500	1979.60
1750	2029.60
2000	2079.60

3. *(Writing to Learn)* Let x represent Sarah's annual medical expenses and y represent the total annual cost of Plan 1, both in dollars.
   a. Use Table 15 to show that if Sarah's annual expenses are \$1000 or less, the cost of Plan 1 is expressed by the equation
   $$y = x + 879.60$$
   b. Use the continuation of Sarah's table in Exercise 2 to show that if her annual expenses are more than \$1000, the cost of Plan 1 is expressed by the equation
   $$y = 0.20x + 1679.60$$
4. Exercise 3 shows that the cost of Plan 1 is
   $$f(x) = \begin{cases} x + 879.60, & \text{if } 0 \leq x \leq 1000 \\ 0.20x + 1679.60, & \text{if } x > 1000 \end{cases}$$
   Use these equations and Sarah's table to sketch the graph of f.

THREE VIEWS OF PIECEWISE LINEAR FUNCTIONS

In Exercises 59–60 you can find functions to express the annual cost of Plans 2 and 3 and decide which plan would be best for a given level of medical expenses. In the meantime, let's notice that function f in Exercise 4 is composed of "pieces" of two separate linear functions, and is therefore called **piecewise linear**. Specifically, a piecewise linear function is one that is described by two or more formulas $y = mx + b$, each applied over a specified interval. Exercise 4 illustrates the usual way of writing such functions analytically. Let's also look at piecewise linear functions from numerical and graphical perspectives.

A Numerical View Example 1 illustrates how to evaluate a piecewise linear function at a particular value of the independent variable.

EXAMPLE 1 *Evaluating a piecewise linear function*

Suppose $f(x) = \begin{cases} 2x - 1, & \text{if } x \leq 3 \\ 8 - 1, & \text{if } x > 3 \end{cases}$

The equation defines $f(x)$ by separate rules in the intervals $(-\infty, 3]$ and $(3, \infty)$, respectively. For example,

Since $2 \leq 3$, $f(-4) = 2(2) - 1 = 3$

Since $3 \leq 3$, $f(3) = 2(3) - 1 = 5$

Since $4 \geq 3$, $f(4) = 8 - (4) = 4$ ■

EXERCISES **5.** *(Making Observations)*

a. Make a table showing the values of $f(x)$ for $x = 2.5, 2.6, 2.7, \ldots, 3.3, 3.4, 3.5$.

b. Make a table showing the values of $f(x)$ for $x = 2.95, 2.96, 2.97, \ldots, 3.03, 3.04, 3.05$.

c. If x is near 3, do your tables indicate that y is near $f(3)$? Explain.

6. *(Making Observations)* Repeat Exercise 5 for

$$g(x) = \begin{cases} x - 2, & \text{if } x < 3 \\ 3x - 1, & \text{if } x \geq 3 \end{cases}$$

A Graphical View Example 2 illustrates a method for graphing a piecewise linear function.

EXAMPLE 2 *The graph of a piecewise linear function*

Let's sketch the graph of the function from Example 1.

$$f(x) = \begin{cases} 2x - 1, & \text{if } x \leq 3 \\ 8 - x, & \text{if } x > 3 \end{cases}$$

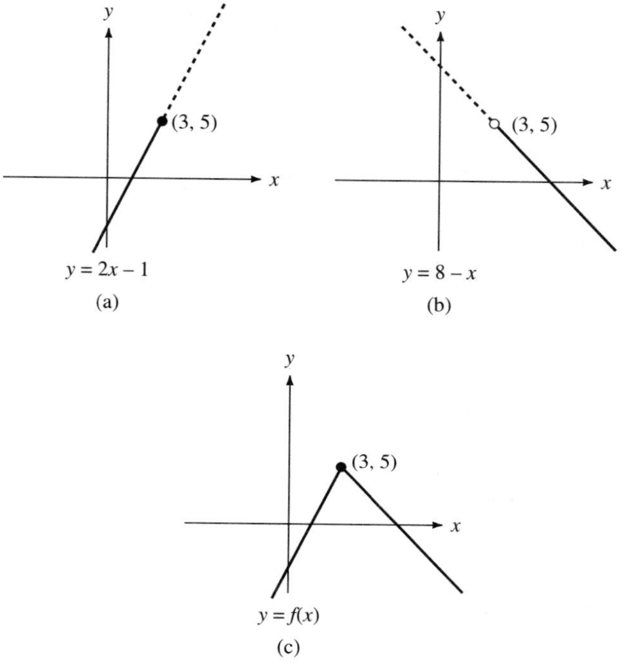

Figure 3–9

Since the expression $2x - 1$ defines $f(x)$ for $x \leqslant 3$, the graph of $y = f(x)$ coincides with the graph of $y = 2x - 1$ on and to the left of its intersection with the line $x = 3$. Similarly, the graph of $y = f(x)$ coincides with the graph of $y = 8 - x$ to the right of $x = 3$. Figure 3-9 indicates how pieces of the two linear graphs combine to form the graph of $f(x)$. ■

EXERCISES

7. Use the method of Example 2 to sketch the graph of the function

$$g(x) = \begin{cases} x - 2, & \text{if } x < 3 \\ 3x - 1, & \text{if } x \geqslant 3 \end{cases}$$

8. *(Writing to Learn)* Describe how the graph in Figure 3-9 and your graph in Exercise 7 support your conclusions in Exercises 5c and 6c.

In Exercises 9–12, sketch the graph of the function. Check your graphs on your calculator.

9. $y = \begin{cases} 3 - x, & \text{if } x \leqslant 1 \\ x + 1, & \text{if } x > 1 \end{cases}$

10. $Z = \begin{cases} 0.5W + 4.5, & \text{if } W < -4 \\ 2.5, & \text{if } W \geqslant -4 \end{cases}$

11. $f(t) = \begin{cases} t - 2, & \text{if } t \leqslant 3 \\ 7 - t, & \text{if } t > 3 \end{cases}$

12. $g(x) = \begin{cases} 1, & \text{if } x < 0 \\ 1 - 0.5x, & \text{if } 0 \leqslant x \leqslant 2 \\ 3x - 2, & \text{if } x > 2 \end{cases}$

13. *(Making Observations)* As you have seen, the "pieces" of a piecewise linear graph may join to form a continuous graph as in Figure 3-9, or they may fail to join as in Exercise 7. How can you tell what will happen by looking at the equation of the function? The functions in Exercises 9–12 can help you answer this question.

THREE VIEWS OF LINEAR ABSOLUTE VALUE FUNCTIONS

Since

$$|x| = \begin{cases} -x, & \text{if } x < 0 \\ x, & \text{if } x \geqslant 0 \end{cases}$$

the function $f(x) = |x|$ is piecewise linear, and is an example of a **linear absolute value function**, that is, a function $f(x) = a|x - h| + k$ for some real numbers a, h, and k. Linear absolute value functions occur in physical situations when we are concerned about the *size* of the difference between two quantities, without regard to the *sign*. The following Mathematical Looking Glass provides an example.

A MATHEMATICAL LOOKING GLASS
Salt Container

Morton International, Inc., produces cylindrical 26-ounce salt containers, designed to have a height of 13.90 cm and a radius of 4.280 cm. The heights and radii of the actual containers deviate slightly from these values, so that the actual capacity deviates slightly from the designed capacity of 800 cm^3.

Morton employees responsible for quality control need to know what deviations in radius and height produce acceptably small deviations in capacity.

We will not attempt to deal with the complexities of considering deviations in radius and height simultaneously. Instead, we will adopt the strategy of *Examining a Simpler Problem*. Let's assume that each container has a radius of exactly 4.280 cm. Then the deviation in capacity results entirely from the deviation between the actual height of the container and the designed height of 13.90 cm. ■

EXERCISES

14. Use the formula $V = \pi r^2 h$ for the volume of a cylinder to verify that a container with the designed height and radius has a capacity of 800 cm^3.

15. a. By how much does a container deviate from its designed height if its actual height is 14.00 cm? 13.95 cm? 13.85 cm? 13.80 cm?

b. *(Writing to Learn)* If the actual height of a container is h cm, explain why the deviation in centimeters from its designed height is $|h - 13.90|$.

c. *(Writing to Learn)* Explain why the resulting deviation in cubic centimeters from its designed capacity is $\pi(4.280)^2|h - 13.90| \cong 57.55|h - 13.90|$.

The acceptable variations in the heights of Morton salt containers are closely related to the linear absolute value function $E(h) = 57.55|h - 13.90|$ in Exercise 15. To prepare ourselves to solve problems involving such functions, let's consider them from numerical, analytical, and graphical perspectives.

A numerical view fails to provide much insight by itself. Without additional information it is usually difficult to tell whether a table was generated by a linear absolute value function. However, a table can help explain the presence of certain features in a linear absolute value graph.

An Analytical View Example 3 illustrates that the equation of every linear absolute value function can be written in piecewise linear form.

EXAMPLE 3 *Writing a linear absolute value function in piecewise linear form*

To write

$$f(x) = 3|x - 2| - 4$$

in piecewise linear form, recall that

$$|x - 2| = \begin{cases} -(x - 2), & \text{if } x - 2 < 0 \text{ (that is, if } x < 2) \\ (x - 2), & \text{if } x - 2 \geqslant 0 \text{ (that is, if } x \geqslant 2) \end{cases}$$

Thus

$$f(x) = \begin{cases} -3(x - 2) - 4, & \text{if } x < 2 \\ 3(x - 2) - 4, & \text{if } x \geqslant 2 \end{cases}$$

In Example 4 you will discover why it is usually better to leave the equations written as they are here, as opposed to multiplying and combining terms in each expression. ■

EXERCISES *In Exercises* 16–19, *write the function in piecewise linear form.*

16. $f(x) = 2|x - 1|$ **17.** $g(x) = 3|x| + 4$

18. $F(t) = 0.5|t + 4| - 1$ **19.** $H(z) = 5 - |z|$

A Graphical View As Exercises 16–19 suggest, every linear absolute value graph is V-shaped. The vertical line through its **vertex** (the point of the V) is an **axis of symmetry** for the graph. That is, the pieces of the graph to the left and right of the line mirror each other as in Figure 3-10. The piecewise linear form of the equation provides some clues as to why this is true and allows us to find the vertex, the axis of symmetry, and the range of the function.

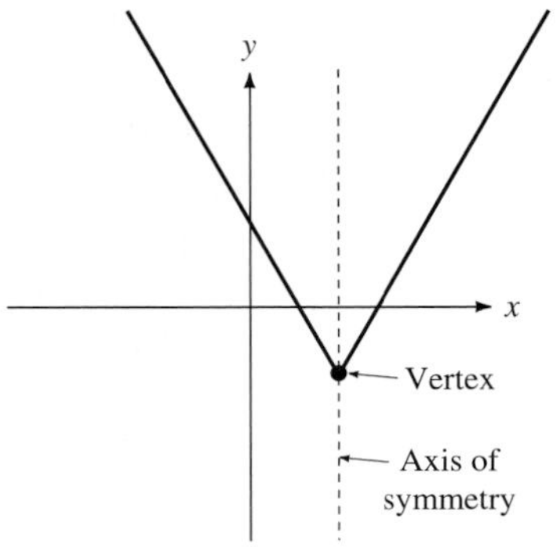

Figure 3–10

EXAMPLE 4 *Graphing a linear absolute value function in piecewise linear form*

To sketch the graph of

$$f(x) = 3|x - 2| - 4$$

write the equation in piecewise linear form, as in Example 3.

$$f(x) = \begin{cases} -3(x-2)-4, & \text{if } x < 2 \\ 3(x-2)-4, & \text{if } x \geqslant 2 \end{cases}$$

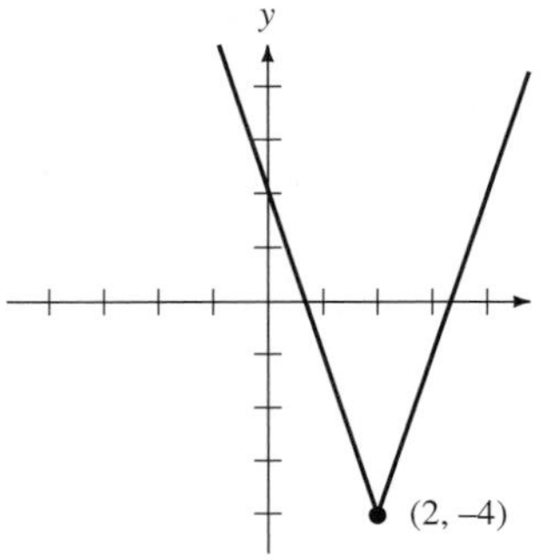

Figure 3–11

Figure 3-11 shows that the graph of $f(x)$ coincides with that of $y = -3(x - 2) - 4$ to the left of $x = 2$ and with that of $y = 3(x - 2) - 4$ to the right of $x = 2$. From their equations, it is easy to verify that the two lines each pass through $(2, -4)$ and that their slopes are -3 and 3, respectively.

The vertex is $(2, -4)$ and the axis of symmetry is the line $x = 2$. The range of the function is $[-4, \infty)$. ■

EXERCISES *In Exercises* 20–23,

a. Write the function in piecewise linear form.

b. Find the slope of each piece and the coordinates of the vertex.

c. Sketch the graph of the function.

d. Identify the axis of symmetry, and find the range of the function.

20. $y = 2|x - 4|$ **21.** $y = 2|x| - 4$

22. $y = \frac{1}{3}|x + 5| + 1$ **23.** $y = -|x - 6| + 3$

24. *(Making Observations)* On the graph of the function $f(x) = a|x - h| + k$, what are the slopes of the two pieces? Where is the vertex? How can you tell whether the graph opens upward or downward? Your results from Exercises 20–23 can help you answer these questions.

25. *(Making Observations)* The functions in parts (a)–(d) are from Exercises 20–23. Make a table for each, using the given values of x.

a. $y = 2|x - 4|$, $x = 2, 3, 4, 5, 6$

b. $y = 2|x| - 4$, $x = -2, -1, 0, 1, 2$

c. $y = \frac{1}{3}|x + 5| + 1$, $x = -7, -6, -5, -4, -3$

d. $y = |x - 6| + 3$, $x = 4, 5, 6, 7, 8$

e. How do the entries in your tables support the appearance of the graphs in Exercises 20–23?

Your observations in Exercise 24 can be stated as follows.

> The graph of $y = a|x - h| + k$ has its vertex at (h, k). It consists of two linear pieces whose slopes are $\pm a$. It opens up if $a > 0$ and down if $a < 0$.

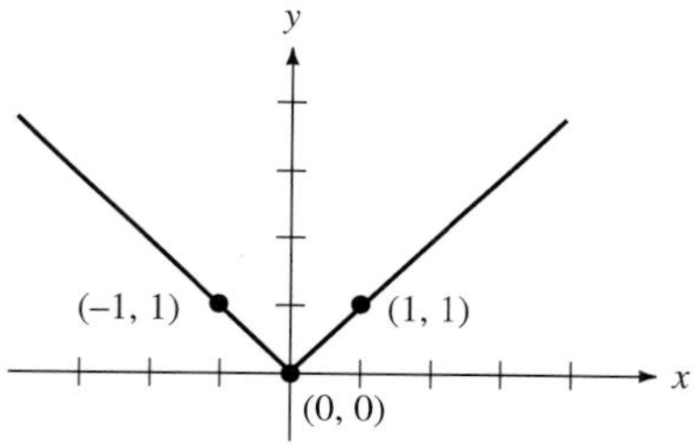

Figure 3–12

A different way to think of the graph of $y = a|x - h| + k$ is outlined in Exercises 26–30. Figure 3-12 shows the graph of the simplest linear absolute value function $y = |x|$. The graph of every other linear absolute value function can be obtained by transforming (stretching, compressing, reflecting, and shifting) this basic graph in appropriate ways.

EXERCISES

26. *(Making Observations)* Graph the functions $y = |x|$ and $y = 2|x|$ on the same set of coordinate axes. Describe how the graph of $y = |x|$ must be transformed to obtain the graph of $y = 2|x|$.

27. Graph the functions in parts (a)–(d) on one set of coordinate axes.

a. $y = |x|$ **b.** $y = 2|x|$

c. $y = 3|x|$ **d.** $y = 4|x|$

e. *(Making Observations)* Describe how the graph of $y = |x|$ must be transformed to obtain the graph of $y = a|x|$ if $a > 1$.

28. Graph the functions in parts (a)–(d) on one set of coordinate axes.

a. $y = |x|$ **b.** $y = 0.5|x|$

c. $y = 0.2|x|$ **d.** $y = 0.1|x|$

e. *(Making Observations)* Describe how the graph of $y = |x|$ must be transformed to obtain the graph of $y = a|x|$ if $0 < a < 1$.

29. Graph the function in parts (a)–(d) on one set of coordinate axes.

a. $y = |x|$ **b.** $y = |x + 1|$

c. $y = |x + 4|$ **d.** $y = |x - 1|$

e. *(Making Observations)* Describe how the graph of $y = |x|$ must be transformed to obtain the graph of $y = |x - h|$.

30. Graph the functions in parts (a)–(d) on one set of coordinate axes.

a. $y = |x|$ **b.** $y = |x| - 1$

c. $y = |x| + 1$ **d.** $y = |x| + 3$

e. *(Making Observations)* Describe how the graph of $y = |x|$ must be transformed to obtain the graph of $y = |x| + k$.

Your observations in Exercise 26–30 are summarized in the following statement and illustrated in Figure 3-13.

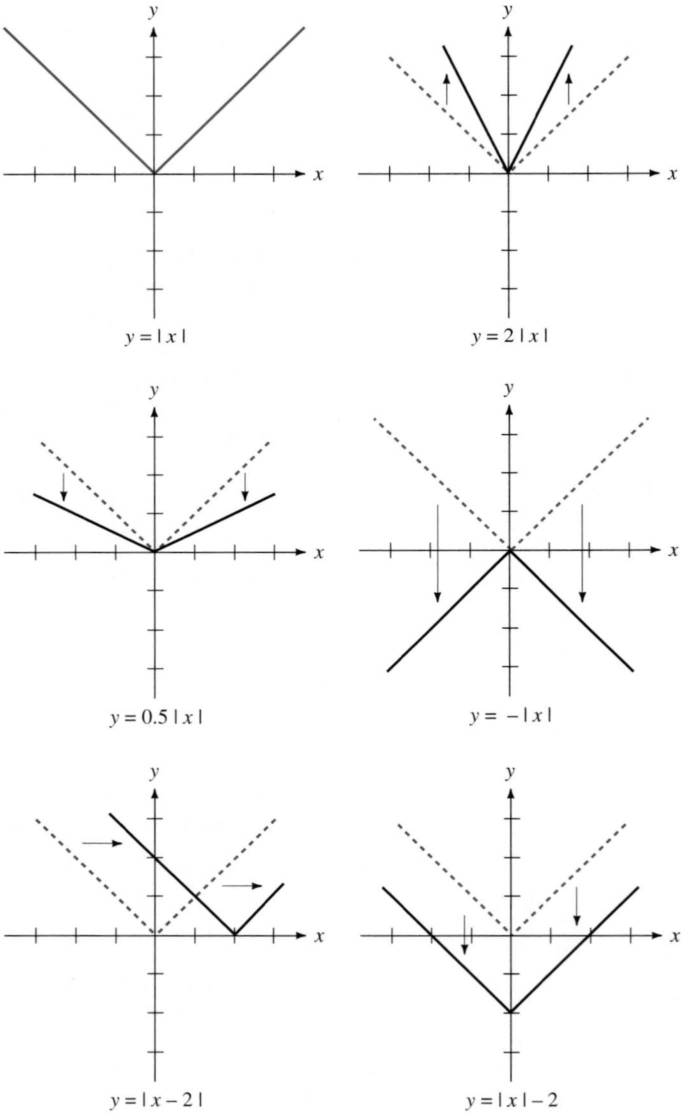

Figure 3–13

To obtain the graph of $y = a|x - h| + k$ by transforming the graph of $y = |x|$:

- Stretch the graph vertically by a factor of $|a|$ if $|a| > 1$, or compress it if $|a| < 1$.
- If $a < 0$, reflect the graph in the x-axis.
- Shift the graph to the right if $h > 0$ or to the left if $h < 0$.
- Shift the graph up if $k > 0$ or down if $k < 0$.

EXERCISE **31.** Describe how the graph of $y = |x|$ must be transformed to obtain the graph of $y = a|x - h| + k$. In particular describe the transformations that must be used to obtain the graphs of the functions in parts (a)–(d), and sketch each graph.

a. $y = 2|x - 2| + 5$

b. $y = 0.2|x + 5| + 2$

c. $y = -|x + 2| - 5$

d. $y = -0.5|x + 6| - 3$

SOLVING LINEAR ABSOLUTE VALUE INEQUALITIES

In ***Salt Container***, Morton employees responsible for quality control need to know what deviations in the height of a container produce acceptably small deviations in its capacity. This question can be answered by solving a linear absolute value inequality. You have already learned how to solve such inequalities both analytically (see Section A-9) and graphically (see Section 2-4). Example 5 reviews the methods.

EXAMPLE 5 *Solving absolute value equations and inequalities*

a. To solve $|3x - 4| < 5$ analytically,

$$-5 < 3x - 4 < 5$$

$$-1 < 3x < 9$$

$$-\frac{1}{3} < x < 3$$

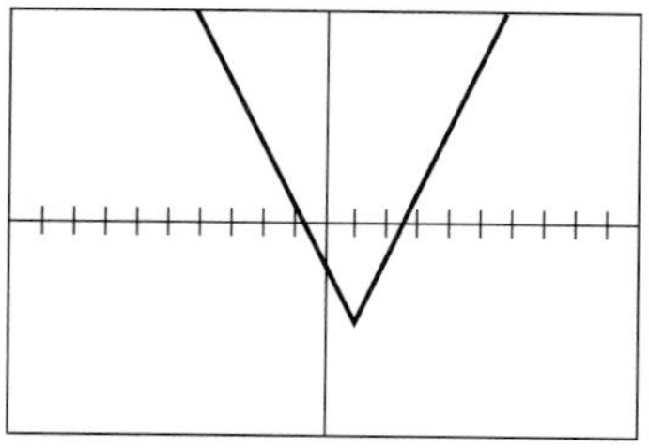

[−10, 10] by [−10, 10]

Figure 3–14

b. To solve $|3x - 4| < 5$ graphically, begin by graphing $y = |3x - 4| - 5$. The graph is shown in Figure 3-14.

The two x-intercepts represent solutions of the equation $|3x - 4| = 5$. By zooming and tracing, you can verify that they occur at $x \cong -0.33$ and $x = 3$.

The inequality is true for all values of x where the graph is below the x-axis. The solution is $(-0.33, 3)$. ■

EXERCISES *In Exercises 32–35 (Review), solve the equation or inequality both analytically and graphically, and make sure your solutions agree.*

32. $|x + 2| = 9$ **33.** $|2t - 1| = 5$

34. $|Q - 7| < 0.1$ **35.** $|5z + 100| \geqslant 12$

36. *(Problem Solving)* In ***Salt Container***, a container whose height is h cm deviates from its designed capacity by $E(h) = 57.55\,|h - 13.90|$ cm^3. What heights will produce a deviation of no more than 10 cm^3?

ADDITIONAL EXERCISES *In Exercises* 37–40,

a. Sketch the graph of the function.

b. Find its domain and range.

c. Find the intervals where the function is increasing and the intervals where it is decreasing.

37. $f(P) = \begin{cases} \dfrac{2P - 5}{3}, & \text{if } P < 2 \\ \dfrac{P}{3} - 1, & \text{if } P \geqslant 2 \end{cases}$

38. $g(x) = \begin{cases} x + 2, & \text{if } -2 \leqslant x \leqslant 0 \\ 2 - x, & \text{if } 0 < x \leqslant 2 \end{cases}$

39. $H(x) = \begin{cases} -2x - 8, & \text{if } -4 \leqslant x < -1 \\ -6, & \text{if } -1 \leqslant x \leqslant 1 \\ 2x - 8, & \text{if } 1 < x \leqslant 4 \end{cases}$

40. $S = \begin{cases} -5, & \text{if } T < -5 \\ T, & \text{if } -5 \leqslant T \leqslant 5 \\ 5, & \text{if } T > 5 \end{cases}$

In Exercises 41–48,

a. Find the slope of each linear piece of the graph and the coordinates of the vertex.

b. Sketch the graph of the function.

c. Identify the axis of symmetry, and find the range of the function.

d. Identify the intervals where the function is increasing and the intervals where it is decreasing.

e. Describe how the basic graph $y = |x|$ must be transformed to produce the given graph.

41. $y = 6\,|x|$ **42.** $y = -0.1\,|x|$

43. $y = |x - 6|$ **44.** $y = |x| + 4$

45. $y = 3\,|x + 2|$ **46.** $y = |x - 4| - 5$

47. $y = 5\,|x + 1| + 7$ **48.** $y = 8 - 2\,|x|$

In Exercises 49–58, *solve the equations and inequalities by any method.*

49. $6\,|x| - 2 = 8$ **50.** $|3 - 0.2x| = 7$

51. $|3 - 0.2x| = 0$ **52.** $|3 - 0.2x| = -7$

53. $|3x + 7| < 19$ **54.** $\dfrac{|x + 4|}{2} \leqslant \dfrac{4}{3}$

55. $2\pi < |\pi x - 2\pi|$

56. $\left|\frac{x}{3} - 2\right| \geqslant 0$

57. $\left|\frac{x}{3} - 2\right|14 \geqslant 0$

58. $\left|\frac{x}{3} - 2\right| - 4 \geqslant 0$

59. *(Creating Models)* Find a piecewise linear function $g(x)$ to describe the cost of Plan 2 in ***Health Insurance***, and find a different piecewise linear function $h(x)$ to describe the cost of Plan 3.

60. *(Problem Solving)* For what levels of medical expenses should Sarah choose Plan 1? Plan 2? Plan 3?

61. *(Writing to Learn)* Which of the three plans in ***Health Insurance*** would you choose for yourself? Explain why.

62. *(Problem Solving)* Dave and his wife Alice buy groceries at the Giant Eagle Supermarket, where ground meat costs less per pound if it is purchased in packages of 3 pounds or more.

a. If the price is \$1.79 per pound for the smaller packages and \$1.39 per pound for the larger ones, express the cost of p pounds as a piecewise linear function $C(p)$.

b. Sketch the graph of $y = C(p)$.

c. What quantities of meat weighing less than 3 pounds cost more than a 3-pound package?

63. *(Creating Models)* In ***Phone Bill*** (page 83), the equation expressing C as a function of M is not quite accurate, since Carrie must pay for a full minute even if she talks for only part of the minute. Using the fact that the first minute (or fraction thereof) costs \$1.15, and each additional minute (or fraction thereof) costs \$0.67, sketch a graph of C as a function of M over the interval $[0, 5]$. (A function such as this one, which is piecewise constant, is referred to as a **step function**.)

64. *(Creating Models)* The 1996 Boston Marathon was won by Kenya's Moses Tanui. Although race officials timed the race with extreme attention to accuracy, Tanui's recorded winning time of 2 hours, 9 minutes, and 16 seconds was almost certainly in error by at least a small fraction of a second. If his actual time was t seconds, express the error as a linear absolute value function $E(t)$.

65. *(Creating Models)* Repeat Exercise 64 using the time of Germany's Uta Pippig, who won the women's race in a recorded time of 2:27:12.

66. *(Problem Solving)* You are driving on Interstate Route I-40, which passes through Arizona's Petrified Forest National Park at milepost 311.

a. Express your distance from milepost 311 as a function of the milepost number you are currently passing.

b. At what mileposts might you see a sign saying "Petrified Forest—15 miles"?

67. *(Problem Solving)* You are restoring a Victorian farmhouse, in which the baseboard in one bedroom was destroyed by the previous owner's pet rabbit. The baseboard must be replaced around the entire perimeter of the room, except for two doorways, which are each 3 feet wide. You will need to allow yourself some slack in ordering the baseboard. However, since it is custom-made and very expensive, you want to allow as little slack as possible. The room is square, and you have measured the length of one side as 10 feet, 9 inches. If the actual length is a slightly different number of inches, say, L, you will make an error in calculating the total length of baseboard required.

a. To help you place your order economically, write the size of the error as a function of L.

b. For what values of L will you have an error of no more than 2 inches in the total length of the baseboard?

68. *(Problem Solving)* Duquesne Light Company provides electric power for customers in the Pittsburgh area. In estimating the cost of cooling and heating a house, the company computes the absolute difference between the average outside temperature and 65° each day. The result is multiplied by a figure that varies according to the size, shape, and amount of insulation for each house. A typical value for this multiplier might be \$0.25 per day.

a. Using these figures, express the estimated cost per day as a function of the average outside temperature.

b. If you can afford up to \$6.00 per day for heating and cooling, what average temperatures will prevent you from paying your entire electric bill?

69. *(Extension)* In ***Salt Container***, assume that each container has a height of exactly 13.90 cm, so that the error in capacity results entirely from the deviation in the radius r.

a. Express the error in capacity as an absolute value function $E(r)$. (It will not be linear.)

b. What radii produce containers with an error in capacity of no more than 10 cm^3? (Use graphical methods to answer.)

70. *(Making Observations)* In parts (a)–(c), solve the equation graphically.

a. $3|x - 5| - 6 = 0$ **b.** $3|x - 5| = 0$

c. $3|x - 5| + 6 = 0$

d. Your experience suggests that every linear absolute value graph is V-shaped. Assuming that this is true, what can you say about the number of solutions to a linear absolute value equation?

Exercise 71 illustrates that any table can be made to fit a piecewise linear function.

71. *(Making Observations)* The following table was generated by a function.

x	y
1	7
2	3
3	5
4	2

Find a piecewise linear function to fit it, as follows.

a. Plot the data on graph paper.

b. Draw line segments between consecutive data points.

c. Find the equation of each segment.

d. Write the equation of a piecewise linear function that fits the data.

72. *(Writing to Learn)* Without additional information, would you believe that the table in Exercise 71 was generated by the function you found? Explain your answer.

CHAPTER REVIEW

Complete Exercises 1–29 (Writing to Learn) before referring to the indicated pages.

WORDS AND PHRASES

In Exercises 1–10, explain the meaning of the words or phrases in your own words.

1. (page 73) **linear functions**

2. (pages 73, 75, 77, 79) **rate of change**

3. (pages 74, 87) **first differences**

4. (page 85) **varies directly**, **directly proportional**, **constant of proportionality**, **varies linearly** (linear variation)

5. (pages 87, 88) **arithmetic sequence**, **arithmetic series**

6. (page 91) **secant line**, **linearization**

7. (page 95) **average rate of change**

8. (page 102) **piecewise linear** function

9. (page 104) **linear absolute value function**

10. (page 106) **vertex**, **axis of symmetry** of a linear absolute value graph

IDEAS

11. (pages 74–75) Describe the test for deciding whether a table fits a linear function. Explain how and why the test can be simplified if the values of the independent variable are equally spaced. Illustrate with specific examples.

12. (page 75) If a table contains only two data points, why should you be wary about concluding that it was generated by a linear function?

13. (pages 75, 77, 79) Linear functions have a constant rate of change in the dependent variable with respect to the independent variable. How is this fact reflected in a table, graph, or equation of a linear function? Illustrate with specific examples.

14. (pages 75, 76, 77, 79) Describe or make up a specific linear function in which the variables represent physical quantities. What physical information is provided by the rate of change?

15. (page 79) What can you say about the slope of an increasing linear function? Of a decreasing linear function?

16. (page 82) Give at least one example of a specific linear function. Then list at least three nonlinear functions, and explain why each fails to be linear.

17. (pages 85, 86) Describe the process for deciding whether a table fits a linear variation, and compare it with the process for deciding whether a table fits a linear function. Explain why, if the ratio $\frac{y}{x}$ is constant for a given table, $\frac{\Delta y}{\Delta x}$ must also be constant.

18. (page 86) In general, you can find the equation of a linear function $y = mx + b$ if you know two data points. You can find the equation of a linear variation $y = mx$ if you know only one data point with $x \neq 0$. Explain why.

19. (pages 87, 88) How does an arithmetic sequence differ from other linear functions? Illustrate the difference with specific examples.

20. (pages 88, 89) Write down the formula for finding the sum of an arithmetic series, and show why the formula works. Illustrate your argument with a specific series.

21. (page 92) What do we gain by using a linear function to approximate a nonlinear function over an interval? What do we lose?

22. (pages 96–98) Compare the ideas of rate of change for a linear function and average rate of change for a nonlinear function. In particular, tell how they are alike and how they are different.

23. (pages 95–98) Describe or make up a nonlinear function in which the variables represent physical quantities. What physical information is provided by the average rate of change?

24. (pages 103, 104) Explain how to sketch the graph of a piecewise linear function. Use a specific function to illustrate the process and then explain the reasoning behind it.

25. (pages 105–107) Why can a linear absolute value function be viewed as a piecewise linear function? Illustrate with a specific example. What graphical information is provided by the piecewise linear form of the equation?

26. (pages 104, 105) Why do linear absolute value functions occur in questions related to errors in measurement?

27. (pages 106, 107) You can graph linear absolute value functions quickly on your calculator. Why do you think it is still helpful to be able to sketch their graphs manually?

28. (pages 107–109) Tell how to sketch the graph of $y = a|x - h| + k$ quickly. In particular, tell what information is provided by each of the parameters a, h, and k.

29. (page 109). The solution to a linear absolute value inequality is typically a finite interval or the union of two semi-infinite intervals. The solution may also be empty or contain a single point. Show how the shape of a linear absolute value graph makes it clear that these are the only possibilities.

CHAPTER 4

LINEAR SYSTEMS

4-1 SYSTEMS OF LINEAR EQUATIONS

PREREQUISITE MAKE SURE YOU ARE FAMILIAR WITH:
Systems of Linear Equations (Section A-10)

According to an old riddle, "A bottle full of wine is worth \$50, and the wine alone is worth \$45 more than the bottle alone. How much is the bottle worth?" The riddle was designed to lure you into giving a quick (and incorrect) answer of \$5. Let's look at two paths to the correct answer.

One path uses the strategy of *Writing an Equation*. If the dollar value of the bottle alone is b, then the dollar value of the wine alone is $b + 45$. Thus

$$b + (b + 45) = 50$$

from which $b = 2.50$, so the bottle is worth \$2.50.

A second path uses the strategy of *Looking for Relationships Among Variables*. Let b and w represent the dollar values of the bottle and the wine. Then the riddle tells you that

$$w + b = 50$$
$$w - b = 45$$

These two equations are an example of a **linear system**; that is, a collection of linear equations or inequalities to be solved simultaneously. Constructing a linear system to solve a problem allows you to name all the important unknown quantities quickly and look for relationships among them later. In the first two sections of this chapter you will study several methods of solving systems of linear equations. (In the meantime you can verify that both of the equations above are true if $b = 2.50$ and $w = 47.50$.) The last section will focus on systems of linear inequalities.

METHODS OF SOLVING 2 × 2 SYSTEMS

A system of m linear equations in n variables is referred to as an $\boldsymbol{m \times n}$ **system of linear equations**. As we will see, systems involving exactly as many equations as variables are usually the easiest to solve. The following Mathematical Looking Glass involves a 2×2 system.

A MATHEMATICAL LOOKING GLASS
Supply and Demand (Fat Cats 4)

The company your economics class formed in ***Fat Cats on Campus*** (page 43) has held its first board meeting. To increase sales, the class has adopted an incentive program, whereby each student will receive 10% of the net profit on each shirt he or she sells.

When you attempted to agree on a selling price, you discovered that financial rewards were more important to some students than to others.

- All 14 students in the class are willing to produce and sell 100 shirts during the semester at a price of $13.50 each.
- Only 4 students are willing to produce and sell 100 shirts during the semester at a price equal to the cost of $3.50 each (that is, no net profit per shirt.)

At this point you were confronted with what economists refer to as a **supply and demand problem**. Although more students were willing to sell shirts at $13.50 each, fewer customers will purchase them at that price. The class realized that their problem was to find the **equilibrium price**, at which the supply exactly equals the demand for shirts.

After the meeting the class divided into three groups, each of which solved the problem in a different way. ■

Numerical Methods At your next board meeting, one group presented a numerical solution. From your first meeting they knew that the supply of shirts would be 1400 (14 students supplying 100 shirts each) at a price of $13.50, and 400 at a price of $3.50. They constructed Table 16, whose first two columns duplicate Table 7 on page 44.

TABLE 16

Price per shirt	Demand	Supply
$3.50	2500	400
5.50	2100	
7.50	1700	
9.50	1300	
11.50	900	
13.50	500	1400

EXERCISES

1. *(Creating Models)* Assuming, as the group did, that price and supply are linearly related, complete Table 16.
2. *(Interpreting Mathematics)* Generate more entries for Table 16 to find the equilibrium price and demand for T-shirts.

Analytical Methods A second group of students had arrived at an analytical solution. Sharing the first group's assumption of linearity, they had constructed a system of linear equations. Exercises 3–5 will help you duplicate their solution.

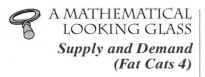

EXERCISES

3. *(Creating Models)* Express the demand for T-shirts during the semester as a function of the price per shirt in dollars.
4. *(Creating Models)* Express the available supply of T-shirts during the semester as a function of the price per shirt in dollars.
5. *(Interpreting Mathematics)* Find the equilibrium price by solving your system of equations from Exercises 3 and 4. Make sure your solution agrees with the one you obtained in Exercise 2.

Graphical Methods The third group had solved the problem graphically. They had plotted the data from Table 16 in a coordinate plane. Also making the assumption of linearity, they sketched the graphs in Figure 4-1.

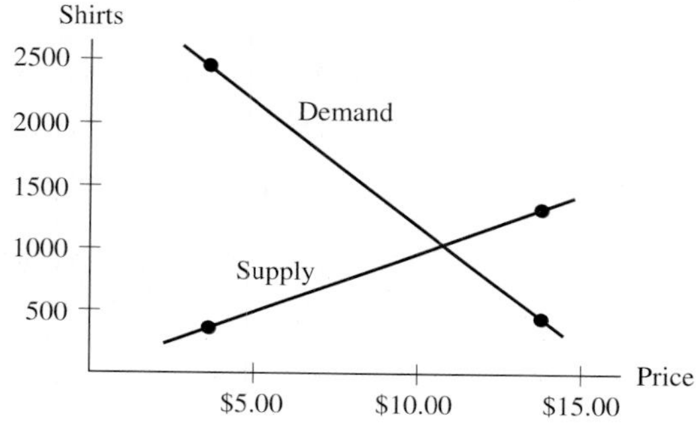

Figure 4–1

The graphs show clearly that supply is an increasing function of price, while demand is a decreasing function, so that their point of intersection represented the equilibrium price and demand, but they could get only a rough estimate with their hand-drawn graphs. In Exercise 6, you can do better with your calculator.

EXERCISES

6. *(Interpreting Mathematics)* Graph your system of equations, and trace the graphs to estimate the solution. Make sure your solution agrees with those in Exercises 2 and 5.

Solve the systems in Exercises 7–10 *graphically.*

7. $y = x + 4$
$y = 6 - x$

8. $L = 1.5W - 0.2$
$L = 3.5W + 6.4$

9. $x + 2y = 7$
$3x - y = 7$

10. $x + 2y = 9$
$3x - y = -6$

11. Solve the systems in Exercises 8 and 9 both numerically and analytically.
12. *(Writing to Learn)* Make up a 2×2 system of linear equations, and solve it both numerically and graphically. Then explain how the information in your table is reflected in your graph, and vice versa.

METHODS OF SOLVING 3×3 SYSTEMS

The following Mathematical Looking Glass leads to a 3×3 system and shows how the methods of solution compare with those used for 2×2 systems.

A MATHEMATICAL LOOKING GLASS
Sunsilk

Sunsilk is a small business run by Paul and Karen Schmitt in Santa Fe, NM. They do three different types of work: landscaping, masonry, and carpentry. Paul likes to schedule the work so that every employee can be fully utilized, but that is not always possible. To analyze Paul's problem, let's make some simplifying assumptions.

- Paul and two helpers do all the manual labor. They can each work 24 days per month.
- Karen manages the paperwork, which includes designing, planning, and billing. She can also work 24 days per month.
- Karen's mother Mary orders and picks up all supplies. Because her grandchildren, Luke and Trevor, keep her busy, she can devote only 12 days per month to the business.

Table 17 shows the division of labor for each type of job they do. For example, on landscaping jobs, 25% of the work is done by Karen, 65% by Paul and the two helpers, and 10% by Mary.

TABLE 17

	Landscaping	**Masonry**	**Carpentry**
Karen	25%	15%	20%
Paul and helpers	65%	80%	65%
Mary	10%	5%	15%

Since Paul and the two helpers can be considered together as one "triple person," Paul has to satisfy time requirements for three people. Since they do three types of work, he feels intuitively that he can produce a work schedule which will utilize everyone to full capacity. (We will see later why this intuition is correct.) Let L, M, and C represent the total days per month spent by all employees on landscaping, masonry, and carpentry. If Paul and his helpers are to work at full capacity, then $0.65L + 0.80M + 0.65C = 72$. ■

EXERCISES

13. *(Interpreting Mathematics)* Explain what the quantities $0.65L$, $0.80M$, and $0.65C$ represent, and then explain what the following equation $0.65L + 0.80M + 0.65C = 72$ says in physical terms.

14. *(Creating Models)* Write a second equation in L, M, and C to describe the use of Karen's time to full capacity. Write a third equation for Mary's time.

Although 2×2 systems can be solved by making a table, this method is usually impractical for 3×3 systems. (If you don't believe this, you are welcome to try.) Therefore, let's consider analytical and graphical methods of solution.

Analytical Methods The analytical methods for 3×3 systems, such as **Gaussian elimination**, discussed in Section A-10, are extensions of those used for 2×2 systems and are usually the easiest path to the solution.

Exercise **15.** *(Interpreting Mathematics)* Solve the linear system you wrote in Exercises 13–14. How many days of each type of work should be scheduled each month?

Graphical Methods Like 2×2 systems, 3×3 systems can be represented graphically. However, the three variables need a coordinate system with three axes, so the representation is three-dimensional. The graph of a linear equation in two variables is a (one-dimensional) line in two-dimensional space, and the graph of a linear equation in three variables is a (two-dimensional) plane in three-dimensional space. Thus, just as solutions of a 2×2 system correspond to points where two lines intersect, solutions of a 3×3 system correspond to points where three planes intersect. The graphical solution to Paul's problem looks something like Figure 4-2.

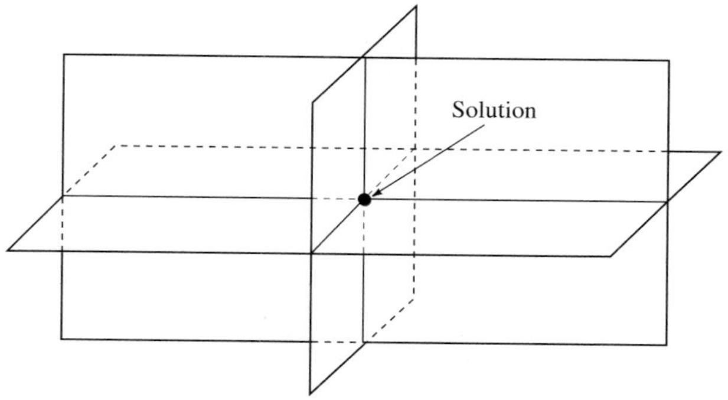

Figure 4–2

We do not expect you to solve 3×3 systems graphically. However, the visualizing of both 2×2 and 3×3 systems graphically will help you to understand how many solutions each type of system can have.

NUMBER OF SOLUTIONS TO 2×2 AND 3×3 SYSTEMS

The number of solutions to any system of linear equations is either zero, one, or infinitely many. (In Figures 4-3 and 4-4, you will see graphical evidence in support of this fact.) You can say which is most likely to occur if you know how many equations and variables the system has. This knowledge is helpful for at least two reasons.

- If you know in advance how many solutions a physical problem has, then any linear system you construct as a model should have the same number of solutions.
- You can conduct a partial check on the accuracy of your solutions by noticing whether you obtained the expected number of solutions.

2×2 Systems Every 2×2 system you have considered so far has a unique solution. Does every 2×2 system have a unique solution? You can discover the answer in Exercises 16–17.

EXERCISES

16. *(Making Observations)* For the system

$$\begin{aligned} x - y &= 1 \\ 2x - 2y &= 6 \end{aligned}$$

complete the following table.

x	y (1st equation)	y (2nd equation)
1		
2		
3		
4		

How many solutions does the system have? Explain how the table indicates this.

17. *(Making Observations)* Repeat Exercise 16 for the system

$$\begin{aligned} x - y &= 1 \\ 2x - 2y &= 2 \end{aligned}$$

Exercises 16 and 17 show that a 2×2 system can have zero or infinitely many solutions, as well as one. Are there any other possibilities? The answer to that question is easier to see graphically.

If two people each draw a line in a plane, the two lines nearly always intersect at a unique point, as in Figure 4-3a. These lines represent a linear system with a unique solution. Occasionally two lines drawn in the plane are parallel, as in Figure 4-3b, and represent a linear system with no solutions. Alternately they may coincide, as in Figure 4-3c, and represent a linear system with infinitely many solutions.

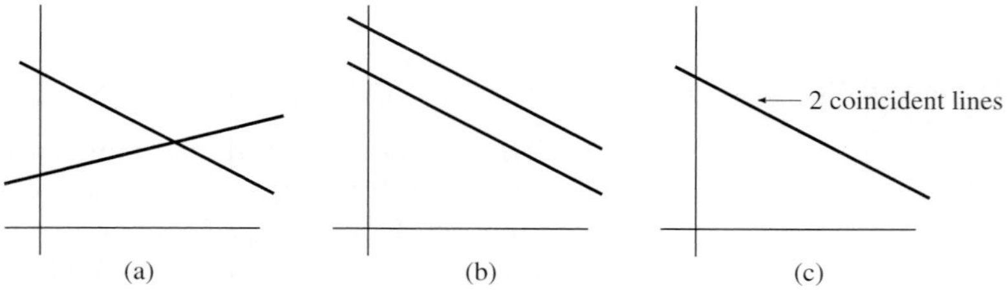

Figure 4–3

Figure 4-3 represents all possible types of 2×2 systems of linear equations. In each case, the number of solutions is equal to the number of points where the lines intersect.

A linear system with at least one solution is called **consistent**, while one with no solutions is called **inconsistent**. A 2×2 linear system with at most one solution is called **independent**, while one with infinitely many is called **dependent**. (For larger linear systems, the meaning of independence and dependence becomes more complicated.)

EXERCISES *Graph the systems in Exercises* 18–21. *Decide which systems are consistent and independent (have unique solutions), which are inconsistent, and which are dependent.*

18. $x + 5y = 13$
$x = 8 - 5y$

19. $3S = 5T - 4$
$9S - 15T = -12$

20. $2A = 3B$
$2A + 3B = 0$

21. $4P - 3Q = 12$
$3Q - 4P = -12$

22. *(Making Observations)* Choose a system in Exercises 18–21 that you identified as inconsistent, and verify your result analytically.

23. *(Making Observations)* Choose a system in Exercises 18–21 that you identified as dependent, and verify your result analytically.

3 × 3 Systems The 3 × 3 linear system you solved in Exercise 15 had a unique solution. In fact, Figure 4-2 illustrates that almost every 3 × 3 linear system has a unique solution, because three planes in three-dimensional space almost always have exactly one point in common. This is why, in ***Sunsilk***, Paul's intuition that he could schedule three types of work to meet the time requirements for three people was correct.

Occasionally, three planes in three-dimensional space do not intersect at a point. All possible ways in which three planes can intersect are indicated in Figure 4-4. As was the case in two dimensions, the corresponding linear system can have zero, one, or infinitely many solutions.

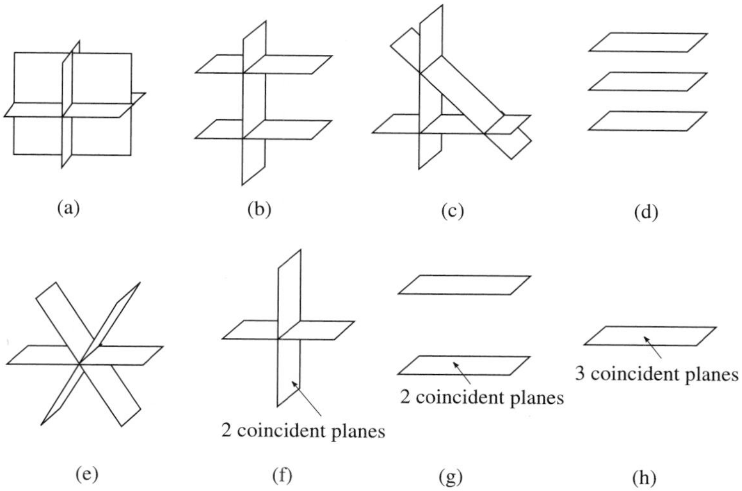

Figure 4–4

EXERCISE **24.** *(Making Observations)* Decide how many solutions each system in Figure 4-4 has. Remember that a solution corresponds to a point lying on all three planes.

ADDITIONAL EXERCISES *Solve the* 2 × 2 *linear systems in Exercises* 25–30 *both graphically and analytically, and make your solutions agree.*

25. $y = 5$
$4x - 3y = 2$

26. $R + 4 = 5S - 3$
$4S + 2R = 0$

27. $2T + 5V = 80$
$3T - 2V = 82$

28. $w + 3z = 0$
$2w - z = 0$

29. $y = 2.5x - 7$
$x = 0.4y + 1$

30. $4x - 6y = 10$
$9y - 6x = -15$

Solve the 3×3 *linear systems in Exercises* 31–34 *(Review) analytically.*

31. $x - 2y = 8$
$2x + 3y - 3z = 8$
$x + 8y - 2z = 8$

32. $3a - 2b - c = 0$
$a + 3b - 5c = 1$
$2a - 7b + 3c = 2$

33. $5p + 2q + r = 0.3$
$2p - q + 3r = 0.1$
$3p - 2q + 6r = 0.3$

34. $x - 0.1y + 30z = 6$
$x + 0.3y - 25z = 3$
$2x + 0.2y - 15z = 5$

Graph the systems in Exercises 35–38. *Decide which are consistent and independent, which are inconsistent, and which are dependent.*

35. $\alpha - 3\beta = 5$
$3\alpha - \beta = -5$

36. $4(5x + 2y) = 15$
$10x + 4y = 11$

37. $x = -1.5y + 1$
$2x + 3y = 2$

38. $4x = 8 - 2y$
$2x = y - 4$

39. Graph each system, and identify the one that is *not* consistent and independent.

a. $x + 2y = 6$
$2x + 3y = 12$

b. $x + 2y = 6$
$2x + 4y = 12$

c. $x + 2y = 6$
$2x + 5y = 12$

40. *(Making Observations)* The system

$$3x - 4y = 7$$
$$ax - 8y = 14$$

is consistent and independent for every value of a except one. What is the exceptional value of a? (*Hint*: Look at your results from Exercise 39.)

41. Graph each system, and identify the one that is *not* inconsistent.

a. $x + 2y = 6$
$2x + 4y = 8$

b. $x + 2y = 6$
$2x + 4y = 12$

c. $x + 2y = 6$
$2x + 4y = 16$

42. *(Making Observations)* The system

$$3x - 4y = 7$$
$$6x - 8y = a$$

is inconsistent for every value of a except one. What is the exceptional value of a? (*Hint*: Look at your results from Exercise 41.)

43. *(Problem Solving)* Dave contributes to each of two retirement plans through Penn State University. TIAA (Teachers Insurance and Annuity Association) has a more stable rate of return, which was 8.7% during a recent 12-month period. CREF (College Retirement Equities Fund) increased at a rate of 10.5% during the same period, but its rate is subject to greater fluctuations. To obtain an optimum combination of income and security, Dave wants to invest as much as possible in TIAA, as long as the interest rate on the total investment remains at least 9%. Assuming that the rates just given remain the same, how much of his $4000 annual contribution should go to TIAA?

44. *(Problem Solving)* Figure 4-5 shows part of the blueprint for a house built for Lynn in 1987. Builder/architect Jack Miller suggested that lengthening the bathroom from 4 feet to 8 feet would provide room for a walk-in closet. Extending into the hallway (shaded area 1) would cost $100 per foot, while extending beyond the outer first-floor wall (shaded area 2) would cost $400 per foot. Lynn wanted to keep the hallway as wide as possible, but could spend no more than $1200 on the bathroom enlargement. How far did she extend the bathroom in each direction?

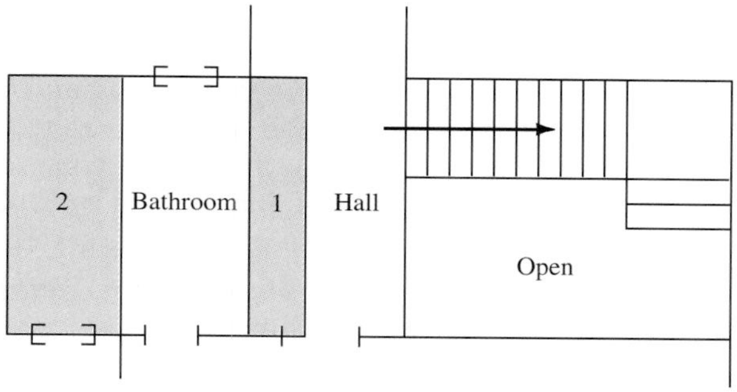

Figure 4–5

45. *(Problem Solving)* On April 4, 1990, millionaire Phil Sokolof placed full-page ads headlined "The Poisoning of America" in several national newspapers. "McDonald's, Your Hamburgers Have Too Much Fat," proclaimed the ad, which went on to ask McDonald's to reduce the fat in its hamburgers by 10%. Shortly afterward, McDonald's introduced the McLean Deluxe hamburger, a 0.25-pound hamburger containing 10 grams of fat. Their other hamburgers contain 90 grams of fat per pound. McDonald's uses about 500,000,000 pounds of ground meat each year. To achieve a 10% reduction by weight in total fat content, how many McLean Deluxes must they sell?

46. *(Problem Solving)* A gasoline's **octane** rating reflects both the chemical composition of the gasoline and the nature of its performance in your car. In general, gasolines with higher octane ratings yield higher mileage and smoother engine performance, but at a higher price. One station near Dave's home offers a choice of 87 (regular), 89 (midgrade), or 93 (premium) octane ratings. These were priced at $1.049, $1.249, and $1.349 per gallon in January, 1996. Dave can buy 10 gallons of midgrade for $12.49, but figures he can get a higher octane rating for his money by mixing regular with premium.

a. How many gallons of each must he buy to get 10 gallons at a cost of $12.49?

b. What is the octane rating of Dave's mixture? The octane rating of a mixture of x gallons of regular and y gallons of premium is

$$\frac{87x + 93y}{x + y}$$

47. *(Problem Solving)* Allegheny Ludlum Steel Corporation produces steel from scrap metal at two locations in western Pennsylvania. The steel's strength and resistance to rust is determined in part by the amount of chromium and nickel it contains. One of Allegheny Ludlum's customers requires steel containing 18% chromium and 8% nickel by weight. A typical batch of 200,000 pounds of scrap steel might contain 15% chromium and 5% nickel. This can be mixed with an alloy that is 50% iron and 50% chromium by weight and a second alloy that is 50% iron and 50% nickel by weight. How much of each alloy must be used to meet the customer's requirements?

48. *(Problem Solving)* Refer to ***Sunsilk***. In the earlier days of the company's operation, Paul and two part-time helpers worked a combined total of 56 days per month. Mary had to spend more time watching the children during those years, and could work for the company only 8 days per month. Karen's hours were unchanged. Under those conditions, was it possible to utilize everyone to full capacity? To find out, make a table similar to Table 17, then write and solve a system of equations. Is your result a valid solution to the system of equations? Is it a valid solution to the problem?

49. *(Making Observations)* Identify several Pólya strategies you used in completing Exercise 48, and describe when and how they were used.

ADDITIONAL TOPIC

Number of Solutions to $m \times n$ Linear Systems

You have seen that most, although not all, 2×2 and 3×3 linear systems have unique solutions. In Section 4-2 you will solve linear systems having more equations than variables, or vice versa. For now, let's see what we can say about the number of solutions to $m \times n$ linear systems (systems of m linear equations in n variables).

Exercises 50–54 explore $m \times 2$ systems.

50. *(Extension)* Suppose three people each draw a line in two-dimensional space. How many points are usually common to all three lines?

51. *(Extension)* What does your answer to Exercise 50 tell you about the usual number of solutions to a 3×2 system of linear equations?

52. *(Extension)* Suppose one person draws a line in two-dimensional space. How many points lie on the line?

53. *(Extension)* What does your answer to Exercise 52 tell you about the usual number of solutions to a 1×2 system of linear equations?

54. *(Extension)* How many solutions does an $m \times 2$ linear system usually have if $m > 2$? If $m < 2$? If $m = 2$?

Exercises 55–59 explore $m \times 3$ systems.

55. *(Extension)* Suppose that four people each draw a plane in three-dimensional space. How many points are usually common to all four planes?

56. *(Extension)* What does your answer to Exercise 55 tell you about the usual number of solutions to a 4×3 system of linear equations?

57. *(Extension)* Suppose two people each draw a plane in three-dimensional space. How many points are usually common to both planes?

58. *(Extension)* What does your answer to Exercise 57 tell you about the usual number of solutions to a 2×3 system of linear equations?

59. *(Extension)* How many solutions does an $m \times 3$ linear system usually have if $m > 3$? If $m < 3$? If $m = 3$?

60. *(Extension)* How many solutions do you expect an $m \times n$ system of linear equations to have if $m = n$? If $m < n$? If $m > n$?

The following are not exercises, but are simply intended to stimulate thought and discussion. Use your imagination, along with your results from Exercises 50–60.

- What do you think the graph of a linear equation in four variables might look like?
- In a two- or three-dimensional coordinate system, all coordinate axes are mutually perpendicular. Do you think it is possible for four axes in a four-dimensional coordinate system to be mutually perpendicular? Why or why not?
- What might the intersection of the graphs of two linear equations in four variables look like? What about the intersection of three such graphs? Four?
- Give a geometric description of a solution to a system of linear equations in four variables.

If these ideas interest you, continue your study of mathematics long enough to take a course in linear algebra.

4–2 MATRIX SOLUTIONS OF SYSTEMS OF LINEAR EQUATIONS

AUGMENTED MATRIX OF A SYSTEM

You solved the following system in Exercise 15 of Section 4-1, probably by using Gaussian elimination.

$$0.25L + 0.15M + 0.20C = 24$$

$$0.65L + 0.80M + 0.65C = 72$$

$$0.10L + 0.05M + 0.15C = 12$$

In this section you will learn a more efficient method that uses the same principles. Begin by writing three columns containing the coefficients of L, M, and C, and a fourth column containing the constant terms.

$$\begin{bmatrix} 0.25 & 0.15 & 0.20 & 24 \\ 0.65 & 0.80 & 0.65 & 72 \\ 0.10 & 0.05 & 0.15 & 12 \end{bmatrix}$$

The result is a **matrix**, that is, a rectangular array of numbers. The term was first used by James Joseph Sylvester (1814–1897) in 1850, although mathematicians had studied rectangular arrays since about a century earlier. The matrix shown here contains all the information in the original system and is called its **augmented matrix**. (The word *augmented* refers to the fact that the matrix containing the coefficients in the system is augmented by a column containing the constant terms.)

EXERCISES *In Exercises* 1–4, *write the augmented matrix of each system of equations.*

1. $\begin{aligned} 3x + 2y &= 6 \\ x + 7y &= 5 \end{aligned}$

2. $\begin{aligned} 2A + 3B + 4C &= 5 \\ 6A - 7B + 8C &= 9 \end{aligned}$

3. $\begin{aligned} 2m + 3n &= 5 \\ -m - 4n &= 0 \\ 3m \quad\quad &= 12 \\ 4n &= -4 \end{aligned}$

4. $\begin{aligned} \frac{1}{3}U + 3V - 5W &= \frac{4}{3} \\ 2U \quad\quad - W &= 7 \\ \frac{1}{5}U - 2V \quad\quad &= 2 \end{aligned}$

In Exercises 5–8, *write a system of linear equations for each augmented matrix, using any symbols you like as variables.*

5. $\begin{bmatrix} 23 & 6 & -2 \\ 7 & 31 & 12 \end{bmatrix}$

6. $\begin{bmatrix} 1 & 4 & -2 & 16 \\ 3 & 0 & 1 & -5 \end{bmatrix}$

7. $\begin{bmatrix} 1 & 3 & 9 & 5 \\ \frac{1}{3} & 4 & -13 & 7 \\ 0 & -2 & 3 & 16 \end{bmatrix}$

8. $\begin{bmatrix} 0 & 0 & 5 & 3 \\ -1 & 4 & 0.6 & 0 \\ 1.5 & -7 & 8 & 2.1 \\ 11 & -3 & -1 & 0.2 \end{bmatrix}$

MATRIX ROW OPERATIONS

The idea of using a matrix to solve a system of linear equations can be found in the Chinese text *K'ui-ch'ang Fuan-shu* (*Arithmetic in Nine Sections*), dating from the Han dynasty (206 B.C.–A.D. 222). In modern times Arthur Cayley (1821–1895) first used matrices to represent linear systems.

The method of solution you are about to learn, developed by the German engineer Wilhelm Jordan (1842–1899), depends on the fact that Gaussian elimination requires only three types of operations on the equations in a system.

- The order of the equations can be changed.
- Any equation can be multiplied by any nonzero constant.
- A multiple of any equation can be added to any other equation.

Each operation produces an equivalent system; that is, one with the same solutions. The corresponding operations on the rows of a matrix can be described as follows.

The matrix **row operations** equivalent to the operations of Gaussian elimination are:

- to interchange two rows,
- to multiply a row by a nonzero constant, and
- to add a multiple of a row to another row.

Example 1 illustrates the use of matrix row operations and shows how they correspond to the operations of Gaussian elimination.

EXAMPLE 1 *Performing row operations on a matrix*

Let's apply each of the three operations of Gaussian elimination to the system at the beginning of this section, and apply the corresponding row operations to its augmented matrix.

a. We can write the third equation first, and vice versa. Equivalently, we can interchange the first and third matrix rows.

System	Matrix
$0.10L + 0.05M + 0.15C = 12$ $0.65L + 0.80M + 0.65C = 72$ $0.25L + 0.15M + 0.20C = 24$	$\begin{bmatrix} 0.10 & 0.05 & 0.15 & 12 \\ 0.65 & 0.80 & 0.65 & 72 \\ 0.25 & 0.15 & 0.20 & 24 \end{bmatrix}$

b. We can multiply the first equation by 10, and the other two by 20. Equivalently, we can multiply the first matrix row by 10 and the other two by 20.

System	Matrix
$L + 0.5M + 1.5C = 120$ $13L + 16M + 13C = 1440$ $5L + 3M + 4C = 480$	$\begin{bmatrix} 1 & 0.5 & 1.5 & 120 \\ 13 & 16 & 13 & 1440 \\ 5 & 3 & 4 & 480 \end{bmatrix}$

c. We can add (-13) times the first equation to the second. Equivalently, we can add (-13) times the first matrix row to the second.

System	Matrix
$L + 0.5M + 1.5C = 120$ $9.5M - 6.5C = -120$ $5L + 3M + 4C = 480$	$\begin{bmatrix} 1 & 0.5 & 1.5 & 120 \\ 0 & 9.5 & -6.5 & -120 \\ 5 & 3 & 4 & 480 \end{bmatrix}$

■

The large numbers in the last system we obtained may seem to make it more cumbersome than the original one, but the first entry in the second row of the matrix is now zero. Thus we have eliminated the variable L from the second equation, so we have simplified the system.

GAUSS–JORDAN ELIMINATION

In performing row operations as we did in Example 1, our objective is to obtain a system and a matrix like this, with L_0, M_0, and C_0 representing numbers.

System	Matrix
$\begin{aligned} L \quad\quad\quad\quad &= L_0 \\ M \quad\quad &= M_0 \\ C &= C_0 \end{aligned}$	$\begin{bmatrix} 1 & 0 & 0 & L_0 \\ 0 & 1 & 0 & M_0 \\ 0 & 0 & 1 & C_0 \end{bmatrix}$

There are several ways we might do this. The following sequence of steps, called **Gauss–Jordan elimination**, is based on Gaussian elimination and always works. To understand the steps more clearly, read them along with Example 2.

STEP 1: Choose a row having a nonzero entry as far to the left as possible. If this is not row 1, interchange it with row 1. Row 1 is now called the **pivot row**, and its first nonzero entry is the **pivot**.

STEP 2: Multiply the pivot row by a constant so that the value of the pivot becomes 1.

STEP 3: Add multiples of the pivot row to each of the other rows so that each number in the column of the pivot (except for the pivot itself) becomes 0.

STEP 4: Choose a row below the previous pivot row having a nonzero entry as far to the left as possible. If this is not row 2, interchange it with row 2. Now row 2 is the pivot row, and its first nonzero entry is the pivot.

STEP 5: Repeat steps 2–4 until you run out of rows.

EXAMPLE 2 *Solving a linear system by Gauss–Jordan elimination*

In Example 1 we performed row operations on the augmented matrix of the system at the beginning of this section, obtaining

System	Matrix
$\begin{aligned} L + 0.5M + 1.5C &= 120 \\ 9.5M - 6.5C &= -120 \\ 5L + 3M + 4C &= 480 \end{aligned}$	$\begin{bmatrix} 1 & 0.5 & 1.5 & 120 \\ 0 & 9.5 & -6.5 & -120 \\ 5 & 3 & 4 & 480 \end{bmatrix}$

Let's continue to solve the system by Gauss–Jordan elimination. Since the first entry in the first row is 1, steps 1 and 2 have been completed. The first row is the pivot row, and the 1 is the pivot.

STEP 3: The first entry in the second row is already 0. To continue, we add (-5) times the first row to the third row.

System	Matrix
$\begin{aligned} L + 0.5M + 1.5C &= 120 \\ 9.5M - 6.5C &= -120 \\ 0.5M - 3.5C &= -120 \end{aligned}$	$\begin{bmatrix} 1 & 0.5 & 1.5 & 120 \\ 0 & 9.5 & -6.5 & -120 \\ 0 & 0.5 & -3.5 & -120 \end{bmatrix}$

STEP 4: We interchange the second and third rows. This is not necessary, since both the second and third rows have nonzero entries in the second column. However, the pivot now becomes 0.5 instead of 9.5, making the next step easier.

System	Matrix
$\begin{aligned} L + 0.5M + 1.5C &= 120 \\ 0.5M - 3.5C &= -120 \\ 9.5M - 6.5C &= -120 \end{aligned}$	$\begin{bmatrix} 1 & 0.5 & 1.5 & 120 \\ 0 & 0.5 & -3.5 & -120 \\ 0 & 9.5 & -6.5 & -120 \end{bmatrix}$

Now we repeat steps 2–4.

STEP 2: Multiply the second row by 2.

System	Matrix
$\begin{aligned} L + 0.5M + 1.5C &= 120 \\ M - 7C &= -240 \\ 9.5M - 6.5C &= -120 \end{aligned}$	$\begin{bmatrix} 1 & 0.5 & 1.5 & 120 \\ 0 & 1 & -7 & -240 \\ 0 & 9.5 & -6.5 & -120 \end{bmatrix}$

STEP 3: Add (-0.5) times the second row to the first and (-9.5) times the second row to the third.

System	Matrix
$\begin{aligned} L + 5C &= 240 \\ M - 7C &= -240 \\ 60C &= 2160 \end{aligned}$	$\begin{bmatrix} 1 & 0 & 5 & 240 \\ 0 & 1 & -7 & -240 \\ 0 & 0 & 60 & 2160 \end{bmatrix}$

STEP 4: Since there is only one row below the pivot row, the 60 in the third row becomes the new pivot.

In Exercises 9–10 you can complete the process of Gauss–Jordan elimination for this system. ■

EXERCISES

9. *(Making Observations)* What row operation was performed on the last matrix in Example 2 to produce the following matrix?

System	Matrix
$L + 5C = 240$ $M - 7C = -240$ $C = 36$	$\begin{bmatrix} 1 & 0 & 5 & 240 \\ 0 & 1 & -7 & -240 \\ 0 & 0 & 1 & 36 \end{bmatrix}$

10. *(Making Observations)* What two row operations were performed on the matrix in Exercise 9 to produce the following matrix?

System	Matrix
$L = 60$ $M = 12$ $C = 36$	$\begin{bmatrix} 1 & 0 & 0 & 60 \\ 0 & 1 & 0 & 12 \\ 0 & 0 & 1 & 36 \end{bmatrix}$

The matrix in Exercise 10 allows us to read the solution to the system directly.

REDUCED ROW-ECHELON MATRICES

The matrix in Exercise 10 is an example of a **reduced row-echelon matrix**, that is, a matrix in which:

- the **leading entry** (first nonzero entry) in each row has a value of 1,
- the leading entry in each row is to the right of the leading entry in any row above it,
- in the column containing the leading entry of each row, the other entries are all zeros, and
- any rows consisting entirely of zeros are at the bottom.

We will say that the corresponding linear system is in **reduced form**. Applying Gauss–Jordan elimination always produces a reduced row-echelon matrix and a system in reduced form.

EXAMPLE 3 *Reduced row-echelon matrices*

a. The matrix

$$\begin{bmatrix} \mathbf{1} & 0 & 2 & 0 & 3 \\ 0 & \mathbf{1} & 4 & 0 & 5 \\ 0 & 0 & 0 & \mathbf{1} & 6 \end{bmatrix}$$

is in reduced row-echelon form. The leading entry in each row is indicated in bold.

b. The matrix

$$\begin{bmatrix} \mathbf{1} & 2 & 2 & 0 & 3 \\ 0 & \mathbf{1} & 4 & 0 & 5 \\ 0 & 0 & 0 & 0 & 0 \\ 0 & 0 & 0 & \mathbf{1} & 6 \end{bmatrix}$$

is not in reduced row-echelon form for two reasons. First, the leading entry in the second row has another nonzero entry above it. Second, the row of zeros is not at the bottom. ■

EXERCISES *In Exercises* 11–14, *decide whether the matrix is in reduced row-echelon form. If it is not, explain why.*

11. $\begin{bmatrix} 1 & 0 & 5 \\ 0 & 1 & 1 \\ 0 & 0 & 1 \end{bmatrix}$ **12.** $\begin{bmatrix} 1 & 0 & 5 \\ 0 & 1 & 1 \\ 0 & 1 & 0 \end{bmatrix}$

13. $\begin{bmatrix} 1 & 0 & 5 \\ 0 & 1 & 1 \\ 0 & 0 & 0 \end{bmatrix}$ **14.** $\begin{bmatrix} 5 & 0 & 1 \\ 0 & 1 & 1 \\ 0 & 0 & 0 \end{bmatrix}$

In Exercises 15–18, *use Gauss–Jordan elimination to obtain reduced row-echelon matrices.*

15. $\begin{bmatrix} 1 & -3 & 5 \\ 4 & 5 & 3 \end{bmatrix}$

16. $\begin{bmatrix} 1 & -1 & 2 & 1 & 3 \\ 4 & -4 & 8 & 6 & 7 \end{bmatrix}$

17. $\begin{bmatrix} 2 & 0 & -1 & -2 \\ 0 & 1 & -2 & 1 \\ 4 & -1 & 1 & 0 \end{bmatrix}$

18. $\begin{bmatrix} 2 & 0 & -1 & -2 \\ 0 & 1 & -2 & 1 \\ 4 & -1 & 0 & 0 \end{bmatrix}$

Use Gauss–Jordan elimination to solve each linear system in Exercises 19–22. *The matrices in Exercises* 19 *and* 21 *are the ones you worked with in Exercises* 15 *and* 17.

19. $\begin{aligned} x - 3y &= 5 \\ 4x + 5y &= 3 \end{aligned}$ **20.** $\begin{aligned} -0.5W + 1.4Z &= 11 \\ 2.5W + 3.2Z &= -4 \end{aligned}$

21. $\begin{aligned} 2x \quad\quad - z &= -2 \\ y - 2z &= 1 \\ 4x - y + z &= 0 \end{aligned}$ **22.** $\begin{aligned} 3p + q + 4r &= 8 \\ 9p - 3q + 6r &= 12 \\ 2q - 4r &= -2 \end{aligned}$

MATRICES AND NUMBERS OF SOLUTIONS

As you saw in Section 4-1, a system of linear equations can have one, zero, or infinitely many solutions. Its reduced row-echelon matrix can tell us how many solutions it has. Let's return to Paul and Karen's landscaping company from ***Sunsilk*** (page 118) to see how.

Matrices of Systems with No Solutions First, let's see what the reduced row-echelon matrix of a system with no solutions looks like.

A MATHEMATICAL LOOKING GLASS
Sunsilk 2

In Exercise 48 of Section 4-1, you discovered that Paul and Karen could not make full use of everyone's available work time. They responded by making their operations more efficient. Since the lumberyard was directly on the route to most of their work, Paul picked up all carpentry supplies, relieving Mary of that responsibility. In return, Mary assumed Paul's responsibility for loading all plants onto the truck for landscaping jobs. This resulted in the following altered version of Table 17 from ***Sunsilk*** (Table 18).

TABLE 18

	Landscaping	**Masonry**	**Carpentry**
Karen	25%	15%	20%
Paul and helpers	60%	80%	70%
Mary	15%	5%	10%

Paul and his helpers still worked a combined total of 56 days per month, and Karen and Mary worked 24 and 8 days per month, respectively. ■

EXERCISES

23. *(Creating Models)* Let L, M, and C represent, respectively, the total days per month spent by all employees on landscaping, masonry, and carpentry. Write a system of linear equations that Paul must solve to make full use of everyone's time.

24. Use Gauss–Jordan elimination to verify that the reduced row-echelon matrix of your system is

$$\begin{bmatrix} 1 & 0 & 0.5 & 0 \\ 0 & 1 & 0.5 & 0 \\ 0 & 0 & 0 & 1 \end{bmatrix}$$

In the reduced form of the system, the third equation is $0 = 1$. Since no values of the variables can make this equation true, the system has no solution. This result can be generalized as follows.

> If a reduced linear system contains an equation $0 = c$ with $c \neq 0$, the system has no solution.
>
> Equivalently, if its reduced row-echelon matrix contains a row with its leading entry in the last column, the system has no solution.

EXERCISES

25. *(Making Observations)*

a. Use Gauss–Jordan elimination to solve

$$\begin{aligned} 2x - 3y &= 10 \\ 4x - 6y &= 15 \end{aligned}$$

b. Use graphical methods to solve the same system.

c. Describe how the graph and the reduced row-echelon matrix provide the same information about the system.

26. Use Gauss–Jordan elimination to solve the following system. The matrix is the one you worked with in Exercise 18.

$$\begin{aligned} 2A - \phantom{B -{}} C &= -2 \\ B - 2C &= 1 \\ 4A - B \phantom{{}-2C} &= 0 \end{aligned}$$

Matrices of Systems with Infinitely Many Solutions Next, let's see what the reduced row-echelon matrix of a system with infinitely many solutions looks like.

A MATHEMATICAL LOOKING GLASS
Sunsilk 3

When we left Paul and Karen in ***Sunsilk 2***, they had discovered that they could not make full use of everyone's available work time. A few months later, Karen became pregnant and decided to work only 16 days per month. Now their system of equations looked like this.

$$\begin{aligned} 0.25L + 0.15M + 0.20C &= 16 \\ 0.60L + 0.80M + 0.70C &= 56 \\ 0.15L + 0.05M + 0.10C &= 8 \end{aligned}$$

■

EXERCISE **27.** For the system in ***Sunsilk 3***, use Gauss–Jordan elimination to obtain the reduced row-echelon matrix

$$\begin{bmatrix} 1 & 0 & 0.5 & 40 \\ 0 & 1 & 0.5 & 40 \\ 0 & 0 & 0 & 0 \end{bmatrix}$$

Writing the system in reduced form, we obtain

$$\begin{aligned} L + \phantom{M +{}} 0.5C &= 40 \\ M + 0.5C &= 40 \\ 0 &= 0 \end{aligned}$$

Each of L and M is a **leading variable** (the first variable in some equation), but C is not. Therefore, we can assign any real value to C and obtain a unique solution for L and M. Thus the system has infinitely many solutions.

To tell which values of L, M, and C are solutions, we can solve for the leading variables L and M in terms of C.

$$L = 40 - 0.5C$$

$$M = 40 - 0.5C$$

Thus if C is assigned any real value, say, c, then

$$L = 40 - \frac{1}{2}c$$

$$M = 40 - \frac{1}{2}c$$

$$C = c$$

is a solution to the system. However, we obtain a solution to the problem only if L, M, and C are all nonnegative. For example, if $c = 20$, then

$$L = 30$$
$$M = 30$$
$$C = 20$$

Thus, one possible way to make full use of everyone's time is to schedule 30 days of landscaping, 30 days of masonry, and 20 days of carpentry each month.

Our observations about the system in ***Sunsilk 3*** can be generalized as follows.

> If a reduced linear system contains no equation $0 = c$ with $c \neq 0$, and has more variables than equations, then the system has infinitely many solutions.
>
> Equivalently, if its reduced row-echelon matrix has no row with its leading entry in the last column, and if (disregarding rows of zeros) it has at least two more columns than rows, the system has infinitely many solutions.

Example 4 further illustrates the process of representing infinitely many solutions to a linear system.

EXAMPLE 4 *Solving a linear system with infinitely many solutions*

Suppose that the reduced row-echelon matrix of a system is

$$\begin{bmatrix} 1 & 0 & 1 & 0 & -1 & 10 \\ 0 & 1 & 1 & 0 & -1 & 1 \\ 0 & 0 & 0 & 1 & 1 & -1 \\ 0 & 0 & 0 & 0 & 0 & 0 \end{bmatrix}$$

The matrix has no row with its leading entry in the last column, and disregarding the row of zeros, it has 3 rows and 6 columns. Therefore, the system has infinitely many solutions. To represent them, we can begin by writing the system in reduced form. If the variables are u, v, x, y, and z, we obtain

$$\begin{aligned} u + \phantom{v +{}} x - \phantom{y +{}} z &= 10 \\ v + x - \phantom{y +{}} z &= 1 \\ y + z &= -1 \end{aligned}$$

Since x and z are not leading variables, they can be assigned any real value. The system then has unique solutions for u, v, and y. If x and z are assigned values of x_0 and z_0, then

$$\begin{aligned} u &= 10 - x_0 + z_0 \\ v &= 1 - x_0 + z_0 \\ x &= x_0 \\ y &= -1 - z_0 \\ z &= z_0 \end{aligned}$$

When all variables are expressed in terms of the nonleading variables, as they are here, the solution is said to be written in **standard form**. ■

EXERCISES **28.** *(Making Observations)*

a. Use Gauss–Jordan elimination to solve

$$2x - 3y = 10$$
$$4x - 6y = 20$$

b. Use graphical methods to solve the system.

c. Describe how the graph and the row-echelon matrix provide the same information about the system.

In Exercises 29–32, write the reduced linear system and write the solutions in standard form. Use any symbols you like as variables.

29. $\begin{bmatrix} 1 & 0 & 2 & 8 \\ 0 & 1 & 4 & 7 \end{bmatrix}$

30. $\begin{bmatrix} 1 & -1 & 0 & 0 & 12 \\ 0 & 0 & 1 & 3 & 17 \end{bmatrix}$

31. $\begin{bmatrix} 1 & 0 & 10 & 0 & 0.5 \\ 0 & 1 & 20 & 0 & 2.5 \\ 0 & 0 & 0 & 1 & 4 \end{bmatrix}$

32. $\begin{bmatrix} 1 & 0 & 0 & 0 & 3 & 0 \\ 0 & 1 & 0 & 0 & 9 & 0 \\ 0 & 0 & 1 & 0 & 5 & 0 \\ 0 & 0 & 0 & 1 & 7 & 0 \end{bmatrix}$

In Exercises 33–36, solve the system and write the solutions in standard form.

33. $\begin{aligned} A + B + C + D &= 1 \\ B + C + D &= 2 \\ C + D &= 3 \end{aligned}$

34. $\begin{aligned} x - 3y + 5z &= 2 \\ 2x \qquad - 4z &= -3 \end{aligned}$

35. $\begin{aligned} p - q + r + s &= 0 \\ p + q + r - s &= 0 \\ q - r - s &= 0 \end{aligned}$

36. $\begin{aligned} p - q + r + s &= 0 \\ p + q + r - s &= 0 \end{aligned}$

Matrices of Systems with Unique Solutions The reduced linear system in Exercise 10 has a unique solution because it contains exactly one equation for each variable, assigning a unique value to that variable. This observation can be generalized as follows.

> If a reduced linear system contains no equation $0 = c$ with $c \neq 0$, and has exactly as many variables as equations, then the system has a unique solution.
>
> Equivalently, if its reduced row-echelon matrix has no row with its leading entry in the last column, and if (disregarding rows of zeros) it has exactly one more column than it has rows, the system has a unique solution.

SUMMARY: MATRIX VERSUS NONMATRIX METHODS

Although matrices are more abstract than linear systems, it is useful to learn matrix methods for several reasons.

- With sufficient practice, most people can solve linear systems more quickly with matrices than without.
- Linear systems with infinitely many solutions are usually awkward to solve without matrices.
- Calculators and computers can deal with linear systems in matrix form. In particular, you can do Gauss–Jordan elimination on your calculator.

ADDITIONAL EXERCISES *In Exercises* 37–40, *write the augmented matrix of the system of equations.*

37. $$\begin{aligned} P + Q + R + S &= 0 \\ 3P - Q + 19R + 6S &= \frac{1}{7} \\ 2P &= 8 \end{aligned}$$

38. $$\begin{aligned} -4A + 7B &= 13 \\ 2A - 15B &= 58 \\ A + B &= 91 \end{aligned}$$

39. $$\begin{aligned} 2J - 3K - 4L - 5M &= 0.001 \\ 2K - 3L - 4M - 5N &= -0.001 \end{aligned}$$

40. $$\begin{aligned} w + x - y &= 3 \\ w - x + z &= 5 \\ w + y + z &= 9 \\ x + y - z &= 1 \end{aligned}$$

In Exercises 41–44, *decide whether the matrices are in reduced row-echelon form.*

41. $$\begin{bmatrix} 1 & 2 & 3 & 4 \\ 0 & 0 & 0 & 1 \\ 0 & 0 & 0 & 0 \\ 0 & 0 & 0 & 0 \end{bmatrix}$$

42. $$\begin{bmatrix} 1 & 0 & 0 & 0 \\ 0 & 0 & 0 & 0 \\ 0 & 1 & 0 & 1 \\ 0 & 0 & 1 & 4 \end{bmatrix}$$

43. $$\begin{bmatrix} 0 & 1 & 0 & 0 \\ 0 & 0 & 1 & 0 \\ 0 & 0 & 0 & 1 \end{bmatrix}$$

44. $$\begin{bmatrix} 1 & 1 & 0 & 1 & 1 & 1 \\ 0 & 0 & 1 & 1 & 1 & 1 \end{bmatrix}$$

In Exercises 45–48, *obtain reduced row-echelon matrices.*

45. $$\begin{bmatrix} 2 & -7 & 6 \\ 5 & -1 & 4 \end{bmatrix}$$

46. $$\begin{bmatrix} 1 & 3 & 4 \\ 10 & 30 & 40 \end{bmatrix}$$

47. $\begin{bmatrix} 1 & 2 & 3 & 0 \\ 1 & 2 & 1 & 0 \\ 2 & 1 & 1 & 0 \end{bmatrix}$

48. $\begin{bmatrix} 1 & 2 & 3 & 0 \\ 1 & 2 & 3 & 1 \\ 1 & 2 & 3 & 2 \end{bmatrix}$

Solve the systems in Exercises 49–56 by Gauss–Jordan elimination.

49. $\begin{aligned} x_1 - 4x_2 &= 6 \\ 4x_1 - 16x_2 &= 18 \end{aligned}$

50. $\begin{aligned} x_1 - 4x_2 &= 6 \\ 4x_1 - 16x_2 &= 24 \end{aligned}$

51. $\begin{aligned} 2x + y + z &= 5 \\ 4x - y + 3z &= 4 \end{aligned}$

52. $\begin{aligned} P - 3Q + 2R &= 4 \\ 3P - 7Q + 5R &= 13 \end{aligned}$

53. $\begin{aligned} R - 2S + T &= 9 \\ R - 2S - T &= 5 \end{aligned}$

54. $\begin{aligned} 3J + 2K - M - N &= 12 \\ J + 2M - N &= 8 \end{aligned}$

55. $\begin{aligned} A + B + C &= 0 \\ 2A + B + C &= 1 \\ A - 2B - 2C &= 0 \\ A + B - C &= 1 \end{aligned}$

56. $\begin{aligned} A + B + C &= 0 \\ A + B - C &= 0 \\ A - B + C &= 0 \\ -A + B + C &= 0 \end{aligned}$

In Exercises 57–60, write the reduced linear system and write the solutions in standard form. Use any symbols you like as variables.

57. $\begin{bmatrix} 1 & 0 & 0 & -2 \\ 0 & 1 & 0.1 & 4 \end{bmatrix}$

58. $\begin{bmatrix} 1 & 0 & 0 & -6 & 1 \\ 0 & 1 & 0 & -6 & 1 \\ 0 & 0 & 1 & -6 & 1 \end{bmatrix}$

59. $\begin{bmatrix} 1 & 0 & 3 & 0 & 0 & 25 \\ 0 & 1 & 1 & 0 & 0 & 0 \\ 0 & 0 & 0 & 1 & 5 & 15 \\ 0 & 0 & 0 & 0 & 0 & 0 \end{bmatrix}$

60. $\begin{bmatrix} 1 & 0 & 0 & -1 & 0 & 0 & 1 \\ 0 & 1 & 0 & 0 & -1 & 0 & 1 \\ 0 & 0 & 1 & 0 & 0 & -1 & 1 \end{bmatrix}$

61. *(Problem Solving)* Figure 4-6 on the following page shows two one-way streets connected by ramps. Eastbound drivers at point A can either continue straight to B or turn north to D. Similarly, northbound drivers at C can either continue straight to D or turn east to B.

To analyze rush-hour traffic flow, the local department of transportation has set up devices to count vehicles passing each of the points A, B, C, and D. The count at each point between 5:00 and 6:00 P.M. is shown. Define the variables.

x_1 = the number of vehicles proceeding from A to B

x_2 = the number of vehicles proceeding from A to D

y_1 = the number of vehicles proceeding from C to B

y_2 = the number of vehicles proceeding from C to D

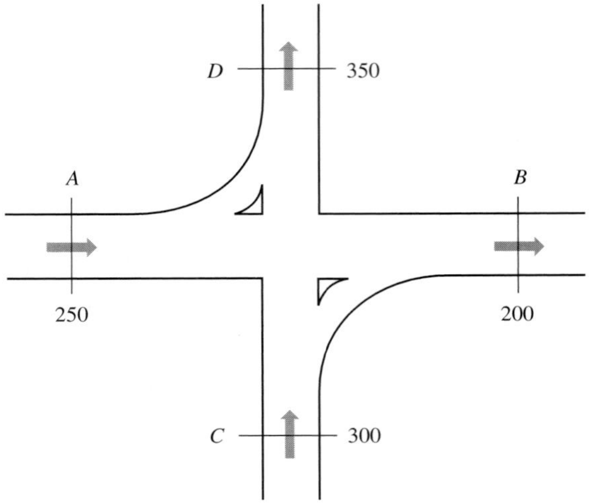

Figure 4–6

a. Use the numbers in Figure 4-6 to write a system of four linear equations in x_1, x_2, y_1, and y_2.

b. Show that the data is not sufficient to determine the values of the variables.

c. List several feasible combinations of values of the variables.

d. Could you find the values of all the variables by setting up one additional counter? If so, where?

e. Could you find the values of all the variables by moving one of the existing counters? If so, which one and to where?

62. (*Problem Solving*) Nutritionists often need to plan meals for persons with special dietary needs. The process involves analyzing the nutritional content of a large number of combinations of foods. In this simplified example, you can plan a meal using the foods listed in the following table.

	Grams fat	Grams carbohydrates	Grams protein	Milligrams cholesterol	Calories
1 baked potato	0.2	32.8	4.0	0	145
1 stalk broccoli	0.5	8.1	5.6	0	47
4 oz. salmon	6.4	0	30.8	40	192
Small slice apple pie	5.5	22.0	1.1	0	139
1 cup 2% milk	2.6	11.7	8.0	10	102

Adapted from Dean Ornish, *Eat More, Weigh Less,* Harper Perennial, 1993.

Define the variables

$P =$ the number of baked potatoes in the meal

$B =$ the number of stalks of broccoli

$S =$ the number of 4-ounce portions of salmon

$A =$ the number of slices of apple pie

$M =$ the number of cups of 2% milk

a. Write a system of linear equations that must be satisfied if the meal is to contain 15 grams of fat and 50 milligrams of cholesterol.

b. Solve the system and write the solution in standard form.

c. Verify that two solutions to the system are $(P, B, S, A, M) = (0, 1, 1, 1, 1)$ and $(P, B, S, A, M) = (27.5, 1, 1, 0, 1)$. Describe the meals represented by each solution.

d. By trial and error, find several other solutions with nonnegative values for all the variables. Would any of these solutions make reasonable meals?

e. Identify a few reasons why a nutritionist would probably not plan a meal by solving a system of linear equations.

63. *(Writing to Learn)* In ***Sunsilk 3***, C is not a leading variable in the system of equations.

a. Can you rewrite the system in such a way that C is a leading variable? If so, how? If not, why not?

b. Can you rewrite the system in such a way that all the variables are leading variables? If so, how? If not, why not?

64. *(Extension)* Consult the owner's manual for your calculator to learn the proper keystrokes for performing matrix operations. Then solve each of the systems in Exercises 19–22 on your calculator.

65. *(Extension)* Using the five foods listed in Exercise 62, is it possible to plan a meal having 15 grams of fat, 100 grams of carbohydrates, 50 grams of protein, 50 milligrams of cholesterol, and 600 calories? (*Hint*: Do the matrix row operations on your calculator.)

4–3 SYSTEMS OF LINEAR INEQUALITIES AND LINEAR PROGRAMMING

GRAPHS OF LINEAR INEQUALITIES IN TWO VARIABLES

In this section you will encounter some relationships that can be more accurately modeled by inequalities in two variables, rather than equations. Just as the solution to an inequality in one variable can be conveniently represented by a graph on a one-dimensional number line, the solution to an inequality in two variables can be conveniently represented by a graph in a two-dimensional coordinate plane. The following Mathematical Looking Glass explores both the use of an inequality to represent a relationship and the use of a graph to represent its solution.

A MATHEMATICAL LOOKING GLASS

Horse Feed

Lynn's daughter Carrie owns a 16-hand Lippizan-Percheron cross named Illiad. His diet includes a mix of timothy hay and alfalfa. During the fall of 1991, Carrie could buy timothy for $1.50 per bale. Alfalfa was harder to obtain and cost $2.50 per bale. Carrie had set aside $165 to spend on Illiad, and wanted to know what combinations of timothy and alfalfa she could buy. However, she realized that she might want to spend only part of the $165 on feed and buy other horse supplies with the rest. ■

EXERCISES **1.** *(Writing to Learn)* The combinations of timothy and alfalfa Carrie can buy are the solutions to the equation $1.50x + 2.50y = 165$, where

$$x = \text{ the number of bales of timothy she buys}$$

and

$$y = \text{ the number of bales of alfalfa she buys}$$

Explain how this equation was obtained.

2. *(Creating Models)* Write an inequality to express the combinations of timothy and alfalfa Carrie can buy for no more than \$165.

The equation in Exercise 1 and the inequality in Exercise 2 each have infinitely many solutions. In both cases, we can represent the solutions graphically. The equation can be solved for y to obtain $y = 66 - 0.6x$. Its graph is shown in Figure 4-7a.

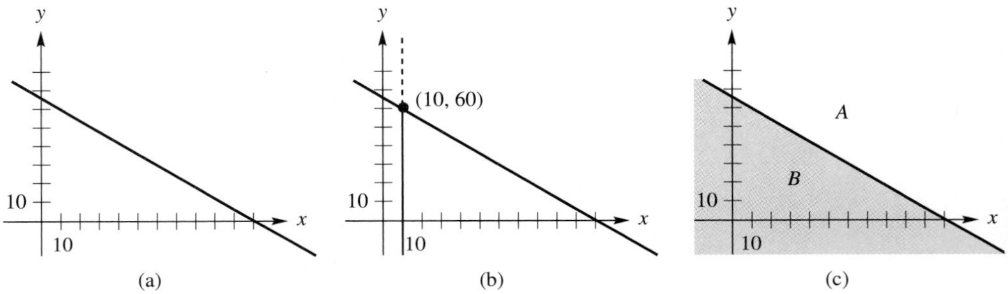

Figure 4–7

The inequality in Exercise 2 can be written $y \leq 66 - 0.6x$. To see what its graph looks like, let's focus on one value of x, say $x = 10$. The inequality then becomes $y \leq 60$. Figure 4-7b shows the points on the line $x = 10$ with $y \leq 60$. Similar calculations show that for every value of x, the inequality is true at points on or below the line $y = 66 - 0.6x$. The graph of the inequality is the shaded portion of the plane in Figure 4-7c. The boundary line $y = 66 - 0.6x$ is included in the graph. It is common practice to draw a boundary as a solid line if it is included in the graph and as a dotted line if it is not.

EXERCISE **3.** *(Writing to Learn)* The graph of a linear equation in two variables is always a line that divides the plane into two regions. The solution to every linear inequality in two variables can be represented as one such region. Explain why.

EXAMPLE 1 *Graphing a linear inequality*

To graph the inequality $2x + y < 6$, begin by graphing the equation $2x + y = 6$, as in Figure 4-8a. The solution to the inequality is one of the two regions A and B. To see which region is the solution, choose a test value in either region. For example, $(0, 0)$ is in region B, and the inequality is true there. Figure 4-8b shows how to indicate that region B is the solution.

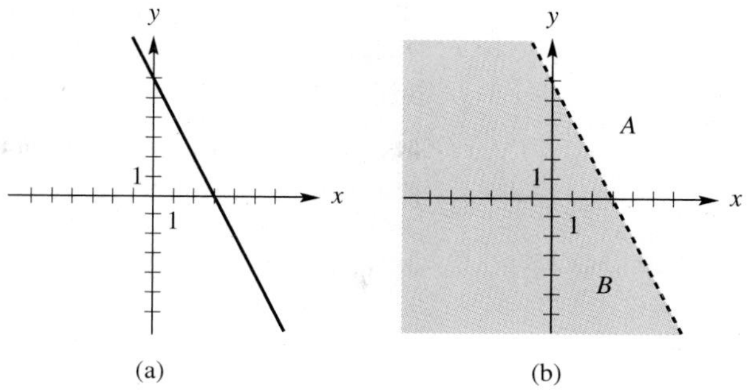

Figure 4–8 ■

Exercises *Graph the inequalities in Exercises* 4–7.

4. $3x + 2y < 5$ **5.** $3x - 2y < 5$

6. $y - 7x \geqslant 3$ **7.** $x + 3y \geqslant 0$

GRAPHS OF SYSTEMS OF LINEAR INEQUALITIES IN TWO VARIABLES

In Section 4-1 you saw that some problems can be modeled by a system of equations in two or more variables. Graphically, the solutions to the system are those points common to the graphs of all the equations in the system. Example 2 describes a problem that can be modeled by a system of linear inequalities in two variables. Graphically, the solutions to the system are those points common to the graphs of all the inequalities in the system.

EXAMPLE 2 *Graphing a system of linear inequalities*

Carrie's problem in ***Horse Feed*** involves the following system of linear inequalities.

$$\begin{aligned} 1.50x + 2.50y &\leqslant 165 \\ x &\geqslant 0 \\ y &\geqslant 0 \end{aligned}$$

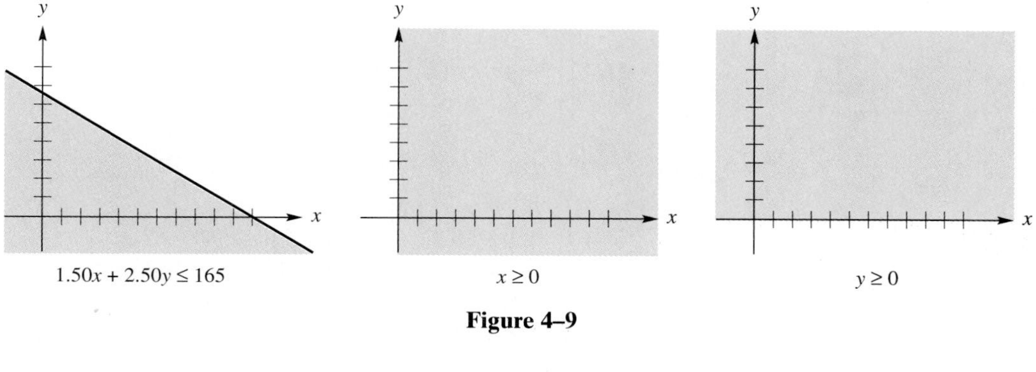

Figure 4–9

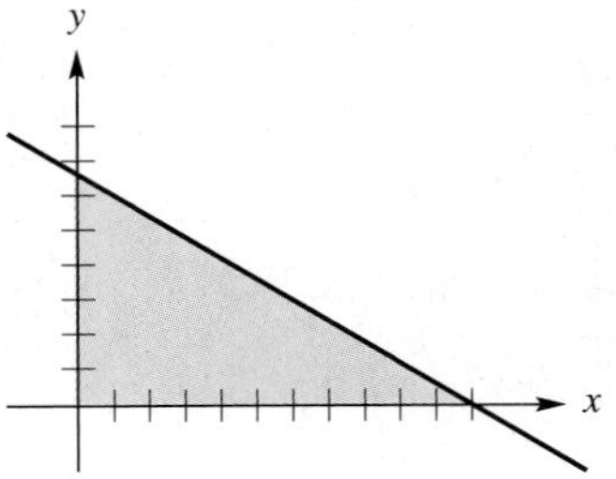

Figure 4–10

The first inequality represents her budget constraint, and the other two represent the constraints that quantities of timothy and alfalfa must be nonnegative. Figure 4-9 shows the graph of each inequality in the system.

A point (x, y) represents a solution to the system if and only if it makes all three inequalities true. Therefore, the graph of the system is the region common to the graphs of the inequalities in the system. Figure 4-10 shows the solution.

■

The solution to a system of inequalities is called its **feasible region**. The feasible regions of larger systems can be obtained in a similar manner. For example, Carrie must buy at least twice as much timothy as alfalfa, because alfalfa is too rich for Illiad to digest unless it is mixed with large quantities of timothy. This additional constraint is expressed by the inequality $x \geq 2y$.

Exercises

8. Graph the inequality $x \geq 2y$.

9. Find the feasible region for the system of inequalities

$$\begin{aligned} 1.50x + 2.50y &\leq 165 \\ x &\geq 0 \\ y &\geq 0 \\ x &\geq 2y \end{aligned}$$

10. *(Interpreting Mathematics)* What does the feasible region in Exercise 9 represent in physical terms?

11. An additional constraint in ***Horse Feed*** is that Carrie's barn holds only 100 bales of hay. Graph the resulting inequality $x + y \leq 100$.

12. Find the feasible region for the system of inequalities

$$\begin{aligned} 1.50x + 2.50y &\leq 165 \\ x &\geq 0 \\ y &\geq 0 \\ x &\geq 2y \\ x + y &\leq 100 \end{aligned}$$

13. *(Interpreting Mathematics)* What does the feasible region in Exercise 12 represent in physical terms?

Graph the solution to each system of inequalities in Exercises 14–17.

14. $\begin{aligned} y &< 2x + 2 \\ 2y &\geq x - 2 \end{aligned}$

15. $\begin{aligned} y &\leq x + 1 \\ 2x + y &\leq -2 \end{aligned}$

16. $\begin{aligned} 2x + 3y &< 12 \\ 2x + 3y &> 6 \\ x - y &< 3 \end{aligned}$

17. $\begin{aligned} 2x + 3y &< 12 \\ 2x + 3y &> 6 \\ x - y &< 3 \\ x - y &> 0 \end{aligned}$

LINEAR PROGRAMMING

As you have seen, systems of linear inequalities in two variables typically have infinitely many solutions. In most situations it is natural to look for one solution that is optimal in some respect. One method of finding optimal solutions, called **linear programming**, was developed during World War II to determine the optimal use of resources in military operations. Linear programming is also used today by businesses and government agencies to solve problems in such diverse areas as economics, personnel, and transportation. Read on to see how Carrie can use linear programming to find an optimal solution to her system of inequalities in ***Horse Feed***.

A MATHEMATICAL LOOKING GLASS
Horse Feed 2

Carrie's system of inequalities from Exercise 12 is

$$1.50x + 2.50y \leqslant 165$$
$$x \geqslant 0$$
$$y \geqslant 0$$
$$x \geqslant 2y$$
$$x + y \leqslant 100$$

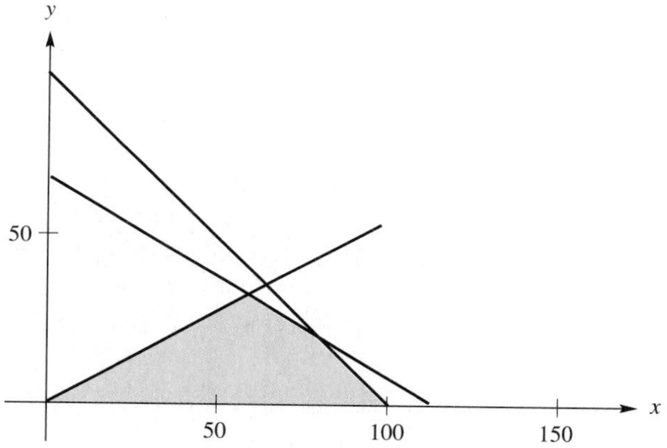

Figure 4–11

Her feasible region, shown in Figure 4-11, contains infinitely many solutions. Her problem is to decide on a "best" solution. She will probably do this by seeking an optimal combination of many factors, such as nutritive value and resistance to spoilage. Let's keep life simple, and assume that she needs to consider only one factor. Suppose that Carrie plans to ride Illiad in a strenuous cross-country event and wants to feed him a protein-rich diet in preparation. Her problem is to choose the solution with the most digestible protein.

Carrie knows that timothy is 6.1% digestible protein, alfalfa is 12.9% digestible protein, and an average bale weighs 50 pounds. Therefore the total amount of digestible protein in the feed shipment is

$$(0.061)(50x) + (0.129)(50y)$$

or

$$3.05x + 6.45y$$

This expression is Carrie's **objective function**. She must now find the point (x, y) in the feasible region where the value of the objective function is largest.

Let's consider several values for the objective function. Any equation $3.05x + 6.45y = k$ represents a feed shipment containing k pounds of digestible protein. The dashed lines in Figure 4-12 represent feed shipments with $k = 100, 200$, and 300 pounds of digestible protein. Their equations are $3.05x + 6.45y = 100$, $3.05x + 6.45y = 200$, and $3.05x + 6.45y = 300$.

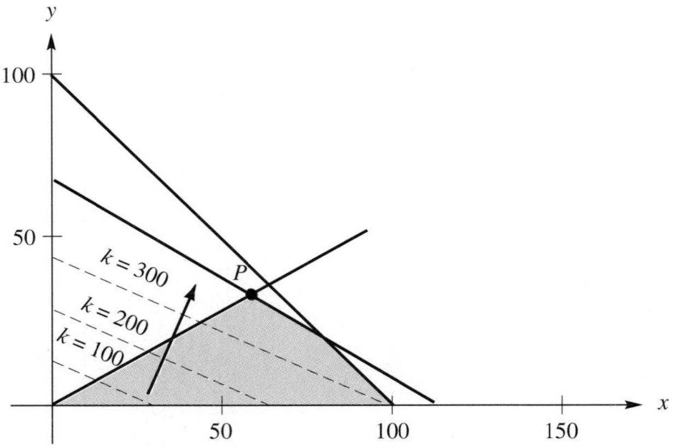

Figure 4–12

The graph of the objective function moves in the direction of the arrow as k increases. Carrie needs to find the largest value of k for which the line $3.05x + 6.45y = k$ intersects the feasible region. This happens when the graph of the objective function passes through P. Since P is the intersection of the lines $x = 2y$ and $1.50x + 2.50y = 165$, $P = (60, 30)$. Thus the optimal line passes through (60, 30) and has equation $3.05x + 6.45y = k$, so

$$(3.05)(60) + (6.45)(30) = k$$

$$k = 376.5$$

Carrie will buy a shipment of 60 bales of timothy and 30 bales of alfalfa, having 376.5 pounds of digestible protein. ■

As Figure 4-12 illustrates, the maximum (or minimum) value of the objective function in a linear programming problem always occurs at an **extreme point** (a corner point) of the system's feasible region. This observation leads to the following method of solution.

To solve a linear programming problem,

- identify the feasible region for the system of inequalities representing the constraints,
- solve systems of linear equations to identify the extreme points of the feasible region, and
- evaluate the objective function at each extreme point.

EXERCISES

18. *(Making Observations) For the linear programming problem in* ***Horse Feed 2***,

a. Find the coordinates of the three other extreme points of the feasible region.

b. Evaluate the objective function at each extreme point. Do your results support the conclusion that the maximum value occurs at $(60, 30)$?

Complete Exercises 19–21 to find the minimum value of $50 - 2x - 3y$ *subject to the constraints*

$$\begin{aligned} x &\geqslant 0 \\ y &\geqslant 0 \\ x + y &\leqslant 12 \\ x + 2y &\leqslant 17 \end{aligned}$$

19. Find the feasible region of the system of inequalities.

20. The feasible region in Exercise 19 has four extreme points, representing the solutions to the systems of equations shown. Find their coordinates.

a. $x = 0$, $y = 0$

b. $y = 0$, $x + 2y = 17$

c. $x = 0$, $x + y = 12$

d. $x + y = 12$, $x + 2y = 17$

21. Evaluate the objective function $50 - 2x - 3y$ at each extreme point. Which value is the solution?

Solve the linear programming problems in Exercises 22–23.

22. Maximize $x + 3y$, subject to:

$$\begin{aligned} y &\leqslant 2x + 2 \\ 2y &\geqslant x - 2 \\ x + y &\leqslant 14 \end{aligned}$$

23. Minimize $2x + 3y$, subject to:

$$\begin{aligned} y &\leqslant x + 1 \\ 2x + y &\leqslant -2 \\ y &\geqslant -5 \end{aligned}$$

ADDITIONAL EXERCISES

Graph the inequalities in Exercises 24–31.

24. $y - 2x \leqslant 10$

25. $y - 2x \leqslant -10$

26. $x + y < 2$

27. $5x - 3y \geqslant 0$

28. $x > 2y$

29. $2x + 3y \geqslant -12$

30. $0.1x + 0.3y < -3$

31. $0.4x - 0.2y > -1$

Graph the systems of inequalities in Exercises 32–37.

32. $2x + 3y > 12$, $2x + 3y < 6$

33. $x > -2 - y$, $y < x + 2$, $x \geqslant 2$

34. $4y \leqslant 12 + 3x$, $x + y \leqslant 5$, $x < 4y - 4$

35. $4y < 12 + 3x$, $x + y < 5$, $x \leqslant 4y - 4$

36. $y < 3x - 3$
$y > x - 7$
$y > -3$
$y < 0$

37. $x > y$
$x \leqslant 3y$
$y \geqslant 5 - x$

Solve the linear programming problems in Exercises 38–41.

38. Maximize $2S - T$, subject to:
$S \geqslant -2 - T$
$T \leqslant S + 2$
$S \geqslant 2$
$S \leqslant 10$

39. Minimize $3Z_1 - Z_2$, subject to:
$4Z_2 \leqslant 12 + 3Z_1$
$Z_1 + Z_2 \leqslant 5$
$Z_1 + 2Z_2 \leqslant 20$
$Z_1 \leqslant 4Z_2 - 4$
$Z_1 \geqslant -1$

40. Maximize $5x + 3y$, subject to:
$y \leqslant 3x - 3$
$y \geqslant x - 7$
$y \leqslant 6 - x$
$y \geqslant -3$
$y \leqslant 0$

41. Maximize $4x_1 + 3x_2$, subject to:
$2x_1 + 3x_2 \leqslant 6$
$-3x_1 + 2x_2 \leqslant 3$
$2x_2 \leqslant 5$
$2x_1 + x_2 \leqslant 4$
$x_1 \geqslant 0$
$x_2 \geqslant 0$

42. *(Creating Models)* You are running for governor of Delaware and would like to get a majority of the votes in each of the state's three counties to demonstrate broad support. Current polls show that you need an additional 9000 votes in New Castle county, 10,000 in Kent County, and 3000 in Sussex County. Your strategy is to buy newspaper ads in the *Delaware State News* and spots on TV station KYW in nearby Philadelphia. You make an educated guess that each newspaper ad will gain you 20 additional votes in New Castle County, 70 in Kent County, and 10 in Sussex County, while each TV spot will gain 400, 100, and 50 votes, respectively. Write and graph a system of inequalities to show all combinations of newspaper ads and TV spots that will accomplish your goal.

43. *(Problem Solving)* Refer to Exercise 42. If TV spots cost 10 times as much as newspaper ads, how many of each should you purchase to obtain a majority of the votes in each county at minimum expense?

44. *(Problem Solving)* (Continuation of Exercise 62 of Section 4-2)

	Grams fat	Grams carbohydrates	Grams protein	Milligrams cholesterol	Calories
1 baked potato	0.2	32.8	4.0	0	145
4 oz. salmon	6.4	0	30.8	40	192

Adapted from Dean Ornish, *Eat More, Weigh Less,* HarperPerennial, 1993.

a. Write and graph a system of inequalities to show all possible combinations of potatoes and salmon having at most 400 calories, and at most 8 grams of fat.

b. Among all such combinations, which has the highest protein content?

c. In Exercise 62 of Section 4-2 you saw that a system of linear equations is not a practical model for most dietary problems. Why can a linear programming problem provide a more realistic model?

CHAPTER REVIEW

Complete Exercises 1–29 (Writing to Learn) before referring to the indicated pages.

WORDS AND PHRASES

In Exercises 1–13, explain the meaning of the words or phrases in your own words.

1. (page 115) **linear system**, **$m \times n$ system of linear equations**
2. (page 120) **consistent**, **inconsistent** system of linear equations
3. (page 120) **dependent**, **independent** system of linear equations
4. (pages 125, 126) **matrix**, **augmented matrix**
5. (page 127) **matrix row operations**
6. (page 128) **Gauss–Jordan elimination**
7. (page 128) **pivot row**, **pivot**
8. (page 130) **reduced row-echelon matrix**, **leading entry**, **reduced form**
9. (page 133) **leading variable** in a system of linear equations
10. (pages 134, 135) **standard form** of the solution to a system of linear equations
11. (page 142) **feasible region** of a system of linear inequalities
12. (pages 143, 144) **linear programming**
13. (pages 143, 144) **objective function**, **extreme point**

IDEAS

14. (page 115) Any problem that can be modeled by a system of linear equations can also be modeled by a single linear equation. Discuss some of the advantages and disadvantages of each type of model.
15. (pages 116, 117) Compare the ease, speed, and accuracy of numerical, analytical, and graphical methods of solving systems of linear equations. Illustrate your points with a particular system.
16. (pages 118, 119) Why is it usually impractical to solve a 3×3 system of linear equations by numerical or graphical methods?
17. (pages 119–121) Most 2×2 and 3×3 systems of linear equations have exactly one solution. How do the graphs of such systems support this statement?
18. (pages 119–121) Explain why it is impossible for a system of linear equations in two or three variables to have more than one solution unless it has infinitely many.
19. (pages 125, 126, 136) In what respect is it efficient to represent a system of linear equations in matrix form?
20. (page 127) Describe the operations on a system of linear equations corresponding to the matrix row operations, and explain why none of these operations change the solutions to the system.
21. (pages 128, 129) Explain why the process of Gauss–Jordan elimination always leads to a reduced row-echelon matrix.
22. (page 130) Why is the reduced row-echelon form of the augmented matrix especially helpful in solving a system of linear equations?
23. (page 132) If a system of linear equations has no solutions, what does its reduced row-echelon matrix look like? Why does the form of the matrix guarantee that the system has no solution? Give an example of such a matrix, and write down the corresponding system.
24. (pages 133, 134) If a system of linear equations has infinitely many solutions, what does its reduced row-echelon matrix look like? Why does the form of the matrix guarantee that the system has infinitely many solutions? Give an example of such a matrix, write down the corresponding system, and list several solutions.
25. (page 135) If a system of linear equations has a unique solution, what does its reduced row-echelon matrix look like? Why does the form of the matrix guarantee that the system has a unique solution? Give

an example of such a matrix, and write down the corresponding system.

26. (page 136) What are some of the advantages of using matrix methods to solve a system of linear equations? When might you prefer to use nonmatrix methods?

27. (pages 139–141) Describe the process of solving a linear inequality in two variables graphically, and illustrate with a particular inequality.

28. (pages 141, 142) The solution of a system of linear inequalities in two variables is usually represented graphically, rather than analytically. Explain why.

29. (pages 142–145) Describe the process of solving a linear programming problem in two variables. In particular, explain why the optimum value of the objective function always occurs at a corner point of the feasible region.

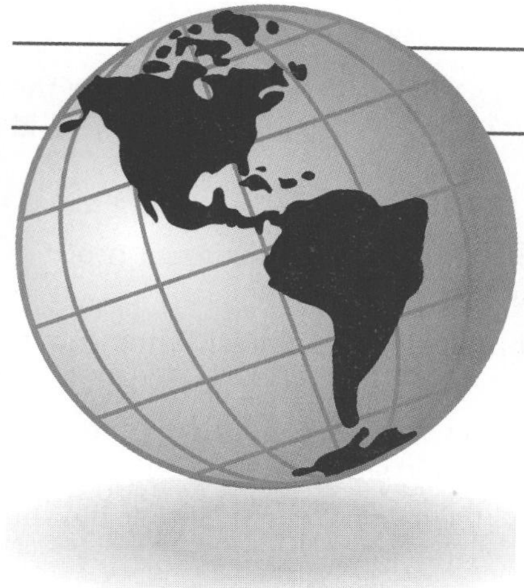

CHAPTER 5

QUADRATIC FUNCTIONS

5–1 THREE VIEWS OF QUADRATIC FUNCTIONS

PREREQUISITES MAKE SURE YOU ARE FAMILIAR WITH:

Operations with Complex Numbers (Section A-14)

In Chapters 3 and 4 you viewed linear functions from several perspectives and learned how to use them in modeling and problem solving. Along the way you learned that their graphs have a distinctive shape, their equations have a distinctive form, and their tables have a distinctive property by which they can be identified.

In this chapter you will study quadratic functions, which are the simplest nonlinear functions. You will see how they appear in situations related to calculating areas, analyzing the motion of a falling object, and maximizing a company's profit. Although quadratic functions are radically different from linear functions, their graphs, equations, and tables also have distinctive shapes, forms, and properties.

The following Mathematical Looking Glass illustrates how a quadratic function can be of interest in a physical context.

A MATHEMATICAL LOOKING GLASS

Maximum Profit (Fat Cats 5)

The T-shirt company your class started in ***Fat Cats on Campus*** (page 43) is about to begin production. A group of students from the class has pointed out that without an intensive marketing campaign, higher sales will not necessarily mean higher profits. Using the profit function

$$P(x) = -0.005x^2 + 12.50x - 612.50$$

that you developed in Exercise 12 of Section 2-2, they present the graph in Figure 5-1 at your next board meeting.

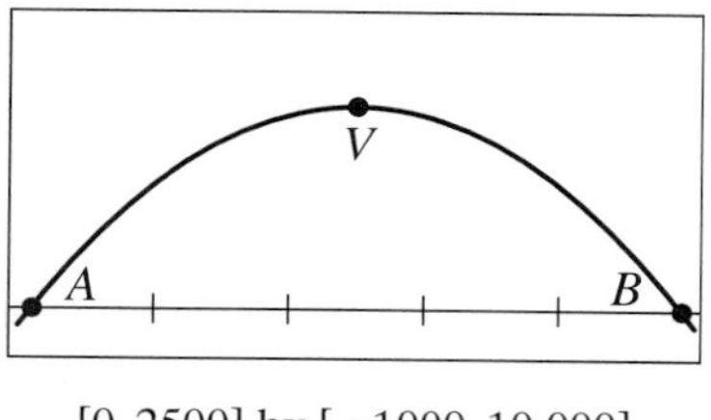

[0, 2500] by [−1000, 10,000]

Figure 5–1

The horizontal axis measures the number of shirts sold, and the vertical axis measures profit in dollars. The group explains that if your level of production is between A and V on the graph, increasing production will cause profits to rise. If your level of production is between V and B, increasing production will cause profits to fall. This is because to sell a large number of shirts, you will have to lower the price.

The class must now identify the target sales level that will generate the largest profit. ■

EXERCISES

1. *(Interpreting Mathematics)*
   a. Explain why the sales level that will generate the maximum profit is represented by the point V in Figure 5-1.
   b. Use Figure 5-1 to estimate the number of shirts you must sell to generate the maximum profit.
2. *(Review)* In ***Breakeven Analysis*** (page 66) you discovered that your company's breakeven sales levels were 50 and 2450 shirts.
   a. Explain why the breakeven sales levels are represented by the x-intercepts A and B in Figure 5-1.
   b. Explain why the breakeven sales levels are of interest to your company.
3. *(Making Observations)* Explain how identifying the x-intercepts in Figure 5-1 can help you estimate the sales level that will generate the maximum profit.

The profit function explained in ***Maximum Profit*** is an example of a **quadratic function**, that is, one that can be written

$$f(x) = ax^2 + bx + c$$

for some real numbers a, b, and c with $a \neq 0$. By studying these functions you will learn systematic methods of solving your company's problem and other similar problems.

A GRAPHICAL VIEW OF QUADRATIC FUNCTIONS

Features of Quadratic Graphs Figure 5-2a shows the graph of the simplest quadratic function $y = x^2$. It shares several striking features with the graph of $y = P(x)$ from ***Maximum Profit***, which is reproduced in Figure 5-2b.

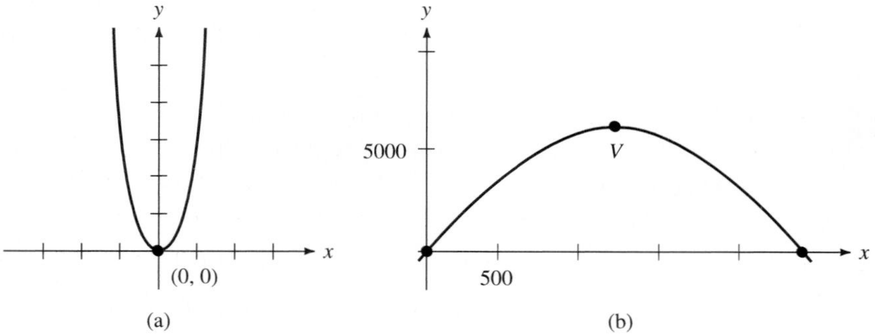

Figure 5–2

Every quadratic graph has the following features.

- It is a **parabola**, a U-shaped curve opening upward as in Figure 5-2a or downward as in Figure 5-2b. In contrast to linear graphs, quadratic graphs do not exhibit a constant rate of change. You can verify this in Exercise 63.

- It has a **vertex**, that is, its extreme point (the origin in Figure 5-2a and the point V in Figure 5-2b). Therefore, the range of a quadratic function does not include all values of the dependent variable, but is an interval extending infinitely in only one direction. From their graphs, the functions in Figure 5-2 appear to have ranges of $[0, \infty)$ and approximately $(-\infty, 7200]$. Later in this section you will learn how to identify the range of a quadratic function analytically.
- The vertical line through the vertex is an **axis of symmetry**. The parts of the graph to the left and right of this line mirror each other. You will learn how to identify the axis of symmetry later in this section.
- It has either 0, 1, or 2 ***x*-intercepts**. You can explain why in Exercises 21–23.

FYI *Parabola* is not a generic name for any U-shaped curve. Parabolas have some distinctive properties that you will study in Section 6-4.

Graphing Quadratic Functions Exercises 1–3 indicate that the vertex and the upward or downward orientation of a parabola are often of physical importance. Let's see if we can read these graphical features quickly from the equation of a quadratic function. We can proceed by *Examining a Special Case* (actually several special cases).

EXERCISES

4. Graph all of the following functions on one set of coordinate axes.

a. $y = x^2$ **b.** $y = 4x^2$

c. $y = 0.5x^2$ **d.** $y = -3x^2$

5. *(Making Observations)* Based on your graphs from Exercise 4, guess the answers to the following questions.

a. Where is the vertex on the graph of $y = ax^2$?

b. How does the steepness of the graph compare with that of $y = x^2$ if $|a| > 1$? If $|a| < 1$?

c. What is the orientation (up or down) of the graph if $a > 0$? If $a < 0$?

6. *(Making Observations)* Use your answers from Exercise 5 to sketch each of the following graphs. Use a calculator graph to support your results.

a. $y = 2x^2$ **b.** $y = 0.2x^2$

c. $y = -x^2$ **d.** $y = -0.5x^2$

7. Graph all of the following functions on one set of coordinate axes.

a. $y = x^2$ **b.** $y = (x + 1)^2$

c. $y = (x + 4)^2$ **d.** $y = (x - 1)^2$

8. *(Making Observations)* Based on your graphs from Exercise 7, guess the location of the vertex on the graph of $y = (x - h)^2$ for any real value of h. Then sketch each of the following graphs and use a calculator graph to support your results.

a. $y = (x - 2)^2$ **b.** $y = (x - 5)^2$

c. $y = (x + 2)^2$ **d.** $y = (x + 6)^2$

9. Graph all of the following functions on one set of coordinate axes.

a. $y = x^2$ **b.** $y = x^2 - 1$

c. $y = x^2 + 1$ **d.** $y = x^2 + 3$

10. *(Making Observations)* Based on your graphs from Exercise 9, guess the location of the vertex on the graph of $y = x^2 + k$ for any real value of k. Then sketch each of the following graphs and use a calculator graph to support your results.

a. $y = x^2 + 5$ **b.** $y = x^2 + 2$

c. $y = x^2 - 5$ **d.** $y = x^2 - 3$

Your observations in Exercises 4–10 can be generalized as follows.

- The graph of $y = a(x - h)^2 + k$ is a parabola with its vertex at (h, k).
- The graph opens up if $a > 0$ and down if $a < 0$.
- The graph is steeper than that of $y = x^2$ if $|a| > 1$ and flatter if $|a| < 1$.

EXAMPLE 1 *Sketching the graph of a function $y = a(x - h)^2 + k$*

The graph of

$$y = -0.5(x + 4)^2 + 2$$

is a parabola with its vertex at $(-4, 2)$. It opens down since $a = -0.5 < 0$. Since $|a| = 0.5 < 1$, the graph is flatter than the basic graph $y = x^2$. Specifically, it is half as steep as the basic graph, so the points 1 unit to the left and right of the vertex have coordinates $(-5, 1.5)$ and $(-3, 1.5)$. Its graph is shown in Figure 5-3.

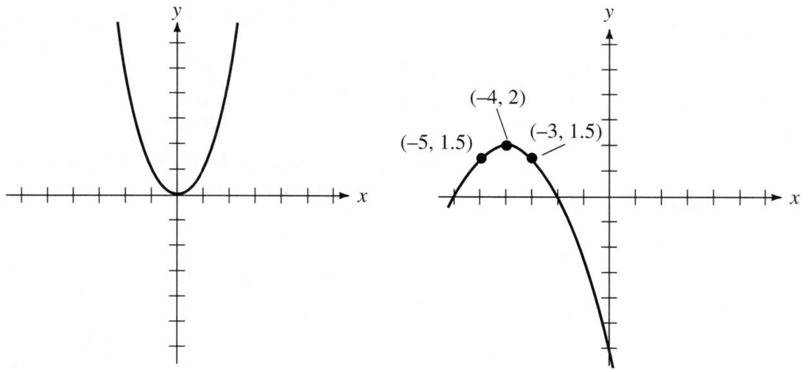

Figure 5–3 ■

EXERCISE **11.** *(Making Observations)* Sketch each of the following graphs, and use a calculator graph to support your results.

a. $y = 2(x - 2)^2 + 5$

b. $y = 0.2(x - 5)^2 + 2$

c. $y = -(x + 2)^2 - 5$

d. $y = -0.5(x + 6)^2 - 3$

Your observations in Exercises 4–10 are important because, as you will soon see, you can write the equation of every quadratic function in the form $y = a(x - h)^2 + k$. When the equation of a quadratic function is written in this form, you can sketch its graph quickly, as you saw in Exercise 11. (We hope you noticed the resemblance to the standard graphing form $y = a|x - h| + k$ for the linear absolute value functions you encountered in Section 3-4.)

EXERCISE **12.** *(Making Observations)* Sketch the graph of $y = -0.5|x + 4| + 2$. In what respects is it like the graph of $y = -0.5(x + 4)^2 + 2$ in Example 1? In what respects is it different?

The graph of a quadratic function, like that of a linear absolute value function, can be described in terms of transformations applied to a basic graph.

> To obtain the graph of $y = a(x - h)^2 + k$ from that of $y = x^2$:
>
> - Stretch the graph vertically by a factor of $|a|$ if $|a| > 1$, or compress it if $|a| < 1$.
> - If $a < 0$, reflect the graph in the x-axis.
> - Shift the graph to the right if $h > 0$ or to the left if $h < 0$.
> - Shift the graph up if $k > 0$ or down if $k < 0$.

EXAMPLE 2 *Sketching the graph of a quadratic function in the form* $y = a(x - h)^2 + k$

The function

$$g(x) = 2(x - 3)^2 - 5$$

has the form $y = a(x - h)^2 + k$ with $a = 2, h = 3, k = -5$. Its graph is that of $y = x^2$ stretched vertically by a factor of 2, then shifted 3 units to the right and 5 units down. As shown in Figure 5-4, its vertex is $(3, -5)$. The points 1 unit to the left and right of the vertex are 2 units above the vertex, at $(2, -3)$ and $(4, -3)$.

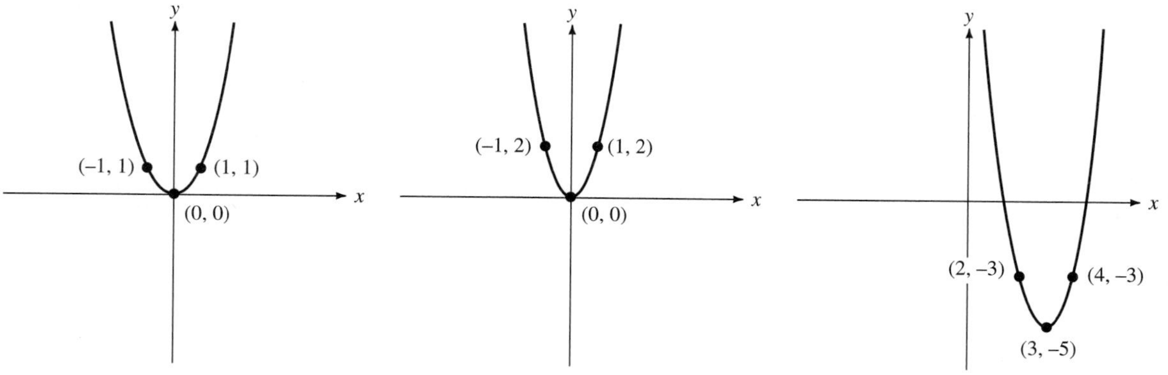

Figure 5–4 ■

EXERCISES *In Exercises* 13–16,

a. Describe the ways in which the graph of $y = x^2$ must be stretched, compressed, reflected, and shifted to produce the graph of the given function.

b. Sketch the graph. Identify the vertex and the points one unit to the left and right of the vertex.

13. $g(x) = 2(x - 4)^2 + 1$ **14.** $g(x) = -3(x + 1)^2 + 3$

15. $g(x) = -\frac{3}{4}x^2 - 3$ **16.** $g(x) = 0.1(x - 2)^2$

Examples 1 and 2 and Exercises 4–16 indicate some of the strong connections between the graphs and the equations of quadratic functions. To establish these connections more clearly, let's look at quadratic functions analytically.

AN ANALYTICAL VIEW OF QUADRATIC FUNCTIONS

The equation of every quadratic function can be written in the following three forms.

$$f(x) = ax^2 + bx + c \qquad \textbf{(expanded form)}$$

$$f(x) = a(x - r_1)(x - r_2) \qquad \textbf{(factored form)}$$

$$f(x) = a(x - h)^2 + k \qquad \textbf{(standard graphing form)}$$

EXAMPLE 3 *The three analytical forms of a quadratic function*

The function

$$f(x) = -0.5(x + 4)^2 + 2$$

is written in standard graphing form. Perform the multiplication and combine like terms to obtain the expanded form

$$f(x) = -0.5x^2 - 4x - 6$$

Factor the expanded form to obtain the factored form

$$f(x) = -0.5(x + 6)(x + 2)$$ ■

Each form provides useful information about the function it represents. Example 3 shows that the value of the parameter a is the same for all three forms, so that it always provides information about the direction and steepness of the graph. You can use each form to find the x-intercepts, vertex, and axis of symmetry, as well as the range of the function.

Expanded Form The x-intercepts, vertex, and axis of symmetry cannot be read directly from the expanded form $f(x) = ax^2 + bx + c$. However, we can identify them with the help of a few formulas.

In Exercise 93 you will show that the vertex on the graph of $f(x) = ax^2 + bx + c$ has an x-coordinate of $-\frac{b}{2a}$. For now you can use this fact to find the vertex. Knowing the vertex, you can also identify the axis of symmetry and the range of $f(x)$.

EXAMPLE 4 *Finding the vertex on the graph of $f(x) = ax^2 + bx + c$*

The vertex on the graph of $f(x) = 2x^2 - 6x + 5$ occurs when

$$x = -\frac{-6}{2(2)} = \frac{3}{2}$$
$$y = 2\left(\frac{3}{2}\right)^2 - 6\left(\frac{3}{2}\right) + 5 = \frac{1}{2}$$

As shown in Figure 5-5, the axis of symmetry is the vertical line through the vertex, and thus has the equation $x = \frac{3}{2}$. The range of $f(x)$ extends upward from the y-coordinate of the vertex, so the range is $\left[\frac{1}{2}, \infty\right)$.

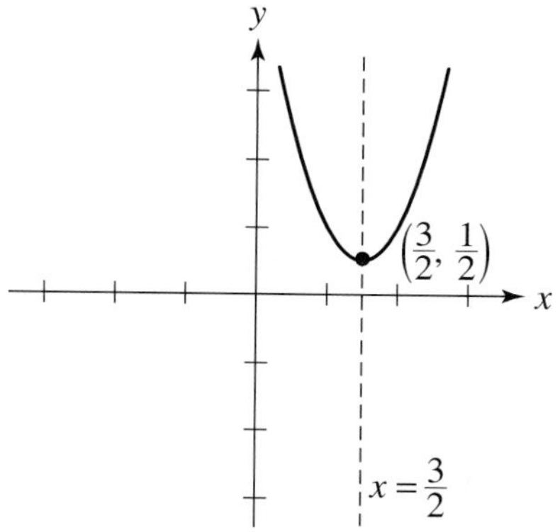

Figure 5–5

■

EXERCISES *In Exercises 17–20, identify the vertex and the axis of symmetry on the graph of $y = H(x)$ and the range of $H(x)$.*

17. $H(x) = x^2 + 8x + 9$ **18.** $H(x) = 3x^2 - 6x + 2$

19. $H(x) = 9 - x^2$ **20.** $H(x) = 9x - x^2$

The quadratic formula allows you to locate the x-intercepts and also shows how the intercepts are related to the vertex.

EXERCISES **21.** *(Review)* Use the quadratic formula $x = \frac{-b \pm \sqrt{b^2 - 4ac}}{2a}$ to solve each equation.

a. $x^2 - 5x + 2.25 = 0$

b. $x^2 - 5x + 6.25 = 0$

c. $x^2 - 5x + 10.25 = 0$

22. *(Making Observations)*

a. How many real solutions does each equation in Exercise 21 have?

b. Graph the functions $y = x^2 - 5x + 2.25$, $y = x^2 - 5x + 6.25$, and $y = x^2 - 5x + 10.25$ on the same set of coordinate axes. How many x-intercepts does each graph have?

c. How can you decide whether a solution to a quadratic equation $f(x) = 0$ corresponds to an x-intercept on the graph of $y = f(x)$?

23. *(Making Observations)* How do your observations in Exercise 22 support the conclusion that a quadratic graph can have 0, 1, or 2 x-intercepts?

24. *(Making Observations)* The quadratic formula can be written

$$x = -\frac{b}{2a} \pm \frac{\sqrt{b^2 - 4ac}}{2a}$$

How does this observation support the conclusion that when a quadratic graph has two x-intercepts, they are at equal distances to the left and right of the vertex?

The discriminant $b^2 - 4ac$ of a quadratic expression $f(x) = ax^2 + bx + c$ can be used to determine the number of real solutions to the equation $f(x) = 0$. (See Section A-13.)

In Exercises 25–28, use the discriminant to determine the number of x-intercepts on the graph of $y = f(x)$. Use a calculator graph to support your results.

25. $f(x) = x^2 + 4$ **26.** $f(x) = x^2 + 4x$

27. $f(x) = x^2 + 4x + 4$ **28.** $f(x) = -x^2 + 4x + 4$

Factored Form You may have learned in a previous algebra course that the quadratic expression $2x^2 - 5x + 2$ is factorable, while $x^2 - 2$ and $x^2 + 2x + 5$ are not. It is more accurate to say that $x^2 - 2$ and $x^2 + 2x + 5$ cannot be written in the form $a(x - r_1)(x - r_2)$ *if r_1 and r_2 are required to be rational numbers*. By contrast, *every* quadratic expression can be written in that form if r_1 and r_2 can take on any complex value. When r_1 and r_2 are real as in Example 5, the factored form of the function is closely connected with the x-intercepts on its graph.

EXAMPLE 5 *Writing quadratic functions in factored form*

a. The function

$$h(x) = 2x^2 - 5x + 2$$

can be written

$$h(x) = (2x - 1)(x - 2)$$

We can then continue by writing $2x - 1$ as $2\left(x - \frac{1}{2}\right)$ to obtain the factored form

$$h(x) = 2\left(x - \frac{1}{2}\right)(x - 2)$$

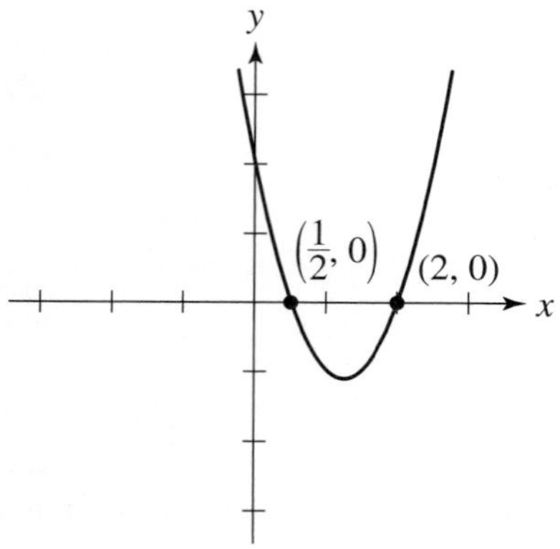

Figure 5–6

The factored form makes it clear that $h\left(\frac{1}{2}\right) = 0$ and $h(2) = 0$, so the x-intercepts on the graph of $y = h(x)$ are $\left(\frac{1}{2}, 0\right)$ and (2, 0). The graph in Figure 5-6 supports these conclusions.

b. To write the factored form of

$$g(x) = x^2 - 2$$

factor $g(x)$ as a difference of squares.

$$g(x) = (x + \sqrt{2})(x - \sqrt{2})$$

The factored form makes it clear that $g(\sqrt{2}) = 0$ and $g(-\sqrt{2}) = 0$, so the x-intercepts on the graph of $y = g(x)$ are $(-\sqrt{2}, 0)$ and $(\sqrt{2}, 0)$. The graph in Figure 5-7 supports this conclusion.

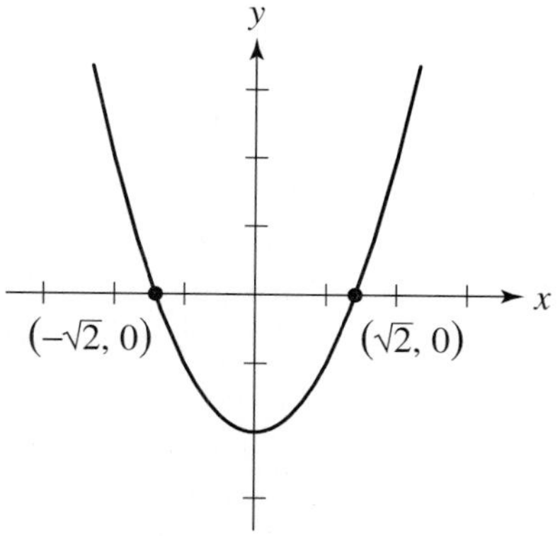

Figure 5–7

■

EXERCISES *In Exercises 29–32, write the function in factored form.*

29. $f(x) = 3x^2 + 8x$ **30.** $Q(x) = -x^2 - 14x - 45$

31. $P(t) = 6t^2 + 7t - 5$ **32.** $G(v) = 13 - v^2$

33. The equation $15x^2 + 41x + 12 = 0$ has solutions $x = -\frac{12}{5}$ and $x = -\frac{1}{3}$. Use this fact to help you write the function $f(x) = 15x^2 + 41x + 12$ in factored form.

In Exercises 34–37, obtain the factored form of $Q(x)$ by solving the equation $Q(x) = 0$ first.

34. $Q(x) = x^2 + 2x - 10$ **35.** $Q(x) = 2x^2 - 4x - 1$

36. $Q(x) = 3 + x - x^2$ **37.** $Q(x) = 0.5x^2 + 4x + 5$

38. *(Making Observations)* Solve $x^2 + 2x + 5 = 0$. How does the solution show that the graph of $y = x^2 + 2x + 5$ has no x-intercepts? Support this conclusion by graphing the function.

The observations on the preceding few pages can be generalized into a statement that allows you to read the x-intercepts on a quadratic graph directly from the factored form of the function.

> If r_1 and r_2 are real, the graph of $f(x) = a(x - r_1)(x - r_2)$ has x-intercepts at $(r_1, 0)$ and $(r_2, 0)$. [If $r_1 = r_2$, there is a single x-intercept at $(r_1, 0)$.]
> If r_1 and r_2 are not real, the graph of $f(x) = a(x - r_1)(x - r_2)$ has no x-intercepts.

EXERCISE **39.** *(Interpreting Mathematics)* Write the factored form of the function $P(x) = -0.005x^2 + 12.50x - 612.50$ from ***Maximum Profit***. Use the factored form to find the breakeven points A and B in Figure 5-1.

Can we also read the vertex of a quadratic graph directly from the factored form of the function? You can explore this question in Exercises 40–41.

EXERCISES **40.** *(Making Observations)* Find the vertex on the graph of each function in Exercises 29–32. How is the vertex related to the x-intercepts in each case?

41. Suppose that the graph of a quadratic function $y = f(x)$ has x-intercepts at $(-2, 0)$ and $(8, 0)$. Use your observations from Exercise 40 to find the x-coordinate of the vertex.

Your observations from Exercise 40 can be stated as a formula that relates the vertex of a quadratic graph to the factored form of its equation. The formula is valid even when the factors involve complex numbers. You can learn why in Exercise 94.

> If $f(x) = a(x - r_1)(x - r_2)$, the vertex on the graph of $f(x)$ occurs when $x = \dfrac{r_1 + r_2}{2}$.

EXAMPLE 6 *Finding the vertex on the graph of $f(x) = a(x - r_1)(x - r_2)$*

The vertex on the graph of

$$f(x) = -4[x - (5 + 2i)][x - (5 - 2i)]$$

occurs when

$$x = \frac{(5 + 2i) + (5 - 2i)}{2} = \frac{10}{2} = 5$$

To find the y-coordinate of the vertex, substitute 5 for x in the equation for $f(x)$.

$$\begin{aligned} y &= -4[5 - (5 + 2i)][5 - (5 - 2i)] \\ &= -4(-2i)(2i) \\ &= 16i^2 \\ &= -16 \end{aligned}$$

The vertex is $(5, -16)$. As usual, the location of the vertex determines the axis of symmetry and the range of f. The axis of symmetry is the line $x = 5$, and the range of f is $(-\infty, -16]$. See Figure 5-8. ■

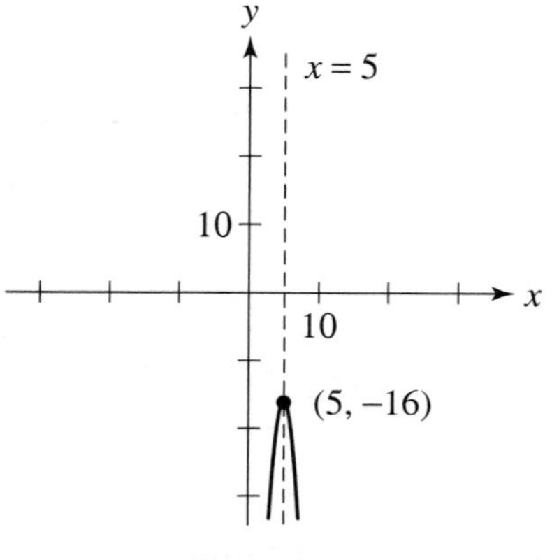

Figure 5–8

EXERCISES *In Exercises 42–45, find the vertex and axis of symmetry on the graph of $y = f(x)$ and the range of f.*

42. $f(x) = 3\left(x - \frac{5}{3}\right)(x - 8)$

43. $f(x) = -(x + \sqrt{7})(x - \sqrt{7})$

44. $f(x) = -5[x - (10 + \sqrt{13})][x - (10 - \sqrt{13})]$

45. $f(x) = 2[x - (1 + 3i)][x - (1 - 3i)]$

Standard Graphing Form By completing the square you can write every quadratic function in the form $f(x) = a(x - h)^2 + k$. The technique of completing the square is discussed in Section A-13. Let's review it in Example 7.

EXAMPLE 7 *Writing a quadratic function in standard graphing form*

To write

$$g(x) = 2x^2 - 12x + 13$$

in standard graphing form, first group the terms containing x^2 and x, and factor out the coefficient of x^2.

$$g(x) = 2(x^2 - 6x) + 13$$

Square half of the x-coefficient [half of -6 is -3, and $(-3)^2 = 9$]. Then add and subtract that quantity inside the parentheses.

$$g(x) = 2(x^2 - 6x + 9 - 9) + 13$$

Remove the -9 from the parentheses, remembering to multiply it by the 2 in front.

$$g(x) = 2(x^2 - 6x + 9) - 18 + 13$$

Finally, write the expression inside the parentheses as the square of a binomial, and combine the constants outside.

$$g(x) = 2(x - 3)^2 - 5$$ ■

When a quadratic function is written in standard graphing form, you can easily find its vertex, its axis of symmetry, and its range. For the function $g(x)$ in Example 7, these features were found in Example 2.

EXERCISES *In Exercises* 46–49,

a. Write the equation in standard graphing form.

b. Sketch the graph of the function. Identify the vertex and the axis of symmetry, and find the range of the function.

46. $y = 4x^2 + 8x - 3$ **47.** $y = 0.5x^2 + 4x + 9$

48. $s = -t^2 - 2t - 4$ **49.** $w = -3z^2 + 6z$

50. *(Interpreting Mathematics)* Write the standard graphing form of $P(x) = -0.005x^2 + 12.50x - 612.50$ from ***Maximum Profit***. Then find the maximum profit and the target sales level that will generate it.

It is possible to read the x-intercepts on the graph of

$$f(x) = a(x - h)^2 + k$$

directly from the equation (see Exercise 56). However, it is more practical to find them by solving the equation $a(x - h)^2 + k = 0$.

EXERCISES *In Exercises* 51–54, *find the t-intercepts on the graph of* $y = F(t)$.

51. $F(t) = (t + 6)^2 - 4$ **52.** $F(t) = 9(t - 1)^2 - 16$

53. $F(t) = -2(t + 3)^2 + 1$ **54.** $F(t) = 2(t + 3)^2 + 1$

55. *(Writing to Learn)* Explain why the graph of $y = a(x - h)^2 + k$ has no x-intercepts if a and k have the same sign.

56. *(Making Observations)* Solve $a(x - h)^2 + k = 0$ for x.

a. How does the result verify that the x-intercepts on the graph of $y = a(x - h)^2 + k$ are $\left(h \pm \sqrt{-\frac{k}{a}}, 0\right)$ if a and k have opposite signs?

b. How does the result show that the vertex is halfway between the x-intercepts?

Comparisons Among the Three Forms Each analytical form of a quadratic function is useful in its own way. The vertex of the graph is obtained most easily from the standard graphing form, while the x-intercepts are obtained most easily from the factored form. In Section 5-2 you will discover that a quadratic function to fit a table can be constructed more easily in expanded form.

A NUMERICAL VIEW OF QUADRATIC FUNCTIONS

The following Mathematical Looking Glass illustrates that it can be useful to know that a table was generated by a quadratic function.

A MATHEMATICAL LOOKING GLASS
Stopping Distance

The data in Table 19 was gathered by General Motors. The left column represents the speed of a car at the instant the driver recognizes a need to stop. The right column represents the distance traveled from that instant until the car comes to a stop. The data assumes that the pavement is dry.

TABLE 19

Initial speed, in mph (v)	Stopping distance, in feet [$S(v)$]
30	77
40	128
50	190
60	263

The relationship of the car's initial speed v to the required stopping distance $S(v)$ is important in many contexts.

- A Department of Transportation employee responsible for timing traffic lights needs to know the stopping distance required by a driver traveling toward an intersection.
- A police officer investigating an accident may need to calculate a driver's speed based on evidence such as the distance of skid marks from the point of impact. ■

To fit a function to Table 19, let's adopt the strategy of *Looking for a Pattern*. In Table 20 we have added a third column for the first differences in

$S(v)$ and a fourth column for the **second differences** (the first differences of the first differences).

TABLE 20

v	$S(v)$	First differences	Second differences
30	77		
		51	
40	128		11
		62	
50	190		11
		73	
60	263		

Are the constant second differences significant? Let's explore a little further.

EXERCISE **57.** *(Making Observations)* Construct a table for each function, using the indicated x-values. Calculate the first and second differences for the y-values.

a. $y = 3x^2$; $x = 1.0, 1.5, 2.0, 2.5, 3.0$

b. $y = ax^2$; $x = -1, 0, 1, 2, 3$

c. $y = ax^2$; $x = x_0, x_0 + h, x_0 + 2h, x_0 + 3h, x_0 + 4h$

In Exercise 57 you have shown that for any function $y = ax^2$, any table with equally spaced x-values has y-values with constant second differences. Let's see whether the same is true for any function $y = ax^2 + bx + c$.

EXERCISE **58.** *(Making Observations)* Construct a table for each function, using the indicated x-values. Calculate the first and second differences for the y-values.

a. $y = 3x^2 + 2x - 5$
$x = 1.0, 1.5, 2.0, 2.5, 3.0$

b. $y = ax^2 + bx + c$
$x = -1, 0, 1, 2, 3$

c. $y = ax^2 + bx + c$
$x = x_0, x_0 + h, x_0 + 2h, x_0 + 3h, x_0 + 4h$

In Exercise 58 you have verified the first of the following statements. The second will be illustrated in Section 5-2, where you will find a quadratic function to fit the table in ***Stopping Distance***.

For a quadratic function $y = f(x)$, a table with equally spaced x-values has y-values with constant second differences.

Conversely, if a table with equally spaced x-values has y-values with constant second differences, it fits a quadratic function.

FYI As in linear functions, we cannot say that a function is quadratic on the basis of a table without additional information. We can only say that the table fits a quadratic function. However, if the second differences are not constant, then the function is not quadratic.

EXERCISES *In Exercises 59–62, decide whether the table fits a linear function, a quadratic function, or neither.*

59.

x	y
3	23
4	16
5	11
6	8
7	7

60.

x	y
−2	−1
−1.5	7
−1	25
−0.5	62
0	123

61.

x	y
−1	−4
1	−6
3	−6
5	−4
7	0
9	6

62.

x	y
−1	−4
1	0
3	4
5	8
7	12
9	16

63. *(Making Observations)*

a. Use Table 19 to find the average rate of change in $S(v)$ with respect to v over the interval [30, 40]. What physical information is provided by this result?

b. Verify that the average rate of change in $S(v)$ with respect to v is larger over the interval [40, 50] than over [30, 40]. What physical information is provided by this result?

ADDITIONAL EXERCISES *In Exercises 64–69,*

a. Describe how the graph of $y = x^2$ must be transformed to obtain the graph of the given function.

b. Identify the vertex and the points one unit to the left and right of the vertex.

c. Sketch the graph.

d. Find the axis of symmetry and the range of the function.

e. Identify the intervals where the function increases and decreases.

64. $y = (x - 6)^2 + 1$

65. $f(x) = (x - 2.5)^2 - 6.8$

66. $z = 2(x + 10)^2$

67. $A(r) = 2\pi(r + 1)^2 - 2\pi$

68. $P(x) = -4(x - 200)^2 + 9000$

69. $q = 100(p + 3)^2 - 5$

In Exercises 70–81,

a. Find the x-intercepts if there are any, and identify the vertex. If there are fewer than two x-intercepts, identify the points one unit to the left and right of the vertex.

b. Sketch the graph.

c. Find the axis of symmetry and the range of the function.

d. Identify the intervals where the function increases and decreases.

70. $y = 3x^2 + 4x + 1$

71. $y = 6 + x - x^2$

72. $g(x) = -0.3x^2 + 7x$

73. $s(t) = 400 - 16t^2$

74. $y = x^2 + 4x + 5$

75. $y = x^2 - 3x + 1$

76. $y = (x + 7)(x - 4)$

77. $V(u) = 2u(u + 6)$

78. $w = -(s + i)(s - i)$

79. $z = (x + \sqrt{5})(x - \sqrt{5})$

80. $y = -2[x - (4 + \sqrt{3})][x - (4 - \sqrt{3})]$

81. $\beta = 5[\alpha - (2 + 4i)][\alpha - (2 - 4i)]$

In Exercises 82–85, find an equation to fit the graph. (Hint: Write the equation in factored form, and use the point that is not an x-intercept to find the value of a.)

82.

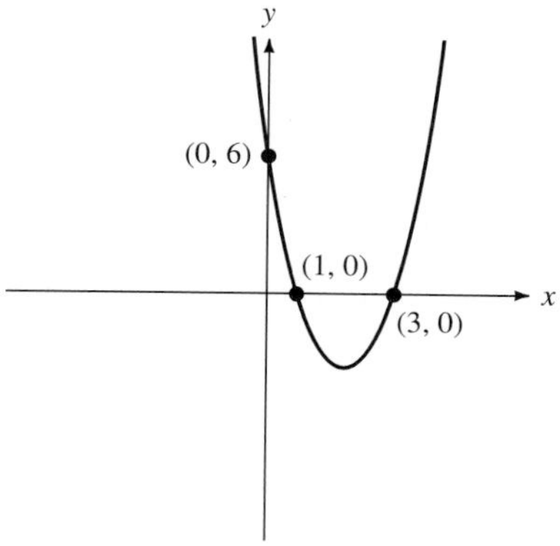

83.

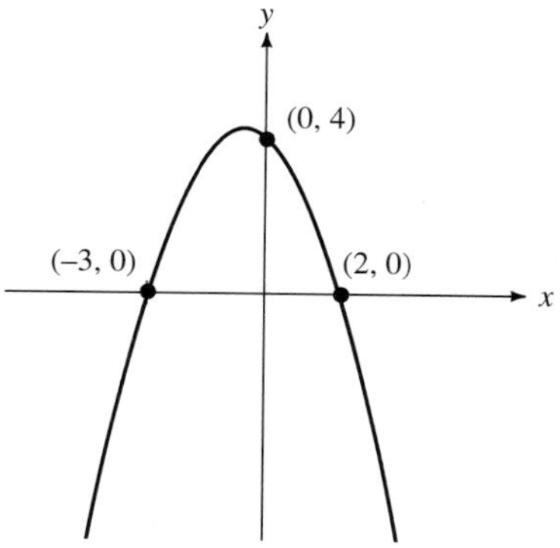

84.

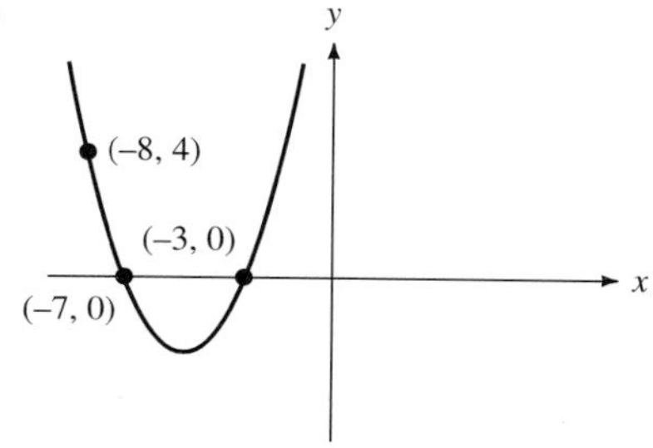

85.

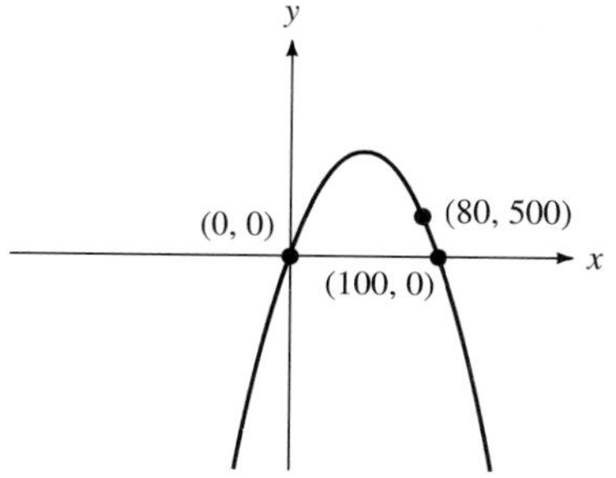

In Exercises 86–89, find an equation to fit the graph. (Hint: Write the equation in standard graphing form, and use the point that is not the vertex to find the value of a.)

86.

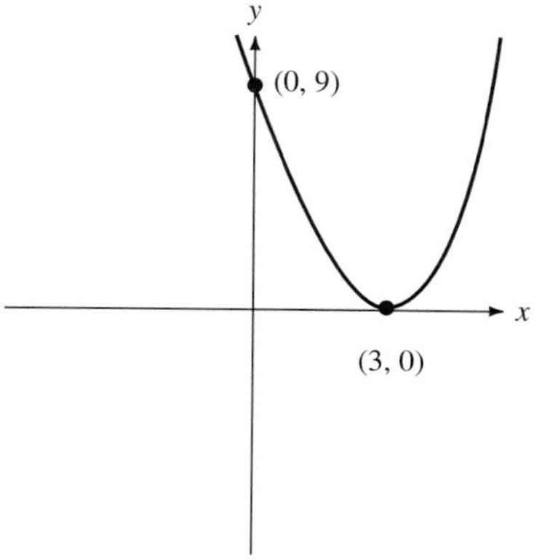

87.

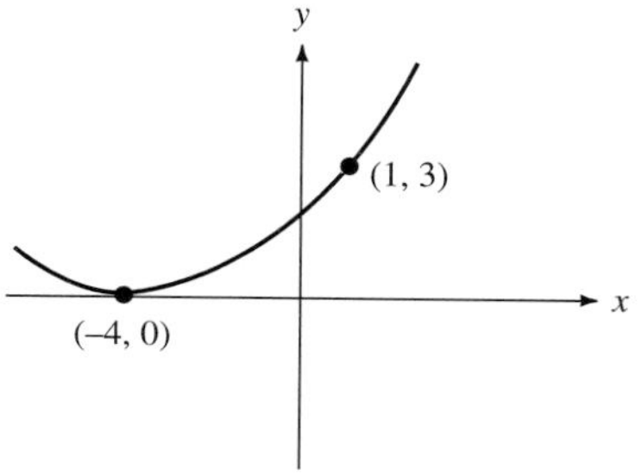

88.

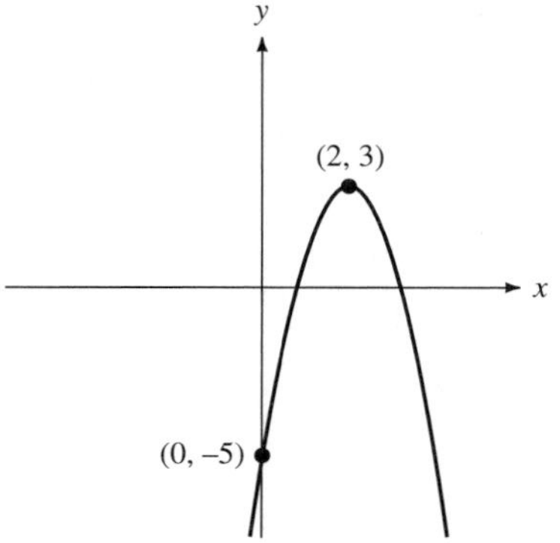

89.

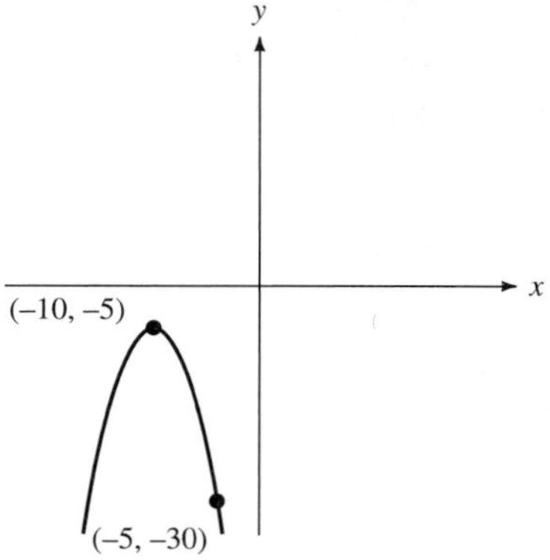

90. *(Interpreting Mathematics)* Assume that each relationship is quadratic. Decide whether the function first increases and then decreases or vice versa.

a. T = elapsed time after a ball is thrown upward
H = height of the ball

b. L = length of a rectangle whose perimeter is 20
A = area of the rectangle

c. D = distance traveled as you ride up, then down a hill on the American Eagle roller coaster at Great America, IL
S = speed of the roller coaster

d. T = elapsed time from the onset of a recession until the economy recovers
P = price of a typical share of stock

e. On May 30, 1986, several creeks northeast of Pittsburgh overflowed, flooding our favorite ice cream stand to a depth of 3 feet.
T = elapsed time from the beginning of the flood
L = water level around Ice Cream World

91. *(Making Observations)* What name is given to the point at which a quadratic function stops increasing and starts to decrease or vice versa?

92. *(Making Observations)* Compare the process of graphing

$$y = a|x - h| + k$$

(see Section 3-4) with that of graphing $y = a(x - h)^2 + k$. In particular, describe the information provided by a, h, and k in each case.

93. *(Making Observations)*

a. Complete the square to verify that $f(x) = ax^2 + bx + c$ has the standard graphing form $f(x) = a\left(x + \frac{b}{2a}\right)^2 - \frac{b^2 - 4ac}{4a}$.

b. Explain why this shows that the vertex on the graph of $y = f(x)$ occurs when $x = -\frac{b}{2a}$.

94. *(Making Observations)* Write $g(x) = a(x - r_1)(x - r_2)$ in expanded form. Explain why the result shows that the vertex on the graph of $y = g(x)$ occurs when $x = \frac{r_1 + r_2}{2}$.

SUPPLEMENTARY TOPICS

Effect of Viewing Window on Apparent Steepness of Graphs

Exercises 95–96 illustrate that the steepness of a parabola can be distorted by the viewing window on your calculator. (Your calculator may have a key or menu item, such as "square," that automatically defines viewing windows in which steepness is not distorted. Consult your owner's manual for details.)

95. *(Making Observations)*

a. Which of the functions $f(x) = x^2$, $g(x) = 4x^2$, or $h(x) = 0.25x^2$ has the steepest graph? The flattest graph?

b. Graph $f(x)$ in the window $[-10, 10]$ by $[-100, 100]$, graph $g(x)$ in $[-5, 5]$ by $[-100, 100]$, and graph $h(x)$ in $[-20, 20]$ by $[-100, 100]$. Do your graphs support what you said in part (a)? Explain any apparent contradiction.

c. Graph all three functions in the window $[-10, 10]$ by $[-100, 100]$, and indicate which graph is which. Do these graphs support what you said in part (a)?

96. *(Making Observations)*

a. Graph $f(x) = 6x - x^2$ in the window $[-10, 10]$ by $[-10, 10]$.

b. By trial and error, find a viewing window in which the graph of $g(x) = 3f(x) = 18x - 3x^2$ appears identical to your graph in part (a).

c. Graph both functions in a window that allows you to compare the graphs accurately.

Complex Factors The factored form of a quadratic function $f(x)$ may involve numbers with imaginary components. In this case the factors of $f(x)$ do not correspond to x-intercepts on the graph, but they still correspond to solutions of the equation $f(x) = 0$.

EXAMPLE 8 *A quadratic function with complex factors*

To write the factored form of

$$P(x) = x^2 + 2x + 5$$

it is easiest to solve the equation $P(x) = 0$ first.

$$x^2 + 2x + 5 = 0$$

$$x = \frac{-2 \pm \sqrt{2^2 - 4(1)(5)}}{2(1)}$$

$$= \frac{-2 \pm \sqrt{-16}}{2}$$

$$= \frac{-2 + 4i}{2}$$

$$= -1 \pm 2i$$

The solutions are $-1 + 2i$ and $-1 - 2i$. Therefore, in factored form,

$$P(x) = [x - (-1 + 2i)][x - (-1 - 2i)]$$ ■

In Exercises 97–100, obtain the factored form of $F(x)$ by solving $F(x) = 0$ first.

97. $F(x) = x^2 - 4x + 13$ **98.** $F(x) = 4x^2 + 4x + 5$

99. $F(x) = 2x^2 + 8x + 9$ **100.** $F(x) = -x^2 + 2x - 2$

5-2 MODELING AND PROBLEM SOLVING WITH QUADRATIC FUNCTIONS

The Mathematical Looking Glasses you encountered in Section 5-1 illustrated that being familiar with quadratic functions can help you solve physical problems. In this section we will focus on methods for solving some common types of problems. Specifically, we will look at problems involving maximum and minimum values of quadratic functions, the solution of quadratic inequalities, and the construction of a quadratic function to fit a table.

OPTIMIZATION PROBLEMS

An **optimization problem** asks for the extreme values of a function, that is, its largest or smallest values. We encounter optimization problems when we want to know a company's most economical production schedule or the point of a comet's closest approach to the earth.

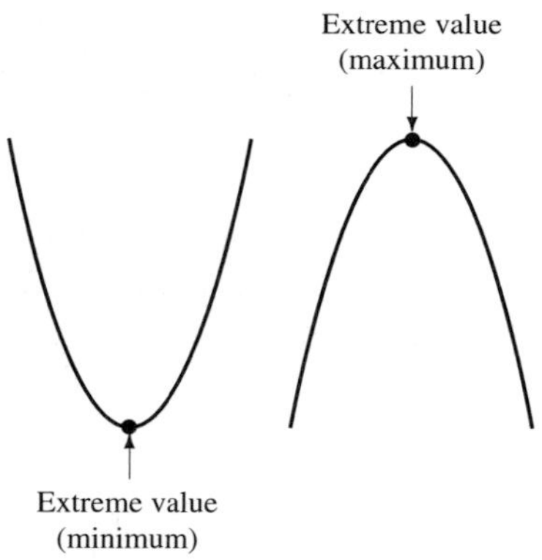

Figure 5–9

Figure 5-9 illustrates that every quadratic function has an extreme value, which occurs at the vertex on its graph. Thus in most optimization problems involving quadratic functions, the most difficult work in the solution is to find an appropriate function. Once that is done, you need only to find the vertex on its graph.

You have already solved an optimization problem. In ***Maximum Profit*** (page 149), your company needed to find the largest value of its profit function. You solved that problem in Exercise 50 of Section 5-1. Here is another.

A MATHEMATICAL LOOKING GLASS

Gravity

If you fire a toy rocket into the air, you might wonder how high it goes before falling back to earth. By using both physics and calculus, it can be shown that if you launch the rocket from a height of s_0 feet at v_0 feet per second, its height in feet after t seconds later is

$$s(t) = -16t^2 + v_0t + s_0$$

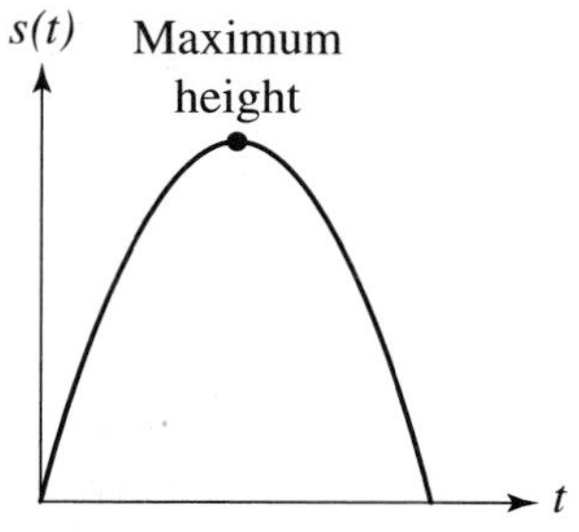

Figure 5–10

As shown in Figure 5-10, this formula considers only the effect of gravity on the rocket and ignores other forces such as thrust and air resistance. If you continue your study of mathematics long enough to learn differential equations, you can construct a more accurate model.

If you launch the rocket from ground level, then $s_0 = 0$. You probably don't know the value of v_0 offhand, but you might observe that the rocket's flight lasts 10 seconds. This is enough information for you to construct an equation from which you can calculate the rocket's maximum height. ■

EXERCISE **1.** *(Problem Solving)*

a. Explain why $s(10) = 0$.

b. Use this fact to construct the equation for $s(t)$ and find the value of v_0.

c. Find the vertex on the graph of $y = s(t)$. How high does the rocket go?

To solve a quadratic optimization problem is simply to find the extreme value (the y-coordinate of the vertex) on its graph. You have learned how to do this in Section 5-1.

EXAMPLE 1 *Finding the extreme value of a quadratic function*

The vertex on the graph of $f(x) = -0.5(x + 4)^2 + 2$ is $(-4, 2)$, as shown in Figure 5-11, so the extreme value of f is 2. Since the graph opens downward, the extreme value is a maximum.

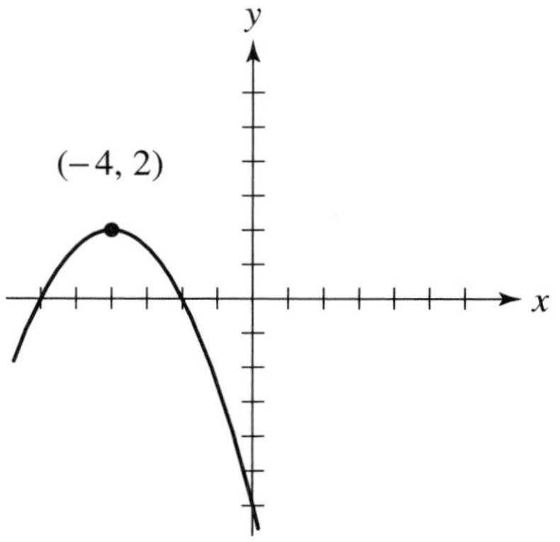

Figure 5–11

If $f(x)$ is not written in standard graphing form, you can still find its extreme value.

If $f(x)$ is written in its expanded form $f(x) = -0.5x^2 - 4x - 6$, we can calculate the x-coordinate of the vertex as $-\dfrac{b}{2a} = -4$. The y-coordinate is then $f(-4) = 2$.

If $f(x)$ is written in its factored form $f(x) = -0.5(x + 6)(x + 2)$, we can read its x-intercepts $(-6, 0)$ and $(-2, 0)$, so the x-coordinate of the vertex is halfway between, at -4. The y-coordinate is then $f(-4) = 2$. ■

EXERCISES *In Exercises 2–5, find the extreme value of the function and state whether it is a maximum or a minimum.*

2. $f(x) = 7 + 4x - x^2$

3. $g(x) = 2x^2 + 4x + 5$

4. $F(t) = 0.1t(t - 100)$

5. $G(t) = -6(t + 20)(t - 8)$

SOLVING QUADRATIC INEQUALITIES

Sometimes you can solve a problem by solving a quadratic inequality. In such a case you can use the method you learned in Section 2-4 to solve quadratic inequalities graphically. Let's use the following Mathematical Looking Glass to learn an analytical method of solution.

A MATHEMATICAL LOOKING GLASS
Dance Floor

In March 1992, Pittsburgh's Community Dance Network (CDN) sponsored a Vintage Dance Weekend featuring Richard Powers, founder of the Flying Cloud Academy of Vintage Dance in Cincinnati. The weekend included an evening ball, preceded by instruction by Mr. Powers in dances from the era 1890–1920.

Prior to the weekend Lisa Tamres of CDN arranged to rent space at the Fox Chapel Yacht Club. The rented space was to be a rectangular area bounded on two sides by a permanent glass wall and on the remaining sides by a number of movable panels, as shown in Figure 5-12.

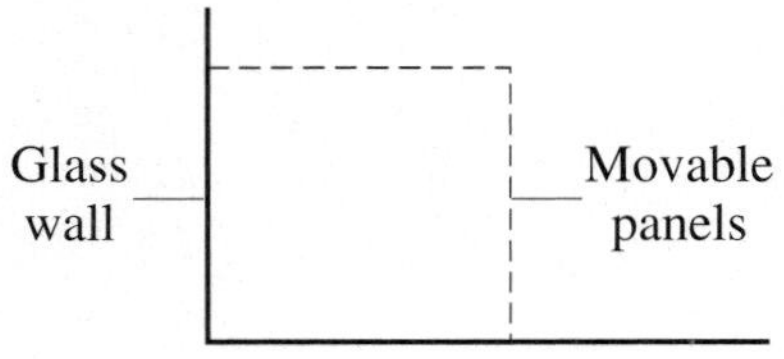

Figure 5–12

Mr. Powers required 30 square feet of floor space per person for the daytime instruction and 20 square feet per person for the evening ball. CDN expected 50 to 70 people for instruction and 100 to 130 people for the ball. Lisa wanted to allow enough floor space for each event. ■

EXERCISE **6.** *(Problem Solving)* The yacht club had 105 linear feet of movable panels, and the room was large enough to arrange them in any rectangular shape Lisa might want. Let W represent the width of the rented space in feet.

a. Express the number of square feet of rented floor space as a function of W.

b. To satisfy Mr. Powers' requirements, explain why Lisa needs to solve the inequality $105W - W^2 \geq 2600$.

c. Use graphical methods to solve the inequality, and describe all acceptable arrangements of paneling.

To solve the inequality in Exercise 6, we needed to

- find the W-intercepts on the graph of $S(W) = -W^2 + 105W - 2600$,
- observe that the intercepts divide the W-axis into the intervals $(-\infty, 40)$, $(40, 65)$, and $(65, \infty)$, and
- observe that the function is positive (so that the inequality is true) only in the interval $(40, 65)$.

Let's see how we can obtain the same information by analytical methods. The process can be carried out in three steps.

EXAMPLE 2 *Solving a quadratic inequality analytically*

STEP 1: To solve the inequality

$$-W^2 + 105W - 2600 \geqslant 0$$

begin by solving the equation

$$\begin{aligned} -W^2 + 105W - 2600 &= 0 \\ W^2 - 105W + 2600 &= 0 \\ (W - 40)(W - 65) &= 0 \\ W &= 40, 65 \end{aligned}$$

This step finds the W-intercepts on the graph of

$$S(W) = -W^2 + 105W - 2600.$$

STEP 2: Next indicate the x-intercepts 40 and 65 on a number line.

This step identifies the intervals $(-\infty, 40)$, $(40, 65)$, and $(65, \infty)$, where the sign of $S(W)$ cannot change.

STEP 3: Choose a "test value" of W in $(-\infty, 40)$ and evaluate $S(W)$ there to see whether the inequality is true or false. For example, $S(0) = -2600$, so the inequality $S(W) \geqslant 0$ is false there. Since the sign of $S(W)$ cannot change in $(-\infty, 40)$, the inequality is false for *every* value of W in that interval.

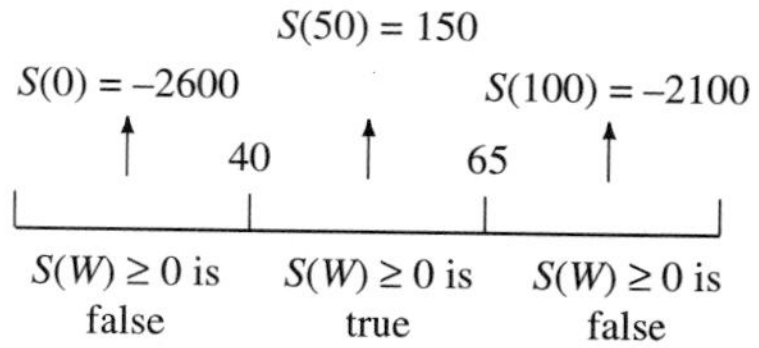

Continue by choosing a test value in $(40, 65)$ and one in $(65, \infty)$. In each case the inequality is true or false over the entire interval according to whether it is true or false at the test value.

Since the inequality is true in (40, 65) and also at the endpoints 40 and 65, the solution is [40, 65]. ■

EXAMPLE 3 *Solving a quadratic inequality analytically*

STEP 1: To solve the inequality

$$2x^2 + 6x + 5 > 0$$

begin by solving the equation

$$2x^2 + 6x + 5 = 0$$

$$x = \frac{-6 \pm \sqrt{6^2 - 4(2)(5)}}{2(2)}$$

$$x = \frac{-6 \pm \sqrt{-4}}{4}$$

Since there are no real solutions, the graph of $f(x) = 2x^2 + 6x + 5$ has no x-intercepts, so the function never changes sign.

STEP 2: There is no need to draw a number line. The only interval to consider is $(-\infty, \infty)$.

STEP 3: Any real number can be used as a test value. For example, $f(0) = 5$. Since the inequality is true there, it must be true on the entire interval $(-\infty, \infty)$. ■

EXERCISES

7. *(Writing to Learn)* Graph the function $f(x)$ from Example 3. How does the graph support the conclusion of the example?

In Exercises 8–11, *solve the inequality analytically.*

8. $x^2 + {}^{x} - 12 < 0$ **9.** $2x^2 + x - 1 > 0$

10. $5x - x^2 \leq 0$ **11.** $100 + 15x - x^2 \geq 0$

12. *(Interpreting Mathematics)* In ***Maximum Profit*** (page 149) your company's profit function is $P(x) = -0.005x^2 + 12.50x - 612.50$.

a. Use graphical methods to solve the inequality $P(x) \geq 0$.

b. Use analytical methods to solve $P(x) \geq 0$. Make sure your solution agrees with the one you obtained in part (a).

c. What useful information does the solution provide?

FITTING A QUADRATIC FUNCTION TO A TABLE

Table 19 from ***Stopping Distance*** (page 160) suggests that a car's stopping distance is a quadratic function of its initial velocity. If that is the case, can we use the table to construct an equation for the function? Let's find out.

EXAMPLE 4 *Fitting a quadratic function to a table*

Table 19 is reproduced on the following page.

Initial speed, in mph (v)	Stopping distance, in feet [$S(v)$]
30	77
40	128
50	190
60	263

If $S(v)$ is a quadratic function of v, then

$$S(v) = av^2 + bv + c$$

for some choice of real numbers a, b, and c. Furthermore,

$$\begin{aligned} S(30) &= 77, \text{ so} & a(30)^2 + b(30) + c &= 77 \\ S(40) &= 128, \text{ so} & a(40)^2 + b(40) + c &= 128 \\ S(50) &= 190, \text{ so} & a(50)^2 + b(50) + c &= 190 \\ S(60) &= 263, \text{ so} & a(60)^2 + b(60) + c &= 263 \end{aligned}$$

Aha! We have constructed a system of linear equations in the variables a, b, and c. In Chapter 4 you learned that a system of three linear equations in three variables usually has a unique solution. The first three equations in the system can be rewritten as

$$\begin{aligned} 900a + 30b + c &= 77 \\ 1600a + 40b + c &= 128 \\ 2500a + 50b + c &= 190 \end{aligned}$$

In Exercise 13 you can solve this system and verify that the solution is $a = 0.055$, $b = 1.25$, $c = -10$. Thus the function is

$$S(v) = 0.055v^2 + 1.25v - 10$$

In Exercises 51–53 you can decide how accurately this function models stopping distances. ■

EXERCISES

13. *(Review)* Solve the 3×3 system of linear equations in Example 4, and confirm that your solution fits all four rows of Table 19.

In Exercises 14–17, *find the equation of a quadratic function that fits the table.*

14.

x	y
0	19
1	12
2	7
3	4

15.

x	y
2	−6
4	6
6	14
8	18

16.

x	y
300	520
400	530
500	520
600	490

17.

x	y
0.5	4
1.0	1
1.5	0
2.0	1

18. *(Writing to Learn)*

a. Why does the technique of Example 4 suggest that nearly every table containing exactly three data points fits a quadratic function?

b. Why does your answer to part (a) suggest that you should look for additional evidence before concluding that a table with three data points was generated by a quadratic function?

In the following Mathematical Looking Glass you can put the skills you just developed to good use.

A MATHEMATICAL LOOKING GLASS
Gravity 2

Imagine that you are an astronaut on the moon, performing an experiment to confirm accepted information about the strength of the moon's gravitational field. You launch a ball upward from ground level, record its flight with a videocamera, and use the images to determine its height at several instants during its flight. Using the strategy of *Making a Table*, you put the information in tabular form.

TABLE 21

Time, in seconds	Height, in feet
5	383.5
10	634
15	751.5
20	736
25	587.5
30	306

You would like to know the total time of the ball's flight and the maximum height it attains. As is often the case with tables, Table 21 allows only crude estimates. The total time of the flight is slightly more than 30 seconds and the maximum height is slightly more than 751.5 feet, but you need an equation for more precise answers. The formula in ***Gravity*** (page 167) does not apply on the moon. ■

EXERCISES

19. *(Creating Models)* Verify that Table 21 fits a quadratic function, and find an equation for the function.

20. *(Interpreting Mathematics)* Find the ball's total flight time and its maximum height.

ADDITIONAL EXERCISES

In Exercises 21–28, find the extreme value of the function, and state whether it is a maximum or a minimum.

21. $y = 3(x - 1)^2$

22. $y = -3(x - 1)^2$

23. $y = x^2 + 4x + 5$

24. $y = x^2 + 4x + 3$

25. $y = -x^2 - 12x + 2$

26. $y = 0.4x^2 + 0.8$

27. $y = -3(x + 3)(x - 9)$

28. $y = (1 - x)(2 - x)$

In Exercises 29–36, solve the inequality analytically. Show a calculator graph to support your answer.

29. $(x + 5)(x - 13) \leq 0$

30. $-3(x + 7)(x - 7) \geq 0$

31. $2x^2 + 7x - 9 > 0$

32. $14 - x - 3x^2 < 0$

33. $x^2 + 16 \geq 8x$

34. $4x^2 + 4x \leq -1$

35. $x^2 + 5x + 8 < 0$

36. $2x^2 + 3x + 2 > 0$

In Exercises 37–42, decide whether the table fits a quadratic function. If so, find its equation.

37.

x	y
−3	11
−2	6
−1	3
0	2

38.

x	y
4	16
5	18
6	18
7	16

39.

x	y
4	18
6	21
8	23
10	26

40.

x	y
2	37
5	35
8	32
11	28

41.

x	y
−6	100
−2	94
2	92
6	94

42.

x	y
0	3
1	4
2	6
3	10

43. *(Problem Solving)* In ***Dance Floor***, what dimensions should the rented space have to provide the maximum floor space?

44. *(Problem Solving)* In ***Fat Cats on Campus*** (page 43) you saw that to sell x T-shirts during the semester, your company would need to charge a price of $16 - 0.005x$ dollars per shirt.

a. What target sales level will generate the maximum revenue?

b. Is this the same sales level that will generate the maximum profit (see Exercise 50 of Section 5-1)?

45. *(Problem Solving)* In Exercise 44, what sales levels will generate more than $10,000 revenue for your company?

46. *(Problem Solving)* A charter bus service has offered a deluxe 7-day tour from Atlanta to Florida's Disney World at a special rate of $800 per person for groups of 50 people. The company will also give a discount of $15 per person for every additional two people in the group. What number of people will generate the greatest revenue for the company?

47. *(Problem Solving)* In Exercise 46, what numbers of people will generate at least $45,000 revenue for the company?

48. *(Problem Solving)* (This exercise was adapted from Jeanne Agnew and Marvin Keener, *Industry Related Problems for Mathematics Students*, Oklahoma State University.) A stretch of Interstate Highway I-35 in Oklahoma consists of straight downhill and uphill segments connected by a parabolic transitional segment. Figure 5-13 shows the required elevations in feet above sea level at three points on the transitional segment.

In a coordinate system with the x-axis at sea level and the y-axis passing through point A of Figure 5-13, find the equation of the transitional segment.

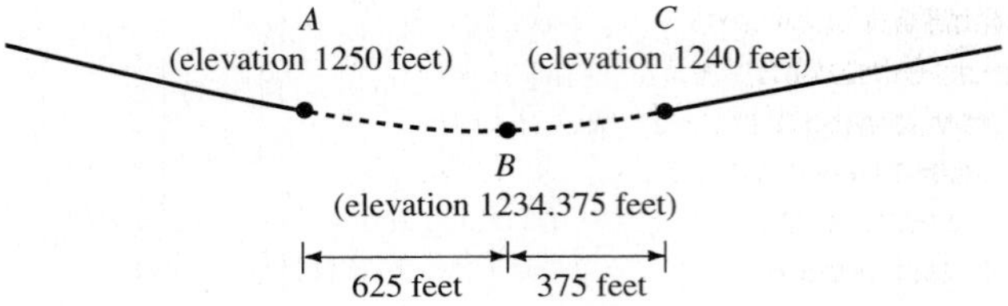

Figure 5–13

49. *(Problem Solving)* A drain is located at point B in Figure 5-13. If a heavy rain causes the roadway to flood to a depth of 6 inches above the drain, how many horizontal feet of the road will be under water?

50. Your car probably gets its best gas mileage at a speed of about 45 miles per hour. Gas mileage is close to optimum at speeds between about 40 and 50 mph, but gets noticeably worse at lower or higher speeds. Of course, at a speed of 0 mph, you get 0 miles per gallon. Let v and M represent your speed in miles per hour and your gas mileage in miles per gallon.

a. *(Writing to Learn)* Explain why the relationship between v and M might be closely approximated by a quadratic function over an interval of normal driving speeds.

b. *(Creating Models)* For a car whose maximum gas mileage is 30 miles per gallon, express M as a quadratic function of v. (*Hint*: Write the function as $M = a(v - h)^2 + k$ and use the given information to find a, h, and k.)

c. *(Writing to Learn)* Explain why your equation in part (b) is not a good model for gas mileage over all possible driving speeds. For what speeds do you think your model is reasonably accurate?

Additional Topic
Validity of a Quadratic Model

In Example 4 we constructed $S(v) = 0.055v^2 + 1.25v - 10$ to fit Table 19 in ***Stopping Distance***.

51. *(Interpreting Mathematics)* Evaluate $S(5)$. What does $S(5)$ mean in physical terms? What does your result say about the validity of $S(v)$ as a model for stopping distances from slow initial speeds?

52. *(Writing to Learn)* Explain why a good model $Q(v)$ for stopping distance should have $Q(0) = 0$ and $Q(v) > 0$ whenever $v > 0$.

53. *(Making Observations)* By *Guessing and Checking*, you might choose $Q(v) = 0.06v^2 + 0.8v$ as a model. It does not quite fit Table 19, but let's make some observations about it.

a. Evaluate $Q(30)$, $Q(40)$, $Q(50)$, and $Q(60)$. How do the values compare with the entries in Table 19?

b. Graph both $y = S(v)$ and $y = Q(v)$ in the window $[25, 65]$ by $[50, 300]$. How do the stopping distances predicted by Q compare with those predicted by S?

c. Graph both $y = S(v)$ and $y = Q(v)$ in the window $[0, 10]$ by $[-10, 20]$. How do the stopping distances predicted by Q compare with those predicted by S?

d. Based on your observations in parts (a) through (c), which function do you think is a better model for predicting stopping distance? Explain why.

CHAPTER REVIEW

Complete Exercises 1–20 (Writing to Learn) before referring to the indicated pages.

WORDS AND PHRASES

In Exercises 1–7, explain the meaning of the words or phrases in your own words.

1. (page 150) **quadratic function**

2. (page 150) **parabola**

3. (page 151) **vertex**

4. (page 151) **axis of symmetry**

5. (page 154) **expanded form, factored form, standard graphing form**

6. (page 161) **second differences**

7. (page 166) **optimization problem**

IDEAS

8. (pages 150–151) In what ways are all quadratic graphs alike?

9. (pages 151–153) What information do the values of a, h, and k provide about the graph of $y = a(x - h)^2 + k$?

10. (page 153) Describe how the basic graph $y = x^2$ is transformed to obtain the graph of $y = a(x - h)^2 + k$. In particular, describe how the graph is affected by the values of a, h, and k.

11. (pages 154–155) Describe the process of finding the vertex and the x-intercepts on a quadratic graph from the expanded form of its equation. Illustrate the process with a specific example.

12. (page 155) If you know the coordinates of the vertex on a quadratic graph, how can you identify its axis of symmetry and find the range of the function?

13. (pages 156–157) Describe the process of finding the vertex and the x-intercepts on a quadratic graph from the factored form of its equation. Illustrate with a specific example.

14. (pages 159–160) Describe the process of finding the vertex and the x-intercepts on a quadratic graph from the standard graphing form of its equation. Illustrate with a specific example.

15. (page 160) Name at least one way in which each of the three analytical forms of a quadratic function is more useful than the other two.

16. (pages 160–162) How can you decide whether a table fits a quadratic function? Make up a table and use it to illustrate.

17. (page 166) Why do optimization problems often arise in a physical context?

18. (pages 169–170) Describe how each step in the analytical solution of a quadratic inequality corresponds to a step in the graphical solution. Illustrate your description by solving a specific inequality both ways.

19. (pages 170–171) Describe the process of fitting a quadratic function to a table. To illustrate your description, make up a table that fits a quadratic function and find the equation of the function.

20. (page 171) If a table fits a quadratic function, why are three data points enough to find its equation? Why are two data points not enough?

CHAPTER 6

QUADRATIC RELATIONS

6–1 RELATIONS

By now you are aware that relationships between two variables often occur in everyday life, as well as in academic fields as diverse as physics, environmental science, economics, and law. Because so many useful relationships are functions, nearly every chapter in this book focuses on a particular type of function. The present chapter is one of the exceptions. Here you will encounter relationships of a more general type. Read on for a simple but important example of a relationship in which neither variable is a function of the other.

A MATHEMATICAL LOOKING GLASS

Fermilab

According to the current theory of atomic structure, every atom of every substance in the universe is made up of two kinds of elementary particles, called **quarks** and **leptons**. The theory holds that there are six types of each. With the recent discovery of the so-called **top quark**, the existence of each type has now been confirmed. Prior to its discovery, the importance of the search for the top quark was summarized by James Trefil in the December 1989 issue of *Discover*.

> With the top quark, the most fundamental theory in all of science will be complete. Without the top quark, the theory is in deep trouble.

The search was made difficult by the extreme elusiveness of quarks. They are normally housed inside larger subatomic particles called **protons** and become observable only when a proton is broken apart by a collision with an **antiproton**, at a speed near the speed of light. Thus an important piece of equipment in the search is the **synchrotron**, a circular tunnel in which magnetic and electrical fields are used to accelerate protons to the required speeds.

The first synchrotron capable of accelerating protons to near light speed was the **Tevatron**, built in 1983 at the Fermi National Accelerator Laboratory in Batavia, IL. Four smaller accelerators first take the protons to a speed of about 45,000 km/sec. The Tevatron then accelerates them to 99.99% of light speed (300,000 km/sec).

A diagram of the Tevatron is shown in Figure 6-1a. Its radius is one kilometer. Figure 6-1b shows a graphical model of the Tevatron centered at $(0, 0)$.

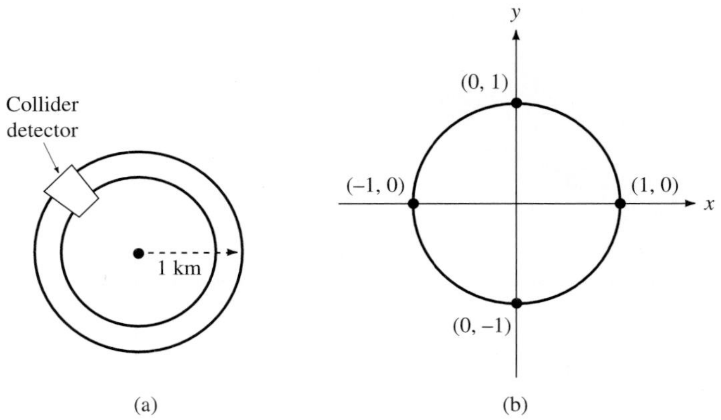

Figure 6–1

The process of accelerating protons and antiprotons to produce a collision is referred to as a "shot." Quarks produced by a shot can be detected only if the collision takes place within the collider detector in Figure 6-1a. To ensure that this occurs, the scientist in charge (the "shotmaster") must be able to calculate the position of each particle at any time. The correspondence between a particle's two coordinates is thus of great interest. ■

The circle in Figure 6-1b is referred to as the **unit circle**. It plays an important role in the development of trigonometry and in a variety of physical problems. In Exercises 1–5 you can verify that neither variable is a function of the other.

EXERCISES

1. *(Review)* How does the graph in Figure 6-1b show that y is not a function of x?
2. *(Review)* Use Figure 6-1b to complete the following table by using two different values of y.

x	y
-1	0
0	
0	
1	0

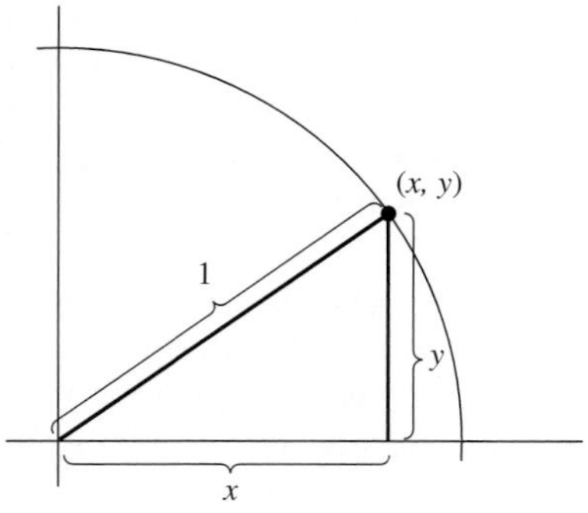

How does the table show that y is not a function of x?

3. *(Writing to Learn)* Explain how the diagram at the left shows that the equation of the unit circle is $x^2 + y^2 = 1$.
4. *(Review)* Solve the equation in Exercise 3 for y. How does your result show that y is not a function of x?

5. *(Review)* Use each of the following to show that x is not a function of y.
   a. the graph in Figure 6-1b
   b. the table in Exercise 2
   c. the equation in Exercise 3

Although the correspondence between x and y in ***Fermilab*** is not a function, it is a **relation**. A relation is *any* correspondence between two variables, so that every function is a relation, although not all relations are functions. The correspondence in Fermilab is a **quadratic relation**, one whose equation is either linear or quadratic in both variables. Most of the relations in this chapter are quadratic.

Let's define relations more precisely from numerical, graphical, and analytical perspectives, and contrast the definition of a relation with that of a function.

THREE VIEWS OF RELATIONS

A Numerical View Any table giving values of two variables defines a relation. The table fits a function if and only if no two rows have the same first entry and different second entries. (See pages 31, 32 to review this idea.)

EXERCISES *In Exercises 6–9 (Review), assume that each table is complete, and decide whether it represents y as a function of x.*

6.

x	y
−1	0
0	0
1	0
−1	1
0	1

7.

x	y
1	7
2	7
3	7
4	7
5	7

8.

x	y
1.3	−8
2.1	−2
3.9	0
4.0	0
5.7	2

9.

x	y
$\sqrt{2}$	0
$\sqrt{3}$	π
$\sqrt{5}$	2π
$\sqrt{5}$	3π
$\sqrt{6}$	4π

A Graphical View Any curve or set of points in a coordinate plane defines a relation. The relation defines y as a function of x if and only if no vertical line intersects the graph more than once. (See pages 33–35 to review this idea.)

EXERCISES *In Exercises* 10–13, *decide whether the graph defines y as a function of x.*

10. *(Review)*

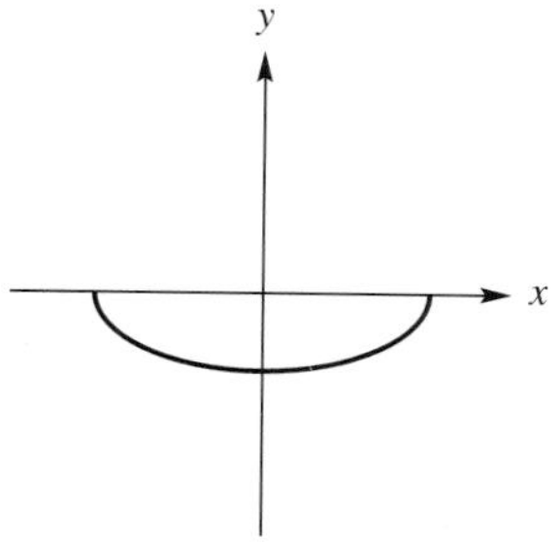

11. *(Review)*

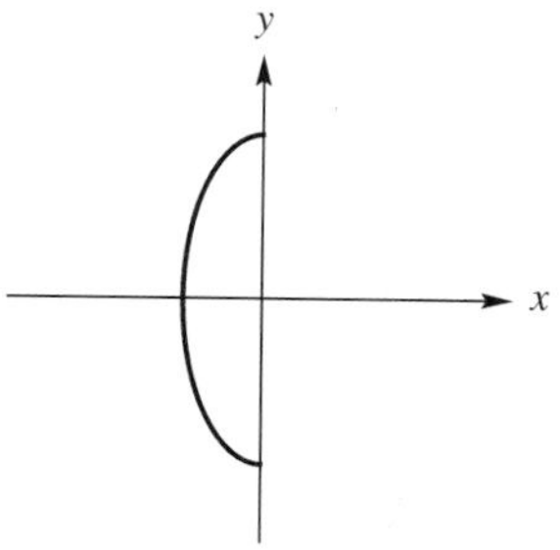

12. *(Review)*

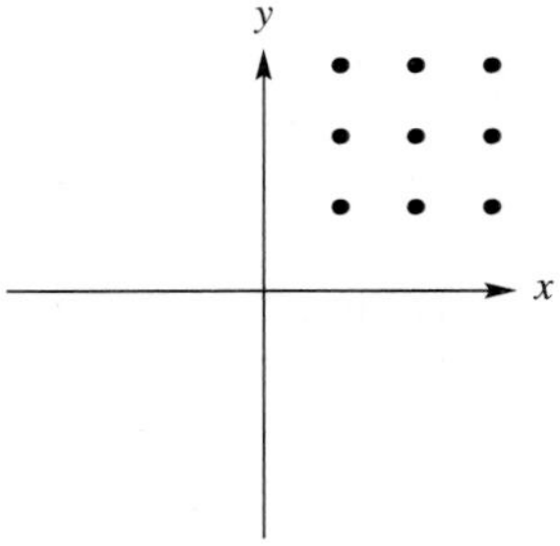

13. *(Review)*

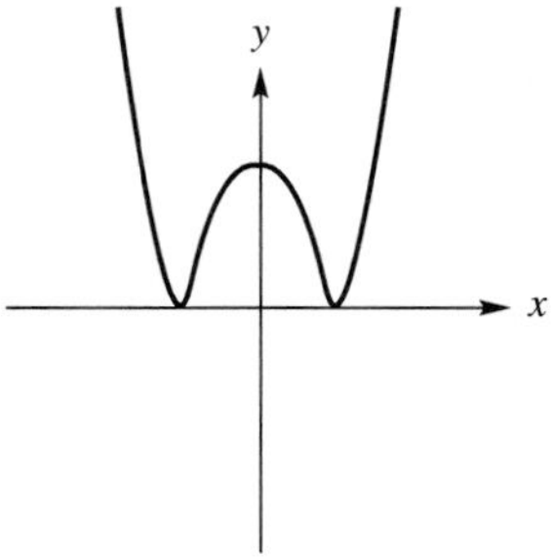

In general, a vertical line can intersect the graph of a relation any number of times. However, no vertical line can intersect the graph of a quadratic relation more than twice. (You can explain why in Exercise 27.) Thus its graph can be broken into two pieces, each of which represents y as a function of x. Example 1 illustrates this property for the unit circle. In Exercise 22 you will find the equation of each piece of its graph. You will then be able to use your knowledge of functions to study quadratic relations.

EXAMPLE 1 *The graph of a quadratic relation*

Figure 6-2 shows the unit circle $x^2 + y^2 = 1$.

Figure 6-3 shows that the upper and lower semicircles each represent y as a function of x.

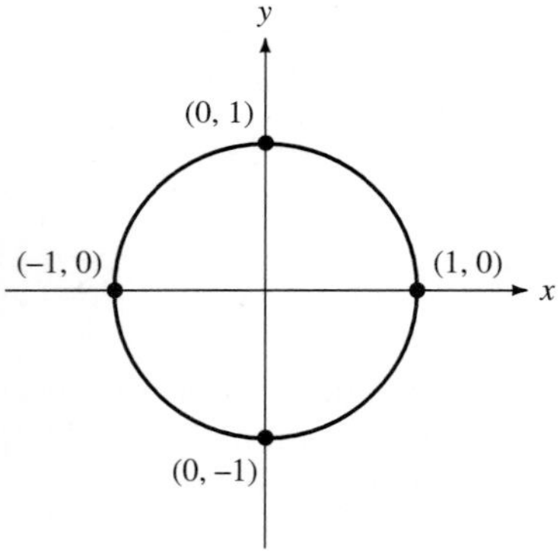

Figure 6–2

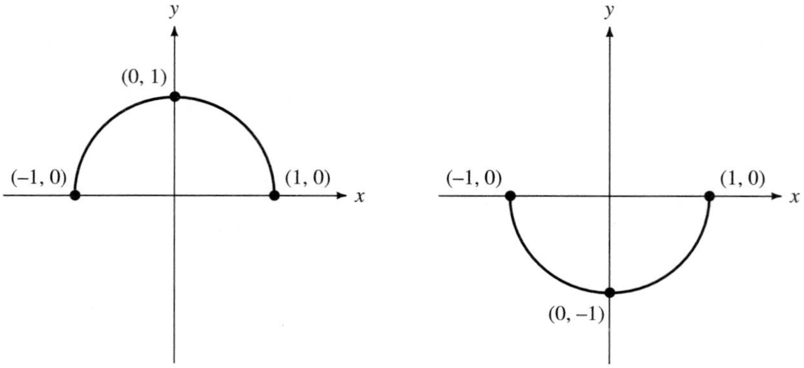

(a) (b)

Figure 6–3

■

EXERCISES *In Exercises* 14–17, *break the graph of the relation into two pieces, each of which represents y as a function of x.*

14.

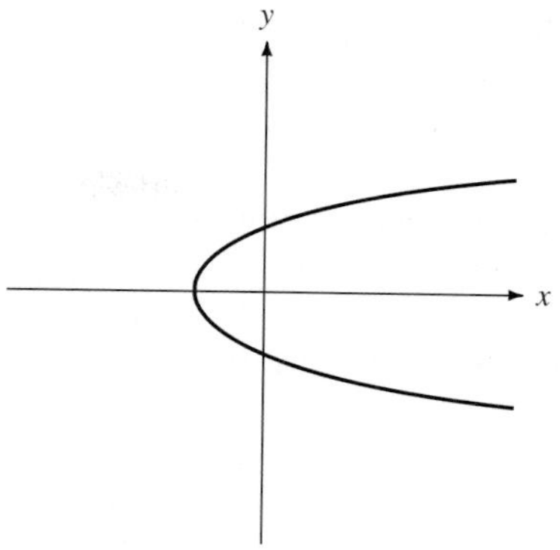

15.

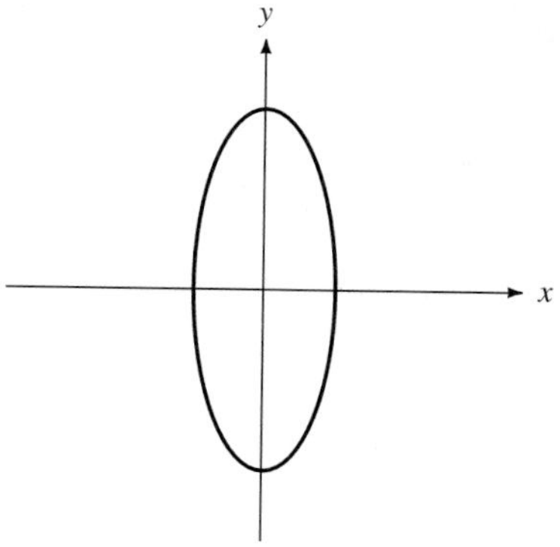

16.

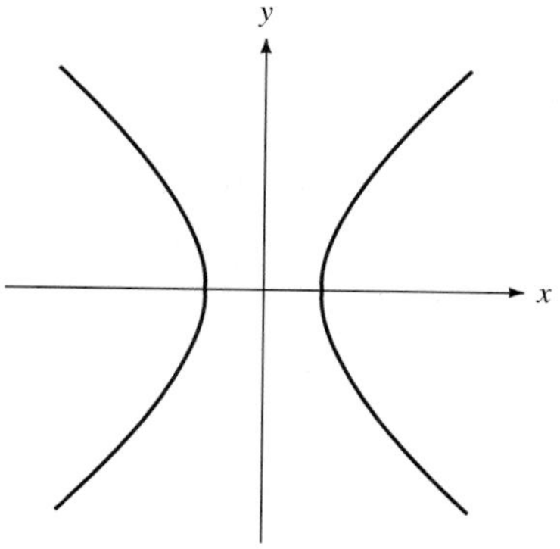

17.

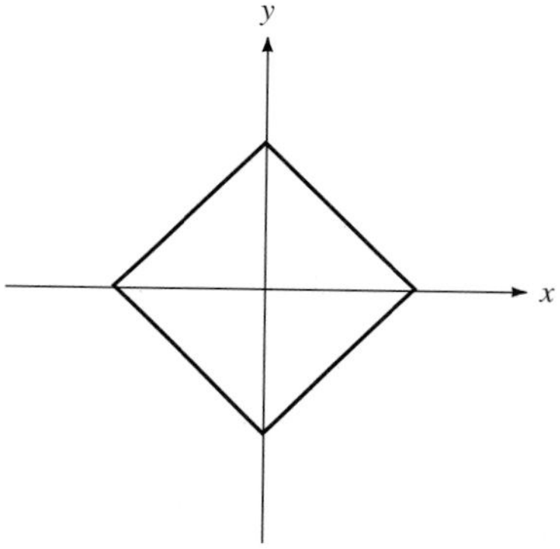

An Analytical View Any equation in two variables defines a relation. The relation is a function if and only if there is at most one value of the dependent variable for each value of the independent variable. There is no general procedure for deciding whether an equation represents y as a function of x, but solving for y sometimes works. (See pages 32, 33 to review this idea.)

EXERCISES *In Exercises* 18–21, *solve for y to decide whether y is a function of x.*

18. *(Review)* $x = y^2$

19. *(Review)* $x^2 + y = 4$

20. *(Review)* $x^2 + y^2 = 2x$

21. *(Review)* $(y - 1)^2 - x^2 = 1$

When a quadratic relation is not a function, solving for y produces an equation involving the square root of a linear or quadratic expression in x, preceded by a $\pm$ sign. The relation can thus be described by two functions, differing only by the sign in front of the radical. The equations of the functions are useful for many purposes, such as graphing quadratic relations on your calculator. Example 2 shows how to associate each function with an appropriate piece of the relation's graph.

EXAMPLE 2 *Finding functions to describe the graph of a relation*

When you solved the equation

$$(y - 1)^2 - x^2 = 1$$

for y in Exercise 21, you obtained

$$y = 1 \pm \sqrt{1 + x^2}$$

The relation can thus be described by the two functions

$$f_1(x) = 1 + \sqrt{1 + x^2} \quad \text{and} \quad f_2(x) = 1 - \sqrt{1 + x^2}$$

The graph of the relation is shown in Figure 6-4. If (x, y) is any point on the graph, then either $y = f_1(x)$ or $y = f_2(x)$.

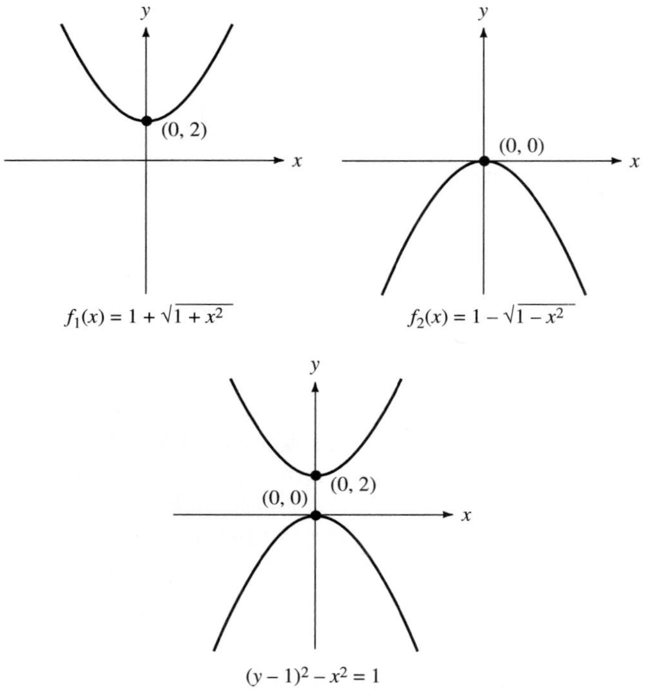

Figure 6–4

To decide what piece of the graph is described by f_1, observe that $f_1(x) \geqslant 1$ throughout its domain while $f_2(x) \leqslant 1$ throughout its domain. The graphs of $y = f_1(x)$ and $y = f_2(x)$ are shown in Figure 6-4. ■

EXERCISES

22. What portion of the unit circle is described by each of the functions $f_1(x) = \sqrt{1 - x^2}$, $f_2(x) = -\sqrt{1 - x^2}$?

Each of Exercises 23–26 *shows the graph of a quadratic relation. Solve for y to obtain two functions of x, and sketch the portion of the graph represented by each function.*

23. $x^2 + 4y^2 = 4$

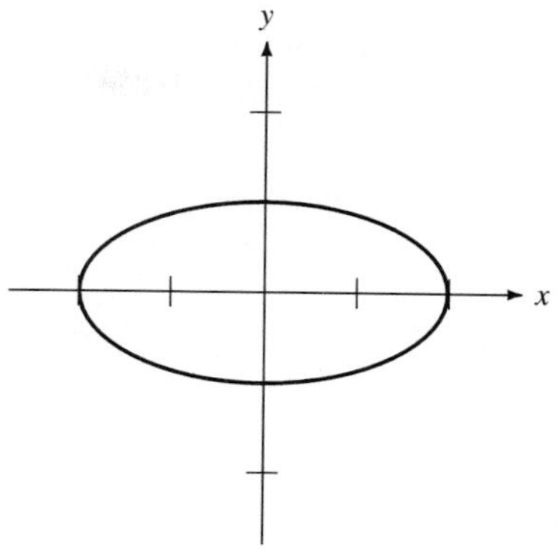

24. $x^2 + y^2 = 2y$

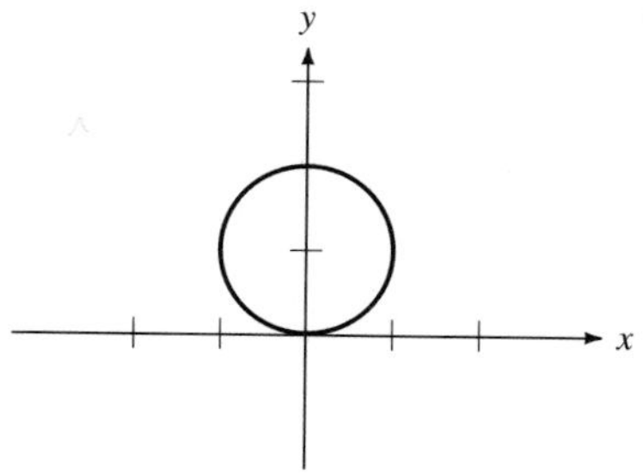

25. $4y^2 - x^2 = 4$

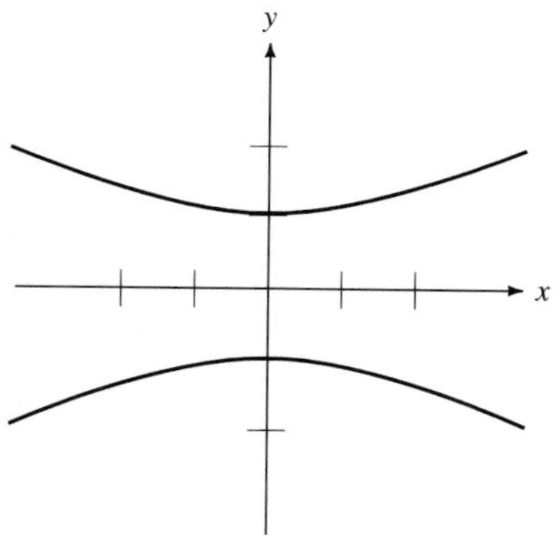

26. $x^2 - 4y^2 = 4$

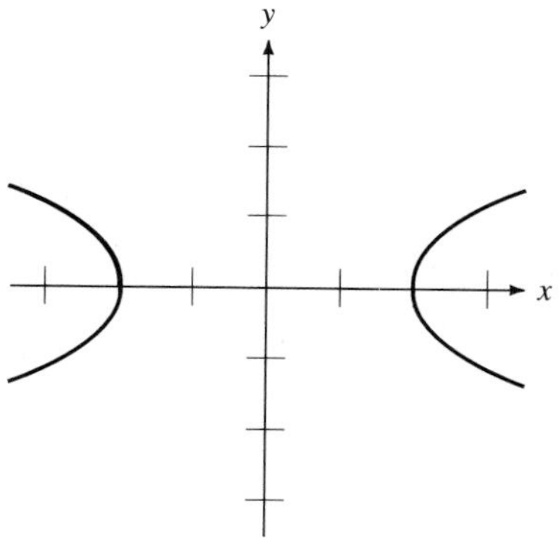

27. *(Writing to Learn)* Why can no vertical line intersect the graph of a quadratic relation more than twice?

Table 22 summarizes our discussion of relations up to this point.

TABLE 22

	Relation	Function
Tables	Any table of paired values of two variables	A table of paired values in which no two rows have the same first entry and different second entries
Graphs	Any curve or set of points in the coordinate plane	A curve or set of points that no vertical line intersects more than once
Equations	Any equation in two variables	An equation for which there is at most one value of the dependent variable for each value of the independent variable

IMPLICIT FUNCTIONS

Relations do not always have unique outputs for each input. However, in a physical situation one particular output, among the many possible, is often of interest to us. Furthermore, additional information about the situation often allows us to find the unique output of interest. For example, in ***Fermilab***, a proton's y-coordinate is $f_1(x) = \sqrt{1 - x^2}$ or $f_2(x) = -\sqrt{1 - x^2}$. We can decide which is the case if we know whether the proton is above or below the x-axis.

The equation $x^2 + y^2 = 1$ is true whenever $y = f_1(x)$. That is,

$$x^2 + [f_1(x)]^2 = 1$$

for all x in the domain of f_1, as you will see in Example 3. A similar statement applies to f_2. The functions f_1 and f_2 are therefore said to be defined **implicitly** by the relation $x^2 + y^2 = 1$. (*The American Heritage Dictionary* defines *implicit* to mean "contained in the nature of something although not readily apparent.") This is in contrast to the equation $y = \sqrt{1 - x^2}$, which defines $f_1(x)$ **explicitly**.

EXAMPLE 3 *Verifying that a function is defined implicitly by a relation*

One way to decide whether the function $f_1(x) = \sqrt{1 - x^2}$ is defined implicitly by the relation $x^2 + y^2 = 1$ is to solve the equation of the relation for y, as you did in Exercise 4. Another way is to substitute $f_1(x)$ for y in the equation of the relation and reduce the equation to an identity.

$$\begin{aligned} x^2 + [f_1(x)]^2 &= 1 \\ x^2 + (\sqrt{1 - x^2})^2 &= 1 \\ x^2 + (1 - x^2) &= 1 \\ 1 &= 1 \end{aligned}$$

Since the equation reduces to an identity, the calculation verifies that $f_1(x)$ is defined implicitly by the relation. ■

EXERCISES

28. Use the method given in Example 3 to verify that the function $f_2(x) = -\sqrt{1 - x^2}$ is defined implicitly by the relation $x^2 + y^2 = 1$.

In Exercises 29–32, use the method of Example 3 to verify that the function $f(x)$ is defined implicitly by the given relation.

29. $x = y^2$; $f(x) = \sqrt{x}$

30. $x^2 + 4y^2 = 16$; $f(x) = -\dfrac{1}{2}\sqrt{16 - x^2}$

31. $9x^2 + (y - 5)^2 = 36$; $f(x) = 5 - 3\sqrt{4 - x^2}$

32. $x^2 - 25(y + 2)^2 = 1$; $f(x) = \dfrac{1}{5}\sqrt{x^2 - 1} - 2$

Obtaining equations for functions defined implicitly by a relation can be useful for several reasons. Specifically:

- Often, as in ***Fermilab***, an implicitly defined function identifies the output of interest from among the many possible outputs.
- Many graphing calculators cannot graph relations unless they are expressed as functions.

If a function f is defined implicitly by a relation, then of all the points (x, y) on the graph of the relation, the points $(x, f(x))$ comprise the graph of $y = f(x)$. For this reason, the graph of the function coincides with part (or all) of the graph of the relation. For example, you have seen that the graphs of $f_1(x) = \sqrt{1 - x^2}$ and $f_2(x) = -\sqrt{1 - x^2}$ are semicircles that together make up the graph of $x^2 + y^2 = 1$.

EXERCISES *In Exercises* 33–36,

a. Find a pair of functions defined implicitly by the relation.

b. Graph each function, and combine the graphs to obtain the graph of the relation.

33. $4x^2 - 9y^2 = 36$

34. $4x^2 + 9y^2 = 36$

35. $x = 2y^2 - 6$

36. $x - x^2 = 2y^2 - 6$

Finding equations for implicitly defined functions is often difficult or impossible. However, you can use the quadratic formula to obtain equations for functions defined implicitly by quadratic relations. The technique is discussed in Section B-4. You can apply it in Exercises 54–57.

In the remainder of this chapter you will explore equations and graphs of quadratic relations. Analyzing tables generated by quadratic relations usually requires more advanced mathematics than you will learn in this book. As evidence to support this statement, we can point to the German astronomer and mathematician Johannes Kepler (1571–1630). Kepler spent years poring over data from observations made by Danish astronomer Tycho Brahe (1546–1601) before discovering that the relations describing planetary orbits are quadratic.

ADDITIONAL EXERCISES *In Exercises* 37–40, *break the graph of the relation into two pieces, each of which represents y as a function of x.*

37.

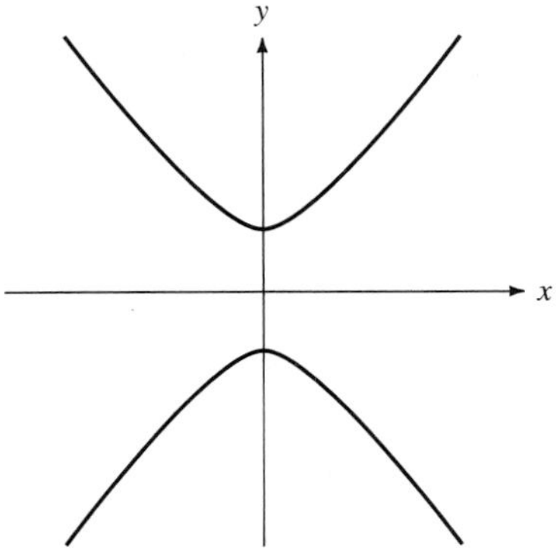

38.

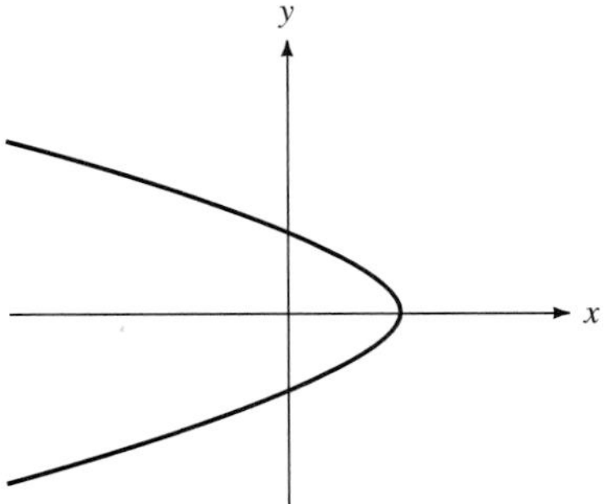

39.

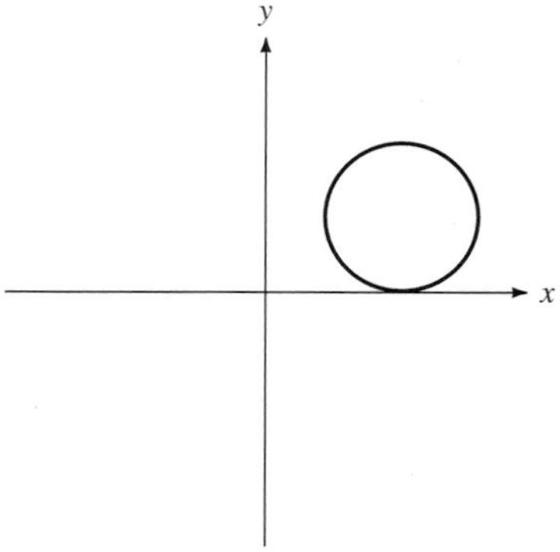

40.

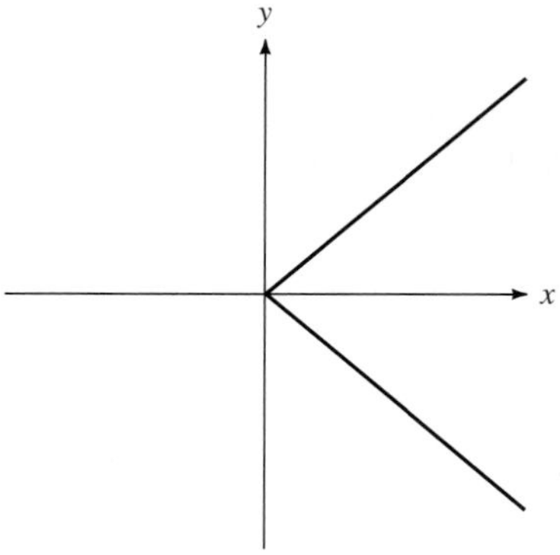

In Exercises 41–48,

a. Solve for y to decide whether y is a function of x. If it is, support your conclusion by graphing the function.

b. If y is not a function of x, find two functions defined implicitly by the relation. Graph each function, and combine the graphs to obtain the graph of the relation.

41. $x^2 + y^2 = 100$

42. $x^2 - y^2 = 100$

43. $4x^2 + y = 4x + 15$

44. $4x^2 + y^2 = 4x + 15$

45. $x^2 + y^2 = 6x + 6y$

46. $x^2 + y^2 = 6x - 6y$

47. $x^2 = y^3 + 1$

48. $x^2 = y^4 + 1$

In Exercises 49–52, *use the method of Example* 3 *to verify that the function* $f(x)$ *is defined implicitly by the given relation.*

49. $3x - y^2 = 6;\ f(x) = \sqrt{3x - 6}$

50. $9x^2 + 25y^2 = 225;\ f(x) = -\frac{3}{5}\sqrt{25 - x^2}$

51. $x^2 - 4(y - 1) = 4x;\ f(x) = \frac{1}{4}(x - 2)^2$

52. $x^2 - 4(y + 1)^2 = 4;\ f(x) = \frac{1}{2}\sqrt{x^2 - 4} - 1$

53. *(Writing to Learn)* Why does the method of Example 3 guarantee that a given function is defined implicitly by a given relation?

In Exercises 54–57,

a. Use the quadratic formula to find two functions defined implicitly by the relation.

b. Graph each function, and combine the two graphs to obtain the graph of the relation.

54. *(Extension)* $\frac{1}{2}x^2 + xy + \frac{1}{2}y^2 + 4x + 5y + 10 = 0$

55. *(Extension)* $x^2 + xy + y^2 + 4x + 5y + 10 = 0$

56. *(Extension)* $x^2 - xy + y^2 - 12 = 0$

57. *(Extension)* $x^2 + 10\sqrt{3}\,xy + 11y^2 - 64 = 0$

Exercise 58 illustrates that there is more than one way to choose a pair of implicitly defined functions to express a quadratic relationship.

58. *(Making Observations)*

a. Verify that the functions $f_1(x) = x$, $f_2(x) = -x$, $g_1(x) = |x|$, and $g_2(x) = -|x|$ are all defined implicitly by the relation $x^2 - y^2 = 0$.

b. Sketch the graphs of $y = f_1(x)$ and $y = f_2(x)$, and combine the graphs to obtain the graph of the relation.

c. Sketch the graphs of $y = g_1(x)$ and $y = g_2(x)$, and combine the graphs to obtain the graph of the relation.

6–2 A GRAPHICAL VIEW OF CONIC SECTIONS

To begin exploring graphs of quadratic relations, take a break and perform a simple experiment.

Get a flashlight, a marker, and three sheets of poster paper, at least 2 by 3 feet each. Tape one sheet of paper to a vertical wall, and shine the flashlight so that the beam falls entirely within the paper, as in Figure 6-5a. Hold the flashlight steady, and have a friend trace around the edge of the beam. The curve, shown in Figure 6-5b, is an **ellipse**.

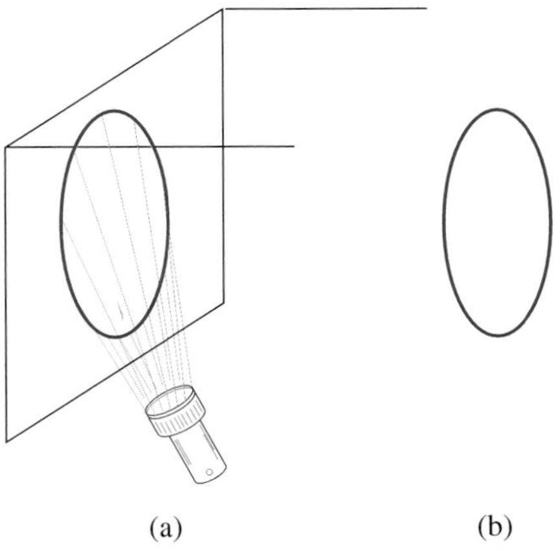

Figure 6–5

Next, tape a second sheet of paper to the wall, hold the flashlight a few inches from the wall near the bottom of the paper, and shine it so that one edge of the beam hits a point on the ceiling directly above the flashlight, as in Figure 6-6a. Again, have a friend trace around the edge of the beam. The curve on the wall, shown in Figure 6-6b, is a **parabola**.

Finally, tape the third sheet of paper to the wall, hold the flashlight a few inches from the wall near the bottom of the paper, and shine the flashlight straight up, as in Figure 6-7a. Once more, have a friend trace around the edge of the beam. The curve on the wall, shown in Figure 6-7b, is one branch of a **hyperbola**. The entire hyperbola, shown in Figure 6-7c, would be traced by the beam of a flashlight with a bulb at each end.

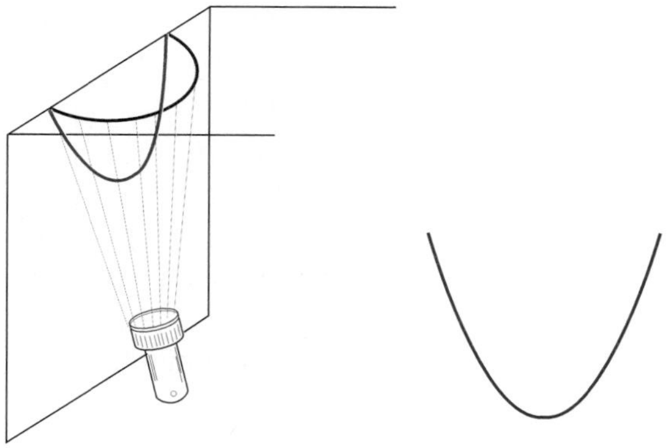

(a) (b)

Figure 6–6

The curves you have traced are called **conic sections**, because each is the intersection of a cone (the light beam) with a plane (the wall). We will develop formal definitions for all three types in Section 6-4, but for now you can think of them as describing the curves traced out by your flashlight. In this section you will discover that the graphs of many second degree equations in two variables resemble the curves you traced with the flashlight beam.

The equations in this chapter have the form

$$Ax^2 + Cy^2 + Dx + Ey + F = 0$$

A quadratic relation can also contain a term Bxy, but a systematic study of such equations and their graphs requires trigonometry. In this chapter it will always be the case that both x and y appear in the equation, and that the equation is actually second degree (A and C are not both 0).

We can explore the graphs of such equations most easily by completing the square in each variable, if possible. This leads to one of several standard graphing forms, depending on whether the product AC is zero, positive, or negative. Let's explore each case in turn.

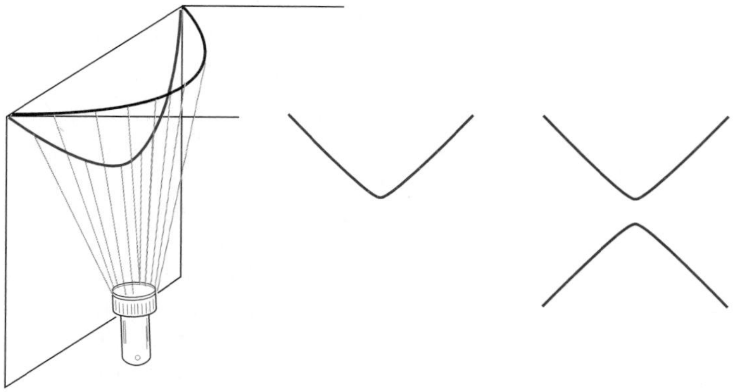

(a) (b) (c)

Figure 6–7

A GRAPHICAL VIEW OF PARABOLAS

The condition $AC = 0$ means that either $A = 0$ or $C = 0$. When $C = 0$ the equation contains no term involving y^2, so y is a quadratic function of x. In this case completing the square produces the standard graphing form of a quadratic function

$$y = a(x - h)^2 + k$$

In Chapter 5 you learned a few distinctive graphical features of such a function. They are restated here.

The graph of $y = a(x - h)^2 + k$ is a parabola.

- Its vertex is (h, k).
- Its axis of symmetry is the line $x = h$.
- It opens up if $a > 0$ and down if $a < 0$.
- The points 1 unit to the left and right of the vertex on the parabola have a y-coordinate of $k + a$.

Similarly, when $A = 0$, x is a quadratic function of y. In this case completing the square produces the standard graphing form

$$x = a(y - k)^2 + h$$

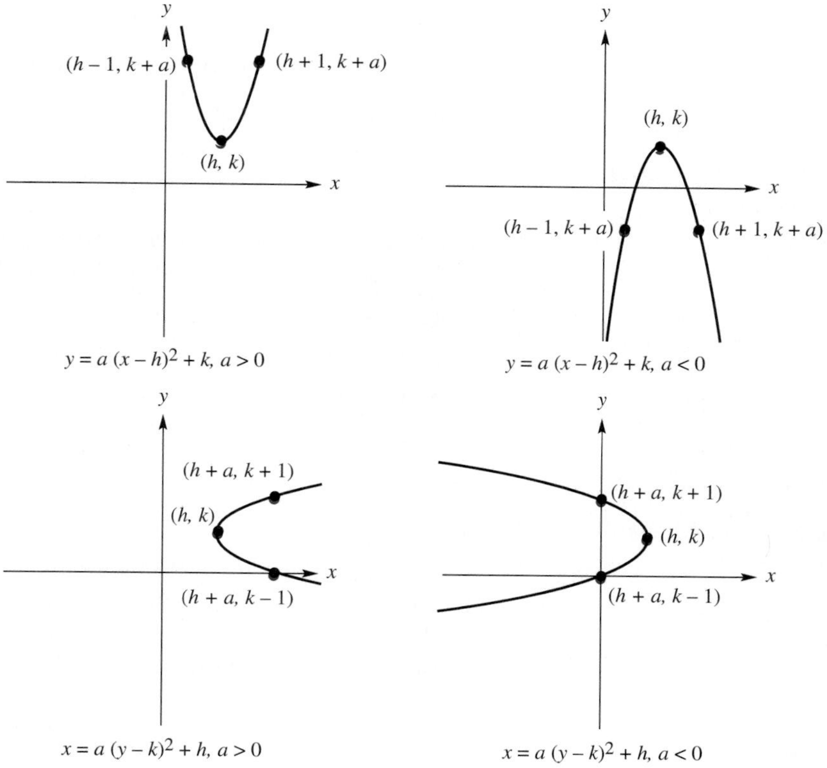

Figure 6–8

In Section 6-3 you will learn the graphing principles needed to establish the following graphical features of this equation.

> The graph of $x = a(y - k)^2 + h$ is a parabola.
>
> - Its vertex is (h, k).
> - Its axis of symmetry is the line $y - k$.
> - It opens right if $a > 0$ and left if $a < 0$.
> - The points 1 unit above and below the vertex on the parabola have an x-coordinate of $h + a$.

Figure 6-8 (see page 189) illustrates the preceding statements.

EXAMPLE 1 *Sketching the graph of an equation* $x = a(y - k)^2 + h$

To sketch the graph of

$$y^2 + 2x - 4y + 6 = 0$$

first write it in the standard graphing form

$$x = -\frac{1}{2}(y - 2)^2 - 1$$

Since the quadratic term involves y, the graph is a parabola opening horizontally. Its vertex is $(-1, 2)$, and it opens to the left since $a = -\frac{1}{2} < 0$. The axis of symmetry is the line $y = 2$. The points 1 unit above and below the vertex have an x-coordinate of $-1 - \frac{1}{2} = -\frac{3}{2}$. The graph is shown in Figure 6-9. ■

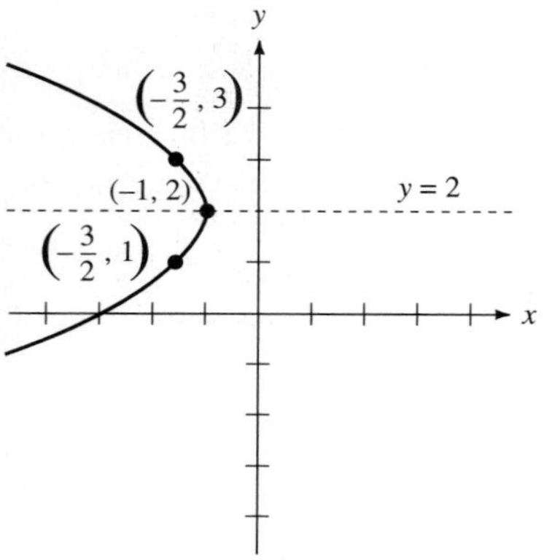

Figure 6–9

EXERCISES *In Exercises 1–4, sketch the graph of the equation. Identify the vertex and two other points on the graph, and the axis of symmetry.*

1. $x^2 + 8x - y + 12 = 0$

2. $x^2 - 2x + 2y + 3 = 0$

3. $y^2 + x - 10y + 18 = 0$

4. $2y^2 - x + 8y + 7 = 0$

A GRAPHICAL VIEW OF ELLIPSES

The condition $AC > 0$ means that A and C have the same sign. Completing the square in both x and y produces an equation of the form $A(x - h)^2 + C(y - k)^2 = G$. Dividing by G on both sides then produces the **standard graphing form**

$$\left(\frac{x - h}{\text{number}}\right)^2 + \left(\frac{y - k}{\text{number}}\right)^2 = 1$$

as long as $G > 0$. Later in this section we will see what happens when $G \leq 0$.

To explore the graphs of these equations, let's begin with a simple example. Figure 6-10 shows the graph of

$$\left(\frac{x}{3}\right)^2 + \left(\frac{y}{2}\right)^2 = 1$$

The graph has the shape of an ellipse. You can explain some of its distinctive graphical features in Exercises 5–6.

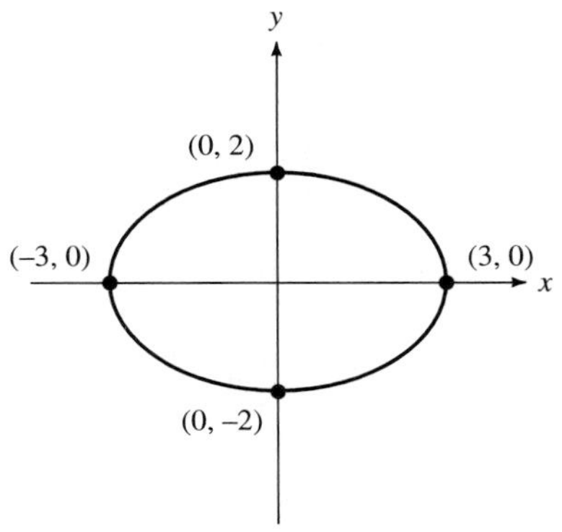

Figure 6–10

EXERCISES 5. *(Writing to Learn)*

a. Show that if $|x| > 3$, the equation has no real solutions for y. Why does this imply that the graph lies on or between the lines $x = -3$ and $x = 3$?

b. Explain why the graph also lies between the lines $y = -2$ and $y = 2$.

c. Show that the points $(\pm 3, 0)$ and $(0, \pm 2)$ are on the graph.

6. *(Writing to Learn)*

a. If a point (a, b) is on the graph, show that $(-a, b)$ is also on the graph. Why does this imply that the y-axis is an axis of symmetry for the graph?

b. If a point (a, b) is on the graph, show that $(a, -b)$ is also on the graph. Why does this imply that the x-axis is an axis of symmetry for the graph?

The features described in Exercises 5 and 6 can be found in every ellipse. Specifically, an ellipse has two axes of symmetry, one horizontal and one vertical, as indicated in Figure 6-11 on page 192. The two axes intersect at the **center** of the ellipse. The longer of the two chords along the axes of symmetry is the **major axis** of the ellipse, and the shorter is its **minor axis**. (The terms *major axis* and *minor axis* can also be used to refer to the entire lines containing the two chords.) The endpoints of the major axis are the **vertices** of the ellipse, and the endpoints of the minor axis are its **covertices**.

Let's see how to sketch the graph of a general equation $Ax^2 + Cy^2 + Dx + Ey + F = 0$ with $AC > 0$. Example 2 shows how to obtain a standard graphing form.

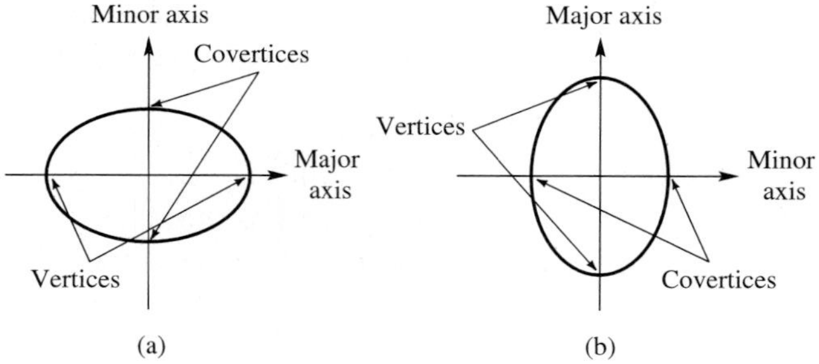

Figure 6–11

EXAMPLE 2 *Standard graphing form for $Ax^2 + Cy^2 + Dx + Ey + F = 0$ when $AC > 0$*

To write $9x^2 + 4y^2 - 180x - 56y + 1060 = 0$ in standard graphing form, first complete the square in x and y.

$$9(x^2 - 20x) + 4(y^2 - 14y) + 1060 = 0$$

$$9(x^2 - 20x + 100 - 100) + 4(y^2 - 14y + 49 - 49) + 1060 = 0$$

$$9(x^2 - 20x + 100) - 900 + 4(y^2 - 14y + 49) - 196 + 1060 = 0$$

$$9(x - 10)^2 + 4(y - 7)^2 = 36$$

Then divide by 36 on both sides.

$$\frac{(x - 10)^2}{4} + \frac{(y - 7)^2}{9} = 1$$

$$\left(\frac{x - 10}{2}\right)^2 + \left(\frac{y - 7}{3}\right)^2 = 1$$

■

EXERCISE 7. *(Writing to Learn)*

a. Show that if $\left|\frac{x - 10}{2}\right| > 1$, the equation in Example 2 has no real solutions for y. Why does this imply that the graph lies on or between the lines $x = 8$ and $x = 12$?

b. Explain why the graph also lies between the lines $y = 4$ and $y = 10$.

c. Show that the points $(8, 7), (12, 7), (10, 4)$, and $(10, 10)$ are on the graph.

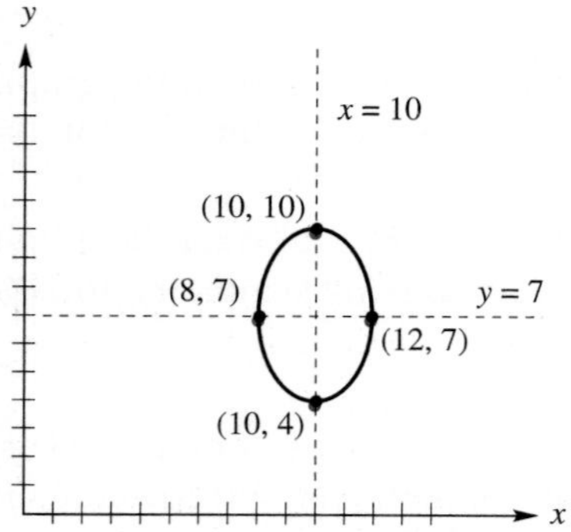

Figure 6–12

In Section 6-3 you will see why the lines $x = 10$ and $y = 7$ are the axes of symmetry for the ellipse in Example 2. The graph is shown in Figure 6-12.

It is traditional to denote the two denominators in the standard graphing form by a and b, with $a > b$. Thus the graphical features of ellipses can be summarized in the following statements.

If $a > b$, the graph of $\left(\frac{x - h}{a}\right)^2 + \left(\frac{y - k}{b}\right)^2 = 1$ is an ellipse.

- Its center is (h, k).
- Its vertices are a units to the left and right of the center, at $(h \pm a, k)$.
- Its covertices are b units above and below the center, at $(h, k \pm b)$.
- Its major axis is the line $y = k$, and its minor axis is the line $x = h$.

If $a > b$, the graph of $\left(\frac{x-h}{b}\right)^2 + \left(\frac{y-k}{a}\right)^2 = 1$ is an ellipse.

- Its center is (h, k).
- Its vertices are a units above and below the center, at $(h, k \pm a)$.
- Its covertices are b units to the left and right of the center, at $(h \pm b, k)$.
- Its major axis is the line $x = h$, and its minor axis is the line $y = k$.

Figure 6-13 illustrates the preceding statements.

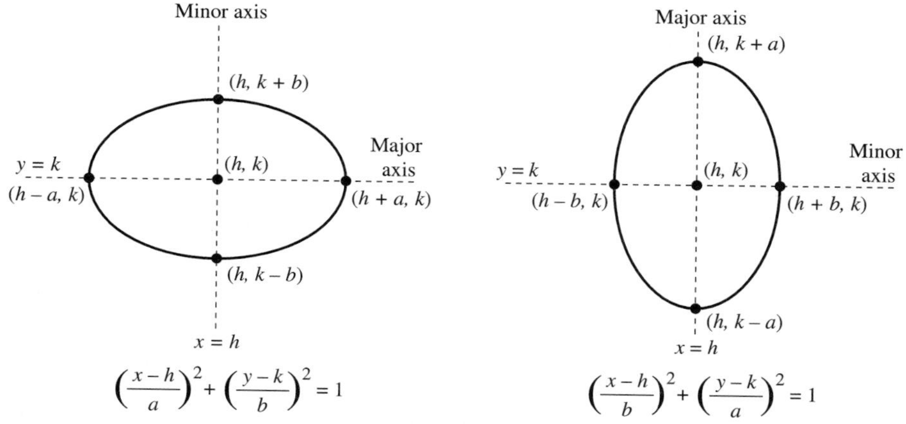

Figure 6–13

Example 3 shows how to sketch an ellipse quickly from its standard graphing form.

EXAMPLE 3 *Sketching an ellipse quickly from the standard graphing form of its equation*

To sketch the graph of

$$\left(\frac{x-1}{5}\right)^2 + \left(\frac{y-3}{2}\right)^2 = 1$$

begin by plotting the center at $(1, 3)$. The major axis is the horizontal line $y = 3$. The vertices are 5 units to the left and right of the center, at $(-4, 3)$ and

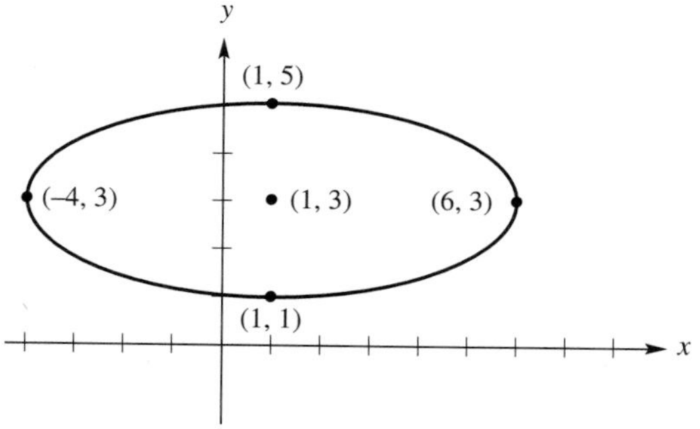

Figure 6–14

(6, 3). The minor axis is the vertical line $x = 1$. The covertices are 2 units above and below the center, at (1, 1) and (1, 5). The graph is shown in Figure 6-14 (see page 193). ■

EXERCISES *In Exercises 8–13, write the equation in standard graphing form and sketch the graph. Identify the center, vertices, covertices, and major and minor axes.*

8. $9x^2 + 25y^2 = 225$

9. $4x^2 + y^2 = 64$

10. $(x - 3)^2 + 5y^2 = 20$

11. $(x + 4)^2 + 4(y - 2)^2 = 16$

12. $49x^2 + 9y^2 + 18y - 432 = 0$

13. $x^2 + 6y^2 - 60x + 300y + 4644 = 0$

Circles If $A = C$ in an equation

$$Ax^2 + Cy^2 + Dx + Ey + F = 0$$

then also $a = b$ in its standard graphing form. In this case, the major and minor axes have equal lengths, and the ellipse is a circle. Equations of circles are usually written in the form

$$(x - h)^2 + (y - k)^2 = r^2$$

In Exercise 14 you can explain why the value of r is the radius of the circle.

EXERCISE **14.** *(Writing to Learn)* Use the figure below and the distance formula to explain why a point (x, y) is on a circle of radius r centered at (h, k) if and only if $(x - h)^2 + (y - k)^2 = r^2$.

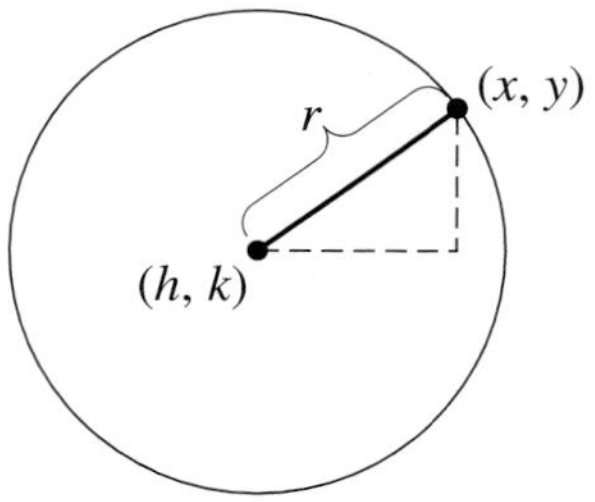

EXAMPLE 4 *Sketching a circle*

To obtain the standard graphing form of the equation

$$4x^2 + 4y^2 + 4x - 8y - 31 = 0$$

begin by completing the square in both x and y to obtain

$$4(x + 0.5)^2 + 4(y - 1)^2 = 36$$

Since $A = C$, do not divide both sides of the equation by 36. Instead, divide both sides by 4 to obtain

$$(x + 0.5)^2 + (y - 1)^2 = 9$$

The graph is a circle with radius 3 centered at $(-0.5, 1)$. Its graph is shown in Figure 6-15.

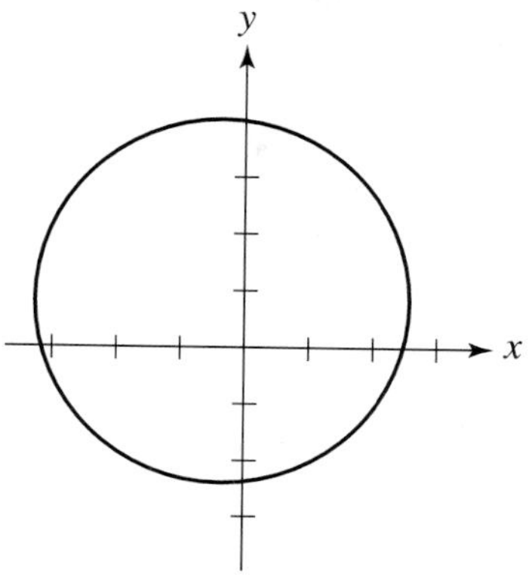

Figure 6–15

■

EXERCISES *In Exercises* 15–20, *sketch the circle and identify its center and radius.*

15. $x^2 + y^2 = 25$

16. $x^2 + y^2 - 4x = 0$

17. $(x + 6)^2 + (y - 2)^2 = 4$

18. $9(x + 1)^2 + 9(y + 3)^2 = 1$

19. $x^2 + y^2 - 8x - 6y + 23 = 0$

20. $4x^2 + 4y^2 - 12x + 4y - 15 = 0$

A GRAPHICAL VIEW OF HYPERBOLAS

Standard Graphing Form The condition $AC < 0$ means that A and C have opposite signs. Completing the square in both x and y produces an equation of the form $A(x - h)^2 + C(y - k)^2 = G$. Dividing by G on both sides then produces a standard graphing form

$$\left(\frac{x - h}{a}\right)^2 - \left(\frac{y - k}{b}\right)^2 = 1$$

or

$$\left(\frac{y - k}{a}\right)^2 - \left(\frac{x - h}{b}\right)^2 = 1$$

as long as $G \neq 0$. Later in this section we will see what happens when $G = 0$.

FYI Although these standard graphing forms closely resemble that of an ellipse, there are two important differences.

- The terms on the left side have opposite signs.
- The denominator of the positive term is denoted by a, regardless of which denominator is larger.

EXAMPLE 5 *Standard graphing form for $Ax^2 + Cy^2 + Dx + Ey + F = 0$ when $AC < 0$*

To write $9x^2 - 4y^2 - 180x + 56y + 668 = 0$ in standard graphing form, first complete the square in x and y.

$$\begin{aligned} 9(x^2 - 20x) - 4(y^2 - 14y) + 668 &= 0 \\ 9(x^2 - 20x + 100 - 100) - 4(y^2 - 14y + 49 - 49) + 668 &= 0 \\ 9(x^2 - 20x + 100) - 900 - 4(y^2 - 14y + 49) + 196 + 668 &= 0 \\ 9(x - 10)^2 - 4(y - 7)^2 &= 36 \end{aligned}$$

Then divide by 36 on both sides.

$$\frac{(x-10)^2}{4} - \frac{(y-7)^2}{9} = 1$$

$$\left(\frac{x-10}{2}\right)^2 - \left(\frac{y-7}{3}\right)^2 = 1$$

■

The simplest equations having the standard graphing form just discussed are $x^2 - y^2 = 1$ and $y^2 - x^2 = 1$. Exercises 21–25 will give you some insight into their graphs.

EXERCISES

21. *(Writing to Learn)* In the first quadrant, show that the equation $x^2 - y^2 = 1$ is equivalent to $y = \sqrt{x^2 - 1}$. Why are the two equations not equivalent in general?

22. *(Review)* What positive values of x are in the domain of the function $y = \sqrt{x^2 - 1}$?

23. Complete the following table for $y = \sqrt{x^2 - 1}$.

x	y
1	
2	
10	
100	

24. *(Writing to Learn)* How does Exercise 23 suggest that if x is large, the graph of $y = \sqrt{x^2 - 1}$ is near the graph of $y = x$?

In the first quadrant the graph of $x^2 - y^2 = 1$ coincides with that of $y = \sqrt{x^2 - 1}$, shown in Figure 6-16a.

EXERCISE

25. *(Making Observations)* Show that if a point (a, b) is on the graph of $x^2 - y^2 = 1$, so are both the points $(-a, b)$ and $(a, -b)$.

Exercise 25 shows that both the x-axis and the y-axis are axes of symmetry for the graph of $x^2 - y^2 = 1$. The entire graph is shown in Figure 6-16b. The lines $y = x$ and $y = -x$, indicated by the dashed lines, are **asymptotes**. In general an asymptote is a line approached arbitrarily closely by a graph as one of the variables becomes arbitrarily large. In Exercise 64 you can show analytically why the lines $y = \pm x$ must be asymptotes for the graph of

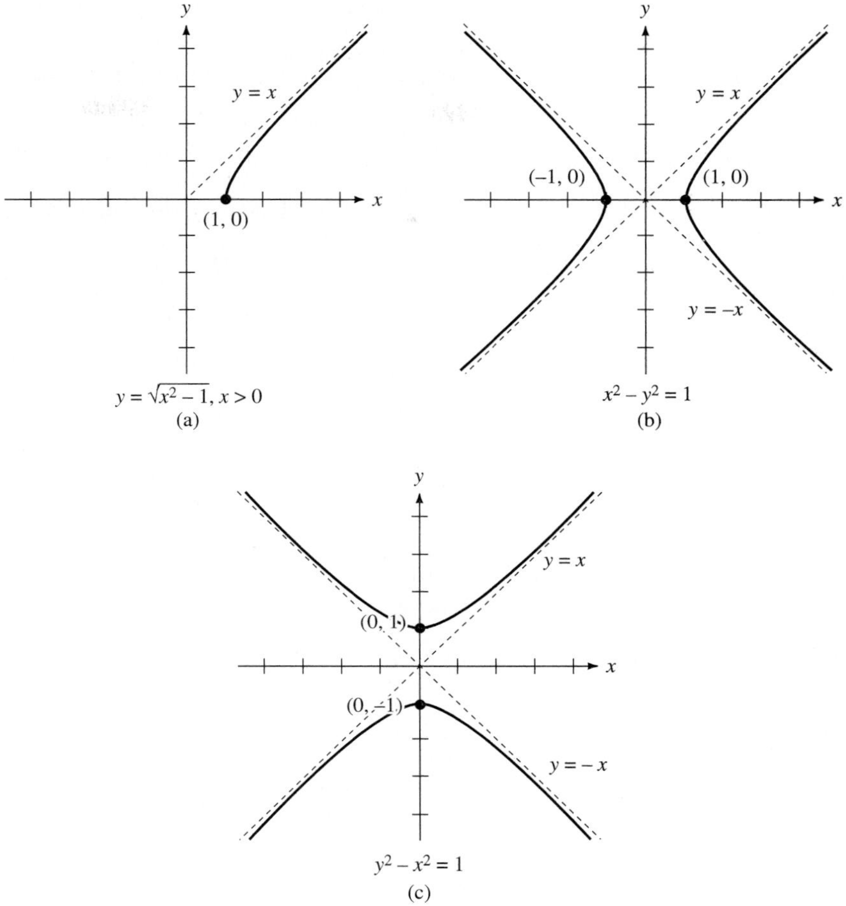

Figure 6–16

$x^2 - y^2 = 1$. Figure 6-16c shows the graph of $y^2 - x^2 = 1$, whose shape can be established by observations similar to those in Exercises 21–25. The graphs of both equations are hyperbolas.

The hyperbolas in Figure 6-16b and c each have two axes of symmetry, intersecting at the **center** of the hyperbola. One axis of symmetry, the **transverse axis**, intersects the hyperbola in two points, called the **vertices**. The other axis of symmetry is the **conjugate axis**.

Observations like those in Exercises 21–25 can be used to establish the following facts.

The graph of $\left(\frac{x-h}{a}\right)^2 - \left(\frac{y-k}{b}\right)^2 = 1$ is a hyperbola.

- Its center is (h, k).
- Its vertices are a units to the left and right of the vertex, at $(h \pm a, k)$.
- Its asymptotes are the lines $\frac{x-h}{a} = \pm\frac{y-k}{b}$ or, equivalently, $y = \pm\frac{b}{a}(x - h) + k$.

The graph of $\left(\frac{y-k}{a}\right)^2 - \left(\frac{x-h}{b}\right)^2 = 1$ is a hyperbola.

- Its center is (h, k).
- Its vertices are a units above and below the vertex, at $(h, k \pm a)$.
- Its asymptotes are the lines $\frac{y-k}{a} = \pm\frac{x-h}{b}$ or, equivalently, $y = \pm\frac{a}{b}(x-h) + k$.

Figure 6-17 illustrates the preceding statements.

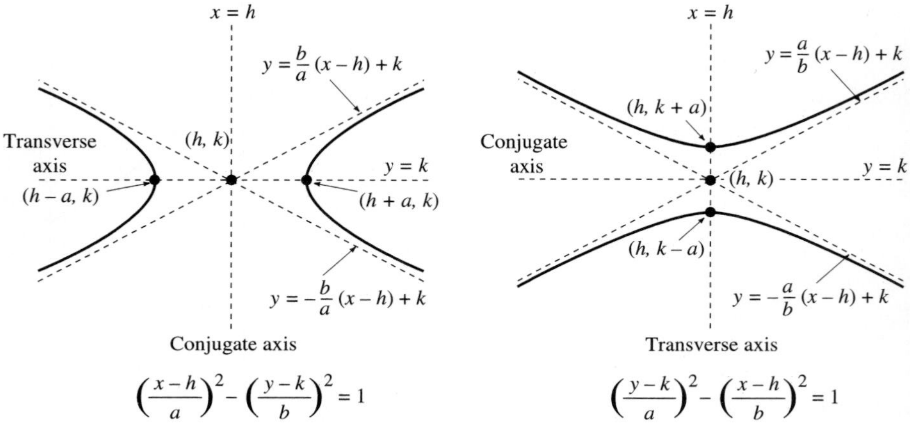

Figure 6–17

EXERCISE **26.** *(Writing to Learn)* Adapt the argument in Exercises 21–25 to show that the hyperbola $\left(\frac{x-h}{a}\right)^2 - \left(\frac{y-k}{b}\right)^2 = 1$ has asymptotes $\frac{x-h}{a} = \pm\frac{y-k}{b}$.

In the standard graphing form of a hyperbola, the positive denominator is the distance from the center to each vertex. To sketch the asymptotes, locate the vertices and the two points b units from the center along the conjugate axis. Sketch a box through these points, as shown in Figure 6-18. The asymptotes are the diagonals of this box.

$$\frac{(x-h)^2}{a^2} - \frac{(y-k)^2}{b^2} = 1 \qquad \frac{(y-k)^2}{a^2} - \frac{(x-h)^2}{b^2} = 1$$

To find the equations of the asymptotes, observe that if x and y are large, the terms on the left of an equation

$$\left(\frac{x-h}{a}\right)^2 - \left(\frac{y-k}{b}\right)^2 = 1$$

are large in comparison to the 1 on the right. Thus the asymptotes have equations

$$\left(\frac{x-h}{a}\right)^2 - \left(\frac{y-k}{b}\right)^2 = 0$$

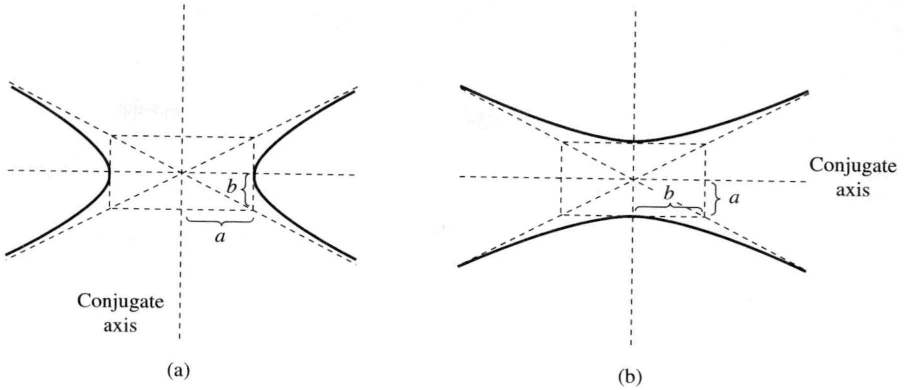

Figure 6–18

Simplifying,

$$\left(\frac{x-h}{a}\right)^2 = \left(\frac{y-k}{b}\right)^2$$

$$\frac{x-h}{a} = \pm\frac{y-k}{b}$$

$$y - k = \pm\frac{b}{a}(x-h)$$

Similarly, a hyperbola

$$\left(\frac{y-k}{a}\right)^2 - \left(\frac{x-h}{b}\right)^2 = 1$$

has asymptotes $y - k = \pm\frac{a}{b}(x-h)$.

You can now sketch any hyperbola quickly from the standard graphing form of its equation.

EXAMPLE 6 *Sketching a hyperbola quickly from the standard graphing form of its equation*

To sketch the graph of

$$\left(\frac{y-3}{2}\right)^2 - \left(\frac{x-1}{5}\right)^2 = 1$$

begin by plotting the center at (1, 3). The transverse axis is the vertical line $x = 1$, and the conjugate axis is the horizontal line $y = 3$. The vertices are 2 units above and below the center, at (1, 1) and (1, 5). The equations of the asymptotes are

$$\left(\frac{y-3}{2}\right)^2 = \left(\frac{x-1}{5}\right)^2$$

which can be simplified to

$$y - 3 = \pm\frac{2}{5}(x-1)$$

The graph is shown in Figure 6-19.

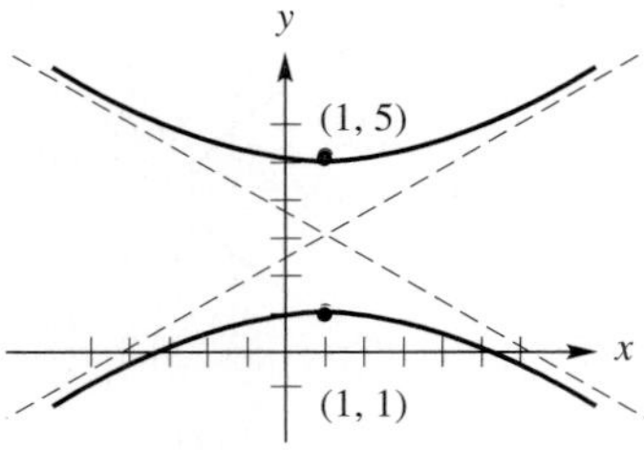

Figure 6–19

■

EXERCISES *In Exercises 27–30, sketch the graph of the hyperbola. Identify the coordinates of the center and vertices and the equations of the transverse axis, conjugate axis, and asymptotes.*

27. $\dfrac{(x+5)^2}{9} - \dfrac{(y+6)^2}{16} = 1$

28. $\dfrac{(x-2)^2}{7} - \dfrac{(y+1.5)^2}{4} = 1$

29. $\dfrac{(y-85)^2}{1225} - \dfrac{(x+129)^2}{2500} = 1$

30. $\dfrac{(x-15)^2}{144} - \dfrac{(y-13)^2}{144} = 1$

EXCEPTIONAL GRAPHS

You know now that the graph of an equation

$$A(x-h)^2 + C(y-k)^2 = G$$

is an ellipse when $AC > 0$, *as long as* $G > 0$, and a hyperbola when $AC < 0$, *as long as* $G \neq 0$. In Exercises 31–34 you can see what happens when the conditions G are not met.

EXERCISES **31.** *(Writing to Learn)* Explain why the graph of $x^2 + y^2 = 0$ is a single point. What is that point?

32. *(Writing to Learn)* Explain why the graph of $x^2 + y^2 = -1$ is empty (contains no points).

33. *(Writing to Learn)* Explain why the graph of $x^2 - y^2 = 0$ consists of two intersecting lines. What are the equations of the two lines?

34. *(Writing to Learn)* Based on your observations in Exercises 31–33, explain why each of the following statements is true about the graph of an equation $A(x-h)^2 + C(y-k)^2 = G$.

a. If $A > 0, C > 0$, and $G = 0$, the graph is the single point (h, k).

b. If $AC > 0$ and $G < 0$, the graph is empty.

c. If $AC < 0$ and $G = 0$, the graph consists of the two intersecting lines

$$y - k = \pm\sqrt{\frac{A}{C}}(x-h)$$

In Exercises 35–38, describe the graph of the equation.

35. $x^2 + 4y^2 - 6x + 10 = 0$

36. $x^2 + 4y^2 - 6x + 9 = 0$

37. $x^2 - 4y^2 - 6x + 9 = 0$

38. $4x^2 - y^2 + 12x + 9 = 0$

ADDITIONAL EXERCISES

In Exercises 39–62, sketch the graph of each equation and identify the following graphical features.

> *For a parabola, coordinates of the vertex and equation of the axis of symmetry*
>
> *For an ellipse, coordinates of the center, vertices, and covertices and equations of the major and minor axes*
>
> *For a hyperbola, coordinates of the center and vertex and equations of the transverse axis, conjugate axis, and asymptotes*

If the graph is one of the exceptional cases discussed in Exercises 31–34, describe the graph.

39. $x = 3y^2$

40. $x^2 - 4y^2 + 16 = 0$

41. $2x^2 + 7y^2 = 28$

42. $4x^2 + 9y^2 = 1$

43. $4(x + 1)^2 + 4(y - 2)^2 = 9$

44. $4(x + 1)^2 - 4(y - 2)^2 = 9$

45. $4(x + 1)^2 - 4(y - 2) = 9$

46. $100x^2 - 4(y - 1)^2 = -400$

47. $5x^2 + y^2 - 70x + 225 = 0$

48. $9x^2 - 36y^2 - 90x + 360y - 999 = 0$

49. $2x^2 - 12x - y + 10 = 0$

50. $4x^2 + y^2 + 24x + 8y + 52 = 0$

51. $4x^2 - y^2 + 24x - 8y + 20 = 0$

52. $y^2 - 2x + 8y + 12 = 0$

53. $100x^2 + 20x + 400y^2 - 480y - 9855 = 0$

54. $9x^2 + y^2 - 72x + 4y + 157 = 0$

55. $4x^2 + 25y^2 + 24x - 50y - 39 = 0$

56. $-x^2 - 4y^2 + 8x - 8y - 4 = 0$

57. $x^2 + 5y^2 + 6x + 20y + 34 = 0$

58. $3x^2 + 2y^2 - 4y - 22 = 0$

59. $4x^2 - 25y^2 + 24x + 50y - 89 = 0$

60. $-x^2 + 4y^2 + 8x + 8y - 12 = 0$

61. $x^2 - 5y^2 + 6x - 20y - 6 = 0$

62. $3x^2 - 2y^2 + 4y - 26 = 0$

63. *(Creating Models)* Imagine that you are an automotive engineer and your company is designing a new model of sports car. One of your responsibilities is to design the headlamp unit. You have been given the specifications shown in Figure 6-20.

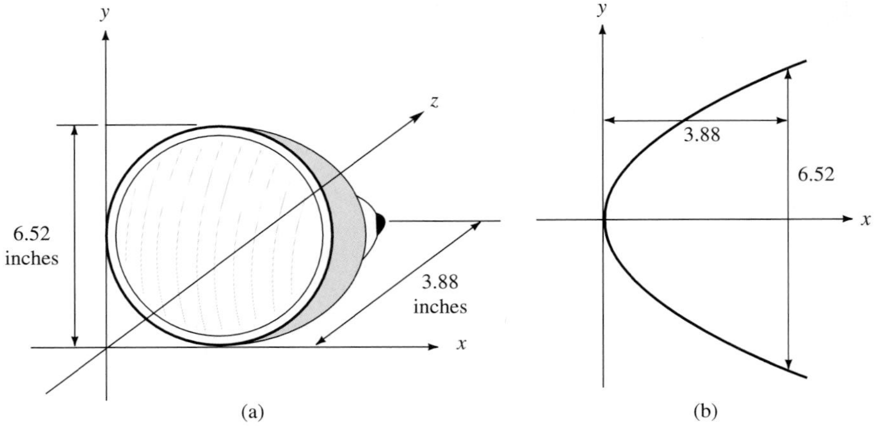

Figure 6–20

In Section 6-4 you will discover why the reflective surface should have parabolic cross sections. For now, find the equation of the cross section shown.

64. *(Extension)* Follow these steps to explain why the portion of the hyperbola $x^2 - y^2 = 1$ in the first quadrant becomes and remains arbitrarily close to the line $y = x$.

a. For a fixed value of x, why is the difference of the y-coordinates on the line and hyperbola equal to $x - \sqrt{x^2 - 1}$?

b. Rationalize the numerator to show that the difference can also be expressed as $\dfrac{1}{x + \sqrt{x^2 - 1}}$. (The process of rationalizing the numerator is discussed in Section A-18.)

c. Why is the difference always less than $1/x$ and therefore as small as you please if x is sufficiently large?

6–3 GRAPHICAL TRANSFORMATIONS

In Section 6-2 you learned how to sketch the graph of a quadratic relation from the standard graphing form of its equation. In this section you will learn an alternate method that uses the idea of transforming a basic graph. You have seen this idea before in Sections 3-4 and 5-1, where you sketched the graphs of linear absolute value functions and quadratic functions by transforming the basic graphs $y = |x|$ and $y = x^2$. Now we will take a closer look at how changes in an equation produce changes in a graph. Our objective is to develop a unified set of principles that we can use to make quick and accurate sketches of a wide variety of graphs. Let's begin our investigation in the context of a situation in which both the equation and the graph of a quadratic relation are of interest.

A MATHEMATICAL LOOKING GLASS
The Blue Planets

Our knowledge of the solar system that is our home has increased greatly over the last two decades, largely because of the development of unmanned spacecraft capable of transmitting visual images back to earth. The images can often be quite dramatic as well as informative. In *The Planets: Portraits of New Worlds*, Nigel Henbest describes humanity's first close glimpse of Neptune.

> When the spacecraft *Voyager 2* swept past its final planet in 1989, astronomers viewing the pictures on Earth were in for a nostalgic surprise. *Voyager*'s cameras showed Neptune to be a blue planet with wispy white clouds, accompanied by a large, rocky-looking moon. No one had expected to find, in the outer fringes of the solar system, a world so reminiscent of our own "Blue Planet." Up till then, we knew virtually nothing about Neptune. The planet is so far away that we need a telescope to see it at all. Astronomers did not know of its existence until the middle of the last century, when they noticed that some unexpected force was pulling Uranus as it moved round its orbit. The best explanation was the gravitational tug of an unknown planet, further from the Sun. Mathematicians in England and France calculated where this planet should lie, and in 1846 observers at Berlin Observatory picked it out as a faint moving "star."

Voyager's extraordinary success was a triumph for mathematics as much as for astronomy. To observe details such as clouds on Neptune, the spacecraft traveled nearly 3 billion miles along an extremely precise path. To calculate that path mathematicians needed the equations and graphs of the orbits of both Earth and Neptune.

Since the time of Johannes Kepler, whom you met in Section 6-1, it has been known that the orbit of each planet is an ellipse, a type of curve closely related to a circle. Before returning to planetary orbits in Exercise 38, let's proceed by *Examining a Related Problem*. Over the next few pages you can discover the changes in the unit circle (see ***Fermilab***, page 177) that are produced by various changes in its equation. ■

STRETCHES AND COMPRESSIONS

First let's see what happens to the unit circle if we replace x by $\frac{x}{a}$ or y by $\frac{y}{b}$ in its equation. Table 23 was generated by the equation $x^2 + y^2 = 1$ of the unit circle.

TABLE 23

x	y
-1	0
0	1
0	-1
1	0

EXERCISES

1. *(Making Observations)*

a. Verify that the following table fits the relation $\left(\frac{x}{2}\right)^2 + y^2 = 1$.

x	y
-2	0
0	1
0	-1
2	0

b. We have *divided* x by 2 in the equation of the unit circle, but to obtain the table in part (a) from Table 23 we have *multiplied* each x-coordinate by 2. Explain why this occurs.

c. Sketch the graph of $\left(\frac{x}{2}\right)^2 + y^2 = 1$. How does the graph support your observations in part (b)?

d. Describe how the unit circle was changed to produce the graph in part (c).

e. Support your result in part (c) by finding and graphing two functions defined implicitly by the relation.

2. Use your results from Exercise 1 to sketch the graph of each relation.

a. $\left(\frac{x}{3}\right)^2 + y^2 = 1$ **b.** $\left(\frac{x}{5}\right)^2 + y^2 = 1$

c. $(2x)^2 + y^2 = 1$ **d.** $(4x)^2 + y^2 = 1$

3. *(Making Observations)*

a. Sketch the graph of $x^2 + \left(\frac{y}{2}\right)^2 = 1$.

b. Describe how the unit circle was changed to produce the graph in part (a).

c. Support your result in part (a) by finding and graphing two functions defined implicitly by the relation.

4. Use your results from Exercise 3 to sketch the graph of each relation.

a. $x^2 + \left(\frac{y}{3}\right)^2 = 1$ **b.** $x^2 + \left(\frac{y}{5}\right)^2 = 1$

c. $x^2 + (2y)^2 = 1$ **d.** $x^2 + (4y)^2 = 1$

Your observations in Exercises 1–4 can be generalized as follows.

> If x is replaced by $\frac{x}{a}$ with $a > 0$ in the equation of a relation, each x-coordinate on the graph is multiplied by a. The graph is stretched horizontally if $a > 1$ and compressed horizontally if $0 < a < 1$.
>
> Similarly, if y is replaced by $\frac{y}{b}$ with $b > 0$ in the equation, each y-coordinate on the graph is multiplied by b. The graph is stretched vertically if $b > 1$ and compressed vertically if $0 < b < 1$.

EXERCISES *In Exercises 5–10, describe how the first graph must be transformed to produce the second, and sketch both graphs on the same set of axes.*

5. $y = x^2;\ y = (3x)^2$

6. $x = y^2;\ x = \left(\frac{y}{4}\right)^2$

7. $x^2 - y^2 = 1;\ \left(\frac{x}{2}\right)^2 - y^2 = 1$

8. $y^2 - x^2 = 1;\ (2y)^2 - x^2 = 1$

9. $x^2 + y^2 = 1;\ \left(\frac{x}{3}\right)^2 + \left(\frac{y}{4}\right)^2 = 1$

10. $x^2 - y^2 = 1;\ \left(\frac{x}{3}\right)^2 - \left(\frac{y}{4}\right)^2 = 1$

SHIFTS

Next let's see what happens to the unit circle if we replace x by $x - h$ or y by $y - k$ in its equation.

EXERCISES **11.** *(Making Observations)*

a. Verify that the following table fits the relation $(x - 3)^2 + y^2 = 1$.

x	y
2	0
3	1
3	−1
4	0

b. We have *subtracted* 3 from x in the equation of the unit circle, but to obtain the table in part (a) from Table 23 we have *added* 3 to each x-coordinate. Explain why this occurs.

c. Sketch the graph of $(x - 3)^2 + y^2 = 1$. How does the graph support your observations in part (b)?

d. Describe how the unit circle was changed to produce the graph in part (c).

e. Support your result in part (c) by finding and graphing two functions defined implicitly by the relation.

12. Use your results from Exercise 11 to sketch the graph of each relation.

a. $(x - 4)^2 + y^2 = 1$

b. $(x - 1)^2 + y^2 = 1$

c. $(x + 2)^2 + y^2 = 1$

d. $(x + 5)^2 + y^2 = 1$

13. *(Making Observations)*

a. Sketch the graph of $x^2 + (y - 3)^2 = 1$.

b. Describe how the unit circle was changed to produce the graph in part (a).

c. Support your result in part (a) by finding and graphing two functions defined implicitly by the relation.

14. Use your results from Exercise 13 to sketch the graph of each relation.

a. $x^2 + (y - 4)^2 = 1$

b. $x^2 + (y - 1)^2 = 1$

c. $x^2 + (y + 2)^2 = 1$

d. $x^2 + (y + 5)^2 = 1$

Your observations in Exercises 11–14 can be generalized as follows.

If x is replaced by $x - h$ in the equation of a relation, h is added to each x-coordinate on the graph. The graph is shifted to the right if $h > 0$, and to the left if $h < 0$.

Similarly, if y is replaced by $y - k$ in the equation, k is added to each y-coordinate on the graph. The graph is shifted up if $k > 0$, and down if $k < 0$.

EXERCISES *In Exercises 15–20, describe how the first graph must be transformed to produce the second, and sketch both graphs on the same set of axes.*

15. $y = x^2$; $y = (x - 3)^2$

16. $x = y^2$; $x = (y + 4)^2$

17. $x^2 - y^2 = 1$; $(x + 2)^2 - y^2 = 1$

18. $y^2 - x^2 = 1$; $(y - 2)^2 - x^2 = 1$

19. $x^2 + y^2 = 1$; $(x + 3)^2 + (y + 4)^2 = 1$

20. $x^2 - y^2 = 1$; $(x + 3)^2 - (y + 4)^2 = 1$

REFLECTIONS

Finally, let's see what happens to the unit circle if we replace x by $-x$ or y by $-y$ in its equation.

EXERCISE **21.** *(Making Observations)* In the equation $x^2 + y^2 = 1$, replace x by $-x$ and simplify. Then replace y by $-y$ and simplify. What do your results say about the graphs of the resulting equations?

Surprise! For this particular relation, there is no effect at all on either the equation or the graph. Let's consider some different relations to see how the replacements of x by $-x$ and y by $-y$ affect graphs in general, and why they do not affect the unit circle.

22. *(Making Observations)*

a. In the equation $(x - 2)^2 + y^2 = 1$, replace x by $-x$, and verify that the resulting equation can be simplified to $(x + 2)^2 + y^2 = 1$.

b. Sketch the graphs of both equations in part (a). How do the two graphs relate to each other?

23. *(Making Observations)*

a. In the equation $x^2 + (y - 2)^2 = 1$, replace y by $-y$ and verify that the resulting equation can be simplified to $x^2 + (y + 2)^2 = 1$.

b. Sketch the graphs of both equations in part (a). How do the graphs relate to each other?

24. *(Writing to Learn)* Use your results from Exercises 22 and 23 to explain why the unit circle is not affected when we replace x by $-x$ or y by $-y$ in its equation.

Your observations in Exercises 22 and 23 can be generalized as follows.

If x is multiplied by -1 in the equation of a relation, each x-coordinate on the graph is multiplied by -1, and the graph is reflected in the y-axis.

Similarly, if y is multiplied by -1 in the equation, each y-coordinate on the graph is multiplied by -1, and the graph is reflected in the x-axis.

If the replacement of x by $-x$ produces an equation equivalent to the original one, as with the unit circle, the graph of the relation is its own reflection in the y-axis. Such a graph is said to be **symmetric about the *y*-axis**. Similarly, if the replacement of y by $-y$ produces an equation equivalent to the original one, the graph of the relation is **symmetric about the *x*-axis**. We will return to the idea of symmetry in Section 7-3.

EXAMPLE 1 *Detecting symmetry analytically*

In the equation $\left(\frac{x-6}{2}\right)^2 + \left(\frac{y}{3}\right)^2 = 1$, if y is replaced by $-y$, the equation becomes

$$\left(\frac{x-6}{2}\right)^2 + \left(\frac{-y}{3}\right)^2 = 1$$

or, equivalently,

$$\left(\frac{x-6}{2}\right)^2 + \left(\frac{y}{3}\right)^2 = 1$$

Thus its graph is symmetric about the x-axis. If x is replaced by $-x$, the equation becomes

$$\left(\frac{-x-6}{2}\right)^2 + \left(\frac{y}{3}\right)^2 = 1$$

or, equivalently,

$$\left(\frac{x+6}{2}\right)^2 + \left(\frac{y}{3}\right)^2 = 1$$

Thus the graph is not symmetric about the y-axis. The graph in Figure 6-21 supports these conclusions.

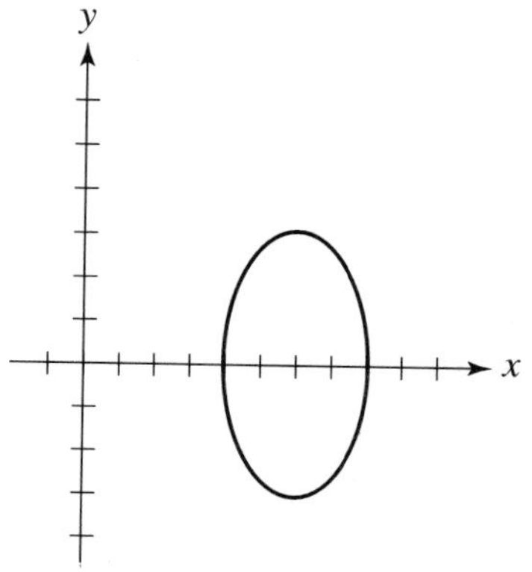

Figure 6–21

■

EXERCISES *In Exercises 25–28, decide whether the graph is symmetric about the x-axis, the y-axis, both, or neither. Then use the equation to confirm your conclusion.*

25. $x - y^2 = 1$

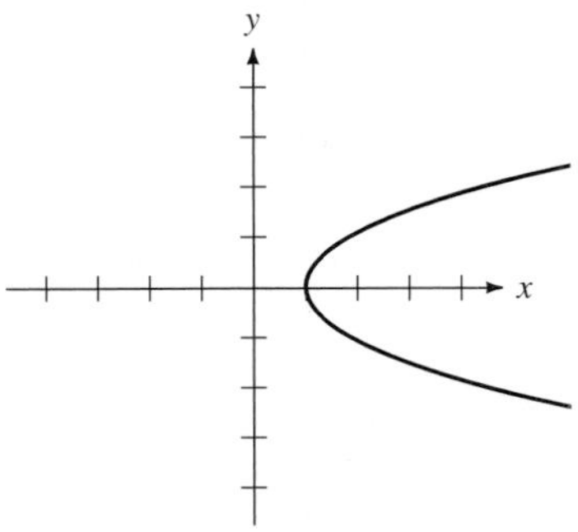

26. $x^2 - y^2 = 1$

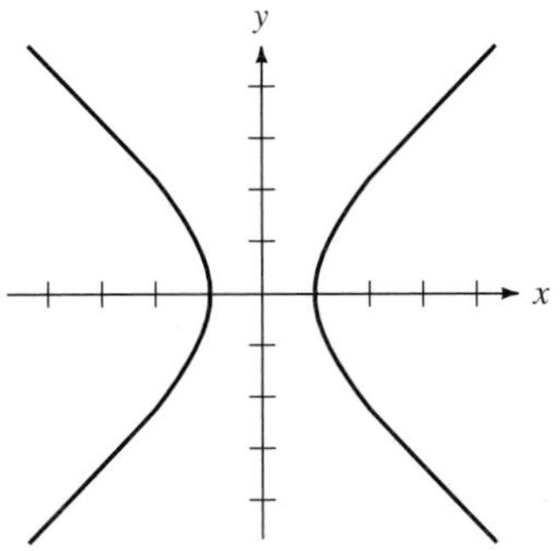

27. $x^2 + \left(\frac{y}{4}\right)^2 = 1$

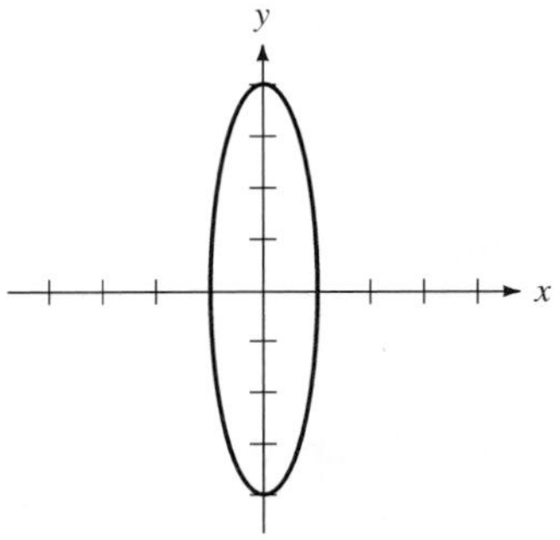

28. $(x + 1)^2 + \left(\frac{y}{4}\right)^2 = 1$

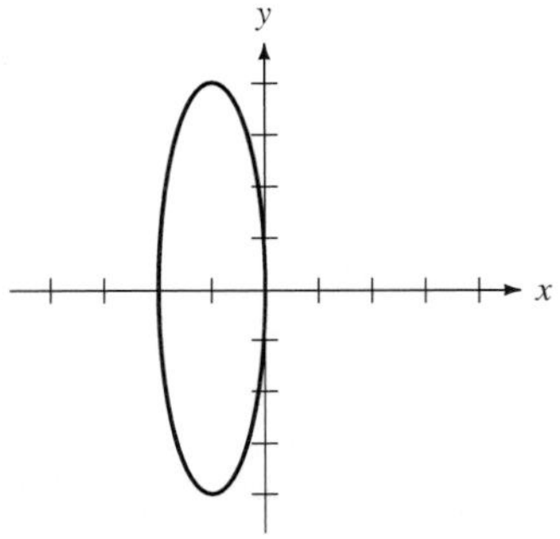

APPLYING A SEQUENCE OF TRANSFORMATIONS

The value of the graphical transformation principles you have just learned is that they allow you to sketch a variety of graphs quickly and accurately by memorizing just a few basic graphs and equations, such as the unit circle and its equation. If a basic equation undergoes several changes of the types just described, you can determine the graphical effects by applying the appropriate transformations in sequence.

EXAMPLE 2 *Graphing a relation by transforming a basic graph*

To sketch the graph of

$$x^2 - \left(\frac{y-3}{2}\right)^2 = 1$$

begin with the basic equation

$$x^2 - y^2 = 1$$

whose graph is shown in Figure 6-22a. Then divide y by 2 to obtain

$$x^2 - \left(\frac{y}{2}\right)^2 = 1$$

whose graph is shown in Figure 6-22b. Finally subtract 3 from y to obtain

$$x^2 - \left(\frac{y-3}{2}\right)^2 = 1$$

whose graph is shown in Figure 6-22c.

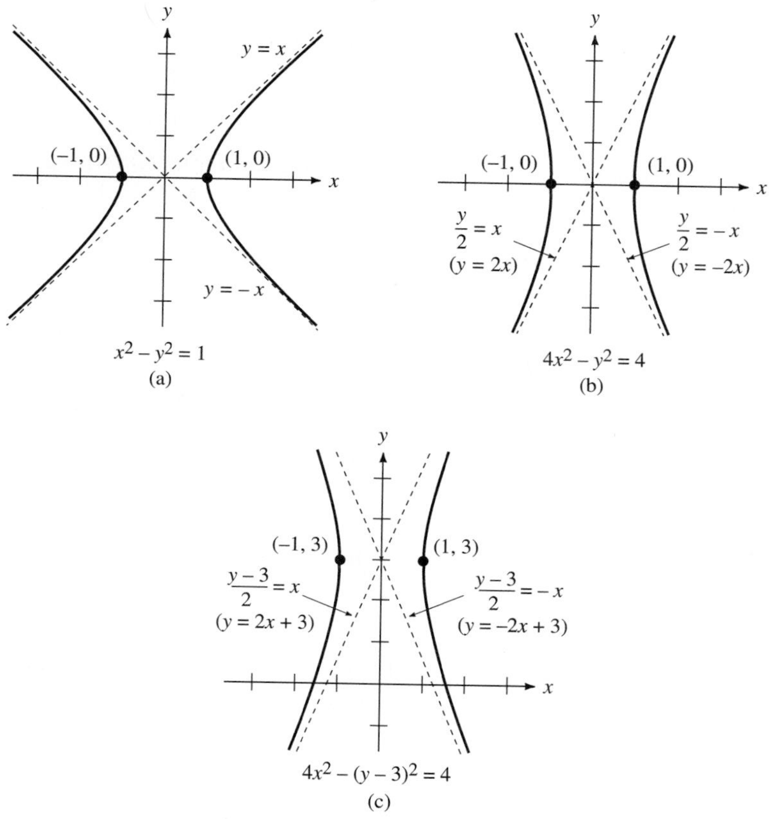

Figure 6–22

Figure 6-22 shows that the basic graph is first stretched vertically by a factor of 2, then shifted 3 units up. ■

FYI When you are applying a sequence of graphical transformations, order matters! In Exercise 29 you can demonstrate that applying the two transformations in Example 2 in the opposite order yields a different graph with a different equation.

EXERCISES **29.** *(Making Observations)*

a. Beginning with the unit circle, shift the graph 3 units up; then stretch it vertically by a factor of 2. Sketch each graph in the sequence and confirm that the final graph is different from the one in Figure 6-22c.

b. Beginning with the equation of the unit circle, first subtract 3 from y; then divide y by 2. Confirm that the resulting equation is different from the last equation in Example 2.

In Exercises 30–33, sketch the ellipse. Identify the coordinates of the center, vertices, and covertices and the equations of the major and minor axes.

30. $\left(\frac{x+5}{3}\right)^2 + \left(\frac{y+6}{4}\right)^2 = 1$

31. $\left(\frac{x+129}{50}\right)^2 + \left(\frac{y-85}{35}\right)^2 = 1$

32. $\left(\frac{x-2}{\sqrt{7}}\right)^2 + \left(\frac{y+1.5}{2}\right)^2 = 1$

33. $(x-0.5)^2 + \left(\frac{y-1.5}{2}\right)^2 = 1$

In Exercises 34–37, sketch the hyperbola. Identify the coordinates of the center and vertices and the equations of the transverse axis, conjugate axis, and asymptotes.

34. $\left(\frac{x+5}{3}\right)^2 - \left(\frac{y+6}{4}\right)^2 = 1$

35. $\left(\frac{x+129}{50}\right)^2 - \left(\frac{y-85}{35}\right)^2 = 1$

36. $\left(\frac{y+1.5}{2}\right)^2 - \left(\frac{x-2}{\sqrt{7}}\right)^2 = 1$

37. $\left(\frac{y-1.5}{2}\right)^2 - (x-0.5)^2 = 1$

38. Sketch the graph of each planet's orbit. Distances are measured in astronomical units (1 astronomical unit = 92.9 million miles).

a. Earth: $(x-0.02)^2 + \left(\frac{y}{0.9998}\right)^2 = 1$

b. Mercury: $\left(\frac{x-0.08}{0.39}\right)^2 + \left(\frac{y}{0.38}\right)^2 = 1$

SUMMARY

The graphical transformation principles of Section 6-3 are summarized in Table 24.

TABLE 24

Change applied to equation	Effect on coordinates	Effect on graph
$x \to \frac{x}{a}$	Multiply x by a	Horizontal stretch ($a > 1$) or compression ($0 < a < 1$)
$y \to \frac{y}{b}$	Multiply y by b	Vertical stretch ($b > 1$) or compression ($0 < b < 1$)
$x \to x - h$	Add h to x	Horizontal shift
$y \to y - k$	Add k to y	Vertical shift
$x \to -x$	Multiply x by -1	Reflection in y-axis
$y \to -y$	Multiply y by -1	Reflection in x-axis

ADDITIONAL EXERCISES

In Exercises 39–50, sketch the graph of each equation by transforming one of the basic graphs $y = x^2$, $x = y^2$, $x^2 + y^2 = 1$, $x^2 - y^2 = 1$, *or* $y^2 - x^2 = 1$. *In each case describe how the basic graph is transformed, and identify the following graphical features.*

For a parabola, coordinates of the vertex and equation of the axis of symmetry

For an ellipse, coordinates of the center, vertices, and covertices and equations of the major and minor axes

For a circle, radius and coordinates of the center

For a hyperbola, coordinates of the center and vertex and equations of the transverse axis, conjugate axis, and asymptotes

39. $x = -5y^2$

40. $4x^2 - y^2 + 64 = 0$

41. $5x^2 + 6y^2 = 30$

42. $9x^2 + 4y^2 = 1$

43. $(x + 10)^2 + (y - 20)^2 = 25$

44. $(x + 10)^2 - (y - 20)^2 = 25$

45. $(x + 10)^2 - (y - 20) = 25$

46. $36x^2 - (y + 7)^2 = -144$

47. $\left(\frac{x + 2}{4}\right)^2 + \left(\frac{y + 3}{6}\right)^2 = 1$

48. $\left(\frac{x - 4}{7}\right)^2 + \left(\frac{y - 1}{3}\right)^2 = 1$

49. $\left(\frac{x + 2}{4}\right)^2 - \left(\frac{y + 3}{6}\right)^2 = 1$

50. $\left(\frac{y - 1}{3}\right)^2 - \left(\frac{x - 4}{7}\right)^2 = 1$

In Exercises 51–56, decide whether the graph is symmetric about the x-axis, the y-axis, both, or neither. Then use the equation to confirm your conclusion.

51. $y = x^2 - 4$

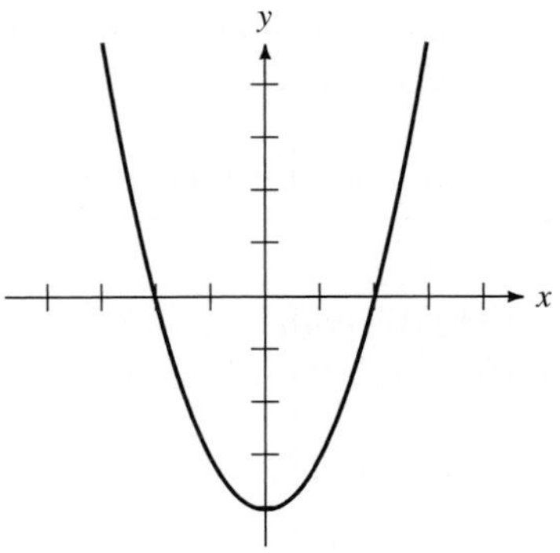

52. $y = x^3 - 4x$

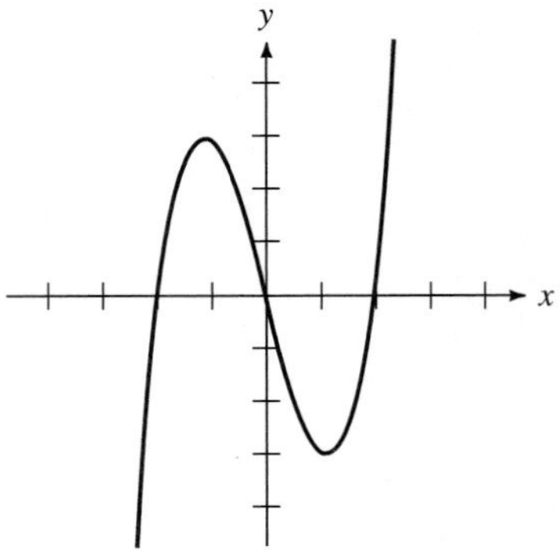

53. $y = x^3 - 4x^2$

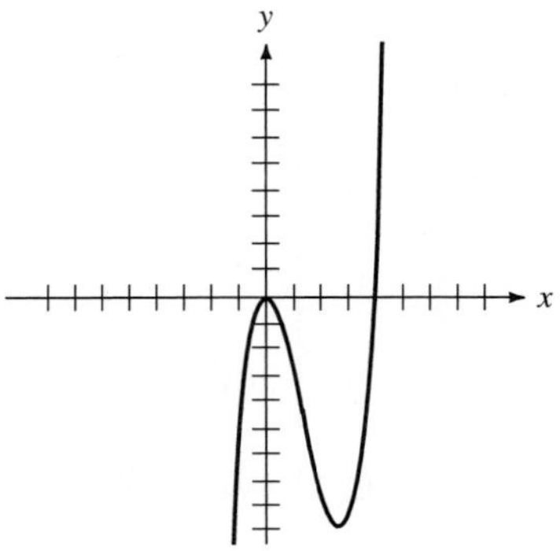

54. $y = \dfrac{x^2}{x^2 + 1}$

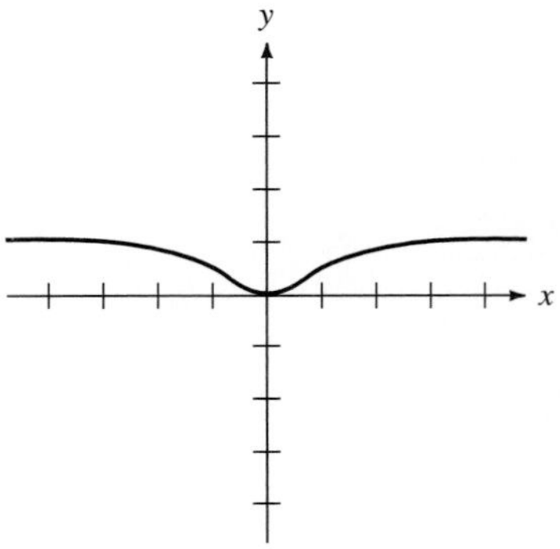

55. $x^3 - 3x^2 - y^2 = 0$

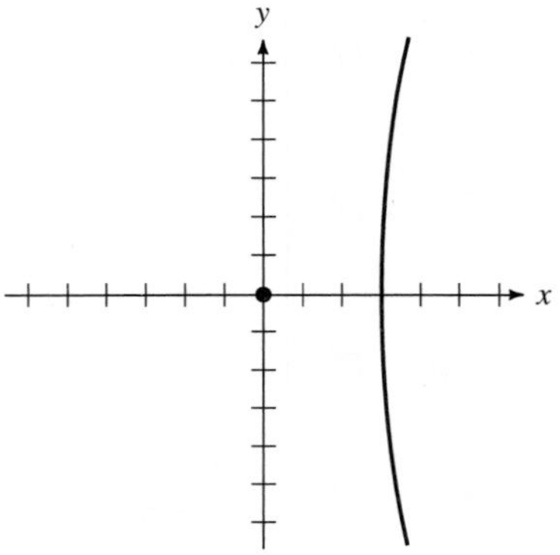

56. $x^6 + y^6 = 1$

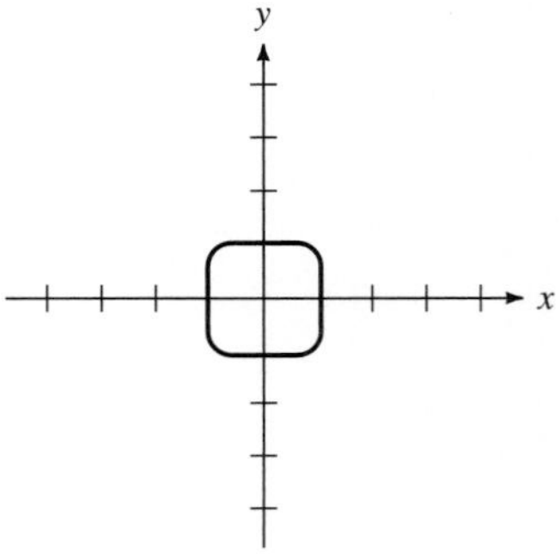

57. What equation results from each sequence of transformations?

a. The unit circle is shifted 5 units to the right and 2 units down.

b. The unit circle is stretched horizontally by a factor of 2 and compressed vertically by a factor of 2.

c. The unit circle is stretched horizontally by a factor of 3, then shifted 1 unit to the left.

d. The unit circle is shifted 1 unit to the left, then stretched horizontally by a factor of 3.

58. What equation results from each sequence of transformations?

a. The graph of $x = y^2$ is compressed horizontally by a factor of 2, then reflected in the y-axis.

b. The graph of $y^2 - x^2 = 1$ is shifted 4 units left, then stretched vertically by a factor of 5.

c. The graph of $y = x^2$ is reflected in the x-axis, then shifted 3 units down.

d. The graph of $x^2 - y^2 = 1$ is stretched horizontally by a factor of 3, then shifted 1 unit right.

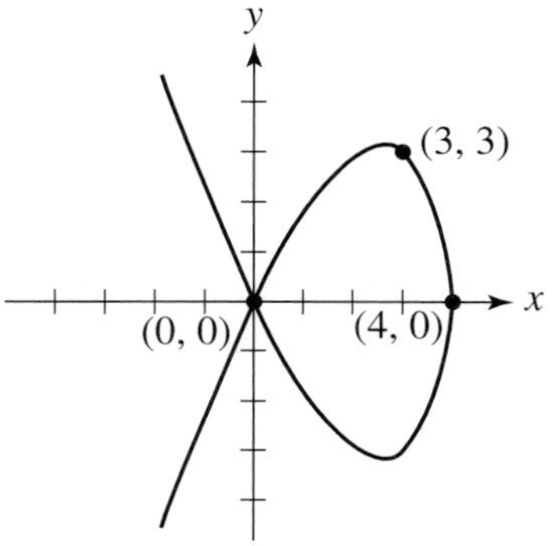

The graph of the relation $x^3 - 4x^2 + y^2 = 0$ *is shown at the left. In Exercises 59–62, sketch the graph of the given relation, and identify three points on the graph.*

59. $x^3 - 4x^2 + \left(\frac{y+1}{3}\right)^2 = 0$

60. $(-4x)^3 - 4(-4x)^2 + y^2 = 0$

61. $(x + 3)^3 - 4(x + 3)^2 + (y - 4)^2 = 0$

62. $\left(\frac{x+3}{2}\right)^3 - 4\left(\frac{x+3}{2}\right)^2 + [2(y - 1)]^2 = 0$

63. *(Creating Models)* Imagine that you are a civil engineer working on a highway construction project. A portion of the roadbed is shown in Figure 6-23a with the coordinate system that one of your colleagues has imposed on it.

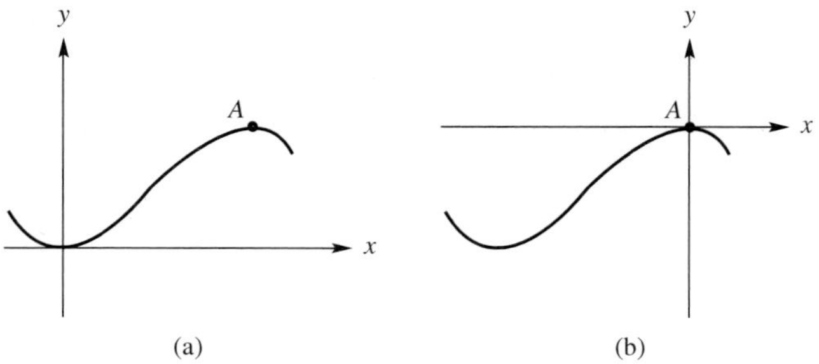

Figure 6–23

You would rather work with the coordinate system in Figure 6-23b since you will be working near the crest of the hill at A. In your colleague's coordinate system, the coordinates of A are (5000, 200) and the equation of the roadbed is

$$y = (2.4 \times 10^{-5})x^2 - (3.2 \times 10^{-9})x^3$$

with both x and y given in feet. What is the equation of the roadbed in your coordinate system?

64. *(Writing to Learn)* The information in the first four lines of Table 24 can be characterized loosely by the statement, "Whatever you do in the equation, do the opposite on the graph." Explain why the information in the last two lines can be characterized in the same way.

SUPPLEMENTARY TOPIC

Graphical Transformations in the Context of Functions

In Sections 3-4 and 5-1 you saw how the graphs of $y = a|x - h| + k$ and $y = a(x - h)^2 + k$ could be obtained by transforming the basic graphs $y = |x|$ and $y = x^2$. Are those transformations compatible with the principles we just developed? They should be, since the equations $y = |x|$ and $y = x^2$ define relations as well as functions. Let's find out.

65. a. *(Review)* Use the methods of Section 3-4 to sketch the graph of $y = 3|x - 4| - 2$.

b. Use the methods of this section to sketch the graph of $\dfrac{y + 2}{3} = |x - 4|$.

c. Your graphs should appear to be identical. Verify that their equations are equivalent.

Your observations in Exercise 65 indicate that if $y = f(x)$, the transformations affecting y can be summarized as in Table 25.

66. *(Writing to Learn)* If $y = f(x)$, explain why the information in Table 25 is consistent with that in Table 24.

In Exercises 67–70, sketch the graph of the first function by transforming the graph of the second, shown below the equations. Identify three points on each graph.

67. $y = 2\sqrt{x + 5}$
$y = \sqrt{x}$

68. $y = (x - 4)^4 + 1$
$y = x^4$

TABLE 25

New function	Effect on coordinates	Effect on graph
$y = bf(x)$	Multiply y by b	Vertical stretch ($b > 1$) or compression ($0 < b < 1$)
$y = f(x) + k$	Add k to y	Vertical shift
$y = -f(x)$	Multiply y by -1	Reflection in x-axis

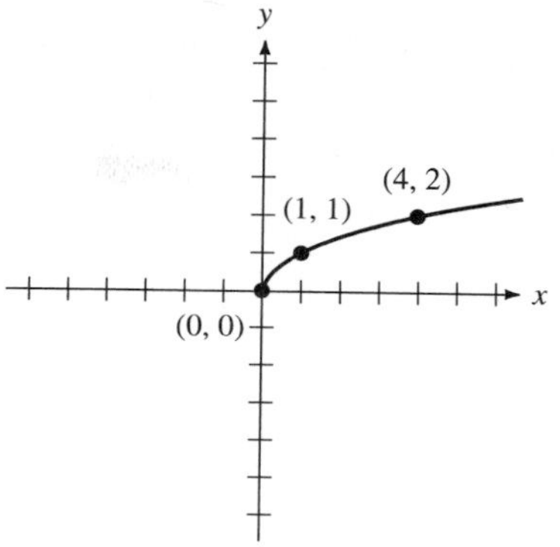

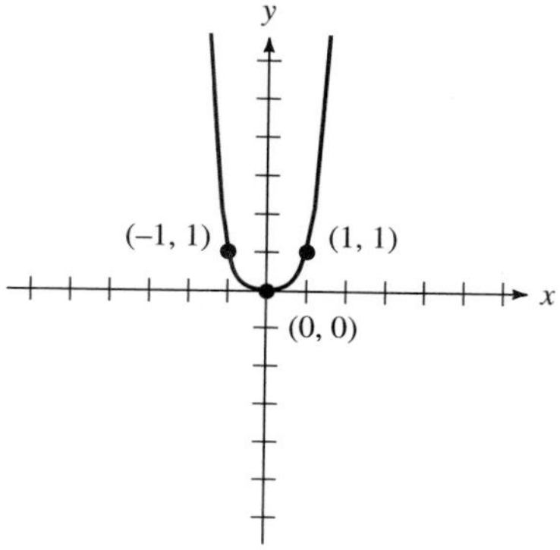

69. $y = 4 - 2x^3$
$y = x^3$

70. $y = 3^{2x} - 5$
$y = 3^x$

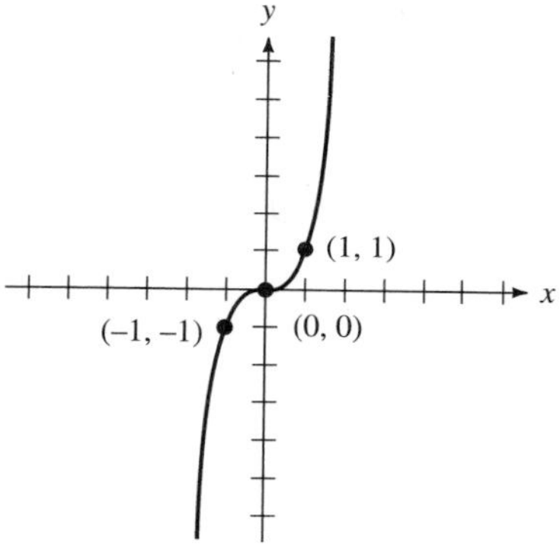

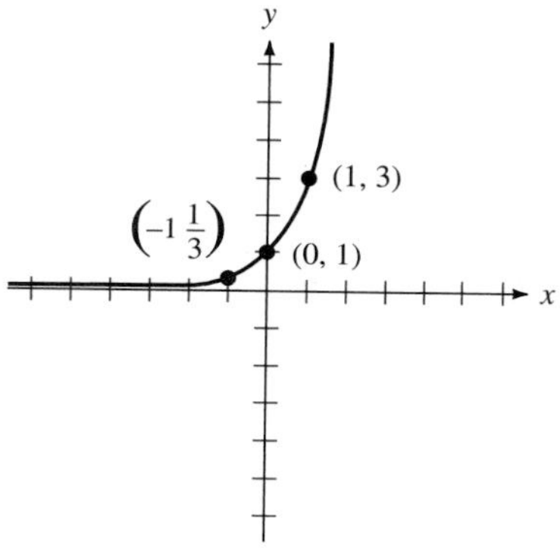

In Exercises 71–74, sketch the graph of each function by transforming the graph of the basic function shown. Identify the points obtained by moving the three identified points on the basic graph.

71. a. $y = -3x^5$

b. $y = (x + 3)^5 + 4$

c. $y = 0.1(x - 2)^5$

d. $y = 2(x - 1)^5 + 3$

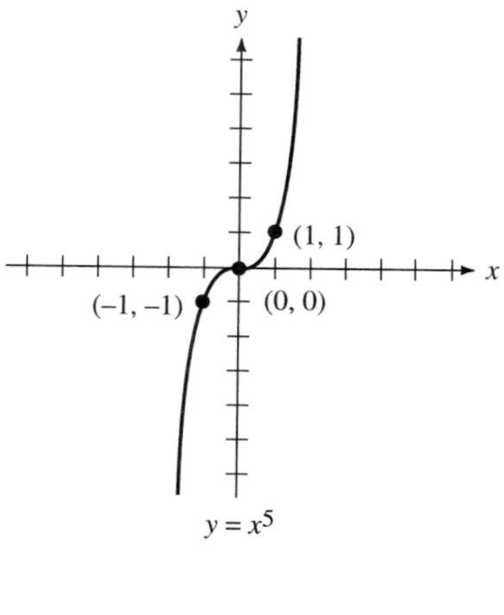

$y = x^5$

72. **a.** $y = -\sqrt{x}$ **b.** $y = \sqrt{-x}$
c. $y = -\sqrt{-x}$ **d.** $y = 2\sqrt{x-1} + 3$

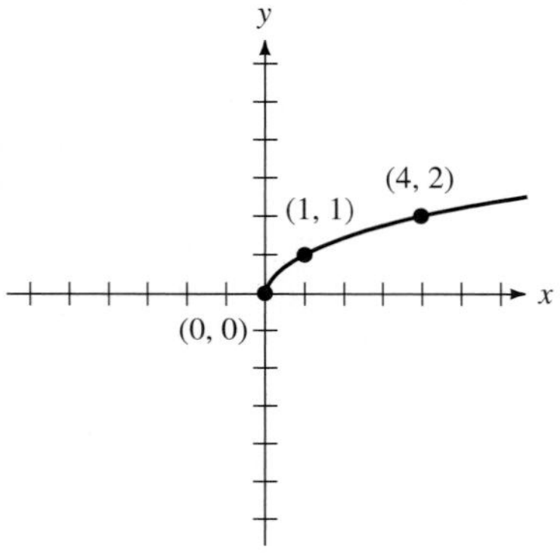

$y = \sqrt{x}$

73. **a.** $y = 2^{x+3} - 5$ **b.** $y = 0.5(2^{3x})$
c. $y = 7 + 2^{-x}$ **d.** $y = 2(2^{x-1}) + 3$

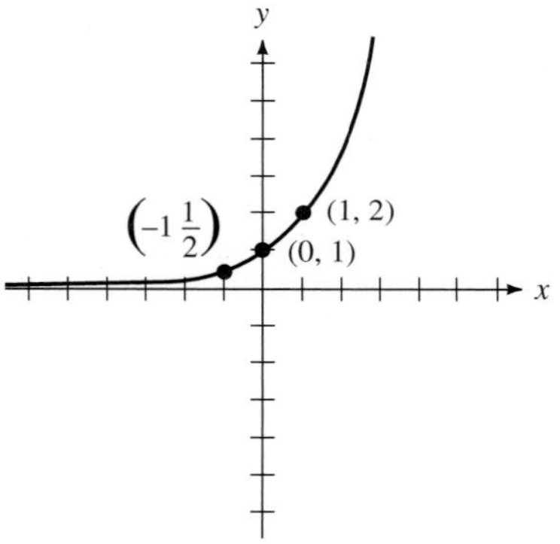

$y = 2^x$

74. **a.** $y = \dfrac{4}{x} - 2$ **b.** $y = -\dfrac{1}{x+5}$
c. $y = 6 - \dfrac{1}{x}$ **d.** $y = \dfrac{2}{x-1} + 3$

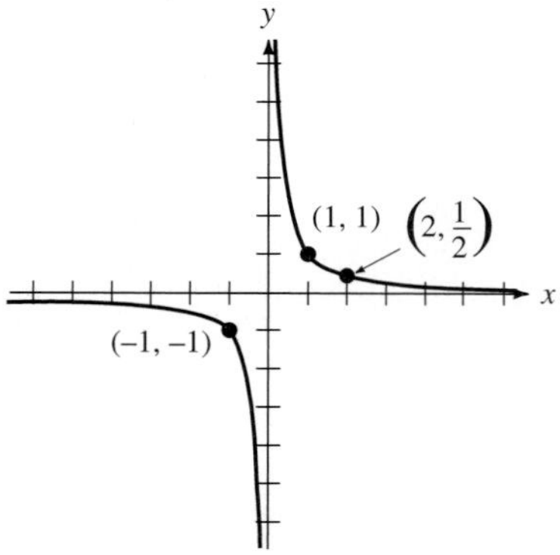

$y = \dfrac{1}{x}$

6–4 AN ANALYTICAL VIEW OF CONIC SECTIONS

PREREQUISITES MAKE SURE YOU ARE FAMILIAR WITH:

Radicals and Rational Exponents (Section A-18)
Equations with Radicals (Section A-19)

The universe abounds with conic sections. The orbits of planets and comets and the cross sections of optical lenses are parabolic, elliptical, or hyperbolic. In this section you will learn that conic sections appear so often in the physical world because they possess some interesting geometric properties related to their equations. You will use the methods of analytic geometry (geometry in the framework of a coordinate system) to discover that the equations of the curves with the required properties are quadratic relations.

AN ANALYTICAL VIEW OF PARABOLAS

Try this experiment. You will need a sheet of wax paper about a foot square and a marker. Make a mark about 1 inch up from the middle of the bottom edge, as in Figure 6-24a. Next, fold up part of the bottom edge so that it just touches your mark, as in Figure 6-24b.

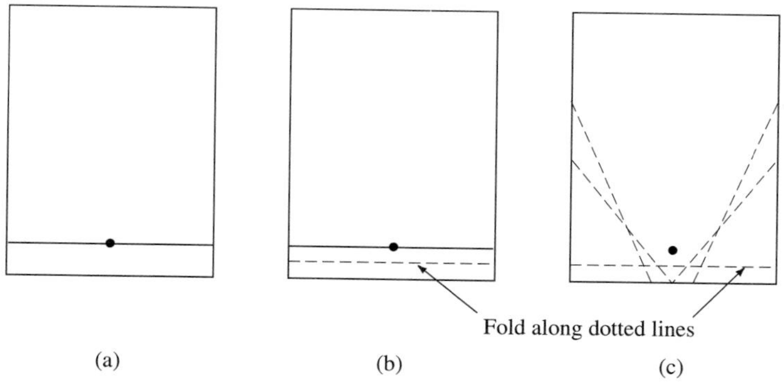

Figure 6–24

Continue as in Figure 6-24c to make about 15 to 20 such folds at different angles. Now spread the paper flat. Do you see a curve on the paper? What type of curve does it look like?

Let's find an equation to describe the wax paper curve. To do this, we need to place the paper into a convenient coordinate system. We choose one in which the y-axis passes through the mark, labeled F in Figure 6-25a, and the origin is midway between the mark and the bottom edge. Thus F has coordinates $(0, p)$, and the bottom edge has the equation $y = -p$ for some $p > 0$. Every point $P = (x, y)$ on the curve was produced by a fold which superimposed a point $D = (x, -p)$ on F, as in Figure 6-25b (see page 218).

EXERCISES **1.** *(Writing to Learn)* Explain why $\overline{PF} = \overline{PD}$ in Figure 6-25.

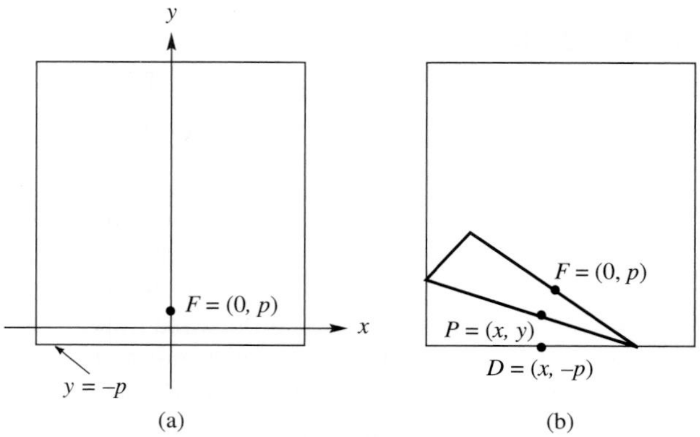

Figure 6–25

2. *(Extension)* Use the distance formula to show that x and y must satisfy the equation

$$\sqrt{x^2 + (y - p)^2} = y + p.$$

3. *(Extension)* Show that the equation in Exercise 2 can be simplified to $y = \dfrac{1}{4p}x^2$.

So the wax paper curve is a parabola! In fact, your experiment provides a formal definition of the word *parabola* at last. A **parabola** is the set of all points in the plane that are equidistant from a fixed line and a fixed point not on the line. The fixed line is called the **directrix** of the parabola (the bottom edge of the wax paper), and the fixed point is its **focus** (the mark on the paper). See Figure 6-26.

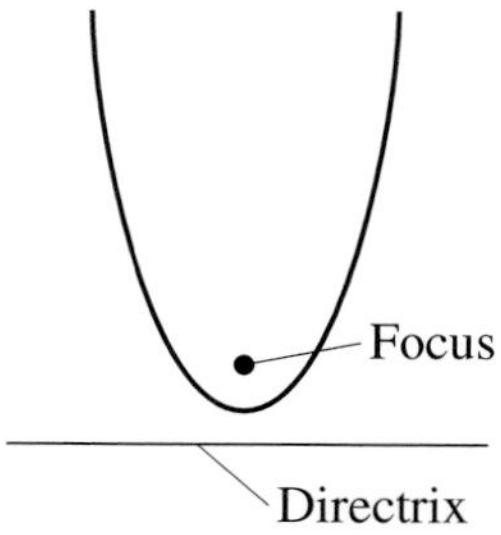

Figure 6–26

Your work in Exercises 1–3 shows that every equation $y = ax^2$ describes a parabola according to our new definition. More generally, in Exercise 45 you will establish the following facts.

- Every equation $y = a(x - h)^2 + k$ or $x = a(y - k)^2 + h$ describes a parabola, so that our new definition coincides with the one given in Section 6-2.
- If $p = \dfrac{1}{4a}$ $\left(\text{equivalently, if } a = \dfrac{1}{4p}\right)$, the focus and directrix are each located p units from the vertex.

- The focus is always on the "inside" of the parabola, and the directrix is always on the "outside." The possible arrangements of focus and directrix are shown in Figure 6-27.

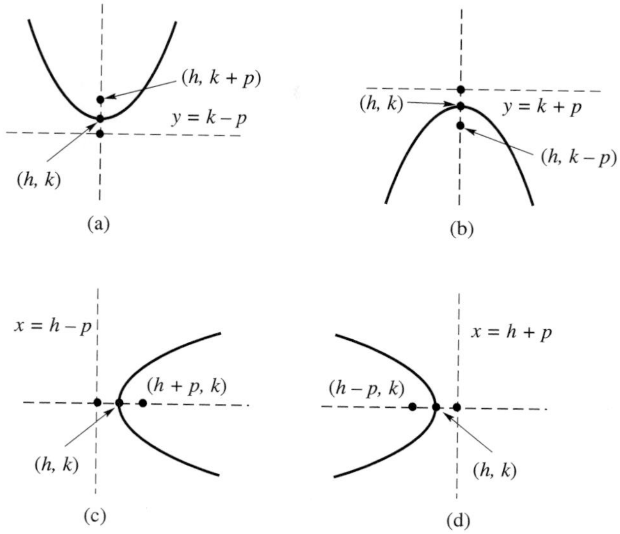

Figure 6–27

Example 1 shows how to find the focus and directrix of a parabola whose equation is known.

EXAMPLE 1 *Using the equation of a parabola to locate its focus and directrix*

Figure 6-28 shows the graph of $x = 0.1(y - 1)^2 - 2$. Its vertex is $(-2, 1)$. Since the quadratic term involves y and has a positive coefficient, the graph opens toward the right. To locate its focus and directrix, first find the value of p.

$$p = \frac{1}{4(0.1)} = 2.5$$

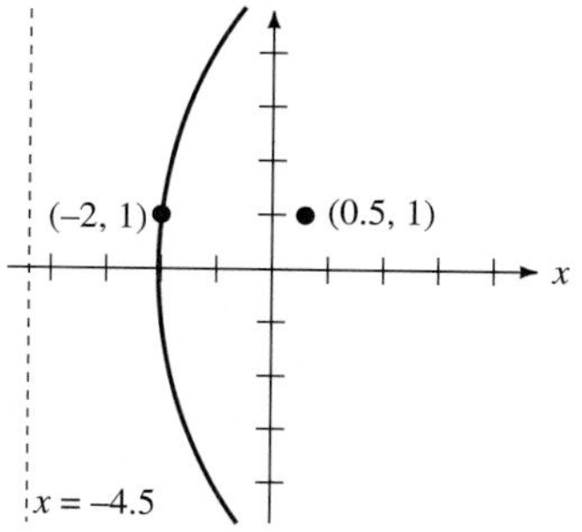

Figure 6–28

The focus is (0.5, 1), 2.5 units to the right of the vertex. The directrix is $x = -4.5$, the vertical line 2.5 units to the left of the vertex. ■

EXERCISES *In Exercises 4–7, find the focus and directrix of the parabola, and include them in a sketch of the graph.*

4. $y = 8 - x^2$

5. $y^2 + 8x - 2y = 15$

6. $x = 0.1(y + 5)^2 - 10$

7. $y = 5(x - 0.4)^2 + 0.03$

Your wax paper experiment illustrated another geometric property of parabolas. The line formed by each of your folds was tangent to the parabola, as indicated in Figure 6-29. (If you take calculus, you will learn the mathematical definition of a line tangent to an arbitrary curve. For now, let's just use the intuitive idea that the line and the curve have the same direction at the point of tangency.)

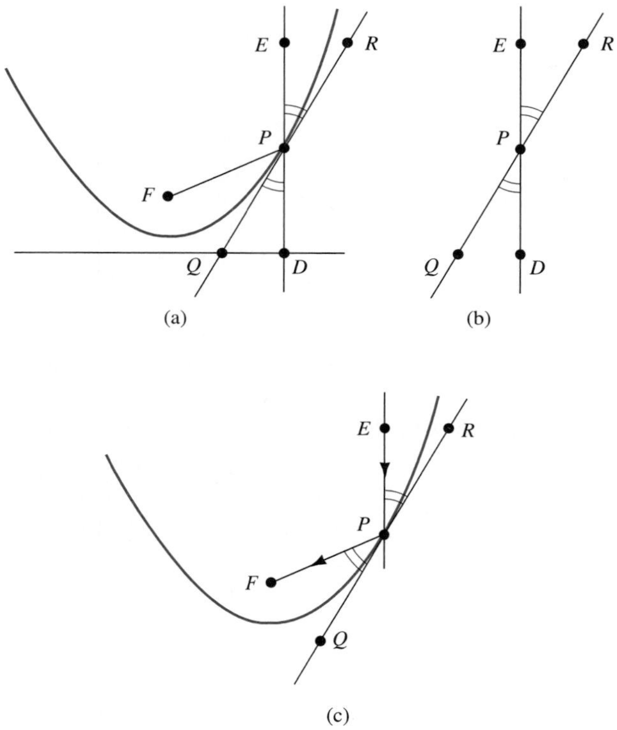

Figure 6–29

In the act of folding as in Figure 6-29a, you verified that the angles FPQ and DPQ are equal. Since DPQ and EPR are vertical angles, as shown in Figure 6-29b, they are also equal. The resulting equality of angles FPQ and EPR has some physical consequences. If you imagine the parabola as a mirror, any light ray traveling along EP will strike the mirror and be reflected toward F as illustrated in Figure 6-29c. In general, rays traveling parallel to the axis of symmetry and striking the parabola will be reflected toward the focus. Conversely, rays directed toward the parabola from the focus will be reflected parallel to the axis of symmetry.

A MATHEMATICAL LOOKING GLASS

Satellite Dish

Lynn's father, Ernie Harvey, lives in Hawthorne, NV, a remote area that had no access to cable television services until 1985. To improve television reception in his home, Dr. Harvey decided in 1983 to buy a satellite dish. The dish, which has parabolic cross sections, collects signals beamed in from a satellite and concentrates them by reflecting them to a receiver at the focus, as shown in Figure 6-30.

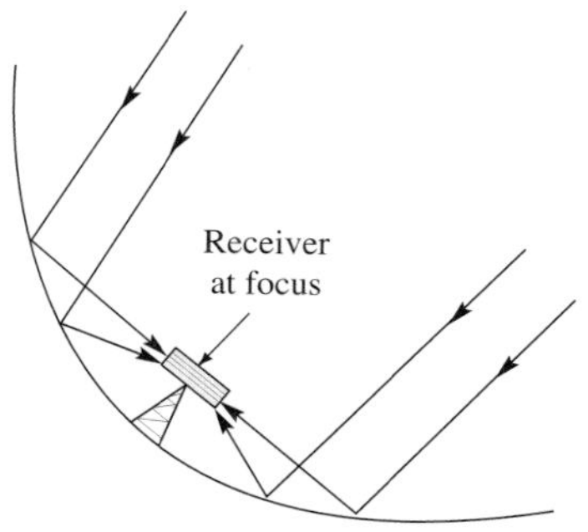

Figure 6–30

After learning that a dish would cost $9000, Dr. Harvey decided to construct his own. He knew that its power would be proportional to the square of the diameter at the edge, so he decided to devote his entire backyard to a dish with a diameter of 20 feet. To minimize the amount of material used in construction, he wanted the dish to be as shallow as possible. However, for a shallow dish he must place the receiver farther from the vertex than for a deep dish. ■

EXERCISES

8. *(Writing to Learn)* Explain why the receiver for a shallow dish must be farther from the vertex than the receiver for a deeper dish of the same diameter.

9. *(Problem Solving)* Dr. Harvey wanted his receiver to be only 6 feet from the vertex of his dish. How deep should the dish be?

AN ANALYTICAL VIEW OF ELLIPSES

Try this experiment. You will need a 1-foot-square piece of corkboard, enough construction paper to cover it, two thumbtacks, a loop of string at least 15 inches long, and a pencil. Secure the paper on top of the corkboard with the thumbtacks, placed about 2 inches in from opposite edges. Tie the string to each thumbtack, leaving about 10 inches of string between the knots. See Figure 6-31a (page 222).

Place the pencil below the string so that the string is taut, and trace an arc above the tacks by moving the pencil so the string remains taut, as in Figure 6-31b. Then place the pencil above the string and trace an arc below the tacks. Your two arcs should join to form a single closed curve. What type of curve does it look like?

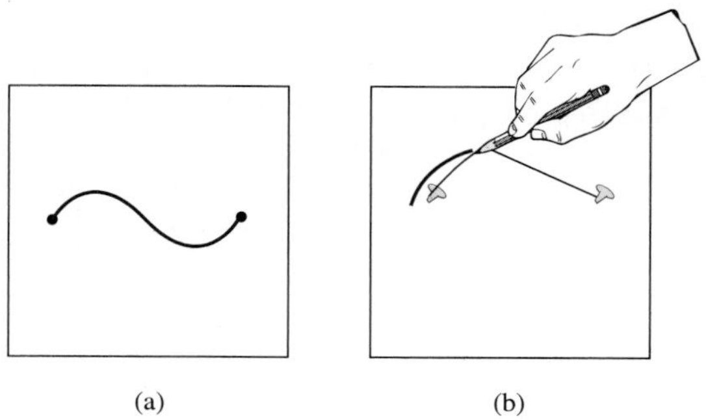

Figure 6–31

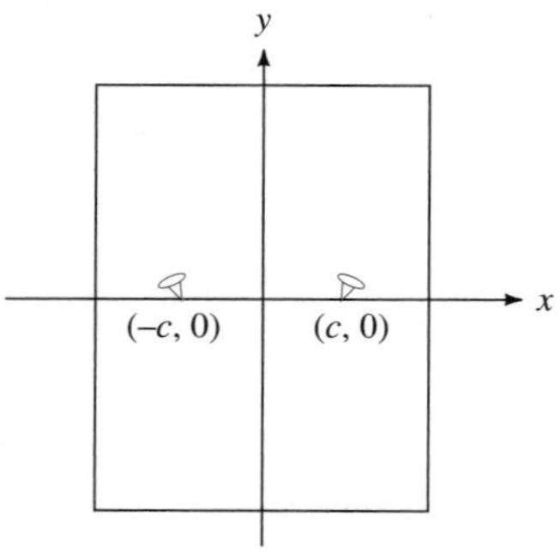

Figure 6–32

Let's find an equation to describe the pencil-and-string curve. To do this, let's place the paper into a coordinate system with the thumbtacks on the x-axis and equidistant from the origin. The tacks, at F_1 and F_2 in Figure 6-32, have coordinates $(c, 0)$ and $(-c, 0)$ for some $c > 0$.

Exercises

10. *(Writing to Learn)* If the length of the string between the knots is $2a$ and $P = (x, y)$ is any point on the curve, explain why $\overline{PF_1} + \overline{PF_2} = 2a$.

11. *(Extension)* Use the distance formula to show that x and y must satisfy the equation $\sqrt{(x + c)^2 + y^2} + \sqrt{(x - c)^2 + y^2} = 2a$.

12. *(Extension)* Show that the equation in Exercise 11 can be simplified to $(a^2 - c^2)x^2 + a^2y^2 = a^2(a^2 - c^2)$.

Figure 6-33 illustrates that $a > c$. Therefore $a^2 - c^2 > 0$, and the equation in Exercise 12 describes an ellipse. In fact, your experiment provides a formal definition of the word *ellipse*. An **ellipse** is the set of all points in the plane, the sum of whose distances from two fixed points is constant. The fixed points are called the **foci** of the ellipse.

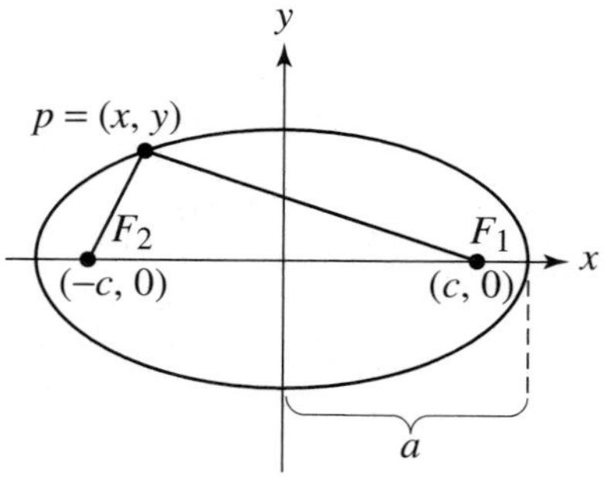

Figure 6–33

If we denote the number $a^2 - c^2$ by b^2, the standard graphing form of the equation in Exercise 12 is

$$\frac{x^2}{a^2} + \frac{y^2}{b^2} = 1$$

and every such equation describes an ellipse according to our new definition. More generally, in Exercise 46 you will establish the following facts.

- Every equation $\frac{(x-h)^2}{a^2} + \frac{(y-k)^2}{b^2} = 1$ or $\frac{(x-h)^2}{b^2} + \frac{(y-k)^2}{a^2} = 1$ describes an ellipse, so that our new definition coincides with the one given in Section 6-2.
- If $c^2 = a^2 - b^2$ (equivalently, $b^2 = a^2 - c^2$), the foci are located c units on each side of the center along the major axis. The possible arrangements of the foci with respect to the vertices and covertices are shown in Figure 6-34.

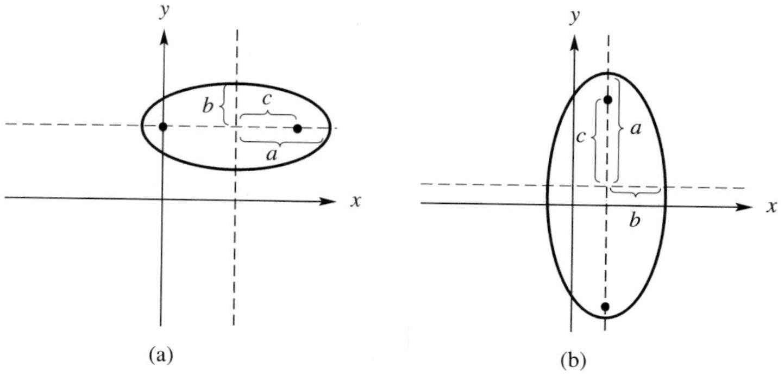

Figure 6–34

Example 2 shows how to find the foci of an ellipse whose equation is known.

EXAMPLE 2 *Finding the foci of an ellipse*

To find the foci of the ellipse $\frac{(x-2)^2}{9} + \frac{(y+5)^2}{16} = 1$, first locate its center at $(2, -5)$. Its major axis is the vertical line $x = 2$ since the y-term has the larger denominator. The value of c can be calculated from the equation $c^2 = 16 - 9$, so $c = \sqrt{7}$. Thus the foci are $\sqrt{7}$ units above and below the center at $(2, -5 + \sqrt{7})$ and $(2, -5 - \sqrt{7})$. From Section 6-2 you also know that the vertices are 4 units above and below the center at $(2, -1)$ and $(2, -9)$ and the covertices are 3 units to the left and right of the center at $(-1, -5)$ and $(5, -5)$. See Figure 6-35.

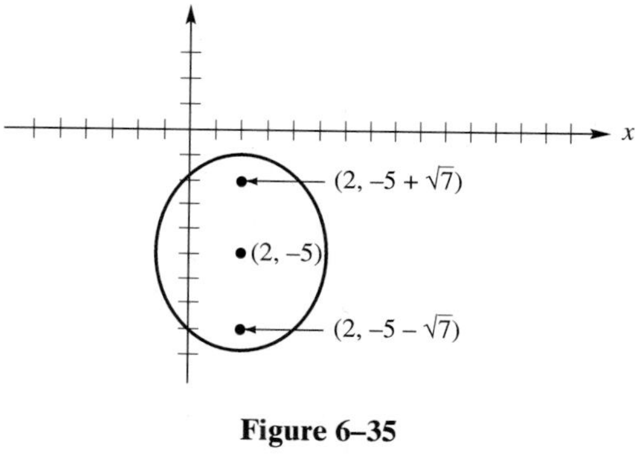

Figure 6–35

■

Exercise 13 suggests that the relative size of a and c determines the shape of an ellipse.

EXERCISE **13.** *(Making Observations)* Repeat the string-and-pencil experiment shown in Figure 6-31, but place the thumbtacks about 2 inches on either side of the origin, and leave about 12 inches of string between the knots.

a. How does the shape of your new ellipse compare with the one you constructed before?

b. What would the curve look like if you placed both thumbtacks at the origin?

Exercise 13 supports the conclusion that a circle is a special case of an ellipse. It also illustrates that an ellipse whose foci are near the vertices is very elongated, while one whose foci are near the center is more nearly circular. The degree of elongation of an ellipse is called its **eccentricity**, denoted by e and is measured by the quantity c/a. See Figure 6-36.

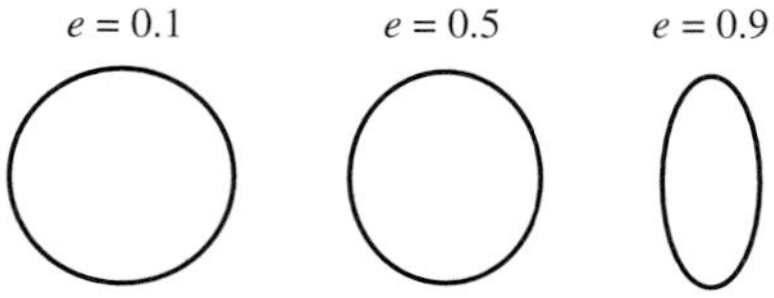

Figure 6–36

EXERCISES **14.** *(Writing to Learn)*

a. Explain why the eccentricity of a circle is 0.

b. Explain why the eccentricity of a very elongated ellipse is close to 1. Is it ever 1?

c. Explain why $0 \leq e < 1$ for any ellipse.

In Exercises 15–18, *find the foci of the ellipse, and calculate its eccentricity. Then sketch the graph, and include the foci.*

15. $16x^2 + 25y^2 = 400$

16. $x^2 + \dfrac{(y-10)^2}{36} = 1$

17. $9x^2 + 4y^2 + 18x + 24y - 99 = 0$

18. $3x^2 + 6y^2 - 24x - 24y = 0$

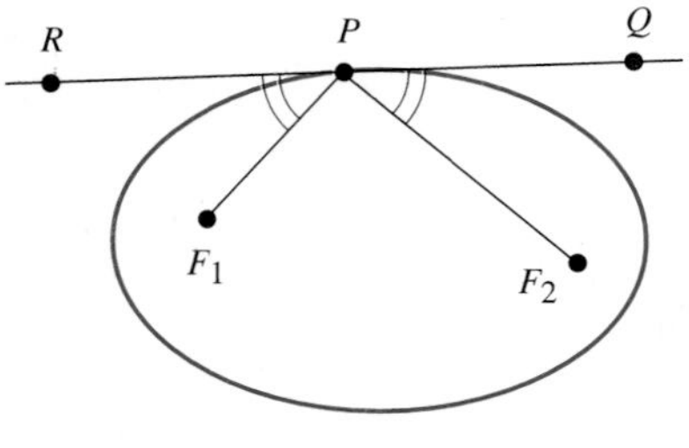

Figure 6–37

Like parabolas, ellipses have a reflective property, illustrated in Figure 6-37. Using calculus it can be shown that the angles F_1PR and F_2PQ are equal. If you imagine the ellipse as a mirror, any light ray traveling along F_1P will strike the mirror and be reflected toward F_2, and vice versa. That is, it will follow the path of the string in your experiment. Furthermore, since $\overline{PF_1} + \overline{PF_2} = 2a$ no matter where P is on the ellipse, all rays leaving F_1 at the same time will arrive at F_2 at the same time because they travel the same distance.

A MATHEMATICAL LOOKING GLASS
Ultrasound

The reflective property of ellipses provides physicians with a noninvasive procedure (that is, one not requiring surgery) for removing kidney stones. The patient is placed within a device with elliptical walls, so that the kidney stone is at one focus. A source of high-frequency sound waves, called ultrasound, is at the other focus, as in Figure 6-38.

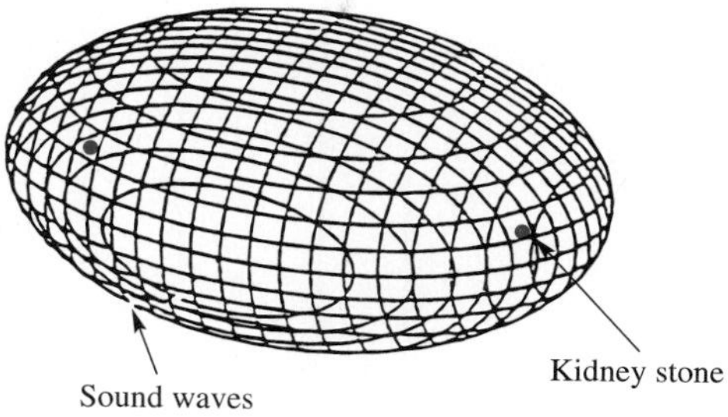

Figure 6–38

The ultrasound waves converge at the kidney stone, causing vibrations that shatter it. The small pieces can then be passed painlessly in the patient's urine. ■

EXERCISE **19.** *(Problem Solving)* If the ellipse is 15 feet long and 5 feet wide, where should the patient and the ultrasound source be placed?

AN ANALYTICAL VIEW OF HYPERBOLAS

We know you are expecting to perform an activity that will generate a hyperbola, but first you will need some information about how radar works in air traffic control systems. A transmitter at an airport sends out electromagnetic signals to an aircraft. These signals travel at the speed of light, about 328 yards every microsecond (1/1,000,000 of a second). When two or more transmitters send out signals simultaneously, the pilot can identify the aircraft's location by comparing arrival times of the signals.

Now try this experiment! You will need a commercial pilot's license and a Boeing 747. Take off from a major airport near your home, and fly in the general direction of Washington, DC. When you are sufficiently close, you will pick up radar signals from both Washington National and Baltimore–Washington International Airports, about 30 miles apart. Adjust your course so that you always receive the signals from Washington National 100 microseconds sooner than those from Baltimore–Washington. This means that you are always (100)(328) = 32,800 yards, or about 18.6 miles, closer to Washington National, as in Figure 6-39a (see page 226). What kind of curve does your airplane appear to trace?

Let's find an equation to describe the airplane's curve. To do this, let's place the Washington area into a coordinate system with the two airports on the x-axis and equidistant from the origin. In Figure 6-39b, they are at $F_1 = (-c, 0)$ and $F_2 = (c, 0)$. Denote the difference in your distances from the two airports as $2a$.

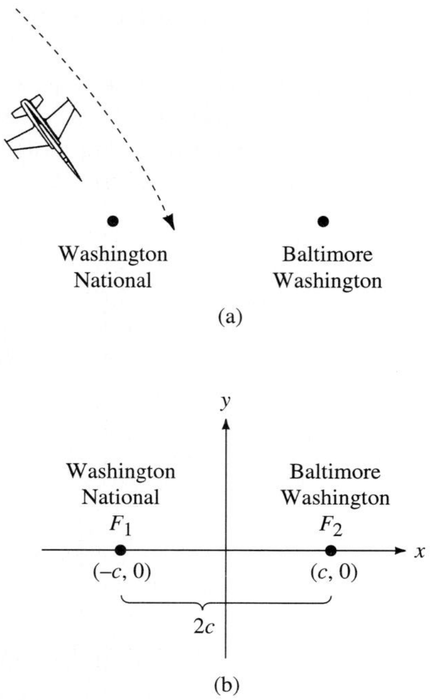

Figure 6–39

EXERCISES

20. *(Writing to Learn)* If $P = (x, y)$ is any point on the curve, explain why $\overline{PF_2} - \overline{PF_1} = 2a$.

21. *(Extension)* Use the distance formula to show that x and y must satisfy the equation $\sqrt{(x - c)^2 + y^2} - \sqrt{(x + c)^2 + y^2} = 2a$.

22. *(Extension)* Show that the equation in Exercise 21 can be simplified to $(c^2 - a^2)x^2 - a^2y^2 = a^2(c^2 - a^2)$.

Figure 6-40 shows that the difference in your distances from the airports ($2a$) must be less than their distance from each other ($2c$). Therefore, $c > a$, so $c^2 - a^2 > 0$ and the curve is a hyperbola. In fact, your experiment provides a formal definition of the word *hyperbola*. A **hyperbola** is the set of all points in the plane, the difference of whose distances from two fixed points is constant. The fixed points are called the **foci** of the hyperbola.

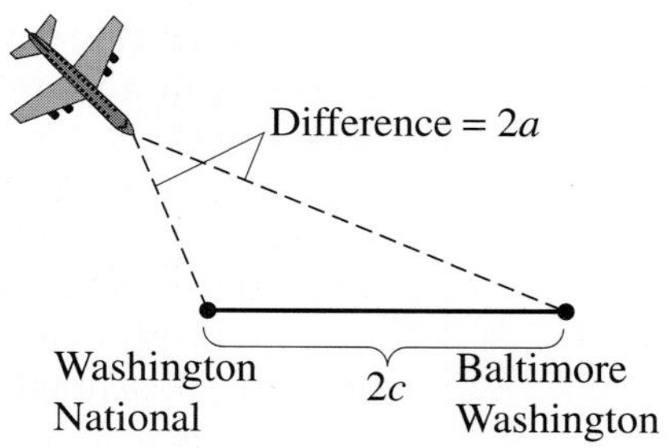

Figure 6–40

As you may have noticed, your airplane traces only the left half of the hyperbola. You would have traced the right half if you had adjusted your course to receive the signals from Washington National 100 microseconds *later* than those from Baltimore–Washington.

If we denote the number $c^2 - a^2$ by b^2, the standard graphing form of the equation in Exercise 22 is

$$\frac{x^2}{a^2} - \frac{y^2}{b^2} = 1$$

and every such equation describes a hyperbola according to our new definition. More generally, in Exercise 47 you will establish the following facts.

- Every equation $\frac{(x-h)^2}{a^2} - \frac{(y-k)^2}{b^2} = 1$ or $\frac{(y-k)^2}{a^2} - \frac{(x-h)^2}{b^2} = 1$ describes a hyperbola, so that our new definition coincides with the one given in Section 6-2.
- If $c^2 = a^2 + b^2$ (equivalently, $b^2 = c^2 - a^2$), the foci are located c units on each side of the center along the transverse axis. The possible arrangements of the foci with respect to the vertices are shown in Figure 6-41.

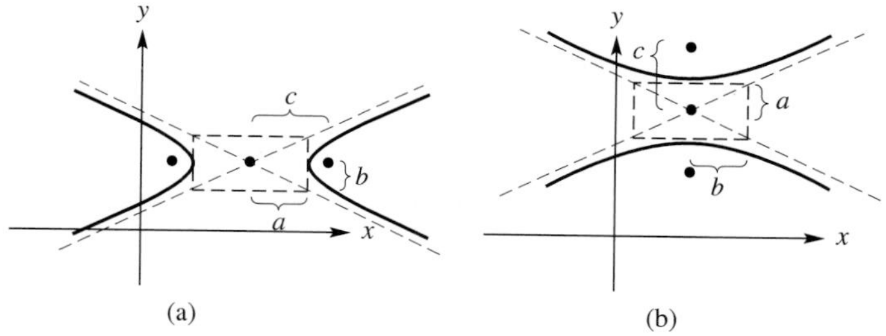

Figure 6–41

Example 3 shows how to find the foci of a hyperbola whose equation is known.

EXAMPLE 3 *Locating the foci of a hyperbola*

To find the foci of the hyperbola $\frac{(x-2)^2}{9} - \frac{(y-5)^2}{16} = 1$, first locate its center at (2, 5). Its transverse axis is the horizontal line $y = 5$, since the x-term has a positive coefficient. The value of c can be calculated from the equation $c^2 = 16 + 9$, so $c = 5$. Thus the foci are 5 units to the left and right of the center, at $(-3, 5)$ and $(7, 5)$. From Section 6-2 you know that the vertices are 3 units to the left and right of the center at $(-1, 5)$ and $(5, 5)$. The asymptotes are the lines $\frac{y-5}{16} = \pm\frac{x-2}{9}$. See Figure 6-42 (page 228).

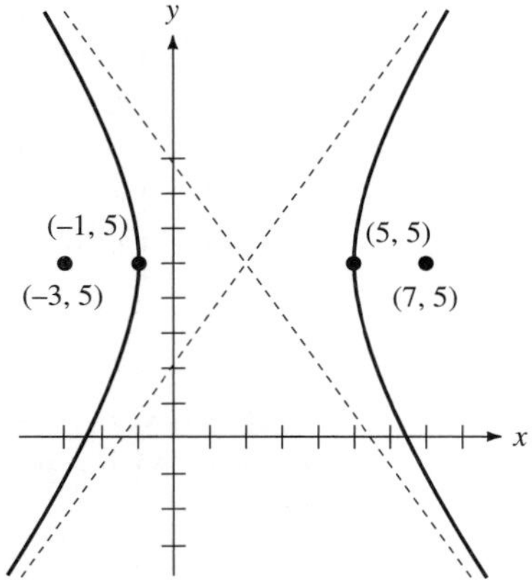

Figure 6–42

■

FYI The equations relating the parameters a, b, and c for ellipses and hyperbolas are similar, but different.

For an ellipse, $c^2 = a^2 - b^2$
For a hyperbola, $c^2 = a^2 + b^2$

EXERCISES *In Exercises 23–26, find the foci of the hyperbola, and include them in a sketch of the graph.*

23. $25x^2 - 16y^2 = 400$

24. $x^2 - \frac{(y-10)^2}{36} = 1$

25. $9x^2 - 4y^2 + 18x - 24y + 171 = 0$

26. $3x^2 - 3y^2 - 24x + 24y + 96 = 0$

Like parabolas and ellipses, hyperbolas have a reflective property, illustrated in Figure 6-43.

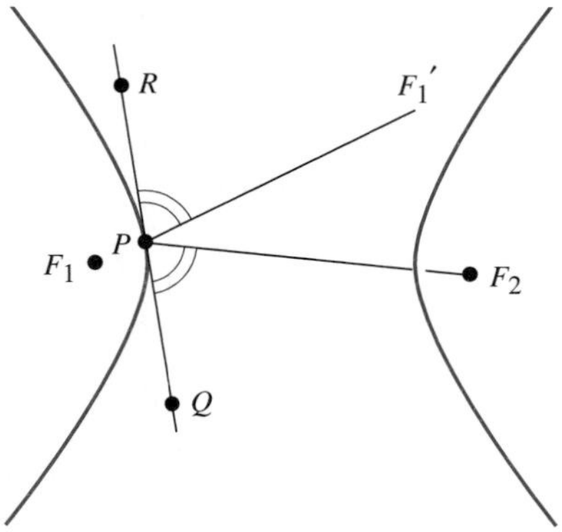

Figure 6–43

Using calculus, it can be shown that the angles $F_1'PR$ and F_2PQ are equal. If you imagine the hyperbola as a mirror, any light ray traveling along $F_1'P$ will strike the mirror and be reflected toward F_2, and vice versa.

A MATHEMATICAL LOOKING GLASS

Earthquakes

When an earthquake occurs, seismologists immediately try to locate its epicenter, or point of origin. Let's see how data from different seismograph stations might have been used to locate the epicenter of the Loma Prieta earthquake of October 17, 1989, also called the San Francisco World Series earthquake because it occurred during a world series game in that city. This earthquake registered 7.1 on the Richter scale and was responsible for \$5.6 billion in damage and 62 deaths. According to Lee Row III of the National Oceanic and Atmospheric Administration in Boulder, CO, "The most deadly structural failure of the earthquake occurred when the upper deck of the Interstate 880 (Nimitz Freeway) in Oakland fell onto the lower roadway, causing an official death toll of 41." The report continues, "It is interesting to compare this earthquake with the slightly smaller (magnitude 6.9) Armenian earthquake of December, 1988, that killed 25,000 people and destroyed entire towns. Good construction and engineering practices in the Loma Prieta area obviously contributed to the preservation of property and human lives."

For our discussion, let's make the following assumptions.

- The epicenter is located near the surface of the earth. (It was actually 19 kilometers below the surface.)
- The primary shock wave travels from the epicenter at a uniform rate of 25 kilometers per second. (In reality, the shock wave's speed varies as it passes through different types of rock.)
- The portion of the earth's surface containing the seismograph positions can be modeled by a portion of a coordinate plane. (A more accurate model requires spherical geometry.)

The primary wave from the World Series earthquake was detected by a seismograph near Sacramento at 22.3 seconds after 3:04 P.M., Pacific Daylight Time. It was detected in Petaluma at 22.8 seconds after 3:04 P.M. Based on the difference between these times, the epicenter must be located somewhere on one branch of a hyperbola with foci at Sacramento and Petaluma. To describe the hyperbola by means of an equation, let's introduce the coordinate system in Figure 6-44. Units are measured in kilometers, Sacramento is at the origin, and Petaluma is 110 kilometers west at $P = (-110, 0)$.

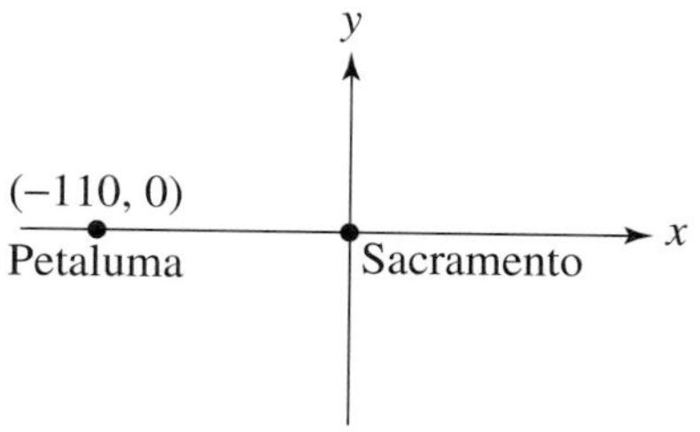

Figure 6–44

■

EXERCISES

27. *(Writing to Learn)* Explain why the value of c for this hyperbola is half the distance between the seismograph positions and the value of a is 12.5 times the difference in the arrival times.

28. *(Problem Solving)* Find an equation for the hyperbola, sketch its graph, and identify the branch on which the epicenter lies.

In Section 6-6, you will learn how data from a third seismograph location can be used to locate the epicenter.

ADDITIONAL EXERCISES *In Exercises 29–38, sketch the graph of each conic section, and identify the following features.*

For each parabola, vertex, focus, directrix
For each ellipse, center, vertices, covertices, foci
For each hyperbola, center, vertices, asymptotes, foci

If an equation leads to one of the exceptional graphs discussed in Section 6-2, these features will not be present, so just sketch the graph.

29. $(x - 5)^2 - 7y^2 = 14$

30. $(x - 5)^2 - 7y = 14$

31. $(x - 5)^2 + 7y^2 = 14$

32. $4(x + 0.05)^2 + (y - 0.1)^2 = 0.02$

33. $9(x - 500)^2 = 4(y - 300)^2$

34. $9(x - 500)^2 = -4(y - 300)^2$

35. $x^2 + 9y^2 - x + 9y - 38 = 0$

36. $3y^2 + 4x - 12y + 168 = 0$

37. $4x^2 + 4y^2 - 556x - 372y = 0$

38. $8x^2 - y^2 + 16x + 4y = 0$

39. *(Problem Solving)* In Exercise 63 of Section 6-2 you analyzed the automobile headlamp unit shown in Figure 6-45.

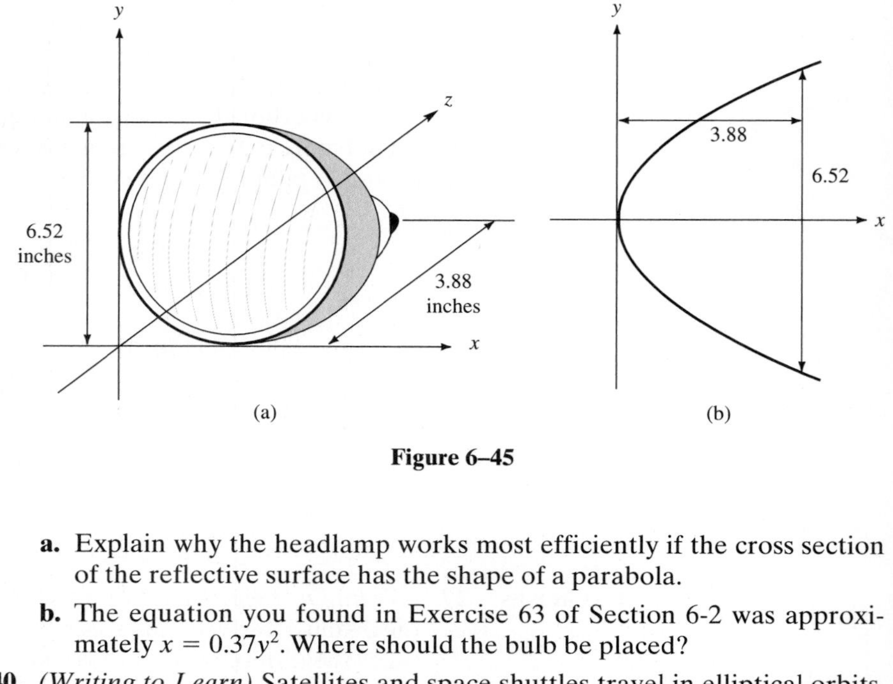

Figure 6–45

a. Explain why the headlamp works most efficiently if the cross section of the reflective surface has the shape of a parabola.

b. The equation you found in Exercise 63 of Section 6-2 was approximately $x = 0.37y^2$. Where should the bulb be placed?

40. *(Writing to Learn)* Satellites and space shuttles travel in elliptical orbits, with the earth at one focus. For such orbits, the distances a and c are inconvenient to measure. It is easier to measure the maximum and minimum distances M and m from the space vehicle to the earth. The orbit's eccentricity can be expressed as

$$e = \frac{M - m}{M + m}$$

Explain why this formula is valid.

41. *(Creating Models)* The formula in Exercise 40 can also be used to calculate the eccentricity of planetary orbits if M and m represent the planet's maximum and minimum distance from the sun. For the planet Neptune, $M = 30.33$ and $m = 29.79$ if distances are measured in astronomical units. (One astronomical unit is the mean distance from the earth to the sun, about 149.6 km.)

a. Calculate the eccentricity of Neptune's orbit.

b. Use the given information to find the values of a and c for Neptune's orbit. (*Hint*: The sun is at one focus.)

c. Find the equation of Neptune's orbit in a coordinate system with the sun at the origin, the orbit's major axis on the x-axis, and the sun at the left-hand focus.

42. *(Creating Models)* Repeat Exercise 41 for the planet Pluto ($M = 49.3$, $m = 29.6$).

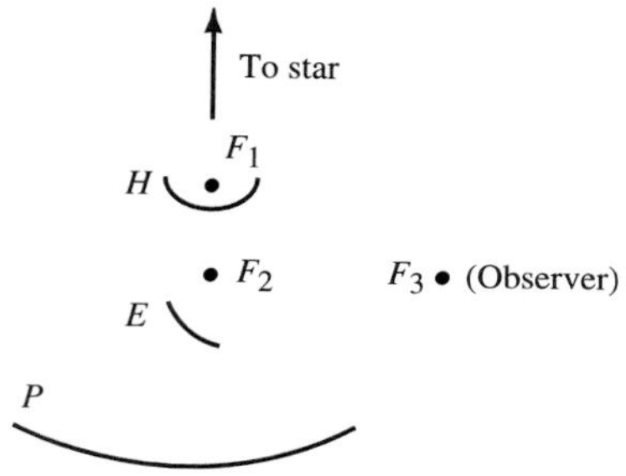

Figure 6–46

43. *(Interpreting Mathematics)* Graph your equations from Exercises 41 and 42 on the same screen. By zooming in on appropriate portions of the graphs, estimate the coordinates of the points where the orbits cross. About what fraction of Pluto's orbit is inside Neptune's orbit?

44. *(Writing to Learn)* Explain how the telescope in Figure 6-46 works. P is an arc of a parabola whose axis of symmetry points directly toward the star, and whose focus is F_1. H is an arc of a hyperbola with foci F_1 and F_2. E is an arc of an ellipse with foci F_2 and F_3.

45. *(Extension)* Show that the parabola in Figure 6-47 has the equation $y = \frac{1}{4p}(x - h)^2 + k$. (*Hint*: Apply the strategy of *Looking for Relationships Among Variables* by using the distance formula, and the fact that $PF = PD$.)

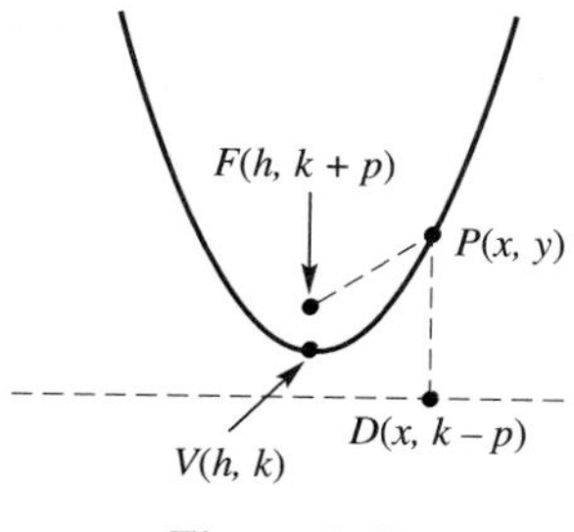

Figure 6–47

46. *(Extension)* Show that the ellipse in Figure 6-48 has the equation $\frac{(x - h)^2}{a^2} + \frac{(y - k)^2}{b^2} = 1$. (*Hint*: Use the distance formula and the fact that $\overline{PF_1} + \overline{PF_2} = 2a$.)

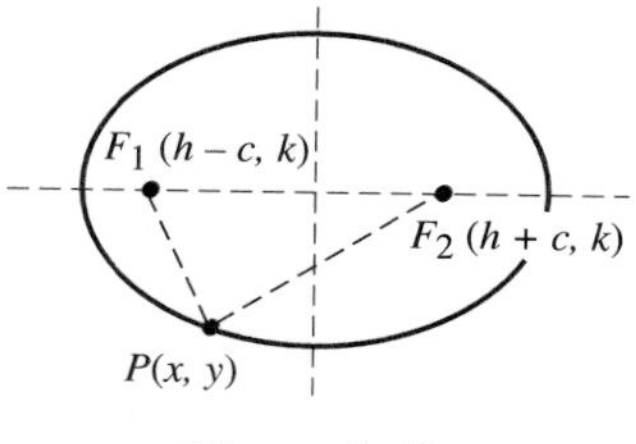

Figure 6–48

47. *(Extension)* Show that the hyperbola in Figure 6-49 has the equation $\dfrac{(x-h)^2}{a^2} - \dfrac{(y-k)^2}{b^2} = 1$. (*Hint*: Apply the strategy of *Examining a Related Problem* by adapting your solution from Exercise 46.)

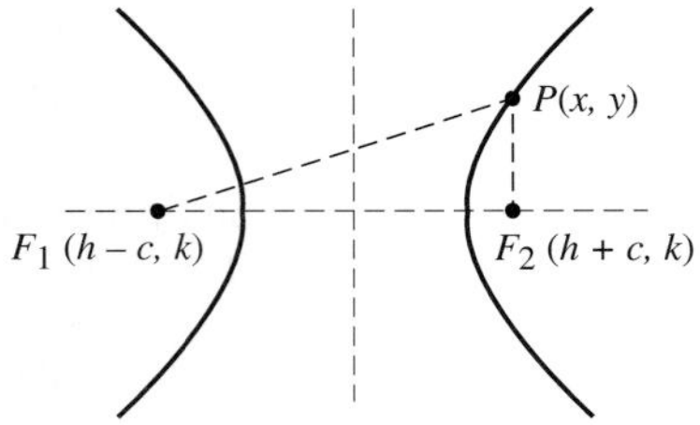

Figure 6–49

6-5 SQUARE ROOT FUNCTIONS

It often happens that a relationship is best modeled not by a quadratic relation, but by one of its implicitly defined functions. In Section 6-1 you saw that those functions usually involve square roots. In this section we will look more closely at square root functions and their graphs. Let's begin by noticing how a relationship between two physical variables produces a square root function.

A MATHEMATICAL LOOKING GLASS
Tsunami

Tsunamis are commonly called seismic sea waves or, incorrectly, tidal waves. The National Oceanic and Atmospheric Administration describes them in their pamphlet *Tsunami Watch and Warning* (1975).

> The phenomenon we call "tsunami" is a series of traveling ocean waves of extremely long length and period, generated by disturbances associated with earthquakes occurring below or near the ocean floor. As the tsunami crosses the deep ocean, its length from crest to crest may be 100 miles or more, its height from trough to crest only a few feet. It cannot be felt aboard ships in deep water, and cannot be seen from the air. But in deep water, tsunami waves may reach forward speeds exceeding 600 miles per hour.
>
> As the tsunami enters the shoaling water of coastlines in its path, the velocity of its waves diminishes and wave height increases. It is in these shallow waters that tsunamis become a threat to life and property, for they can crest to heights of more than 100 feet, and strike with devastating force.

If you live in Hawaii or on the Pacific Coast of the United States or Canada, you may receive an occasional warning that a tsunami is approaching. To tell you how much time you have to evacuate the area, the National Weather Service must predict the tsunami's speed. *Tsunami Watch and Warning* says,

> Tsunami speed is determined solely by water depth, and this fixed relationship makes it possible to forecast tsunami arrival times for distant

locations. The tsunami illustrated here, although somewhat exaggerated in the vertical dimension, is characteristic.

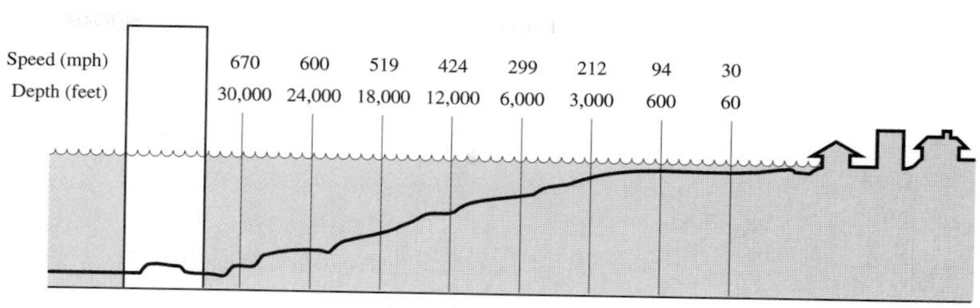

Figure 6–50

Source: *Tsunami Watch and Warning,* National Oceanic and Atmospheric Administration, 1975. ■

EXERCISES

1. *(Making Observations)* For the data in Figure 6-50, show that the ratio $\frac{d}{s^2}$ is constant.
2. *(Writing to Learn)* Explain why the data in Figure 6-50 fits an equation $d = ks^2$. What is the value of k?
3. *(Writing to Learn)* Tell why you could not use the method of Section 5-1 to discover that the data fits a quadratic function.
4. The National Weather Service needs to use ocean depths as inputs and obtain predicted tsunami speeds as outputs. For this purpose, it is more convenient to express s as a function of d.
   - **a.** *(Creating Models)* Use your equation from Exercise 2 to express the speed of the tsunami as a function of water depth.
   - **b.** *(Writing to Learn)* Explain why s is not a function of d if s and d are abstract variables and $d = ks^2$.
5. *(Interpreting Mathematics)* How does your equation reflect the fact that a tsunami slows down as the water becomes shallower?
6. *(Interpreting Mathematics)* Use the data in Figure 6-50 to decide which statement best describes the speed of a tsunami as it goes from deep to shallow water. Explain your conclusions.
   - **a.** It slows down at a constant rate.
   - **b.** It slows down gradually in deep water and more sharply as it enters shallow water.
   - **c.** It slows down rapidly in deep water and more gradually as it enters shallow water.
7. *(Making Observations)* Graph your function from Exercise 4a on your calculator. How does the graph support your conclusion from Exercise 6?

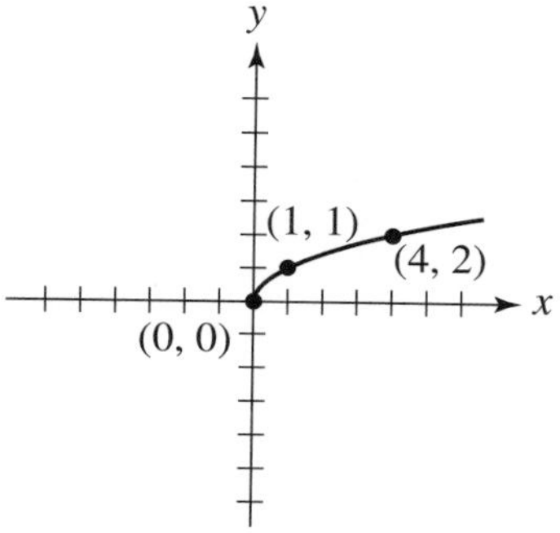

Figure 6–51

Your function in Exercise 4a involves a square root. To see how the equations of such functions relate to their graph, let's proceed by *Examining a Related Problem*. In Sections 3-4 and 5-1 you saw how the graphs of the basic functions $y = |x|$ and $y = x^2$ can be transformed to produce the graph of any function $y = a|x - h| + k$ or $y = a(x - h)^2 + k$. Therefore, let's begin to study square root functions by looking at the graph of the basic function $y = \sqrt{x}$ in Figure 6-51.

EXERCISES

8. *(Writing to Learn)* The graph in Figure 6-51 appears to be half of a parabola. Is it really? How do you know?
9. *(Review)* (See Section 2-3 to review the ideas needed for this exercise.) Both the domain and range of the function $y = \sqrt{x}$ are $[0, \infty)$.
   a. How does the graph of $y = \sqrt{x}$ show this?
   b. How does the equation show this?

You can use your knowledge of the basic graph $y = \sqrt{x}$ and graphing transformations to sketch the graphs of a variety of square root functions.

EXAMPLE 1 *Sketching the graph of a square root function*

To sketch the graph of $y = 3\sqrt{5 - x}$, begin with the basic graph $y = \sqrt{x}$ in Figure 6-52a and replace x by $5 + x$ to obtain

$$y = \sqrt{5 + x}$$

This subtracts 5 from each x-coordinate on the graph, and the graph is shifted 5 units to the left as in Figure 6-52b.

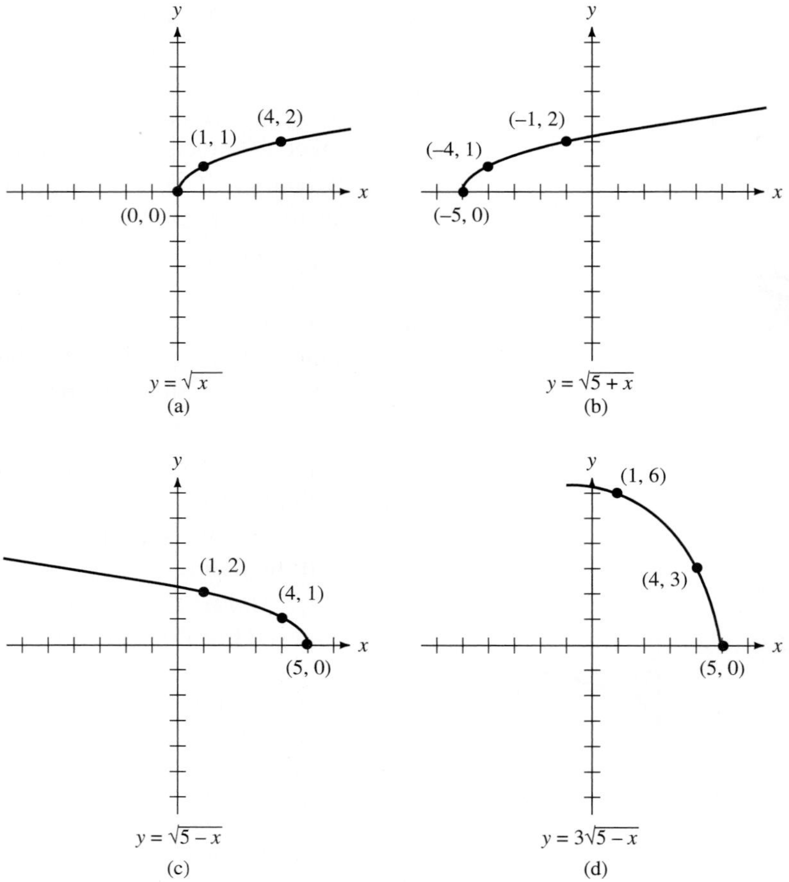

Figure 6–52

Next, replace x by $-x$ to obtain

$$y = \sqrt{5 - x}$$

This multiplies each x-coordinate by -1, and the graph is reflected in the y-axis as in Figure 6-52c. Finally, replace y by $\frac{y}{3}$ to obtain

$$\frac{y}{3} = \sqrt{5 - x}$$

or, equivalently,

$$y = 3\sqrt{5 - x}$$

This multiplies each y-coordinate by 3, and the graph is stretched vertically as in Figure 6-52d. ■

EXERCISES

10. Show how the graph of the function $f(d) = \sqrt{15d}$ in ***Tsunami*** can be obtained by transforming the basic square root graph.

In Exercises 11–14,

a. Sketch the graph of the equation by transforming the basic graph $y = \sqrt{x}$. Identify the points obtained by moving the points $(0, 0)$ and $(1, 1)$ on the basic graph.

b. Find the domain and range of the function.

11. $y = 1 + \sqrt{x + 3}$ **12.** $y = 1 - \sqrt{x + 3}$

13. $y = 3\sqrt{x} - 2$ **14.** $y = \sqrt{3x} - 2$

So far we have encountered only square root functions that are defined implicitly by equations of parabolas. Let's look at some that are defined implicitly by the other conic sections.

A MATHEMATICAL LOOKING GLASS
Kinzua Bridge

Engineers who design bridges must include small spaces between metal components, to allow for expansion and contraction as the surrounding temperature changes. Failure to do so can have surprising consequences. To illustrate, let's take a walk on the Kinzua Railroad Bridge near Marienville, PA. Built in 1882 at a height of 301 feet above Kinzua Creek Valley, it was the world's highest railroad bridge at the time, and is the fourth highest even today.

The horizontal span of the bridge is 2053 feet. If no allowance were made for the rails to expand during hot weather, they would buckle in the middle, as in Figure 6-53.

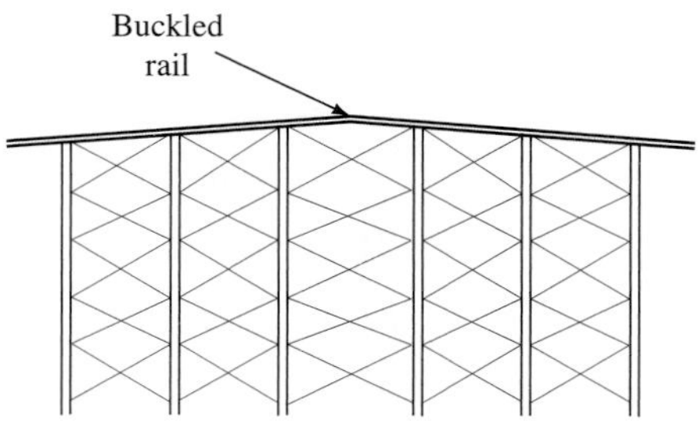

Figure 6–53

■

EXERCISES

15. *(Creating Models)* Let x represent the expansion of each half of the rail, and let y represent the vertical displacement at the center, both in feet. Express y as a function $f(x)$.

16. *(Review)*
    a. What is the domain of f if x and y are abstract variables?
    b. What is the domain of f if x and y are given their physical meanings?

17. *(Making Observations)* Square both sides of the equation $y = f(x)$ to obtain the equation of a conic section. Then sketch the graph of
    a. the conic section
    b. the function $y = f(x)$ with x and y considered as abstract variables
    c. the function $y = f(x)$ with x and y given their physical interpretations

18. *(Interpreting Mathematics)* If each half of the rail were to expand 2 feet, what would be the resulting vertical displacement?

In Exercises 19–22,

a. Eliminate the radical to obtain the equation of a conic section that defines the function implicitly.
b. Sketch the graph of both the conic section and the function.
c. Find the domain and range of the function.

19. $y = \sqrt{16 - 4x^2}$ **20.** $y = \sqrt{4x^2 - 16}$
21. $y = 2\sqrt{25 + x^2}$ **22.** $y = 1 - \sqrt{9 - x^2}$

ADDITIONAL EXERCISES

In Exercises 23–30,

a. Sketch the graph of the equation by transforming the basic graph $y = \sqrt{x}$. Identify the points obtained by moving the points $(0, 0)$ and $(1, 1)$ on the basic graph.
b. Find the domain and range of the function.

23. $y = -2\sqrt{x}$ **24.** $y = \sqrt{-2x}$
25. $y = -2\sqrt{-2x}$ **26.** $y = 4 + \sqrt{x - 1}$
27. $y = 4 + \sqrt{1 - x}$ **28.** $y = 4 - \sqrt{1 - x}$
29. $y = \sqrt{2x + 8} - 1$
[*Hint*: Think of $2x + 8$ as $2(x + 4)$.]
30. $y = 3\sqrt{0.5x - 2}$

In Exercises 31–38,

a. Eliminate the radical to obtain the equation of a conic section that defines the function implicitly.
b. Sketch the graph of both the conic section and the function.
c. Find the domain and range of the function.

31. $y = 3\sqrt{1 - x^2}$ **32.** $y = 3\sqrt{x^2 - 1}$
33. $y = 3\sqrt{x^2 + 1}$ **34.** $y = -3\sqrt{x^2 + 1}$
35. $y = 5 + \sqrt{9x^2 - 36}$ **36.** $y = 5 - \sqrt{9x^2 - 36}$
37. $y = \sqrt{4 - (x - 5)^2} - 6$ **38.** $y = 2 - 4\sqrt{25 - (x + 1)^2}$

Exercises 39 and 40 refer to ***Tsunami***. If a tsunami's point of origin and time of origin are known, calculus can be used to predict its arrival time at any coastal point. In these exercises let's assume that the ocean's depth x miles east of Honolulu is $15x + 135$ feet, as in Figure 6-54.

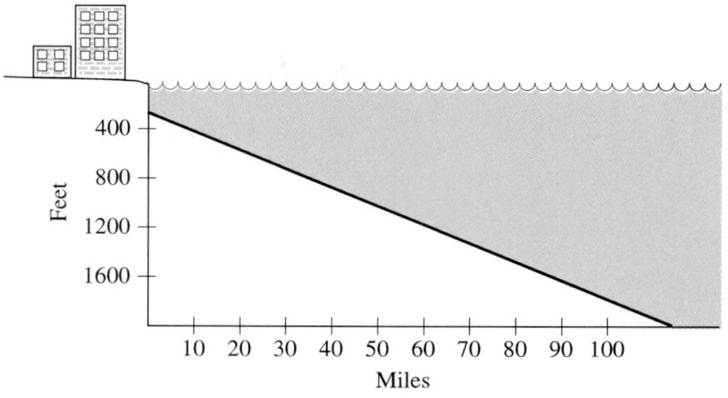

Figure 6–54

If a ***Tsunami*** *is generated x miles east of Honolulu, it can be shown that it will take* $h(x) = \dfrac{\sqrt{x+9}-3}{15}$ *hours to strike the coast.*

39. *(Review)*

a. What are the domain and range of $h(x)$ if x is regarded as an abstract variable?

b. What are the domain and range of $h(x)$ if x is given its physical meaning?

40. *(Interpreting Mathematics)*

a. Calculate the average rate of change in $h(x)$ with respect to x over $[160, 280]$.

b. What physical information does your answer provide?

c. Calculate the average rate of change in $h(x)$ with respect to x over $[0, 16]$.

d. Which of your answers to parts (a) and (c) is larger? What physical information does that observation provide?

Exercises 41–44 *refer to Figure* 6-55, *in which a police car equipped with a radar gun is positioned* 20 *feet from the shoulder of a highway.*

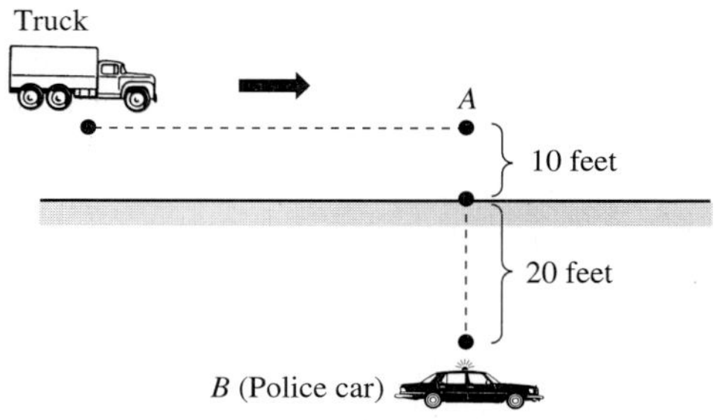

Figure 6–55

41. *(Creating Models)* A truck is traveling 10 feet from the shoulder of the road and is approaching the police car. If x represents the truck's distance in feet from point A in Figure 6-55 and y represents its distance in feet from the police car at point B, express y as a function $f(x)$.

42. *(Creating Models)* The truck is traveling at 90 feet per second and is 450 feet from point A at time $t = 0$ seconds. Express its distance y from the police car as a function $g(t)$.

43. Eliminate the radical from your equation in Exercise 42. Sketch the graph of the resulting conic section and the graph of $y = g(t)$ over the interval $[-10, 10]$.

44. *(Interpreting Mathematics)* When the radar gun points at the truck, it displays a number. This number does not represent the truck's speed along the road, but the rate at which the truck is approaching the radar gun at point B.

a. Is this speed greater than 90 feet per second or less? Explain how you arrived at your answer.

b. As the truck approaches point A, will the displayed number increase or decrease? Explain how you arrived at your answer.

c. When does the number on the radar gun approximate the truck's speed most accurately?

6-6 SYSTEMS OF QUADRATIC EQUATIONS AND INEQUALITIES

In Sections 6-4 and 6-5 you have seen physical situations in which two variables satisfy a quadratic relation. It sometimes happens that two variables must satisfy two quadratic relations simultaneously. The relations then form a **system of quadratic equations** that can be solved either analytically or graphically, using methods similar to those used for linear systems in Chapter 4.

A MATHEMATICAL LOOKING GLASS
Earthquakes 2

In ***Earthquakes*** (page 229) you used data from two seismographs to place the epicenter of the San Francisco World Series earthquake on one branch of a hyperbola with foci at Sacramento and Petaluma, as in Figure 6-56a.

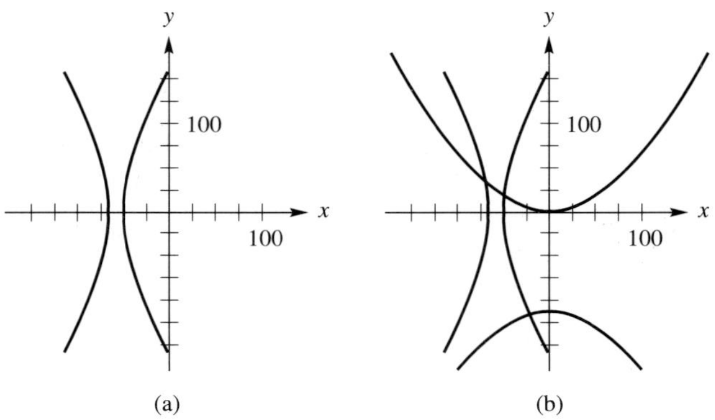

Figure 6–56

Now let's see how a third seismograph might have been used to locate the epicenter exactly. Recall our assumptions that the primary shock wave traveled at 25 kilometers per second and was detected in Sacramento at 22.3 seconds after 3:04 P.M. Pacific Daylight Time. Suppose that a seismograph in Modesto detected it at 18.3 seconds after 3:04 P.M. Once again, the difference in arrival times can be used to place the epicenter on one branch of a hyperbola. We have sketched its graph for you in Figure 6-56b. ■

EXERCISE **1.** *(Problem Solving)* Modesto is located at $(0, -110)$ in the coordinate system of Figure 6-56. Find an equation for the hyperbola, and identify the branch on which the epicenter lies. (Refer to Exercises 27–28 of Section 6-4 for help.)

Since the earthquake's epicenter lies on both hyperbolas in Figure 6-56, it is located at a point where they intersect. At such a point the equations of both hyperbolas are true. Thus to find the points of intersection, you must solve a system of equations. This system differs from those in Chapter 4 in that its equations are quadratic rather than linear. Before solving it, you will need to learn some commonly used techniques for solving quadratic systems.

It is usually impractical to solve systems of quadratic equations in two variables numerically, so we will consider analytical and graphical methods in turn.

SOLVING SYSTEMS OF QUADRATIC EQUATIONS ANALYTICALLY

No simple analytical method will solve every system of quadratic equations. Let's look at some methods of solving the types of systems that occur most often.

Systems with One Quadratic and One Linear Equation If one of the equations in the system is linear, rather than quadratic, you can solve the system by substitution. Since you have used substitution to solve linear systems, we are applying the strategy of *Examining a Related Problem.*

EXAMPLE 1 *A system with one quadratic and one linear equation*

To solve the system

$$x^2 + y^2 = 25$$
$$x - 2y = -5$$

first solve the linear equation for either x or y. In this case it is easier to solve for x.

$$x = 2y - 5$$

Now replace x by $2y - 5$ in the quadratic equation.

$$(2y - 5)^2 + y^2 = 25$$
$$4y^2 - 20y + 25 + y^2 = 25$$
$$5y^2 - 20y = 0$$
$$5y(y - 4) = 0$$
$$y = 0 \quad \text{or} \quad y = 4$$

You can now use the equation $x = 2y - 5$ to find x.

$$\text{If } y = 0, \text{ then } x = 2(0) - 5 = -5$$

$$\text{If } y = 4, \text{ then } x = 2(4) - 5 = 3$$

The solutions are $(-5, 0)$ and $(3, 4)$. ■

EXERCISES *Solve the systems in Exercises 2–5 analytically.*

2. $2x^2 - 3y^2 = -3$
$x - y = 1$

3. $x + y^2 = 1$
$x + 2y = 2$

4. $x^2 + 3xy = 13$
$y = x + 5$

5. $4x^2 + y^2 = 100$
$3x - 2y = 0$

Systems Involving Only x^2 and y^2 If the variable terms in a quadratic system contain only x^2 or y^2, you can solve the system by treating it as linear in the variables x^2 and y^2.

EXAMPLE 2 *A quadratic system involving only x^2 and y^2*

To solve the system

$$3x^2 - y^2 = 6$$

$$2x^2 + 3y^2 = 37$$

you can proceed by elimination, as in Section 4-1, regarding x^2 and y^2 as the variables. Multiply both sides of the first equation by 3.

$$9x^2 - 3y^2 = 18$$

$$2x^2 + 3y^2 = 37$$

Then add the equations.

$$11x^2 = 55$$

$$x^2 = 5$$

$$x = \pm\sqrt{5}$$

Replace x^2 by 5 in the first equation in the system.

$$3(5) - y^2 = 6$$

$$y^2 = 9$$

$$y = \pm 3$$

The solutions are $(\sqrt{5}, 3)$, $(-\sqrt{5}, 3)$, $(\sqrt{5}, -3)$, and $(-\sqrt{5}, -3)$. ■

EXERCISES *Solve the systems in Exercises 6–9 analytically.*

6. $7x^2 + 2y^2 = 13$
$2x^2 - 3y^2 = -16$

7. $x^2 = 2y^2 - 14$
$y^2 = 2x^2 - 5$

8. $5x^2 + 4y^2 = 2$
$6x^2 + 7y^2 = 9$

9. $x^2 + 2y^2 = 11$
$2x^2 - y^2 = 7$

Systems of the Form $Ax^2 + Cy^2 = G$, $xy = K$ An equation $xy = K$ in a system can be solved for either variable. Substitution into the other equation then produces a single equation in one variable. If the other equation has the form $Ax^2 + Cy^2 = G$, the resulting equation can be solved.

EXAMPLE 3 *A quadratic system of the form $Ax^2 + Cy^2 = G, xy = K$*

To solve the system

$$4x^2 + y^2 = 26$$
$$xy = 6$$

you can begin by solving the second equation for either variable. Solving for y,

$$y = \frac{6}{x}$$

Substitution into the first equation yields

$$4x^2 + \left(\frac{6}{x}\right)^2 = 26$$
$$4x^2 + \frac{36}{x^2} = 26$$
$$4x^4 + 36 = 26x^2$$
$$4x^4 - 26x^2 + 36 = 0$$
$$2(x^2 - 2)(2x^2 - 9) = 0$$
$$x^2 = 2 \quad \text{or} \quad x^2 = \frac{9}{2}$$
$$x = \pm\sqrt{2} \quad \text{or} \quad x = \pm\sqrt{\frac{9}{2}} = \pm\frac{3\sqrt{2}}{2}$$

Since $y = \dfrac{6}{x}$, the corresponding values of y are

$$y = \pm\frac{6}{\sqrt{2}} = \pm 3\sqrt{2} \quad \text{or} \quad y = \pm\frac{12}{3\sqrt{2}} = \pm 2\sqrt{2}$$

The solutions to the system are

$$(\sqrt{2}, 3\sqrt{2}), (-\sqrt{2}, -3\sqrt{2}), \left(\frac{3\sqrt{2}}{2}, 2\sqrt{2}\right), \text{ and } \left(-\frac{3\sqrt{2}}{2}, -2\sqrt{2}\right)$$ ■

EXERCISES *Solve the systems in Exercises* 10–13 *analytically.*

10. $x^2 + y^2 = 25$
$xy = -12$

11. $x^2 + y^2 = 25$
$xy = 156$

12. $2x^2 + 3y^2 = 10$
$xy = -2$

13. $x^2 + 4y^2 = 40$
$xy = -6$

SOLVING SYSTEMS OF QUADRATIC EQUATIONS GRAPHICALLY

The analytical methods you have just learned apply only to a few special classes of systems of quadratic equations in two variables. By contrast, you can solve any such system graphically. Furthermore, although a graph does not provide exact solutions, it can indicate the number of solutions and their approximate values at a glance.

Number of Solutions Graphical methods depend on the idea that solutions to a system are represented by points where the graphs of all the equations in the system intersect. You encountered this idea before in connection with linear systems. However, in contrast to linear systems, quadratic systems can have more than one solution without having infinitely many.

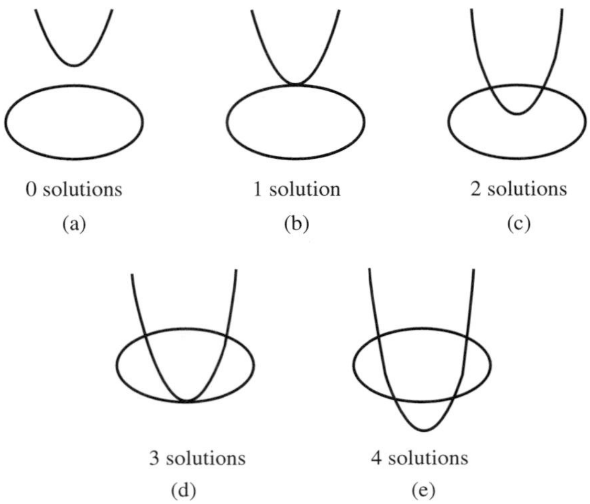

Figure 6–57

Figure 6-57 suggests that two distinct conic sections cannot intersect at more than four points, so that a system of quadratic equations cannot have any more than four solutions without having infinitely many. You can provide additional evidence in Exercises 14–15.

EXERCISES

14. *(Making Observations)* Draw a five-part figure, similar to Figure 6-57, to illustrate that two ellipses can intersect at 0, 1, 2, 3, or 4 points.

15. *(Making Observations)* Repeat Exercise 14 for

a. an ellipse and a hyperbola

b. a parabola and a hyperbola

Finding Points of Intersection You can solve a system of quadratic equations graphically by the method described for linear systems in Section 4-1, but you need to make sure you have found all points of intersection on the graphs or, equivalently, all solutions to the system of equations.

EXAMPLE 4 *Solving a quadratic system graphically*

Figure 6-58 shows the graph of the system

$$x^2 - y^2 = 4$$
$$x = y^2 - 2$$

(To graph the system, solve each equation for y to obtain $y = \pm\sqrt{x^2 - 4}$ and $y = \pm\sqrt{x + 2}$.) It appears to show three solutions, labeled A, B, and C.

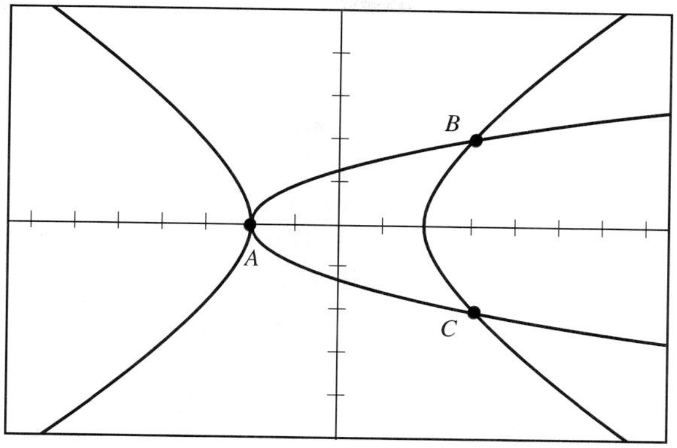

Figure 6–58

You can't be sure from the figure whether the two graphs cross, touch, or do not quite touch in the vicinity of A. However, you might guess that $(-2, 0)$ is a solution, and substitution into each equation confirms the guess. Furthermore, by looking at the equations in the system you can verify that $(-2, 0)$ is the vertex of the parabola, and also a vertex of the hyperbola. Therefore, the graphs simply touch at that point and do not cross, so there is no other solution close to $(-2, 0)$.

The solutions at B and C, estimated by zooming in, are $(3.00, \pm 2.24)$ with an error of no more than 0.01. [The exact solutions, obtained analytically, are $(3, \pm\sqrt{5})$.] ■

Since a quadratic system can have more than one solution, you may need to decide which solution is of interest when the system models a physical problem. The situation often provides clues, as in ***Earthquakes 2***. In Exercise 16, you will use those clues to locate the epicenter of the San Francisco World Series earthquake.

EXERCISES

16. *(Problem Solving)* In Exercises 27–28 of Section 6-4, and Exercise 1 of this section, you placed the earthquake's epicenter on a particular branch of each of two hyperbolas. Find all points of intersection of the hyperbolas, and tell which point is the epicenter.

In Exercises 17–20, *solve the system in the indicated exercise graphically.*

17. Exercise 4

18. Exercise 6

19. Exercise 8

20. Exercise 12

SOLVING SYSTEMS OF QUADRATIC INEQUALITIES

The following Mathematical Looking Glass leads to a **system of quadratic inequalities**, that is, a collection of quadratic inequalities in two variables to be solved simultaneously. Over the next few pages you will discover that the solution of a quadratic inequality, like that of a linear inequality, is represented most conveniently as a region of the plane. The solution of a system of quadratic inequalities is then the region common to the solutions of all inequalities in the system.

A MATHEMATICAL LOOKING GLASS
Bermuda Triangle

The Bermuda Triangle, a region of ocean bounded by Bermuda, Miami, and Puerto Rico, is a notorious area where ships and airplanes disappear mysteriously. In fact, the number of ships lost in the Bermuda Triangle is due largely to its unusually large volume of boating traffic. As Charles J. Cazeau and Stuart D. Scott point out in *Exploring the Unknown* (Da Capo Press, 1980),

> During 1975 the U.S. Coast Guard came to the rescue of 140,000 people. The Coast Guard aided 5600 boats that sank or capsized, 1300 with fires or explosions, and 1000 boats that did not know where they were. . . . We hasten to point out that these statistics apply to *all* of the United States coastline and not just the Bermuda Triangle. However, of the 37,000 U.S. Coast Guardsmen, the busiest are those rescuing people in the Bermuda Triangle, one of the great resort areas of the world. . . . We think that it is a remarkable tribute to the Coast Guard that they let only 2.5 to 6(?) ships and/or planes slip away from them each year.

To assist in their rescue work, the Coast Guard maintains helicopters at several stations in the Caribbean. Each can operate within a 100-mile radius of its base. That is, it can fly to a specific location up to 100 miles away, spend up to one half-hour there, and fly back without refueling. Let's follow a hypothetical helicopter rescue mission.

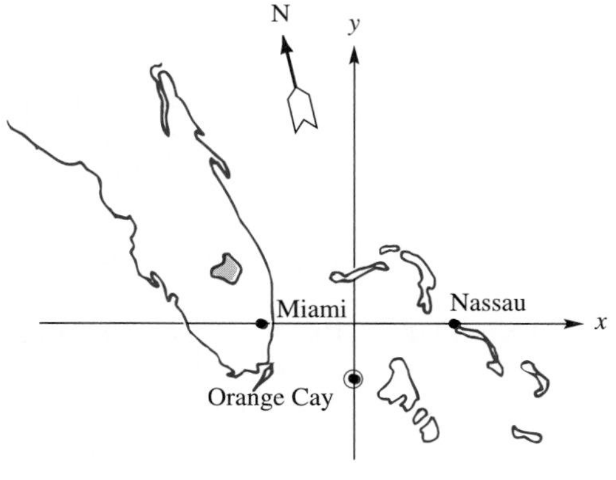

Figure 6–59

Figure 6-59 shows a portion of the Bermuda Triangle between Miami and Nassau. It is placed in a coordinate system with units measured in miles, so that Miami is at $(-80, 0)$, and Nassau is at $(100, 0)$.

A ship is lost in a storm shortly after leaving Orange Cay, at $(0, -40)$. Although it may have been blown in any direction, the Coast Guard is reasonably certain that the ship is still within 20 miles of Orange Cay. ■

EXERCISES

21. *(Creating Models)*

a. Copy Figure 6-59 onto a sheet of graph paper, and shade in the region where the ship might be.

b. Write the equation of the boundary of the region.

22. a. Use a different style of shading to indicate the region that could be covered by a helicopter based in Miami.

b. Write the equation of the boundary of the region.

23. *(Interpreting Mathematics)* Can the helicopter in Exercise 22 search the entire area where the ship might be located? How can you tell?

Your shaded areas in Exercises 21 and 22 are solutions of quadratic inequalities in two variables. Such inequalities often model physical problems. Like linear inequalities, they are solved most conveniently by graphical methods. After a general discussion of systems of quadratic inequalities, we will return to ***Bermuda Triangle*** to see how the Coast Guard might improve their chances of finding the lost ship. Figure 6-60 indicates how the graph of a quadratic relation divides the coordinate plane into two regions. (For our present purpose, consider the two regions "outside" the hyperbola in Figure 6-60c as one region.)

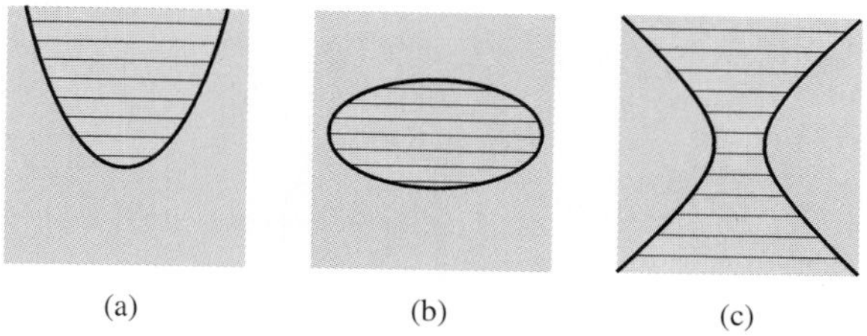

Figure 6–60

Any quadratic inequality in two variables is true at all points in one region, and false at all points in the other. The solution of the inequality can be found by evaluating the inequality at a test value.

EXAMPLE 5 *Solving a quadratic inequality in two variables*

To solve the inequality

$$4x^2 - 9y^2 \geqslant 36$$

first graph $4x^2 - 9y^2 = 36$, as in Figure 6-61a (see page 246).

You can use any point not on the hyperbola as a test value. We have chosen $(0, 0)$. At that point, the inequality asserts

$$4(0)^2 - 9(0)^2 \geqslant 36$$

which is false. The solution of the inequality is therefore the region not containing $(0, 0)$, indicated in Figure 6-61b. The hyperbola itself is included in the solution, since the inequality is true there.

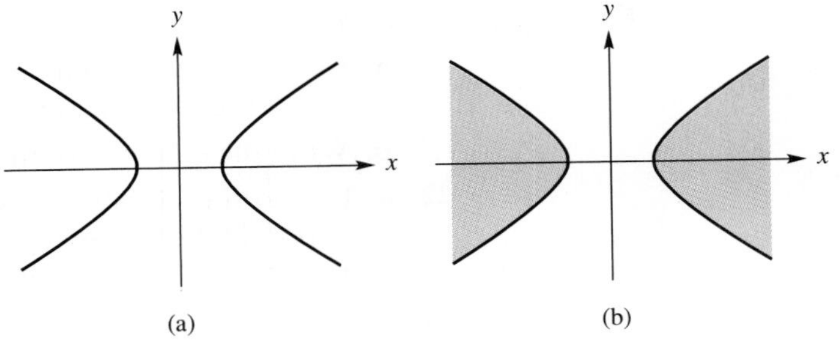

Figure 6–61

The solution of a system of quadratic inequalities is the region common to the solutions of the inequalities in the system. For example, the solution of the system

$$4x^2 - 9y^2 \geqslant 36$$
$$4x^2 + 25y^2 < 100$$

is shown in Figure 6-62.

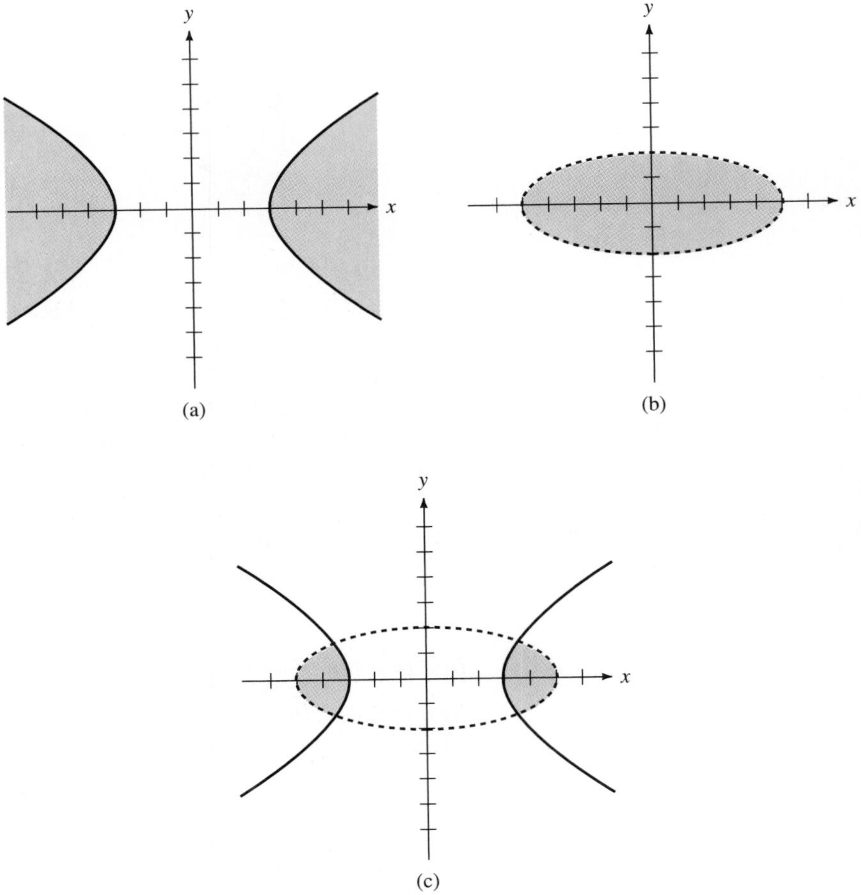

Figure 6–62

EXERCISES *Solve the inequalities in Exercises 24–27.*

24. $x^2 + y^2 > 64$

25. $3x - y^2 \leq 6$

26. $y^2 - 16x^2 < 96$

27. $y^2 - 16x^2 \geq 96$

Solve the systems of inequalities in Exercises 28–31.

28. $x^2 + y^2 > 64$
$3x - y^2 \leq 6$

29. $3x - y^2 \leq 6$
$y^2 - 16x^2 < 96$

30. $x^2 + y^2 > 64$
$3x - y^2 \leq 6$
$y^2 - 16x^2 < 96$

31. $x^2 + y^2 > 64$
$x^2 + y^2 < 100$
$y - x^2 \geq 8$

32. *(Creating Models)* Write inequalities to describe the regions you shaded in Exercises 21 and 22.

33. *(Problem Solving)* Refer to ***Bermuda Triangle***. Write and solve graphically a system of inequalities to represent the overlap of the lost ship's possible locations with the area that can be covered by a helicopter based in Nassau.

34. *(Problem Solving)*

a. Suppose that a helicopter takes off from Nassau, conducts a search, and lands in Miami, flying a total of 200 miles. Explain why the region it can cover lies inside an ellipse with foci at Nassau and Miami.

b. Write an inequality to represent the area that can be covered by the helicopter in part (a).

c. Write and solve graphically a system of inequalities to represent the overlap of the lost ship's possible locations with the area that can be covered by this helicopter.

35. *(Interpreting Mathematics)* Suppose that you are in charge of helicopter operations out of Nassau, and you have received a call from Miami, saying that a helicopter is already searching for the lost ship, as in Exercise 22. You are to assist them by sending a second helicopter, which will either return to Nassau, as in Exercise 33, or fly on to Miami, as in Exercise 34. Which option do you choose? Why?

ADDITIONAL EXERCISES *Solve the systems of equations in Exercises 36–43 analytically.*

36. $y = x^2 - 1$
$y = x + 1$

37. $x^2 + 5y^2 = 24$
$xy = -4$

38. $3x^2 + 8y^2 = 139$
$2x^2 - 3y^2 = 1$

39. $x^2 + y^2 = 4$
$9x^2 + 16y^2 = 144$

40. $x^2 - y^2 = 9$
$5x - 4y = 9$

41. $x^2 + y^2 = 36$
$y = x^2 - 6$

42. $3(x - 1)^2 + 2(y + 2)^2 = 216$
$x + 4y = 7$

43. $x^2 + 4y^2 - 8x + 16y + 16 = 0$
$xy = 0$

44. Use graphical methods to find the real solutions to the system in the indicated exercise.

a. Exercise 36 **b.** Exercise 37
c. Exercise 38 **d.** Exercise 39
e. Exercise 40 **f.** Exercise 41

Solve each system of inequalities in Exercises 45–48.

45. $3x^2 + 5y^2 \geq 15$
$3x^2 + 5y^2 < 90$

46. $x^2 - y^2 < 4$
$y^2 - x^2 < 4$

47. $25x^2 + 4y^2 \leq 100$
$y^2 - 9x^2 \geq 9$
$xy < 4$

48. $y > x^2 - 16$
$y > 4x - x^2$
$y < 2x + 1$

49. *(Problem Solving)* In an appropriate coordinate system, the orbit of the earth around the sun has the equation

$$\frac{x^2}{1.03} + \frac{y^2}{0.97} = 1$$

Distances are measured in astronomical units (an astronomical unit is the mean distance from the earth to the sun, about 92.9 million miles). In the same coordinate system, a newly discovered comet might have an orbit whose equation is

$$xy + x + y = 0$$

a. Comets with elliptic orbits return periodically to the same position, while comets with parabolic or hyperbolic orbits visit the solar system only once. Will this comet ever be seen again? (*Hint*: Solve the comet's equation for y, and graph it.)

b. How many times will the comet cross the earth's orbit? At what points? (*Hint*: Solve the earth's equation for y, and answer the question by graphical methods.)

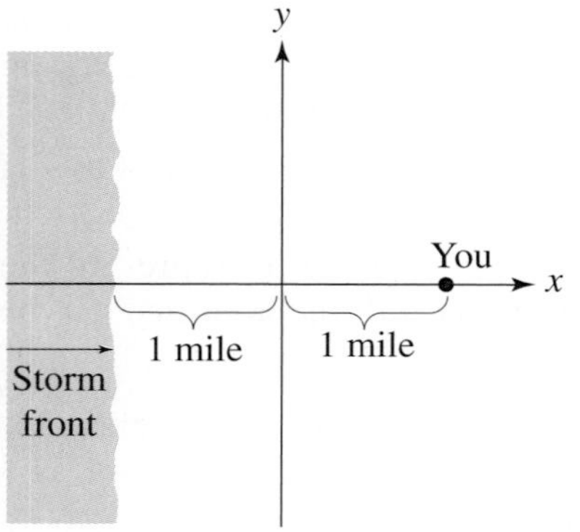

Figure 6–63

50. *(Problem Solving)* Imagine that as you are taking your morning jog, you hear over your portable radio that a storm front is approaching at 8 mph. Looking up and seeing the front about 2 miles ahead, you run for shelter at 8 mph.

a. Explain why the points you can reach without getting wet lie on one side of a parabola with its focus at your present position and its directrix along the storm front.

b. In the coordinate system of Figure 6-63, write and solve graphically an inequality to show all such points.

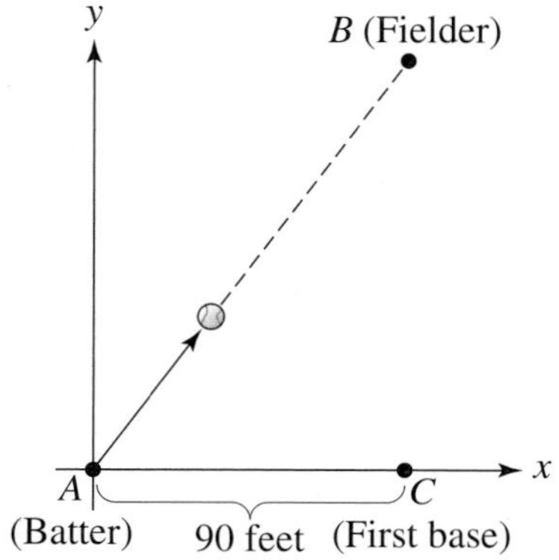

Figure 6–64

51. *(Problem Solving)* Figure 6-64 depicts a critical moment in a baseball game. The batter at A has just hit a ground ball to the fielder at B. The fielder must throw the ball to first base (C) before the batter gets there.

Assume that the batted ball and thrown ball both travel in a straight line at 120 feet per second, that the batter can run to first base in 4 seconds, and that the fielder must take 1 second after catching the ball before making the throw.

a. Explain why the fielder's throw will reach first base before the batter arrives if the fielder is positioned inside an ellipse with foci at A and C.

b. Write and solve graphically a system of inequalities to show all points in fair territory (above the x-axis and to the right of the y-axis) where the fielder can play.

CHAPTER REVIEW

Complete Exercises 1–48 (Writing to Learn) before referring to the indicated pages.

WORDS AND PHRASES

In Exercises 1–19, explain the meaning of the words or phrases in your own words.

1. (page 178) **unit circle**
2. (page 179) **relation, quadratic relation**
3. (page 184) **implicitly defined function**
4. (pages 187, 188) **conic section**
5. (page 190) **standard graphing form** of an ellipse
6. (page 191) **center, major axis, minor axis, vertices, covertices** of an ellipse
7. (page 195) **standard graphing form** of a hyperbola
8. (page 196) **asymptote**
9. (page 197) **center, transverse axis, vertices, conjugate axis** of a hyperbola
10. (page 207) **symmetric about the y-axis, symmetric about the x-axis**
11. (page 218) **parabola**
12. (page 218) **directrix, focus** of a parabola
13. (page 222) **ellipse**
14. (page 222) **foci** of an ellipse
15. (page 224) **eccentricity** of an ellipse
16. (page 226) **hyperbola**
17. (page 226) **foci** of a hyperbola
18. (page 238) **system of quadratic equations**
19. (page 243) **system of quadratic inequalities**

IDEAS

20. (page 179) In what respects are the concepts of function and relation alike? In what respects are they different? Give an example of a correspondence that is both a function and a relation, and an example of a correspondence that is a relation but not a function.
21. (pages 179–183) A relation may be represented by a table, by a graph, or by an equation. In each case, explain how to decide whether the relation is a function.
22. (pages 184–185) Why might we want to obtain equations for functions defined implicitly by a relation?
23. (page 184) Describe a method for verifying that a given function is defined implicitly by a given relation.
24. (pages 188–201) In an equation $Ax^2 + Cy^2 + Dx + Ey + F = 0$, what do the values of A and C tell you about the graph?
25. (pages 189–190) In an equation $y = a(x - h)^2 + k$ or $x = a(y - k)^2 + h$, what do the values of a, h, and k tell you about the graph?
26. (pages 192–193) In an equation $\left(\frac{x-h}{a}\right)^2 + \left(\frac{y-k}{b}\right)^2 = 1$ or $\left(\frac{x-h}{b}\right)^2 + \left(\frac{y-k}{a}\right)^2 = 1$, what do the values of $a, b, h,$ and k tell you about the graph?
27. (page 194) What condition on an equation $Ax^2 + Cy^2 + Dx + Ey + F = 0$ distinguishes a circle from other ellipses?
28. (pages 194–195) In an equation $(x - h)^2 + (y - k)^2 = r^2$, what do the values of h, k, and r tell you about the graph?
29. (pages 195–198) In an equation $\left(\frac{x-h}{a}\right)^2 - \left(\frac{y-k}{b}\right)^2 = 1$ or $\left(\frac{y-k}{a}\right)^2 - \left(\frac{x-h}{b}\right)^2 = 1$, what do the values of $a, b, h,$ and k tell you about the graph?
30. (pages 196–199) How does the equation of a hyperbola relate to the equations of its asymptotes?
31. (page 200) The graph of a quadratic relation is not always a parabola, ellipse, or hyperbola. Describe all the exceptional graphs. How can you identify each exceptional graph from its equation?
32. (pages 203–207) How does the graph of a relation change if x is replaced by $-x$, by $\frac{x}{a}$, or by $x - h$ in its equation? How does the graph change if y is replaced by $-y$, by $\frac{y}{b}$, or by $y - k$?
33. (pages 208–210) When you are applying a sequence of graphical transformations, explain why order is important. Specifically, compare the effects of apply-

ing a reflection in the x-axis before or after a vertical shift.

34. (pages 217–220) How are the points on a parabola determined by its focus and directrix? In an equation $y = a(x - h)^2 + k$ or $x = a(y - k)^2 + h$, what do the values of a, h, and k tell you about the location of the focus and directrix?

35. (pages 220–221) Describe the reflective property of parabolas, and give an example of its use in a physical context.

36. (pages 221–223) In an equation $\left(\frac{x-h}{a}\right)^2 + \left(\frac{y-k}{b}\right)^2 = 1$ or $\left(\frac{x-h}{b}\right)^2 + \left(\frac{y-k}{a}\right)^2 = 1$, what do the values of a, b, h, and k tell you about the location of the foci?

37. (page 224) What are the possible values for the eccentricity of an ellipse? What does the eccentricity tell you about the shape of the ellipse?

38. (pages 224–225) Describe the reflective property of ellipses, and give an example of its use in a physical context.

39. (pages 225–228) In an equation $\left(\frac{x-h}{a}\right)^2 - \left(\frac{y-k}{b}\right)^2 = 1$ or $\left(\frac{y-k}{a}\right)^2 - \left(\frac{x-h}{b}\right)^2 = 1$, what do the values of a, b, h, and k tell you about the location of the foci?

40. (pages 228–229) Describe the reflective property of hyperbolas, and give an example of its use in a physical context.

41. (pages 232–238) Even when a quadratic relation does not represent a function in the abstract, it may represent a function if the variables represent physical quantities. Explain how this can occur.

42. (pages 234–240) Describe an analytical method of solution for a system consisting of one quadratic equation and one linear equation. Illustrate your description with a specific system.

43. (page 240) Describe an analytical method of solution for a system of quadratic equations involving only x^2 and y^2. Illustrate your description with a specific system.

44. (pages 240–241) Describe an analytical method of solution for a system of quadratic equations of the form $Ax^2 + Cy^2 = G, xy = K$. Illustrate your description with a specific system.

45. (pages 241–242) Why is it useful to know both analytical and graphical methods for solving systems of quadratic equations?

46. (page 242) How many solutions can a system of quadratic equations have? How do the graphs of conic sections support your answer?

47. (pages 239–243) In what respects is the process of solving a system of quadratic equations like that of solving a system of linear equations? In what respects is it different?

48. (pages 245–247) Describe the process of solving a system of quadratic inequalities. In what respects is it like that of solving a system of linear inequalities? In what respects is it different?

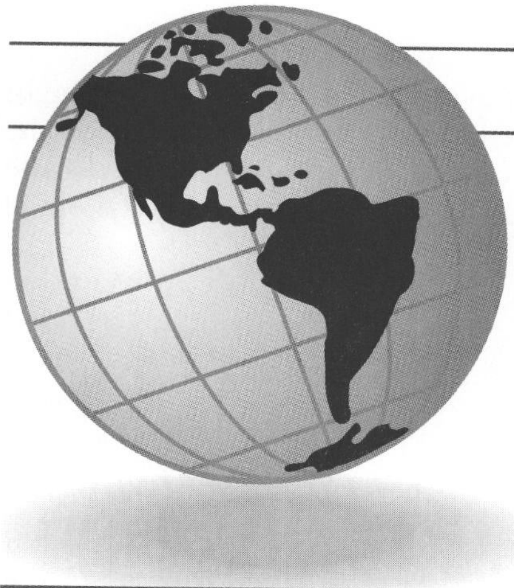

CHAPTER 7

POLYNOMIAL FUNCTIONS

7-1 POWER FUNCTIONS

BASIC POWER FUNCTIONS AND THEIR GRAPHS

In Chapter 3 you studied linear functions, whose equations have the form $f(x) = mx + b$. In Chapter 5 you studied quadratic functions, whose equations have the form $f(x) = ax^2 + bx + c$. Both of these are special cases of **polynomial functions**, whose equations have the form

$$f(x) = a_nx^n + a_{n-1x}{}^{n-1+}\cdots + a_1x + a_0$$

for some nonnegative integer n and choice of real numbers $a_n, a_{n-1}, \ldots, a_1, a_0$. The value of n is the **degree** of the function. Thus the equations

$$f(x) = 3x - 4$$

$$f(x) = x^2 + \pi x + \sqrt{2}$$

$$f(x) = x^7 - \frac{2}{3}x^4 - 97x$$

$$f(x) = 3$$

define polynomial functions of degrees 1, 2, 7, and 0. The equation

$$f(x) = (x - 1)(x + 1)(x - 4)$$

defines a polynomial function of degree 3 since it can be rewritten in the required form as

$$f(x) = x^3 - 4x^2 - x + 4$$

The equations

$$f(x) = 2\sqrt{x} - 5$$

$$f(x) = 1 + \frac{1}{x}$$

do not define polynomial functions, since they cannot be rewritten $f(x) = a_nx^n + a_{n-1x}{}^{n-1+}\cdots + a_1x + a_0$.

You already know that the graph of every linear function is a straight line, and the graph of every quadratic function is a parabola. By contrast, the graphs of polynomial functions of higher degree appear at first to have little in common with each other. Figure 7-1 shows some typical polynomial graphs.

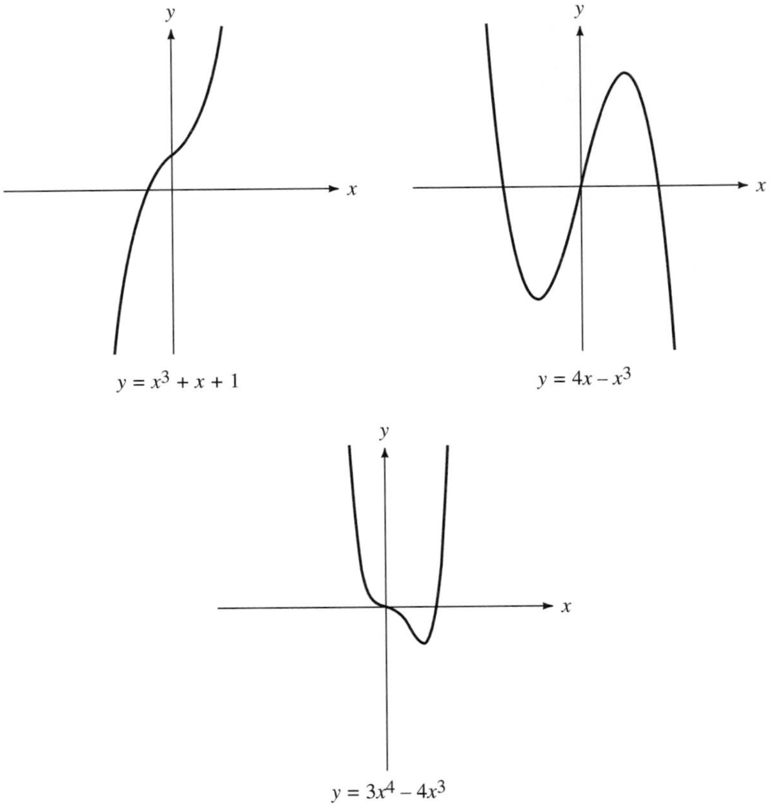

Figure 7–1

In Sections 7-2 to 7-4 you will discover some of the properties shared by all polynomial functions. In this section we will adopt the strategy of *Examining a Simpler Problem* and confine our attention to a simple but important type of polynomial function. These are the **power functions** $y = x^n$, and their graphs can be regarded as basic graphs. You are already familiar with the graphs of $y = x$ and $y = x^2$. Figure 7-2 shows the next few basic graphs.

EXERCISES

1. *(Making Observations)* Graph the power functions $y = x^3$, $y = x^5$, $y = x^7$, and $y = x^9$ together, first in the window $[-2, 2]$ by $[-10, 10]$, then in the window $[-1, 1]$ by $[-1, 1]$. Describe the similarities and differences among the graphs.
2. *(Making Observations)* Repeat Exercise 1 for the power functions $y = x^4$, $y = x^6$, $y = x^8$, and $y = x^{10}$.
3. *(Making Observations)* Guess what the graphs of $y = x^{49}$ and $y = x^{50}$ look like, and sketch them. Use calculator graphs to check your guesses.

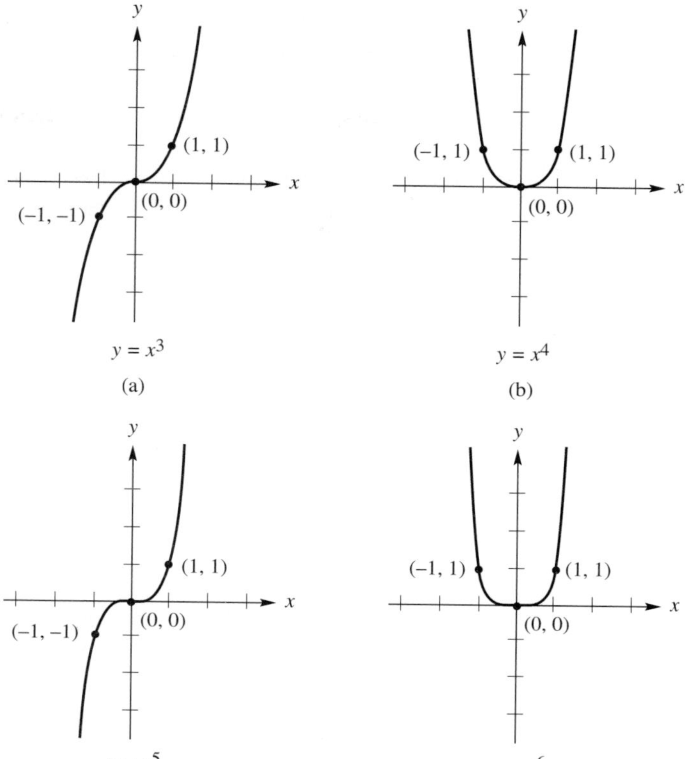

$y = x^3$ (a)

$y = x^4$ (b)

$y = x^5$ (c)

$y = x^6$ (d)

Figure 7–2

Your observations in Exercises 1–3 can be summarized as follows.

If n is odd, the function $f(x) = x^n$

- has domain $(-\infty, \infty)$ and range $(-\infty, \infty)$,
- is increasing on $(-\infty, \infty)$, and
- passes through $(-1, -1)$, $(0, 0)$, and $(1, 1)$.

If n is even, the function $f(x) = x^n$

- has domain $(-\infty, \infty)$ and range $[0, \infty)$,
- is decreasing on $(-\infty, 0)$, increasing on $(0, \infty)$, and
- passes through $(-1, 1)$, $(0, 0)$, and $(1, 1)$.

EXERCISE **4.** *(Making Observations)* Explain why each of the preceding statements in the box is true.

GRAPHICAL TRANSFORMATIONS

In Section 5-1, you discovered that you could sketch the graph of every quadratic polynomial function by transforming the basic graph $y = x^2$. No similar statement can be made for polynomials of higher degree. However, many polynomial graphs can be obtained by transforming the graphs of power functions, as in Example 1.

EXAMPLE 1 *Using transformations to graph a polynomial function*

To graph $y = 2(x + 3)^6$, begin with the basic graph of $y = x^6$, and make the changes

$$y = x^6 \rightarrow y = 2x^6 \rightarrow y = 2(x + 3)^6$$

These changes stretch the graph of $y = x^6$ vertically by a factor of 2, then shift it 3 units left, as in Figure 7-3.

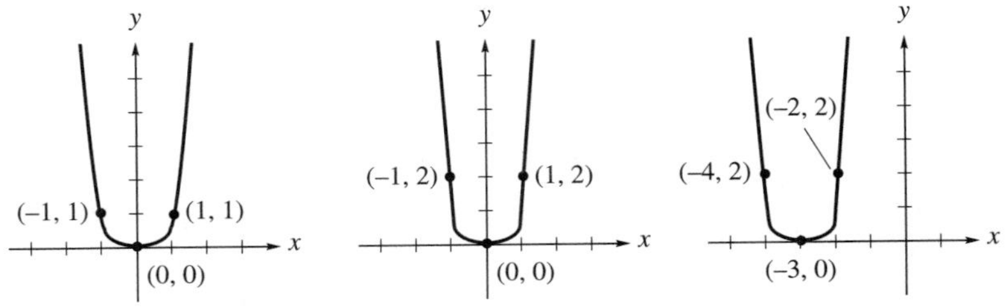

Figure 7–3

■

EXERCISES *In Exercises 5–8, sketch each graph. Support your results with a calculator graph.*

5. $y = (x - 6)^5$ **6.** $y = -x^6$

7. $y = \frac{1}{2}(x - 5)^4$ **8.** $y = 6 - 2x^3$

In Exercises 9–12,

a. Sketch the graph of the function whose graph is that of $f(x) = x^3$ transformed as described.

b. Write the equation of the function.

9. reflected in the y-axis.

10. reflected in the x-axis and compressed horizontally by a factor of 2.

11. stretched vertically by a factor of 4, then shifted 2 units up.

12. shifted 2 units up, then stretched vertically by a factor of 4.

POLYNOMIAL VARIATION

If a relationship is described by an equation $y = ax^n$, we say that **y varies directly with the nth power of x**. For example, the volume V of a sphere varies directly with the cube of its radius r, since $V = \frac{4}{3}\pi r^3$. Linear variation, discussed in Section 3-2, is a special case of polynomial variation. Polynomial variation has the following property.

If $y = ax^n$, then whenever x is multiplied by a constant k, y is multiplied by k^n.

Informally,

For nth power polynomial variations, doubling x multiplies y by 2^n.

For example, if V varies directly with r^3, then doubling r multiplies V by $2^3 = 8$.

EXERCISE 13. *(Writing to Learn)* Section 3-2 contains special cases of the preceding statements. Find them, and explain why the statements in Section 3-2 are consistent with those given here.

It is easy to tell whether a table fits a polynomial variation of a given degree, and to find its equation when it does.

> In any table generated by $y = ax^n$, the ratio $\dfrac{y}{x^n} = a$.
>
> *Exception*: A row with $x = 0$ fits if and only if $y = 0$ also.

EXAMPLE 2 *Deciding whether a table fits a polynomial variation of a given degree*

To see whether the table

x	y
1	3
2	48
3	243
4	768

fits a polynomial variation of degree 4, evaluate the ratio $\dfrac{y}{x^4}$ for each row of the table.

$$\frac{3}{1^4} = 3, \qquad \frac{48}{2^4} = 3, \qquad \frac{243}{3^4} = 3, \qquad \frac{768}{4^4} = 3$$

The table fits the equation $y = 3x^4$. ■

EXERCISES *In Exercises 14–17, decide whether the table fits a polynomial variation $y = ax^n$ for the given value of n. If so, write an equation relating x and y.*

14.

x	$y\ (n = 5)$
1	0.1
2	3.2
3	24.3
5	312.5
6	777.6

15.

x	$y\ (n = 3)$
3	54
4	128
6	432
9	1458
12	3456

16.

x	$y\ (n = 2)$
2	4
4	16
5	32
7	128
9	512

17.

x	$y\ (n = 4)$
1	0.5
2	8
3	40.5
4	128
5	312.5

18. In Exercises 14–17, if the table fits a polynomial variation, find the value of y when $x = 20$.

In the following Mathematical Looking Glass, polynomial variation provides insight into a biological question.

A MATHEMATICAL LOOKING GLASS

Weighing Dinosaurs

How large were the largest dinosaurs? We may never be able to answer this question with more than rough accuracy. The largest for which a nearly complete skeleton has been found is *Brachiosaurus*, which lived over 100 million years ago and despite its immense size was probably a vegetarian. From its skeleton we can be reasonably sure that *Brachiosaurus* stood about 43 feet high. Its weight is more open to question, since we do not have access to the flesh that once surrounded the skeleton. In *Dynamics of Dinosaurs and Other Extinct Giants* (Columbia University Press, 1989), R. McNeill Alexander describes several methods for estimating dinosaur weights. In his preface he states,

> I have used the methods of physics and engineering to try to discover how extinct animals could have lived and moved. I like to think about animals in the kinds of ways that engineers think about machines and vehicles.

Let's estimate *Brachiosaurus*' weight by a method that is not highly accurate, but is easily carried out. The plan is to estimate the amount of weight that its skeleton could have supported. The strategy for implementing the plan is to compare the *Brachiosaurus* skeleton to that of a modern large animal, say a hippopotamus, whose weight is known.

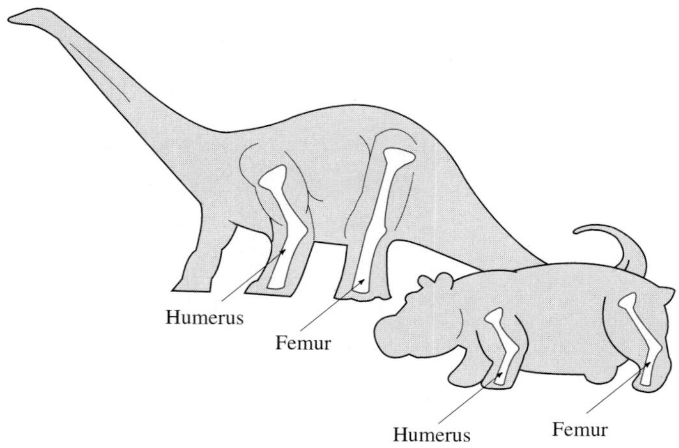

Figure 7–4

When either animal is standing most of its weight is supported by its upper leg bones, the femur and humerus, as shown in Figure 7-4. In *Dynamics of Dinosaurs,* Professor Alexander describes the work of an international team of scientists who measured and added together the circumference of the two *Brachiosaurus* leg bones. Repeating this for a hippopotamus, they found that the sum was about 400 mm for the hippopotamus, which has a weight of about 5500 pounds, and about 1400 mm for *Brachiosaurus*.

In Exercise 19 you can use this data to estimate *Brachiosaurus*' weight. You will need to make the following assumptions.

- For each animal, the femur and humerus bones have circular cross sections of equal size.
- The amount of weight any animal bone can support is proportional to its cross-sectional area.
- The weight of each animal is equal to the greatest weight its bones could support. ■

EXERCISES

19. *(Problem Solving)*

a. Use the given data to express an animal's weight W in pounds as a function of the circumference C in millimeters of its femur or humerus bone.

b. Use your result from part (a) to estimate the *Brachiosaurus* weight. (Researchers' estimates quoted in *Dynamics of Dinosaurs* range from about 69,500 to 191,400 pounds. How does your estimate compare with those?)

20. *(Problem Solving)* Professor Alexander's own estimate of the weight of the *Brachiosaurus* is based on an experiment he performed using a model *Brachiosaurus* from London's Natural History Museum. By submerging the model in water, he determined that its volume was 728 cm^3.

a. If the model was $\frac{1}{40}$ life size, explain why the dinosaur's volume was $(40)^3(728)\ cm^3$.

b. The bodies of most large land animals have about the same density as water (1 kg for each 1000 cm^3). Use this fact and your result from part (a) to estimate *Brachiosaurus*' weight in pounds. How does this estimate compare with the one in Exercise 19?

ADDITIONAL EXERCISES

In Exercises 21–24,

a. Use graphical transformations to sketch each graph.

b. State the domain and range of the function.

21. $y = (x - 1)^3 - 5$ **22.** $y = 10 - 0.2x^7$

23. $y = \frac{1}{3}x^4 + 12$ **24.** $y = \frac{1}{3}(x^4 + 12)$

25. *(Writing to Learn)* Describe how the domain and range of a power function $y = x^n$ are affected by each of the following transformations.

a. a vertical stretch or compression

b. a reflection in the x-axis

c. a horizontal shift

d. a vertical shift

In Exercises 26–31,

a. Sketch the graph of the function whose graph is that of $f(x) = x^4$ transformed as described.

b. Write the equation of the function.

26. reflected in the y-axis.

27. reflected in the y-axis, then shifted 1 unit right.

28. shifted 1 unit right, then reflected in the y-axis.

29. stretched vertically by a factor of 3.

30. stretched vertically by a factor of 3, then shifted 2 units up.

31. shifted 2 units up, then stretched vertically by a factor of 3.

In Exercises 32–35, decide whether the table fits a polynomial variation $y = ax^n$ for the given value of n. If it does, write an equation relating x and y.

32.

x	y ($n = 4$)
0.2	48
0.3	243
0.5	1,875
0.8	12,288
1.0	30,000

33.

x	y ($n = 1$)
3	45
5	75
10	150
12	180
18	270

34.

x	y ($n = 2$)
−20	−8
10	−2
30	−18
40	−32
100	−200

35.

x	y ($n = 3$)
2	1
4	16
6	54
8	128
10	250

36. In Exercises 32–35, if the table fits a polynomial variation, find the value of y when $x = 1$.

37. *(Problem Solving)* Imagine you are an aeronautical engineer, employed by NASA. You have been assigned to a project whose objective is to put a satellite into orbit around Mars. A similar satellite is currently orbiting the earth at an altitude of 230 km, making one revolution every 89 minutes. You are to calculate an altitude for the Mars satellite so that its period of revolution will also be 89 minutes.

You know that the radius r of a satellite's orbit is related to the mass M of the central body by the equation

$$M = \left(\frac{4\pi}{GT^2}\right)r^3$$

where T is the period of revolution and G is a universal gravitational constant. The value of r is the distance of the satellite from the center of the central body. For example, since the earth has a radius of 6370 km, the value of r for the earth satellite is $6370 + 230 = 6600$ km. The radius of Mars is 3398 km, and its mass is 0.11 times that of the earth. What should you tell your colleagues at NASA about the altitude of the Mars satellite's orbit?

38. *(Problem Solving)* Physiologist J. L. Poiseuille (1779–1869) discovered that the volume of blood flowing through a small artery (arteriole) per unit time varies directly as the fourth power of the arteriole's radius. If a heart attack victim is given a medicine, such as Cardizem, to dilate his blood vessels by 10%, by what factor will the flow of blood be increased?

7-2 AN ANALYTICAL VIEW OF POLYNOMIAL FUNCTIONS

PREREQUISITES MAKE SURE YOU ARE FAMILIAR WITH:

Algebra of Complex Numbers (Section A-14)

Division and Synthetic Division of Polynomials (Section A-15)

You have seen that polynomial functions of higher degree are studied as models of physical problems, although they occur less often than linear or quadratic functions. Another reason to study them is simply for their own sake. This reason is not as flimsy as it might seem. Great people often do things out of sheer passion and curiosity. Sir George Mallory, whose disappearance on the slopes of Mount Everest in 1924 made him a legend among climbers, was once asked why he was so determined to conquer the mountain. His reply was, "Because it is there!"

Mathematicians of the sixteenth century were climbing mountains of their own as they attempted to find solutions to polynomial equations. But sometimes even they were overwhelmed by its difficulty, as Howard Eves noted in *An Introduction to the History of Mathematics* (Holt, Rinehart, and Winston, 1976). He refers to François Viète's laborious method for solving equations of higher degree by noting that ". . . one seventeenth century mathematician described it as 'work unfit for a Christian'." Eves presents a delightful account of this struggle.

> Probably the most spectacular mathematical achievement of the sixteenth century was the discovery, by Italian mathematicians, of the algebraic solution of cubic and quartic equations. The story of this discovery, when told in its most colorful version, rivals any page written by Benvenuto Cellini. Briefly told, the facts seem to be these. About 1515, Scipione del Ferro (1465–1526), a professor of mathematics at the University of Bologna, solved algebraically the cubic equation $x^3 + mx = n$, probably basing his work on earlier Arabic sources. He did not publish his result but revealed the secret to his pupil Antonio Fior. Now about 1535, Nicolo Fontana of Brescia, commonly referred to as Tartaglia (the stammerer) because of a childhood injury that affected his speech, claimed to have discovered an algebraic solution of the cubic equation $x^3 + px^2 = n$. Believing that this claim was a bluff, Fior challenged Tartaglia to a public contest of solving cubic equations, whereupon the latter exerted himself and only a few days before the contest found an algebraic solution for cubics lacking a quadratic term. Entering the contest equipped to solve two types of cubic equations, whereas Fior could solve but one type, Tartaglia triumphed completely. Later Girolamo Cardano, an unprincipled genius who taught mathematics and practiced medicine in Milan, upon giving a solemn pledge of

secrecy, wheedled the key to the cubic from Tartaglia. In 1545, Cardano published his *Ars magna*, a great Latin treatise on algebra, at Nuremberg, Germany, and in it appeared Tartaglia's solution of the cubic. Tartaglia's vehement protests were met by Ludovico Ferrari, Cardano's most capable pupil, who argued that Cardano had received his information from del Ferro through a third party and accused Tartaglia of plagiarism from the same source. There ensued an acrimonious dispute from which Tartaglia was perhaps lucky to escape alive.

Since the actors in the above drama seem not always to have had the highest regard for truth, one finds a number of variations in the details of the plot.

Benvenuto Cellini (1500–1571), mentioned in Eves' account above, was an artist, craftsman, and sculptor of great significance. His fame, however, is due mainly to his autobiography, which despite its manifest exaggerations and boastful tone is surprisingly frank and authentic.

In this section we will study polynomial functions from an historical perspective, focusing on three questions that occupied mathematicians until the early part of the nineteenth century.

- What is the largest number of zeros that a polynomial function can have? [A complex number r is a **zero** of a polynomial function P if and only if $P(r) = 0$. Our discussion will focus primarily on real zeros.]
- What is the smallest number of zeros that a polynomial function can have?
- Can we find all zeros of a polynomial function?

FACTORS AND ZEROS

Let's focus on the first of the preceding questions. Exercises 1–4 suggest that the number of zeros of a polynomial function is related to its degree.

EXERCISES

1. *(Review)* Find all the zeros of each function P by solving the equation $P(x) = 0$.
   a. $P(x) = x^2 + x - 6$
   b. $P(x) = x^3 - 5x^2 + 6x$
   c. $P(x) = 2x^3 + 4x^2 - 2x - 4$
   d. $P(x) = x^4 - 10x^2 + 9$
2. *(Making Observations)* How is the number of zeros for each function in Exercise 1 related to the degree of the function?
3. *(Review)* Find all the zeros of each function Q by solving the equation $Q(x) = 0$.
   a. $Q(x) = x^2 - 8x + 16$
   b. $Q(x) = x^5 - 9x^3$
   c. $Q(x) = x^3 - x^2 + x - 1$
   d. $Q(x) = x^4 - 4x^2 + 4$
4. *(Making Observations)* Exercise 3 illustrates that the number of zeros of a polynomial function can be less than its degree. Do you think the number of zeros can ever exceed the degree? Why or why not?

By completing Exercises 1–4 you have duplicated the reasoning outlined by René Descartes in his *Discours de la méthode pour bien conduire sa raison et chercher la vérité dans les sciences* (Discourse of the method for conducting reason well and seeking the truth in the sciences), published in 1637. His conclusion, stated here in more modern language, has come to be known as the Factor Theorem.

The Factor Theorem

A polynomial function P has a zero at a complex number r if and only if $P(x)$ has a factor of $x - r$.

Example 1 illustrates the computational power of the Factor Theorem.

EXAMPLE 1 *Using the Factor Theorem to find the zeros of a polynomial function*

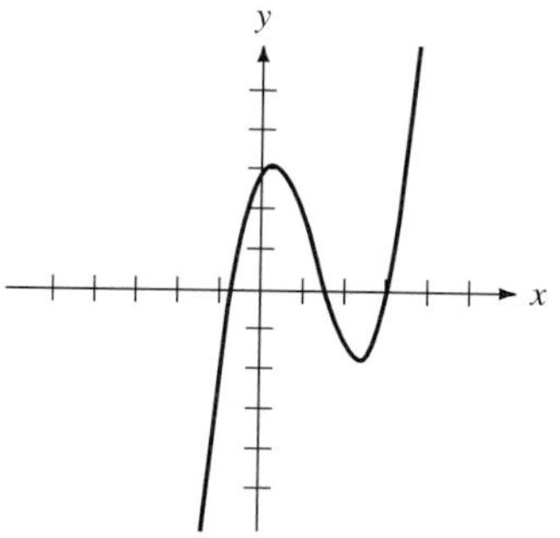

Figure 7–5

Figure 7-5 shows a graph of $P(x) = x^3 - 4x^2 + 2x + 3$. The real zeros of P correspond to the x-intercepts on the graph. One x-intercept appears to be at about $(3, 0)$. A quick check shows that

$$P(3) = (3)^3 - 4(3)^2 + 2(3) + 3 = 0$$

Because $x = 3$ is a zero of P, $x - 3$ must be a factor of $P(x)$. Dividing $P(x)$ by $x - 3$ yields

$$P(x) = (x - 3)(x^2 - x - 1)$$

The other zeros must be solutions of

$$x^2 - x - 1 = 0$$

By the quadratic formula, they are

$$x = \frac{1 \pm \sqrt{5}}{2} \cong -0.62, 1.62$$

■

EXERCISES *In Exercises 5–8, graph $y = f(x)$ in the indicated window. Use the graph to guess the value of one of the zeros of f; then use division to find the others, as in Example 1.*

5. $f(x) = x^3 + 2x^2 - 5x - 6$; $[-10, 10]$ by $[-10, 10]$

6. $f(x) = 2x^3 - 7x^2 - 5x + 4$; $[-10, 10]$ by $[-30, 30]$

7. $f(x) = -x^3 - x^2 + 3x + 2$; $[-10, 10]$ by $[-10, 10]$

8. $f(x) = 3x^3 + x^2 - 7x - 5$; $[-10, 10]$ by $[-10, 10]$

The Factor Theorem answers the first of our questions at the beginning of this section. Since a polynomial of degree n has at most n linear factors, the Factor Theorem implies that a polynomial function of degree n has at most n zeros. This fact was stated as early as 1600 by the German mathematician Peter Rothe in *Arithmetica philosophic*, and later by Descartes and Albert Girard (1593–1632), both Frenchmen working in Holland. Girard's statement appeared in *L'invention nouvelle en l'algèbre* (New results in algebra), a 1629 publication based on a 1585 treatise by Simon Stevin. Girard and Descartes recognized the nature of polynomial zeros because they were among the first mathematicians to take imaginary numbers seriously.

Multiplicity of Zeros Let's turn now to the second of our three questions from the beginning of the section. What is the smallest number of zeros that a polynomial function can have? The Factor Theorem suggests that we begin by asking whether a polynomial function of degree n must always have n zeros. The answer to this question depends on how we count the zeros. If we confine our attention to real zeros, Exercise 9 shows that the answer is no.

EXERCISE **9.** *(Review)*

a. Sketch the graph of $f(x) = x^2 + 1$, and verify that it has no x-intercepts.

b. Explain why the graph suggests that f has no real zeros.

c. Confirm that f has no real zeros by solving the equation $x^2 + 1 = 0$.

Although the function in Exercise 9 has no real zeros, it has two complex zeros. This suggests that we ask whether a polynomial function of degree n must always have n complex zeros. The answer still depends on how we count the zeros. If a number can only be counted once, Example 2 shows that the answer is still no.

EXAMPLE 2 *A cubic polynomial function with fewer than three distinct zeros*

In Descartes' *Discours*, he discusses the function

$$P(y) = y^3 - 10y^2 + 17y - 8$$

which can be written in the factored form

$$P(y) = (y - 1)^2(y - 8)$$

Its only zeros are therefore $y = 1$ and $y = 8$. ■

Although the function in Example 2 has only two distinct zeros, it has three linear factors (two factors of $y - 1$ and one factor of $y - 8$). Descartes' *Discours* describes this situation by saying that the polynomial $P(y)$ "... has only three roots, two of them, 1 and 8, being true roots, and the third, also 1, being false; ...." A more modern view is to say that $P(y)$ has a **double zero** at $y = 1$, corresponding to the factor $(y - 1)^2$. More generally, if k is a positive integer, a zero $x = r$ corresponding to a factor $(x - r)^k$ is called a **simple zero** if $k = 1$ and a **multiple zero** if $k > 1$. The number k is called the **multiplicity** of the zero. In our last example, the zero $y = 1$ has multiplicity 2, and the zero $y = 8$ has multiplicity 1.

EXERCISE **10.** As parts (a)–(d) of this exercise, state the multiplicity of each zero for each function in Exercise 3.

THE FUNDAMENTAL THEOREM OF ALGEBRA

Now let's consider the question of whether a polynomial function of degree n must always have n zeros. If we mean complex zeros, and if we count them according to their multiplicity, the answer is yes. This statement is referred to as the Fundamental Theorem of Algebra.

The Fundamental Theorem of Algebra

Every polynomial function P of degree n can be written in factored form as

$$P(x) = a_n(x - r_1)(x - r_2)\cdots(x - r_{n-1})(x - r_n)$$

for some choice of complex numbers $r_1, r_2, \ldots, r_{n-1}, r_n$, not necessarily distinct.

Equivalently, every polynomial function of degree n has exactly n complex zeros, counted according to their multiplicity.

Both Girard and Descartes stated the Fundamental Theorem. Its proof, however, was the Mount Everest of seventeenth-century mathematics. Such mathematical giants as Newton, Euler, D'Alembert, and Lagrange all attempted unsuccessfully to establish its truth. The Fundamental Theorem was not conquered until 1799, when Carl Friedrich Gauss proved it in his doctoral dissertation, written at the age of 22 at the University of Helmstädt. (You have seen another example of Gauss' ingenuity in Section 1-3.) Gauss thus answered our second question from the beginning of this section.

Exercise **11.** *(Making Observations)* Show that the Fundamental Theorem of Algebra is true for each of the functions in Exercise 3 by verifying that the number of complex zeros is equal to its degree.

Now let's turn to our third question from the beginning of this section. Can we find all the zeros of a polynomial function? The Fundamental Theorem of Algebra tells how many zeros there are, but does not tell how to find them. The task of finding zeros becomes increasingly more complex as the degree of the polynomial increases.

Your expertise in solving linear and quadratic equations may have been shared by your $(\text{great})^{200}$-grandparents. Methods for solving linear equations were known in Egypt as far back as 3500 B.C. The quadratic formula was known to the Babylonians of 1700 B.C. A similar formula for cubic equations was first published by Girolamo Cardano (1501–1576) in his *Ars magna* in 1545. Cardano himself rejected negative and imaginary solutions, because he saw no physical meaning in them. The acceptance of such solutions was delayed until 1732, when Leonhard Euler (1707–1783), writing in the journal *Commentarii Academiae Scientiarum Petropolitanae*, showed that Cardano's formula can be used to obtain three complex zeros of every cubic function. *Ars magna* also contained a formula of Ludovico Ferrari (1522–1565) for solving a general quartic equation. You can find the details of Cardano's and Ferrari's formulas in Howard Eves' *An Introduction to the History of Mathematics.*

For over 250 years following the appearance of *Ars magna*, European mathematicians searched in vain for a formula for the zeros of general quintic functions. François Viète (1540–1603), James Gregory (1638–1675), Ehrenfried Walter von Tschirnhausen (1651–1708), Joseph Louis Lagrange (1736–1813), and Leonhard Euler all tried and failed. Gregory, through his work on calculus, was led to believe that no such formula existed. In 1824 Norwegian Niels Henrik Abel (1802–1829), building on the work of

Lagrange, proved Gregory right. There is no general formula for finding zeros of polynomials of degree 5 or greater. Abel thus answered our third question from the beginning of this section.

Which equations of degree 5 or greater can be solved by algebraic formulas? This question was answered in 1831 by the Frenchman Evariste Galois. Any discussion of his work is far too ambitious for this book, but it is interesting to note how it came to the attention of the rest of the world. Galois was jailed as a political activist several times while still in his teens. At age 20 he was challenged to a duel, probably provoked for political reasons, over a woman of dubious reputation. The only surviving original record of his work was hastily jotted down the night before the duel, in which he was mortally wounded. In Leopold Infeld's biography of Galois, *Whom the Gods Love*, it is told how the manuscript was entrusted to his brother Alfred as he lay dying.

> "I must tell you something. Important. Mathematical manuscripts are in my room. On the table. A letter to Chevalier. It is for you, too. Both. Alfred and Auguste. Take care of my work. Make it known. Important."
>
> He felt relieved and listened to Alfred's violent words:
>
> "I swear to you that I will do it. I shall do everything. I swear to you that they will be published and recognized. I swear to you that if necessary I shall devote my whole life to it."
>
> Now he broke out with a new, more violent torrent of tears. Evariste looked at him with pity and said very slowly, tearing the words out of his body with increasing pain, "Don't cry, Alfred. I need all my courage—to die at twenty."

Alfred did indeed devote his life to obtaining proper recognition for his brother's work. A copy of the manuscripts eventually found its way into the hands of Joseph Liouville, who published them in *Journal de mathématiques pures et appliquées* in 1846.

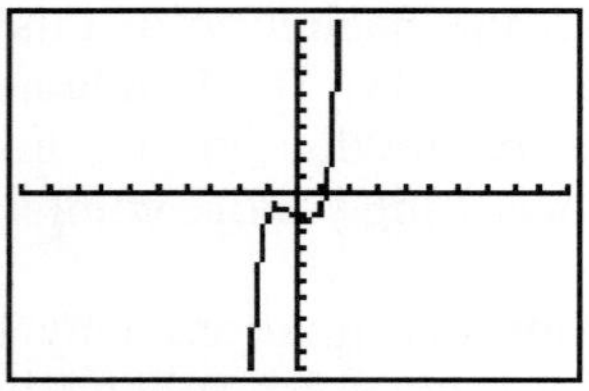

[−10, 10] by [−10, 10]

Figure 7–6

Galois' work shows that no algebraic formula will find the zeros of $P(x) = x^5 - x - 1$. Yet the graph in Figure 7-6 reveals a zero in $[1, 2]$. Although we cannot find this zero analytically, we can approximate it graphically. In Section 7-4 you will also learn how to approximate it numerically.

ADDITIONAL EXERCISES

In Exercises 12–15, *find all the zeros of the polynomial function.*

12. *(Review)* $f(x) = 16x - x^3$

13. *(Review)* $h(t) = t^4 + t^3 - 42t^2$

14. *(Review)* $Q(z) = 2z^5 - 10z^3 + 8z$

15. *(Review)* $r(v) = v^3 - 3v^2 - 2v + 6$

In Exercises 16–19,

a. Decide whether $x = -1$ is a zero of g by evaluating $g(-1)$.

b. Write $g(x)$ in its factored form, and list all of the zeros of g. Make sure the result agrees with your conclusion from part (a).

16. $g(x) = x^3 + 2x^2$

17. $g(x) = x^7 + x^5 - 2x^3$

18. $g(x) = x^6 - 1$

19. $g(x) = 2x^4 + 5x^2 + 2$

In Exercises 20–25, identify the zeros of the function and state the multiplicity of each zero.

20. $f(x) = (x + 10)(x - 20)^6$

21. $A = (b - 4)^3(b + \pi)^5$

22. $S(x) = x^4(x^2 - 16)^4$

23. $h(r) = (r^2 - 2r - 8)^3(r^2 + 2r - 15)^2$

24. $y = x^3 + 2x^2 - 4x - 8$

25. $z = x^6 - 2x^4 - 4x^2 + 8$

26. *(Making Observations)* Verify that the Fundamental Theorem of Algebra is true for each function in Exercises 20–25.

The functions in Exercises 27–30 each have some nonreal zeros. Find the zeros of each function by factoring and verify that the Fundamental Theorem of Algebra is satisfied.

27. $y = 2x^2 + 4x + 10$

28. $C(x) = x^4 + 7x^2 + 12$

29. $F(r) = r^5 + 6r^3 + 9r$

30. $w = z^5 - 2z^4 + 4z^3 - 8z^2$

31. In Exercise 5 of Section 2-4 you solved the following equation: $x^3 - 34x^2 + 408x - 1728 = 0$ graphically, obtaining $x = 12$.

a. *(Writing to Learn)* Why does the *Fundamental Theorem of Algebra* guarantee that the equation has two other solutions?

b. Use division to find the other solutions, and thus verify that there are no other real solutions.

32. *(Writing to Learn)* Suppose that f and g are polynomials, with $f(3) = 7$ and $g(3) = 0$. What does the Factor Theorem tell you about f and g?

33. *(Making Observations)* You can use the Factor Theorem to find polynomial functions having specified zeros with specified multiplicities. For example, a function having zeros $x = \frac{1}{2}$ with multiplicity 2 and $x = \pm\sqrt{3}$ each with multiplicity 1 is

$$P(x) = \left(x - \frac{1}{2}\right)^2 (x + \sqrt{3})(x - \sqrt{3})$$

a. Verify that in expanded form,

$$P(x) = x^4 - x^3 - \frac{11}{4}x^2 + 3x - \frac{3}{4}$$

b. Graph $y = P(x)$ in the window $[-3, 3]$ by $[-10, 10]$. Then graph

$$y = 4P(x) = 4x^4 - 4x^3 - 11x^2 + 12x - 3$$

on the same screen. How do the graphs suggest that the two functions have the same zeros?

c. How could you draw the same conclusion without seeing the graphs?

d. If the zeros of a polynomial function Q are the same as those of P with the same multiplicities, explain why Q is a constant multiple of P.

In Exercises 34–39, *use the idea in Exercise* 33 *to find a polynomial function with integer coefficients having the given zeros with the indicated multiplicities.*

34. $x = 1$, multiplicity 3

35. $x = 3$, multiplicity 2; $x = \frac{1}{2}$, multiplicity 1

36. $x = \pm\sqrt{5}$, each with multiplicity 1; $x = 0$, multiplicity 3

37. $x = -5$, multiplicity 1; $x = -\frac{3}{2}$, multiplicity 2

38. $x = 0$, multiplicity 4; $x = \frac{1}{3}$, multiplicity 2

39. $x = \pm\sqrt{2}$, each with multiplicity 2

40. *(Writing to Learn)* Write a few pages on the life of Tartaglia, Cardano, Descartes, Gauss, or Galois. You will need biographical information beyond what we have provided. One of the sources we cited might be a good starting point.

SUPPLEMENTARY TOPIC
The Rational Root Theorem

As you saw in Example 1, one strategy for finding the zeros of a polynomial function is to graph the function, use the graph to guess the value of one or more zeros, then use division to find the remaining zeros. You can often reduce the amount of guesswork in the process by using the following theorem, developed by Descartes.

The Rational Root Theorem

If a polynomial function

$$f(x) = a_nx^n + a_{n-1x}{}^{n-1} + a_{n-2}x^{n-2} + \ldots + a_1x + a_0$$

with integer coefficients $a_n, a_{n-1}, \cdots, a_1, a_0$, has a zero $x = \frac{p}{q}$, with p and q integers and $\frac{p}{q}$ in lowest terms, then a_0 is evenly divisible by p, and a_n is evenly divisible by q.

The Rational Root Theorem allows you to make a list of all possible rational zeros of a given polynomial function with integer coefficients. For example, suppose

$$P(x) = 3x^4 + 8x^3 - 72x^2 - 127x + 50$$

The possible values of p are the integer factors of 50, which are $\pm 1, 2, 5, 10, 25, 50$. The possible values of q are the integer factors of 3, which are $\pm 1, 3$. The possible rational zeros $x = \frac{p}{q}$ are

$$x = \pm 1, 2, 5, 10, 25, 50, \frac{1}{3}, \frac{2}{3}, \frac{5}{3}, \frac{10}{3}, \frac{25}{3}, \frac{50}{3}$$

Example 3 illustrates the use of the Rational Root Theorem in identifying the zeros of this polynomial function.

EXAMPLE 3 *Using the Rational Root Theorem to find the zeros of a polynomial function*

Figure 7-7 shows the graph of $P(x) = 3x^4 + 8x^3 - 72x^2 - 127x + 50$.

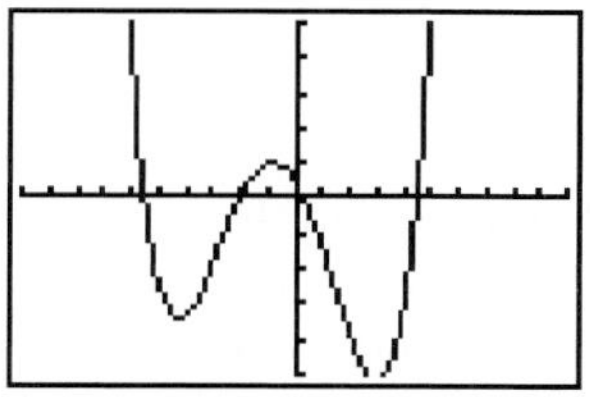

$[-10, 10]$ by $[-500, 500]$

Figure 7–7

We know that the window has captured all the zeros of P, because it shows four x-intercepts, and the degree of P is 4. The intercepts in the intervals $[-6, -5]$ and $[4, 5]$ must be irrational, because none of our listed possible rational zeros lies in either of these intervals. The other two intercepts appear to be near $(-2, 0)$ and $\left(\frac{1}{3}, 0\right)$.

You can use synthetic division to verify that $x = -2$ is a zero of P, so

$$P(x) = (x + 2)(3x^3 + 2x^2 - 76x + 25)$$

Similarly, you can verify that $x = \frac{1}{3}$ is a zero of P, so

$$P(x) = (x + 2)\left(x - \frac{1}{3}\right)(3x^2 + 3x - 75)$$

The remaining zeros must satisfy the quadratic equation

$$3x^2 + 3x - 75 = 0$$

You can find them through the quadratic formula, so that altogether the zeros are

$$x = -2, \frac{1}{3}, \frac{-1 \pm \sqrt{101}}{2}$$ ■

In Exercises 41–48, *use the Rational Root Theorem together with a calculator graph to find all rational zeros of the polynomial. If possible, use synthetic division and the quadratic formula to find the remaining zeros. Otherwise, estimate them graphically.*

41. $P(x) = x^3 - x^2 - 17x + 20$

42. $G(x) = 2x^4 - 3x^3 + 2x^2 - x - 3$

43. $s(t) = 6t^3 - 7t^2 - 40t + 21$

44. $w = 3x^3 - 10x + 6$

45. $f(x) = x^3 - 2x^2 - 12x + 16$

46. $r = x^3 + x^2 - 20x + 12$

47. $P(w) = w^4 + 2w^3 + w - 24$

48. $Y(x) = 4x^3 + 12x^2 - \frac{7}{2}$

7–3 A GRAPHICAL VIEW OF POLYNOMIAL FUNCTIONS

Sixteenth-century mathematicians could not view polynomial functions graphically, because the coordinate plane had not yet been invented. The first steps in its development were taken by Descartes, who assigned equations to curves in a plane, and by Pierre de Fermat (1601?–1665), who began with equations and studied the corresponding curves. According to Howard Eves, in *An Introduction to the History of Mathematics*,

> There are a couple of legends describing the initial flash that led Descartes to the contemplation of analytic geometry. According to one story, it came to him in a dream. On St. Martin's Eve, November 10, 1616, while encamped in the army's winter quarters on the banks of the Danube, Descartes experienced three singularly vivid and coherent dreams that, he claimed, changed the whole course of his life. The dreams, he said, clarified his purpose in life and determined his future endeavors by revealing to him "a marvelous science" and "a wonderful discovery." Descartes never explicitly disclosed just what were the marvelous science and the wonderful discovery, but some believe them to have been analytical geometry, or the application of algebra to geometry, and then the reduction of all science to geometry. It was eighteen years later that he expounded some of his ideas in his *Discours*.

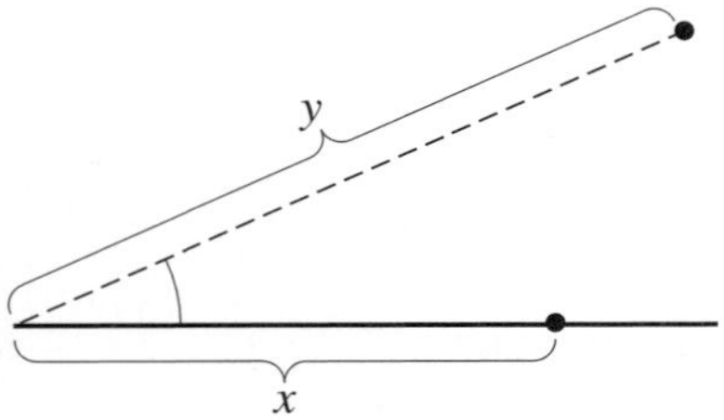

Figure 7–8

The coordinate plane is often called the Cartesian plane after Cartesius (the Latinized spelling of Descartes), but he did not in fact develop it as we know it. In particular, he did not use negative coordinates or mutually perpendicular axes. As Eves describes, he simply ". . . marked off x on a given axis and then a length y at a fixed angle to this axis and endeavored to construct points whose x's and y's satisfy a given relation," as in Figure 7-8. The coordinate plane as we now know it evolved gradually, and it was not until the eighteenth century that mutually perpendicular axes became standard.

With the coordinate plane and graphing technology, you can gain graphical insights into polynomial functions more easily than the best mathematicians of Descartes' day. Let's begin by building on our analytical work from Section 7-2.

POINTS OF INTEREST

***x*-intercepts** Of all the points on the graph of a polynomial function P, the x-intercepts have a special significance. The graph has an x-intercept at $x = r$ if and only if $P(r) = 0$. According to the Factor Theorem, this happens if and only if $x - r$ is a factor of $P(x)$. Since a polynomial of degree n has at most n real linear factors, we can draw a graphical conclusion from our observations.

> The graph of a polynomial function of degree n has at most n x-intercepts.

EXERCISES

1. *(Review)* List all real zeros of P according to multiplicity, and list all x-intercepts on the graph of $y = P(x)$.
   - **a.** $P(x) = (x + 4)(x + 1)(x - 3)^3$
   - **b.** $P(x) = x^3 + 4x$
   - **c.** $P(x) = x^2(x + 2\sqrt{7})(x - 2\sqrt{7})$
   - **d.** $P(x) = x^4 - 16$
2. *(Review)* Write a polynomial function with integer coefficients having the given graph. Assume that all zeros are real and simple (have multiplicity 1).

a.

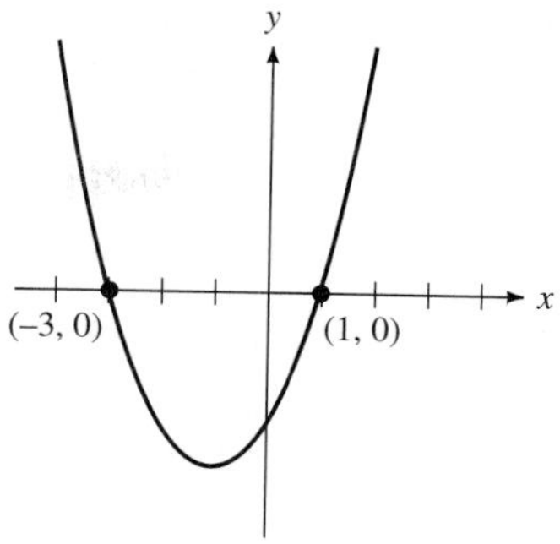

b.

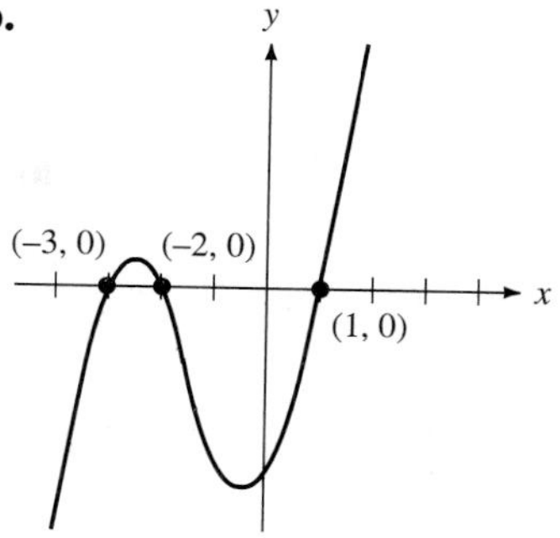

c.

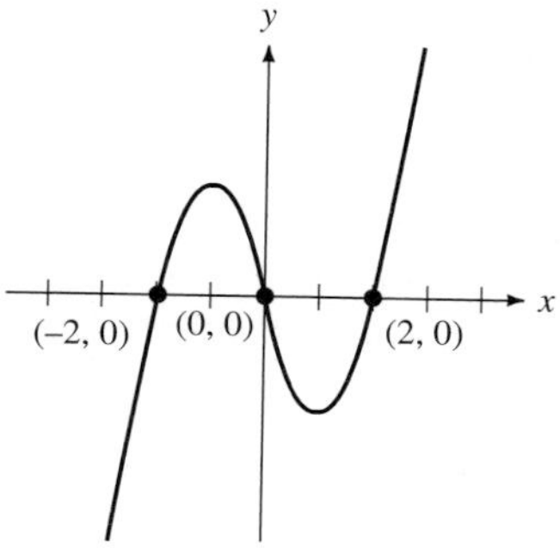

d.

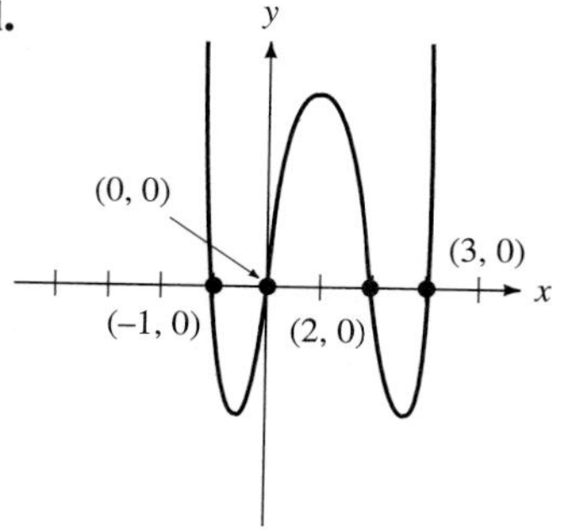

3. *(Review)* Write a polynomial function with integer coefficients having the given zeros with the indicated multiplicities, and list the x-intercepts on its graph.

a. $-2, 1, 3$ (each multiplicity 2)

b. $0, \sqrt{6}, -\sqrt{6}$ (each multiplicity 1)

c. $\frac{3}{2}$ (multiplicity 2), $-\frac{1}{3}$ (multiplicity 1)

d. $\pm\sqrt{2}$ (each multiplicity 2)
$\pm\sqrt{3}$ (each multiplicity 1)

***x*-intercepts at Multiple Zeros** The shape of a polynomial graph around an x-intercept is determined by the multiplicity of the corresponding zero. You can use this relationship to sketch graphs more accurately and quickly. In Section 7-5 you will also use it to solve polynomial inequalities.

EXAMPLE 1 *The shape of a polynomial graph around an x-intercept*

Figure 7-9 shows the graph of $P(x) = (x + 4)^2(x - 1)$.

One striking feature of this graph is that it crosses the x-axis at $(1, 0)$, but not at $(-4, 0)$. Stated differently, the sign of $P(x)$ changes at $x = 1$, but not at $x = -4$.

What could account for this difference? Since $P(x)$ is the product of the factors $(x + 4)^2$ and $(x - 1)$, the sign of $P(x)$ at each x-value depends on the sign of each factor. Let's follow the graph from left to right, and analyze the sign of each factor along the way.

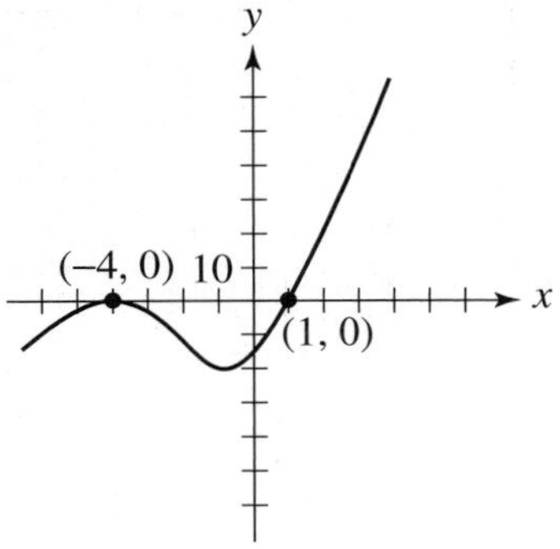

Figure 7–9

To the left of $x = -4$, $(x + 4)^2$ is positive while $(x - 1)$ is negative. As we pass $x = -4$ neither factor changes sign, so that $P(x)$ remains negative and the graph does not cross the x-axis. As we pass $x = 1$ the sign of $(x + 4)^2$ does not change, but $(x - 1)$ changes from negative to positive. Thus the sign of $P(x)$ changes and the graph crosses the x-axis. ■

Example 1 illustrates a general principle.

> If a polynomial function P has a zero of multiplicity k at $x = a$, then the graph of $y = P(x)$
>
> - crosses the x-axis at $(a, 0)$ if k is odd.
> - touches, but does not cross, the x-axis at $(a, 0)$ if k is even.

Using calculus, it can be shown that at an x-intercept corresponding to a zero of multiplicity $k > 1$ a polynomial graph flattens out as in Figure 7-10b and 7-10c. Therefore the three possible shapes of a polynomial graph around an x-intercept of multiplicity k are as shown in Figure 7-10.

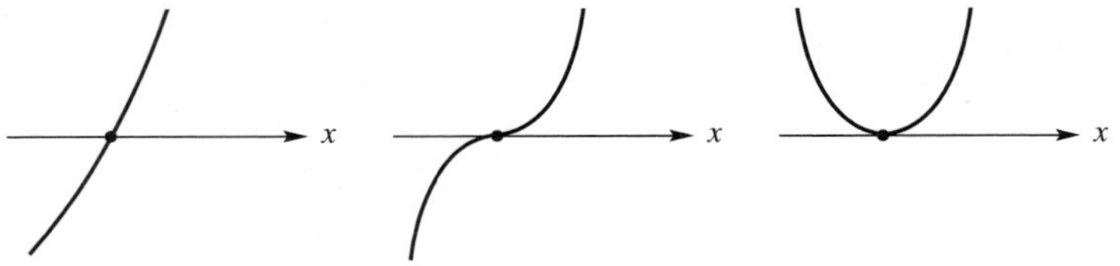

$k = 1$ (a) k odd, $k > 1$ (b) k even (c)

Figure 7–10

EXERCISES *In Exercises 4–7, write an equation in factored form for a sixth degree polynomial whose graph resembles the one shown.*

4.

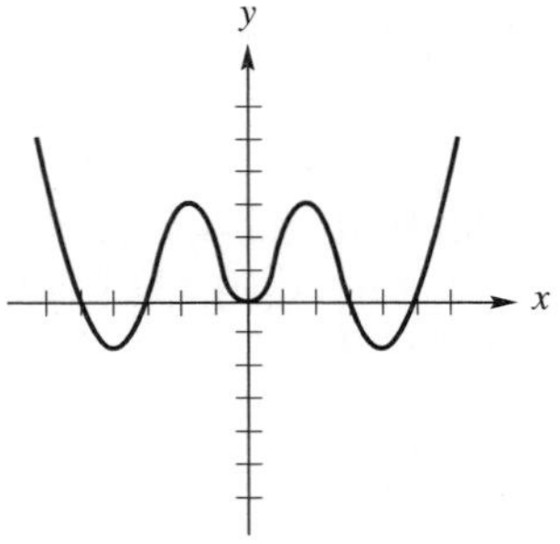

5.

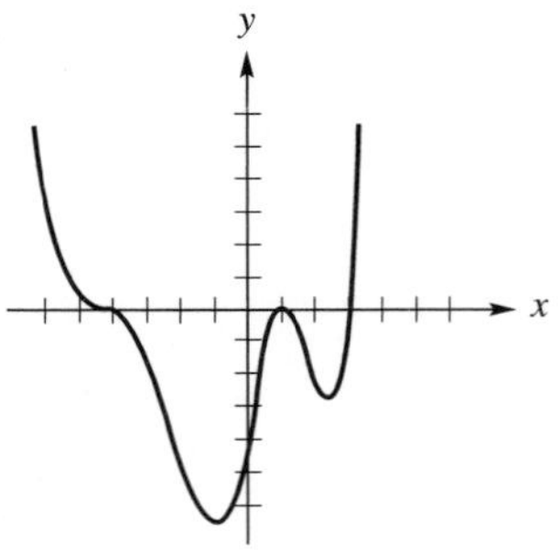

6.

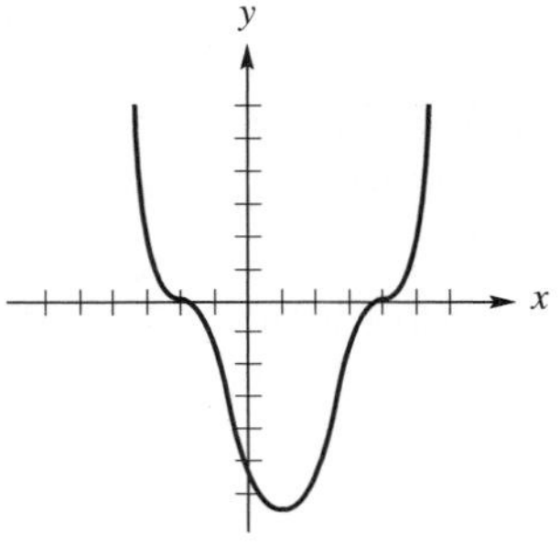

7.

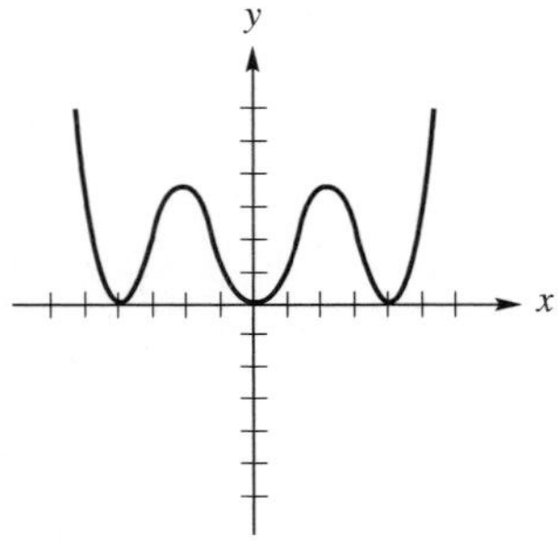

Extreme Values Most polynomial graphs have one or more **turning points** (points where the function changes from increasing to decreasing, or vice versa). Using calculus, it can be shown that a polynomial function of degree n has at most $n - 1$ turning points. The y-coordinates of polynomial functions are called **extreme values** of the function. The following terms describe specific types of extreme values shown in Figure 7-11.

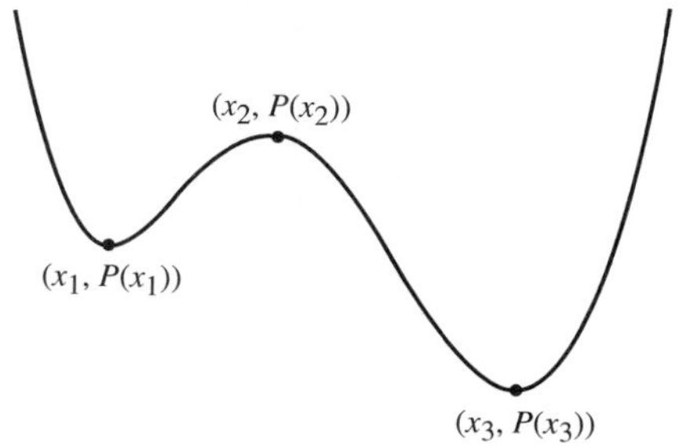

Figure 7–11

- A **local minimum** is the smallest value of the function with some open interval.
  Example: $P(x_1)$, $P(x_3)$.
- A **local maximum** is the largest value of the function within some open interval.
  Example: $P(x_2)$.
- An **absolute minimum** is the smallest value of the function over its entire domain. An absolute minimum is also a local minimum.
  Example: $P(x_3)$.
- An **absolute maximum** is the largest value of the function over its entire domain. An absolute maximum is also a local maximum. None is present in Figure 7-11.

Johannes Kepler, whom you met in Section 6-1, was among the first mathematicians to study the problem of locating extreme values of polynomial functions. His concern came from the practical problem of finding an optimal shape for a wine cask. Kepler calculated y-values for a large number of x-values and estimated the extreme value by noticing where the function increased and decreased. Since Kepler did not have a graphing calculator, or even the concept of a graph, he computed each y-value by hand. In 1629 Fermat devised a method of locating extreme values exactly, using the just-emerging ideas of analytic geometry. Fermat's work was criticized by Descartes and other mathematicians who questioned the logical soundness of his methods. Always reluctant to publish his work, Fermat waited until 1637 to describe his method in a manuscript, which was published by his son in 1679, 14 years after his death. The theory to justify his method was finally developed about 150 years later, by Augustin Louis Cauchy (1789–1857).

If you take calculus, you will learn Fermat's method of finding exact extreme values, with the benefit of a few centuries of refinements. For now, you can follow the spirit of Kepler by finding approximate extreme values, but with the aid of your graphing calculator you will be able to accomplish your task in minutes, rather than months.

FYI In abstract problems, estimates of extreme values should have an error of less than 0.01 in both coordinates. Calculator techniques for estimating extreme values are discussed in Section B-5.

EXAMPLE 2 *Finding extreme values of a polynomial function*

Figure 7-12 shows the graph of $f(x) = 2x^3 - 12x + 6$ and indicates that f has two turning points. By zooming and tracing, you can estimate that they occur at $(-1.41, 17.31)$ and $(1.41, -5.31)$. See Figures 7-13a and b, respectively.

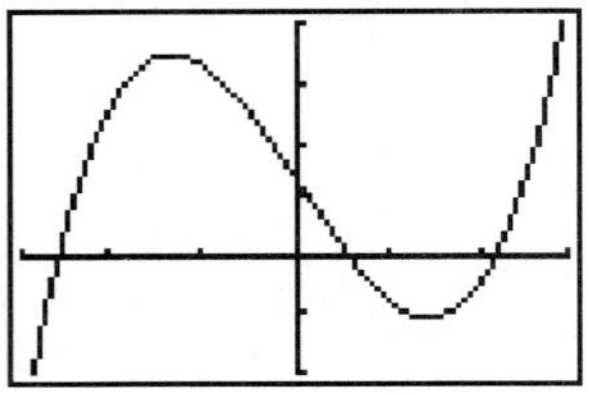

$[-3, 3]$ by $[-10, 20]$

Figure 7–12

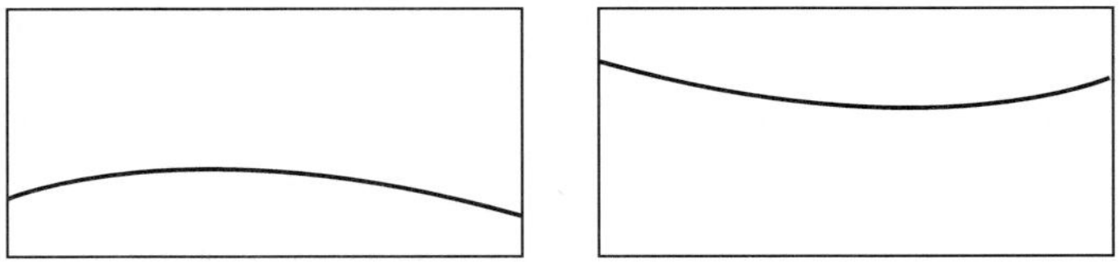

(a)

$[-1.42, -1.40]$ by $[17.31, 17.32]$

(b)

$[1.40, 1.42]$ by $[-5.32, -5.31]$

Figure 7–13 ■

From the turning points of a polynomial graph you can identify the intervals where it is increasing or decreasing. For example, the function in Example 2 is increasing on $(-\infty, -1.41)$ and $(1.41, \infty)$ and decreasing on $(-1.41, 1.41)$.

EXERCISES *In Exercises* 8–11,

a. Find all extreme values of the polynomial function, and identify any which are absolute.

b. Identify the intervals where the function is increasing and decreasing.

c. Use the graph of the function to estimate its range.

8. $f(x) = x^2 - 6x + 8$ **9.** $G(x) = 12x - x^3$

10. $y = x^4 - 4x^3$ **11.** $w = 3x^3 - 10x + 6$

A MATHEMATICAL LOOKING GLASS
Popcorn

Paul White is a transplanted Englishman who lives in a remote part of Tanzania. When he was recently visited by Dave's wife, Alice, she offered to send him a package of gourmet caramel popcorn from the United States, since he is fond of popcorn and cannot obtain it near his home. Paul warned Alice that food items sent to Tanzania may not reach their intended recipient, but she planned to label her package "Contents: Styrofoam" to fool the postal inspectors.

At her local post office Alice discovered that postal regulations limit a package to a combined length and girth of no more than 108 inches. ■

EXERCISES **12.** *(Creating Models)* If Alice's package has a combined length and girth of 108 inches, as in Figure 7-14, express its volume in cubic inches as a function of its radius.

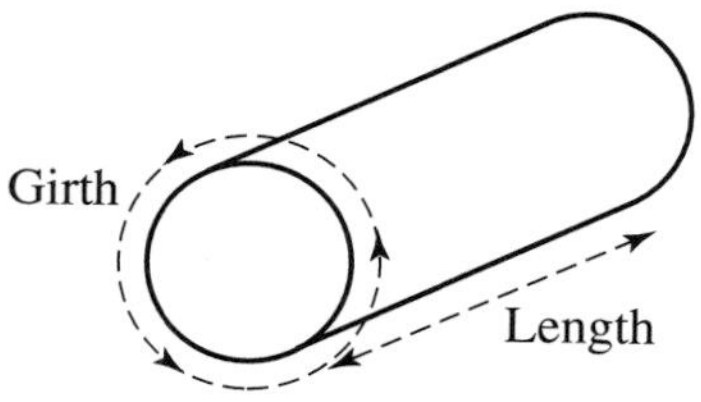

Figure 7–14

13. *(Interpreting Mathematics)*

a. What are the dimensions, in cubic inches, of the container that will hold the most popcorn?

b. Alice must send the popcorn already popped with the caramel coating applied. How many ounces can she send? Popped popcorn weighs 1.3 ounces per cubic foot.

14. a. Identify the intervals where the function in Exercise 12 is increasing and decreasing.

b. *(Writing to Learn)* How could your response to part (a) help you find the dimensions of the optimal container?

15. *(Problem Solving)* Suppose that Alice decides to send the popcorn in a package with a square cross section, as in Figure 7-15. How many ounces can she send? Is this more or less than she could send in a cylindrical package?

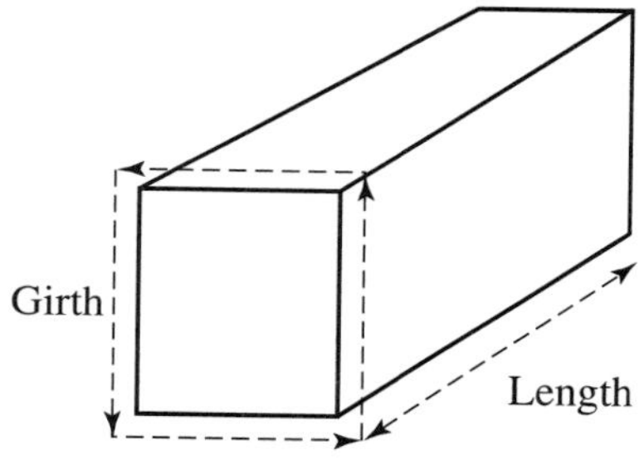

Figure 7–15

END BEHAVIOR

The x-intercepts and extreme values on a polynomial graph occur in the graph's "midsection," sometimes referred to as its region of interest. The graph's behavior outside the region of interest is referred to as its **end behavior**. For an example of physical information provided by a function's end behavior, see ***Robert Arnold*** (page 312).

The simplest question about end behavior is whether the "tails" of the graph point up or down. Let's look at an example to see how to answer that question.

EXAMPLE 3 *End behavior of a polynomial graph*

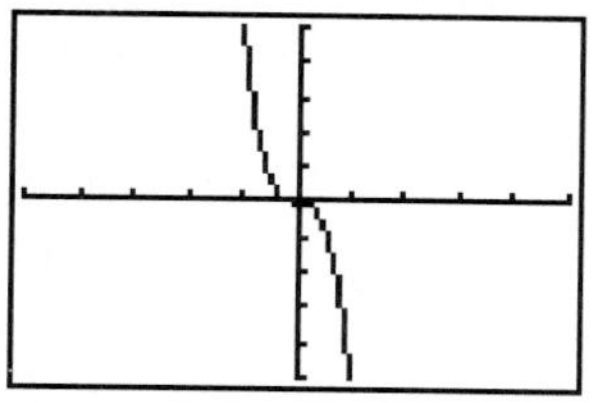

$[-5, 5]$ by $[-100, 100]$

Figure 7–16

Figure 7-16 shows a graph of $P(x) = x^4 - 100x^3$.

Does Figure 7-16 present an accurate picture of the graph's end behavior? We can check by looking at the factored form

$$P(x) = x^3(x - 100)$$

If $x < 0$, both x^3 and $x - 100$ are negative, so their product is positive. Thus the left tail of the graph actually does point up. When $x > 100$, both x^3 and $x - 100$ are positive, so the right tail of the graph also points up! Figure 7-17 shows a more accurate picture.

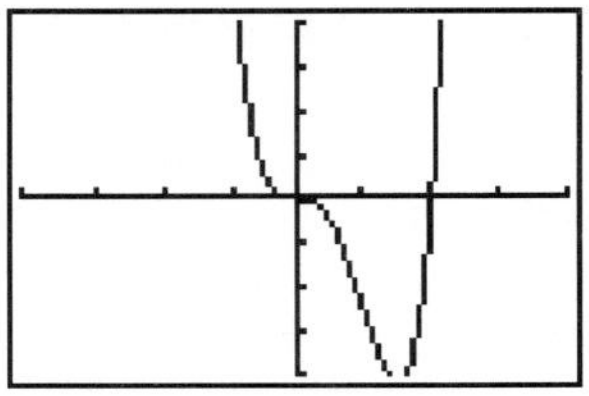

[−200, 200] by
[−10 000 000, 10 000 000]

Figure 7–17

■

Exercises 16–17 illustrate a way to analyze the end behavior of P by looking at the terms of $P(x)$, rather than the factors.

EXERCISES **16.** Complete the following table for the function P in Example 3.

Value of x	Value of x^4	Value of $-100x^3$	Value of $P(x)$
−10,000			
−1,000			
−100			
0			
100			
1,000			
10,000			

17. *(Making Observations)* How does the table in Exercise 16 suggest that when $|x|$ is large, the sign of $P(x)$ is determined by the sign of x^4?

Your observation in Exercise 17 can be generalized. If we go far enough to the left or right along the graph of any polynomial function

$$P(x) = a_nx^n + a_{n-1}x^{n-1} + a_{n-2}x^{n-2} + \cdots + a_1x + a_0$$

the size (absolute value) of the term a_nx^n of largest degree eventually exceeds that of all the other terms combined. This means that on the tails of the graph, the sign of $P(x)$ will agree with that of a_nx^n. This conclusion can be stated as

The Highest Degree Theorem (Polynomial Version)

The end behavior of the graph of any polynomial function

$$P(x) = a_nx^n + a_{n-1}x^{n-1} + a_{n-2}x^{n-2} + \cdots + a_1x + a_0$$

is the same as that of $y = a_nx^n$. (In end behavior, only the term of highest degree counts.)

The graph of $y = a_nx^n$ is always that of a basic power function, possibly stretched or compressed vertically, and reflected in the x-axis if

$a_n < 0$. Thus the end behavior of all polynomial functions is summarized in Table 26.

TABLE 26

	Conditions on n and a_n	**End behavior**	**Graph**
Example: $y = x^3$	n odd, $a_n > 0$	Left tail down, right tail up	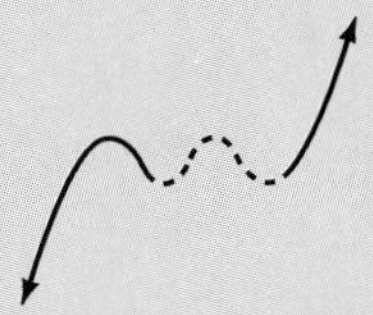
Example: $y = -x^3$	n odd, $a_n < 0$	Left tail up, right tail down	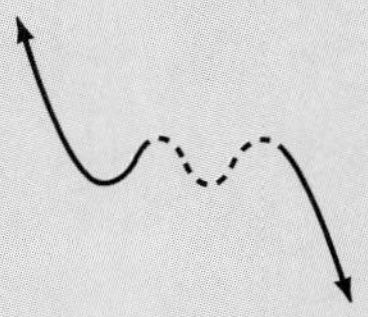
Example: $y = x^2$	n even, $a_n > 0$	Left tail up, right tail up	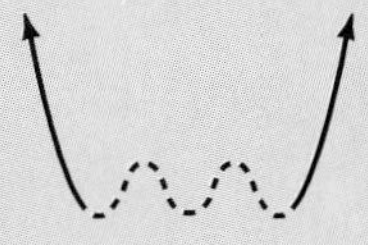
Example: $y = -x^2$	n even, $a_n < 0$	Left tail down, right tail down	

You do not need to memorize Table 26. It is easier to apply the principles that generated it, as in Example 4.

EXAMPLE 4 *Describing the end behavior of a polynomial function*

The end behavior of

$$P(x) = -4x^5 + 6x^4 + x^3 - 10x^2 + 5x - 1$$

is the same as that of $y = -4x^5$. The graph of $y = -4x^5$ is that of $y = x^5$ stretched vertically by a factor of 4 and reflected in the x-axis. On the graph of $y = x^5$, the left tail points down and the right tail points up as in Figure 7-18a. The reflection reverses the direction of the tails so that the left tail on the graph of $y = -4x^5$ points up and the right tail points down as in Figure 7-18b. Figure 7-18c shows that the graph of $y = P(x)$ exhibits the same end behavior.

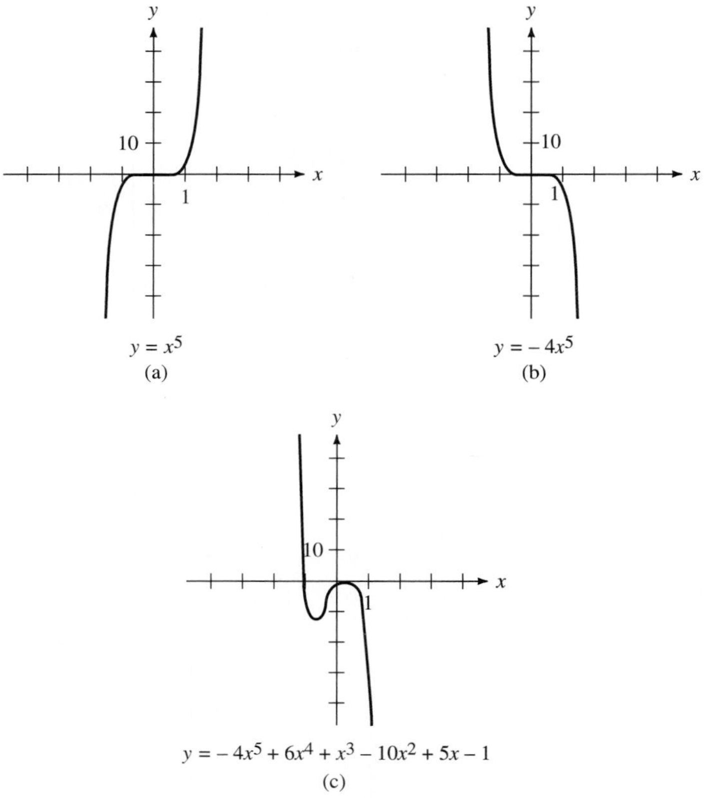

Figure 7–18

EXERCISES *In each of Exercises* 18–21, *describe the end behavior of the graph of the function.*

18. $g(t) = t^4 - 5t^3 + 7$

19. $P(x) = 2x^5 - 9x^4 + 12x^3$

20. $f(x) = 20 - x^2 - 5x^4$

21. $y = 1{,}000{,}000x - x^5$

Match the equations in Exercises 22–25 *with the graphs in Figure* 7–19.

22. $y = x^3 - 4x$ **23.** $y = 4x^3 - 3x^4$

24. $y = -2x^5 + 2x + 3$ **25.** $y = x^6 + 3x^4 - 4x^2$

26. *(Writing to Learn)* Explain how to generate Table 26 from your knowledge of power function graphs and transformation principles.

SYMMETRY

You may have noticed the symmetry of the graphs in Figure 7-19a, b, and c. You can detect symmetry analytically, using ideas similar to those in Section 6-3.

One type of symmetry is illustrated in Figure 7-20a, which shows that $f(a) = b$ and $f(-a) = b$.

More generally in Figure 7-20a, $f(-x) = f(x)$ for any x, so that replacing x by $-x$ leaves the equation unchanged. This means that reflecting the graph

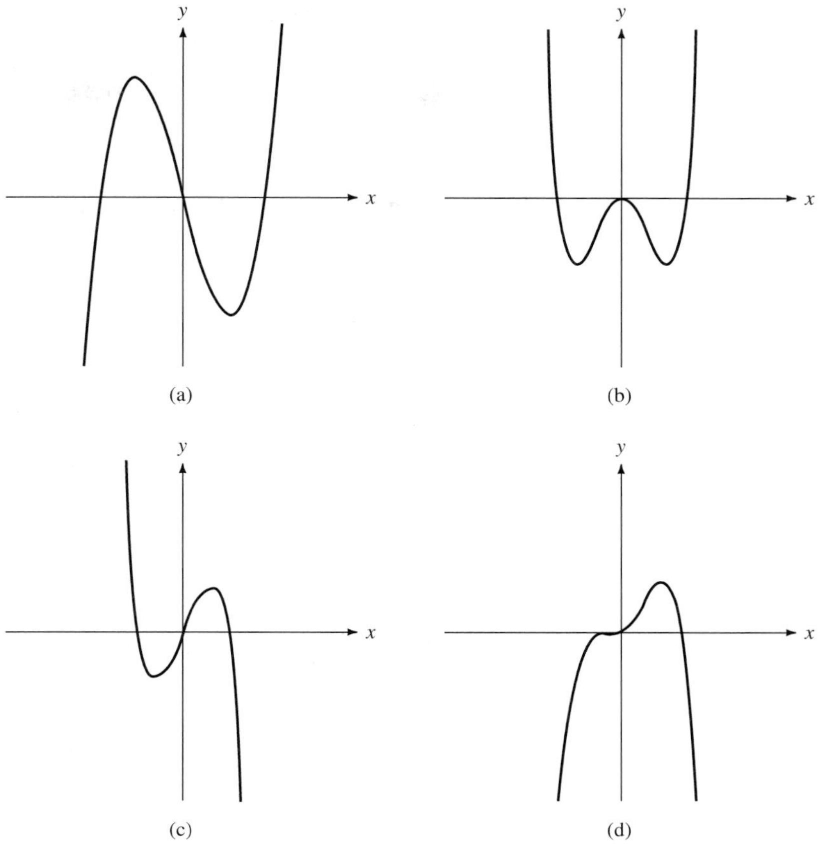

Figure 7–19

about the y-axis leaves the graph unchanged, so the graph is symmetric about the y-axis.

The graph in Figure 7-20b is not symmetric about any axis, but it is **symmetric about the origin**. The figure shows that $g(a) = b$ and $g(-a) = -b$. The symmetry results from the more general fact that $g(-x) = -g(x)$ for any x. Thus the combination of replacing x by $-x$ and y by $-y$ leaves the equation unchanged. [Remember that on the graph y and $g(x)$ are the same.]

These results can be summarized in the following analytical tests for symmetry.

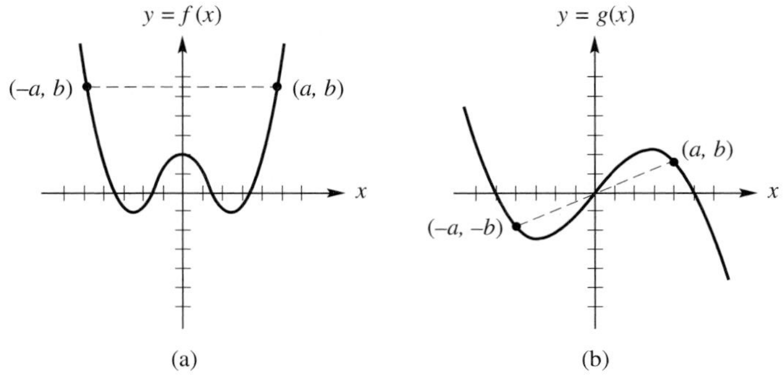

Figure 7–20

If $f(-x) = f(x)$ for each x in the domain of f, then the graph of $y = f(x)$ is symmetric about the y-axis.

If $f(-x) = -f(x)$ for each x in the domain of f, then the graph of $y = f(x)$ is symmetric about the origin.

EXERCISES *In Exercises 27–32, decide whether the graph of $y = f(x)$ is symmetric about the y-axis, the origin, or neither by evaluating $f(-x)$.*

27. $f(x) = 4 - x^2$ **28.** $f(x) = x^3 + x^2$

29. $f(x) = 3x$ **30.** $f(x) = x^7 - 4x^5 + 10x^3$

31. $f(x) = 2x^6 - x^4 + 3$ **32.** $f(x) = 2x^5 - x^3 + 3$

33. *(Writing to Learn)* An analytical method of detecting symmetry about the y-axis was described in Section 6-3. In what respects is it equivalent to the one in this section? In what respects is it different?

34. *(Writing to Learn)* Explain why you never need to test the graph of a function for symmetry about the x-axis.

It is no coincidence that those functions in Exercises 27–32 exhibiting symmetry about the y-axis involve only even powers of x, and that those exhibiting symmetry about the origin involve only odd powers. (A constant term may be regarded as a coefficient of x^0.) A function whose graph is symmetric about the y-axis is often called **even**, while one whose graph is symmetric about the origin is called **odd**. Most polynomial functions involve both even and odd powers of x, and are therefore neither even nor odd.

FYI The terms "even" and "odd" can be applied to functions other than polynomials. In this wider context, functions with odd powers of x can be even, and vice versa. For example, $f(x) = |x|$ is even, and $g(x) = \dfrac{x^4 + 1}{x}$ is odd.

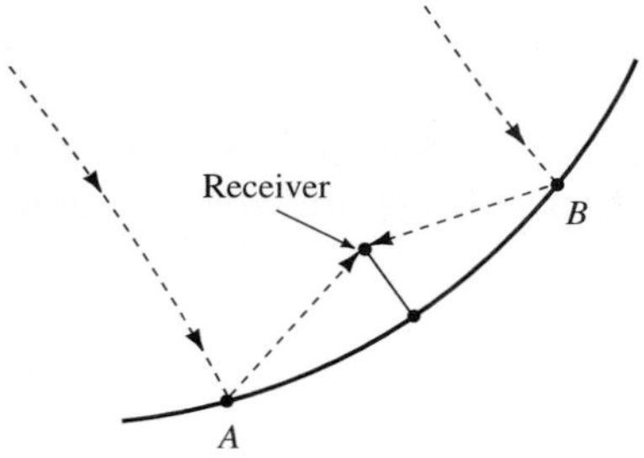

Figure 7–21

Symmetry in a graph that models a physical situation often results from symmetry in the situation. For example, in ***Satellite Dish*** (page 221), the path of a signal to the receiver from point A in Figure 7-21 mirrors the path from point B. The graph of the resulting equation of the dish, $y = \dfrac{x^2}{24}$, is symmetric about the y-axis. Here, symmetry provides a partial check on the accuracy of our model. If we see symmetry in the physical situation, but not in the model, we need to modify the model.

Conversely, symmetry in a mathematical model often leads to the discovery of symmetry in a physical situation. For example, during the 1860s, James Clerk Maxwell rewrote the equations of electromagnetism to make them symmetric. Although no experimental evidence suggested that his altered equations had any physical meaning. Maxwell used them to predict, among other things, the existence of radio waves. Some years later, Heinrich Hertz confirmed Maxwell's predictions by detecting radio waves experimentally. Perhaps it is for this reason that, according to A. Zee in *Fearful Symmetry*, physicists today ". . . search their own minds for that which constitutes symmetry and beauty. In the silence of the night, they listen for voices telling them about yet-undreamed-of symmetries."

ADDITIONAL EXERCISES

In Exercises 35–38, *write the equation of a polynomial function with only simple real zeros, having the indicated x-intercepts.*

35. $(1, 0), (4, 0)$

36. $(-5, 0), (0, 0), (3, 0)$

37. $(0.1, 0), (1, 0), (10, 0), (100, 0)$

38. $(-\sqrt{2}, 0), (\sqrt{2}, 0), (4, 0)$

In Exercises 39–42, *write an equation in factored form for a sixth degree polynomial whose graph resembles the one shown.*

39.

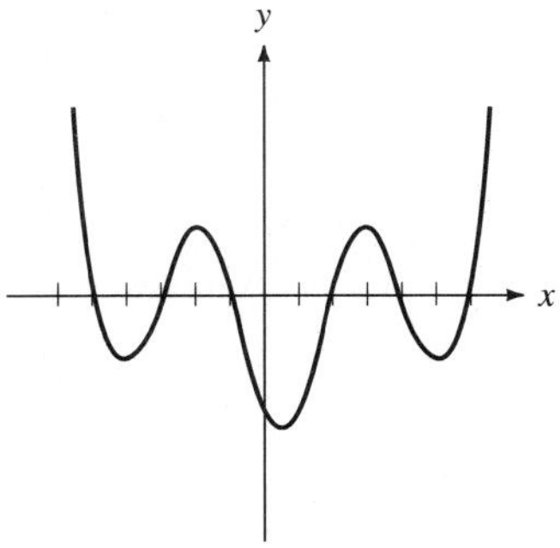

40.

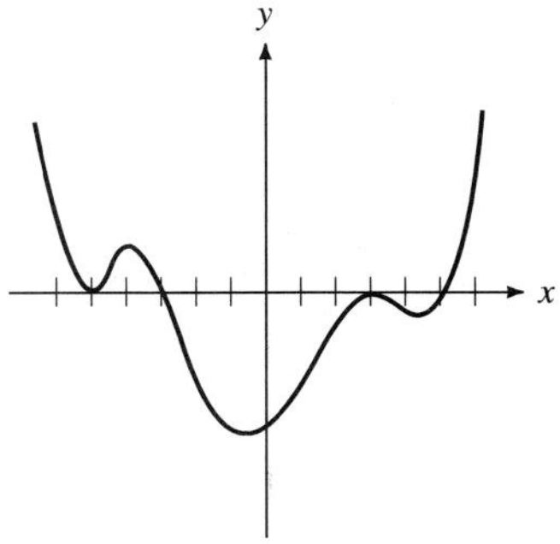

41.

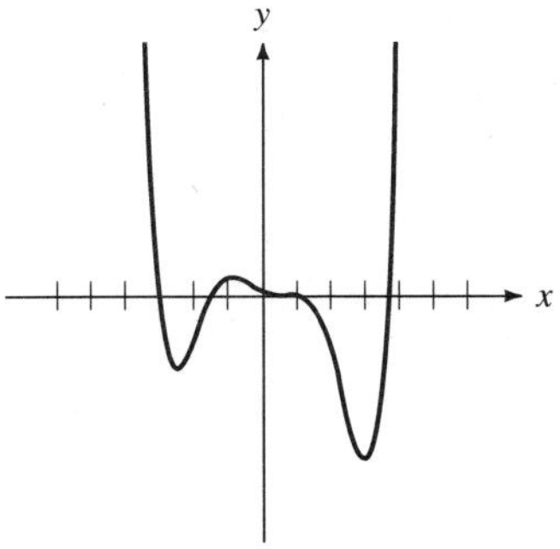

42.

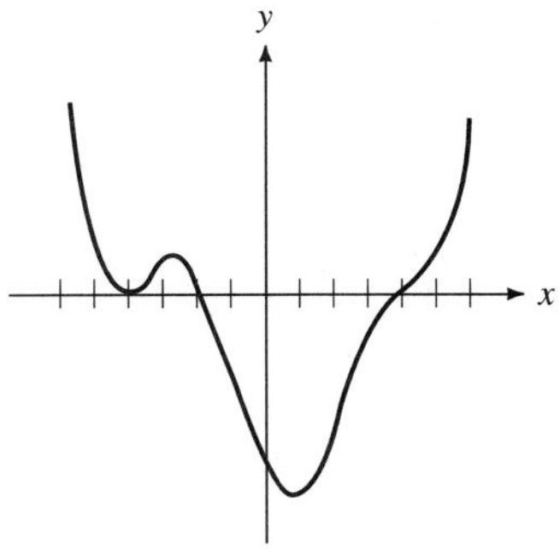

In Exercises 43–48,

a. Find the exact or approximate value of each real zero of the function.

b. Find all extreme values, and identify any which are absolute.

c. Describe the end behavior of the graph.

d. Identify a viewing window which accurately depicts the intercepts, turning points, and end behavior of the graph. State the information you needed in order to choose the window.

e. Determine the intervals over which the function is increasing and decreasing.

43. $y = x^4 - 6x^2 + 4$

44. $y = 4x^5 - 5x^4$

45. $y = 49 - 97x + 23x^2 - x^3$

46. $y = 6 + x + 0.3x^2 - 0.01x^5$

47. $y = -x^4 - x^3 + 3x^2 - 4x + 28$

48. $y = x^5 - 42x^3 + 6x^2 + 185x - 30$

Classify the functions in Exercises 49–52 as even, odd, or neither, using an analytical test. Support your answer with a calculator graph.

49. $U(x) = x^4 - 1$

50. $V(x) = x(x^4 - 1)$

51. $W(x) = (x - 1)^4$

52. $Z(x) = (x - 1)^4(x + 1)^4$

53. *(Making Observations)* Carefully sketch the graph of $y = \frac{1}{3}x^3 - 3x$ over the interval $[-4, 4]$, making each unit about half an inch.

a. Calculate the average rate of change in y with respect to x over each of the intervals $[-2.5, -2]$ and $[3, 3.5]$. Over which interval does the average rate of change have the larger absolute value?

b. Draw secant lines (lines joining points on the graph) over each of the intervals $[-2.5, -2]$ and $[3, 3.5]$. Over which interval is the secant line steeper?

c. Calculate the average rate of change in y with respect to x over each of the intervals $[-1.5, -0.5]$ and $[-0.5, 0.5]$. Over which interval does the average rate of change have the larger absolute value?

d. Draw secant lines on the graph over each of the intervals $[-1.5, -0.5]$ and $[-0.5, 0.5]$. Over which interval is the secant line steeper?

54. *(Writing to Learn)* The range of a polynomial function is $(-\infty, \infty)$ if and only if its degree is odd. Explain how your knowledge of end behavior leads to this conclusion.

Find the domain and range of each function in Exercises 55–60.

55. *(Review)* $g(x) = 400x - x^3$

56. *(Review)* $f(x) = x^6 - 7x^4 + 9x^2 - 12$

57. *(Review)* $f(x) = -2x^6 + 24x^4 - 96x^2$

58. *(Review)* $f(x) = x^4 - 14x^3 + 24x^2$

59. *(Review)* $f(x) = x^6 - 12x^3 + 100$

60. *(Review)* $f(x) = x^7 - 1000x$

61. "From a distance, the world looks blue and green,
And the snow-capped mountains white."

From a Distance
Julie Gold

(Writing to Learn) Graph each of the following functions: $f(x) = x^7$ and $g(x) = x^7 - 3x^6 + x^5$, first in the window $[-2, 4]$ by $[-50, 50]$ and then in the window $[-100, 100]$ by $[-10^{14}, 10^{14}]$. These graphs provide visual support for the Highest Degree Theorem. Explain why. (What does the line from the song have to do with the exercise? You may want to use it in your explanation.)

62. *(Writing to Learn)* In examining extreme values of polynomial functions, Johannes Kepler adopted the strategies of *Making a Table* and *Looking for a Pattern*. In *Nova stereometria doliorum vinariorum*, published in 1615, he stated that "near a maximum the decrements on both sides are in the beginning only imperceptible." What do you think he meant by this? (Before attempting to answer, graph a polynomial function that has a maximum. Trace along the graph, and notice how the y-coordinates change as you trace both near the maximum and farther away from it.)

7–4 A NUMERICAL VIEW OF POLYNOMIAL FUNCTIONS

*n*TH-ORDER DIFFERENCES

What tables fit polynomial functions? This question is of considerable practical importance. In this section you will see how scientists can check the accuracy of data transmitted from a space probe by fitting a polynomial function to the data.

Let's use the strategy of *Guessing and Checking* to explore the question. Since tables from linear and quadratic functions have constant first and second differences, respectively, we might guess that tables from polynomial functions of degree n have constant nth-order differences. We can begin to check our guess by *Examining a Special Case*. Table 27 was generated by the simplest cubic function $f(x) = x^3$.

TABLE 27

x	$f(x) = x^3$	First differences	Second differences	Third differences
−3	−27			
		19		
−2	−8		−12	
		7		6
−1	−1		−6	
		1		6
0	0		0	
		1		6
1	1		6	
		7		
2	8			

EXERCISES

1. *(Making Observations)* Construct a table for each cubic function using the indicated x-values. Calculate the first, second, and third differences for the y-values.

   a. $y = 3x^3; x = 1.0, 1.5, 2.0, 2.5, 3.0, 3.5$

   b. $y = ax^3; x = -1, 0, 1, 2, 3, 4$

   c. $y = ax^3; x = x_0, x_0 + h, x_0 + 2h, x_0 + 3h, x_0 + 4h, x_0 + 5h$

In Exercise 1 you have shown that for any function $y = ax^3$, any table with equally spaced x-values has y-values with constant third differences. Now let's see whether the same thing is true of any function $y = ax^3 + bx^2 + cx + d$.

2. *(Making Observations)* Construct a table for each cubic function using the indicated x-values. Calculate the first, second, and third differences for the y-values.

   a. $y = 3x^3 + x^2 + 2x - 5$
   $x = 1.0, 1.5, 2.0, 2.5, 3.0, 3.5$

   b. $y = ax^3 + bx^2 + cx + d$
   $x = -1, 0, 1, 2, 3, 4$

Exercise 2 suggests that for any cubic function, a table with equally spaced x-values always has y-values with constant third differences.

Conversely, if a table with equally spaced x-values has y-values with constant third differences, it fits a cubic function. It can be shown that similar statements apply to polynomial functions of any degree.

> Any table with equally spaced x-values fits a polynomial function of degree n if and only if it has y-values with constant nth-order differences.

FYI As with linear and quadratic functions, we cannot say that a function is a polynomial on the basis of a table without additional information. We can only say that the table fits a polynomial function. However, if the nth-order differences are not constant, then the function is not a polynomial of degree n.

EXERCISES *In Exercises* 3–6, *find the smallest degree of a polynomial function that fits the table.*

3.

x	y
2	0
3	15
4	48
5	105
6	192
7	315
8	480

4.

x	y
−2	−45
−1.5	35.5
−1	17
−0.5	−1.5
0	−5
0.5	−2.5
1.0	33

5.

x	y
−3	−1
−1	−3
1	−3
3	−1
5	3
7	9
9	17

6.

x	y
0	1
1	0
2	−1
3	−2
4	−1
5	0
6	1

FYI A table with only $n + 1$ points always fits a polynomial function of degree n. For this reason, you should look for additional evidence before concluding that a polynomial function of degree n actually generated the table.

FITTING A POLYNOMIAL FUNCTION TO A TABLE

The following Mathematical Looking Glass illustrates that it is sometimes useful to find the equation of a polynomial function that fits a given table, even if that function did not generate the table.

A MATHEMATICAL LOOKING GLASS

Error-Correcting Codes

Recent advances in technology have enabled us to produce photographlike images of objects that are inaccessible to standard photographs. We see a CT scan of a human brain or a detailed view of a distant planet so often that we may not stop to think about how the images are generated. The first step in "photographing" such an object is to gather data and store it in a computer. For example, a CT scan sends x-rays through a person's head from many directions. The intensity of each ray decreases sharply as it passes through the skull, and less sharply as it passes through the less dense material of the brain. As it exits the head its intensity is recorded as a number that indicates the average density of the material it has passed through. A visual image is then reconstructed from the collection of numbers, using techniques that depend heavily on calculus. Any area of abnormal density in the brain, such as a tumor, appears on the reconstructed image.

To obtain images of distant planets, a space probe has sensors that record the intensity of light at many points. The intensities are beamed back to Earth as numbers from which a visual image can be mathematically reconstructed. A portion of the data sent by the probe might be

$$(0, 6), (1, 5), (2, 4), (3, 9)$$

indicating that the intensity level was 6 at position 0, 5 at position 1, 4 at position 2, and 9 at position 3. However, the data may become garbled as it travels through space, and may arrive on Earth as

$$(0, 6), (1, 5), (2, 9), (3, 4)$$

The accuracy of the reconstructed visual image can be affected by such errors. Fitting the data to a polynomial function is used to check the accuracy of the transmissions. In Exercise 8 you can find a cubic function to fit the four ordered pairs sent from the probe. The next step is to generate and transmit many more ordered pairs that fit the same function. These "extra" ordered pairs, which are not used to reconstruct the visual image, are referred to as **error-correcting codes**. Even if a few of the many ordered pairs become altered in transmission, it is generally easy to identify the function that generated the collection. The process is described in more detail by Linda Kurz in "Error-Correcting Codes," *The AMATYC Review* (Vol. 17, no. 1, 1995).

By the way, scientists at NASA are thankful that they have computers to do their number-crunching for them. The data from *Voyager 2*, whose images of Neptune were described in ***The Blue Planets*** (page 203), had to be fit to a polynomial of degree 222. ■

EXERCISES

7. *(Writing to Learn)* Why is it reasonable to look for a cubic function to fit the data in ***Error-Correcting Codes***, rather than a polynomial function of some other degree?

8. *(Creating Models)* Find a cubic function to fit the ordered pairs (0, 6), (1, 5), (2, 4), (3, 9). (*Hint*: The function has the form $f(x) = ax^3 + bx^2 + cx + d$. You know $f(0) = 6$, so $a(0)^3 + b(0)^2 + c(0) + d = 6$. Use the other three ordered pairs to create a system of four equations in the variables a, b, c, and d, and solve the system.)

9. Find six other appropriate ordered pairs to transmit back to Earth along with the original four.

In Exercises 10–13, *find a polynomial function of the smallest possible degree to fit the given ordered pairs.*

10. (2, 1), (3, 4), (4, 5)

11. (0, 1), (1, 3), (2, 1), (3, 3)

12. (1, 7), (2, 1), (3, −1), (4, −1)

13. (0, 1), (1, −3), (2, −5), (3, 19), (4, 117)

A NUMERICAL METHOD FOR FINDING ZEROS

Analytical and graphical methods for finding zeros of polynomial functions are discussed in Sections 7-2 and 2-4. You are about to see a numerical method, called the **bisection method**, that approximates zeros by repeatedly refining an initial guess. The method evolved over the course of centuries, rather than being developed by a single person.

To use the method, you must first find an interval containing a zero. You might find the interval by using a graph or a table. Example 1 applies the method to $P(x) = x^5 - x - 1$.

EXAMPLE 1 *Using the bisection method to approximate a zero*

The following table was generated by $P(x) = x^5 - x - 1$.

$$P(0) = -1$$
$$P(1) = -1$$
$$P(2) = 29$$
$$P(3) = 239$$
$$P(4) = 1019$$
$$P(5) = 3119$$

Since $P(1)$ and $P(2)$ have opposite signs, it is reasonable to expect that P has a zero in the interval $(1, 2)$.

To find a smaller interval around the zero, evaluate P at the midpoint of $(1, 2)$. Now your table becomes

$$P(1) = -1$$
$$P(1.5) \cong 5.09$$
$$P(2) = 29$$

Since $P(1)$ and $P(1.5)$ have opposite signs, P has a zero in $(1, 1.5)$. The point 1.5 bisects the interval $(1, 2)$ and thus cuts the size of the interval containing the zero in half. This explains why the technique is called the bisection method.

You can bisect the interval again by evaluating P at the midpoint of $(1, 1.5)$, so that your table becomes

$$P(1) = -1$$
$$P(1.25) \cong 0.80$$
$$P(1.5) \cong 5.09$$

and there is a zero in $(1, 1.25)$.

At the next step, the table becomes

$$P(1) = -1$$
$$P(1.125) \cong -0.32$$
$$P(1.25) \cong 0.80$$

and there is a zero in $(1.125, 1.25)$.

To approximate the zero with an error of less than 0.01, continue until the width of the interval around the zero is less than 0.01. The work is summarized in Table 28. Since the width of the last interval is less than 0.01, you can choose any number within it as your estimate. One possible choice is $x = 1.17$. ■

TABLE 28

Step	Interval	Midpoint	Value of P at left endpoint	Value of P at midpoint	Value of P at right endpoint
1	(1, 2)	1.5	−1	5.09	29
2	(1, 1.5)	1.25	−1	0.80	5.09
3	(1, 1.25)	1.125	−1	−0.32	0.80
4	(1.125, 1.25)	1.1875	−0.32	0.17	0.80
5	(1.125, 1.1875)	1.15625	−0.32	−0.09	0.17
6	(1.15625, 1.1875)	1.171875	−0.09	0.04	0.17
7	(1.15625, 1.171875)	1.1640625	−0.09	−0.03	0.04
8	(1.1640625, 1.171875)				

FYI If $P(x_1)$ and $P(x_2)$ have opposite signs, then P has at least one zero in (x_1, x_2). However, if $P(x_1)$ and $P(x_2)$ have the same sign, there still may be zeros in (x_1, x_2), as in Figure 7-22. A graph can help you decide whether this is the case.

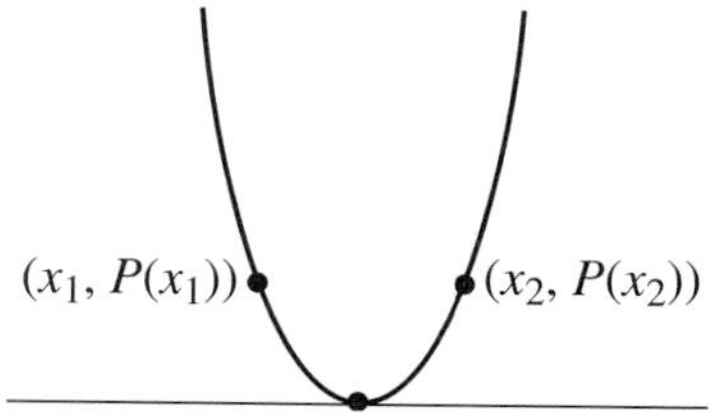

Figure 7–22

EXERCISES *In Exercises 14–17, use the bisection method to approximate the largest real zero of the function with an error of no more than 0.01. Use a graph to establish initial intervals in which the zero can be found.*

14. $f(x) = x^3 + 4x - 8$

15. $w = 3x^3 - 10x + 6$

16. $G(x) = 2x^4 - 3x^3 + 2x^2 - x - 3$

17. $y = x^7 - 3x^5 + 2x^2 - 1$

Now that you understand how the bisection method works, you might program it into your calculator for future use.

A COMPARISON OF METHODS FOR FINDING ZEROS

The method you choose to find the zeros of a polynomial function depends on the nature of the function and the required degree of precision of your answer. A comparison of analytical, graphical, and numerical methods reveals several differences in effectiveness and convenience.

- The analytical method of factoring is convenient only for polynomials that factor easily.
- The graphical method of "zooming-in" can be used with any polynomial function, but several iterations are often required to produce the desired accuracy. Using the Rational Root Theorem can enhance the efficiency of these methods.
- The numerical bisection method can be programmed into your calculator and can easily provide a high degree of accuracy. However, it requires prior knowledge of an interval containing a zero. A graph can reveal such an interval.
- Analytical methods yield exact results. Graphical and numerical methods usually yield only approximations.
- Although neither graphical nor numerical methods can identify multiple zeros, a graph indicates when multiple zeros may be present.

As you have seen, it is impractical to rely exclusively on any one method for all cases. The most painless "route to the roots" usually involves a combination of two or more. Combining methods is an application of the Pólya strategy of *Breaking the Problem into Parts*.

EXERCISES *In Exercises* 18–21, *find all real zeros of the function, exact if possible, by any combination of methods.*

18. $P(x) = 3x^3 + 8x^2 - 3x$

19. $Q(t) = t^4 - 40t + 39$

20. $H(z) = z^4 - 20z^3 + 126z^2 - 260z + 25$

21. $y = x^7 - 3x^5 + 2x^2 + 1$ (Compare the methods needed here to those needed for Exercise 17.)

ADDITIONAL EXERCISES *In Exercises* 22–25, *determine the smallest degree of a polynomial function that fits the table.*

22.

x	y
−2	18
−1	−15
0	−20
1	−21
2	−30
3	−47
4	−60

23.

x	y
1	0
2	9
3	13
4	14
5	14
6	15
7	18

24.

x	y
−3	−41
−2	−5
−1	5
0	1
1	−5
2	−1
3	25
4	85

25.

x	y
−0.9	−3.12
−0.8	−2.28
−0.7	−1.48
−0.6	−0.72
−0.5	0
−0.4	0.68
−0.3	1.32
−0.2	1.92

In Exercises 26–29, *find a polynomial function of the smallest possible degree to fit the data.*

26. $(2, -16), (4, -8), (6, 48), (8, 176)$

27. $(0, 0), (1, 10), (2, 4), (3, 7)$

28. $(1, 2), (2, 17), (3, 82), (4, 257), (5, 626)$

29. $(-1, 25), (0, 18), (1, 12), (2, 7), (3, 3)$

In Exercises 30–33, use the bisection method to approximate the largest real zero of the polynomial. Use a graph to establish an initial interval in which the zero can be found.

30. $y = 2 - x^5$

31. $y = x^4 - 6x^2 + 4$

32. $y = 98 - 194x + 46x^2 - 2x^3$

33. $y = 6 + x + 0.3x^2 - 0.01x^5$

In Exercises 34–37, find all real zeros of the function, exact if possible, by any combination of methods.

34. $y = x^5 + x^3 - 3x^2$

35. $H(z) = z^5 + z^3 - 2z - 4$

36. $S(a) = a^6 + a^3 - 2$

37. $P(x) = 3x^4 - 4x^3 - 16$

38. *(Problem Solving)* The following table shows the population of the United States according to censuses taken in the given years.

Year	Population (millions)
1950	151
1960	179
1970	203
1980	227

a. Find a cubic function to fit the data.

b. Use your function to predict the population in 1990, and compare it to the actual population of 249 million.

c. Use your function to predict the population in each of the years 2000, 2050, and 2100. Which of these predictions, if any, do you trust to be accurate to a reasonable degree?

d. Do you trust the method used to obtain the estimates in parts (b) and (c)? Why or why not?

7–5 SOLVING POLYNOMIAL INEQUALITIES

The graphical and analytical methods you learned in Chapter 5 for solving quadratic inequalities can be applied to polynomials of higher degree. As with quadratic inequalities, the first step in solving a polynomial inequality is to solve a polynomial equation. For most polynomials of higher degree, this first step is the most difficult part of the solution. Because of this, mathematicians of past centuries concentrated on solving polynomial equations. The precise historical origins of the methods used to solve polynomial inequalities are unknown.

SOLVING POLYNOMIAL INEQUALITIES GRAPHICALLY

In the following Mathematical Looking Glass, your t-shirt company from ***Fat Cats on Campus*** (page 43) encounters a polynomial inequality.

A MATHEMATICAL LOOKING GLASS
Growth of Cost (Fat Cats 6)

As a special assignment, your economics professor has asked your class to see how an increase in the size of your company would affect your pricing strategies and your profit. The cost and revenue functions you calculated in ***Fat Cats on Campus*** are

$$C(x) = 3.50x + 612.50$$

$$R(x) = -0.005x^2 + 16x$$

If your company grows, it will be more convenient to express sales in units of 100 shirts. In Exercise 5 you can verify that if x represents hundreds of shirts, the cost and revenue functions should be

$$C(x) = 350x + 612.50$$

$$R(x) = -50x^2 + 1600x$$

Your professor has suggested that a more accurate cost function might be

$$C(x) = 0.08x^3 - 6x^2 + 350x + 612.50$$

In Exercise 6 you can discover why costs for a larger company might be modeled accurately by a cubic function. Because your profit is the difference between revenue and cost, your profit function will also be cubic. To discover your profitable target sales levels, you need to solve the cubic inequality $P(x) \geq 0$. ■

The inequality in ***Growth of Cost*** can be solved either graphically or analytically. You already know how to solve polynomial inequalities graphically. Let's look at Example 1 to review the method.

EXAMPLE 1 *Solving a polynomial inequality graphically*

To solve $x^3 + 2x^2 - 9x < 0$, begin by graphing $f(x) = x^3 + 2x^2 - 9x$, as in Figure 7-23.

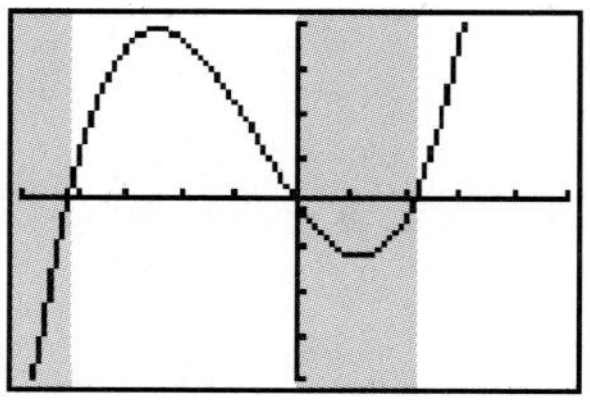

$[-5, 5]$ by $[-20, 20]$

Figure 7–23

The inequality is true at those values of x where $f(x) < 0$, or equivalently, where the graph is below the x-axis. The solution is shaded in Figure 7-23. You can identify it by finding the x-intercepts, or equivalently, solving $f(x) = 0$. You can verify that the intercepts are at $x = 0$ and $x = -1 \pm \sqrt{10} \cong -4.16, 2.16$. The inequality is false at each intercept, but true to the left of $-1-\sqrt{10}$ and between 0 and $-1 + \sqrt{10}$. The solution is

$$(-\infty, -1 - \sqrt{10}) \cup (0, -1 + \sqrt{10}) \cong (-\infty, -4.16) \cup (0, 2.16)$$ ■

EXERCISES *In Exercises 1–4, solve the inequality graphically.*

1. $x^3 - 8x^2 + 19x - 13 > 0$

2. $3 + 2x^2 - x^4 > 0$

3. $x^4 - 4x^3 - 2x^2 + 12x + 9 \leqslant 0$

4. $x^5 + x^4 + 2x^3 \geqslant 0$

5. *(Writing to Learn)* In ***Growth of Cost***, if x represents hundreds of shirts, explain why your former cost and revenue functions should be modified to

$$C(x) = 350x + 612.50$$
$$R(x) = -50x^2 + 1600x$$

[*Hint*: Begin with the former functions

$$C(x) = 3.50x + 612.50$$
$$R(x) = -0.005x^2 + 16x$$

and write expressions for $C(100x)$ and $R(100x)$.]

6. *(Interpreting Mathematics)* As your company grows, the average cost of producing additional shirts may decrease for a while. For example, you may be able to buy larger quantities of shirts at a cheaper rate. Economists refer to such savings as **economies of scale**. Eventually, however, the average cost of producing additional shirts may increase again due to size-related costs, called **diseconomies of scale**. For example, you may need to hire new employees and rent additional space.

a. For your professor's suggested cost function

$$C(x) = 0.08x^3 - 6x^2 + 350x + 612.50$$

calculate the average rate of change in $C(x)$ with respect to x over each of the intervals $[0, 10]$, $[20, 30]$, and $[50, 60]$.

b. Interpret the results from part (a) in physical language. (Remember that x is expressed in units of 100 shirts.)

c. Over which interval in part (a) do you see evidence of economies of scale? Over which interval do you see evidence of diseconomies of scale? Explain your answers.

7. *(Problem Solving)*

a. Use the cost and revenue functions

$$C(x) = 0.08x^3 - 6x^2 + 350x + 612.50$$
$$R(x) = -50x^2 + 1600x$$

to find your company's profit function P.

b. Graph $y = P(x)$ in the window $[0, 50]$ by $[-10{,}000,\ 10{,}000]$ and solve $P(x) \geqslant 0$ to find your profitable target sales levels.

c. Graph $y = P(x)$ in the window $[-1000, 1000]$ by $[-3{,}000{,}000, 3{,}000{,}000]$ and solve $P(x) \geqslant 0$ as an abstract inequality. (*Hint*: The graph has three x-intercepts.) How do your solutions in parts (b) and (c) differ? Why?

SOLVING POLYNOMIAL INEQUALITIES ANALYTICALLY

The Test-Value Method Polynomial inequalities can be solved analytically by using the test-value method you learned in Section 5-2. Example 2 reviews the method.

EXAMPLE 2 *Solving a polynomial inequality analytically*

STEP 1: To solve $x^3 + 2x^2 - 9x < 0$, begin by solving $x^3 + 2x^2 - 9x = 0$ to obtain $x = 0$ and $x = -1 \pm \sqrt{10} \cong -4.16, 2.16$.

STEP 2: Indicate $-4.16, 0$, and 2.16 on a number line.

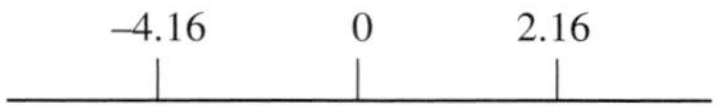

STEP 3: Evaluate $f(x) = x^3 + 2x^2 - 9x$ at a test value in each of the intervals $(-\infty, -4.16), (-4.16, 0), (0, 2.16)$, and $(2.16, \infty)$.

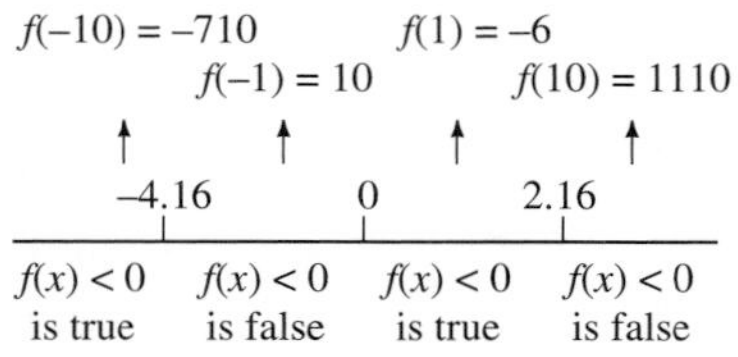

The solution is $(-\infty, -1 - \sqrt{10}) \cup (0, -1 + \sqrt{10}) \cong (-\infty, -4.16) \cup (0, 2.16)$. ■

EXERCISES *In Exercises* 8–11, *solve the inequality analytically.*

8. $x^3 - 9x > 0$ **9.** $8x^2 - x^4 \leqslant 0$

10. $x^5 + x^4 - 2x^3 \geqslant 0$ **11.** $x^5 + x^4 + 2x^3 \geqslant 0$

12. *(Writing to Learn)* Compare your graphical solution to Exercise 4 with your analytical solution to Exercise 11. In what respects, if any, was each solution easier?

The Scan Method The following method is especially useful for inequalities involving factored polynomials. It uses some graphical observations from Section 7-3 to create a streamlined analytical procedure we will call the **scan** method (our acronym for Sign Change ANalysis).

EXAMPLE 3 *Solving a polynomial inequality by the scan method*

Let $P(x) = (x + 5)(2x + 1)^2(x - 3)^3$. To solve $P(x) \geqslant 0$,

STEP 1: Solve $P(x) = 0$ to obtain $x = -5, -\frac{1}{2}$, and 3.

STEP 2: Indicate these solutions on a number line.

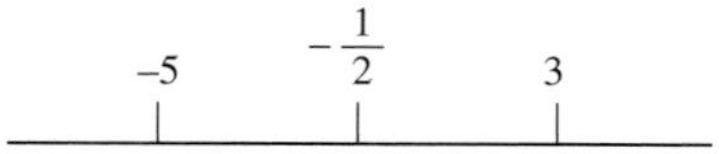

STEP 3: To find the sign of $P(x)$ on $(-\infty, -5)$, we need to determine the left-end behavior of P. The leading term of $P(x)$ in its expanded form is $(x)(2x)^2(x)^3 = 4x^6$. [This is the product of the terms of highest

degree in each factor. You do not need to multiply $P(x)$ out.] The end behavior of $P(x)$ agrees with that of $4x^6$, so $P(x) > 0$ on $(-\infty, -5)$.

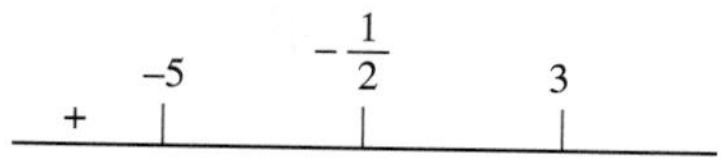

STEP 4: To find the sign of $P(x)$ in the intervals $\left(-5, -\frac{1}{2}\right)$, $\left(-\frac{1}{2}, 3\right)$, and $(3, \infty)$, *scan* the zeros, remembering that the graph of P crosses the x-axis only at a zero of odd multiplicity. That is, the sign of $P(x)$ changes only at a zero of odd multiplicity.

The zero at $x = -5$ has multiplicity 1, so the sign of $P(x)$ changes there. Thus $P(x)$ becomes negative in $\left(-5, -\frac{1}{2}\right)$.

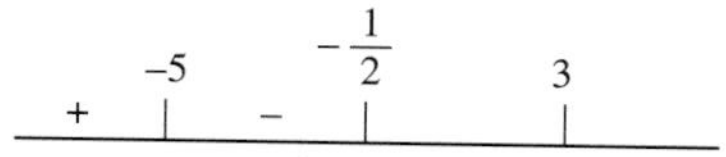

The zero at $x = -\frac{1}{2}$ has multiplicity 2, so the sign of $P(x)$ does not change there. Thus $P(x)$ remains negative in $\left(-\frac{1}{2}, 3\right)$.

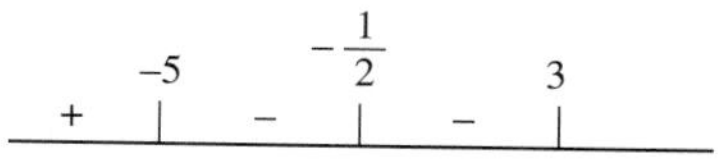

The zero at $x = 3$ has multiplicity 3, so the sign of $P(x)$ changes there. Thus $P(x)$ becomes positive in $(3, \infty)$.

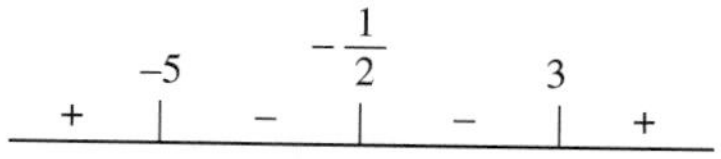

We can now see that $P(x) > 0$ in $(-\infty, -5) \cup (3, \infty)$. Since $P(x) = 0$ at each of the zeros, the solution to $P(x) \geq 0$ is

$$(-\infty, -5] \cup \left\{-\frac{1}{2}\right\} \cup [3, \infty)$$

■

EXERCISES *In Exercises* 13–16, *solve the inequality using the scan method.*

13. $(x + 7)(3x + 4) \geq 0$

14. $(x + 7)^2(3x + 4)^3 \geq 0$

15. $(1 - x)(7 - 2x)^2(5 - x)^2 < 0$

16. $(x + 2)^3(x - 5)^2(x^2 + 4) > 0$ (*Hint:* The factor of $x^2 + 4$ does not correspond to any real zeros.)

17. *(Problem Solving)* Alice Wells (see ***Popcorn***, page 272) soon realized that she could not find a cylindrical container of optimal size to send popcorn to Paul. She therefore decided to look for a cylindrical container that holds at least 10 ounces of popcorn weighing 1.3 ounces per cubic foot. What radii can her package have if it has a combined length and girth of exactly 108 inches?

18. *(Writing to Learn)* Compare the convenience of the test-value and scan methods of solving polynomial inequalities. How might the form of the inequality influence your choice of methods?

ADDITIONAL EXERCISES

In Exercises 19–22, *solve the inequality graphically.*

19. $x^3 + 2x^2 - 15x < 0$

20. $4x^3 - x^4 > 0$

21. $x^6 - 12x^4 + 48x^2 - 64 \leq 0$

22. $x^4 - 25x^2 + 2x + 3 \geq 0$

In Exercises 23–30, *solve the inequality analytically, using either the test-value method or the scan method.*

23. $x^3 + 2x^2 - 15x < 0$

24. $4x^3 - x^4 > 0$

25. $x^6 - x^4 - 16x^2 + 16 \leq 0$

26. $4x - 3x^3 - x^5 \geq 0$

27. $(2x + 5)(3x - 1)(x - 7) < 0$

28. $(x + 4)^2(x - 12)^2 > 0$

29. $(x + 4)^3(x - 12)^3 > 0$

30. $(x + 79)^3(x + 0.3)^4(x - 140) \leq 0$

SUPPLEMENTARY TOPIC

Using Scan Method to Graph Polynomial Functions

The scan method relies on knowledge of the end behavior of polynomial graphs, as well as the behavior around multiple zeros. The processes of solving an inequality by the scan method, and of graphing a polynomial function, thus complement each other. In particular, the reasoning of the scan method can be used to obtain a quick rough sketch of a polynomial graph.

EXAMPLE 4 *Using the scan method to graph a polynomial function*

The information contained in Example 3 can help you to sketch a graph of $P(x) = (x + 5)(2x + 1)^2(x - 3)^3$. Since $P(x) > 0$ in $(-\infty, -5) \cup (3, \infty)$, the graph is above the x-axis there and below the x-axis in

$$\left(-5, -\frac{1}{2}\right) \cup \left(-\frac{1}{2}, 3\right)$$

Furthermore, the graph flattens out at $x = -\dfrac{1}{2}$ and 3, since the multiplicity of each zero is greater than 1. Figure 7-24 shows a rough sketch of the graph. (You cannot show extreme values accurately without plotting some additional points.)

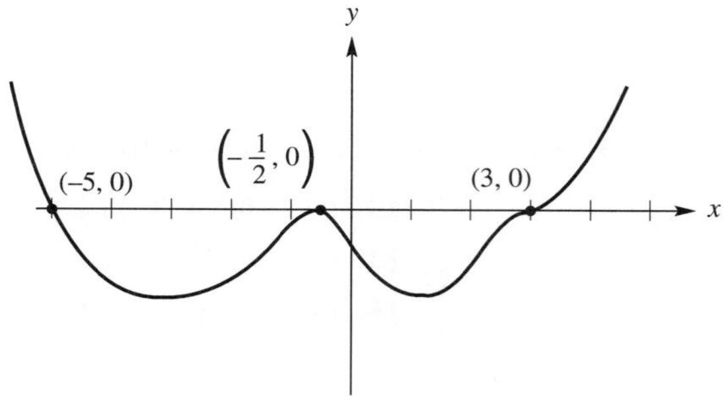

Figure 7–24 ■

In Exercises 31–34, sketch a graph of the function by using information from Exercises 13–16. Your sketch should accurately depict end behavior and behavior around the zeros. It does not need to show extreme values to scale. Support your sketch with a calculator graph.

31. $y = (x + 7)(3x + 4)$

32. $y = (x + 7)^2(3x + 4)^3$

33. $y = (1 - x)(7 - 2x)^2(5 - x)^2$

34. $y = (x + 2)^3(x - 5)^2\,(x^2 + 4)$

35. *(Writing to Learn)* Describe how your graph in Exercise 33 supports your solution in Exercise 15, and vice versa.

In Exercises 36–39, sketch a graph of the function using information from Exercises 27–30. Your sketch should accurately depict end behavior and behavior around the zeros. It does not need to show extreme values to scale. Support your sketch with a calculator graph.

36. $y = (2x + 5)(3x - 1)(x - 7)$

37. $y = (x + 4)^2(x - 12)^2$

38. $y = (x + 4)^3(x - 12)^3$

39. $y = (x + 79)^3(x + 0.3)^4(x - 140)$

CHAPTER REVIEW

Complete Exercises 1–34 (Writing to Learn) before referring to the indicated pages.

WORDS AND PHRASES

In Exercises 1–11, explain the meaning of the words or phrases in your own words.

1. (page 251) **polynomial function**

2. (page 251) **degree** of a polynomial function

3. (page 253) **power function**

4. (page 254) ***y* varies directly with the *n*th power of *x***

5. (page 260) **zero** of a polynomial function

6. (page 262) **simple zero, multiple zero, multiplicity** of a zero

7. (page 271) **turning point, extreme value**

8. (page 271) **local minimum, local maximum, absolute minimum, absolute maximum**

9. (page 273) **end behavior** of a polynomial graph

10. (page 277) **symmetry about the origin**

11. (page 278) **even** function, **odd** function

IDEAS

12. (pages 253, 254) Describe the graph of $y = x^n$ for all integers $n \geq 2$, and describe how to obtain the graph of $y = a(x - h)^n + k$ from the basic graph $y = x^n$. Illustrate with a specific example.

13. (page 255) How can you tell whether a table fits a polynomial variation of a given degree? Make up a table and use it to illustrate the process.

14. (page 254) Why can you regard linear variation as a special case of polynomial variation?

15. (page 261) What does the Factor Theorem say? How can it be helpful in finding the zeros of a polynomial function?

16. (page 261) How does the Factor Theorem guarantee that a polynomial function of degree n has at most n zeros?

17. (page 262) Why is it sometimes reasonable to refer to a zero of a polynomial function as a double zero?

18. (pages 262, 263) What does the Fundamental Theorem of Algebra tell us about the number of zeros of a polynomial function? In the statement of the theorem, are we to count all complex zeros or only the real ones? Are we to count multiple zeros more than once?

19. (page 263) For what values of n are there formulas for solving all polynomial equations of degree n? When were these formulas discovered? When was it discovered that there are degrees for which no formula exists?

20. (page 268) What is the maximum number of x-intercepts on the graph of a polynomial function of degree n? How can this result be established analytically?

21. (pages 269, 270) Describe the behavior of a polynomial graph around a zero of multiplicity k for all integers $k \geqslant 1$.

22. (page 271) What are the similarities and differences between the phrases *local minimum* and *absolute minimum*? Use a sketch to illustrate your explanation.

23. (pages 273–275) Why is the end behavior of a polynomial graph determined by the term of highest degree in its equation?

24. (page 275) Describe the four possible end behaviors for a polynomial graph. Under what conditions does each occur?

25. (pages 276–278) How can you use the equation of a function to decide whether its graph is symmetric about the y-axis? about the origin? Use specific polynomial functions to illustrate.

26. (page 278) What is the origin of the terms *even function* and *odd function*? Must every polynomial function be either even or odd? Exhibit specific functions in support of your answer.

27. (pages 278, 279) Why is it important to consider symmetry in constructing mathematical models of physical situations?

28. (pages 281, 282) How can you tell whether a table fits a polynomial function of a given degree? Make up a table and use it to illustrate.

29. (page 282) Why is it true that every finite table fits a polynomial function of some degree?

30. (page 283) Describe the process of fitting a polynomial function to data. In particular, explain why we can fit a polynomial function of at most degree n to any collection of $n + 1$ data points.

31. (pages 284, 285) Explain how the bisection method works. Make up a polynomial equation and use it to illustrate the method.

32. (pages 285, 286) What are some of the advantages and limitations of the various methods of finding zeros of polynomial functions? How can the effectiveness of the methods be improved by using them in combinations?

33. (pages 289, 290) Describe the steps in the test value method of solving polynomial inequalities, and explain why the method works.

34. (pages 290, 291) Describe the steps in the scan method of solving polynomial inequalities, and explain why the method works.

CHAPTER 8

RATIONAL FUNCTIONS

8-1 RECIPROCAL POWER FUNCTIONS

PREREQUISITES MAKE SURE YOU ARE FAMILIAR WITH:

Algebraic Fractions (Section A-16)

Equations with Algebraic Fractions (Section A-17)

The category of polynomial functions (Chapter 7) includes all linear functions (Chapter 3) and quadratic functions (Chapter 5). The present chapter explores a still more inclusive category. A **rational function** is one that can be written as a ratio (quotient) of two polynomial functions. That is, f is rational if and only if $f(x) = \dfrac{N(x)}{D(x)}$ for some polynomial functions N and D.

Every polynomial function is a rational function. For example, the function $f(x) = x^3 + 5$ can be thought of as $f(x) = \dfrac{x^3 + 5}{1}$. However, as you will see in this chapter, rational functions do not always behave like polynomial functions. In particular,

- Rational functions are not always continuous.
- The end behavior of some rational graphs is different from that of polynomial graphs.

These differences will be explored in depth in Section 8-2. However, they are illustrated by the simplest rational functions, the **reciprocal power functions** $y = \dfrac{1}{x^n}$. These, and other rational functions closely related to them, will be the focus of the present section.

A MATHEMATICAL LOOKING GLASS
Speed Limits

Shortly after Dave graduated from college in 1966, he and his friend Dick Sills drove to Everglades National Park. At that time the speed limit on many interstate highways in southern states was 75 mph, so that Dave and Dick were able to drive a stretch of 450 miles from Atlanta, GA, to Orlando, FL, in 6 hours without breaking any laws. Along the way they were passed by many drivers who were less observant of the speed limit. As one car whizzed by, Dick wondered aloud why anyone would feel the need to exceed a limit of 75 mph. Dave observed that people probably speed in a 75-mph zone for the same reasons that they speed in a 45-mph zone. Speeding cuts time off the trip, and it gives the speeder a feeling of "getting away with something."

Recently Dave recalled this incident and decided to look more critically at his observation. Specifically, he asked whether speeding saves the same amount of time in a 45-mph zone as in a 75-mph zone. You can duplicate his calculations in Exercises 1–3. ■

EXERCISES

1. *(Creating Models)* Let t represent the time, in hours, required to drive from Atlanta to Orlando at a speed of s miles per hour. Express t as a function $g(s)$.
2. *(Interpreting Mathematics)*
   a. Calculate the average rate of change in $g(s)$ with respect to s in each of the intervals $[45, 60]$ and $[75, 90]$.
   b. What does the average rate of change in $g(s)$ with respect to s represent physically?
   c. How does your result from part (a) enable you to answer Dave's question in ***Speed Limits***?
3. *(Problem Solving)* Today the speed limit between Atlanta and Orlando is 65 mph.
   a. If you drive 80 mph, how much time will you save?
   b. If you drive 80 mph and are stopped for speeding, you will be delayed for about 20 minutes and the ticket will cost about \$150. What is your net savings in time? What is the cost of that time in dollars per minute?

Although your function in Exercise 1 is not a reciprocal power function, it is closely related to $f(x) = \frac{1}{x}$. Over the next few pages you will discover several distinctive properties shared by reciprocal power functions and certain related functions. You will also see what information those properties can provide in a physical context like that of ***Speed Limits***.

THREE VIEWS OF RECIPROCAL POWER FUNCTIONS

An Analytical View A reciprocal power function is described analytically by an equation $f(x) = \frac{1}{x^n}$ for some positive integer n. We can identify several properties of f by looking at its equation. In Exercise 4 you can explain why each of the following statements is true.

- The domain of f is $(-\infty, 0) \cup (0, \infty)$.
- The value of $f(x)$ is never 0, but it can be made arbitrarily close to 0 by making $|x|$ sufficiently large. Similarly, $|f(x)|$ can be made arbitrarily large by making x sufficiently close to 0.
- The range of f is $(-\infty, 0) \cup (0, \infty)$ if n is odd, and $(0, \infty)$ if n is even.

EXERCISE **4.** *(Making Observations)*

a. Explain how each of the three preceding statements can be derived from the equation $f(x) = \frac{1}{x^n}$.

b. Describe how each of the statements is reflected in the graphs of $y = \frac{1}{x}$ and $y = \frac{1}{x^2}$, shown in Figure 8-1.

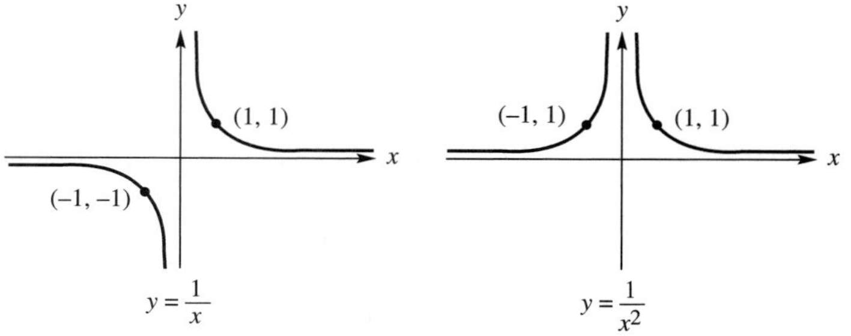

Figure 8–1

A Numerical View The preceding analytical observations can be supported by appropriate tables.

EXERCISES **5.** For $f(x) = \frac{1}{x}$, complete each table.

a.

x	$f(x)$
1	
10	
100	
1000	

b.

x	$f(x)$
−1	
−10	
−100	
−1000	

c.

x	$f(x)$
1	
0.1	
0.01	
0.001	

d.

x	$f(x)$
−1	
−0.1	
−0.01	
−0.001	

6. Repeat Exercise 5 for $g(x) = \frac{1}{x^2}$.

7. *(Writing to Learn)* How do your tables in Exercises 5 and 6 support the statements preceding Exercise 4?

A Graphical View Figure 8-2 shows the graphs of the next few reciprocal power functions after $y = \frac{1}{x}$ and $y = \frac{1}{x^2}$.

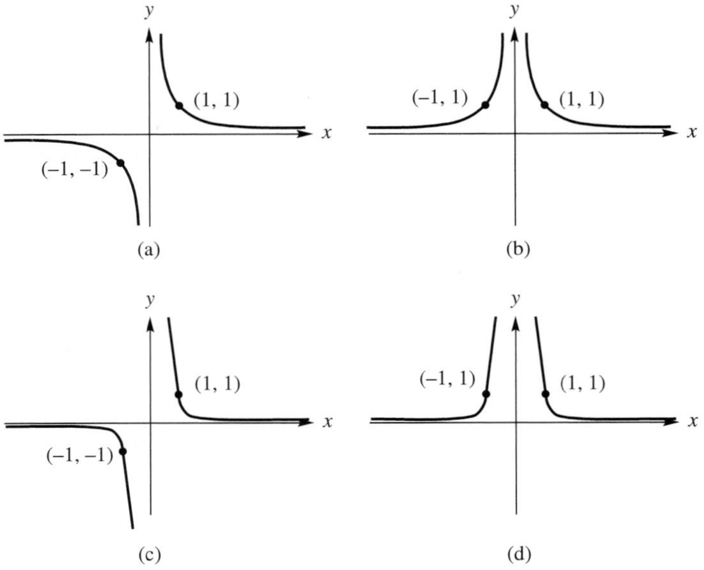

Figure 8–2

EXERCISE **8.** *(Writing to Learn)* How do the graphs in Figure 8-2 support the statements preceding Exercise 4? How do they support your tables in Exercises 5 and 6?

Recall from Section 6-2 that an **asymptote** is a line that a graph approaches as one of the variables becomes arbitrarily large. As $|x|$ becomes large, the graph of $f(x) = \frac{1}{x^n}$ comes arbitrarily close to the x-axis, so that the x-axis is a **horizontal asymptote** for the graph. Similarly, the y-axis is a **vertical asymptote**.

INVERSE VARIATION

Now let's see what properties of reciprocal power functions are shared by functions of the form $y = \frac{a}{x^n}$. If a relationship is described by such an equation, we say that **y varies inversely with the nth power of x**. For example, Newton's law of gravitation states that the gravitational force between any two bodies in the universe is described by an equation $F = \frac{k}{d^2}$, where k is a constant depending on the mass of the two bodies and d is the distance between them. Because the force varies inversely with the square of the distance, Newton's law of gravitation is sometimes called an inverse square law. The attractive force between two magnets and the intensity of light from a fixed source are also described by inverse square laws.

In Exercise 1 you showed that the number of hours required to drive from Atlanta to Orlando at a speed of s miles per hour is $g(s) = \frac{450}{s}$. Thus the required time varies inversely with your speed. In Exercises 9–12 you can verify that this function has the properties just described for reciprocal power functions.

EXERCISES

9. How many hours would it take each of the following to go from Atlanta to Orlando?

a. a jet at 600 mph

b. a spaceship at 10,000 mph

c. a light ray at 186,000 miles per *second*

10. *(Writing to Learn)* What feature on the graph of $g(s) = \frac{450}{s}$ is suggested by your result from Exercise 9?

11. How many hours would it take each of the following to go from Atlanta to Orlando?

a. a pedestrian at 2.5 mph

b. a snail at 0.01 mph

c. a glacier at 0.1 miles per *year*

12. *(Writing to Learn)* What feature on the graph of $g(s) = \frac{450}{s}$ is suggested by your result from Exercise 11?

Inverse variations behave in a manner similar to other types of variations. In particular,

If $y = \frac{a}{x^n}$, then whenever x is multiplied by a constant k, y is divided by k^n.

Informally,

For inverse nth power variations, doubling x divides y by 2^n.

For example, if y varies inversely with x^3, then doubling x divides y by $2^3 = 8$.

It is easy to tell whether a table fits an inverse nth power variation, and to find its equation when it does.

In any table generated by $y = \frac{a}{x^n}$, the product $x^n y = a$ for every row.

EXAMPLE 1 *Deciding whether a table fits an inverse variation of a given degree*

To see whether the table

x	y
1	8100
2	506.25
3	100
5	12.96

fits an inverse variation of degree 4, evaluate the product x^4y for each row of the table.

$$(1)^4(8100) = 8100, \quad (2)^4(506.25) = 8100,$$
$$(3)^4(100) = 8100, \quad (5)^4(12.96) = 8100$$

The table fits the equation $y = \dfrac{8100}{x^4}$. ■

EXERCISES *In Exercises* 13–16, *decide whether the table fits an inverse variation* $y = \dfrac{a}{x^n}$ *for the given value of n. If it does, write an equation relating x and y.*

13.

x	$y\ (n = 1)$
3	28
4	21
6	14
7	12
12	7

14.

x	$y\ (n = 1)$
10	240
50	200
100	150
125	125
150	100

15.

x	$y\ (n = 2)$
2	225
3	100
5	36
6	25
10	9

16.

x	$y\ (n = 3)$
2	25
4	3.125
5	1.6
10	0.2
20	0.025

17. If P varies inversely with V and $P = 20$ when $V = 6$, what is the value of P when $V = 10$?

18. If N varies inversely with M and $N = 7.2$ when $M = 3$, what is the value of N when $M = 10.8$?

19. If s varies inversely with r^3 and $s = 250$ when $r = 1$, what is the value of s when $r = 5$?

20. If y varies inversely with x^4 and $y = 13$ when $x = 2$, what is the value of y when $x = 3$?

GRAPHICAL TRANSFORMATIONS

Not all rational graphs have shapes similar to those in Figure 8-2. In particular, not all have vertical or horizontal asymptotes. In Section 8-2 you will learn how to decide whether a rational graph has an asymptote of either type.

For now, let's consider a class of rational functions whose graphs always have exactly one asymptote of each type. These rational functions are the **linear rational functions**. Every linear rational function has the form

$$f(x) = \frac{Ax + B}{Cx + D}$$

for some real numbers A, B, C, and D with $C \neq 0$. You can always obtain the graph of a linear rational function by transforming the basic graph $y = \frac{1}{x}$ as in Example 2.

EXAMPLE 2 *Using transformations to graph a rational function*

Before sketching the graph of $f(x) = \frac{2x + 7}{x + 1}$, perform the division.

$$\begin{array}{r|l} & \quad\;\; 2 \\ \hline x + 1 & 2x + 7 \\ & \underline{2x + 2} \\ & \quad\;\; 5 \end{array}$$

Thus $f(x) = \frac{5}{x + 1} + 2$. To sketch its graph, begin with the basic graph $y = \frac{1}{x}$, and make the changes

$$y = \frac{1}{x} \rightarrow y = \frac{5}{x} \rightarrow y = \frac{5}{x + 1} \rightarrow y = \frac{5}{x + 1} + 2$$

These changes stretch the graph vertically by a factor of 5, then shift it 1 unit left and 2 units up, as in Figure 8-3.

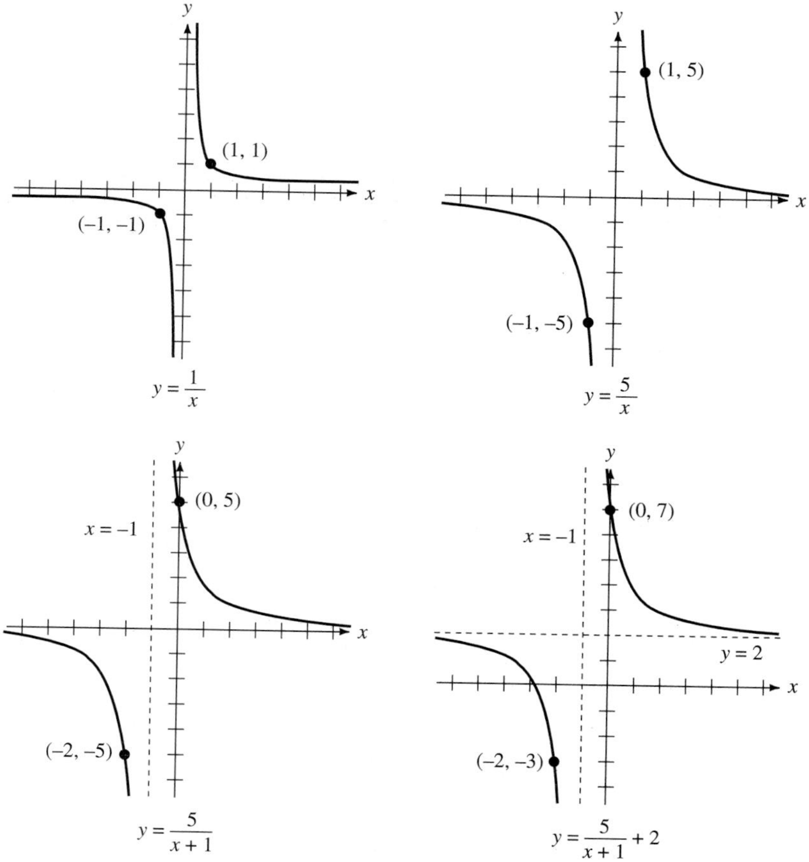

Figure 8–3

■

EXERCISES

21. *(Writing to Learn)* In Example 2, what tells us that the graph of f has a vertical asymptote at $x = -1$ and a horizontal asymptote at $y = 2$?

In Exercises 22–25, sketch the graph of each function by transforming the basic graph $y = \frac{1}{x}$. *Identify the points obtained by moving* $(1, 1)$ *and* $(-1, -1)$, *the vertical and horizontal asymptotes, and the domain and range of the function.*

22. $B(x) = -\dfrac{1}{x - 4}$ **23.** $q = \dfrac{1}{x + 3} - 2$

24. $y = \dfrac{5}{x} + 1$ **25.** $R(x) = 7 - \dfrac{3}{x}$

It is possible to sketch the graph of a linear rational function and identify important graphical features without performing the division. In Exercise 26 you can verify the following facts.

The graph of $f(x) = \dfrac{Ax + B}{Cx + D}$ has

- a y-intercept at $\left(0, \dfrac{B}{D}\right)$ if $D \neq 0$
- an x-intercept at $\left(-\dfrac{B}{A}, 0\right)$ if $A \neq 0$
- a vertical asymptote at $x = -\dfrac{D}{C}$
- a horizontal asymptote at $y = \dfrac{A}{C}$

EXERCISES

26. *(Writing to Learn)* Explain why the graph of $f(x) = \dfrac{Ax + B}{Cx + D}$ must have the features described above. (*Hint*: Perform the division to locate the horizontal asymptote.)

In Exercises 27–30, sketch the graph of the function by locating its intercepts and asymptotes.

27. $y = \dfrac{3x + 7}{x + 1}$ **28.** $y = \dfrac{2x}{x - 4}$

29. $y = \dfrac{3 - x}{x - 2}$ **30.** $y = \dfrac{x}{3x + 6}$

ADDITIONAL EXERCISES

In Exercises 31–34, decide whether the table fits an inverse variation $y = \dfrac{a}{x^n}$ *for the given value of n. If it does, write an equation relating x and y.*

31.

x	$y\ (n = 2)$
0.01	1000
0.1	10
1	0.1
10	0.001

32.

x	$y\ (n = 1)$
2	105
3	70
5	42
6	35

33.

x	$y\ (n = 2)$
1	4096
2	1024
4	256
8	64

34.

x	$y\ (n = 3)$
1	1728
2	216
3	64
4	27

35. If b varies inversely with a and $b = 54$ when $a = 7$, what is the value of b when $a = 21$?

36. If H varies inversely with G and $H = 2$ when $G = \pi$, what is the value of H when $G = 2$?

37. If v varies inversely with t^3 and $v = 4$ when $t = 3$, what is the value of v when $t = 6$?

38. If y varies inversely with x^2 and $y = 25$ when $x = \sqrt{3}$, what is the value of y when $x = \sqrt{5}$?

39. *(Problem Solving)* Under certain conditions, the pressure exerted by a gas in a closed container varies inversely with the volume of the container. Suppose that the air in a balloon exerts an outward pressure of 20 lb/in^2.

a. If the volume of the balloon is reduced to one-third of its present volume, how much outward pressure will be exerted by the air?

b. If the balloon is designed to withstand an outward pressure of 100 lb/in^2, to what fraction of its present volume can it be reduced without breaking?

40. *(Problem Solving)* Your weight varies inversely with the square of your distance from the center of the earth. Suppose that you weigh 150 lb at sea level, 3960 miles from the center of the earth.

a. How much would you weigh on the top of Mt. Everest, about 5.5 miles above sea level?

b. At what altitude above sea level would you weigh 1 pound?

41. *(Problem Solving)* The attractive force between two magnets varies inversely with the square of the distance between them. If the attractive force between two particular magnets is 5 newtons at a distance of 3 meters, what is the force at a distance of 10 meters?

42. *(Problem Solving)* Tides on the earth's oceans are caused by the gravitational pull of external bodies, primarily the moon and the sun. The tide-raising force of an external body varies directly with its mass and inversely with the cube of its distance from the earth. That is, the tide-raising force of a body of mass m at a distance d from the earth is $F = \dfrac{km}{d^3}$ for some constant k.

a. If the moon were twice as far from the earth as it actually is, by what factor would its mass have to be increased to exert the same tide-raising force?

b. The sun is 391 times as far from the earth as the moon, and its tide-raising force is 0.45 times that of the moon. What is the ratio of the mass of the sun to the mass of the moon?

In Exercises 43–46, sketch the graph of each function by transforming the basic graph $y = \dfrac{1}{x}$. *Identify the points obtained by moving* $(1, 1)$ *and* $(-1, -1)$, *the vertical and horizontal asymptotes, and the domain and range of the function.*

43. $f(x) = \dfrac{10}{x - 2}$ **44.** $g(x) = -\dfrac{3}{x}$

45. $y = \dfrac{2}{x + 7}$ **46.** $z = 4 - \dfrac{1}{x + 3}$

In Exercises 47–50, sketch the graph of the function by locating its intercepts and asymptotes.

47. $y = \dfrac{x + 7}{x + 5}$ **48.** $y = \dfrac{4x - 9}{x - 3}$

49. $y = \dfrac{2x + 4}{x}$ **50.** $y = \dfrac{x}{2x + 4}$

In Exercises 51–54, sketch the graph of each function by transforming a reciprocal power graph $y = \dfrac{1}{x^n}$. *Identify the points obtained by moving* $(1, 1)$ *and either* $(-1, 1)$ *or* $(-1, -1)$, *identify the vertical and horizontal asymptotes, and find the domain and range of the function.*

51. $f(x) = \dfrac{6}{x^3}$ **52.** $T(x) = \dfrac{1}{x^4} + 5$

53. $u = 3 - \dfrac{1}{x^2}$ **54.** $K = \dfrac{5}{(x + 10)^4}$

8-2 DISCONTINUITIES AND END BEHAVIOR OF RATIONAL FUNCTIONS

In Section 8-1 you observed that the graphs of reciprocal power functions have two striking features not found in polynomial graphs.

- The graph consists of two pieces separated by a vertical asymptote. Thus there is a break, or **discontinuity**, in the graph, and the function is said to be **discontinuous**.
- The "tails" of the graph approach a horizontal asymptote.

In this section you will discover that not all rational functions have these graphical features. As we begin to explore the graphs of rational functions, it will be helpful to frame a few questions to guide our exploration.

- Is every rational function discontinuous at some point?
- Is every discontinuity a vertical asymptote?
- How does a table or an equation indicate the behavior of a rational function near a discontinuity?
- When a rational function models a physical problem, what physical information is provided by a discontinuity?
- Does the graph of every rational function have a horizontal asymptote?
- How does a table or an equation indicate the end behavior of a rational graph?
- When a rational function models a physical problem, what physical information is provided by the end behavior of its graph?

DISCONTINUITIES AND VERTICAL ASYMPTOTES

Let's focus on discontinuities first. Your experience in Section 8-1 suggests that a rational function is discontinuous only where its denominator is zero. Using calculus, it can be shown that this is actually the case. You can use this observation to answer our first two questions, and in each case the answer is no. The function in Exercise 1 is not discontinuous at any point, and the function in Exercise 2 has a discontinuity that is not a vertical asymptote.

EXERCISES

1. *(Making Observations)* For $f(x) = \dfrac{10}{x^2 + 1}$,

a. Verify that the denominator is never 0, so that f is continuous for all values of x.

b. Graph $y = f(x)$ in the window $[-10, 10]$ by $[-10, 10]$. How does the graph support your conclusion in part (a)?

2. *(Making Observations)* For $g(x) = \dfrac{x^2 - 1}{x - 1}$,

a. Use the equation to verify that g is discontinuous at $x = 1$.

b. Graph $y = g(x)$ in the window $[-10, 10]$ by $[-10, 10]$. Does the graph have a vertical asymptote at $x = 1$?

You may have been surprised by the graph in Exercise 2. The function g is discontinuous at $x = 1$, and yet the graph appears to be an unbroken line. To resolve this apparent paradox, let's consider our third question from the beginning of the section. How does an equation or a table indicate the behavior of a rational function near a discontinuity? In Exercises 3–6 you can explore this question for $h(x) = \dfrac{1}{x - 1}$ and $g(x) = \dfrac{x^2 - 1}{x - 1}$. From Section 8-1 you know that $h(x)$ has a vertical asymptote at $x = 1$, and from Exercise 2 you know that $g(x)$ does not.

EXERCISES

3. *(Review)*

   a. Complete the following table for $h(x) = \dfrac{1}{x - 1}$.

x	$h(x)$
0	
0.9	
0.99	
0.999	
1	
1.001	
1.01	
1.1	
2	

   b. How does the table indicate that the graph of $y = h(x)$ has a vertical asymptote at $x = 1$?

4. *(Making Observations)*

   a. Repeat Exercise 3a for $g(x) = \dfrac{x^2 - 1}{x - 1}$.

   b. How does the table indicate that the graph of $y = g(x)$ has no vertical asymptote at $x = 1$?

Your tables in Exercises 3 and 4 reveal something about the behavior of h and g near their discontinuities. When x is near 1, $|h(x)|$ is large, while $g(x)$ is near 2. In Exercises 5 and 6 you can see how the equations of the two functions support this observation.

5. *(Writing to Learn)* For $h(x) = \dfrac{1}{x - 1}$, explain without using any specific values of x why $|h(x)|$ must be large if x is near 1.

6. *(Making Observations)*

   a. Simplify the expression $g(x) = \dfrac{x^2 - 1}{x - 1}$, and verify that $g(x) = x + 1$ if $x \neq 1$.

   b. Why is $g(x)$ not equal to $x + 1$ when $x = 1$?

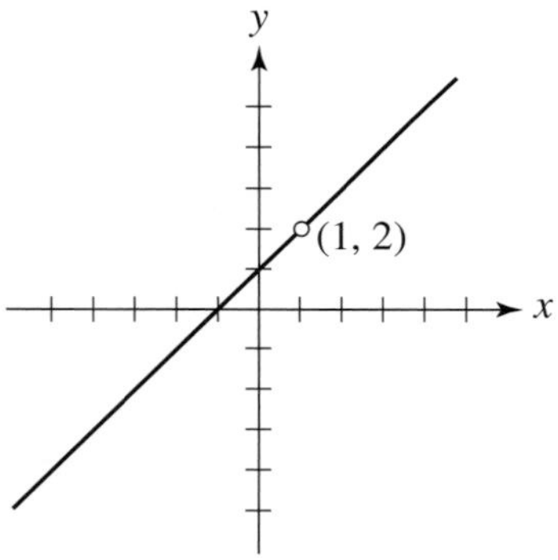

Figure 8–4

Both the table in Exercise 4 and the equation in Exercise 6 indicate that the graph of $y = g(x)$ coincides with that of $y = x + 1$ except that the point $(1, 2)$ is missing. Thus your calculator graph in Exercise 2 was not quite accurate. A more accurate graph is shown in Figure 8-4. A discontinuity such as the one in Figure 8-4 is called a **missing-point discontinuity**.

Your observations in Exercises 3–6 can be generalized.

A rational function $f(x) = \dfrac{N(x)}{D(x)}$ has a discontinuity at $x = a$ if $D(a) = 0$.

The discontinuity is

- a vertical asymptote as in Figure 8-5a if $\dfrac{N(x)}{D(x)}$ is in lowest terms.
- a missing point discontinuity as in Figure 8-5b if $f(x)$ can be reduced to an expression whose denominator is not 0 at $x = a$.

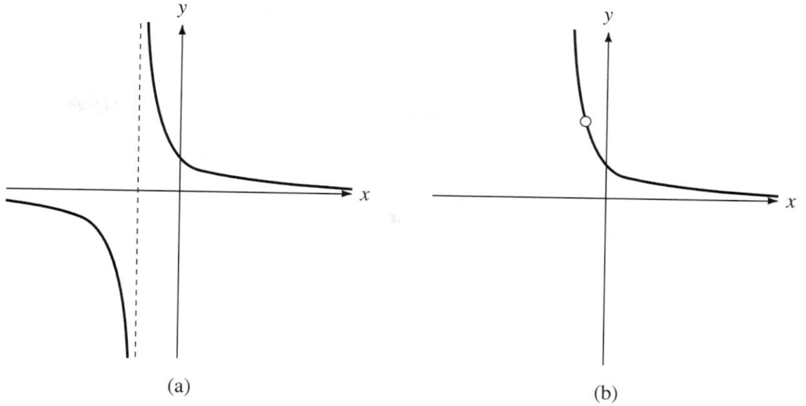

Figure 8–5

EXERCISES *In Exercises* 7–10,

a. List all discontinuities for each function.

b. Decide whether each discontinuity is a vertical asymptote or a missing point.

7. $y = \dfrac{x}{x + 4}$ **8.** $y = \dfrac{x}{x^2 - 16}$

9. $y = \dfrac{2x - 10}{x - 5}$ **10.** $y = \dfrac{2x - 10}{x^2 - 5x}$

Let's turn now to our fourth question from the beginning of the section. When a rational function models a physical problem, what physical information is provided by a discontinuity?

A MATHEMATICAL LOOKING GLASS

Average Cost (Fat Cats 7)

The T-shirt company started by your economics class in ***Fat Cats on Campus*** (page 43) has done so well that the class has decided to expand its operations. You plan to open a store near your school, in competition with several other companies. You must therefore sell your shirts at or near the prevailing local price of \$9 each. You can make a profit as long as the selling price exceeds the average cost of producing each shirt. To determine profitable production levels you need to know your anticipated average cost, which will be a function of your production level.

Your profits from the semester were used to pay off the heat press you purchased to produce the shirts, so it is no longer considered a cost. However, you need to rent production and office space at \$700 per month. Materials are still \$3.50 per shirt. You estimate the advertising cost of selling x shirts per month to be $\$0.01x^2$. The average cost per shirt is the total cost of producing and selling the shirts, divided by the number of shirts sold. ■

EXERCISES **11.** *(Interpreting Mathematics)* Near your campus you can advertise inexpensively through posters and word of mouth. To reach a wider market you must use more expensive forms of advertising, such as newspaper ads or radio commercials. Why does this indicate that a quadratic expression is a better model for your advertising costs than a linear expression?

12. *(Writing to Learn)*

a. Explain why the cost in dollars of producing x shirts per month is

$$C(x) = 700 + 3.50x + 0.01x^2$$

b. Explain why the average cost per shirt is

$$A(x) = \frac{700 + 3.50x + 0.01x^2}{x} \text{ dollars.}$$

c. At what value of x is A discontinuous? Is the discontinuity a vertical asymptote or a missing point?

13. *(Making Observations)*

a. Complete the following table for $y = A(x)$.

x	y
100	
10	
1	
0.1	

(Think of 0.1 shirts per month as 1 shirt every 10 months.) What physical information is provided by each row in the table?

b. What does the table in part (a) tell you about the average cost at low production levels? How does your answer reflect the type of discontinuity in the average cost function?

14. Suppose that you are currently producing 300 shirts per month, and you are thinking about increasing your level of production to s shirts per month.

a. *(Writing to Learn)* Explain why the average cost of producing each additional shirt will be

$$f(s) = \frac{C(s) - C(300)}{s - 300}$$

b. Show that $f(s)$ can be written

$$f(s) = \frac{0.01s^2 + 3.50s - 1950}{s - 300}$$

c. At what value of s is f discontinuous? Is the discontinuity a vertical asymptote or a missing point?

15. *(Interpreting Mathematics)* If you increase production by a small amount, the value of s will be near 300.

a. The value of $f(s)$ will be near what number? [*Hint*: Simplify the expression for $f(s)$ by factoring the numerator.]

b. If you are currently selling shirts for \$9.00 each, does your result in part (a) indicate that you should increase production? Explain.

END BEHAVIOR AND HORIZONTAL ASYMPTOTES

Now let's explore the end behavior of rational graphs, starting with our fifth question from the beginning of the section. Does the graph of every rational function have a horizontal asymptote? In Exercises 16–18 you can confirm that the answer is no.

EXERCISES

16. *(Review)* Use the methods of Section 8-1 to sketch the graph of $f(x) = \dfrac{-6}{x - 1}$ and identify its horizontal asymptote.

17. *(Review)*

a. For $g(x) = \dfrac{x - 6}{x - 1}$, perform the division to verify that

$$g(x) = 1 - \frac{5}{x - 1}$$

b. Sketch the graph of $y = g(x)$ and identify its horizontal asymptote.

18. *(Making Observations)*

a. For $h(x) = \dfrac{x^2 + x - 6}{x - 1}$, perform the division to verify that

$$h(x) = x + 2 - \frac{4}{x - 1}$$

b. Graph $y = x + 2$ along with $y = h(x)$ in the window $[-10, 10]$ by $[-10, 10]$. What happens when $|x|$ is large?

19. *(Writing to Learn)*

a. As $|x|$ becomes large, explain why any expression $\dfrac{\text{constant}}{x - 1}$ becomes small.

b. Why does this observation guarantee that when $|x|$ is large, the graphs of the functions in Exercises 16–18 approach the graphs $y = 0$, $y = 1$, and $y = x + 2$, respectively?

In Exercises 20–22 *you can see how tables support your conclusions in Exercises* 16–19.

20. *(Making Observations)* Complete the following table for $f(x) = \dfrac{-6}{x - 1}$.

x	$f(x)$
10	
100	
1,000	
10,000	

How does the table indicate the end behavior of the graph of $y = f(x)$?

21. *(Making Observations)* Repeat Exercise 20 for $g(x) = \dfrac{x - 6}{x - 1}$.

22. *(Making Observations)* Complete the following table for

$$h(x) = \frac{x^2 + x - 6}{x - 1}$$

x	$x + 2$	$h(x)$
10		
100		
1,000		
10,000		

How does the table indicate the end behavior of the graph of $y = h(x)$?

Exercises 16–22 suggest the answer to our sixth question from the beginning of the section. How does a table or an equation indicate the end behavior of a rational graph? By performing division, you can express every rational function as the sum of a polynomial function and a remainder. The polynomial part is 0 if the degree of the numerator in the rational function is less than that of the denominator, as in Exercise 16. This case is illustrated by Figure 8-6a. The polynomial part is a nonzero constant if the degrees are equal, as in Exercise 17, and has degree at least 1 if the degree of the numerator is larger than that of the denominator, as in Exercise 18. These cases are illustrated by Figure 8-6b and c. In each case, the remainder is a **proper rational function**. That is, its numerator has a smaller degree than its denominator.

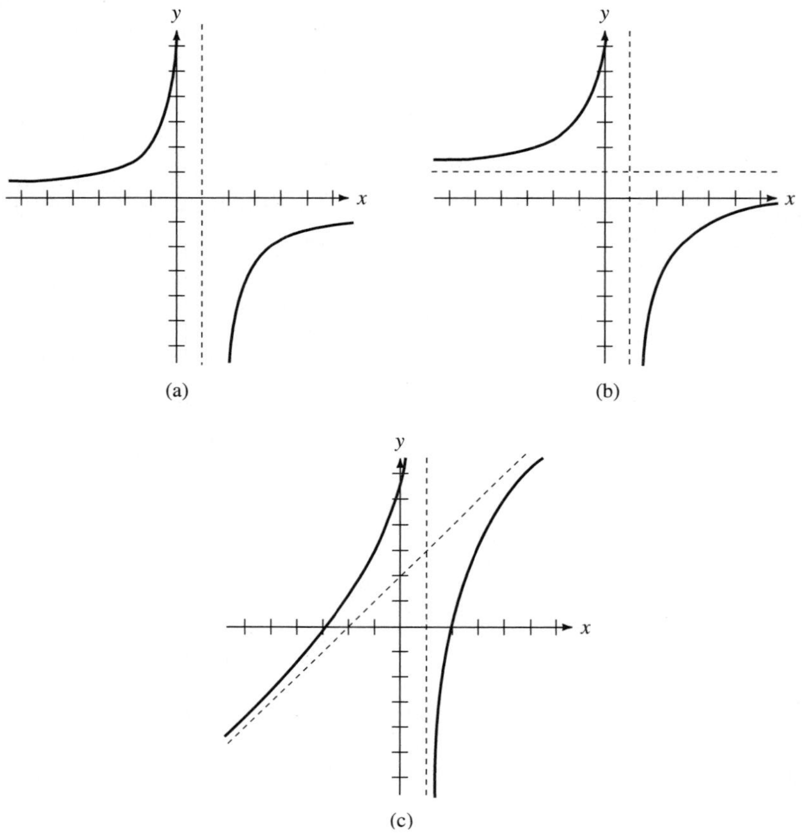

Figure 8–6

In Exercises 48–49 you can show that every proper rational function becomes small as $|x|$ becomes large, so that the end behavior of a rational function depends only on its polynomial part. Your observations are summarized in the following statement.

The Highest Degree Theorem (Rational Function Version)

The end behavior of the graph of any rational function

$$f(x) = \frac{N(x)}{D(x)} = \frac{a_n x^n + a_{n-1}x^{n-1} + \cdots + a_1 x + a_0}{b_d x^d + b_{d-1}x^{d-1} + \cdots + b_1 x + b_0}$$

is the same as that of $y = \frac{a_n x^n}{b_d x^d}$. (In end behavior, only the terms of highest degree count.) In particular, the graph has

- a horizontal asymptote at $y = 0$ if $n < d$
- a horizontal asymptote at $y = \frac{a_n}{b_d}$ if $n = d$
- no horizontal asymptote if $n > d$

EXAMPLE 1 *Identifying a horizontal asymptote by using the Highest Degree Theorem*

a. The Highest Degree Theorem implies that the end behavior of

$$g(x) = \frac{8x^2 + 3}{4x^2 - 2x + 1}$$

is the same as that of $y = \frac{8x^2}{4x^2} = 2$. Thus the graph of $y = g(x)$ has a horizontal asymptote at $y = 2$.

b. The end behavior of

$$h(x) = \frac{8x^3 + 3}{4x^2 - 2x + 1}$$

is the same as that of $y = \frac{8x^3}{4x^2} = 2x$. Thus the left tail of the graph points down, and the right tail points up. In particular, there is no horizontal asymptote. ■

EXERCISES *For each function in Exercises 23–26, use the Highest Degree Theorem to decide whether the graph has a horizontal asymptote. If there is one, find its equation. If not, decide whether each tail of the graph points up or down.*

23. $f(x) = \frac{x^2 + 4x}{3x^2 - 5}$

24. $y = \frac{2x^3 - 16}{x^3 + 3x^2}$

25. $A(t) = \frac{4t - t^5}{6t^3 - 4t + 10}$

26. $Z = \frac{100s^2 + 37s + 2751}{s^4 - s - 1}$

Now let's turn to our seventh and final question from the beginning of the section. When a rational function models a physical problem, what physical information is provided by the end behavior of its graph?

A MATHEMATICAL LOOKING GLASS
Robert Arnold

Taos, NM, is a famous ski resort and artist's colony. One of Taos' nationally known resident artists is Robert Arnold, a painter of southwestern scenes. To make his works accessible to a wide audience, he has concentrated on marketing his art since 1987. By 1990, he had produced about 500 original and limited-edition pieces. Since then his reputation has grown, and he has expanded his market by producing affordable open-edition items such as note cards, prints, and images on ceramic tiles, which are sold by several distributors. Since 1990 he has produced about 50 original and limited-edition pieces and about 2000 open-edition pieces each year. ■

EXERCISES

27. *(Creating Models)* Let t denote time in years since 1990.

a. Write a function $P(t)$ to express the total number of open-edition pieces produced prior to time t.

b. Write a function $Q(t)$ to express the total number of original and limited-edition pieces produced by Robert Arnold prior to time t, including those produced before 1990.

c. Write an expression for the rational function $f(t) = \dfrac{P(t)}{Q(t)}$. What does $f(t)$ represent in physical terms?

28. *(Interpreting Mathematics)* The graph of the function f in Exercise 27c has a horizontal asymptote.

a. Find its equation.

b. What physical information does it provide?

ADDITIONAL EXERCISES

In Exercises 29–44,

a. List the values of x where each function is discontinuous.

b. Decide whether each discontinuity is a vertical asymptote or a missing point.

c. Decide whether the graph has a horizontal asymptote. If there is one, find its equation. If not, decide whether each tail of the graph points up or down.

29. $y = \dfrac{x+8}{x-4}$

30. $y = \dfrac{2x+3}{3x+2}$

31. $y = \dfrac{x^2-4x}{x}$

32. $y = \dfrac{x^2-4x}{x-4}$

33. $y = \dfrac{6-2x}{x+3} + 1$

34. $y = \dfrac{6-2x}{x-3} - 1$

35. $y = \dfrac{1}{x^2-3}$

36. $y = \dfrac{1}{x^2+3}$

37. $y = \dfrac{x-1}{x^2-8x+7}$

38. $y = \dfrac{x-2}{x^2-8x+7}$

39. $y = \dfrac{x^2+4x+4}{x+2}$

40. $y = \dfrac{x+2}{x^2+4x+4}$

41. $f(x) = \dfrac{x^4-x^3}{2-x^4}$

42. $J(x) = \dfrac{x^2-1}{x}$

43. $R(x) = \dfrac{10x^2}{x^4 + 16}$ **44.** $Q(x) = \dfrac{6x^3 + 4x - 3}{2x^3 - x^2 - x}$

45. *(Interpreting Mathematics)* The function

$$f(t) = \frac{2000t}{50t + 500}$$

in Exercise 27 has a discontinuity. Where does it occur, and what information, if any, does it provide?

46. *(Interpreting Mathematics)* Determine the end behavior of the average cost function

$$A(x) = \frac{700 + 3.50x + 0.01x^2}{x}$$

in Exercise 12. What physical information, if any, does it provide?

47. A cylindrical 12-ounce soft drink can must have a volume of 21.7 in^3. Let r denote its radius in inches.

a. *(Creating Models)* Express the height of the container in terms of r.

b. *(Creating Models)* Express the surface area of the container as a function $S(r)$.

c. *(Interpreting Mathematics)* Where is S discontinuous? What physical information, if any, does it provide?

d. *(Interpreting Mathematics)* Describe the end behavior of the graph of S. What physical information, if any, does it provide?

48. *(Writing to Learn)*

a. Verify that the function

$$f(x) = \frac{2x^4 + x^3 + 5x - 10}{2x^2 + x - 4}$$

can be written

$$f(x) = x^2 + 2 + \frac{3x - 2}{2x^2 + x - 4}$$

b. Explain why the remainder can be written

$$\frac{\dfrac{3}{x} - \dfrac{2}{x^2}}{2 + \dfrac{1}{x} - \dfrac{4}{x^2}}$$

c. Explain why the expression in part (b) is small if $|x|$ is large. (*Hint*: Which terms in the numerator and denominator become small as $|x|$ becomes large?)

49. *(Writing to Learn)*

a. Graph $y = f(x)$ and $y = x^2 + 2$ in a window that shows the end behavior of both graphs. Why does Exercise 48 suggest that the two graphs are close when $|x|$ is large?

b. Let $P(x)$ denote the polynomial obtained when the numerator of the rational expression $f(x)$ is divided by its denominator. Explain why the graph of $y = f(x)$ approaches the graph of $y = P(x)$ when $|x|$ is large.

SUPPLEMENTARY TOPIC
Slant Asymptotes

In Exercise 18 you showed that the graph of $h(x) = \dfrac{x^2 + x - 6}{x - 1}$ approaches that of $y = x + 2$ when $|x|$ is large. The line $y = x + 2$ is called a **slant asymptote** for the graph of $y = h(x)$, shown in Figure 8-7.

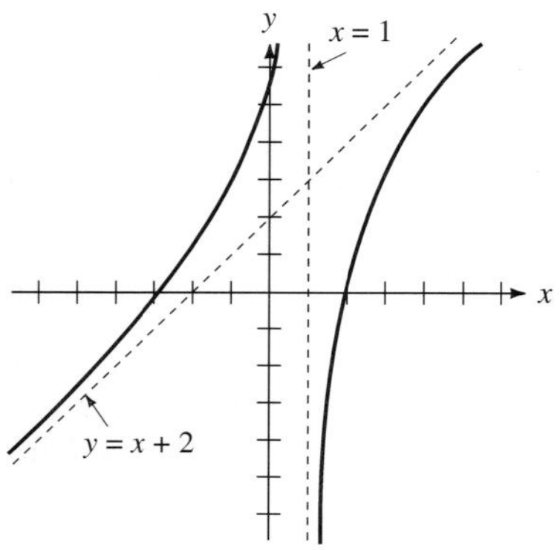

Figure 8–7

50. *(Writing to Learn)* If $f(x) = \dfrac{N(x)}{D(x)}$ and the degree of N exceeds the degree of D by 1, explain why the graph of $y = f(x)$ has a slant asymptote.

In Exercises 51–58, find the slant asymptote for each function. Graph the function and its asymptote together in the window $[-10, 10]$ *by* $[-10, 10]$.

51. $y = \dfrac{x^2 + 1}{x}$ **52.** $y = \dfrac{x^2 + 2x}{x - 2}$

53. $y = \dfrac{6x^2}{2x - 3}$ **54.** $y = \dfrac{6x^2}{3 - 2x}$

55. $y = \dfrac{2x^3}{x^2 - x + 1}$ **56.** $y = \dfrac{x^3 + x^2 - 4x}{4 - x^2}$

57. $y = \dfrac{x^4 - 2x^2 + 1}{2x^3 + 8x}$ **58.** $y = \dfrac{0.1x^6}{1 - x^5}$

8–3 SOLVING RATIONAL INEQUALITIES

In ***Average Cost*** (page 307) you discovered that it costs your T-shirt company an average of

$$A(x) = \frac{700 + 3.50x + 0.01x^2}{x}$$

dollars per shirt to produce x shirts per month. You can make a profit as long as you can sell the shirts for more than it costs to produce them. Since your selling price is fixed at \$9, your profitable levels of production are the solutions of the inequality

$$\frac{700 + 3.50x + 0.01x^2}{x} < 9$$

In this section you will learn how to solve inequalities involving rational functions. Since the first step in solving any inequality is to solve an equation, let's begin by recalling that rational equations can be solved graphically, as in Section 2-4, or analytically, as in Section A-17. Example 1 reviews the analytical method.

EXAMPLE 1 *Solving a rational equation*

To solve

$$\frac{5x + 1}{x^2 - 1} = 2$$

begin by multiplying both sides by $x^2 - 1$.

$$5x + 1 = 2(x^2 - 1)$$

$$2x^2 - 5x - 3 = 0$$

$$(2x + 1)(x - 3) = 0$$

$$x = -\frac{1}{2}, 3$$

In solving the equation, you multiplied both sides by the variable quantity $x^2 - 1$. Therefore, your solutions are valid if and only if they do not produce a value of 0 in the denominator of the original equation. A quick check shows that both solutions are valid. ■

EXERCISES *In Exercises* 1–4, *solve the equation.*

1. *(Review)* $\dfrac{2x + 7}{x - 1} = 5$

2. *(Review)* $\dfrac{x - 2}{x} = \dfrac{x + 4}{2x + 3}$

3. *(Review)* $\dfrac{2x - 1}{x - 5} - 3 = \dfrac{9}{x - 5}$

4. *(Review)* $\dfrac{2x - 1}{x - 5} - x = \dfrac{9}{x - 5}$

5. a. Solve the equation $\dfrac{700 + 3.50x + 0.01x^2}{x} = 9$ from ***Average Cost***.

b. *(Interpreting Mathematics)* What physical information is provided by the solution?

As with polynomial inequalities, the next step in solving a rational inequality is to rewrite it in one of the forms $f(x) \geqslant 0$, $f(x) > 0$, $f(x) \leqslant 0$, $f(x) < 0$. When f is a rational function, however, you also need to find the discontinuities of f. This is done in Example 2. In Example 3 you will see why the extra step is needed.

EXAMPLE 2 *Preparing to solve a rational inequality*

To prepare to solve

$$\frac{5x + 1}{x^2 - 1} \geqslant 2$$

begin by rewriting it

$$\frac{5x + 1}{x^2 - 1} - 2 \geqslant 0$$

The function

$$f(x) = \frac{5x + 1}{x^2 - 1} - 2$$

is discontinuous when the denominator $x^2 - 1$ is zero, that is, when $x = \pm 1$. ■

The information from Examples 1 and 2 can be used to solve the inequality either graphically or analytically. Example 3 shows the graphical solution, and Examples 4 and 5 show two analytical methods of solution.

A GRAPHICAL METHOD

EXAMPLE 3 *Solving a rational inequality graphically*

To solve

$$\frac{5x+1}{x^2-1} \geqslant 2$$

begin by rewriting it as in Example 2. Then graph $f(x) = \dfrac{5x+1}{x^2-1} - 2$, as in Figure 8-8.

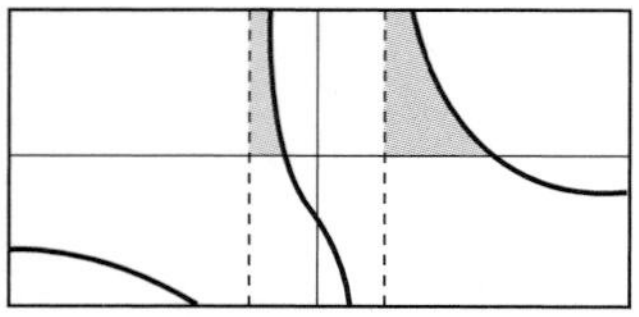

[−5, 5] by [−5, 5]

Figure 8–8

The inequality is true at all x-values where the graph is on or above the x-axis. Figure 8-8 indicates that the endpoints of the solution intervals correspond to the x-intercepts and the discontinuities on the graph. The x-intercepts are the solutions to $f(x) = 0$. From Example 1, these are $x = -\dfrac{1}{2}$ and $x = 3$. From Example 2, the discontinuities occur at $x = \pm 1$. The inequality is true at both x-intercepts, but false at both discontinuities. The solution is $\left(-1, -\dfrac{1}{2}\right] \cup (1, 3]$. ■

EXERCISES *In Exercises 6–9, solve the inequality graphically.*

6. $\dfrac{x+1}{x^2-9} \leqslant 0$ **7.** $\dfrac{(x+3)(x-7)}{x-2} \geqslant 0$

8. $\dfrac{1}{(x+4)^2} < 1$ **9.** $\dfrac{1}{x} > \dfrac{1}{x^2}$

10. a. Solve the inequality $\dfrac{700 + 3.50x + 0.01x^2}{x} < 9$ from ***Average Cost***.

b. *(Interpreting Mathematics)* What physical information is provided by the solution?

THE TEST-VALUE METHOD

Like polynomial inequalities, rational inequalities can be solved by the test-value method, but because of the possible presence of discontinuities, the method must be modified as shown in Example 4.

EXAMPLE 4 *Solving a rational inequality by the test-value method*

To solve

$$\frac{5x+1}{x^2-1} \geqslant 2$$

define $f(x) = \dfrac{5x+1}{x^2-1} - 2$, so that the inequality is $f(x) \geqslant 0$.

STEP 1: Solve $f(x) = 0$ as in Example 1. The solutions are $x = -\dfrac{1}{2}$ and $x = 3$.

STEP 2: Determine where f is discontinuous as in Example 2. The discontinuities occur when $x = \pm 1$.

STEP 3: As Figure 8-8 showed, the sign of $f(x)$ can change at a value of x where $f(x)$ is either zero or discontinuous. Indicate these values on a number line.

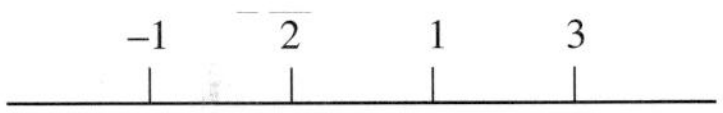

STEP 4: Evaluate $f(x)$ at a test value in each interval.

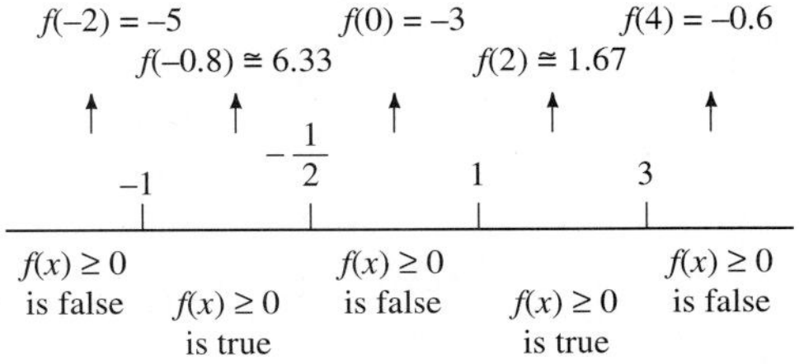

As in Example 3, the inequality is true when $f(x) = 0$ and false when $f(x)$ is discontinuous. The solution is $\left(-1, -\frac{1}{2}\right] \cup (1, 3]$. ■

EXERCISES

11. As parts (a)–(d) of this exercise, solve the inequalities in Exercises 6–9 by the test-value method.

12. *(Writing to Learn)* Compare the graphical and test-value methods of solving rational inequalities. Which method do you prefer to use, and why? Does your choice depend on the inequality you are solving? If so, how?

THE SCAN METHOD

Knowing the end behavior of rational functions, you can use the scan method to solve rational inequalities. Example 5 indicates how the method must be modified.

EXAMPLE 5 *Solving a rational inequality by the scan method*

Suppose $f(x) = \dfrac{x - 4}{x^2(x - 6)}$. To solve $f(x) \leq 0$,

STEP 1: Solve $f(x) = 0$ to obtain $x = 4$.

STEP 2: Determine that f is discontinuous when $x = 0$ or 6.

STEP 3: Indicate these x-values on a number line.

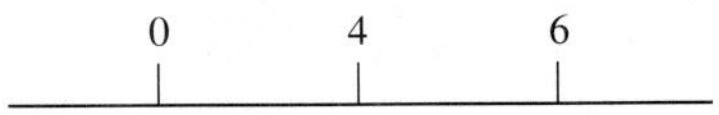

STEP 4: Now you must find the sign of $f(x)$ in each of the four intervals shown. Find the sign of $f(x)$ on $(-\infty, 0)$ by looking at the behavior of the left side of its graph. If the denominator of $f(x)$ is multiplied out, the terms of highest degree in its numerator and denominator are x and x^3, respectively. Thus the graph of $f(x)$ has the same left-end behavior as that of $\frac{x}{x^3} = \frac{1}{x^2}$, so that $f(x) > 0$ on $(-\infty, 0)$.

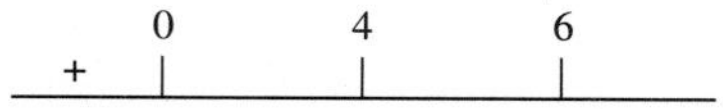

STEP 5: Next, find the sign of $f(x)$ in the intervals $(0, 4)$, $(4, 6)$, and $(6, \infty)$ by *scan*ning the zeros and discontinuities.

The discontinuity at $x = 0$ corresponds to the factor x^2 in the denominator of f. This factor has multiplicity 2, so the sign of $f(x)$ does not change there and remains positive in $(0, 4)$.

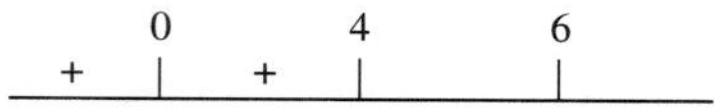

Since the zero at $x = 4$ has multiplicity 1, the sign of $f(x)$ changes there and becomes negative in $(4, 6)$.

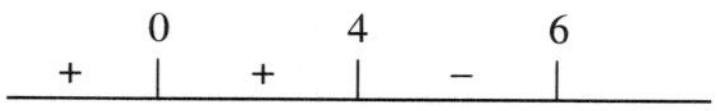

The discontinuity at $x = 6$ corresponds to the factor of $x - 6$ in the denominator of f. This factor has multiplicity 1, so the sign of $f(x)$ changes there and becomes positive in $(6, \infty)$.

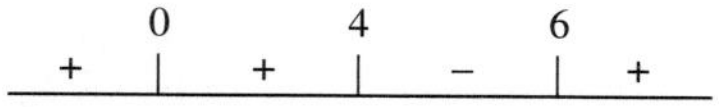

Thus $f(x) < 0$ in $(4, 6)$. Since $f(x) = 0$ at the zero $x = 4$, the solution to $f(x) \leq 0$ is $[4, 6)$.

The solution is supported by the graph of $y = f(x)$, shown in Figure 8-9.

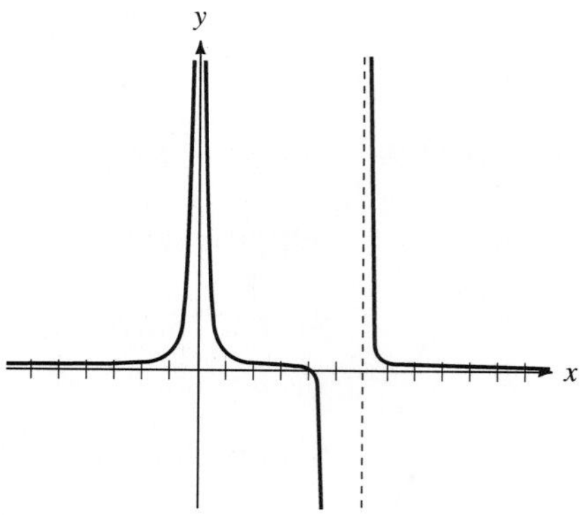

Figure 8–9

■

EXERCISES

13. As parts (a)–(d) of this exercise, solve the inequalities in Exercises 6–9 by the scan method. (You also solved them by the test-value method in Exercise 11.)

14. *(Writing to Learn)* Compare the test-value and scan methods of solving rational inequalities. Which method do you prefer to use, and why? Does your choice depend on the inequality you are solving? If so, how?

ADDITIONAL EXERCISES

In Exercises 15–32, solve the inequality by any method.

15. $\dfrac{x^2 + 3x}{x - 4} > 0$

16. $\dfrac{(x - 1)^2}{x^2 - 25} < 0$

17. $\dfrac{x^2 + 3x - 1}{x + 2} \geqslant 1$

18. $\dfrac{x^2 + 3x - 1}{x + 2} \geqslant x$

19. $\dfrac{x}{x - 1} < 0$

20. $\dfrac{x}{(x - 1)^2} < 0$

21. $\dfrac{x^2 + x - 6}{x} \geqslant 1$

22. $\dfrac{x^2 + x - 6}{x} \geqslant 2$

23. $\dfrac{(x + 20)^2(x - 10)^3}{(x + 10)(x - 30)^4} > 0$

24. $\dfrac{1}{x} \leqslant 0$

25. $\dfrac{x}{x^2 + 1} \geqslant 0$

26. $\dfrac{x}{x^2 + 1} \geqslant 2$

27. $\dfrac{x^3 + 2x - 1}{x^3} \geqslant 1$

28. $\dfrac{x^3 + x^2 - 2x}{x^2 - 16} \leqslant 0$

29. $\dfrac{(x - 1)(x - 7)}{(x - 3)^4} > 0$

30. $\dfrac{x^3}{9 - x^2} < 0$

31. $\dfrac{x^2(x + 5)^2}{x + 2} \leqslant 0$

32. $\dfrac{(x + 7)(x - 4)}{x(x - 8)^3} \geqslant 0$

33. *(Problem Solving)* In Exercise 47 of Section 8-2 you found that a cylindrical 12-ounce soft drink can of radius r inches has a surface area of

$$S(r) = \pi r^2 + \frac{43.4\pi}{r}$$

square inches.

a. For what values of r is the surface area of the can less than 100 in^2?

b. For what value of r is the surface area the smallest?

c. The can in part (b) costs the least to produce, since it uses the least material. How does its shape differ from that of most soft drink cans? List several reasons why soft drink companies do not use cans of this optimal shape.

34. *(Problem Solving)* The intensity I of light from a fixed source at a distance d from the source is $\dfrac{I_0}{d^2}$, where I_0 is a constant reference intensity. In this problem, assume that units of measurement have been chosen so that $I_0 = 1$.

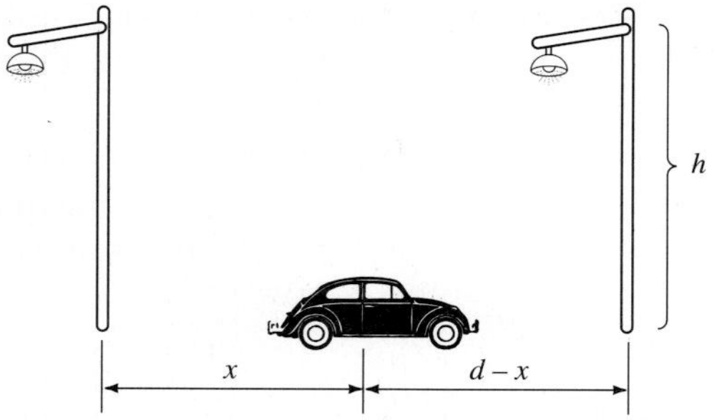

Figure 8–10

Figure 8-10 shows part of a line of lamps designed to illuminate a stretch of highway near an interchange.

a. Show that the intensity of light reaching the car from the two lamps in Figure 8-10 is

$$f(x) = \frac{1}{x^2 + h^2} + \frac{1}{(d - x)^2 + h^2}$$

(For simplicity, ignore the light reaching the car from the more distant lamps.)

b. Suppose the lamp poles are planned to be 40 feet high and 200 feet apart. Graph $y = f(x)$ in the window [0, 200] by [0, 0.001]. Your graph suggests that the light intensity is lowest halfway between the poles. Show that in general the intensity halfway between the poles is

$$\frac{8}{4h^2 + d^2}$$

c. Suppose $d = 200$ and $h = 40$ as in part (b). At what points between the two middle poles is the intensity of light reaching the car less than 0.0002? (*Hint*: Solve the inequality graphically.)

d. The intensity of light along the road can be increased by lowering the poles and/or placing them closer together. If the intensity must be at least 0.0002 at each point for safety, explain why lowering the poles alone will not be satisfactory.

CHAPTER REVIEW

Complete Exercises 1–21 (Writing to Learn) before referring to the indicated pages.

WORDS AND PHRASES

In Exercises 1–8, explain the meaning of the words or phrases in your own words.

1. (page 295) **rational function**

2. (page 295) **reciprocal power function**

3. (page 298) **horizontal asymptote, vertical asymptote**

4. (page 298) ***y* varies inversely with the *n*th power of *x***

5. (page 300) **linear rational function**

6. (page 304) **discontinuity, discontinuous** function

7. (page 306) **missing-point discontinuity**

8. (page 310) **proper rational function**

IDEAS

9. (page 295) How do the equations and the graphs of polynomial functions differ from those of rational functions?

10. (page 298) Describe the graph of $y = \frac{1}{x^n}$ for all integers $n \geq 1$. What are the domain and range of the function in each case?

11. (page 298) Why are certain laws of physics referred to as inverse square laws?

12. (pages 299, 300) How can you tell whether a table fits an inverse variation of a given degree? Make up a table and use it to illustrate the process.

13. (pages 300, 301) Describe the process of obtaining the graph of a linear rational function by transforming the basic graph $y = \dfrac{1}{x}$. Illustrate the process with a specific function.

14. (page 302) Describe the process of sketching the graph of a linear rational function by locating its intercepts and asymptotes. Illustrate the process with a specific function.

15. (pages 305, 306) Under what conditions is a rational function discontinuous? Give an example of a rational function $f(x)$ that is continuous for all values of x.

16. (pages 306, 307) What are the two possible types of discontinuity for a rational function? How do tables, equations, and graphs indicate the presence of each type? Make up specific functions to illustrate your explanation.

17. (pages 310, 311) What does the Highest Degree Theorem say about the end behavior of a rational function? Make up three functions to illustrate the three cases described by the theorem.

18. (pages 310, 311) Explain informally why the Highest Degree Theorem is true.

19. (pages 314–317) To solve a rational inequality $f(x) \geqslant 0$, why is it necessary to find the zeros of f? Why is it necessary to find the discontinuities of f?

20. (pages 316, 317) Describe the process of solving a rational inequality by the test value method. Make up an inequality to illustrate your explanation. What is the difference between the test-value method for polynomials and for rational functions?

21. (pages 317, 318) Describe the process of solving a rational inequality by the scan method. Make up an inequality to illustrate your explanation. What steps in the scan method are the same as in the test-value method? What steps are different?

CHAPTER 9

EXPONENTIAL AND LOGARITHMIC FUNCTIONS

9-1 EXPONENTIAL FUNCTIONS

Exponential functions are arguably the most useful mathematical functions. They are used to describe phenomena that grow or decrease at an increasingly rapid rate. (You may have heard the phrase *exponential growth.*) Exponential functions are used to calculate the interest on a bond, estimate the age of an artifact, and place the frets on the neck of a guitar. In this chapter you will learn how they are used to model a variety of problems. Let's begin with the problem of global overpopulation.

A MATHEMATICAL LOOKING GLASS
Standing Room Only

Rudy Rucker observes in *Mind Tools* (Houghton Mifflin, 1987)

> Overpopulation can only lead to famine, war, and plague. . . . As we seem unable to change our patterns of reproduction, shouldn't we just resign ourselves to a lot more plagues, wars, and famines in the future? Why get so upset every time a lot of people die? . . . no one is upset if a tenth of the crabs in Chesapeake Bay are steamed to death every year. . . . Why don't we feel this way about famine in Ethiopia, typhoons in Bangladesh, or an earthquake in Mexico? Most people are about the same, so there's no great global information loss if a few million disappear. . . . The argument seems perfectly logical, but if I accept it, I'm put in a position of having to accept the unacceptable statement: "It makes no real difference if all my friends and relatives drop dead tomorrow." The difference for me between people and crabs seems to be that *I* am a person, and I care about people as individuals. Perhaps a crab sees things differently.

Professor Rucker points out a serious moral dilemma that continued population growth forces upon us. Since we do not want to steam people like crabs, we must learn to control the population and/or generate new sources of food and air. A first step toward solving the population problem is to understand both its scope and its urgency. A mathematical model can provide the needed insight. Let's see if we can construct one, and use it to figure out how long it will take the present population to double. You will need to know that for the last several decades world population has grown at the rate of about 2% per year, reaching 6 billion in 1992. ■

To answer the question about doubling time, it will be convenient to express the population P in billions as a function of time t in years since 1992. As a first step in constructing the function, we can make a table, such as Table 29. An annual increase of 2% means that each year the population is multiplied by 1.02. Thus each value of P in Table 29 is 1.02 times the previous value.

TABLE 29

t	P
0	6
1	6.12
2	6.2424
3	6.367248
4	6.49459296

Intuition might tell you that a 2% increase each year means a 100% increase (doubling the population) in 50 years. However, Table 29 shows that each successive year's increase is 2% *of an increasingly larger number*, so that the doubling time should be less than 50 years. Since it would be tedious to continue the table until the value of P reached 12, let's construct an equation to represent the population function.

EXERCISE **1.** *(Writing to Learn)*

a. Explain why the entry opposite $t = 4$ in Table 29 can be written as $6(1.02)^4$.

b. Explain why the relationship of time to population can be described by the equation $P = 6(1.02)^t$.

To find the doubling time, we need to solve the equation $6(1.02)^t = 12$. You will learn how to solve this equation analytically in Section 9-4. For now, let's solve it graphically.

EXERCISE **2.** *(Interpreting Mathematics)* Graph the function $y = 6(1.02)^t$ in the window $[0, 50]$ by $[0, 18]$.

a. How does the graph support the observation that each successive year's increase is 2% of an increasingly larger number?

b. Estimate, to the nearest year, the number of years required for the population to double.

The function $f(t) = 6(1.02)^t$ is an example of an **exponential function**, that is, a function for which increases of equal *size* in the independent variable produce increases (or decreases) of equal *percentage* in the dependent variable. Let's explore exponential functions from numerical, graphical, and analytical perspectives.

A NUMERICAL VIEW OF EXPONENTIAL FUNCTIONS

Tables of Exponential Functions The definition just given tells you how to decide whether a table with equally spaced x-values fits an exponential function $y = f(x)$.

A table with equally spaced x-values fits an exponential function if and only if the ratio of consecutive y-values is constant.

EXERCISES *In Exercises 3–6, decide whether the table fits an exponential function. If it does, specify the ratio between consecutive values of y.*

3.

x	y
0	$\frac{2}{3}$
1	2
2	6
3	18

4.

x	y
0	64
2	16
4	4
6	1

5.

x	y
−3	0
0	16
3	256
6	1296

6.

x	y
1	8
2	−12
3	18
4	−27

FYI A table with only two rows always fits an exponential function. Therefore, look for more evidence before concluding that it was generated by an exponential function.

Predicting Graphical Features from Tables A table generated by an exponential function can suggest some graphical features that you can confirm later in this section. Table 30 was generated by an exponential function $y = f(x)$ for which $f(0) = 1$ and y triples (increases by 200%) whenever x increases by 1.

TABLE 30

x	y
0	1
1	3
2	9
3	27
4	81

EXERCISES **7.** *(Making Observations)*

a. Extend Table 30 to show values of y for $x = 5, 6, 7, 8, 9$, and 10. What does your table indicate about values of y when x is large?

b. Extend Table 30 to show values of y for $x = -1, -2, \ldots, -10$. What does your table indicate about values of y when $x < 0$ and $|x|$ is large?

8. *(Making Observations)* Suppose that $y = g(x)$ is a function for which $g(0) = 1$ and y is cut in half (decreases by 50%) whenever x increases by 1. Make a table to show values of y for $x = -10, -9, -8, \ldots, 8, 9, 10$. (*Hint*: Start at $x = 0$ and work in both directions.) What does your table indicate about values of y when x is large? When $x < 0$ and $|x|$ is large?

9. *(Making Observations)* Based on your results from Exercises 7 and 8, make guesses about

a. the range of an exponential function

b. the end behavior on the graph of a decreasing exponential function

c. the end behavior on the graph of an increasing exponential function

Over the next few pages you can see how accurate your guesses are.

AN ANALYTICAL VIEW OF EXPONENTIAL FUNCTIONS

The Equation of an Exponential Function Exercise 1 suggests that every exponential function can be described by an equation $y = Ab^x$. To see if that is the case, let's see if we can find such an equation for a table that fits an exponential function. Example 1 illustrates the method.

EXAMPLE 1 *Fitting an exponential function to a table*

The following table fits an exponential function $y = f(x)$.

x	y
2	4
4	36
6	324
8	2916

To find an equation $f(x) = Ab^x$ to fit the table, use the first two rows of the table to obtain

$$Ab^2 = 4$$

$$Ab^4 = 36$$

You now have a system of two equations in the two variables A and b. If the equations were linear, you could solve the system by the method of substitution. Although this system is not linear, the method still works. (By looking at linear systems to gain insight into nonlinear systems, you are applying the strategy of *Examining a Related Problem.*)

Solve the first equation for A.

$$A = \frac{4}{b^2}$$

Replace A by $\frac{4}{b^2}$ in the second equation.

$$\left(\frac{4}{b^2}\right)(b^4) = 36$$

$$b^2 = 9$$

$$b = 3$$

Now $A = \frac{4}{b^2} = \frac{4}{9}$, so $f(x) = \frac{4}{9}(3^x)$. ■

EXERCISES

10. Verify that the table in Example 1 fits an exponential function.

11. In Exercises 3–6, if the table fits an exponential function, find its equation.

The method of Example 1 can be shown to produce an equation $y = Ab^x$ for every exponential table. Thus another definition of an **exponential function** is one that can be written $f(x) = Ab^x$ for some constants A and b with $A \neq 0, b \neq 1$, and $b > 0$.

These restrictions are imposed on A and b because

- If $A = 0$ or $b = 0$, then $f(x) = 0$ for all x, so that f is a constant function.
- If $b = 1$, then $f(x) = A \cdot 1^x = A$ for all x, so that f is again a constant function.
- If $b < 0$, then $f(x)$ is not always real. For example, if $A = 1$ and $b = -4$, then $f\left(\frac{1}{2}\right) = (-4)^{1/2} = \sqrt{-4} = 2i$, which is not a real number.

EXERCISES

12. *(Making Observations)* For each function, evaluate $f(0)$, $f(1)$, $f(2)$, and $f(3)$, and confirm that the ratio of consecutive values of f is constant.

a. $f(x) = 3(4^x)$ **b.** $f(x) = -2(0.4)^x$

13. *(Writing to Learn)* How does an equation $f(x) = Ab^x$ indicate that increases of equal size in x produce increases of equal percentage in $f(x)$? If x increases by 1, what is the ratio of the new value of f to the previous value?

Irrational Exponents By now it may have occurred to you to ask what b^x means if x is irrational. For example, your calculator tells you that $2^\pi \cong 8.82497782708\ldots$, but where does that value come from? Let's explore.

Figure 9-1a shows the points obtained from a table of the function $f(x) = 2^x$ for the values $x = 0, 1, 2,$ and 3.

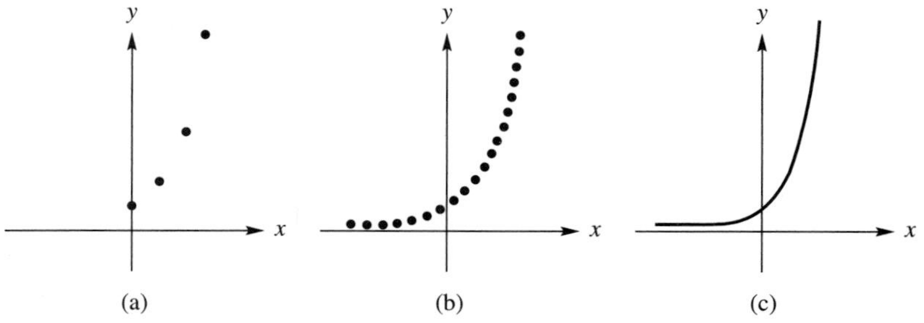

Figure 9–1

You already know that $f(x)$ can be defined for all rational values of x, since, for example, $2^{-3} = \frac{1}{8}$ and $2^{1/2} = \sqrt{2}$. Figure 9-1b suggests the graph of $f(x) = 2^x$ with the set of all rational numbers as its domain. To obtain the unbroken curve in Figure 9-1c, we must define 2^x for irrational values of x. Let's see how to define 2^π so that the graph of $f(x) = 2^x$ is unbroken when $x = \pi$.

A graphing calculator can illustrate the definition. As Figure 9-2a shows, π is in the interval $(3, 4)$, so we should define 2^{π} to be in the interval $(2^3, 2^4) = (8, 16)$. Zooming in, Figure 9-2b shows that π is in $(3.1, 3.2)$, so 2^{π} should be in $(2^{3.1}, 2^{3.2}) \cong (8.57, 9.19)$. Similarly, Figure 9-2c shows that π is in $(3.14, 3.15)$, so 2^{π} should be in $(2^{3.14}, 2^{3.15}) \cong (8.82, 8.88)$.

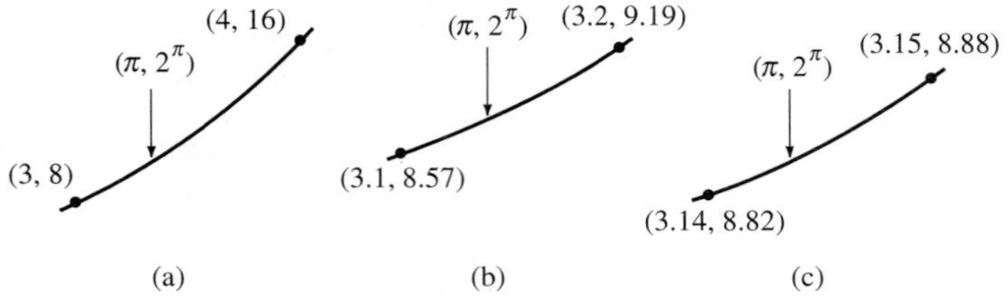

Figure 9–2

Repeated zooming-in makes it appear that there is a number (approximately 8.82497782708) smaller than 2^x for any rational $x > \pi$ and larger than 2^x for any rational $x < \pi$. In a more advanced course, it can be proved that exactly one such number exists, and can therefore be defined as 2^{π}. Values of b^x for other bases and other irrational exponents are defined in a similar manner.

EXERCISES

14. *(Extension)* Estimate the value of $5^{\sqrt{3}}$ by continuing the following table, using $\sqrt{3} = 1.732050\ldots$.

$5^1 =$	$5^2 =$
$5^{1.7} =$	$5^{1.8} =$
$5^{1.73} =$	$5^{1.74} =$

Continue until you can estimate $5^{\sqrt{3}}$ with an error of no more than 0.01. Confirm your estimate by evaluating $5^{\sqrt{3}}$ on your calculator.

15. *(Extension)* Use tables similar to the one in Exercise 14 to estimate each of the following with an error of no more than 0.01.

a. $3^{\sqrt{6}}$ **b.** π^{π}

Predicting Graphical Features from Equations In Exercise 9 you made guesses about several graphical features of functions $f(x) = b^x$. Let's see if we can predict the same features analytically. Specifically, let's see if the equation supports the following statements about an exponential function $f(x) = b^x$.

- The domain of f is $(-\infty, \infty)$.
- The range of f is $(0, \infty)$.
- If $b > 1$, f is increasing. The value of f becomes arbitrarily large when x is large and becomes arbitrarily close to 0 when $x < 0$ and $|x|$ is large.
- If $0 < b < 1$, f is decreasing. The value of f becomes arbitrarily close to 0 when x is large and becomes arbitrarily large when $x < 0$ and $|x|$ is large.

Based on our discussion of irrational exponents, the expression b^x is defined for all real values of x, so the first of the preceding statements is true. You can explore the rest in Exercises 16–17.

EXERCISES **16.** *(Writing to Learn)* Suppose $f(x) = 2^x$.

a. What can you say about the value of $f(x)$ if x is large? Explain.

b. What can you say about the value of $f(x)$ if $x < 0$ and $|x|$ is large? Explain.

c. Which of the statements preceding this exercise are supported by your observations in parts (a) and (b)?

17. *(Writing to Learn)* Repeat Exercise 16 for $g(x) = \left(\frac{1}{2}\right)^x$.

A GRAPHICAL VIEW OF EXPONENTIAL FUNCTIONS

Our preceding discussion suggests that the graphical features of a function $f(x) = b^x$ depend on whether $b > 1$ or $0 < b < 1$. Figures 9-3a and 9-3b show the graphs of $y = 2^x$ and $y = \left(\frac{1}{2}\right)^x$, respectively.

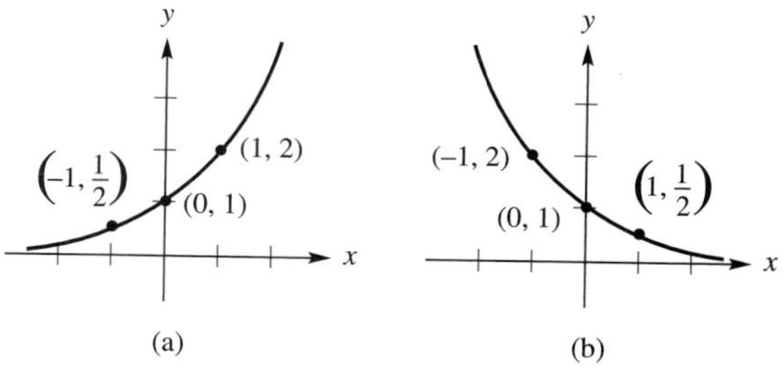

Figure 9–3

Exercises 18–19 show that these graphs illustrate behavior typical of exponential functions.

EXERCISES **18.** *(Making Observations)*

a. Graph the following functions together in the window [−2, 2] by [−1, 5]: $y = 2^x$, $y = 5^x$, $y = 20^x$.

b. Graph the following functions together in the window [−2, 2] by [−1, 5]: $y = 2^x$, $y = 1.5^x$, $y = 1.1^x$.

Based on your graphs, guess the answers to the following questions about a function $f(x) = b^x$ with $b > 1$.

c. What is the domain of f? What is its range?

d. Does the graph have any x-intercepts? y-intercepts? If so, where are they?

e. In terms of b, what is the value of $f(1)$? $f(-1)$?

f. Does the graph have any vertical asymptotes? Horizontal asymptotes? If so, where are they?

g. Is f increasing or decreasing?

19. *(Making Observations)*

a. Repeat Exercise 18a for the functions $y = \left(\frac{1}{2}\right)^x$, $y = \left(\frac{1}{5}\right)^x$, $y = \left(\frac{1}{20}\right)^x$.

b. Repeat Exercise 18b for the functions $y = \left(\frac{1}{2}\right)^x$, $y = \left(\frac{2}{3}\right)^x$, $y = \left(\frac{9}{10}\right)^x$.

c–g. Repeat Exercises 18c–18g for functions $f(x) = b^x$ with $0 < b < 1$.

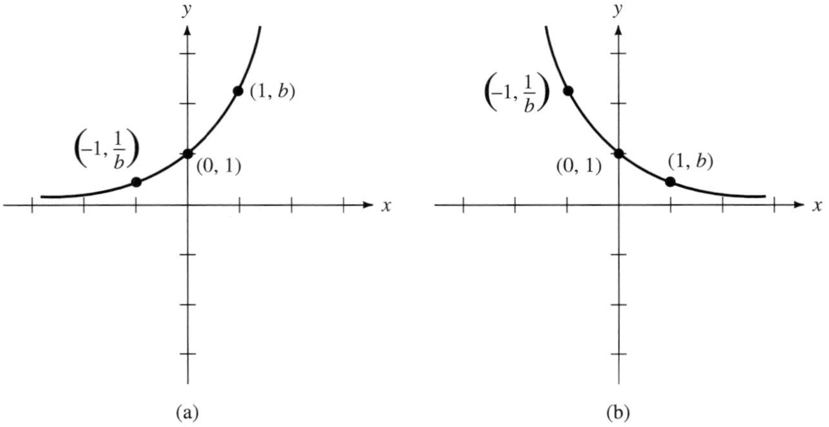

Figure 9–4

Exercises 18–19 illustrate that the graph of $y = b^x$ resembles Figure 9-4a if $b > 1$ and Figure 9-4b if $0 < b < 1$. To sketch it quickly, begin by plotting the y-intercept at $(0, 1)$. Then plot $(1, b)$ and $\left(-1, \frac{1}{b}\right)$, which are always on the graph. Finally, draw a smooth curve through those points with one of the two basic shapes.

The graphs of many functions with similar equations can be obtained by transforming an appropriate graph $y = b^x$.

EXAMPLE 2 *Transforming the graph of $y = b^x$*

To sketch the graph of $y = 3 \cdot 5^{-x}$, begin with the basic graph $y = 5^x$, and make the changes

$$y = 5^x \to y = 5^{-x} \to y = 3 \cdot 5^{-x}$$

Replacing x by $-x$ reflects the graph in the y-axis, and multiplying the function by 3 stretches the graph vertically by a factor of 3. Figure 9-5 shows how the points $\left(-1, \frac{1}{5}\right)$, $(0, 1)$, and $(1, 5)$ on the basic graph move as the transformations are applied.

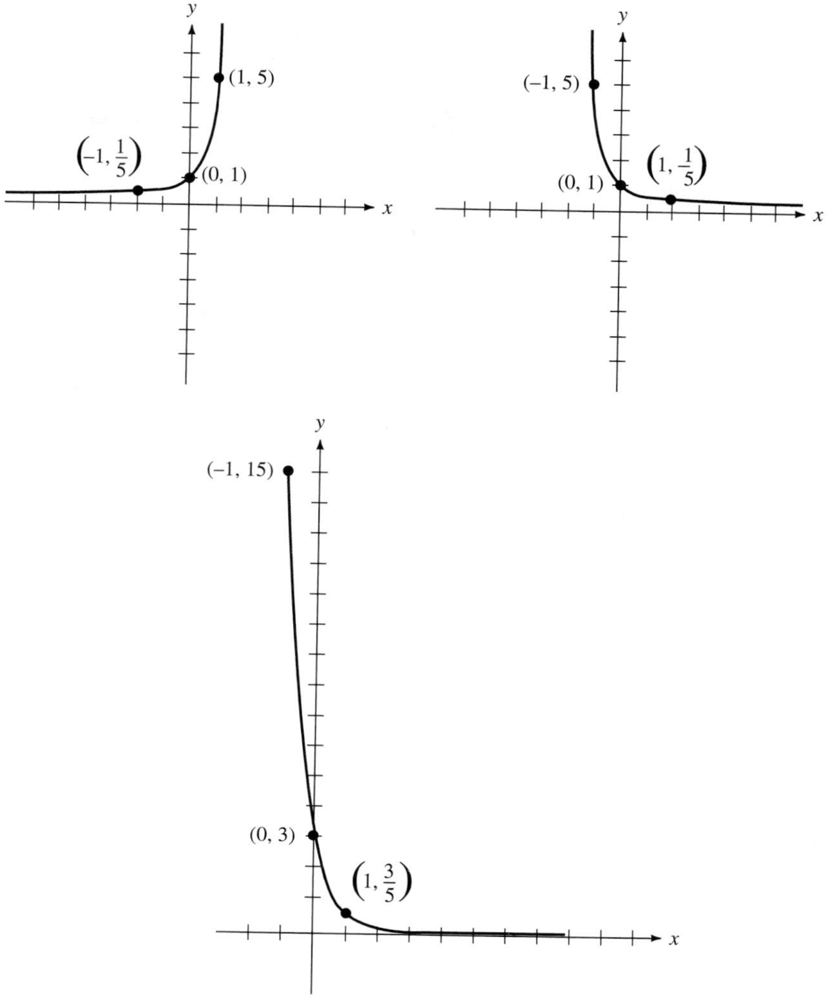

Figure 9–5

■

EXERCISES *In Exercises 20–23, sketch the graph of each function by transforming an appropriate graph $y = b^x$. In each case,*

a. State the domain and range of the function.

b. Decide whether the function is increasing or decreasing.

c. Plot the points obtained by moving $\left(-1, \dfrac{1}{b}\right)$, $(0, 1)$, and $(1, b)$.

20. $y = 4^{x+3}$ **21.** $y = 0.1^{2x} - 2$

22. $y = 7\left(\dfrac{1}{2}\right)^{x-1}$ **23.** $y = \left(-\dfrac{1}{2}\right)(3^{-x})$

24. *(Writing to Learn)* Describe how each transformation in parts (a)–(f) affects the following graphical features of $y = b^x$.

- domain and range
- asymptotes

- intercepts
- increasing/decreasing behavior

a. a horizontal stretch or compression
b. a vertical stretch or compression
c. a reflection in the x-axis
d. a reflection in the y-axis
e. a horizontal shift
f. a vertical shift

Exercises 25–28 illustrate that every exponential function $f(x) = Ab^x$ *has a constant* **doubling time**. *That is, for each such function there is a number* h *such that* $f(x + h) = 2f(x)$ *for every value of* x. *In each exercise, estimate the doubling time by graphing* $f(x)$ *and estimating the values of* x *for which* $f(x) = 1, f(x) = 2, f(x) = 4$, *and* $f(x) = 8$.

25. $f(x) = 1.5^x$ **26.** $f(x) = 4(3^x)$
27. $f(x) = 5(2^x)$ **28.** $f(x) = 0.8^x$ (*Hint*: The doubling time is negative.)

GEOMETRIC SEQUENCES AND SERIES

Many problems can be modeled by an exponential function whose domain is a set $\{K, K + 1, K + 2 \ldots\}$ for some integer K. Such a function is called a **geometric sequence**. (In Section 2-3 you learned that a sequence is a function whose domain is such a set.) The following situation can be modeled by a geometric sequence.

A MATHEMATICAL LOOKING GLASS
20 Questions

As a child, you probably played the game of "20 Questions," in which your opponent thinks of an object, and you try to identify it by asking up to 20 yes/no questions. Did it ever surprise you that among the millions of objects your opponent could choose, you were often able to identify the right one? You are about to discover that your ability is less amazing than it might appear. .

Let's play a simplified version of "20 Questions," in which we will think of a symbol, such as a letter of the alphabet, and you must guess it. To identify it with certainty, the number of questions you need to ask depends on the size of our pool of symbols.

- If our pool contains only one symbol, then you can identify it without asking any questions.
- If our pool contains two symbols, say A and B, then you can identify the chosen letter by asking the single question, "Is it A?" Our answer will tell you whether it is A or B.
- If our pool contains the four symbols A, B, C, and D, you can identify the chosen letter by asking at most two questions. For example, you can first ask, "Is it one of the symbols A or B?" If our answer is yes, you know that the symbol is either A or B. If the answer is no, you know it is either C or D. In either case, you have reduced the size of the pool to two symbols, and you can guess the correct one with one more question.
- If our pool contains the eight symbols A, B, C, D, E, F, G, and H, you can identify the chosen letter by asking at most three questions. For example,

you can first ask, "Is it one of the symbols A, B, C, or D?" Regardless of our answer, you have reduced the size of the pool to four symbols, and you can guess the correct one with two more questions.

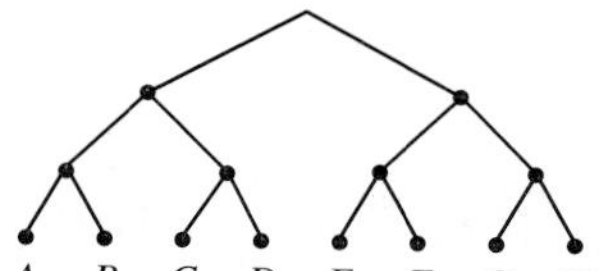

Figure 9–6

With this strategy, each additional question allows you to identify the correct symbol from an original pool twice as large. For example, by asking four questions, you can identify the correct symbol from a pool of 16.

This strategy, called a **binary search**, is illustrated in Figure 9-6. Starting at the top vertex, ask a question about one half of the pool of symbols. If the answer is "yes," proceed on the path toward that half. If the answer is "no," proceed toward the other half. When you arrive at the next vertex, repeat the procedure. ■

EXERCISES

29. *(Creating Models)* If n represents the number of questions needed to guess the correct symbol, write a function of n to represent the maximum size of the pool of symbols.

30. *(Interpreting Mathematics)* If you can ask 20 questions, how large can the pool of symbols be?

31. *(Writing to Learn)* Why do you think children playing "20 Questions" succeed so often in identifying the correct object? In particular, how does your answer to Exercise 30 provide insight? What other factors might be involved?

The function in Exercise 30 is a geometric sequence, and its values 1, 2, 4, 8, . . . are the terms of the sequence. We can give the sequence a functional name such as a and denote its terms by $a_0 = 1, a_1 = 2, a_2 = 4, a_3 = 8, \ldots$. Geometric sequences can also be described by the set of values of the dependent variable, so that this sequence may be written

$$a = \{1, 2, 4, 8, \ldots\}$$

The geometric sequence in Exercise 30 can also be described by the equation $f(n) = 2^n$. An equation has the advantage of giving a formula for each term in the sequence.

EXERCISES

In Exercises 32–35, write out the terms a_0, a_1, a_2, a_3, and a_4, and decide whether the sequence is geometric.

32. $a_n = 4^n$ **33.** $a_n = \dfrac{2}{(-3)^n}$

34. $a_n = \dfrac{1}{5}n^2$ **35.** $a_n = \dfrac{1}{5}2^n$

In Exercises 36–39, decide whether the given sequence is geometric. If it is, find a formula for a_n, and use your formula to find a_{10}.

36. $a = \{1, 3, 9, 27, \ldots\}$

37. $a = \{4, 8, 12, 16, \ldots\}$

38. $a = \{1000, 200, 40, 8, \ldots\}$

39. $a = \{1, 1.1, 1.21, 1.331, \ldots\}$

In solving problems we sometimes need to add a large number of consecutive terms in a geometric sequence. Such a sum is referred to as a **geometric series**. The addition is usually cumbersome, but in Exercises 40–41 you can develop a formula for the sum.

EXERCISES

40. For the sequence $a_n = \left(\frac{1}{4}\right)^n$, let $S_4 = 1 + \frac{1}{4} + \left(\frac{1}{4}\right)^2 + \left(\frac{1}{4}\right)^3 + \left(\frac{1}{4}\right)^4$.

a. Multiply each term by $\frac{1}{4}$ to obtain an expression for $\frac{1}{4}S_4$.

b. Subtract this expression from the one for S_4, and simplify to obtain a shorter expression for $\frac{3}{4}S_4$.

c. Divide by $\frac{3}{4}$ to obtain the value of S_4.

Now, in Exercise 41, *generalize your work from Exercise* 40 *to obtain the formula*

$$S_n = \frac{A(1 - r^{n+1})}{1 - r}$$

41. For the sequence $a_n = Ar^n$, let $S_n = A + Ar + Ar^2 + \cdots + Ar^{n-1} + Ar^n$.

a. Write an expression for rS_n in terms of A, r, and n.

b. Subtract this from the expression for S_n, and simplify to obtain a shorter expression for $(1 - r)S_n$.

c. Divide by $1 - r$ to obtain a formula for S_n.

Use the formula in Exercise 41 *to find* S_5, S_{10}, *and* S_{20} *for each sequence in Exercises* 42–45.

42. $a_n = 6\left(\frac{1}{3}\right)^n$ **43.** $\left\{16, -4, 1, -\frac{1}{4}, \ldots\right\}$

44. $a_n = (-2)^n$ **45.** $\{2, 2.2, 2.42, 2.662, \ldots\}$

46. *(Making Observations)* In two of the sequences in Exercises 42–45, the values of S_n appear to approach a particular value as n becomes large. Which two sequences are they, and what is the value approached by S_n in each case?

If S_n approaches a particular value as n becomes large, that value is called the **sum of the infinite geometric series** $a_1 + a_2 + a_3 + \cdots$. Every geometric sequence $a_n = Ar^n$ with $0 < r < 1$ has a sum, such as the ones you observed in Exercise 46. By using the formula

$$S_n = \frac{A(1 - r^{n+1})}{1 - r}$$

you can derive an expression for the sum in terms of A and r. If $0 < r < 1$ and n becomes sufficiently large, r^{n+1} becomes arbitrarily small, so that S_n becomes arbitrarily close to $\frac{A}{1 - r}$. This number is the sum of the series whenever $0 < r < 1$. (If $r > 1$ and n becomes sufficiently large, r^{n+1} and S_n become arbitrarily large as well, so that the series has no sum.)

47. *(Making Observations)* Check your conclusions about the sums in Exercise 46 by evaluating $\frac{A}{1 - r}$.

ADDITIONAL EXERCISES

In Exercises 48–51, decide whether the table fits an exponential function. If it does, find its equation.

48.

x	y
1	100,000
2	10,000
3	1,000
4	100

49.

x	y
−0.5	0.007
0	0.021
0.5	0.063
1	0.189

50.

x	y
1.5	8
2.5	32
3.5	128
4.5	512

51.

x	y
2	90
3	99
4	99.9
5	99.99

In Exercises 52–55, find an equation for an exponential function $f(x) = A \cdot b^x$ *passing through the given points.*

52. $(0, 6)$ and $(2, 24)$ **53.** $(0, 7)$ and $(2, 49)$

54. $(3, 1)$ and $(4, 10)$ **55.** $(3, 40)$ and $(8, 30)$

In Exercises 56–59, sketch the graph of each function by transforming an appropriate graph $y = b^x$. *In each case,*

a. State the domain and range of the function.

b. Decide whether the function is increasing or decreasing.

c. Plot the points obtained by moving $\left(-1, \frac{1}{b}\right)$, $(0, 1)$, and $(1, b)$.

56. $y = (0.1)(3^{2x})$ **57.** $y = 5^{x-2} + 1$

58. $y = -4^{-x}$ **59.** $y = 8 - 2^x$

60. *(Making Observations)*

a. Use transformations to sketch the graphs of $y = 3^{x+2}$ and $y = 9 \cdot 3^x$. The two graphs should appear to be identical. Support this observation with calculator graphs.

b. Part (a) suggests that any function $y = A \cdot b^{x-h}$ can be written $y = A_1 \cdot b^x$ for some value of A_1. Find a value of A_1 for which $y = 3 \cdot 4^{x+1}$ is equal to $y = A_1 \cdot 4^x$. (*Hint*: Use the laws of exponents to rewrite 4^{x+1}.)

61. *(Making Observations)*

a. Use the transformations to sketch the graphs of $y = 3^{2x}$ and $y = 9^x$. The two graphs should appear to be identical. Support this observation with calculator graphs.

b. Part (a) suggests that any function $y = A \cdot b^{cx}$ can be written $y = A(b_1)^x$ for some value of b_1. Find a value of b_1 for which $y = 4^{3x}$ is equal to $y = (b_1)^x$. (*Hint*: Use the laws of exponents to rewrite 4^{3x}.)

In Exercises 62–69,

a. Decide whether the sequence is geometric.

b. If the sequence is geometric, find a formula for a_n, and use your formula to find a_8.

c. If the sequence is geometric, decide whether the infinite series $a_1 + a_2 + a_3 + \cdots$ has a sum. If the sum exists, find it.

62. $a = \left\{1, \frac{1}{3}, \frac{1}{9}, \frac{1}{27}, \ldots\right\}$

63. $a = \{-1, -0.9, -0.81, -0.729, \ldots\}$

64. $a = \left\{\frac{1}{2}, \frac{1}{4}, \frac{1}{6}, \frac{1}{8}, \ldots\right\}$

65. $a = \{10, 50, 250, 1250, \ldots\}$

66. $a = \left\{1, -\frac{4}{3}, \frac{16}{9}, -\frac{64}{27}, \ldots\right\}$

67. $a = \left\{1, -\frac{3}{4}, \frac{9}{16}, -\frac{27}{64}, \ldots\right\}$

68. $a = \left\{9, 6, 4, \frac{8}{3}, \ldots\right\}$

69. $a = \{2, 8, 18, 32, \ldots\}$

70. *(Problem Solving)* The Occupational Health and Safety Act of 1971 established permissible levels of occupational noise in the United States. In the table below, t is the maximum acceptable duration of a noise level of A decibels.

A	t
95	4
100	2
105	1
110	0.5
115	0.25

a. Write an equation for t as a function of A.

b. Graph the function in part (a), and estimate the maximum acceptable duration for a noise level of 103 decibels.

c. What is the maximum noise level that can be tolerated through an 8-hour workday?

71. *(Creating Models)* The **pitch**, or frequency, of a musical note determines whether we hear it as a high sound like that produced by a flute or violin or a low sound like that produced by a tuba or bass viol.

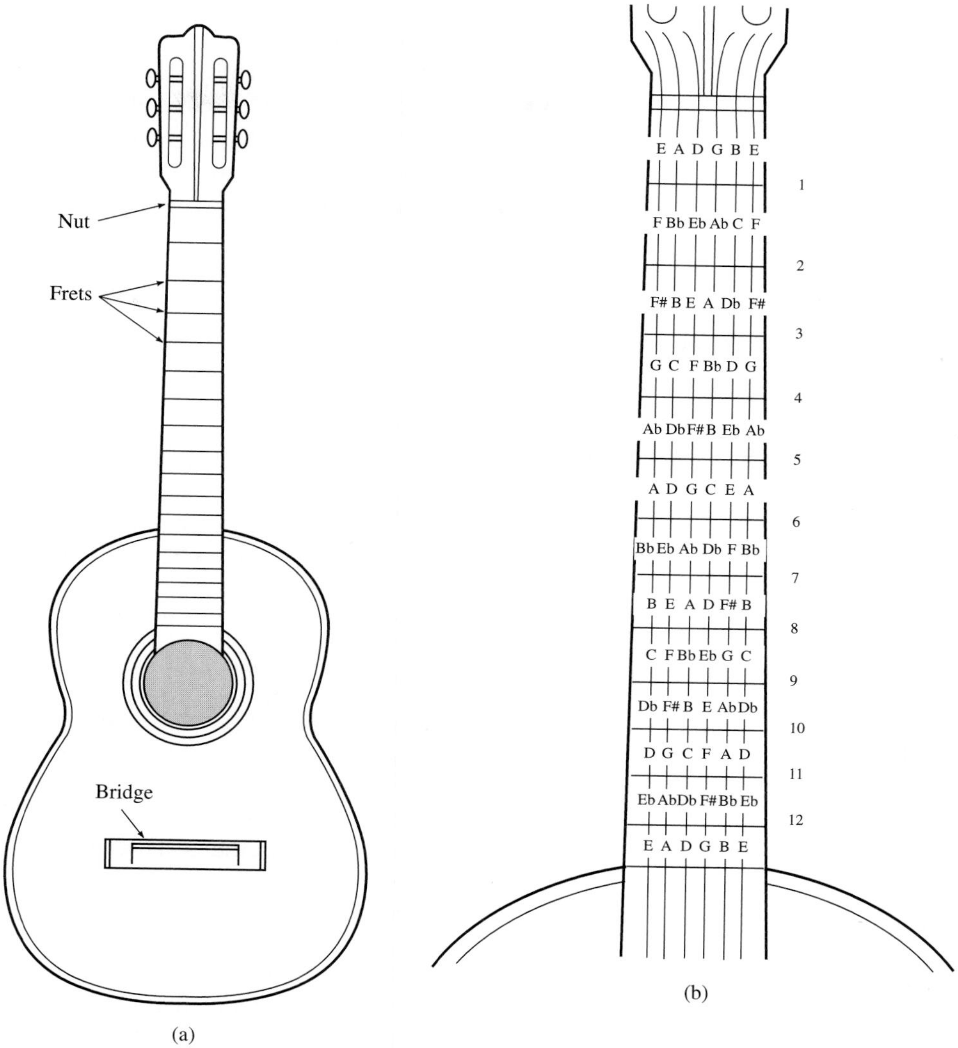

Figure 9–7

Figure 9-7 shows the neck of a guitar. It must be constructed so that moving your finger from one fret to the next higher fret always produces the same percentage increase in the frequency of the note. The frequency produced at the 12th fret is twice the frequency produced at the nut (the 0th fret). For the fifth string, the frequency at the nut is 220 cycles per second. Find an equation to express the frequency F of the fifth string at the nth fret as a function of n.

72. *(Problem Solving)* The frequency of the note produced by a guitar string at a given fret is inversely proportional to the distance D from the fret to the bridge (see Figure 9-7). On a Gibson ES-335 guitar the distance from the nut to the bridge is 24.75 inches.

a. If F is the frequency of the note produced by the fifth string when the distance from fret to bridge is D inches, write an equation to express D as a function of F.

b. Combine your functions in Exercises 71 and 72a to express D as a function of n.

c. At what distance from the bridge should each of the first 12 frets be placed on a Gibson ES-335 guitar?

73. *(Problem Solving)* A diabolical mathematician has imprisoned a pregnant fruit fly in a jar. She notices that the number of flies doubles each day. At the end of the 30th day, the jar is full. After how many days was it half full?

9-2 THE SPECIAL NUMBER *e*

THE DEFINITION OF *e*

Exponential functions are used to calculate interest on an investment or a loan. In this section you will discover that continuously compounded interest is calculated by using an exponential function with a particular irrational base called e. You will also see how the same exponential function is used to model other physical phenomena. In the following Mathematical Looking Glass you will learn how the number e is defined, and you will find its approximate value.

A MATHEMATICAL LOOKING GLASS
Dealing with e

On December 19, 1993, the *Pittsburgh Post-Gazette* ran an advertisement placed by Barden-McKain Ford, offering automobile loans at 5.9% annual interest over a period of 5 years. Such a statement, typical in automobile ads, appears to define a specific interest rate. However, the amount of interest you pay depends on whether it is compounded once a year, once a month, once a day, or continuously.

To demonstrate the effect of compounding in a simple setting, suppose that you are going to borrow \$10,000 at 5.9% annual interest, keep it for 5 years, and pay it back all at once. How much will you pay altogether? (You would pay back a real loan by making monthly payments. Determining these is much more difficult, and we leave the problem to students of accounting and/or differential equations.) ■

EXERCISES

1. Suppose that the interest is compounded annually (once a year). Then the amount you owe is increased by 5.9% once a year.

a. *(Writing to Learn)* Explain why you owe \$10,000(1 + 0.059) after 1 year.

b. Complete the following table to find the amount you owe after 5 years.

Interest periods	Amount owed
0	\$10,000
1	\$10,000(1 + 0.059) = \$10,590
2	\$10,590(1 + 0.059) = \$11,214.81
3	
4	
5	

c. *(Creating Models)* The table in part (b) defines a geometric sequence. Write an expression Ab^t for the amount owed after t years.

2. Suppose that the interest is compounded semiannually (twice a year). Then the amount you owe is increased by half of 5.9% twice a year.

a. *(Writing to Learn)* Explain why after half a year you owe $\$10{,}000\left(1 + \frac{0.059}{2}\right)$.

b. Complete the following table to find the amount you owe after 5 years.

Interest periods	Amount owed
0	\$10,000
1	$\$10{,}000\left(1 + \frac{0.059}{2}\right) = \$10{,}295$
2	$\$10{,}295\left(1 + \frac{0.059}{2}\right) = \$10{,}598.70$
3	
4	
.	
.	
.	
10	

c. *(Creating Models)* The table in part (b) defines a geometric sequence. Write an expression Ab^k for the amount owed after k interest periods. Then rewrite the expression to denote the amount owed after t years. (*Hint*: There are $2t$ interest periods in t years.)

3. *(Creating Models)* Suppose that the interest is compounded three times a year. Then the amount you owe is increased by one-third of 5.9% three times a year.

a. Find the amount you owe after one-third of a year.

b. Complete a table to find the amount you owe after 5 years (15 interest periods).

c. Write an expression Ab^k for the amount owed after k interest periods. Then rewrite the expression to denote the amount owed after t years.

Exercises 1–3 illustrate that if \$$P$ is borrowed for t years at an interest rate of r, compounded m times per year, the total amount owed is

$$A = P\left(1 + \frac{r}{m}\right)^{mt}$$

As stated in ***Dealing with e***, the amount you owe depends on how often the interest is compounded. However, this is not the end of the story. If the interest is compounded ridiculously often, you should now be asking whether your debt could be ridiculously large. You can answer this question in Exercises 4–5.

EXERCISES

4. Using the formula just developed, compute the amount you owe after 5 years if you borrow \$10,000 at 5.9% interest compounded

a. once a month **b.** once a day

c. once an hour **d.** once a minute

5. *(Making Observations)* Exercise 4 suggests the presence of a number above which the amount will not rise, no matter how often the interest is compounded. To the nearest dollar, what is this number?

You might now guess that if you borrow any amount $\$P$ at any interest rate r for any period of t years, there is a ceiling on the amount you owe, no matter how often the interest is compounded. In Exercises 6–8 you can describe that ceiling in terms of P, r, and t only.

EXERCISE

6. *(Writing to Learn)* Explain why the formula

$$A = P\left(1 + \frac{r}{m}\right)^{mt}$$

can be written in each of the following ways.

$$A = P\left(1 + \frac{r}{m}\right)^{(m/r)rt}$$

$$A = P\left[\left(1 + \frac{r}{m}\right)^{(m/r)}\right]^{rt}$$

Exercises 4 and 5 suggest that we should look for the ceiling by exploring the behavior of $\left(1 + \frac{r}{m}\right)^{(m/r)}$ as the number of compoundings m becomes large. If $n = \frac{m}{r}$, then the above expression becomes $\left(1 + \frac{1}{n}\right)^n$. Since $n = \frac{m}{r}$, if m is large, then n will be large also, so let's ask whether the value of $\left(1 + \frac{1}{n}\right)^n$ approaches a ceiling as n becomes large.

EXERCISES

7. *(Making Observations)* Complete the following table to estimate the ceiling on the values of $\left(1 + \frac{1}{n}\right)^n$ as n becomes large.

n	$\left(1 + \frac{1}{n}\right)^n$
1,000	
1,000,000	
1,000,000,000	

8. *(Making Observations)* Estimate the ceiling on the values of $\left(1 + \frac{1}{n}\right)^n$ as n becomes large by looking at the end behavior of the graph of $y = \left(1 + \frac{1}{x}\right)^x$.

Using calculus, it can be shown conclusively that if n is large, the value of $\left(1 + \frac{1}{n}\right)^n$ approaches 2.71828182846. . . . This special number is called e to honor Swiss mathematician Leonhard Euler, whom you met in Section 7-2.

In summary, if interest is compounded many times each year, our interest formula becomes approximately

$$A = Pe^{rt}$$

This equation is used to define what is called **continuous compounding**. (This does not mean that the interest is actually compounded in infinitesimally small time periods. It simply means that the amount owed is calculated according to this formula.)

FYI From now on, unless otherwise specified, *interest* means continuously compounded interest. This assumption is made in many everyday situations, such as the automobile loan advertisement in ***Dealing with e***.

Don't be confused by the number e. Like π, e is an irrational number which has been given a letter name because it is used in so many situations. Its decimal expansion neither terminates nor repeats, but for many purposes, such as sketching graphs, you may approximate e by 2.7. For most other purposes in this book, you may approximate it by 2.72. If greater accuracy is desired, your calculator can provide it. (Actually, you can do better than your calculator. The first 15 decimal places of e can be grouped conveniently as $e = 2.7\ 1828\ 1828\ 45\ 90\ 45\ldots$. This memory device is due to George Simmons, *Calculus with Analytic Geometry,* McGraw-Hill, 1985, p. 214.)

EXERCISES

9. In each case, find the amount owed when $\$P$ is borrowed for t years at an annual interest rate r compounded annually. Repeat the calculation for interest compounded monthly, then continuously.

a. $P = 4000, t = 3, r = 14.9\%$

b. $P = 50{,}000, t = 30, r = 8.6\%$

c. $P = 500, t = 0.5, r = 200\%$

10. In each case, find the amount obtained when $\$P$ is invested for t years at an annual interest rate r compounded annually. Repeat the calculation for interest compounded monthly, then continuously.

a. $P = 2000, t = 2, r = 2.5\%$

b. $P = 8200, t = 1, r = 20.85\%$

c. $P = 5, t = 0.25, r = 7.5\%$

11. Sketch the graphs of $y = 2^x$, $y = e^x$, and $y = 3^x$ on the same set of axes, using $e \cong 2.7$. Support your results with calculator graphs.

USING THE BASE e TO EXPRESS EXPONENTIAL FUNCTIONS

The number e is often used as a base for exponential functions to model physical relationships. Exercises 12 and 13 suggest one reason why.

EXERCISES

12. *(Making Observations)* In each case, calculate the average rate of change in $f(x) = e^x$ over the interval $[a, b]$, and verify that your result is a number between e^a and e^b.

a. $a = 1, b = 3$ **b.** $a = 1.9, b = 2.1$

c. $a = 3, b = 5$ **d.** $a = 3.99, b = 4.01$

13. *(Writing to Learn)* Your results in Exercise 12 suggest that the average rate of change in $f(x) = e^x$ over any interval is always equal to the value of $f(x)$ at some point within the interval. Explain why this in turn suggests that $f(x)$ is an appropriate model for physical quantities, such as compound interest and world population, that grow at a rate proportional to their size.

If you take a course in calculus, you will expand on the ideas in Exercises 12 and 13, and you will discover that the operations of calculus are performed more easily on exponential functions when the base is e. The following Mathematical Looking Glass illustrates that every exponential function can be written with a base of e.

A MATHEMATICAL LOOKING GLASS
Crater Lake

Crater Lake is a circular lake in southwestern Oregon, about 6 miles in diameter and almost 2000 feet deep. It is contained in a crater formed about 6700 years ago when the volcano Mount Mazama erupted with such violence that its top was blown off (much like Washington's Mount St. Helens in 1980). The lake is remarkable for its dramatic origin and its deep blue color. Its level remains surprisingly constant, even though it is fed entirely by snowfall and rainfall and drained entirely by evaporation.

Have you ever wondered how scientists can determine when a prehistoric event, such as the formation of Crater Lake, occurred? One method that can determine age up to 50,000 years with reasonable accuracy is known as radiocarbon dating. It is based on two facts.

- Among the carbon atoms contained in all living organisms, the portion of radioactive carbon-14 atoms is constant. When an organism dies, its carbon-14 begins to decay into nonradioactive carbon-12, so that the portion of carbon-14 atoms among all carbon atoms decreases.
- Every radioactive substance has a **half-life**, a fixed length of time during which half of the atoms in any sample of the substance will decay. The half-life of carbon-14 is about 5730 years.

This means that a sample of bone from an animal or wood from a tree 5730 years old contains 50% of the carbon-14 it contained when the animal or tree was alive. A sample 11,460 (twice 5730) years old contains 25% (50% of 50%) of the original portion of carbon-14. ■

If the half-life of a radioactive substance is known, we can write a function $f(t) = 2^{-kt}$ for the portion of the substance present after t years. The base 2 occurs naturally, because when the value of t is equal to the half-life of the substance, the value of f is $\frac{1}{2} = 2^{-1}$.

EXERCISES

14. *(Creating Models)* Write a function $f(t) = 2^{-kt}$ for the portion of carbon-14 in a sample of material from an organism that died t years ago. [*Hint*: To find k, use the fact that $f(5730) = 2^{-1}$.]

15. *(Interpreting Mathematics)* The estimate of Crater Lake's age is based in part on a sample of charcoal from a tree that burned in the eruption of Mount Mazama when the lake was formed. Given that the lake is 6700 years old, what percentage of the carbon-14 found in living wood was in the sample?

Scientists often write exponential functions in the form $f(t) = e^{kt}$. In Section 9-4, you will learn how to find k analytically. In the meantime, you can use a graphical approach in Exercise 16.

EXERCISES

16. **a.** Zoom in on the graph of $y = e^x$ to find a value $x = c$ with $e^c = 2$.
    b. Rewrite your function from Exercise 14 as $f(t) = e^{kt}$ by substituting e^c for 2.
    c. Use your function from part (b) to answer the question in Exercise 15. Make sure this answer agrees with the one you obtained there.

In Exercises 17–20, *use the method of Exercise* 16 *to write each function in the form* $y = Ae^{kt}$.

17. $y = 3^t$ **18.** $y = 5^{2t}$

19. $y = 3\left(\frac{1}{2}\right)^t$ **20.** $y = 7(0.8)^{3t}$

ADDITIONAL EXERCISES

In Exercises 21–24, *sketch the graph of each function by transforming the graph of* $y = e^x$. *In each case, plot the points obtained by moving* $\left(-1, \frac{1}{e}\right)$, $(0, 1)$, *and* $(1, e)$.

21. *(Review)* $y = e^{0.5x}$ **22.** *(Review)* $y = e^{-0.5x}$

23. *(Review)* $y = 2e^{-0.5x}$ **24.** *(Review)* $y = 2e^{-0.5x} - 3$

25. *(Writing to Learn)* Describe how each graph in Exercises 21–24 differs from the preceding one.

26. *(Problem Solving)* A few years ago Lynn bought a 6-month CD with Atlantic Financial Savings and Loan in the amount of \$10,000 at an annual interest rate of 6.2%.

a. Explain why the amount of money in her account after t months is

$$A(t) = 10{,}000e^{0.062t/12}$$

b. After 2.2 months Lynn's bank was bought by Ellwood Federal Savings. How much was her account worth at that time?

27. *(Problem Solving)* (continuation of Exercise 26) Ellwood Federal Savings lowered the annual interest rate on Lynn's CD to 4.85%.

a. Write a function for the amount of money in Lynn's account after t months at 4.85% annual interest. (*Hint*: When $t = 0$, the amount is your answer to Exercise 26b.)

b. If she leaves her money in the account at 4.85% for the remaining 3.8 months of its term, what will her account be worth?

28. *(Problem Solving)* Lynn wanted to know whether to leave her money in the account, as in Exercise 27, or pay a penalty for early withdrawal and reinvest elsewhere at a higher rate. Head teller Madeline Kelley told her that the penalty would be \$119.59. (She also said that if Lynn had acted more promptly after the takeover, there would have been no penalty.)

a. If she pays the penalty, how much will she have left to reinvest?

b. If she reinvests at an annual interest rate of r, write a function of r for the amount of money in her account after 3.8 months.

c. Graph your function from part (a). Use the graph and your answer from Exercise 27 to determine what interest rates would make it profitable to pay the penalty.

29. *(Problem Solving)* Six months ago, you received a high school graduation gift of \$10,000 from your grandfather. Since then you have had it in a CD while making long-term plans for it. Your grandfather has cautioned you that your return on a long-term investment will be very sensitive to the interest rate.

a. Six months ago, you could have invested the money at 6.2% at a bank 20 miles away. You chose instead to put it into the bank around the corner at 6%. How much did you lose?

You point out that taking a lower interest rate for the sake of convenience didn't cost you very much. Your grandfather smiles knowingly and says, "Yes, but that was only for six months, and you will keep this investment until you retire."

b. Assuming that you will keep the money for 50 years before passing it on to your grandchildren, what will be the difference between your returns at 6.2% and at 6%?

30. *(Problem Solving)* How old is a sequoia tree if its (nonliving) heartwood contains 65% of its original carbon-14? (*Hint*: Solve graphically.)

31. *(Problem Solving)* Radiocarbon dating, as described in ***Crater Lake***, can be used reliably to date only organic objects up to about 50,000 years old. Dating older objects requires other radioactive elements with longer half-lives. For example, rocks are sometimes dated by analyzing their uranium 238, which has a half-life of 4.5 billion years and decays into lead 206. The ratio of uranium to lead indicates the portion of uranium which has decayed since the rock was formed.

a. Write a function $f(t) = 2^{-kt}$ for the portion of uranium 238 remaining in a rock t years after it is formed.

b. What portion of the original uranium 238 is left in a rock whose age is 200 million years? [*Hint*: Write your function from part (a) so that the base is e.]

c. How old is a rock that contains 73% of its original uranium 238?

32. *(Problem Solving)* The following elements are among the by-products of a nuclear explosion. How long does it take for 99% of each one to decay?

a. polonium 212 (half-life 3×10^{-7} seconds)

b. krypton 91 (half-life 10 seconds)

c. strontium 90 (half-life 28 years)

In calculus, you will learn another way to construct the number *e*. Exercises 33–34 demonstrate the construction, while leaving the theoretical justification to calculus.

33. *(Extension)* If n is any positive integer, the symbol $n!$ denotes $1 \cdot 2 \cdot 3 \cdot \cdots \cdot (n-1) \cdot n$. Thus

$$\begin{aligned} 1! &= 1 \\ 2! &= 1 \cdot 2 = 2 \\ 3! &= 1 \cdot 2 \cdot 3 = 6 \end{aligned}$$

and so on. By convention, $0! = 1$. Evaluate each of the numbers $6!$, $7!$, $10!$, $25!$.

34. *(Extension)* Define a sequence $\{c_n\}$ by

$$\begin{aligned} c_0 &= \frac{1}{0!} \\ c_1 &= \frac{1}{0!} + \frac{1}{1!} \\ c_2 &= \frac{1}{0!} + \frac{1}{1!} + \frac{1}{2!} \end{aligned}$$

etc.

Find the values of the terms c_0 through c_{10}. Do the values of c_n appear to approach a ceiling as n becomes large? What does that ceiling appear to be? You're right!

9–3 INVERSE FUNCTIONS

In ***Standing Room Only*** (page 323) we asked how long it would take the world's population to double. To answer the question, you constructed the equation

$$P = 6(1.02)^t$$

to describe population as a function of time. This equation is essentially of the form $y = b^x$. Up to this point, we have regarded x as the input and y as the output in such equations. By contrast, the question in ***Standing Room Only*** provides a value of y and asks you to solve for x. In Section 9-1 you solved the equation by graphical methods. In Section 9-4 you will learn an analytical method utilizing the concept of an inverse function, which you will learn in the present section.

You have seen the word *inverse* before in mathematics. For example, the additive inverse of 2 is -2, and the multiplicative inverse of 2 is $\frac{1}{2}$. The word can also be applied to mathematical operations. For example, the inverse of adding 2 is subtracting 2, and the inverse of multiplying by 2 is dividing by 2.

When a pair of inverse operations is applied in succession, the net result is no change. For example, if we multiply 16 by 2, then divide the result by 2, our final result is our starting point 16. When we speak of inverse functions, we have the same idea in mind. To illustrate, suppose that f is the "multiplying by 2" function defined by $f(x) = 2x$, and g is the "dividing by 2" function defined by $g(x) = \frac{1}{2}x$. If these two functions are applied in succession in either order, the net result is no change. Recall from Section 2-2 that applying f and g in succession produces one of the composed functions $f \circ g$ or $g \circ f$.

$$(f \circ g)(x) = f[g(x)] = 2[g(x)] = 2\left[\frac{1}{2}x\right] = x$$

$$(g \circ f)(x) = g[f(x)] = \frac{1}{2}[f(x)] = \frac{1}{2}[2x] = x$$

Similarly, the equations $F(x) = x + 2$ and $G(x) = x - 2$ define a pair of inverse functions. If they are applied in succession in either order, our final output is our original input x.

EXERCISE **1.** *(Review)* For the functions F and G that were just mentioned, verify that $(F \circ G)(x) = x$ and $(G \circ F)(x) = x$ for all real values of x.

The following definition is made with these examples in mind.

Two functions f and g are called **inverse functions** if and only if $(f \circ g)(x) = x$ for all x in the domain of g and $(g \circ f)(x) = x$ for all x in the domain of f. In this case it is customary to write $g = f^{-1}$ and $f = g^{-1}$.

FYI The notation f^{-1} denotes a compositional inverse, and not a multiplicative inverse. In general, $f^{-1}(x)$ and $\frac{1}{f(x)}$ are not the same.

ONE-TO-ONE FUNCTIONS

Not all functions have inverses. A prime example is the squaring function $f(x) = x^2$. For this function, $f(2)$ and $f(-2)$ are both equal to 4. If f^{-1} existed, $f^{-1}(4)$ would have to be both 2 and -2, violating the requirement that a function must have only one output for each input. See Figure 9-8.

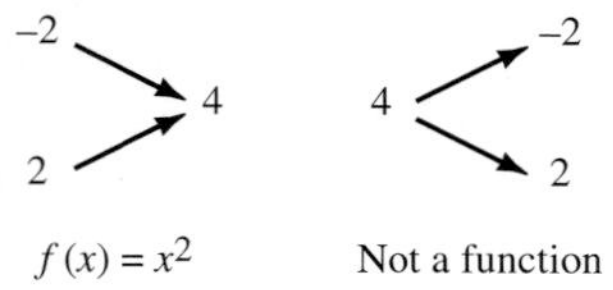

Figure 9–8

Which functions have inverses? We can begin to answer this question by focusing on what prevents the squaring function from having an inverse. It is the fact that two input values (2 and -2) produce the same output value (4). Equivalently, the squaring function $y = x^2$ does not have an inverse because x is not a function of y. Conversely, if only one input produces each output, the function has an inverse. Such functions are called **one-to-one**. Since only one-to-one functions have inverses (and only one-to-one functions can *be* inverses), our study of inverse functions begins with a study of one-to-one functions.

The following observation is useful, both in deciding whether a function is one-to-one and in finding the inverse of a one-to-one function.

A function $y = f(x)$ is one-to-one if and only if x is also a function of y.

A Numerical View of One-to-One Functions It is easy to tell whether a function described by a (complete) table is one-to-one. A function is one-to-one if no two distinct inputs lead to the same output. Equivalently, it is one-to-one if no two rows of the table with different first entries have the same second entry. More generally,

A function is one-to-one if and only if no two rows in any table of values have different first entries and the same second entry.

EXERCISES *In Exercises 2–5, decide whether the table represents a one-to-one function. Assume that each is complete.*

2.

x	y
0	0
1	1
2	4
3	9
4	16

3.

x	y
−4	16
−2	4
0	0
2	4
4	16

4.

x	y
2	35
5	7
8.1	−2
19	100
$-\sqrt{3}$	0.05

5.

x	y
10	3
20	3
30	3
40	3
50	3

A Graphical View of One-to-One Functions It is also easy to tell whether a function described by a graph is one-to-one. Figure 9-9a shows the graph of $y = x^2$, which is not one-to-one.

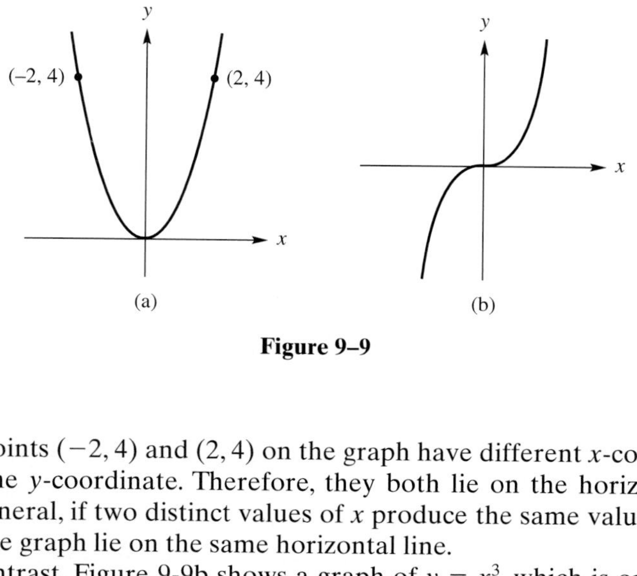

Figure 9–9

The points $(-2, 4)$ and $(2, 4)$ on the graph have different x-coordinates, but the same y-coordinate. Therefore, they both lie on the horizontal line $y = 4$. In general, if two distinct values of x produce the same value of y, the points on the graph lie on the same horizontal line.

By contrast, Figure 9-9b shows a graph of $y = x^3$, which is one-to-one, since no two distinct real numbers have the same cube. Equivalently, no two points on the graph of $y = x^3$ lie on the same horizontal line.

> A function is one-to-one if and only if no horizontal line intersects its graph more than once.

EXERCISES *In Exercises 6–9, decide whether the function is one-to-one.*

6.

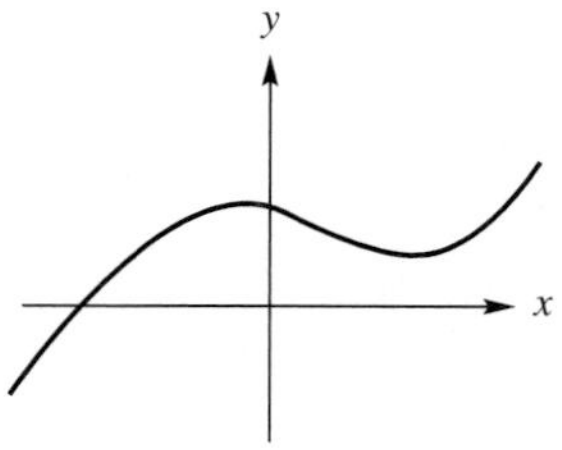

7.

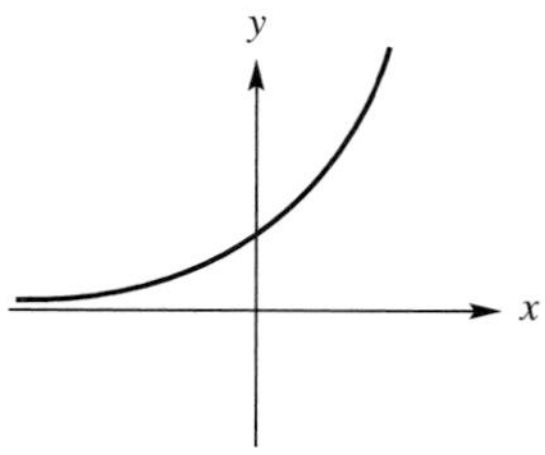

8.

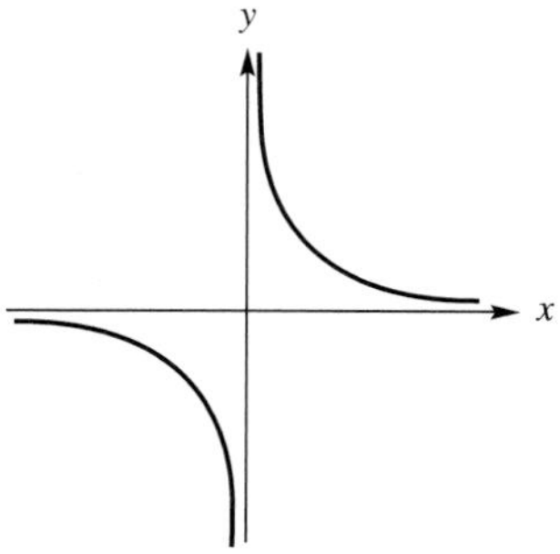

9.

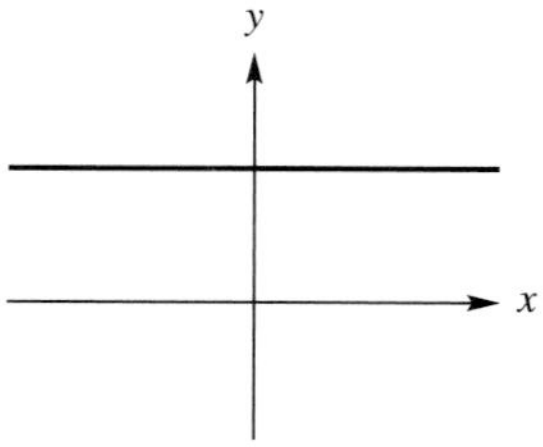

In Exercises 10–13, graph the function and decide whether it is one-to-one.

10. $f(x) = 4 - x^2$ **11.** $f(x) = 2x + 3$

12. $f(x) = \sqrt{x - 3}$ **13.** $f(x) = 2^x$

14. *(Making Observations)*

a. If $f(x) = mx + b$ with $m \neq 0$, does the graph of $y = f(x)$ pass the horizontal line test?

b. If $f(x) = ax^2 + bx + c$ with $a \neq 0$, does the graph of $y = f(x)$ pass the horizontal line test?

c. If $f(x) = Ab^x$ with $A \neq 0, b \neq 1$, and $b > 0$, does the graph of $y = f(x)$ pass the horizontal line test?

Exercise 14 allows you to conclude that nonconstant linear and exponential functions are *always* one-to-one, while quadratic functions *never* are.

Except for a few classes of functions such as those in Exercise 14, it is generally difficult or impossible to tell whether a function described by an equation is one-to-one. One method that sometimes works is to attempt to find the equation of the function's inverse. You will learn this method later in this section.

FINDING INVERSES OF ONE-TO-ONE FUNCTIONS

Once you know that a function is one-to-one, how do you find its inverse? To describe the process, let's consider a pair of inverse functions you have seen earlier in this book.

In ***Temperature Scales*** (page 30) you encountered the function $F = \frac{9}{5}C + 32$, which converts Celsius temperatures to Fahrenheit. In a subsequent exercise you derived the function $C = \frac{5}{9}(F - 32)$ to convert Fahrenheit temperatures to Celsius. If the two are applied in succession, the net result is no change, so that the functions are inverses. You can verify this in Exercise 15.

EXERCISE **15.** Define functions f and g by $f(x) = \frac{9}{5}x + 32$ and $g(x) = \frac{5}{9}(x - 32)$, and simplify the expressions for $f[g(x)]$ and $g[f(x)]$. How does your result confirm that f and g are inverses?

Our two functions from ***Temperature Scales*** provide another way of thinking about inverse functions. A pair of inverse functions looks at the same relationship from two different points of view, by reversing the roles of the independent and dependent variables. This insight suggests the following process for finding the inverse of a given one-to-one function.

- To reverse the roles of the independent and dependent variables, physically interchange them. For example, if the variables are x and y, relabel each x as y, and vice versa.
- Display the new dependent variable explicitly as the output.

Over the next few pages you will apply this process to functions described numerically, graphically, and analytically.

Numerical Methods If a one-to-one function is described by a (complete) table, then to find the inverse function is to describe it by a table.

EXAMPLE 1 *Finding the inverse of a one-to-one function numerically*

Suppose the function f is given by the following table.

x	y
-1	10
2	7
3	4
6	9
9	1

To obtain a table for f^{-1}, first interchange the labels "x" and "y" to reverse the roles of independent and dependent variables. Then interchange the positions of the columns to display the new y column as the output.

x	y
10	-1
7	2
4	3
9	6
1	9

■

EXERCISE **16.** In Exercises 2–5, if the function is one-to-one, write a table to describe the inverse function.

Graphical Methods If a one-to-one function is described by a graph, then to find the inverse is to describe it by a graph.

EXAMPLE 2 *Finding the inverse of a one-to-one function graphically*

The function $f(x) = \sqrt{x}$ is one-to-one. The graph of $y = f(x)$ is shown in Figure 9-10a.

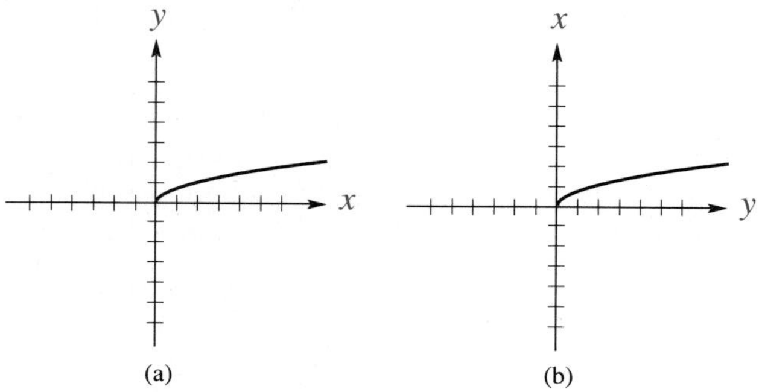

Figure 9–10

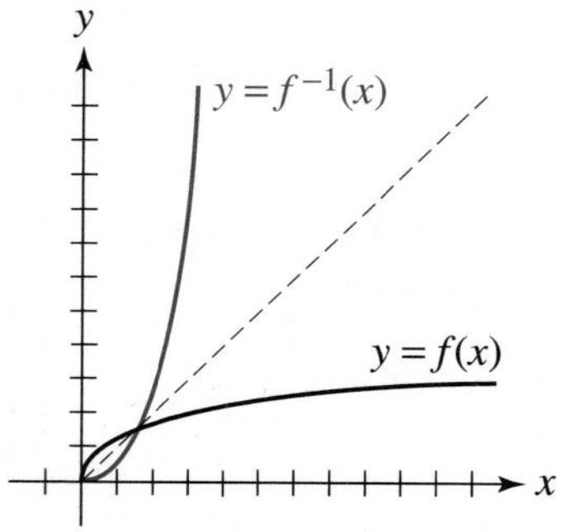

Figure 9–11

To reverse the roles of independent and dependent variables, relabel the axes, as in Figure 9-10b.

To display the new y as output, move the plane so that the positive x-axis points to the right and the positive y-axis points upward, as usual. The required movement is a reflection of the plane across the diagonal line (the graph of $y = x$) shown in Figure 9-11.

Later in this section you will find an equation for the graph in Figure 9-11 by showing analytically that $f^{-1}(x) = x^2$ for $x \geq 0$. ■

The process described in Example 2 can be summarized as follows.

> If f is a one-to-one function, the graph of $y = f^{-1}(x)$ is the reflection of the graph of $y = f(x)$ in the line $y = x$.

EXERCISES **17.** *(Making Observations)*

a. Sketch the graph of $f(x) = 2^x$ and label the points $\left(-1, \frac{1}{2}\right)$, $(0, 1)$, $(1, 2)$, and $(2, 4)$.

b. Explain why the points $\left(\frac{1}{2}, -1\right)$, $(1, 0)$, $(2, 1)$, and $(4, 2)$ must be on the graph of $y = f^{-1}(x)$.

c. Plot the points in part (b). How do the results suggest that the graph of $y = f^{-1}(x)$ is the reflection of the graph of $y = f(x)$ in the line $y = x$?

d. Sketch the graph of $y = f^{-1}(x)$. (You will find its equation in Section 9-4.)

18. In Exercises 6–13, if the function is one-to-one, sketch the graph of its inverse.

Analytical Methods If a one-to-one function is described by an equation, then to find the inverse function is to describe it by an equation. The analytical process requires an extra step, compared to the numerical and graphical processes. Example 3 shows why.

EXAMPLE 3 *Finding the inverse of a one-to-one function analytically*

Suppose that $f(x) = \dfrac{x}{x+2}$. The graph in Figure 9-12 indicates that f is one-to-one and, therefore, has an inverse.

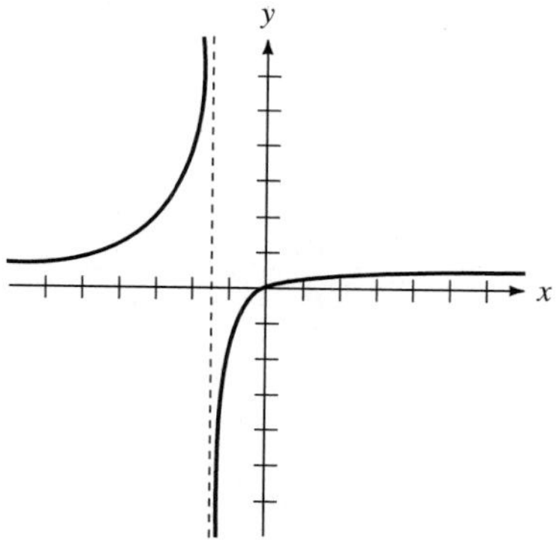

Figure 9–12

To find an equation for $f^{-1}(x)$, first write f using the variables x and y.

$$y = \frac{x}{x+2}$$

Then interchange x and y.

$$x = \frac{y}{y+2}$$

Displaying the new y as output means solving for y.

$$x(y+2) = y$$

$$xy + 2x = y$$

$$xy - y = -2x$$

$$y(x-1) = -2x$$

$$y = -\frac{2x}{x-1}$$

We now have a new function $g(x) = -\dfrac{2x}{x-1}$, but we can't be too hasty in concluding that $g = f^{-1}$. In solving for y we have performed two operations that may not lead to equivalent equations. We have multiplied both sides of the equation by a variable quantity $(y + 2)$ and divided both sides by a variable quantity $(x - 1)$. As a result, we do not know whether g is the same as f with the roles of x and y reversed.

To see whether $g = f^{-1}$, we need to apply the definition of inverse function. If the composition $(g \circ f)(x) = x$ for all x in the domain of f and $(f \circ g)(x) = x$ for all x in the domain of g, then we will know that $g = f^{-1}$.

First let's look at $(g \circ f)(x)$, bearing in mind that the domain of f includes all real numbers $x \neq -2$.

$$
\begin{aligned}
(g \circ f)(x) = g[f(x)] &= -\frac{2f(x)}{f(x) - 1} \\
&= -\frac{2\dfrac{x}{x+2}}{\dfrac{x}{x+2} - 1} \\
&= -\frac{2x}{x - (x+2)}, \quad \text{if } x \neq -2 \\
&= -\frac{2x}{-2} \\
&= x
\end{aligned}
$$

Similarly, you can verify that $(f \circ g)(x) = x$ for all x in the domain of g, so $g = f^{-1}$. ■

EXERCISES

19. For the functions f and g in Example 3,

a. Find the domain of g.

b. Verify that $(f \circ g)(x) = x$ for all x in the domain of g.

In Exercises 20–23,

a. Sketch the graph of $y = f(x)$ to verify that f is one-to-one, and sketch the graph of $y = f^{-1}(x)$.

b. Find f^{-1} analytically, and verify your result by simplifying the compositions $(f \circ f^{-1})(x)$ and $(f^{-1} \circ f)(x)$.

c. Support your results by graphing $y = f(x)$ and $y = f^{-1}(x)$ on your calculator.

20. $f(x) = 3x + 5$ **21.** $f(x) = 6 - 4x$

22. $f(x) = 2x^3 - 7$ **23.** $f(x) = \dfrac{1}{2x - 1}$

Finding the inverse of a function analytically can be difficult for two reasons.

- After interchanging x and y, you may not be able to solve the new equation for y. In this case you can describe the inverse function only numerically or graphically.
- Even if you can solve the new equation explicitly for y, the new function g is not guaranteed to be the inverse of the original function f. That is, it may be that $g[f(x)] \neq x$ for some x in the domain of f or that $f[g(x)] \neq x$ for some x in the domain of g. When this occurs, you can often obtain f^{-1} by restricting the domain of g. You have seen such a restriction in Example 2, where we found graphically that the inverse of $f(x) = \sqrt{x}$ is $g(x) = x^2$ for $x \geqslant 0$. Without the restriction on the domain, $g(x) = x^2$ is not one-to-one, and so could not be f^{-1}. Example 4 provides an analytical illustration.

EXAMPLE 4 *Finding the inverse of a one-to-one function analytically*

Suppose that $f(x) = \sqrt{1 - 2x}$. The graph in Figure 9-13 indicates that f is one-to-one.

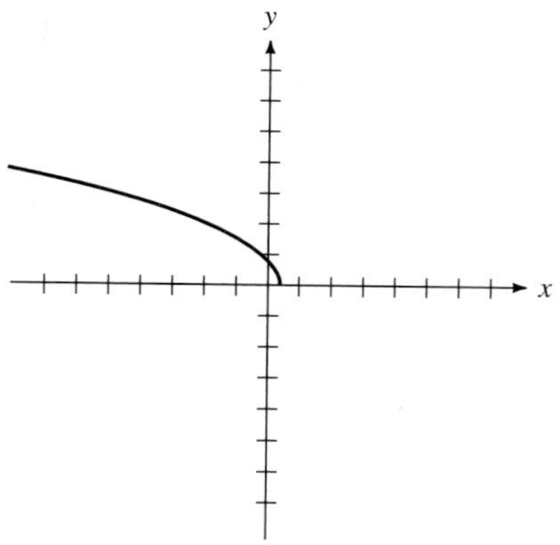

Figure 9–13

To find f^{-1}, first write f using the variables x and y.

$$y = \sqrt{1 - 2x}$$

Then interchange x and y.

$$x = \sqrt{1 - 2y}$$

Next, solve for y to obtain

$$y = \frac{1}{2}(1 - x^2)$$

We now have a new function $g(x) = \frac{1}{2}(1 - x^2)$. To see whether $g = f^{-1}$, let's compose the two. First, let's look at $(g \circ f)(x)$, bearing in mind that the domain of f is $\left(-\infty, \frac{1}{2}\right]$.

$$\begin{aligned}(g \circ f)(x) = g[f(x)] &= \frac{1}{2}[1 - f(x)^2] \\ &= \frac{1}{2}[1 - (\sqrt{1 - 2x})^2] \\ &= \frac{1}{2}[1 - (1 - 2x)] \\ &= \frac{1}{2}(2x) \\ &= x \quad \text{for all } x \text{ in } \left(-\infty, \frac{1}{2}\right]\end{aligned}$$

Next, let's look at $(f \circ g)(x)$, bearing in mind that the domain of g is $(-\infty, \infty)$.

$$\begin{aligned}(f \circ g)(x) = f[g(x)] &= \sqrt{1 - 2g(x)} \\ &= \sqrt{1 - 2\left[\frac{1}{2}(1 - x^2)\right]} \\ &= \sqrt{1 - (1 - x^2)} \\ &= \sqrt{x^2} \\ &= |x|\end{aligned}$$

Thus, $(f \circ g)(x)$ is not equal to x on $(-\infty, \infty)$, but is equal to x on $[0, \infty)$. Therefore, the inverse of f is $f^{-1}(x) = \frac{1}{2}(1 - x^2)$ for $x \geqslant 0$. The restriction on the domain of f^{-1} is critical. As Figure 9-14 shows, the function $y = \frac{1}{2}(1 - x^2)$ is not one-to-one on $(-\infty, \infty)$, but is one-to-one on $[0, \infty)$.

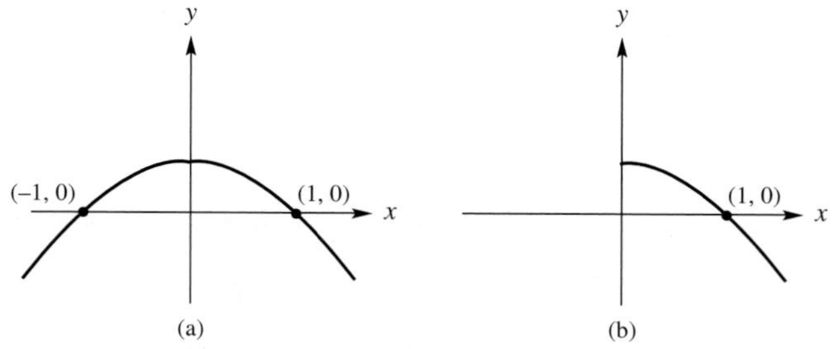

Figure 9–14 ■

EXERCISES *In Exercises* 24–27,

a. Sketch the graph of $y = f(x)$ to verify that f is one-to-one, and sketch the graph of $y = f^{-1}(x)$.

b. Find f^{-1} analytically, and verify your result by simplifying the compositions $(f \circ f^{-1})(x)$ and $(f^{-1} \circ f)(x)$.

c. Support your results by graphing $y = f(x)$ and $y = f^{-1}(x)$ on your calculator.

24. $f(x) = \sqrt{x - 4}$ **25.** $f(x) = \sqrt{5 - x}$

26. $f(x) = \sqrt{3x + 2}$ **27.** $f(x) = \sqrt[4]{x}$

Sometimes a physical problem is modeled by a function that is not one-to-one over its entire domain, but is one-to-one when its domain is suitably restricted. In the following Mathematical Looking Glass, it is useful to find the inverse of such a restricted function.

A MATHEMATICAL LOOKING GLASS

Safe Speeds

The distance d in feet required to stop a car traveling at an initial speed of v miles per hour is approximately

$$d = 0.06v^2 + 0.8v$$

This function was discussed in Exercise 53 of Section 5-2. It is a convenient model of stopping distance as long as speed is the independent variable. For example,

- A member of the National Safety Council might want to construct a table of stopping distances associated with selected initial velocities.
- A Department of Transportation employee who is setting the timing of traffic lights needs to know the stopping distance for a vehicle traveling at the speed limit.

However, there are situations in which it is more convenient to use stopping distance as the independent variable. For example,

- A police officer investigating an accident may need to calculate a driver's initial speed, based on evidence such as the distance from some skid marks to the point of impact.
- A Department of Transportation employee who is setting speed limits needs to know the normal range of visibility on each road so that drivers traveling at the speed limit can stop safely in that distance.

The function $d = 0.06v^2 + 0.8v$ is not one-to-one on $(-\infty, \infty)$, but its domain in a physical context is $[0, \infty)$. In Exercises 28–29 you can verify that the function is one-to-one on this domain, and you can find its inverse. ■

EXERCISES

28. *(Making Observations)* Find the vertex on the graph of $d = 0.06v^2 + 0.8v$. How does your result show that the function is one-to-one on each of the intervals $\left(-\infty, -\frac{20}{3}\right]$ and $\left[-\frac{20}{3}, \infty\right)$?

29. a. To solve the equation $d = 0.06v^2 + 0.8v$ for v, rewrite it as $0.06v^2 + 0.8v - d = 0$ and use the quadratic formula.

b. How does your result indicate that the functions $d = 0.06v^2 + 0.8v$ and $v = \dfrac{5\sqrt{6d + 16} - 20}{3}$ are inverses when v and d are both nonnegative?

In Exercises 30–33, break the domain of f into two intervals so that the restriction of f to each interval is one-to-one. Then find the inverse of each restricted function.

30. $f(x) = (x - 3)^2$

31. $f(x) = 10 - 2x^2$

32. $f(x) = (x + 4)^2 - 5$

33. $f(x) = 3(x - 1)^2 + 2$

34. Lynn's daughter, Carrie, often rides her horse near her home on Old Mill Road. Because many other residents also ride horses, she feels that the speed limit should be no more than 25 mph. She went to a township meeting armed with the two functions from ***Safe Speeds*** and the fact that because of curves, drivers on the road must be able to stop within 50 feet.

a. *(Interpreting Mathematics)* Use the function $v = \dfrac{5\sqrt{6d + 16} - 20}{3}$ to argue her point that the speed limit should be no more than 25 mph.

b. *(Interpreting Mathematics)* Use the function $d = 0.06v^2 + 0.8v$ to argue her point.

c. *(Writing to Learn)* Which function makes it easier for Carrie to argue her point? Why?

35. *(Writing to Learn)* When you don't use material you have learned, you tend to forget much of it. For example, suppose you memorize 50 new vocabulary words for a final exam in Spanish, and thereafter never speak Spanish again. A reasonable model for the number of those words you retain over time might be

$$n = 10 + \frac{30}{t}$$

where

t = number of days after the exam

n = number of words retained

Describe at least one situation in which it is convenient to use the equation in its present form and at least one situation in which you would want to solve for t and use the inverse function.

ADDITIONAL EXERCISES *In Exercises 36–39, assume that each table is complete, and decide whether it describes a one-to-one function. If it does, write a table to describe its inverse.*

36.

x	y
-1	-3
0	-2
8	0
9	-1
23	1

37.

x	y
1	1
2	2
3	3
4	4
5	5

38.

x	y
3	6
5	6
2π	6
7.96	6
$\sqrt{139}$	6

39.

x	y
0	1
1	2
2	4
3	8
4	16

In Exercises 40–41, write a table to represent f^{-1}, and demonstrate that $f^{-1}[f(x)] = x$ for each x in the domain of f and $f[f^{-1}(x)] = x$ for each x in the domain of f^{-1}.

40.

x	$f(x)$
-7	28
-4	13
0	-2
5	25

41.

x	$f(x)$
1	$\sqrt{3}$
2.8	12,946
7π	0
$\frac{2}{3}$	e

In Exercises 42–45, decide whether the function is one-to-one. If it is, sketch the graph of its inverse.

42.

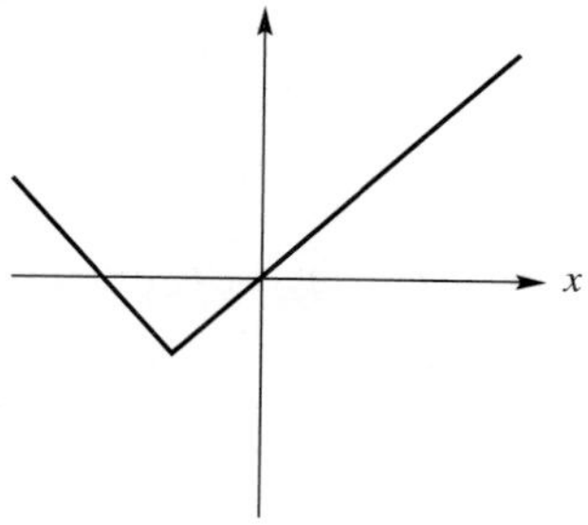

43.

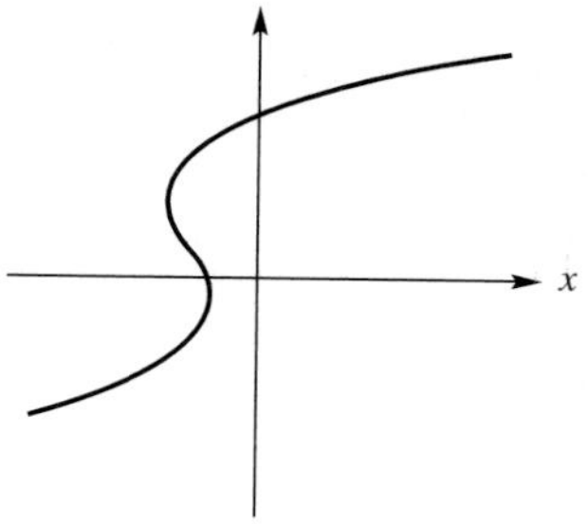

44.

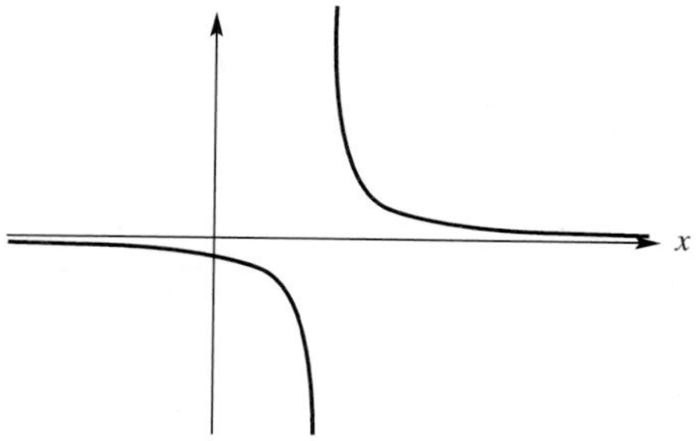

45.

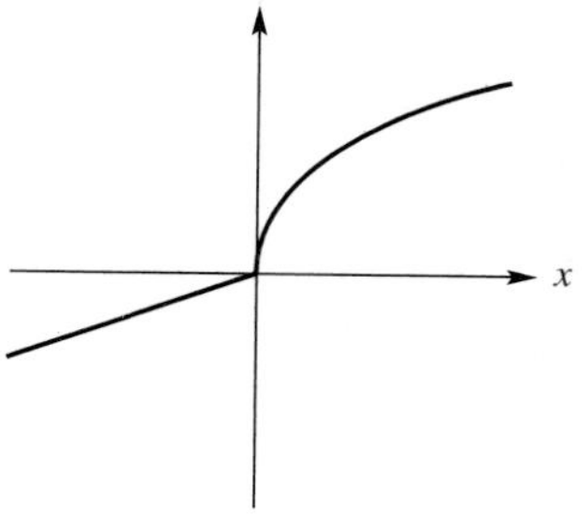

In Exercises 46–53, graph the function and decide whether it is one-to-one. If it is, sketch the graph of its inverse.

46. $f(x) = x^3 - 4x$

47. $f(x) = x^3 + 4x$

48. $f(x) = \dfrac{1}{x^3 + 4x}$

49. $f(x) = 2 \cdot 3^{-x}$

50. $f(x) = \sqrt{1 + x^3}$

51. $f(x) = \sqrt{1 + x^2}$

52. $f(x) = x - \dfrac{1}{x}$ on $(0, \infty)$

53. $f(x) = \dfrac{1}{1 + x^2}$ on $[0, \infty)$

In Exercises 54–61, decide whether f and g are inverses by simplifying the compositions $f[g(x)]$ and $g[f(x)]$.

54. $f(x) = 8x + 16,\ g(x) = \dfrac{1}{8}x - 2$

55. $f(x) = 3 - x,\ g(x) = 3 - x$

56. $f(x) = \dfrac{1}{x},\ g(x) = \dfrac{1}{x}$

57. $f(x) = \dfrac{x}{x + 6},\ g(x) = \dfrac{6x}{1 - x}$

58. $f(x) = (x - 7)^5,\ g(x) = \sqrt[5]{x} + 7$

59. $f(x) = (x - 7)^4,\ g(x) = \sqrt[4]{x} + 7$

60. $f(x) = 2x^3 + 1,\ g(x) = \sqrt[3]{\dfrac{x - 1}{2}}$

61. $f(x) = 4,\ g(x) = \dfrac{1}{4}$

In Exercises 62–69, find f^{-1} analytically, and graph f and f^{-1} on the same set of axes.

62. $f(x) = 0.2x$ **63.** $f(x) = 0.2x - 0.8$

64. $f(x) = x^7$ **65.** $f(x) = \dfrac{x^3 + 8}{3}$

66. $f(x) = \dfrac{x}{x - 5}$ **67.** $f(x) = \dfrac{5}{x - 5}$

68. $f(x) = \sqrt{x}$ **69.** $f(x) = \sqrt{10 - x}$

In Exercises 70–73, break the domain of f into two intervals so that the restriction of f to each interval is one-to-one. Then find the inverse of each restricted function.

70. $f(x) = (2x - 1)^2$ **71.** $f(x) = 7 - 3x^2$

72. $f(x) = 7 - 3x^4$ **73.** $f(x) = x^2 + 4x + 5$

In Exercises 74–77, graph the function, and find at least one interval on which it has an inverse.

74. $h(x) = x^4 - 8x^2$ **75.** $P(t) = \dfrac{1}{t^2 - 1}$

76. $y = x^3 - 12x$ **77.** $A(z) = |z^2 - 9|$

9-4 LOGARITHMIC FUNCTIONS

Section 9-3 began with a promise that our discussion of inverse functions would produce an analytical method for solving exponential equations. Here is a situation in which the solution of such an equation was crucial.

A MATHEMATICAL LOOKING GLASS
A Murder in Ceylon

Newton's law of cooling states that when a hot object cools, the difference between its temperature and that of the surrounding medium decreases by equal percentages over time intervals of equal length. The temperature difference is thus an exponential function of time. In *Clouds in a Glass of Beer* (John Wiley & Sons, Inc., 1987), Craig Bohren tells how Newton's law of cooling played an important role in a murder trial.

> Judged by his superbly written autobiography, *Mostly Murder*, the late Sir Sydney Smith must have been a charming and delightful man. But the field in which he achieved eminence, forensic pathology, could hardly be so described; morbidly fascinating might be more apt. The function of a forensic pathologist is to determine how the dead met their end: murder, suicide, or accident. Sir Sydney chose to recount mostly the murders he investigated during his long, distinguished career, and one of them gives the title to this chapter.
>
> About thirty years ago Sir Sydney was involved in a case of murder in Ceylon (now Sri Lanka). A man was accused of murdering his wife. His whereabouts were known reliably after a certain time, and the victim was known to have been alive before a certain time. Whether or not the accused was to be hanged therefore crucially depended on determining

the time of death as accurately as possible. After death a body cools slowly at a rate depending on its size and clothing and its surroundings (still or windy, for example). From a body's *cooling curve* the time of death can be estimated by extrapolating backwards from when the temperature is known to when it may be assumed to have been normal (about 37° C) . . . .

Here are Sir Sydney's words on what happened:

> The experiments on the cooling of recently dead bodies that I had suggested had been carried out. Professor deSaram [a former pupil of Sir Sydney's] had been fortunate in obtaining three bodies of young murderers immediately after they had been executed. Dressed in petticoats and sarees, the bodies were kept under conditions as close as possible to those in which the body of the murdered woman had been placed, and their temperatures were taken every half-hour. Approximately seven hours were required for these bodies to lose 5.2 degrees.

■

EXERCISES

1. *(Creating Models)* It is reasonable to assume that the victim's temperature at the time of her death was 37° C, that the temperature of her surroundings was 20° C, and that her body, like those of the executed murderers, cooled 5.2° in the first seven hours after death. Find a function $y = 20 + Ae^{-kt}$ to express her body's temperature t hours after death.

2. *(Interpreting Mathematics)*

a. The victim's body was discovered at 6:00 A.M., at which time its temperature was 27°. Use your graph to estimate the time of the murder.

b. Her husband's whereabouts were known at all times since 10:30 A.M. the previous day. What does your graph suggest about his guilt or innocence?

c. Write an equation in t that must be solved to determine the time of the murder more precisely.

We will return to this murder mystery in Exercise 34, where you can solve the equation in Exercise 2. To see how the concept of an inverse function will help, let's consider a simple equation with variable exponents, say, $2^x = 7$.

EXERCISE

3. *(Writing to Learn)*

a. The function $g(x) = 2^x$ has an inverse. How can you tell?

b. Explain why solving the equation $2^x = 7$ is the same as finding $g^{-1}(7)$.

If we could find an equation for the inverse of $g(x) = 2^x$, it would be a simple matter to evaluate $g^{-1}(7)$. Unfortunately, no algebraic operation will solve $y = 2^x$ for x. We are thus in the awkward position of needing to work with the function g^{-1} without having an equation for it. John Napier (1550–1617), the first person to study inverses of exponential functions, provided the name **logarithm** to describe their outputs, such as $g^{-1}(7)$. (The name, meaning "proportionate numbers," was suggested by the computational processes he used.) Mathematicians today, following Napier, call the inverse of the exponential function $y = b^x$ the **logarithmic function** to the base b, and write it as $y = \log_b x$.

Simply naming the inverses of exponential functions provides neither insight into their nature nor help in using them to solve equations. However, let's see what we can discover about logarithmic functions from numerical, graphical, and analytical perspectives.

A NUMERICAL VIEW OF LOGARITHMIC FUNCTIONS

Table 31 was generated by $y = 2^x$.

TABLE 31

x	$y = 2^x$
0	$1 = 2^0$
1	$2 = 2^1$
2	$4 = 2^2$
3	$8 = 2^3$
4	$16 = 2^4$

TABLE 32

x	$y = \log_2 x$
$1 = 2^0$	0
$2 = 2^1$	1
$4 = 2^2$	2
$8 = 2^3$	3
$16 = 2^4$	4

By interchanging x and y, we obtain Table 32, which fits $y = \log_2 x$.

So logarithms are just disguised exponents!

> The logarithm to the base b of a number x is the power to which b must be raised in order to obtain x. That is,
>
> $$y = \log_b x \quad \text{means} \quad x = b^y$$

We can use this observation to obtain certain values of logarithmic functions.

EXAMPLE 1 *Evaluating a logarithm (a disguised exponent)*

To evaluate $\log_8 64$, ask yourself to what power 8 must be raised to obtain 64. Recognizing that $8^2 = 64$, you can write $\log_8 64 = 2$. ■

We can't use the same process to find all values of logarithmic functions. For example, to find $\log_2 7$, we need to find the power to which 2 must be raised in order to obtain 7. Equivalently, we have to solve $2^x = 7$ for x, which we cannot do algebraically. The evaluation of $\log_2 7$ will have to wait until Exercise 46.

EXERCISES

4. *(Making Observations)* Use the x-values 0, 1, 2, 3, and 4 to make a table for $y = 3^x$. Then make a second table for $y = \log_3 x$. Verify that each value of $\log_3 x$ is the power to which 3 must be raised to obtain x.

5. Make a table of values for $y = \log_5 x$, using the x-values 5, 25, 125, and 625.

6. Use the observation that logarithms are disguised exponents to evaluate each of the following.

   a. $\log_4 64$ **b.** $\log_2 128$

   c. $\log_7 7$ **d.** $\log_{12} 1$

   e. $\log_6 \frac{1}{6}$ **f.** $\log_9 3$

7. *(Writing to Learn)* Explain why it must be true that $\log_b 1 = 0$ and $\log_b b = 1$ for any base b.

In Section 9-1 you learned that a table with equally spaced x-values fits an exponential function $y = Ab^x$ if and only if the ratio of consecutive y-values is constant. Interchanging x and y, we can discover how to tell whether a table fits the inverse of a function $y = Ab^x$. In Exercise 41, you will verify that such a function has the form $y = \log_b x + k$. Therefore,

> A table with equally spaced y-values fits a function $y = \log_b x + k$ if and only if the ratio of consecutive x-values is constant.

EXERCISES *In Exercises 8–11, decide whether the table fits a function* $y = \log_b x + k$.

8.

x	y
2	−4
8	−3
32	−2
128	−1
512	0

9.

x	y
162	1
108	3
72	5
48	7
32	9

10.

x	y
1	4
1.1	5
1.11	6
1.111	7
1.1111	8

11.

x	y
1	4
1.1	5
1.21	6
1.331	7
1.4641	8

Until we can verify analytically that functions $y = \log_b x + k$ are inverses of functions $y = Ab^x$, Exercises 12–13 will provide some numerical evidence.

12. *(Making Observations)*

    a. Use the x-values 0, 1, 2, 3, and 4 to make a table for $y = 3^x$.

    b. Make a second table for $y = \log_3 x$.

    c. Use your table in part (b) to make a table for $y = \log_3 x - 2$.

13. **a.** *(Making Observations)* Use the x-values −2, −1, 0, 1, and 2 to make a table for $h(x) = 9 \cdot 3^x$.

    b. Make a second table for the function h^{-1}.

c. *(Writing to Learn)* Explain why your tables in Exercises 12c and 13a suggest that $y = 9 \cdot 3^x$ and $y = \log_3 x - 2$ are inverses.

Evaluating Logarithms on Your Calculator Until the advent of computers, the only convenient way to evaluate logarithmic functions accurately was to use lengthy tables first compiled by Napier. (Napier also invented the decimal point.) According to Isaac Asimov's *Biographical Encyclopedia of Science and Technology* (Doubleday & Company, Inc., 1982),

> Napier spent twenty years working out rather complicated formulas for obtaining exponential expressions for various numbers. . . . Finally, in 1614, Napier published his tables of logarithms, which were not improved on for a century, and they were seized on with avidity. Their impact on the science of the day was something like that of computers on the science of our own time.

Your calculator almost certainly has two keys labeled "log" and "ln." Asimov's *Biographical Encyclopedia of Science and Technology* describes the historical reasons for the presence of those two keys, dating back to John Napier and his contemporary, Henry Briggs (1561–1630).

> [Briggs] is remembered chiefly for his reaction to Napier's publication of logarithms. He was lost in admiration for the beauty of the system and its simplicity (and aghast at his own stupidity in not seeing it until it was shown him).
>
> He went to the considerable trouble of making a trip to Edinburgh to see Napier and talk to him. Napier had written his exponential numbers as e^2, $e^{2.32}$, $e^{3.97}$, and so on, where e is an unending decimal fraction that starts 2.7182818284. . . . There are good mathematical reasons for doing this and such Napierian or "natural" logarithms are still used in calculus. However, Briggs pointed out during his conversation with Napier the convenience of using exponential numbers such as 10^2, $10^{2.32}$, $10^{3.97}$, and so on. Logarithms in this fashion are called Briggsian or "common" logarithms and are almost invariably used for ordinary calculations.
>
> Briggs worked out the first logarithm tables for numbers from 1 to 20,000 and from 90,000 to 100,000 (to fourteen places!) in 1624. Briggs also invented the modern method of long division.

Thus 10 and e have always been, and continue to be, the most common and useful bases for logarithms. The notations $\log_{10} x$ and $\log_e x$ are usually abbreviated as $\log x$ and $\ln x$, respectively. Your two calculator keys will give values of logarithms to the bases 10 and e.

EXERCISE **14.** *(Making Observations)*

a. Guess the value of log 3, using the fact that the value is the power to which 10 must be raised to obtain 3.

b. Use your calculator to evaluate log 3. Confirm your result on your calculator by raising 10 to that power. How does your result compare with your guess in part (a)?

c. Use your calculator to evaluate ln 3. Confirm your result on your calculator by raising e to that power.

A GRAPHICAL VIEW OF LOGARITHMIC FUNCTIONS

Basic Graphs Since the functions $y = b^x$ and $y = \log_b x$ are inverses, their graphs are reflections of each other in the line $y = x$. Furthermore, each point on the graph of $y = \log_b x$ is a point on the graph of $y = b^x$ with its coordinates reversed. For example, Figure 9-15 shows graphs of $g(x)$ and 2^x and $g^{-1}(x) = \log_2 x$.

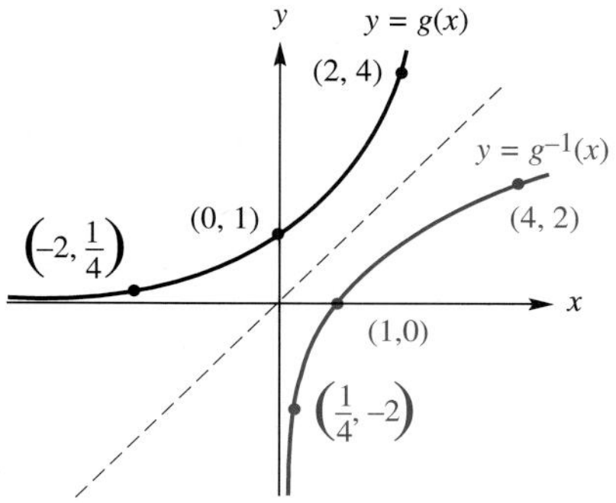

Figure 9–15

Since any function $f(x) = b^x$ with $b > 1$ has a graph similar to that of $g(x) = 2^x$, any function $f^{-1}(x) = \log_b x$ with $b > 1$ has a graph similar to that of $g^{-1}(x) = \log_2 x$.

Exercises

15. Sketch the graph of $y = \log_3 x$ by reflecting the graph of $y = 3^x$ in the line $y = x$. Identify the points obtained by moving $\left(-1, \frac{1}{3}\right)$, $(0, 1)$, and $(1, 3)$.

16. Repeat Exercise 15 for base 5, using the points $\left(-1, \frac{1}{5}\right)$, $(0, 1)$, and $(1, 5)$.

17. a. *(Making Observations)* For any $b > 0$ and any x, it is true that $\left(\frac{1}{b}\right)^x = b^{-x}$. Use this fact to demonstrate that $\log_{1/b} x = -\log_b x$. [*Hint*: If $y = \left(\frac{1}{b}\right)^x$, then also $y = b^{-x}$. Rewrite each equation using logarithms.]

b. Use your result from part (a), together with those from Exercises 15 and 16, to sketch the graphs of $y = \log_{1/3} x$ and $y = \log_{1/5} x$.

Exercises 15–17 illustrate several graphical features of logarithmic functions $g(x) = \log_b x$. These features are the same as those exhibited by exponential functions $f(x) = b^x$, but with x and y interchanged.

- The range of $f(x) = b^x$ is $(0, \infty)$, so the domain of $g(x) = \log_b x$ is $(0, \infty)$. This means that the graph of $y = g(x)$ is entirely to the right of the y-axis.
- The domain of $f(x) = b^x$ is $(-\infty, \infty)$, so the range of $g(x) = \log_b x$ is $(-\infty, \infty)$.
- Since the graph of $f(x) = b^x$ passes through $(0, 1)$, the graph of $g(x) = \log_b x$ passes through $(1, 0)$.
- Since the x-axis is a horizontal asymptote for the graph of $f(x) = b^x$, the y-axis is a vertical asymptote for the graph of $g(x) = \log_b x$.

- Since $f(x) = b^x$ is increasing if $b > 1$, so is $g(x) = \log_b x$. (To say that a function is increasing is to say that as x increases, so does y. Interchanging x and y does not change the statement.) Similarly, $g(x)$ is decreasing if $0 < b < 1$.

To sketch the graph of $y = \log_b x$ quickly for any $b > 1$, plot the x-intercept at (1, 0). Then plot $\left(\frac{1}{b}, -1\right)$ and $(b, 1)$, which are always on the graph. Finally, sketch a smooth curve through those points having the shape of Figure 9-16a. If $b < 1$, use the fact that $\log_{1/b} x = -\log_b x$ (see Exercise 17) to obtain a graph having the shape of Figure 9-16b.

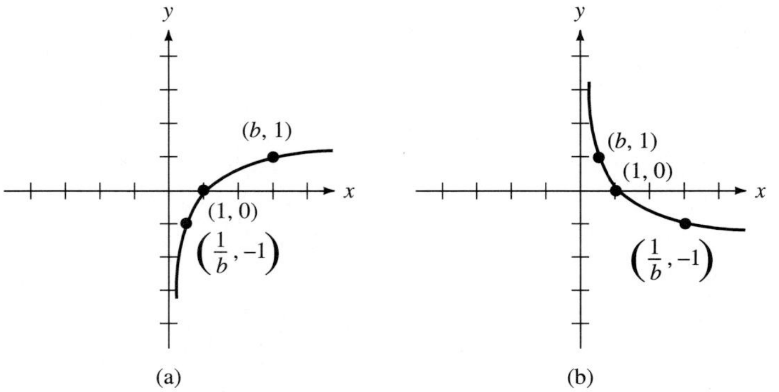

Figure 9–16

Transformations The graphs of many logarithmic functions can be obtained by transforming the graph of $y = \log_b x$.

EXAMPLE 2 *Transforming the graph of* $y = \log_b x$

To sketch the graph of $y = \frac{1}{2}\log_3 x + 1$, begin with the basic graph $y = \log_3 x$ and make the changes

$$y = \log_3 x \to y = \frac{1}{2}\log_3 x \to y = \frac{1}{2}\log_3 x + 1$$

These changes compress the graph vertically by a factor of 2, then shift it 1 unit up, as in Figure 9-17.

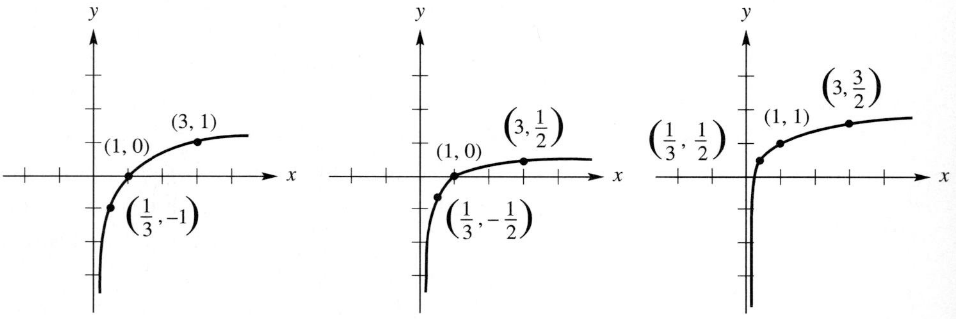

Figure 9–17

The domain of the function is $(0, \infty)$ and the range is $(-\infty, \infty)$. The function is increasing on its domain. ■

EXERCISES *In Exercises 18–21, sketch the graph of each function by transforming an appropriate graph $y = \log_b x$. In each case,*

a. Find the domain and range of the function.

b. Decide whether the function is increasing or decreasing.

c. Identify the points obtained by moving $\left(\frac{1}{b}, -1\right)$, $(1, 0)$, and $(b, 1)$.

18. $y = \log_4 x - 3$ **19.** $y = 0.5 \log_2(x + 2)$

20. $y = \log_7 x + 3$ **21.** $y = \log_3(x - 4) - 1$

22. *(Writing to Learn)* Describe how each transformation below affects the following features of $y = \log_b x$.

- domain and range
- asymptotes
- intercepts
- increasing/decreasing behavior

a. a horizontal stretch or compression

b. a vertical stretch or compression

c. a reflection in the x-axis

d. a reflection in the y-axis

e. a horizontal shift

f. a vertical shift

In Exercises 23–26, sketch the graph of the function and support your results with a calculator graph.

23. $y = \log x$ **24.** $y = \ln x$ (Use $e \cong 2.7$.)

25. $y = 3 \log 2x$ **26.** $y = 4 \ln x - 5$

Later in this section you will discover how to graph logarithmic functions with bases other than 10 or e on your calculator. However, you first need to learn some of the analytical properties of logarithms.

AN ANALYTICAL VIEW OF LOGARITHMIC FUNCTIONS

Basic Analytical Properties The analytical properties of logarithmic functions follow from the fact that they are inverses of exponential functions. In Section 9-3 you learned that pairs of inverse functions have the property that $f^{-1}[f(x)] = x$ for all x in the domain of f, and $f[f^{-1}(x)] = x$ for all x in the domain of f^{-1}. When stated in terms of exponential and logarithmic functions, these statements assert that

$$\log_b b^x = x \quad \text{for all real } x$$

and

$$b^{\log_b x} = x \quad \text{for all } x > 0$$

In particular, two facts you verified in Exercise 7 are worth remembering.

$$\log_b 1 = 0 \quad \text{for any base } b$$

and

$$\log_b b = 1 \quad \text{for any base } b$$

In Section 9-3 you also saw that the inverse of a function reverses the roles of the independent and dependent variables. Applied to logarithmic functions, this says that

$$y = \log_b x \quad \text{means} \quad x = b^y$$

This is another way of seeing that a logarithm is a disguised exponent. You can use this property to solve certain exponential and logarithmic equations, as in Example 3.

EXAMPLE 3 *Solving a logarithmic equation*

To solve

$$3 \log_5(2x + 50) = 7$$

first rewrite the equation in the form

$$\log_b(\text{variable expression}) = (\text{number})$$

$$\log_5(2x + 50) = \frac{7}{3}$$

Next, convert the equation to the exponential form

$$(\text{variable expression}) = b^{(\text{number})}$$

$$2x + 50 = 5^{7/3}$$

$$2x + 50 \cong 42.75$$

$$2x \cong -7.25$$

$$x \cong -3.63$$

■

In Exercises 27–32, solve the equation.

27. $4e^{5t} = 102.8$

28. $e^{s+2} - 27.1 = 1$

29. $2 \cdot 10^x + 0.42 = 3.29$

30. $\log_2 x = 3.5$

31. $\ln(4t - 100) = 0.57$

32. $5 \log(3r) = 13$

33. *(Interpreting Mathematics)* In ***Crater Lake*** (page 342), the charcoal sample used to determine the age of the lake contained 45% of its original carbon-14. Solving the equation $e^{-0.00012t} = 0.45$ gives the age of the lake in years. Solve the equation, and verify that the lake is about 6700 years old.

34. *(Interpreting Mathematics)* In Exercise 2, relating to ***A Murder in Ceylon***, you constructed the equation $27 = 20 + 17e^{-0.052t}$. Its solution represents the number of hours the victim had been dead when her body was discovered. Do the facts provide conclusive evidence of her husband's guilt or innocence? Tell why or why not.

Combining Logarithms Napier, Briggs, and other seventeenth-century mathematicians and scientists promoted the use of logarithms primarily because of their practical usefulness. Their usefulness derives from the following three distinctive properties.

If $m > 0, n > 0, b > 0$, and $b \neq 1$, then

$$\log_b(mn) = \log_b m + \log_b n$$

$$\log_b\left(\frac{m}{n}\right) = \log_b m - \log_b n$$

$$\log_b(m^a) = a \log_b m$$

The three properties can be stated in English as follows.

- To obtain the logarithm of a product, add the logarithms of the factors.
- To obtain the logarithm of a quotient, subtract the logarithms of the factors.
- To obtain the logarithm of a positive number raised to a power, multiply the exponent by the logarithm of the number.

Before discussing the practical importance of these properties, let's see why they are true.

To see why $\log_b(mn) = \log_b m + \log_b n$, let's look at the expressions

$$b^{\log_b(mn)}$$

and

$$b^{(\log_b m + \log_b n)}$$

If these two powers of b are equal, then the exponents must be equal. (This is another way of saying that the function $y = b^x$ is one-to-one.)

Since $b^{\log_b x} = x$ for all $x > 0$, the first expression reduces to mn, and the second can be rewritten as

$$b^{\log_b m}\, b^{\log_b n}$$

which also reduces to mn. This establishes that $\log_b(mn) = \log_b m + \log_b n$.

EXERCISE **35.** *(Making Observations)* Use arguments similar to the one just given to verify that

a. $\log_b\left(\frac{m}{n}\right) = \log_b m - \log_b n$

b. $\log_b(m^a) = a \log_b m$

These three properties lay at the heart of Napier's interest in logarithms. As Asimov says,

> It occurred to Napier that all numbers could be expressed in exponential form. That is, 4 can be written as 2^2, while 8 can be written as 2^3, and 5, 6, and 7 can be written as 2 to some fractional power between 2 and 3. Once numbers were written in such exponential form, multiplication could be carried out by adding exponents, and division by subtracting exponents. Multiplication and division would at once become no more complicated than addition and subtraction.

In Chapter 6 you learned of Johannes Kepler's discovery that the orbits of the planets are ellipses, with the sun at one focus. This fact is the first of three famous planetary laws established by Kepler. News of the invention of logarithms reached him after he had spent almost a decade laboring over calculations based on the astronomical data of Tycho Brahe. The properties of logarithms greatly simplified his remaining work, and the development of Kepler's laws was the first significant application of logarithms. Example 4 both illustrates the properties of logarithms and indicates the manner in which they were used by Kepler to simplify computations.

EXAMPLE 4 *Using the properties of logarithms*

To evaluate an expression of the form

$$\frac{a\sqrt{b}}{c}$$

call the unknown quotient y, so that

$$y = \frac{a\sqrt{b}}{c}$$

Since logarithmic functions are one-to-one, this equation is equivalent to

$$\log y = \log \frac{a\sqrt{b}}{c}$$

The properties of logarithms tell us that multiplying and dividing numbers corresponds to adding and subtracting their logarithms.

$$\log y = \log a + \log \sqrt{b} - \log c$$

Also $\sqrt{b} = b^{1/2}$, and raising a number to a power corresponds to multiplying its logarithm by the exponent.

$$\log y = \log a + \frac{1}{2} \log b - \log c$$

In Exercise 36 you can see how Kepler might have continued from here to evaluate the expression $\frac{a\sqrt{b}}{c}$. ■

EXERCISE **36.** *(Making Observations)*

a. In Example 4, suppose that $a = 12.9$, $b = 535$, and $c = 8.24$. Kepler's tables of logarithms would have told him that

$$\log 12.9 \cong 1.1106$$
$$\log 535 \cong 2.7284$$
$$\log 8.24 \cong 0.9159$$
$$\log 36.2 \cong 1.5587$$

Use this information to evaluate $\dfrac{12.9\sqrt{535}}{8.24}$ without using your calculator.

b. How was the calculation made simpler by the use of logarithms?

In Exercises 37–40, *write the expression in terms of* log a, log b, *and* log c, *as in Example* 4.

37. $\log \dfrac{a}{bc}$ **38.** $\log \sqrt[3]{a}$

39. $\log a^2b$ **40.** $\log \sqrt{\dfrac{b}{c}}$

41. *(Making Observations)* Interchange x and y in the equation $y = Ab^x$ and solve for y, and thus verify that the inverse of the function $y = Ab^x$ has the form $y = \log_b x + k$.

With the computational power of your calculator, you do not need logarithms to simplify cumbersome arithmetic calculations. (Actually, in one sense you do need them, since your calculator does some of its work by using logarithms.) In Section 9-5, you will see an example of why the properties of logarithms are useful today.

In an interesting aside, Asimov, who was also an accomplished science fiction writer, tells about one of Kepler's more unusual works.

> Kepler, by the way, wrote a story, "Somnium," about a man who traveled to the moon in a dream. For the first time the lunar surface was described as it really was, so that "Somnium" may be considered the first piece of authentic science fiction, as opposed to fantasy. It was published after Kepler's death.

The properties of logarithms can be used to solve logarithmic equations more complicated than those in Exercises 27–32. When you combine logarithms in solving an equation, you must check to see whether your solutions yield undefined logarithms in the original equation. Example 5 illustrates.

EXAMPLE 5 *Solving a logarithmic equation*

To solve

$$\ln x + \ln(3x + 2) = 0$$

begin by expressing the left side as a single logarithm.

$$\ln[x(3x + 2)] = 0$$

Then rewrite the equation in an exponential form, remembering that "ln" means "$\log_e$."

$$x(3x + 2) = e^0$$
$$3x^2 + 2x = 1$$
$$3x^2 + 2x - 1 = 0$$
$$(3x - 1)(x + 1) = 0$$
$$x = \frac{1}{3} \quad \text{or} \quad x = -1$$

When $x = \frac{1}{3}$, the original equation becomes

$$\ln \frac{1}{3} + \ln 3 = 0$$

It is easy to verify that this is true. However, when $x = -1$, the equation becomes

$$\ln(-1) + \ln(-1) = 0$$

Since $\ln(-1)$ is not defined, the solution $x = -1$ is extraneous, and the only solution is $x = \frac{1}{3}$. ■

EXERCISES *In Exercises* 42–45, *solve the equation.*

42. $\log(2x - 1) - \log(x + 5) = 0$

43. $2 \log_2 x + \log_2 2x = 1$

44. $\log_x 120 - \log_x 15 = 3$

45. $\dfrac{\ln x}{\ln(x - 12)} = 2$

The Change of Base Formula Now we are ready to return to the problem of using your calculator to graph logarithmic functions with bases other than 10 or e. Such graphs require the following **change of base formula**.

> For any bases $b > 0, b \neq 1$ and $c > 0, c \neq 1$, and any $x > 0$,
>
> $$\log_b x = \frac{\log_c x}{\log_c b}$$

Let's see why the formula is valid. Multiplying both sides of the equation by $\log_c b$, we obtain

$$(\log_c b)(\log_b x) = \log_c x$$

In words, this equation says that the power to which c is raised to obtain b, times the power to which b is raised to obtain x, equals the power to which c is raised to obtain x.

For example, suppose $c = 3, b = 9$, and $x = 6561$, which is both 3^8 and 9^4. Then the power to which 3 is raised to obtain 9 is 2, the power to which 9 is raised to obtain 6561 is 4, and the product $(2)(4) = 8$ is the power to which 3 is raised to obtain 6561. The statement is true in general because you multiply exponents when raising a power to a power.

In practice, c is nearly always chosen to be either 10 or e.

EXERCISES

46. *(Making Observations)* Evaluate $\log_2 7$ on your calculator by evaluating $\dfrac{\log 7}{\log 2}$, then by evaluating $\dfrac{\ln 7}{\ln 2}$. Confirm that the results are the same.

47. Evaluate each of the following on your calculator.

a. $\log_4 7$ **b.** $\log_7 4$
c. $\log_{2/3} 13$ **d.** $\log_{29} 29$
e. $\log_5 \sqrt{11}$ **f.** $\log_{0.4} 16$
g. $\log_3 17.6$ **h.** $\log_{12} \dfrac{16}{9}$

48. Use your calculator to graph each of the functions in Exercises 18–21.

ADDITIONAL EXERCISES

49. Make a table of values for $y = \log x$, using the x-values 1, 10, 100, 1000.

50. Make a table of values for $y = \log_3 x$, using values of x that yield integer values of y.

51. Use the observation that logarithms are disguised exponents to evaluate each of the following.

a. $\log_{36} 6$ **b.** $\log_6 36$
c. $\log_{11} \sqrt{11}$ **d.** $\log_3 \dfrac{1}{81}$
e. $\log_5 5^{-7}$ **f.** $\log_{1/2} 16$
g. $\log_{2/3} \dfrac{4}{9}$ **h.** $\log_{\sqrt{2}} 4$

In Exercises 52–55, decide whether the table fits a function $y = \log_b x + k$.

52.

x	y
−2	1
−1	2
0	4
1	8
2	16

53.

x	y
1	5
2	2
4	−1
8	−4
16	−7

54.

x	y
0.04	4
0.2	5
1	6
5	7
25	8

55.

x	y
1000	1
100	3
10	5
1	7
0.1	9

In Exercises 56–63, sketch the graph of each function by transforming an appropriate graph $y = \log_b x$. In each case,

a. Find the domain and range of the function.

b. Decide whether the function is increasing or decreasing.

c. Identify the points obtained by moving $\left(\dfrac{1}{b}, -1\right)$, $(1, 0)$, $(b, 1)$.

Support your results with a calculator graph.

56. $y = \log(x + 2)$ **57.** $y = 4 \ln x$

58. $y = -2 \log_8(x - 4)$ **59.** $y = \dfrac{1}{3} \log_3 x + 3$

60. $y = 2 \ln x + 6$ **61.** $y = 6 \log 3x$

62. $y = 5 - \log x$ **63.** $y = \ln(x + 1) - 2$

64. a. *(Making Observations)* Graph each of the following pairs of functions on your calculator.

$$y = -\log_2 x \quad \text{and} \quad y = \log_{1/2} x$$
$$y = \log_2(8x) \quad \text{and} \quad y = \log_2 x + 3$$

b. *(Writing to Learn)* Explain what you observe about each pair of graphs in part (a), and support your observations analytically.

In Exercises 65–80, solve the equation.

65. $5^{3t-4} = 9$

66. $0.3^{x/2} - 1 = 2$

67. $3 \cdot 4^{v+5} - 2 = 0.21$

68. $e^{1/x} = 0.1$

69. $e^{4x+5} = 2e^{3x-1}$

70. $5^x = 10(4^x)$

71. $12 \log_8(x-1) = 4$

72. $\log_{0.32}(5x) = 0$

73. $\ln \sqrt{t} = 5$

74. $\log x + \log(x-3) = 1$

75. $2 \ln x = \ln(6-x)$

76. $\log_3 x^2 - \log_3(x-4) = 2$

77. $\log(x-3) + \log(x^2 + 3x + 9) = \log 37$

78. $\log_x 27 + \log_x 8 = 3$

79. $\log_4(x+5) + \log_4\left(\frac{1}{x}\right) = \frac{1}{2}$

80. $\ln x + \ln(x^2 + ex - 1) = 1$

Rewrite each expression in Exercises 81–88 to have no logarithms except $\log_4 x$.

81. $\log_4 \sqrt{x}$

82. $\log_4(16x)$

83. $\log_4 \frac{x}{16}$

84. $\log_4 x^4$

85. $\log_4 \frac{1}{x^3}$

86. $\log_4 x^3 - \log_4 x^2$

87. $\log_4(x^2 + x) - \log_4(x+1)$

88. $\log_4 \frac{x^2}{y^3} + 3 \log_4 y$

89. Identify each statement as true or false.

a. $\log_2(32x) = 5 + \log_2 x$

b. $\log_b(x+3) = \log_b x + \log_b 3$

c. $\log_2(3x) = 3 \log_2 x$

d. $\ln x^4 = 4 \ln x$

e. $-3 \log(x-2)^4 = \log(x-2)^{-12}$

f. $\log_b(bx) = \log_b x + 1$

90. *(Problem Solving)* Use the function $P = 6(1.02)^t$ from ***Standing Room Only*** (page 323)to predict when the world's population will reach "standing room only," one person for each square foot of land. You will need to know that the total land mass of the world is 57,280,000 square miles.

Exercises 91–93 illustrate the use of logarithmic scales to represent physical quantities.

91. *(Interpreting Mathematics)* The apparent brightness of a star is measured by the amount of light energy from the star reaching a unit area of the earth's surface. A star's apparent brightness is described by a number called its **magnitude** on a scale originated by the Greek astronomer Hipparchus (160–127 B.C.). In his textbook, *Foundations of Astronomy* (Wadsworth, 1990), Michael Seeds says, "We must use logarithms to describe the magnitude system because Hipparchus designed it to represent the way stars look to our eyes, and our eyes work logarithmically."

The *difference* between the magnitudes M_1 and M_2 of two stars is related to the *ratio* of their apparent brightnesses I_1 and I_2. Specifically,

$$\frac{I_1}{I_2} = (2.512)^{M_2 - M_1}$$

a. Show that this equation can be rewritten

$$M_2 - M_1 = 2.5(\log I_1 - \log I_2)$$

b. How many times brighter is a star of magnitude 6 (the faintest magnitude visible to the unaided eye) than a star of magnitude 29 (the faintest magnitude visible to the Hubble space telescope)?

c. If one star is 10 times as bright as another, what is the difference in their magnitudes?

92. *(Interpreting Mathematics)* The loudness of sounds is often measured in **decibels**. A sound's decibel level A is related to its intensity I by the equation

$$A = 10 \log I$$

if I is expressed in appropriate units.

a. At close range, the decibel levels of a truck and a jet plane are 100 and 120, respectively. What is the ratio of the intensity of the jet's noise to the intensity of the truck's noise?

b. In Exercise 70 of Section 9-1, you constructed the equation

$$t = 2^{21-0.2A}$$

to describe the maximum number of hours t that workers can safely be exposed to noise at a level of A decibels. Show that t varies approximately inversely with $I^{0.60}$.

93. *(Interpreting Mathematics)* The intensity of earthquakes is sometimes measured on the **Richter scale**. On this scale the magnitude M of an earthquake is related to its released energy E in ergs. Specifically,

$$M = \frac{\log E - 11.8}{1.5}$$

a. By what factor is the released energy of an earthquake multiplied by each increase of one on the Richter scale?

b. In terms of released energy, an earthquake of magnitude 8.0 is how many times as powerful as one of magnitude 4.0?

c. Two of the most famous earthquakes to occur in North America struck San Francisco in 1906 and the Alaskan peninsula in 1964. Their magnitudes were 8.25 and 8.4. How many ergs of energy were released by each earthquake?

94. *(Writing to Learn)* In the quote on page 368, Isaac Asimov says that "5, 6, and 7 can be written as 2 to some fractional power between 2 and 3." What do you think he means by the word *fractional* in this context?

SUPPLEMENTARY TOPIC
Slide Rules

A great advance in the use of logarithms was made in 1622 by William Oughtred (1574–1660). According to Asimov,

> Oughtred, who obtained his master's degree at Cambridge in 1600, was a minister and not a professional mathematician, but that makes little difference since he spent almost all the time he could spare on mathematics, even when it meant sleeping but two or three hours a night. . . .
>
> His greatest invention, however, came in 1622, and consisted of two rulers along which logarithmic scales were laid off. By manipulating the rulers and sliding one against the other, calculations could be performed mechanically by means of logarithms. We know it now as a slide rule, and, for centuries, engineers carried slide rules at least as lovingly as any physician ever carried his stethoscope and tongue depressor.

Figure 9-18 shows a sketch of a slide rule. The distance from the left end of a ruler to a number x on that ruler is proportional to $\log x$.

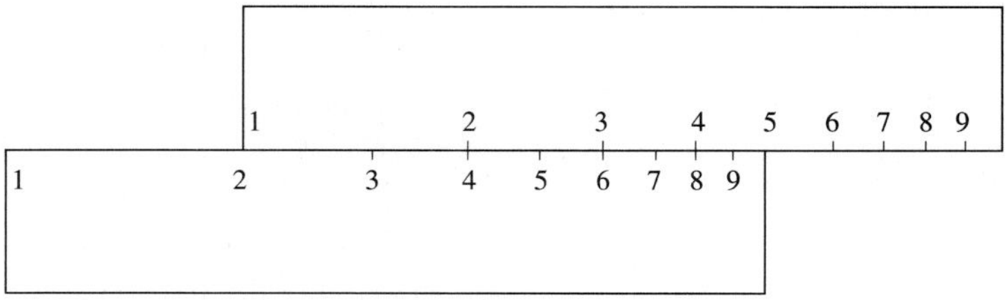

Figure 9–18

95. *(Writing to Learn)* Explain how the position of the slide rule in Figure 9-18 verifies that $2 \cdot 3 = 6$.

96. *(Extension)* Explain how a slide rule could be used to multiply numbers of any size. (*Hint*: Talk to a scientist or engineer who was educated before 1975.)

9–5 CURVE-FITTING

Tables we encounter as we solve problems are seldom cooperative enough to fit one of the functions you have studied in this book. Frequently we might want to construct a function that does not quite fit a table. For example,

- When we conduct an experiment to find the relationship between two physical quantities, we cannot measure the quantities precisely. Therefore, our table does not describe the relationship exactly.
- The table may have come from a function that is too complicated to find, or to be of practical value. We may be forced to, or prefer to, find a simpler function which fits the data approximately.

The process of constructing functions to approximate data is referred to as **curve-fitting**. In this section, you will learn how to find a "best fit" for a table using linear, logarithmic, exponential, and power functions.

FITTING LINEAR FUNCTIONS TO DATA

Fitting a linear function to data is the simplest process, and it can be modified to fit logarithmic, exponential, or power functions to data. The following Mathematical Looking Glass describes a relationship that is approximately linear.

A MATHEMATICAL LOOKING GLASS
Compasses

A navigational compass, such as you might use on a backcountry hike, works because the earth is a huge magnet. The magnetized compass needle points toward the earth's magnetic north pole, a point in arctic Canada about 800 miles from the geographic north pole, as shown in Figure 9-19.

Figure 9–19

By studying magnetic patterns in rocks whose ages are approximately known, scientists have concluded that the polarity (direction) of the earth's magnetic field changes about every 700,000 years. This means that a compass needle would have pointed toward the magnetic south pole 700,000 years ago and toward the magnetic north pole 700,000 years before that.

Near the time of each polarity reversal, the strength of the earth's magnetic field decreases, reaching zero at the time of the reversal. (The units used to measure field strength are the product of electrical current in amperes and area in square meters. We'll let the physicists explain why.) Using past measurements of field strength, we can predict when the next reversal will occur.

Table 33 shows the results of selected past measurements of field strength.

TABLE 33

Year	Strength
1835	8.56
1885	8.36
1910	8.25
1950	8.07
1985	7.87

Adapted from A. C. Fraser-Smith, Centered and Eccentric Geomagnetic Dipoles and Their Poles, 1600–1985, *Reviews of Geophysics,* Vol. 25, no. 1 (1987), pp. 1–16.

When you plot these data points in Exercise 1, you will see that they are almost, but not quite, collinear. Although this discrepancy is partly due to unavoidable errors in the measurements, physicists have proven that the relationship of time to field strength is not quite linear. However, a linear relationship that fits the past data closely can be expected to fit reasonably well for some time to come. In Exercises 1–2 you can find an appropriate linear function and use it to predict the time of the next polarity reversal. Although you will obtain a rough estimate, it will probably be accurate to within about a century.

For a few hundred years before and after the direction change, the magnetic field will be so weak as to render compasses useless, so if you happen to be out in the woods during this time, don't get lost! ■

EXERCISES

1. *(Making Observations)*
   a. Carefully plot the data from Table 33. (*Suggestion*: Scale both axes to intervals which do not extend very far beyond the data points.)
   b. Sketch the line that you think comes closest to passing through all the data points. (Such a line need not actually pass through any of the points.)
2. *(Problem Solving)* Estimate the coordinates of two points on your line, and use those points to find its equation. Then estimate the year in which the next direction reversal will occur.
3. *(Writing to Learn)* List several possible ways to decide what line best fits a given table. Which way makes the most sense to you? Explain why.

The most commonly accepted method of finding a line of best fit, called **linear regression**, is used by your calculator. To find the **regression line** for data $(x_1, y_1), (x_2, y_2), (x_3, y_3), \ldots, (x_n, y_n)$:

- First find the mean (the "average") of the x-values. This is $\bar{x} = \frac{\Sigma x}{n}$, where Σx represents the sum $x_1 + x_2 + x_3 + \cdots + x_n$.
- Next find the mean $\bar{y} = \frac{\Sigma y}{n}$ of the y-values.
- Then calculate the slope of the regression line.

$$m = \frac{n(\Sigma xy) - (\Sigma x)(\Sigma y)}{n(\Sigma x^2) = (\Sigma x)^2}$$

Here $\Sigma xy = x_1y_1 + x_2y_2 + x_3y_3 + \cdots + x_ny_n$, and the other Σ symbols have similar meanings.

- The equation of the regression line is

$$y - \bar{y} = m(x - \bar{x})$$

EXERCISES **4. a.** Carefully plot the following data, estimate a line of best fit, and write the equation of your line.

x	y
6	−2
8	5
9	7
10	10

b. Find the equation of the regression line for the data. How does it compare with your result from part (a)?

5. The following table lists world record times for the mile run in selected years.

Year	Time
1875	4:24.5 (Walter Slade, England)
1895	4:17.0 (Fred Bacon, Scotland)
1915	4:12.6 (Norman Taber, United States)
1934	4:06.8 (Glen Cunningham, United States)
1954	3:59.4 (Roger Bannister, England)
1975	3:51.0 (Filbert Bayi, Tanzania)

a. *(Creating Models)* Carefully plot the data, estimate a line of best fit, and write the equation of your line.

b. *(Interpreting Mathematics)* Use your equation to estimate the record time for the mile run in 1985. The actual record, established that year by Steve Cram of England, was 3:46.3. How close is your estimate?

c. *(Interpreting Mathematics)* Use your equation to estimate the record time for the mile run in the year 1 and in the year 3000.

d. *(Writing to Learn)* For what years do you think your equation is approximately valid? Explain your answer.

Although you can estimate a line of best fit for any table, the fit is better for some relationships than for others. Furthermore, a good fit does not always indicate that the relationship is linear, but a poor fit always indicates that the relationship is not linear.

EXERCISES

6. Carefully plot each of the following tables and assume that the values of x and y are subject to small errors. Decide whether x and y might be linearly related, and explain your answer.

a.

x	y
0.5	7
0.9	6
1.5	5
1.9	4

b.

x	y
−4	230
−2.5	250
0	350
1	460

c.

x	y
1000	10
1500	16
1800	19
2300	23

7. As parts (a)–(c) of this exercise, find the equation of the regression line for each table in Exercise 6. For which of the tables is the line reasonably close to all the data points?

FITTING LOGARITHMIC FUNCTIONS TO DATA

It is not appropriate to fit a linear function to every table. In the following Mathematical Looking Glass, patterns in the data suggest that a logarithmic function is more appropriate. Let's see how the technique of fitting a linear function to a table can be adapted to this situation.

A MATHEMATICAL LOOKING GLASS
Glottochronology

It has been said that the United States and England are two countries separated by the Atlantic Ocean and a common language. Although the English spoken by seventeenth-century American colonists was virtually identical to that spoken in England at the time, the English spoken on the two sides of the Atlantic is now distinctly different. For example, the words "lift" and "elevator" have the same meaning, but are unmistakably British and American, respectively.

Such differences illustrate that when two groups of people with a common language are separated, the language spoken by each will evolve over time. However, the fact that people in London and Denver can still understand one another shows that languages change slowly. Linguistic change is usually measured in millenia (thousands of years). Linguists need information about the rate at which languages evolve to estimate when two related languages descended from a common parent language, or to determine the geographical origin of a family of languages.

Glottochronology is a mathematical method of studying the rates at which languages change. A more detailed description is given by Anthony Lo Bello and Maurice Weir in *Glottochronology: An Application of Calculus to Linguistics* (COMAP, 1983). This analysis begins with a list of words describing basic, time-independent concepts such as "sky," "fly," or "high" that were present in a language at some known benchmark date. The fundamental assumption of glottochronology, based on observations of languages whose history is known, is that equal percentages of the words disappear from the language over equal time intervals. This is equivalent to saying that the number N of words remaining on the list t millenia after the benchmark date is

$$N = Ae^{-kt}$$

In practice, N is often known but t is not. It is then convenient to regard N as the independent variable and write the relationship in the form

$$t = m \ln N + b$$

The constants m and b are determined by curve-fitting. To illustrate how curve-fitting works for approximately logarithmic data, let's begin with Table 34. The table is based on a list of 210 Chinese words in use in the year 950 A.D. and shows the number of words remaining in the language t millenia later. The first and last entries in the table are authentic. We made up the others to create a simple illustration.

TABLE 34

Words (N)	Time in millenia (t)
210	0
200	0.2
186	0.5
174	0.8
167	1

The method you used for fitting linear functions to data is not practical in this situation, because unless you are much better at graphing than we are, you can't sketch an accurate logarithmic curve easily. Exercise 8 indicates what to do instead. ■

EXERCISES

8. *(Creating Models)*

a. Complete the following extension of Table 34.

N	$n = \ln N$	t
210		0
200		0.2
186		0.5
174		0.8
167		1

b. Carefully plot the data points (n, t), estimate a line of best fit, and write the equation of your line.

c. Replace n with $\ln N$ in your equation to obtain a function $t = m \ln N + b$.

d. Graph the function in part (c) in a window large enough to include each pair (N, t) from Table 34.

9. *(Making Observations)* Evaluate your function from Exercise 8c for $N = 200$, 186, and 174. Compare your results with the values of t given in Table 34. How closely do your values fit the data?

10. **a.** Find the equation of the regression line for the data (n, t) from Table 34.

b. Use the regression line to obtain a function $t = m \ln N + b$.

c. Evaluate the function in part (b) for $N = 200$, 186, and 174. How closely do these values fit the data? Is the fit better or worse than what you obtained in Exercise 9?

11. Suppose that an ancient Chinese text has been found, in which half the words do not appear in the modern Chinese language. Use your function in Exercise 10 to estimate the age of the text.

FITTING EXPONENTIAL FUNCTIONS TO DATA

The dynamics of epidemics can be described by mathematical models based on differential equations. According to many such models, the growth of an epidemic during its early stages is approximately exponential. Let's see how to fit an exponential function to data on the HIV virus, which causes AIDS. (Later in this section you will decide whether a different type of function might fit the data better.)

A MATHEMATICAL LOOKING GLASS
Growth of AIDS

During 1991, employees at the Veterans Administration Hospital in Pittsburgh were accidentally punctured on 83 occasions by needles contaminated by blood or body fluids from patients. Four of these patients were infected with the HIV virus. Although none of the employees contracted HIV, the risk of infection is a serious concern. Epidemiologist Angella Goetz gathered data in an attempt to quantify the expected degree of future risk. In making her assessment, she needed to estimate the rate at which the incidence of HIV among patients will increase over the next several years. This estimate requires a mathematical model for the spread of HIV among the entire population. Ms. Goetz consulted with one of your authors to analyze the reliability of several models. In Exercises 12–17 and 52 you will re-create two of the models and explore their reliability.

It has been difficult to collect reliable data on the HIV virus, because people with HIV exhibit no symptoms when they are first infected. However, the spread of HIV can be roughly estimated from the number of recorded deaths from AIDS. The U.S. Department of Health compiled the data in Table 35 for the years 1981–1989.

TABLE 35

Year	Deaths from AIDS	Total deaths from AIDS since 1981
1981	335	335
1982	950	1,285
1983	2,648	3,933
1984	5,082	9,015
1985	9,180	18,195
1986	13,257	31,452
1987	14,601	46,053
1988	9,335	55,388
1989	22,827	78,215

■

Fitting an exponential curve directly to the data will not work well, because it is difficult to sketch exponential graphs accurately. Once again, taking some logarithms produces an approximately linear relationship. Exercises 12–13 will guide you through the process.

EXERCISES

12. *(Writing to Learn)* In Table 35, let t represent time in years since 1981 and y represent the total number of deaths from AIDS since 1981. Assume that the relationship between t and y is described by an equation $y = Ae^{mt}$.

a. Explain why the above equation is equivalent to $\ln y = \ln A + \ln e^{mt}$.

b. Explain why the equation in part (a) is equivalent to $\ln y = \ln A + mt$.

c. Explain why the equation in part (b) shows that t and $\ln y$ are linearly related.

13. *(Problem Solving)*

a. Add a new column to Table 35, labeled $Y = \ln y$. Calculate the value of Y for each of the years 1981–1989.

b. Carefully plot the data points (t, Y), estimate a line of best fit, and write the equation of your line.

c. Your equation in part (b) has the form $Y = mt + b$. Use the fact that $Y = \ln y$ and $b = \ln A$ to obtain an equation $y = Ae^{mt}$.

d. Graph your equation from part (c) in a window large enough to contain each point (t, y) from Table 35.

14. *(Making Observations)* Evaluate your function from Exercise 13c for each year from 1981 through 1989. Compare your results with the values of y given in Table 35. How closely do your values fit the data?

15. a. Find the equation of the regression line for the data (t, Y) from Table 35.

b. Use the regression line to obtain a function $y = Ae^{mt}$.

c. Evaluate the function in part (b) for each year from 1981 through 1989. How closely do these values fit the data? Is the fit better or worse than what you obtained in Exercise 14?

16. Use the function in Exercise 15b to predict the total number of deaths from AIDS at the end of 1991.

17. *(Writing to Learn)* The actual number of deaths from AIDS at the end of 1991 was 133,232. What conclusions can you draw by comparing this figure with your result from Exercise 16?

FYI A good fit between function and data does not always indicate that the function describes the relationship. In fact, because the spread of HIV is affected by so many complex physical and sociological factors, it is unlikely that any simple function can accurately model its long-term growth. In general, models of complex phenomena are accurate only for a short time.

FITTING POWER FUNCTIONS TO DATA

Now let's see how to fit a power function to a table, using astronomical data collected by the sixteenth-century Danish astronomer Tycho Brahe and analyzed by Johannes Kepler.

A MATHEMATICAL LOOKING GLASS
Kepler's Third Law

In Section 9-4 you read about Kepler's three laws of planetary motion. His third law states that the square of a planet's period of revolution is proportional to the cube of its distance from the sun. Table 36 shows the distance of each planet from the sun, along with its period of revolution.

TABLE 36

Planet	Distance (millions of km)	Period (years)
Mercury	57.9	0.241
Venus	108.2	0.615
Earth	149.6	1.000
Mars	227.9	1.881
Jupiter	778.3	11.86
Saturn	1427	29.46
Uranus	2870	84.01
Neptune	4497	164.8
Pluto	5900	247.7

■

EXERCISES

18. *(Writing to Learn)* In Table 36, let x and y represent distance and period, respectively. Explain why Kepler's third law says that x and y are related by a power function $y = Ax^m$. According to the statement of the law, what is the value of m?

19. *(Writing to Learn)*

a. Explain why the equation $y = Ax^m$ is equivalent to

$$\ln y = \ln A + \ln x^m$$

b. Explain why the equation in part (a) is equivalent to

$$\ln y = \ln A + m \ln x$$

c. Explain why the equation in part (b) shows that $\ln x$ and $\ln y$ are linearly related.

20. *(Creating Models)*

a. Use Table 36 to complete the following table.

x	$X = \ln x$	y	$Y = \ln y$
57.9		0.241	
108.2		0.615	
149.6		1.000	
227.9		1.881	
778.3		11.86	
1427		29.46	
2870		84.01	
4497		164.8	
5900		247.7	

b. Carefully plot the data points (X, Y), estimate a line of best fit, and write the equation of your line.

c. Your equation in part (b) has the form $Y = mX + b$. Use the fact that $X = \ln x$, $Y = \ln y$, and $b = \ln A$ to obtain $y = Ax^m$.

d. Graph your equation from part (c) in a window large enough to contain each point (x, y) from Table 36.

21. *(Making Observations)* Evaluate your function from Exercise 20c for each value of x in Table 36. Compare your results with the values of y given there. How closely do your values fit the data?

22. a. Find the equation of the regression line for the data (X, Y) from Table 36.

b. Use the regression line to obtain a function $y = Ax^m$.

c. Evaluate the function in part (b) for each planet in Table 36. How closely do these values fit the data? Is the fit better or worse than what you obtained in Exercise 21?

23. Use the statement of Kepler's third law to write an equation relating x and y, and compare it with your equation from Exercise 22b.

CURVE FITTING ON YOUR CALCULATOR

If you use your calculator to construct a curve of best fit for a table, the calculator will also display a number r, called the **coefficient of linear correlation**. It is calculated by the formula

$$r = \frac{n(\Sigma xy) - (\Sigma x)(\Sigma y)}{\sqrt{n(\Sigma x^2) - (\Sigma x)^2}\sqrt{n(\Sigma y^2) - (\Sigma y)^2}}$$

All you need to know for now is that the value of r is always between -1 and 1 and that the best fit occurs when $|r|$ is close to 1.

EXERCISES

24. *(Creating Models)* The following table gives men's world record times for running selected distances as of September 1992.

Distance, in meters	Record time, in minutes : seconds
1,500	3:28.9 (Noureddine Morceli, Algeria)
2,000	4:50.8 (Said Aouita, Morocco)
3,000	7:29.0 (Moses Kiptanui, Kenya)
5,000	12:58.4 (Said Aouita, Morocco)
10,000	27:08.2 (Arturo Barrios, Mexico)
42,220 (marathon)	126:50 (Belayneh Densimo, Ethiopia)

Rounding all times to the nearest hundredth of a minute, use your calculator to fit each type of curve to the data. Record the value of r for each, and decide which gives the best fit.

a. a linear curve

b. a logarithmic curve

c. an exponential curve

d. a power curve

25. *(Creating Models)* Repeat Exercise 24 using the following women's world record times for the same distances.

Distance, in meters	Record time, in minutes : seconds
1,500	3:52.5 (Tatyana Kazankina, USSR)
2,000	5:28.7 (Maricica Puica, Romania)
3,000	8:22.6 (Tatyana Kazankina, USSR)
5,000	14:37.3 (Ingrid Kristiansen, Norway)
10,000	30:13.7 (Ingrid Kristiansen, Norway)
42,220 (marathon)	141:06 (Ingrid Kristiansen, Norway)

26. Use your calculator to fit the specified type of curve to the data, and compare your results with the ones you obtained in the indicated exercises.

a. Fit a linear curve to Table 33 and compare your results with those in Exercise 2.

b. Fit a linear curve to the table in Exercise 5 and compare your results with those in Exercise 5.

c. Fit a logarithmic curve to Table 34 and compare your results with those in Exercise 8.

d. Fit an exponential curve to Table 35 and compare your results with those in Exercise 13.

e. Fit a power curve to Table 36 and compare your results with those in Exercise 20.

ADDITIONAL EXERCISES *In Exercises* 27–28, *use the method of Exercises* 1 *and* 2 *to fit a linear function to the table.*

27.

x	y
1	13
2	10
3	9
4	7

28.

x	y
0.3	−120
0.5	100
0.8	360
1.0	600

29. Use the method of Exercise 8 to fit a logarithmic curve to the following table.

x	y
1	−5
7	−1
12	0
20	1

30. a. Find the equation of the regression line for the points (x, y) in Exercise 27.

b. Find the equation of the regression line for the points (x, y) in Exercise 28.

c. Find the equation of the regression line for the points ($\ln x$, y) in Exercise 29, and use the regression line to fit a logarithmic curve to the data.

In Exercises 31–34, plot the pairs ($\ln x$, y) and assume that the values of x and y are subject to small errors. Decide whether x and y might be related by a logarithmic function.

31.

x	y
2	0.6
5	1.4
10	2.0
20	2.6

32.

x	y
2	23
5	14
10	7
20	0

33.

x	y
2	0
5	1
10	10
20	50

34.

x	y
2	32
5	68
10	97
20	128

35. *(Creating Models)* During a mission in space, it is vitally important for both crew members and ground controllers to respond quickly and correctly to information displayed on the control panels. For this reason, NASA has conducted studies measuring the time it takes to respond to varying numbers of choices. The results of one such study are shown in the following table, where N represents the number of choices and R is the average response time in seconds.

N	R
1	0.17
2	0.34
3	0.37
4	0.42
5	0.48
6	0.52
7	0.56
8	0.58
9	0.59
10	0.57

Source: Bernice Kastner, *Space Mathematics: A Resource for Secondary School Teachers*, NASA, 1985.

a. Carefully plot the data points (N, R), then plot the points ($\ln N$, R) on a second set of axes. Decide whether the data looks more nearly linear or logarithmic.

b. Estimate the equation of either a line or a logarithmic curve of best fit, depending on your conclusion from part (a).

x	y
1	1
7	3
12	9
20	33

36. Use the method of Exercise 13 to estimate an exponential curve of best fit for the data in the table at the left. Then use the regression line for the points (x, $\ln y$) to find an exponential curve of best fit, and compare your results.

In Exercises 37–40, plot the pairs $(x, \ln y)$ and assume that the values of x and y are subject to small errors. Decide whether x and y might be related by an exponential function.

37.

x	y
3	8
5	9
7	10.5
9	12.5

38.

x	y
−6	1
−3	10
−1	13
0	14

39.

x	y
−1	0.5
1	2.3
2	3.9
4	14.8

40.

x	y
10	6.2
20	17.5
30	51.0
40	149.0

41. Use the method of Exercise 20 to estimate a power curve of best fit for the following data.

x	y
1	2
7	90
12	300
20	750

In Exercises 42–45, plot the pairs $(\ln x, \ln y)$ on graph paper, and assume that the values of x and y are subject to small errors. Decide whether x and y might be related by a power function.

42.

x	y
1	10
2	5
3	3
4	2

43.

x	y
10	1
20	3.5
40	14.1
70	43.2

44.

x	y
3	2
5	10
6	15
8	12

45.

x	y
2	0.7
3	2.5
5	12.0
6	21.0

46. *(Writing to Learn)* Explain what goes wrong when you attempt to fit a power curve to a table when one of the entries is 0.

In Exercises 47–50, use your calculator to decide which type of curve (linear, logarithmic, exponential, or power) fits the table best.

47.

x	y
2.3	9.1
4.1	7.2
5.6	5.7
8.7	2.4

48.

x	y
5	12.1
10	13.3
15	14.6
20	16.0

49.

x	y
6	2.5
20	4.5
50	7.1
80	8.9

50.

x	y
1	20
3	85
4	180
7	1400

51. The following table gives the population of the United States as recorded in each census from 1790 through 1830.

Year	Population to the nearest thousand
1790	3,929,000
1800	5,308,000
1810	7,240,000
1820	9,564,000
1830	12,866,000

a. *(Creating Models)* Use your calculator to decide whether the data most nearly fits a linear, logarithmic, exponential, or power function. Find the equation of the curve that fits best.

b. *(Interpreting Mathematics)* Verify that your result from part (a) predicts the actual 1860 census figure of 31,433,000 within half a million people but overestimates the 1870 figure of 38,558,000 by almost 3 million.

c. *(Making Observations)* What occurred in the United States during the 1860s that would help explain the discrepancy between the predicted and actual populations for 1870?

52. *(Making Observations)* In ***Growth of AIDS***, you discovered that an exponential curve of best fit for data on AIDS deaths through 1989 failed miserably to predict the number of deaths through 1991. Find a power curve of best fit for the same data, use it to predict the number of deaths through 1991, and compare your result with the actual total of 133,232.

53. *(Writing to Learn)* To find the relationship between two variables, scientists sometimes plot data on **semilog paper** (Figure 9-20a), on which the y-axis is scaled logarithmically (the scale mark for $y = y_0$ is at a distance of $\log y_0$ units above the x-axis). They also may plot data on **log-log paper** (Figure 9-20b), on which both axes are scaled logarithmically.

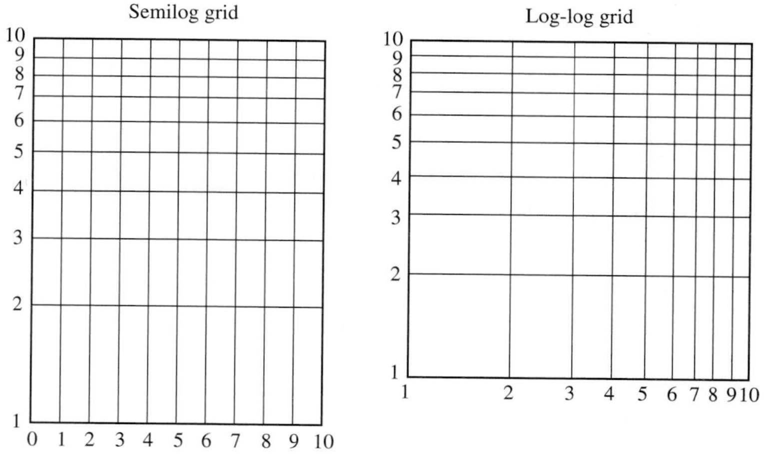

Figure 9–20

Decide what type of data would plot linearly on each type of paper, and explain why.

CHAPTER REVIEW

Complete Exercises 1–44 (Writing to Learn) before referring to the indicated pages.

WORDS AND PHRASES

In Exercises 1–11, explain the meaning of the words or phrases in your own words.

1. (pages 324, 327) **exponential function**
2. (page 332) **doubling time**
3. (page 332) **geometric sequence**
4. (page 334) **geometric series**
5. (page 335) **sum of an infinite geometric series**
6. (page 341) the number ***e***
7. (page 341) **continuous compounding** of interest
8. (page 346) **inverse** of a function
9. (page 346) **one-to-one** function
10. (pages 359, 360) **logarithm**, **logarithmic function**
11. (page 374) **curve-fitting**

IDEAS

12. (pages 324, 325) How do you decide whether a table fits an exponential function? Make up a table to illustrate your explanation.
13. (pages 326, 327) Describe the process of fitting an exponential function to a table. Make up a table that fits an exponential function and find its equation to illustrate your explanation.
14. (page 327) The equation $y = Ab^x$ defines an exponential function as long as $A > 0$, $b > 0$, and $b \neq 1$. For what reasons are those restrictions imposed on A and b?
15. (pages 327, 328) How are irrational exponents defined?
16. (pages 328–330) Describe the graph of a function $f(x) = Ab^x$. In particular, state its domain and range, and tell whether it is increasing or decreasing. Be sure to consider both the cases $b > 1$ and $0 < b < 1$.
17. (pages 330–332) Describe the process of obtaining the graph of a function $y = Ab^{x-h} + k$ from that of $y = b^x$. Illustrate your explanation with a specific function.
18. (pages 332, 333) Explain why a geometric sequence is an exponential function. What is the difference between geometric sequences and other exponential functions?
19. (pages 332–335) What is the difference between a geometric sequence and a geometric series? Make up an example to illustrate the difference.
20. (pages 334, 335) How do you find the sum of a finite geometric series? Under what conditions does an infinite geometric series have a sum? How is the formula for the sum of an infinite geometric series derived from the formula for the sum of a finite geometric series?
21. (pages 338–341) How is the formula for continuously compounded interest derived from the formula for interest compounded m times per year? What expression in n approaches a value of e if n is large? What is the value of e to the nearest tenth?
22. (pages 341–343) Describe the process of rewriting a given exponential function to have a base of e.
23. (pages 345, 346) How does the definition of a pair of inverse functions express the idea that each function in the pair undoes the action of the other? Make up a pair of inverse functions to illustrate your response.
24. (pages 346–348) How can you tell whether a table, a graph, or an equation describes a one-to-one function? Make up specific functions to illustrate your response.
25. (page 346) Why is it that only a one-to-one function can have an inverse?
26. (pages 349, 350) How can you find the inverse of a one-to-one function that is described by a table? Illustrate the process with a specific table.
27. (pages 350, 351) How can you find the inverse of a one-to-one function that is described by a graph? Illustrate the process with a specific graph.
28. (pages 351, 352) How can you find the inverse of a one-to-one function that is described by an equation? Illustrate the process with a specific equation.
29. (pages 353, 354) When you find the inverse of a one-to-one function f analytically, why must you evaluate $f \circ f^{-1}$ and $f^{-1} \circ f$ as part of the process?
30. (pages 355, 356) How can it happen that a function is not one-to-one, but becomes one-to-one when the variables are given physical meanings?
31. (pages 359, 360) How do we know that a function $f(x) = b^x$ $(b > 0, b \neq 1)$ has an inverse? Why do we need the new name *logarithm* to describe the inverse function?
32. (page 360) How can tables be used to show that logarithms are disguised exponents? Make up a table to illustrate your explanation.
33. (pages 361, 362) How can you decide whether a table fits a logarithmic function? Make up a table to illustrate your explanation.
34. (page 362) What two bases for logarithmic functions are used most often?

35. (pages 363, 364) Describe the graphical features of a function $f(x) = \log_b x$ when $b > 1$ and when $0 < b < 1$. What are the corresponding graphical features of the function $g(x) = b^x$?

36. (pages 364, 365) Describe the process of obtaining the graph of $f(x) = a \log_b(x - h) + k$ by transforming a basic graph $y = \log_b x$. Make up a specific function to illustrate your description.

37. (page 365) Explain why for any base b, $\log_b b^x = x$ for all real x, $b^{\log_b x} = x$ for $x > 0$, $\log_b 1 = 0$, and $\log_b b = 1$.

38. (pages 367–369) In what way can we rewrite the logarithm of a product, a quotient, or a number raised to a power? How are these properties of logarithms related to the laws of exponents?

39. (pages 370, 371) What does the change of base formula say? How can it be used to evaluate logarithms on your calculator? Perform a specific calculation to illustrate your explanation.

40. (page 374) Why might we want to construct a function that does not quite fit a given table?

41. (pages 375, 376) Must a line of best fit for a given set of data pass through any of the data points? Why or why not?

42. (pages 378–380) If the points $(\ln x, y)$ are collinear, then x and y are related by a logarithmic function. Show why this is true.

43. (pages 380, 381) If the points $(x, \ln y)$ are collinear, then x and y are related by an exponential function. Show why this is true.

44. (pages 381–383) If the points $(\ln x, \ln y)$ are collinear, then x and y are related by a power function. Show why this is true.

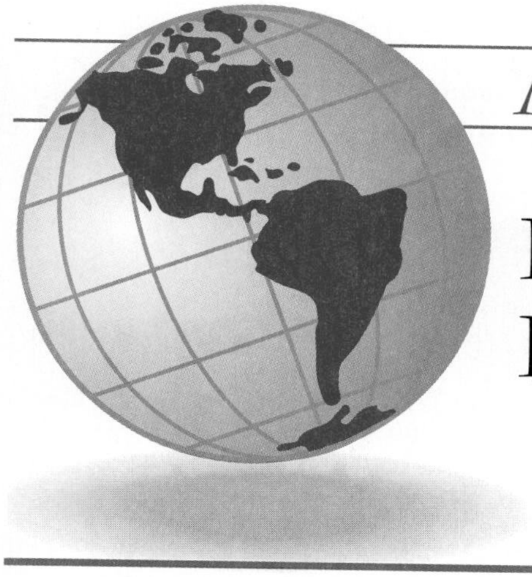

APPENDIX A

BASIC ALGEBRA REFERENCE

A-1 ACCURACY AND PRECISION

Numbers can represent physical quantities either exactly or approximately. For example, suppose that a friend tells you

> "Out of 32 people in my algebra class, I live farthest from campus. I drive 32 miles to school."

The 32 in the first sentence is *exactly* the number of people in your friend's class. By contrast, the 32 in the second sentence is *approximately* the distance she drives to school. She probably means that the distance is closer to 32 miles than to 31 or 33 miles. That is, it is between 31.5 and 32.5 miles. As your friend's statements illustrate, you must often rely on context to decide whether a number is exact or approximate.

When a number is approximate, you often need to know how good the approximation is. Informally, you need to know how many digits you can trust. Furthermore, when you add, subtract, multiply, or divide approximate numbers, you need to know how many digits to trust in your result. The following discussion provides the needed information.

SIGNIFICANT DIGITS

If the population of the United States is reported to be 249,000,000, it is reasonable to believe that the reported number is an approximation to the nearest million, so that the actual population is between 248,500,000 and 249,500,000. Thus only the first three digits in the reported number have meaning, in the sense that their values reflect the size of the population. For this reason, the first three digits are said to be **significant**, while the last six are not.

The following rules determine which digits of an approximate number are significant.

- All nonzero digits are significant.
- All zeros between two nonzero digits are significant.
- All zeros in a string of trailing zeros are significant if the string ends to the right of the decimal point.
- No other zeros are significant.

EXAMPLE 1 *Deciding which digits of an approximate number are significant*

In the table below, the significant digits are underlined.

Approximate number	Actual value of number
3600	closer to 3600 than to 3500 or 3700, that is, between 3550 and 3650
3600.0000	closer to 3600.0000 than to 3599.9999 or 3600.0001, that is, between 3599.99995 and 3600.00005
3600.07	closer to 3600.07 than to 3600.06 or 3600.08, that is, between 3600.065 and 3600.075
0.07	closer to 0.07 than to 0.06 or 0.08, that is, between 0.065 and 0.075
0.0700	closer to 0.0700 than to 0.0699 or 0.0701, that is, between 0.06995 and 0.07005

■

The **accuracy** of an approximate number is its number of significant digits. The **precision** of the number is the decimal place of its rightmost significant digit. Thus the approximate number 3600.07 is accurate to six significant digits, and precise to the nearest hundredth.

SCIENTIFIC NOTATION

Suppose that you are writing a paper in which you want to report the population of the United States, rounded to the nearest million. Suppose, further, that the actual population is 250,341,829, and you write the rounded population as 250,000,000. In this case, the zero in the millions place has meaning in the sense that its value reflects the size of the population. However, a reader of your paper will not regard that zero as a significant digit, and will thus misinterpret your intent.

In order to overcome such ambiguities, scientists have devised a different way to represent approximate numbers. In **scientific notation**, a number is written as the product of two numbers, the first having a value between 1 and 10 and the second being an integer power of 10. *All digits in the first factor of this product are significant.* Thus the approximate population figure in the foregoing paragraph can be written as 2.50×10^8 if three significant digits are intended or as 2.5×10^8 if two significant digits are intended. Here are scientific notation representations of a few other approximate numbers:

Approximate number	Representation in scientific notation
3600	3.6×10^3
3600.0000	3.6000000×10^3
0.07	7×10^{-2}
0.0700	7.00×10^{-2}

ACCURACY OF PRODUCTS AND QUOTIENTS

How should you round off a product or quotient of two or more approximate numbers? The answer depends on the accuracy of the factors. Specifically, the accuracy of the result is equal to that of the least accurate factor.

EXAMPLE 2 *Finding the accuracy of a product*

If we multiply 432.17 by 3.089, the result is 1334.97313. However, if the factors are approximate numbers, then they are accurate to five and four significant digits, respectively. Thus the product should be rounded to four significant digits as 1335. ■

Exact numbers are more accurate than any approximation. Thus, when multiplying 432.17 (approximate) by 14 (exact), round the product 6050.38 to five significant digits as 6050.4.

PRECISION OF SUMS AND DIFFERENCES

How should you round off a sum or difference of two or more approximate numbers? The answer depends on the precision of the terms. Specifically, the precision of the result is equal to that of the least precise term.

EXAMPLE 3 *Finding the precision of a sum*

If we add 432.17 to 3.089, the result is 435.259. However, if the terms are approximate numbers, then they are precise to the nearest hundredth and the nearest thousandth (two and three decimal places), respectively. Thus the sum should be rounded to two decimal places as 435.26. ■

Exact numbers are more precise than any approximation. Thus, when adding 432.17 (approximate) to 14 (exact), round the sum 446.17 to two decimal places. That is, leave it alone.

EXERCISES *In Exercises* 1–8,

a. Underline the significant digits in each approximate number.

b. Find the possible actual values of the number, as in Example 1.

c. Write the number in scientific notation.

1. 8230 **2.** 8230.6

3. 0.0032 **4.** 1.0032

5. 61,400.009 **6.** 61,400.00900

7. 0.010203 **8.** 0.010200

In Exercises 9–16, perform the indicated operations and round the result appropriately. Assume all numbers are approximate.

9. (32.9)(0.0041) **10.** $\dfrac{3300}{780.05}$

11. 5.32761 + 3.5 **12.** 0.000800 − 0.00052

13. 602.987 ÷ 7 **14.** 10,000 + 0.25

15. $\dfrac{(4320)(1.047)}{513.92}$ **16.** 4320 + 1.047 + 513.92

In Exercises 17–20, *repeat Exercises* 13–16, *assuming that all integers are exact.*

17. $602.987 \div 7$

18. $10{,}000 + 0.25$

19. $\dfrac{(4320)(1.047)}{513.92}$

20. $4320 + 1.047 + 513.92$

A–2 LINEAR EQUATIONS

EQUATIONS, SOLUTIONS, AND EQUIVALENCE

An **equation** is a statement that two quantities are equal. Like other statements, an equation may be true or false. For example,

$$2 + 2 = 4$$

is a true equation, while

$$2 + 2 = 5$$

is a false equation.

Some equations involve one or more **variables**, that is, symbols which can be replaced by real numbers. A replacement that produces a true equation is a **solution** of the equation. For example, 2 is a solution of the equation

$$8x - 2 = 2(x + 5)$$

since the equation is true if x is replaced by 2. Furthermore, 2 is the only solution, since the equation is false if x is replaced by any other number.

Two equations are **equivalent** if and only if they have the same solutions. For example, the equations

$$4x - 1 = x + 5$$

$$3x = 6$$

$$x = 2$$

are all equivalent to $8x - 2 = 2(x + 5)$, since they all have 2 as their only solution.

If you perform any of the following operations on an equation, you always produce an equivalent equation.

- Replace quantities by equal quantities anywhere in the equation.
- Add or subtract equal quantities on each side of the equation.
- Multiply or divide each side of the equation by equal nonzero quantities.

SOLVING LINEAR EQUATIONS

An equation in one variable, say x, is called **linear** if the only operations performed on x are addition and subtraction, and multiplication and division by *constant* quantities. Thus $3x + 20 = 7x$ is linear, while $x^2 = 9$ is not. A linear equation in one variable can always be solved by performing only those operations listed above.

EXAMPLE 1 *Solving a linear equation in one variable*

Here is one sequence of steps which can be used to solve

$$8x - 2 = 2(x + 5)$$

Replace $2(x + 5)$ by $2x + 10$.

$$8x - 2 = 2x + 10$$

Add 2 to both sides.

$$8x = 2x + 12$$

Subtract $2x$ from both sides.

$$6x = 12$$

Divide both sides by 6.

$$x = 2$$

■

EXERCISES *In Exercises* 1–8, *solve the equation.*

1. $9 - 4x = 33$

2. $q - 13 = 5q + 7$

3. $5(t - 4) = -(7 - 2t)$

4. $0.01z = 7.86 - 2.61z$

5. $\frac{x}{2} + 4 = \frac{x}{3}$

6. $325(1 - y) = 497y$

7. $8w + 5 = 5$

8. $\sqrt{3}\,p = \frac{1}{\sqrt{3}}$

9. A sequence of steps to solve $8x - 2 = 2(x + 5)$, different from the sequence in the example, is shown on page A4. Identify the operation that was used to obtain each equation from the preceding one.

A–3 THE COORDINATE PLANE

A **coordinate plane** (also called a **Cartesian plane** in honor of René Descartes, whom you will meet in Chapter 7) is a plane containing two number lines at right angles to each other, as in Figure A-1.

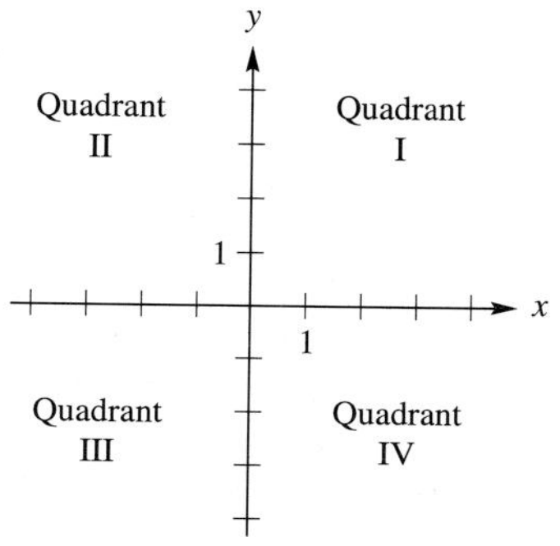

Figure A–1

The two number lines are the **horizontal axis** (***x*-axis**) and the **vertical axis** (***y*-axis**). Their point of intersection is the **origin**. The two axes divide the plane into four **quadrants**, numbered as in Figure A-1.

Every **ordered pair** of real numbers corresponds to a unique point in the coordinate plane, and vice versa. For example, the ordered pair $(3, -2)$ describes point A in Figure A-2, located 3 units to the right of the origin and 2 units below it. The numbers 3 and -2 are the ***x*-coordinate** and ***y*-coordinate** of A.

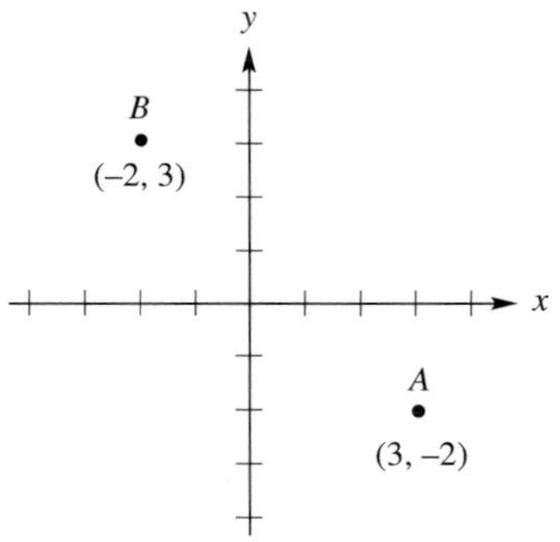

Figure A–2

Points are described by *ordered* pairs of numbers. Thus, the ordered pair $(-2, 3)$ does not describe A in Figure A-2, but instead describes B, located 2 units to the left of the origin and 3 units above it.

EXERCISES **1.** Find the coordinates of each labeled point in the coordinate plane below.

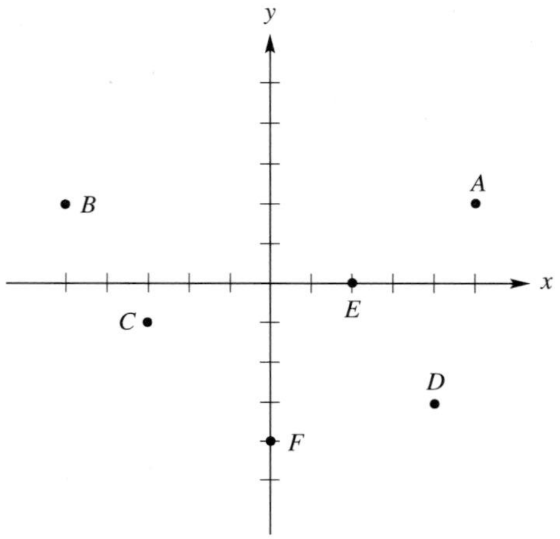

In Exercises 2–9, plot the given point in a coordinate plane, and tell which quadrant or which axis contains it.

2. $(6, 1)$

3. $(-1, -4)$

4. $(-2, 7)$

5. $(3, -1)$

6. $(5, -5)$

7. $(-5, 5)$

8. $(0, 2)$

9. $(3.5, 0)$

A–4 THE PYTHAGOREAN THEOREM AND THE DISTANCE FORMULA

THE PYTHAGOREAN THEOREM

A **right angle** is an angle having a measure of 90 degrees, as in Figure A-3a. A **right triangle** is a triangle containing a right angle, as in Figure A-3b. In a right triangle, the side opposite the 90-degree angle is called the **hypotenuse**.

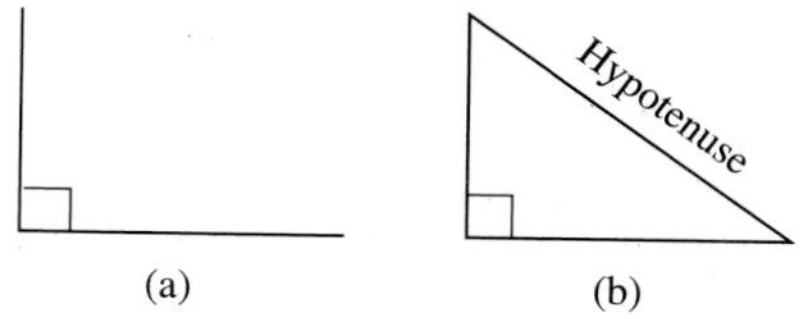

Figure A–3

The **Pythagorean theorem** states:

> If the hypotenuse in a right triangle has length c and the other two sides have lengths a and b, then
>
> $$a^2 + b^2 = c^2$$

EXAMPLE 1 *Using the Pythagorean theorem to find the length of a side of a right triangle*

In Figure A-4, $a = 4$ and $c = 9$. To find b,

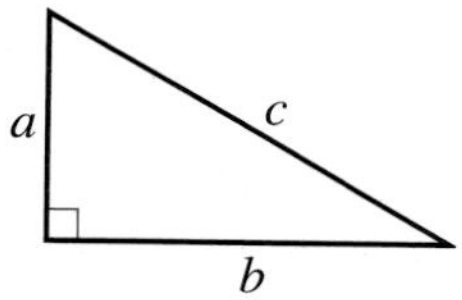

Figure A–4

$$\begin{aligned} a^2 + b^2 &= c^2 \\ (4)^2 + b^2 &= (9)^2 \\ 16 + b^2 &= 81 \\ b^2 &= 65 \\ b = \sqrt{65} &\cong 8.06 \end{aligned}$$

■

THE DISTANCE FORMULA

An important consequence of the Pythagorean theorem is the **distance formula**.

> The distance between two points (x_1, y_1) and (x_2, y_2) in the coordinate plane is
>
> $$\sqrt{(x_2 - x_1)^2 + (y_2 - y_1)^2}$$

Figure A-5, on page A8, shows how the distance formula follows from the Pythagorean theorem. The distance between the points is represented by d. The horizontal distance between them is $|x_2 - x_1|$, so that the square of the horizontal distance is $(x_2 - x_1)^2$. Similarly, the square of the vertical distance is $(y_2 - y_1)^2$, so

$$d^2 = (x_2 - x_1)^2 + (y_2 - y_1)^2$$

Then

$$d = \sqrt{(x_2 - x_1)^2 + (y_2 - y_1)^2}$$

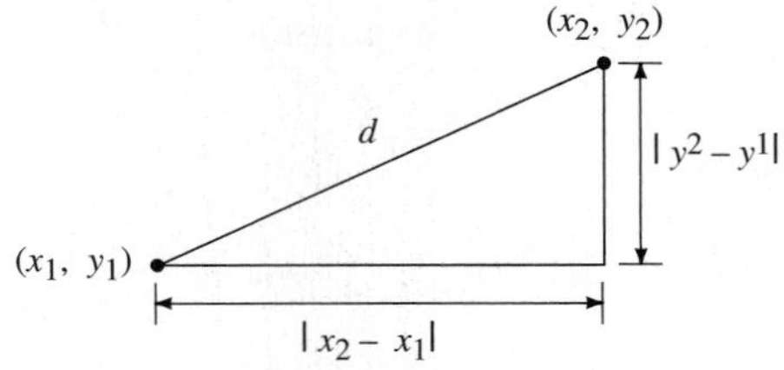

Figure A–5

EXERCISES *In Exercises 1–8, suppose that the hypotenuse of a right triangle has length c and that the other two sides have lengths a and b. Find the length of the side not given.*

1. $a = 6,\ c = 10$

2. $b = 8,\ c = 17$

3. $a = 12,\ b = 5$

4. $a = 7,\ c = 25$

5. $b = 10,\ c = 20$

6. $a = 10,\ b = 10$

7. $a = 1,\ b = 3$

8. $a = 1,\ c = 3$

In Exercises 9–16, find the distance between the points.

9. $(4, -2)$ and $(1, 2)$

10. $(0, 12)$ and $(5, 0)$

11. $(-7, -3)$ and $(-9, -3)$

12. $(6, -3)$ and $(6, 5)$

13. $(-4, -4)$ and $(8, 3)$

14. $(2, 9)$ and $(7, 6)$

15. $(0, 0)$ and $(9, 40)$

16. $(1, 99)$ and $(-3, 97)$

A–5 BASIC GRAPHING TECHNIQUES

The equation

$$y = 2x - 1$$

describes a relationship between the variables x and y. A **solution** to this equation is an ordered pair of real numbers which produce a true equation when substituted for x and y, respectively. The **graph** of the equation consists of all points in the coordinate plane that represent solutions to the equation. The process of obtaining the graph of an equation is referred to as **graphing**. Let's graph $y = 2x - 1$.

EXAMPLE 1 *Graphing an equation*

To graph $y = 2x - 1$, begin by listing several solutions to the equation, choosing any values you please for x, and calculating the values for y.

When $x = 0$, the equation is true if $y = 2(0) - 1 = -1$, so $(0, -1)$ is a solution.

A few other solutions, calculated in the same way, are $(-1, -3)$, $(1, 1)$, $(2, 3)$, and $(3, 5)$. The points representing these solutions are shown in Figure A-6a.

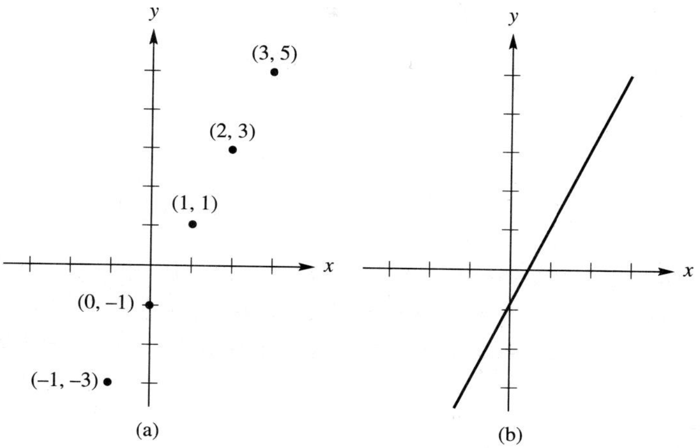

Figure A–6

Since there are infinitely many solutions, you cannot complete the graph by calculating all of them. Eventually, you must do one of two things.

- Make a guess based on the points you have plotted, or
- recall that the equation has the form $y = mx + b$, so its graph is a straight line (see Section A-6).

Figure A-6b shows the graph of $y = 2x - 1$. ■

Of course, the variables in a relationship need not be called x and y, as illustrated by the following example.

EXAMPLE 2 *Graphing an equation*

To graph

$$q = \frac{10}{p^2 + 1}$$

you must first decide whether the horizontal axis is to be the p-axis or the q-axis. Equivalently, you must decide whether the first number in an ordered pair is to represent a value of p or of q. Either choice is correct, but since the equation expresses q in terms of p, it is traditional to make the p-axis horizontal. (We will return to this idea in Section 2-1.)

Several solutions to the equation are $(-3, 1)$, $(-2, 2)$, $(-1, 5)$, $(0, 10)$, $(1, 5)$, $(2, 2)$, and $(3, 1)$, plotted in Figure A-7a (see page A10).

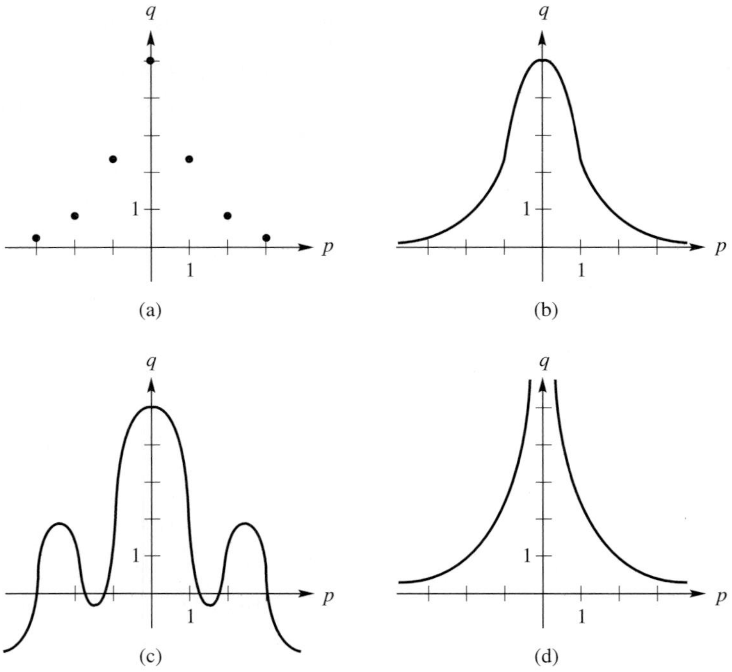

Figure A–7

Until you study equations like this one in Chapter 8, you must guess the overall shape of the graph. A good rule to follow is to sketch as simple a graph as you can, while including all of your plotted points. In particular, don't include any bends, corners, or jumps, unless your plotted points indicate that they are there. The most reasonable guess is shown in Figure A-7b, rather than Figure A-7c or A-7d. (You may need to plot a few more points to convince yourself that the graph never crosses the p-axis.) ■

EXERCISES *In Exercises* 1–4, *decide which points are on the graph of the equation.*

1. $y = x + 5$; $(-1, 4), (8, 3), (5, 10)$

2. $y = x^2 - 3x$ $(0, 0), (4, 4), (8, 8)$

3. $x^2 + y^2 = 25$ $(3, 2), (3, 4), (5, 0)$

4. $s = 2t$ $(2, 4), (3, 6), (100, 200)$

(Regard the first number in each ordered pair as a value of t.)

Sketch the graph of each equation in Exercises 5–12.

5. $y = x - 4$

6. $y = x^2 - 4$

7. $y = 3 - 2x$

8. $y = 3 - 2x^2$

9. $y = x^2 - 4x$

10. $y = x^2 - 4x + 4$

11. $y = x^3$

12. $y = x^3 - 2$

13. Sketch the graph of $z = 3w + 1$:

a. by making the w-axis horizontal

b. by making the z-axis horizontal

14. Sketch the graph of $v = \dfrac{u - 5}{2}$:

a. by making the u-axis horizontal

b. by making the v-axis horizontal

A–6 GRAPHING LINEAR EQUATIONS

An equation that can be written in the form

$$Ax + By = C$$

with A and B not both 0 is a **linear equation in two variables.**

EXAMPLE 1 *Deciding whether an equation in two variables is linear*

a. $y = 3x - 4$ is a linear equation in two variables, since it is equivalent to $3x - y = 4$.

b. $q - 5 = 2(p + 1)$ is a linear equation in two variables, since it is equivalent to $2p - q = -7$.

c. $x = -8$ is a linear equation in two variables, since it is equivalent to $x + 0y = -8$.

d. $x^2 + y^2 = 1$ is not a linear equation in two variables. ■

There are several techniques for graphing linear equations in two variables. Let's review two of the most common ones, the two-point method and the slope-intercept method.

THE TWO-POINT METHOD

You can graph a linear equation by first plotting any two points on the graph, then drawing the line through the plotted points. The **intercepts** (points where the graph intersects the coordinate axes) are convenient points to plot, especially if the equation is written in the form $Ax + By = C$. It is a good idea to plot a third point to check the accuracy of your work.

EXAMPLE 2 *Using the two-point method to graph a linear equation*

To graph

$$2x - 3y = 8$$

begin by plotting the intercepts. Since the x-intercept must have a y-coordinate of 0, you can find it by substituting 0 for y in the equation.

$$2x - 3(0) = 8$$
$$2x = 8$$
$$x = 4$$

Thus the x-intercept is at $(4, 0)$. Similarly, you can find the y-intercept at $\left(0, -\dfrac{8}{3}\right)$ by substituting 0 for x in the equation. The intercepts are shown in

Figure A-8a, along with the point (7, 2), obtained by substituting 7 for x. The graph is shown in Figure A-8b.

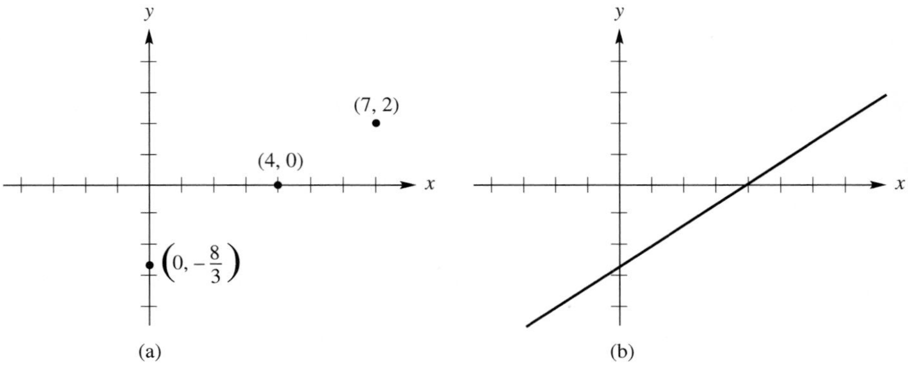

Figure A–8

EXAMPLE 3 *Using the two-point method to graph a linear equation*

To graph

$$2x - 3y = 0$$

begin by plotting the intercepts. This time, both intercepts are at (0, 0), and thus provide only one point on the graph.

To obtain a second point, substitute any value except 0 for one of the variables. For example, substituting 6 for x yields

$$2(6) - 3y = 0$$
$$12 = 3y$$
$$y = 4$$

Thus, a second point on the graph is (6, 4). Similarly, substituting -3 for x yields the point $(-3, -2)$. The graph is shown in Figure A-9.

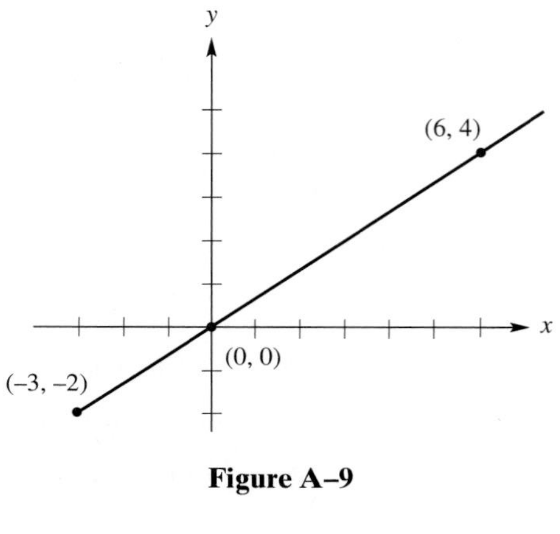

Figure A–9

THE SLOPE-INTERCEPT METHOD

An equation $Ax + By = C$ with $B \neq 0$ can be solved for y to obtain an equation $y = mx + b$. The coefficients m and b provide information about the graph of such an equation.

- The graph of $y = mx + b$ is a nonvertical straight line.
- The value of m is equal to the ratio

$$\frac{y_2 - y_1}{x_2 - x_1}$$

for any pair of points (x_1, y_1) and (x_2, y_2) on the line. This ratio is the **slope** of the line.

- The y-intercept is $(0, b)$.

When a linear equation is written $y = mx + b$, it is said to be in **slope-intercept form**. The following example shows how to use the coefficients m and b to graph the equation.

EXAMPLE 4 *Using the slope-intercept method to graph a linear equation*

To graph

$$y = 5 - 2x$$

begin by rewriting the equation as

$$y = -2x + 5$$

so that $m = -2$ and $b = 5$.

Next, plot the y-intercept $(0, 5)$, as in Figure A-10.

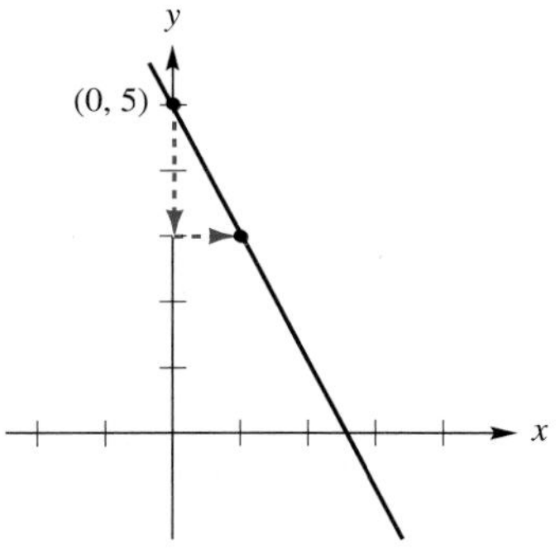

Figure A–10

The slope $-2 = \frac{-2}{1}$ represents the ratio of the change in y to the change in x between any two points on the line. You can therefore obtain a second point by starting at $(0, 5)$ and letting y and x change by -2 and 1, respectively. The second point is 2 units down from $(0, 5)$, and 1 unit to the right, as in Figure A-10. As in the two-point method, you can complete the graph by drawing the line through the two plotted points. ■

HORIZONTAL AND VERTICAL LINES

Equations of the form $x =$ (constant) and $y =$ (constant) graph as vertical and horizontal lines, respectively. The slope of a horizontal line is 0, and the slope of a vertical line is undefined.

Horizontal and vertical lines are graphed most easily by remembering the geometric meanings of their equations.

EXAMPLE 5 *Graphing horizontal and vertical lines*

a. To graph

$$y = 3$$

recall that a point's y-coordinate is its directed vertical distance from the x-axis. Thus, the equation describes all points 3 units above the x-axis. Its graph is shown in Figure A-11a.

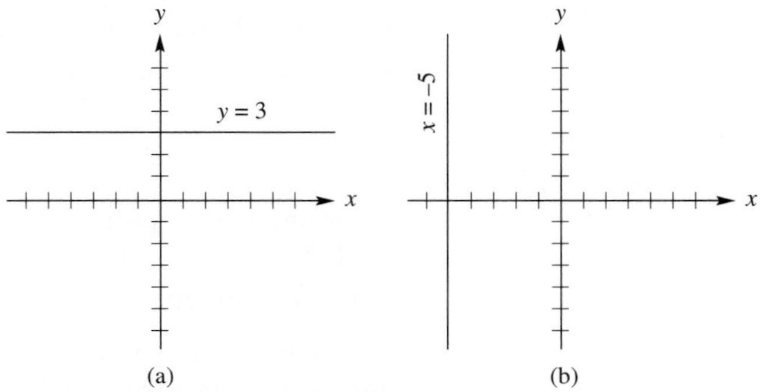

Figure A–11

b. Similarly, the equation

$$x = -5$$

describes all points 5 units to the left of the y-axis. Its graph is shown in Figure A-11b. ■

FINDING THE EQUATION OF A LINE

If a line with slope m passes through (x_1, y_1), then another point (x, y) is on the line if and only if

$$\frac{y - y_1}{x - x_1} = m$$

Multiplying both sides by $x - x_1$ produces the **point-slope form** of the equation

$$y - y_1 = m(x - x_1)$$

The following example shows how to find the equation of a line if you know either its slope and the coordinates of one of its points or the coordinates of two of its points.

EXAMPLE 6 *Finding the equation of a line*

a. The equation of the line through $(-2, 4)$ with slope 3 is

$$y - 4 = 3[x - (-2)]$$

$$y - 4 = 3(x + 2)$$

in point-slope form. Its slope-intercept form is

$$y = 3x + 10$$

b. The line through $(-1, 5)$ and $(3, 7)$ has a slope of

$$\frac{7 - 5}{3 - (-1)} = \frac{1}{2}$$

Using $(-1, 5)$ as (x_1, y_1), we find that the equation in point-slope form is

$$y - 5 = \frac{1}{2}(x + 1)$$

The slope-intercept form is

$$y = \frac{1}{2}x + \frac{11}{2}$$

[If we use $(3, 7)$ as (x_1, y_1), the point-slope equation $y - 7 = \frac{1}{2}(x - 3)$ also simplifies to $y = \frac{1}{2}x + \frac{11}{2}$.] ■

EXERCISES *In Exercises* 1–10, *sketch the graph of the equation.*

1. $4x + y = 12$ **2.** $x - 3y = 6$

3. $2x + 5y = 10$ **4.** $2x + 5y = 0$

5. $y = 3x + 2$ **6.** $y = 2 - x$

7. $y = -5x$ **8.** $x = 2y - 8$

9. $x = 2$ **10.** $y = 0$

In Exercises 11–14, *find the equation of the line through the given point with the given slope.*

11. $(2, -2), m = 1$ **12.** $(5, 4), m = -3$

13. $(-3, -5), m = \frac{2}{3}$ **14.** $(-1, 4), m = 0$

In Exercises 15–18, *find the equation of the line through the given points.*

15. $(6, -1)$ and $(8, 7)$ **16.** $(-2, 9)$ and $(10, 5)$

17. $(1, -1)$ and $(0, 0)$ **18.** $(3, 0)$ and $(3, 4)$

A–7 INTERVALS

If $a < b$, the set of all numbers between a and b is an **interval**. The set of all numbers to the right of a, and the set of all numbers to the left of b, are also intervals. The intervals may or may not include their **endpoints** a and b.

Every interval can be described by an inequality. For example, the set of all numbers between -2 and 3 can be described by one of the inequalities $-2 < x < 3$, $-2 < x \leqslant 3$, $-2 \leqslant x < 3$, or $-2 \leqslant x \leqslant 3$, depending on whether the interval includes either or both of the endpoints -2 and 3.

Interval notation provides a concise way of describing intervals. Table A-1 (page 406) shows the graph of each type of interval, the inequality that describes it, and the corresponding description in interval notation.

TABLE A-1

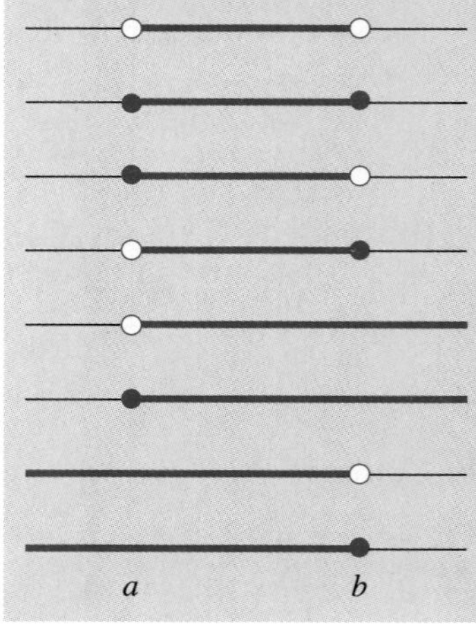

Graph	Inequality	Interval notation
◦——————◦	$a < x < b$	(a, b)
●——————●	$a \leq x \leq b$	$[a, b]$
●——————◦	$a \leq x < b$	$[a, b)$
◦——————●	$a < x \leq b$	$(a, b]$
◦————————	$a < x$	(a, ∞)
●————————	$a \leq x$	$[a, \infty)$
————————◦	$x < b$	$(-\infty, b)$
————————●	$x \leq b$	$(-\infty, b]$

a b

The first four types of intervals are **finite**, and the last four are **semi-infinite**. The interval $(-\infty, \infty)$, which includes all real numbers, is **infinite**. A finite interval which contains both its endpoints is **closed**, and one which contains neither endpoint is **open**.

We often need to describe sets of numbers made up of two or more disjoint intervals. For example, Figure A-12 shows the set of all numbers whose absolute values are between 1 and 2, inclusive.

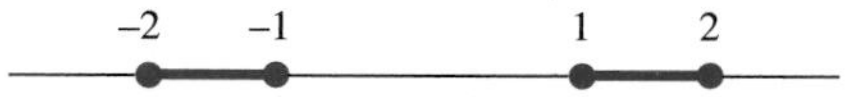

Figure A–12

This set is the **union** of the intervals $[-2, -1]$ and $[1, 2]$ and is written in interval notation as $[-2, -1] \cup [1, 2]$. Similarly, the set of all real numbers except π, shown in Figure A-13, can be written as $(-\infty, \pi) \cup (\pi, \infty)$.

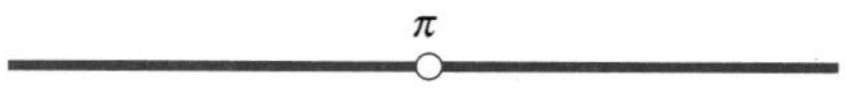

Figure A–13

EXERCISES *In Exercises 1–8, graph the interval described by the given inequality, and describe it in interval notation.*

1. $1 < x \leq 5$ **2.** $-3 \leq x < 0$

3. $-40 < x < 40$ **4.** $4.9 \leq x \leq 5.1$

5. $x < \sqrt{2}$ **6.** $x \geq -10$

7. $x \leq 0$ **8.** $x > 0$

In Exercises 9–16, graph the interval whose interval notation description is given, and describe it by an inequality.

9. $(-2, 2)$

10. $[-2, 2]$

11. $[0, 1.5)$

12. $[1, \infty)$

13. $(-\infty, 1)$

14. $(-10.4, -4.7]$

15. $\left[-\frac{1}{2}, \frac{2}{3}\right)$

16. $\left(-\frac{1}{2}, \frac{2}{3}\right]$

In Exercises 17–24, use both an inequality and an interval notation to describe the interval whose graph is shown.

17. 

18. 

19. 

20. 

21. 

22. 

23. 

24. 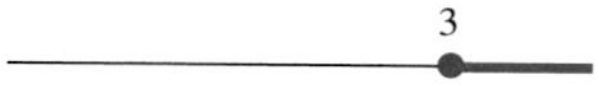

In Exercises 25–28, graph the set of numbers described.

25. $(-4, 1) \cup [3, 5]$

26. $[2, 6) \cup (6, \infty)$

27. $(-\infty, -5] \cup (-\sqrt{3}, \sqrt{3})$

28. $[0, 1] \cup (2, 3) \cup [4, 5]$

In Exercises 29–32, use interval notation to describe the set of numbers whose graph is shown.

29. 

30. 

31. 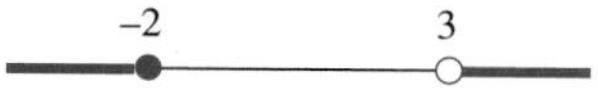

32. 

A–8 LINEAR INEQUALITIES

An **inequality** is a statement having one of the following forms.

$a < b$	a is less than b
$a > b$	a is greater than b
$a \leqslant b$	a is less than or equal to b
$a \geqslant b$	a is greater than or equal to b

Like equations, inequalities can be either true or false. For example,

$$x < 5$$

is true for some values of x, such as 1, and false for others, such as 7.

A **solution** of an inequality is a number that produces a true statement when substituted for the variable. Inequalities typically have infinitely many solutions. These can be represented graphically, as in Figure A-14.

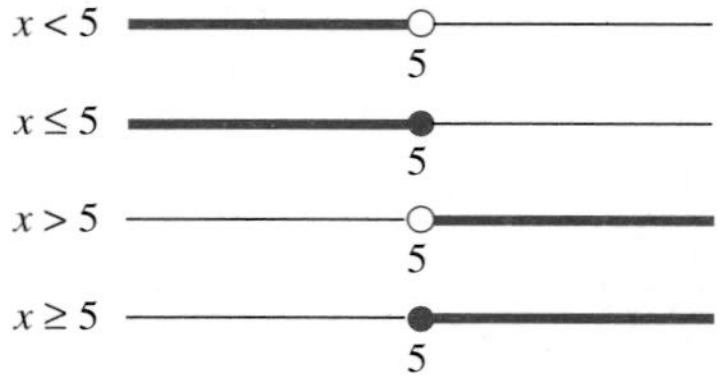

Figure A–14

Two inequalities are **equivalent** if they have the same solutions. If you perform any of the following operations on an inequality, you always produce an equivalent inequality.

- Replace quantities by equal quantities anywhere in the inequality.
- Add or subtract equal quantities on each side of the inequality.
- Multiply or divide each side of the inequality by equal *positive* quantities.
- Multiply or divide each side of the inequality by equal *negative* quantities, *and reverse the direction of the inequality.*

Some care must be taken in applying the last two rules. For example,

$3x < -15$ is equivalent to $x < -5$.
(Dividing both sides by 3 does not reverse the direction of the inequality.)

$-3x < -15$ is equivalent to $x > 5$.
(Dividing both sides by -3 reverses the direction of the inequality.)

An inequality in one variable, say x, is called **linear** if the corresponding equation in x is linear. Thus $3x + 20 \geqslant 7x$ is linear, while $x^2 \leqslant 9$ is not. You can always solve a linear inequality in one variable by performing only those operations listed above. The solution can be written in one of the forms $x < a, x > a, x \leqslant a$, or $x \geqslant a$.

EXAMPLE 1 *Solving a linear inequality in one variable*

Here is one sequence of steps which can be used to solve

$$8x - 2 > 2(x + 5)$$

Replace $2(x + 5)$ by $2x + 10$.

$$8x - 2 > 2x + 10$$

Add 2 to both sides.

$$8x > 2x + 12$$

Subtract $2x$ from both sides.

$$6x > 12$$

Divide both sides by 6.

$$x > 2$$

The solution can also be represented in interval notation as $(2, \infty)$ or represented graphically, as in Figure A-15.

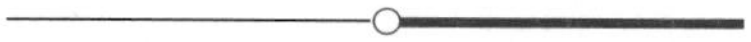

Figure A–15 ■

EXERCISES *In Exercises 1–8, solve the inequality. Represent the solution as an inequality, as a graph, and in interval notation.*

1. $9 - 4x < 33$

2. $q - 13 \leqslant 5q + 7$

3. $5(t - 4) \geqslant -(7 - 2t)$

4. $0.01z > 7.86 - 2.61z$

5. $\dfrac{x}{2} + 4 \leqslant \dfrac{x}{3}$

6. $325(1 - y) < 497y$

7. $8w + 5 > 5$

8. $\sqrt{3}\,p \geqslant \dfrac{1}{\sqrt{3}}$

A–9 ABSOLUTE VALUE EQUATIONS AND INEQUALITIES

The **absolute value** of a real number x, written $|x|$, is defined as follows.

$$\text{If } x \geqslant 0, \text{ then } |x| = x.$$

$$\text{If } x < 0, \text{ then } |x| = -x.$$

According to the definition, $|x|$ is always nonnegative. For example, $|3| = 3$, and $|-4| = 4$. Thus, the absolute value of a number is found by simply eliminating the negative sign, if one is there. However, this simple rule does not apply to expressions involving more than one number or involving variables.

- It is *not* true that $|6 - 8| = 6 + 8$, because the left side is $|6 - 8| = |-2| = 2$ and the right side is $6 + 8 = 14$.
- It is *not always* true that $|-x| = x$. For example, if $x = -5$, then the left side is $|-(-5)| = |5| = 5$, and the right side is -5.

Geometrically, $|x|$ represents the undirected distance of x from 0 on the number line. Figure A-16 shows that $|-5| = 5$ and $|5| = 5$.

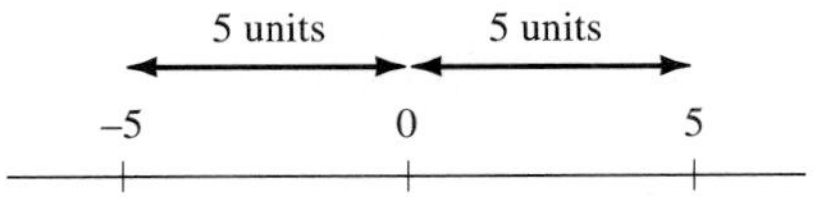

Figure A–16

For any constant a, the equation $|x| = a$ and the inequalities $|x| < a$, $|x| \leqslant a$, $|x| > a$, $|x| \geqslant a$ can all be written as equivalent statements without absolute values.

EXAMPLE 1 *Rewriting absolute value equations and inequalities without absolute values*

a. The equation

$$|x| = 5$$

says that x is 5 units from 0 on the number line and is therefore equivalent to the statement

$$x = -5 \quad \text{or} \quad x = 5$$

b. The inequality

$$|x| < 5$$

says that x is less than 5 units from 0. Figure A-17 shows that this is true if and only if x is in $(-5, 5)$, so the inequality is equivalent to the statement

$$-5 < x < 5$$

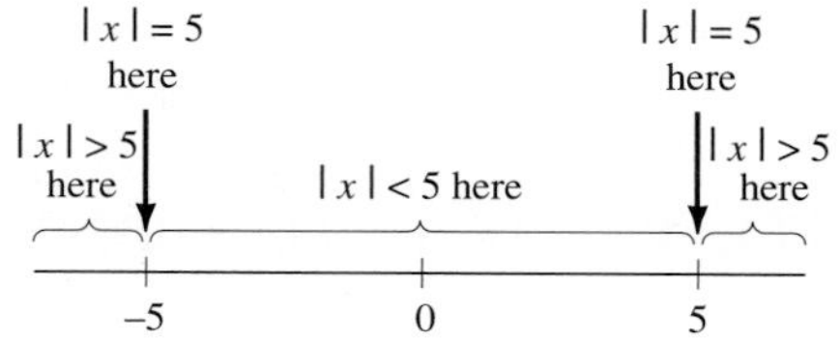

Figure A–17

c. The inequality

$$|x| > 5$$

says that x is more than 5 units from 0. Figure A-17 shows that this is true if and only if x is in $(-\infty, -5) \cup (5, \infty)$, so the inequality is equivalent to the statement

$$x < -5 \quad \text{or} \quad x > 5$$ ■

You can use the observations in Example 1 to solve equations and inequalities involving absolute values of linear expressions in x.

EXAMPLE 2 *Solving a linear absolute value equation*

The equation

$$|3 - 2x| = 5$$

is equivalent to the statement

$$3 - 2x = -5 \quad \text{or} \quad 3 - 2x = 5$$

Solving each equation yields

$$-2x = -8 \quad \text{or} \quad -2x = 2$$

$$x = 4 \quad \text{or} \quad x = -1$$

The equation has two solutions, namely, -1 and 4. ■

EXAMPLE 3 *Solving a linear absolute value inequality*

The inequality

$$|3 - 2x| < 5$$

is equivalent to the statement

$$-5 < 3 - 2x < 5$$

Solving yields

$$-8 < -2x < 2$$

$$4 > x > -1$$

Figure A–18

The solution can also be represented in interval notation as $(-1, 4)$ or graphically as in Figure A-18. ■

EXAMPLE 4 *Solving a linear absolute value inequality*

The inequality

$$|3 - 2x| \geqslant 5$$

is equivalent to the statement

$$3 - 2x \leqslant -5 \quad \text{or} \quad 3 - 2x \geqslant 5$$

Solving each equation yields

$$-2x \leqslant -8 \quad \text{or} \quad -2x \geqslant 2$$

$$x \geqslant 4 \quad \text{or} \quad x \leqslant -1$$

The solution can also be represented in interval notation as $(-\infty, -1] \cup [4, \infty)$ or graphically as in Figure A-19.

Figure A–19 ■

EXAMPLE 5 *Solving linear absolute value equations and inequalities*

a. To solve

$$|4x + 7| = -2$$

notice that the left side is an absolute value and cannot be equal to -2. Thus, the equation has no solution.

b. To solve

$$|2x - 6| > 0$$

notice that the left side must be positive unless the quantity inside the absolute value signs is 0. The inequality is true unless $2x - 6 = 0$, that is, unless $x = 3$. The solution is therefore $(-\infty, 3) \cup (3, \infty)$. ■

EXERCISES *In Exercises* 1–8,

a. Solve the equation.

b. Replace $=$ by $\leqslant$, and solve the inequality.

c. Replace $=$ by $>$, and solve the inequality.

1. $|2x| = 14$

2. $|x + 6| = 0.05$

3. $|2x - 13| = 3$

4. $|13 - 2x| = 3$

5. $|3x + 5| = 13$ **6.** $|5 - 4x| = 7$

7. $|x + 3| = -1$ **8.** $|x + 3| = 0$

A-10 SYSTEMS OF LINEAR EQUATIONS

A **linear equation in *n* variables**, say $x_1, x_2, \ldots, x_n$, is one having the form

$$a_1x_1 + a_2x_2 + \cdots + a_nx_n = c$$

A **system of linear equations** is a set of one or more linear equations to be solved simultaneously. A **solution** to the system is an ordered set of n real numbers that makes every equation in the system true when substituted for $x_1, \ldots, x_n$.

A system of m linear equations in n variables is an $m \times n$ **linear system**. Let's review some commonly used methods for solving 2×2 and 3×3 linear systems.

SOLVING 2 × 2 LINEAR SYSTEMS

Two analytical methods of solving 2×2 linear systems are the **method of substitution** and the **method of elimination** (also called the **Gaussian elimination**). Let's review each with an example.

EXAMPLE 1 *Solving a* 2×2 *linear system by the method of substitution*

To solve the system

$$\begin{aligned} 2x + 5y &= 3 \\ 3x - 4y &= 16 \end{aligned}$$

by the method of substitution, begin by solving either equation for either of the variables. For example, solve the first equation for x.

$$\begin{aligned} 2x &= 3 - 5y \\ x &= \frac{3 - 5y}{2} \end{aligned}$$

Then substitute the resulting expression for x in the other equation.

$$\begin{aligned} 3x - 4y &= 16 \\ 3\left(\frac{3 - 5y}{2}\right) - 4y &= 16 \end{aligned}$$

Next, solve this equation for y.

$$\begin{aligned} 3(3 - 5y) - 8y &= 32 \\ 9 - 15y - 8y &= 32 \\ -23y &= 23 \\ y &= -1 \end{aligned}$$

Finally, replace y by -1 in the expression for x.

$$x = \frac{3 - 5y}{2}$$

$$x = \frac{3 - 5(-1)}{2}$$

$$x = 4$$

The system has one solution, $(x, y) = (4, -1)$. ■

EXAMPLE 2 *Solving a 2×2 linear system by the method of elimination*

To solve

$$\begin{aligned} 2x + 5y &= 3 \\ 3x - 4y &= 16 \end{aligned}$$

by the method of elimination, begin by multiplying each equation by a constant, so that in the resulting equations, the coefficients of one variable are either equal, or opposites of each other. For example, multiply the first equation by 4 and the second by 5.

$$\begin{aligned} 8x + 20y &= 12 \\ 15x - 20y &= 80 \end{aligned}$$

Then add corresponding sides of the two equations, and solve for x.

$$\begin{array}{r} 8x + 20y = 12 \\ 15x - 20y = 80 \\ \hline 23x \qquad = 92 \end{array}$$

$$x \qquad = 4$$

Finally, replace x by 4 in either of the original equations, and solve for y.

$$\begin{aligned} 2x + 5y &= 3 \\ 2(4) + 5y &= 3 \\ 5y &= -5 \\ y &= -1 \end{aligned}$$

The solution is $(4, -1)$. ■

SOLVING 3×3 LINEAR SYSTEMS

The methods of substitution and elimination can also be used to solve 3×3 linear systems.

EXAMPLE 3 *Solving a 3×3 linear system by the method of substitution*

To solve

$$\begin{aligned} x + 2y - z &= 7 \\ 2x + y + 2z &= 5 \\ x - y - 2z &= -2 \end{aligned}$$

using the method of substitution, begin by solving any equation for any variable. For example, solve the last equation for x.

$$x - y - 2z = -2$$

$$x = y + 2z - 2$$

Then substitute the resulting expression for x in each of the other equations.

$$\begin{aligned} x + 2y - z &= 7 \\ 2x + y + 2z &= 5 \end{aligned}$$

$$\begin{aligned} (y + 2z - 2) + 2y - z &= 7 \\ 2(y + 2z - 2) + y + 2z &= 5 \end{aligned}$$

$$\begin{aligned} 3y + z &= 9 \\ 3y + 6z &= 9 \end{aligned}$$

Next, solve the resulting 2×2 system. Subtracting corresponding sides of the two equations yields

$$\begin{aligned} 3y + z &= 9 \\ \underline{3y + 6z} &\underline{= 9} \\ -5z &= 0 \\ z &= 0 \end{aligned}$$

Replacing z by 0 in either equation yields

$$3y = 9$$

$$y = 3$$

Finally, replace y by 3 and z by 0 in the expression for x.

$$\begin{aligned} x &= y + 2z - 2 \\ x &= (3) + 2(0) - 2 \\ x &= 1 \end{aligned}$$

The solution is $(x, y, z) = (1, 3, 0)$. ■

EXAMPLE 4 *Solving a* 3×3 *linear system by the method of elimination*

To solve

$$\begin{aligned} x + 2y - z &= 7 \\ 2x + y + 2z &= 5 \\ x - y - 2z &= -2 \end{aligned}$$

using the method of elimination, begin by choosing any two equations, and multiply each by a constant to eliminate any variable. For example, choose the first two equations, multiply the first by 2 and subtract to eliminate x.

$$\begin{aligned} x + 2y - z &= 7 \\ 2x + y + 2z &= 5 \end{aligned}$$

$$\begin{array}{r} 2x + 4y - 2z = 14 \\ 2x + y + 2z = 5 \\ \hline 3y - 4z = 9 \end{array}$$

Then choose a different pair of equations, and eliminate the same variable. For example, eliminate x from the first and third equations by subtracting them.

$$\begin{array}{r} x + 2y - z = 7 \\ x - y - 2z = -2 \\ \hline 3y + z = 9 \end{array}$$

Next, solve the system

$$3y - 4z = 9$$

$$3y + z = 9$$

to obtain $y = 3, z = 0$. Finally, replace y by 3 and z by 0 in any of the original equations, and solve for x.

$$x + 2y - z = 7$$

$$x + 2(3) - (0) = 7$$

$$x = 1$$

The solution is the ordered triple $(x, y, z) = (1, 3, 0)$. ■

EXERCISES *In Exercises* 1–16, *solve the linear system.*

1. $\begin{aligned} 2x + 5y &= 9 \\ 2x - 5y &= -1 \end{aligned}$

2. $\begin{aligned} s + 4t &= 12 \\ 2s + t &= 10 \end{aligned}$

3. $\begin{aligned} 4x + 3y &= 6 \\ 5x + 4y &= 2 \end{aligned}$

4. $\begin{aligned} 3x - 7y &= 0.3 \\ 2x - 8y &= 1.5 \end{aligned}$

5. $\begin{aligned} 5x - 3y &= 21 \\ y &= x - 5 \end{aligned}$

6. $\begin{aligned} q &= 6p - 9 \\ q &= 4p + 11 \end{aligned}$

7. $\begin{aligned} x &= 8 - 3y \\ y &= 8 - 3x \end{aligned}$

8. $\begin{aligned} 42x - 79y &= 0 \\ 35x + 27y &= 0 \end{aligned}$

9. $\begin{aligned} x + y + z &= 0 \\ 2x + y - z &= -4 \\ -x + 4y + z &= 7 \end{aligned}$

10. $\begin{aligned} 3x - 2y + 4z &= 7 \\ x + 5y - 2z &= 24 \\ 4x + 3y + z &= 29 \end{aligned}$

11. $\begin{aligned} 2x - 3y + z &= 1 \\ 2x + y - 6z &= 2 \\ 2y + z &= 0 \end{aligned}$

12. $\begin{aligned} x - y &= 1 \\ y - z &= 2 \\ x + z &= 5 \end{aligned}$

13. $\begin{aligned} a + b - 3c &= 0 \\ a - 2b + c &= 0 \\ 2a - b - c &= 0 \end{aligned}$

14. $\begin{aligned} a + b - 3c &= 0 \\ a - 2b + c &= 0 \\ 2a - b - c &= 3 \end{aligned}$

15. $\begin{aligned} x - 2y - 3z &= 4 \\ x - 3y - 4z &= 0 \\ x - 4y - 6z &= 3 \end{aligned}$

16. $\begin{aligned} 7x - 6y + 2z &= 5 \\ 4x + 3y - 2z &= 5 \\ 2x - 3y - 3z &= 1 \end{aligned}$

A–11 THE LAWS OF EXPONENTS

In the expression x^p, the number p is an **exponent**. Its meaning is defined as follows.

- If p is a positive integer, then x^p is shorthand for the product $x \cdot x \cdot \cdots \cdot x$, with p factors. For example,

$$3^4 \text{ means } 3 \cdot 3 \cdot 3 \cdot 3 = 81$$

  and

$$x^3y^2 \text{ means } x \cdot x \cdot x \cdot y \cdot y$$

- If $x \neq 0$, then $x^0 = 1$. The expression 0^0 is undefined.
- If p is a positive integer, then $x^{-p} = \dfrac{1}{x^p}$. For example,

$$3^{-4} = \frac{1}{3^4} = \frac{1}{81}$$

  and

$$x^3y^{-2} = x^3\left(\frac{1}{y^2}\right) = \frac{x^3}{y^2}$$

- If p is a rational number $\dfrac{r}{s}$ in lowest terms, then $x^p = x^{r/s}$ is defined to be either $\sqrt[s]{x^r}$ or the equivalent expression $\left(\sqrt[s]{x}\right)^r$, as long as $\sqrt[s]{x}$ is defined. For example,

$$4^{3/2} = \sqrt{4^3} = \sqrt{64} = 8$$

  Equivalently,

$$4^{3/2} = \left(\sqrt{4}\right)^3 = 2^3 = 8$$

$$4^{-3/2} = \frac{1}{4^{3/2}} = \frac{1}{8}$$

$$(-4)^{3/2} = \left(\sqrt{-4}\right)^3$$

  which is not defined as a real number.

When expressions involving exponents are combined, they obey the following **laws of exponents**.

1. $x^px^q = x^{p+q}$

2. $\dfrac{x^p}{x^q} = x^{p-q}$

3. $(xy)^p = x^py^p$

4. $\left(\dfrac{x}{y}\right)^p = \dfrac{x^p}{y^p}$

5. $(x^p)^q = x^{pq}$

In words, the laws say

- To multiply (divide) powers of the same base, add (subtract) exponents (laws 1 and 2).

- Exponents distribute over products and quotients (laws 3 and 4).
- To raise a power to a power, multiply exponents (law 5).

EXAMPLE 1 *Applying the laws of exponents*

a. $x^3(x^4 - 2x^2 + 1)$

$= x^3x^4 - 2x^3x^2 + x^3(1)$

$= x^7 - 2x^5 + x^3$

b. $\dfrac{k}{k^3} = k^{1-3} = k^{-2}$ or $\dfrac{1}{k^2}$

c. $(5y)^3 = 5^3y^3 = 125y^3$

d. $\left(\dfrac{5}{y}\right)^{-3} = \left(\dfrac{1}{5/y}\right)^3 = \left(\dfrac{y}{5}\right)^3 = \dfrac{y^3}{5^3} = \dfrac{y^3}{125}$

e. $(9t^3)^{1/2} = 9^{1/2}(t^3)^{1/2} = \sqrt{9}\,t^{3/2} = 3t^{3/2}$ ■

Although exponents distribute over products and quotients, they do not distribute over sums and differences. For example, it is *not* correct to say that $(u^2 + v^3)^2 = u^4 + v^6$. A correct calculation is

$$\begin{aligned}(u^2 + v^3)^2 &= (u^2 + v^3)(u^2 + v^3)\\ &= u^2u^2 + v^3u^2 + u^2v^3 + v^3v^3\\ &= u^4 + 2u^2v^3 + v^6\end{aligned}$$

EXERCISES *In Exercises 1–8, evaluate without using a calculator.*

1. 2^5 **2.** 6^{-2}

3. $(-1)^6$ **4.** $(-3)^{-3}$

5. 19^0 **6.** $27^{2/3}$

7. $16^{-3/4}$ **8.** $\left(\dfrac{9}{4}\right)^{-1/2}$

In Exercises 9–20, combine powers of the same base where possible, and eliminate parentheses and negative exponents.

9. k^3k^{-2} **10.** $(k^3)^{-2}$

11. $(4b^5)(2b^2)$ **12.** $(4b^5)^2$

13. $\dfrac{3z^7}{12z^3}$ **14.** $\left(\dfrac{w^2}{2w^5}\right)^3$

15. $(2xy^4)^2$ **16.** $(2x - y^4)^2$

17. $c^{-3}(c^4 + 3c^3 - 5c)$ **18.** $\left(\dfrac{x^3}{7y^2}\right)^{-2}$

19. $v^{1/2}v^{1/3}$ **20.** $(64a^{1/2})^{2/3}$

A-12 FACTORING

A **polynomial** is a sum of terms, each of which is a product of a number and a nonnegative integer power of a variable. A few examples of polynomials are

$$x^4y^3 - 4x^2y$$

$$3p - 6q + r$$

$$7$$

Polynomials with one, two, and three terms are called, respectively, **monomials**, **binomials**, and **trinomials**. Thus 7 is a monomial, $x^4y^3 - 4x^2y$ is a binomial, and $3p - 6q + r$ is a trinomial. There are no commonly used names for polynomials with four or more terms.

The **degree** of a term in a polynomial is the number of its variable factors. For example, $2x^3 = 2 \cdot x \cdot x \cdot x$ has degree 3, and the term $4x^2y = 4 \cdot x \cdot x \cdot y$ also has degree 3. The degree of a polynomial is defined as the highest degree of any of its terms, so the three polynomials in the last paragraph have degrees, 0, 7, and 1, respectively.

To **factor** a polynomial is to express it as a product of one or more polynomial factors. Here are factorizations of several polynomials. You can check the correctness of each by performing the multiplications on the right side.

Polynomial	Factorization
$2x^3 + 8x^2$	$2x^2(x + 4)$
$x^2 - 4y^2$	$(x + 2y)(x - 2y)$
$2p^2 - 5p + 2$	$(p - 2)(2p - 1)$
$3z^5 + 3z^4 - 12z - 12$	$3(z + 1)(z^2 + 2)(z^2 - 2)$

A polynomial is **factored completely** if it is expressed as a product of polynomials that cannot be factored again. If the original polynomial has integer coefficients, then the phrase "cannot be factored again" is usually understood to mean "cannot be factored again using only integer coefficients." For example, it is not customary to write $2p - 1$ as $2\left(p - \frac{1}{2}\right)$ or to write $z^2 - 2$ as $\left(z + \sqrt{2}\right)\left(z - \sqrt{2}\right)$. Both of these factorizations are correct, and you will perform similar ones in Chapters 5 and 7, but in the present section, we will use only integer coefficients.

Let's review a few common factoring techniques.

REMOVAL OF COMMON FACTORS

If all terms in a polynomial have a factor in common, you can use the distributive law to write the polynomial as a product of the common factor and another polynomial. The process is referred to as the **removal of common factors**.

EXAMPLE 1 *Removing common factors*

a. $6x^2 + 12x - 3 = 3(2x^2 + 4x - 1)$

b. $x^2y^3z^6 - x^4y^3z^4 + x^2y^5z^4 = x^2y^3z^4(z^2 - x^2 + y^2)$

c. $-2x + 6 = 2(-x + 3)$ or

$-2x + 6 = -2(x - 3)$

Each factorization in part (c) is correct, and each is a complete factorization of $-2x + 6$. They differ by the removal of a common factor of -1. A factor of -1 can be removed from any completely factored polynomial to produce an equally correct complete factorization. ■

Before applying any other factoring techniques to a polynomial, it is good practice to remove all common factors first.

TECHNIQUES FOR FACTORING BINOMIALS

The following formulas are useful for factoring binomials.

$$A^2 - B^2 = (A - B)(A + B)$$
$$A^3 - B^3 = (A - B)(A^2 + AB + B^2)$$
$$A^3 + B^3 = (A + B)(A^2 - AB + B^2)$$

These are the formulas for a **difference of squares**, **difference of cubes**, and **sum of cubes**. You can check the correctness of each by performing the multiplication on the right side. In applying the formulas, you may replace A and B by any polynomial.

EXAMPLE 2 *Factoring binomials*

a. To factor

$$4r^2 - 9s^2$$

think of the polynomial as

$$(2r)^2 - (3s)^2$$

and apply the formula for the difference of squares, replacing A by $2r$ and B by $3s$.

$$A^2 - B^2 = (A - B)(A + B)$$
$$(2r)^2 - (3s)^2 = (2r - 3s)(2r + 3s)$$

b. To factor

$$2z^6 - 54$$

remove common factors first to obtain

$$2(z^6 - 27)$$

Then think of $z^6 - 27$ as $(z^2)^3 - 3^3$, and apply the formula for the difference of cubes, replacing A by z^2 and B by 3.

$$\begin{aligned} A^3 - B^3 &= (A - B)(A^2 + AB + B^2) \\ (z^2)^3 - 3^3 &= (z^2 - 3)[(z^2)^2 + (z^2)(3) + 3^2] \\ &= (z^2 - 3)(z^4 + 3z^2 + 9) \end{aligned}$$

The complete factorization is

$$2z^6 - 54 = 2(z^2 - 3)(z^4 + 3z^2 + 9)$$

c. To factor

$$x^9 + 1$$

think of the polynomial as

$$(x^3)^3 + 1^3$$

and apply the formula for the sum of cubes, replacing A by x^3 and B by 1.

$$\begin{aligned} A^3 + B^3 &= (A + B)(A^2 - AB + B^2) \\ (x^3)^3 + 1^3 &= (x^3 + 1)[(x^3)^2 - (x^3)(1) + 1^2] \\ &= (x^3 + 1)(x^6 - x^3 + 1) \end{aligned}$$

The first factor can be factored again, using the sum of cubes formula and replacing A by x and B by 1.

$$\begin{aligned} &(x^3 + 1)(x^6 - x^3 + 1) \\ &= (x + 1)(x^2 - x + 1)(x^6 - x^3 + 1) \end{aligned}$$

■

TECHNIQUES FOR FACTORING TRINOMIALS

The most common trinomials have the form $ax^2 + bx + c$ and can sometimes be factored as

$$ax^2 + bx + c = (a_1x + s_1)(a_2x + s_2)$$

If the trinomial is **monic** ($a = 1$), then multiplying on the right confirms that $a_1a_2 = 1$, $s_1 + s_2 = b$, and $s_1s_2 = c$. These observations often allow you to factor it quickly.

EXAMPLE 3 *Factoring monic quadratic trinomials*

a. If it is possible to factor

$$x^2 + 3x - 40$$

the factorization must have the form $(x + s_1)(x + s_2)$, with $s_1 + s_2 = 3$ and $s_1s_2 = -40$. A little trial and error produces $s_1 = 8$, $s_2 = -5$ (or vice versa), so

$$x^2 + 3x - 40 = (x + 8)(x - 5)$$

b. If it is possible to factor

$$x^2 + 2x - 40$$

the factorization must have the form $(x + s_1)(x + s_2)$, with $s_1 + s_2 = 2$ and $s_1 s_2 = -40$. The only pairs of integers with a product of -40 and a positive sum are

$$-1 \text{ and } 40 \text{ (sum} = 39)$$
$$-2 \text{ and } 20 \text{ (sum} = 18)$$
$$-4 \text{ and } 10 \text{ (sum} = 6)$$
$$-5 \text{ and } 8 \text{ (sum} = 3)$$

Since no pair has a sum of 2, it is impossible to factor $x^2 + 2x - 40$ using only integer coefficients. ■

If the trinomial is not monic, a modification of the foregoing procedure can be used, as in the following example.

EXAMPLE 4 *Factoring nonmonic quadratic trinomials*

To factor a quadratic trinomial $ax^2 + bx + c$, first express the product ac as a product of two integers with a sum of b, if possible. For the particular polynomial

$$6x^2 - 19x + 15$$

this means expressing $(6)(15) = 90$ as a product of two integers with a sum of -19. The two integers are -9 and -10.

Second, rewrite the term bx using the two integers just obtained. In this case, the trinomial is rewritten as

$$6x^2 - 9x - 10x + 15$$

Third, group the four resulting terms into two pairs, and remove common factors from each pair.

$$(6x^2 - 9x) + (-10x + 15) = 3x(2x - 3) - 5(2x - 3)$$

Fourth, think of the polynomial in its present form as having two terms, $3x(2x - 3)$ and $-5(2x - 3)$, and remove the common factor $(2x - 3)$.

$$3x(2x - 3) - 5(2x - 3) = (2x - 3)(3x - 5)$$ ■

TECHNIQUES FOR FACTORING POLYNOMIALS WITH FOUR TERMS

Polynomials with four terms can sometimes be factored by the process of **grouping**, so named because it begins by factoring groups of terms within the polynomial. The technique was illustrated in the example you just read. The following example illustrates further.

EXAMPLE 5 *Factoring by grouping*

a. To factor

$$x^3 + x^2 - x - 1$$

group the terms in pairs, and remove common factors from each pair.

$$(x^3 + x^2) + (-x - 1) = x^2(x + 1) - (x + 1)$$

Then remove a common factor of $x + 1$ from the two resulting terms.

$$x^2(x + 1) - (x + 1) = (x + 1)(x^2 - 1)$$

Finally, factor $x^2 - 1$ using the formula for a difference of squares.

$$\begin{aligned}(x + 1)(x^2 - 1) &= (x + 1)(x - 1)(x + 1)\\ &= (x - 1)(x + 1)^2\end{aligned}$$

b. To factor

$$p^3q - p^2q - pq^3 + pq^2$$

begin by removing common factors to obtain

$$pq(p^2 - p - q^2 + q)$$

Next, group the terms inside parentheses in pairs and remove common factors from each pair.

$$pq[(p^2 - p) + (-q^2 + q)] = pq[p(p - 1) - q(q - 1)]$$

At this point, the two terms $p(p - 1)$ and $q(q - 1)$ have no common factors, so you cannot continue. Nor can you leave the polynomial in its present form, since the polynomial in the brackets is not factored. Try a different grouping.

$$pq[(p^2 - q^2) + (-p + q)] = pq[(p - q)(p + q) - (p - q)]$$

Here $p^2 - q^2$ has been factored using the formula for a difference of squares. Now continue by removing a common factor of $(p - q)$.

$$\begin{aligned}pq[(p - q)(p + q) - (p - q)] &= pq(p - q)[(p + q) - 1]\\ &= pq(p - q)(p + q - 1)\end{aligned}$$

■

EXERCISES *In Exercises* 1–20, *factor the polynomial, if possible:*

1. $x^2yz + xy^2z + xyz^2$

2. $36m^2 - 54m$

3. $c^2 - 25d^2$

4. $c^2 + 25d^2$

5. $r^2 + 10r + 21$

6. $5z^3 - 40$

7. $x^2 - x - 30$

8. $10 + 3z - z^2$

9. $3x^2 + 8x - 11$

10. $ac + ad + bc + bd$

11. $2p^2 - 5p + 2$

12. $4_{q^2} - 17q + 4$

13. $x^4y + 27xy^4$

14. $s^2 + 7s - 44$

15. $ax^3 - 2a^2x^2 + 5a^3x$

16. $ax^3 - 6a^2x^2 + 5a^3x$

17. $x^4 - 16$

18. $y^4 - 3y^3 + 4y^2 - 12y$

19. $q^6 - 1$

20. $x^4 - 10x^2 + 9$ (*Hint*: Regard this as a quadratic trinomial in the variable x^2.)

A–13 QUADRATIC EQUATIONS

A **quadratic equation** in x is one that can be written in the form $ax^2 + bx + c = 0$ for some real numbers a, b, and c.

TECHNIQUES FOR SOLVING QUADRATIC EQUATIONS

Factoring If $ax^2 + bx + c$ can be factored, the equation $ax^2 + bx + c = 0$ can be solved using the Principle of Zero Products. This principle states that a product is equal to zero if and only if one of the factors is zero.

EXAMPLE 1 *Solving a quadratic equation by factoring*

You can solve

$$2x^2 + 3x - 2 = 0$$

by factoring the left-hand side to obtain

$$(2x - 1)(x + 2) = 0$$

By the Principle of Zero Products, this equation is equivalent to the statement

$$2x - 1 = 0 \quad \text{or} \quad x + 2 = 0$$

The solutions are $x = \frac{1}{2}$ and $x = -2$. ■

The Principle of Zero Products can be used to solve a quadratic equation only when one side is zero. To solve

$$(x - 1)(x + 4) = -6,$$

it is *not* appropriate to solve $x - 1 = -6$ and $x + 4 = -6$. You must obtain zero on one side before factoring, as follows.

$$x^2 + 3x - 4 = -6$$

$$x^2 + 3x + 2 = 0$$

$$(x + 1)(x + 2) = 0$$

$$x = -1 \quad \text{or} \quad x = -2$$

Completing the Square Every expression $ax^2 + bx + c$ can be written in the form $a(x - h)^2 + k$ for some real numbers h and k. This process, called **completing the square**, can be used to solve quadratic equations.

EXAMPLE 2 *Completing the square for a quadratic expression*

To complete the square for $2x^2 + 3x - 2$,

- First, group the first and second degree terms together, and factor out the coefficient of x^2.

$$2\left(x^2 + \frac{3}{2}x\right) - 2$$

- Second, take half of the x-coefficient $\left(\text{half of } \frac{3}{2} \text{ is } \frac{3}{4}\right)$ and square it $\left[\left(\frac{3}{4}\right)^2 = \frac{9}{16}\right]$. Add and subtract this quantity inside the parentheses.

$$2\left(x^2 + \frac{3}{2}x + \frac{9}{16} - \frac{9}{16}\right) - 2$$

- Third, remove the last term from the parentheses, remembering to multiply it by the factor in front.

$$2\left(x^2 + \frac{3}{2}x + \frac{9}{16}\right) - \frac{9}{8} - 2$$

- Fourth, express the quantity inside the parentheses as the square of a binomial, and combine the constants outside the parentheses.

$$2\left(x + \frac{3}{4}\right)^2 - \frac{25}{8}$$

■

EXAMPLE 3 *Solving a quadratic equation by completing the square*

To solve

$$2x^2 + 3x - 2 = 0$$

begin by completing the square as in the last example, to obtain

$$2\left(x + \frac{3}{4}\right)^2 - \frac{25}{8} = 0$$

Continue by isolating the binomial square on one side of the equation.

$$\left(x + \frac{3}{4}\right)^2 = \frac{25}{16}$$

Then take square roots on both sides.

$$x + \frac{3}{4} = \pm\frac{5}{4}$$

Thus

$$x = -\frac{3}{4} + \frac{5}{4} = \frac{1}{2} \quad \text{or} \quad x = -\frac{3}{4} - \frac{5}{4} = -2$$

■

The Quadratic Formula Every equation $ax^2 + bx + c = 0$ can be solved by using the **quadratic formula**, which states that the solutions are

$$x = \frac{-b \pm \sqrt{b^2 - 4ac}}{2a}$$

EXAMPLE 4 *Solving a quadratic equation by using the quadratic formula*

To solve

$$2x^2 + 3x - 2 = 0$$

replace a, b, and c by 2, 3, and -2 in the quadratic formula.

$$x = \frac{-3 \pm \sqrt{3^2 - 4(2)(-2)}}{2(2)}$$

$$x = \frac{-3 \pm 5}{4}$$

$$x = \frac{1}{2} \quad \text{or} \quad x = -2$$

■

COMPLEX SOLUTIONS

If i is the imaginary number $\sqrt{-1}$, then any number $a + bi$ with a and b real is a **complex number**. A complex number is real if $b = 0$ and nonreal if $b \neq 0$. The solutions to some quadratic equations are nonreal complex numbers.

EXAMPLE 5 *A quadratic equation with nonreal complex solutions*

You can solve

$$x^2 - 6x + 13 = 0$$

either by completing the square or by using the quadratic formula. Either method yields the solutions

$$x = 3 \pm \sqrt{-4}$$

which can be rewritten as

$$x = 3 \pm 2i$$

■

THE NATURE OF THE SOLUTIONS OF A QUADRATIC EQUATION

The solutions of $ax^2 + bx + c = 0$ can be described in one of three ways, depending on whether the expression $b^2 - 4ac$ in the quadratic formula is positive, zero, or negative. The three possibilities are summarized in Table A-2.

TABLE A-2

Value of $b^2 - 4ac$	Nature of the solutions
Positive	2 real
Zero	1 real
Negative	2 nonreal

Because the expression $b^2 - 4ac$ *discriminates* among the three possibilities, it is called the **discriminant** of the expression $ax^2 + bx + c$.

EXAMPLE 6 *Information provided by the discriminant*

The discriminant of $x^2 + 4x + 3$ is

$$4^2 - 4(1)(3) = 4$$

so the equation $x^2 + 4x + 3 = 0$ has 2 real solutions. (Check this by solving to obtain $x = -3$ and $x = -1$.)

The discriminant of $x^2 + 4x + 4$ is

$$4^2 - 4(1)(4) = 0$$

so the equation $x^2 + 4x + 4 = 0$ has 1 real solution. (Check this by solving to obtain $x = -2$.)

The discriminant of $x^2 + 4x + 5$ is

$$4^2 - 4(1)(5) = -4$$

so the equation $x^2 + 4x + 5 = 0$ has 2 nonreal solutions. (Check this by solving to obtain $x = -2 \pm i$.) ■

EXERCISES *In Exercises* 1–6, *solve by factoring. In Exercises* 7–12, *solve by completing the square. In Exercises* 13–18, *solve by using the quadratic formula. In Exercises* 19–24, *solve by any method.*

1. $x^2 + 3x - 28 = 0$

2. $x^2 - 5x - 66 = 0$

3. $-4x^2 + x + 14 = 0$

4. $-2x^2 + 7x = -15$

5. $x(x - 6) = 27$

6. $(x - 9)(x + 13) = -117$

7. $x^2 + 5x - 9 = 0$

8. $2x^2 - 3x + 8 = 0$

9. $-x^2 + 8x - 14 = 0$

10. $x^2 - 36 = -5x$

11. $(2x - 1)^2 = 9$

12. $(x - 3)(x - 4) = 20$

13. $x^2 + 6x - 15 = 0$

14. $1 - 4x + 2x^2 = 0$

15. $7x^2 + 5 = 0$

16. $(1 - 2x)(3 + 4x) = 5$

17. $100x^2 - 20x + 1 = 0$

18. $x^2 - x - 1 = 0$

19. $x^2 + 7x + 6 = 0$

20. $x^2 - 5x + 4 = 0$

21. $3x^2 + 7x - 10 = 0$

22. $x^2 = 90 - x$

23. $x(x + 7) = 60$

24. $8 + 6x - 5x^2 = 0$

In Exercises 25–30, *use the discriminant to determine the nature of the solutions. Check by solving, using a method of your choice.*

25. $6x^2 - 10x + 8 = 0$

26. $4x^2 + 25 = 0$

27. $(x - 7)^2 - 79 = 0$

28. $3x^2 = 17x$

29. $4x^2 - 20x + 25 = 0$

30. $(2x + 3)(2x - 3) = -9$

A-14 OPERATIONS WITH COMPLEX NUMBERS

If i is the imaginary number $\sqrt{-1}$, then any number $a + bi$ with a and b real is a **complex number**. Thus the set of complex numbers includes all real numbers, such as -7, $\frac{\sqrt{3}}{2}$, and π, as well as numbers with an imaginary part, such as $3i$ and $1 - \sqrt{2}\,i$.

Operations on complex numbers are performed by treating i as an ordinary algebraic symbol, except that i^2 can be replaced by -1. The following example illustrates addition, subtraction, and multiplication of complex numbers.

EXAMPLE 1 *Adding, subtracting, and multiplying complex numbers*

a. $(5 + 7i) + (4 - i) = (5 + 4) + (7i - i) = 9 + 6i$

b. $(5 + 7i) - (4 - i) = (5 - 4) + [7i - (-i)] = 1 + 8i$

c. $(5 + 7i)(4 - i) = 20 - 5i + 28i - 7i^2$

$$= 20 + 23i - 7i^2$$

$$= 20 + 23i - 7(-1)$$

$$= 27 + 23i$$

■

To divide complex numbers, and express the result in the form $a + bi$, you need to make use of the fact that when a complex number $x + yi$ is multiplied by its **conjugate** $x - yi$, the result is always real. Specifically,

$$\begin{aligned}(x + yi)(x - yi) &= x^2 - xyi + xyi - y^2i^2 \\ &= x^2 - y^2(-1) \\ &= x^2 + y^2\end{aligned}$$

The following example illustrates the use of conjugates in the division of complex numbers.

EXAMPLE 2 *Dividing complex numbers*

$$\begin{aligned}\frac{5 + 7i}{4 - i} &= \frac{5 + 7i}{4 - i} \cdot \frac{4 + i}{4 + i} \\ &= \frac{20 + 5i + 28i + 7i^2}{16 + 4i - 4i - i^2} \\ &= \frac{20 + 33i + 7(-1)}{16 - (-1)} \\ &= \frac{13 + 33i}{17} \\ &= \frac{13}{17} + \frac{33}{17}i\end{aligned}$$

■

By using the fact that $i^2 = -1$, you can express any integer power of i as either ± 1 or $\pm i$.

$$\begin{aligned}
i^1 & & &= i \\
i^2 & & &= -1 \\
i^3 &= i^2 i = (-1)i & &= -i \\
i^4 &= (i^2)^2 = (-1)^2 & &= 1 \\
i^5 &= i^4 i = (1)(i) & &= i \\
i^6 &= i^4 i^2 = (1)(-1) & &= -1 \\
i^7 &= i^4 i^3 = (1)(-i) & &= -i \\
i^8 &= i^4 i^4 = (1)(1) & &= 1
\end{aligned}$$

and so on, repeating in cycles of four.

EXERCISES *In Exercises 1–16, perform the indicated operations and express the result in the form $a + bi$ with a and b real.*

1. $(6 + 2i) + (-4 + 5i)$
2. $-7i - (3 + 8i)$
3. $(5 + \sqrt{2}i) - (5 - \sqrt{2}i)$
4. $(5 + \sqrt{2}\,i) + (5 - \sqrt{2}\,i)$
5. $(6 + 2i)(-4 + 5i)$
6. $-7i(3 + 8i)$
7. $(5 + \sqrt{2}i)(5 - \sqrt{2}i)$
8. $(5 + \sqrt{2}\,i)^2$
9. $\dfrac{6 + 2i}{-4 + 5i}$
10. $\dfrac{-7i}{3 + 8i}$
11. $\dfrac{3 + 8i}{-7i}$
12. $\dfrac{5 + \sqrt{2}\,i}{5 - \sqrt{2}\,i}$
13. $(1 + i)^3$
14. $2i(2 - 3i)(1 + 2i)$
15. i^{100}
16. i^{99}

A–15 DIVISION AND SYNTHETIC DIVISION OF POLYNOMIALS

Division of one polynomial by another produces a quotient that can be expressed in the form

$$Q(x) + \frac{N(x)}{D(x)}$$

where Q, N, and D are polynomials and N has a smaller degree than D. The rational expression $\dfrac{N(x)}{D(x)}$ is the **remainder** of the division.

DIVISION OF MONOMIALS

You can divide a polynomial by a monomial term by term.

EXAMPLE 1 *Dividing a polynomial by a monomial*

a. $$\frac{6x^5 + 8x - 1}{2x} = \frac{6x^5}{2x} + \frac{8x}{2x} - \frac{1}{2x}$$

$$= 3x^4 + 4 - \frac{1}{2x}$$

The polynomial part of the quotient is $3x^4 + 4$, and the remainder is $-\dfrac{1}{2x}$.

b. $$\frac{6x^5 + 8x - 1}{2x^2} = \frac{6x^5}{2x^2} + \frac{8x - 1}{2x^2}$$

$$= 3x^3 + \frac{8x - 1}{2x^2}$$

The polynomial part of the quotient is $3x^3$, and the remainder is $\dfrac{8x - 1}{2x^2}$. ■

DIVISION BY OTHER POLYNOMIALS

When the divisor is not a monomial, polynomial division requires several steps, much like long division of numbers. The following example illustrates the process.

EXAMPLE 2 *Dividing polynomials*

To divide $3x^4 - 4x^3 + 7x - 1$ by $x^2 - 2$, begin by writing the terms of the divisor and dividend in descending order of degree.

$$
\begin{array}{r|l}
 & \\ \hline
x^2 - 2 & 3x^4 - 4x^3 \qquad + 7x - 1
\end{array}
$$

The dividend contains no term involving x^2, but your work will have better visual organization if you leave a space where that term would be.

STEP 1: Divide the term of highest degree in the dividend by the term of highest degree in the divisor. Enter the result as the first term in the x^2 column of the quotient.

$$
\begin{array}{r|l}
 & \qquad\qquad 3x^2 \\ \hline
x^2 - 2 & 3x^4 - 4x^3 \qquad + 7x - 1
\end{array}
$$

STEP 2: Multiply the resulting term in the quotient by the divisor. Enter the result below the dividend, keeping terms with the same degree in the same column.

$$
\begin{array}{r|l}
 & \qquad\qquad 3x^2 \\ \hline
x^2 - 2 & 3x^4 - 4x^3 \qquad + 7x - 1
\end{array}
$$

STEP 3: Subtract.

$$
\begin{array}{r|l}
 & \qquad\qquad 3x^2 \\ \hline
x^2 - 2 & 3x^4 - 4x^3 \qquad + 7x - 1 \\
 & 3x^4 \qquad - 6x^2 \\ \cline{2-2}
\end{array}
$$

STEP 4: Regard the bottom line as a new dividend, and repeat steps 1–3.

$$
\begin{array}{r|l}
 & \qquad\qquad 3x^2 - 4x \\ \hline
x^2 - 2 & 3x^4 - 4x^3 \qquad + 7x - 1 \\
 & 3x^4 \qquad - 6x^2 \\ \cline{2-2}
 & \quad - 4x^3 + 6x^2 + 7x - 1 \\
 & \quad - 4x^3 \qquad + 8x \\ \cline{2-2}
 & \qquad\qquad 6x^2 - x - 1
\end{array}
$$

Continue to repeat steps 1–3 as long as the degree of the dividend is at least as large as that of the divisor.

$$
\begin{array}{r|l}
 & 3x^2 - 4x + 6 \\
\cline{2-2}
x^2 - 2 & 3x^4 - 4x^3 + 7x - 1 \\
 & 3x^4 - 6x^2 \\
\cline{2-2}
 & - 4x^3 + 6x^2 + 7x - 1 \\
 & - 4x^3 + 8x \\
\cline{2-2}
 & 6x^2 - x - 1 \\
 & 6x^2 - 12 \\
\cline{2-2}
 & - x + 11
\end{array}
$$

The final result can be written as

$$3x^2 - 4x + 6 + \frac{-x + 11}{x^2 - 2}$$

■

SYNTHETIC DIVISION

When the divisor has the form $x - a$, the long division process just described can be replaced by a streamlined process called **synthetic division**.

EXAMPLE 3 *Synthetic division of polynomials*

To divide $3x^4 - 4x^3 + 7x - 1$ by $x - 2$, begin with the following arrangement of numbers.

$$
\begin{array}{r|rrrrr}
2 & 3 & -4 & 0 & 7 & -1 \\
 & & & & & \\
\hline
\end{array}
$$

The number to the left of the vertical bar is the value of x for which the divisor is zero. The numbers to the right are the coefficients of the terms in the dividend, in order of decreasing degree.

STEP 1: Copy the first coefficient in the dividend below the horizontal bar.

$$
\begin{array}{r|rrrrr}
2 & 3 & -4 & 0 & 7 & -1 \\
 & & & & & \\
\hline
 & 3 & & & &
\end{array}
$$

STEP 2: Multiply that number by the number to left of the vertical bar, and write the result in the next column to the right.

$$
\begin{array}{r|rrrrr}
2 & 3 & -4 & 0 & 7 & -1 \\
 & & 6 & & & \\
\hline
 & 3 & & & &
\end{array}
$$

STEP 3: Add the numbers in the second column.

$$
\begin{array}{r|rrrrr}
2 & 3 & -4 & 0 & 7 & -1 \\
 & & 6 & & & \\
\hline
 & 3 & 2 & & &
\end{array}
$$

STEP 4: Repeat steps 2–3 until addition has been performed in all columns.

$$\begin{array}{r|rrrrr} 2 & 3 & -4 & 0 & 7 & -1 \\ & & 6 & 4 & 8 & 30 \\ \hline & 3 & 2 & 4 & 15 & 29 \end{array}$$

The quotient is represented synthetically by the numbers below the horizontal bar. The rightmost number is the numerator of the remainder. The others, from left to right, are the coefficients of the polynomial quotient, in order of decreasing degree. (The degree of the polynomial quotient is always one less than the degree of the dividend.) Thus the result of this division is

$$3x^3 + 2x^2 + 4x + 15 + \frac{29}{x - 2}$$

■

EXERCISES *Perform the divisions in Exercises 1–12. Use synthetic division if the divisor has the form $x - a$.*

1. $\dfrac{x^6 - 4x^5 - 7x^3}{2x^3}$

2. $\dfrac{3x^2 + 5}{x^2}$

3. $\dfrac{2x^3 - 7x + 8}{x^2}$

4. $\dfrac{2x^3 - 7x + 8}{x^2 + 1}$

5. $\dfrac{x^7 + 4x^5 - x^2 + 3}{x^3 - 2x + 1}$

6. $\dfrac{x^5 - 6x^3 + x}{x^4 - 1}$

7. $\dfrac{x^3 - 4x^2 + 3x - 5}{x - 4}$

8. $\dfrac{2x^4 - 4x^2 + 3}{x + 2}$

9. $\dfrac{2x^2 - 5x + 2}{x - \frac{1}{2}}$

10. $\dfrac{x^5 + 1}{x + 1}$

11. $\dfrac{x^3 + 3x^2 - 8x + 1}{x^3 + 3x^2 + 4x - 5}$

12. $\dfrac{x^6}{x - 3}$

A–16 ALGEBRAIC FRACTIONS

An **algebraic fraction** is a quotient of two algebraic expressions. A **rational expression** is an algebraic fraction whose numerator and denominator are polynomials. Thus

$$\frac{3y^4 + 6y^2 + 3}{y^2 - 1} \quad \text{and} \quad \frac{12}{x}$$

are rational expressions, while

$$\frac{a - 4}{\sqrt{a} - 2}$$

is not, because its denominator is not a polynomial. All, however, are algebraic fractions. This section will review operations with rational expressions, but the rules given apply to all algebraic fractions.

EQUIVALENT FRACTIONS

Two algebraic fractions $\frac{a}{b}$ and $\frac{c}{d}$ are **equivalent** if and only if

- $ad = bc$

  and

- the equations $b = 0$ and $d = 0$ have the same solutions.

EXAMPLE 1 *Equivalent algebraic fractions*

a. The algebraic fractions

$$\frac{x + 3}{x + 2} \quad \text{and} \quad \frac{4x + 12}{4x + 8}$$

are equivalent, because

- Both $(x + 3)(4x + 8)$ and $(4x + 12)(x + 2)$ are equal to $4x^2 + 20x + 24$.
- The equations $x + 2 = 0$ and $4x + 8 = 0$ each have the single solution $x = -2$.

b. The fractions

$$\frac{x + 3}{x + 2} \quad \text{and} \quad \frac{(x + 3)(x - 5)}{(x + 2)(x - 5)}$$

are not equivalent if x is allowed to assume all real values, because the equations $x + 2 = 0$ and $(x + 2)(x - 5) = 0$ do not have the same solutions. However, if 5 is excluded as a replacement value for x, the two fractions are equivalent. ■

In performing operations on an algebraic fraction such as $\frac{x + 3}{x + 2}$, it may be convenient to write it as $\frac{(x + 3)(x - 5)}{(x + 2)(x - 5)}$, as though the two are equivalent. You will see why in Example 5. However, it is well to remember that the substitution is valid only if the values of x are restricted as in Example 1.

In the remainder of this section, we will assume that the values of all variables are restricted so that no denominators are zero.

REDUCTION OF FRACTIONS

The fractions $\frac{a}{b}$ and $\frac{ca}{cb}$ are equivalent (as long as $c \neq 0$), so that $\frac{ca}{cb}$ can be reduced by canceling a factor of c from both numerator and denominator.

EXAMPLE 2 *Reduction of algebraic fractions*

a. To reduce

$$\frac{x^2 - 9}{x^2 - 4x + 3}$$

begin by factoring the numerator and denominator.

$$\frac{(x + 3)(x - 3)}{(x - 1)(x - 3)}$$

Cancel a factor of $x - 3$ to obtain

$$\frac{x + 3}{x - 1}$$

b. To reduce

$$\frac{12 - 2x}{x^2 - 8x + 12}$$

begin by factoring

$$\frac{2(6 - x)}{(x - 2)(x - 6)}$$

Then recognize that $6 - x$ and $x - 6$ are opposites, so that $2(6 - x) = -2(x - 6)$. Thus the fraction can be reduced by rewriting it as

$$\frac{-2(x - 6)}{(x - 2)(x - 6)}$$

and canceling to obtain

$$\frac{-2}{x - 2} \quad \text{or} \quad \frac{2}{2 - x}$$

c. To reduce

$$\frac{4x - 2(x + 1)}{(x - 1)(x + 1)}$$

begin by multiplying, combining terms, and factoring the numerator.

$$\frac{4x - 2x - 2}{(x - 1)(x + 1)} = \frac{2x - 2}{(x - 1)(x + 1)}$$

$$= \frac{2(x - 1)}{(x - 1)(x + 1)}$$

Cancel to obtain

$$\frac{2}{x + 1}$$ ■

Fractions can be reduced only by canceling *factors* of the numerator and denominator. Thus in part (c) of the foregoing example, you should not make the mistake of attempting to cancel the $x + 1$ immediately, because $x + 1$ is not a factor of the numerator.

MULTIPLICATION AND DIVISION OF FRACTIONS

Multiplication Like numeric fractions, algebraic fractions are multiplied "straight across." That is, $\frac{a}{b} \cdot \frac{c}{d} = \frac{ac}{bd}$. For this reason, a factor in either numerator can be canceled with the same factor in either denominator.

EXAMPLE 3 *Multiplication of algebraic fractions*

$$\frac{x^3 - 8}{5x + 20} \cdot \frac{x^2 - 16}{x^3 - 6x^2 + 8x}$$
$$= \frac{(x - 2)(x^2 + 2x + 4)}{5(x + 4)} \cdot \frac{(x + 4)(x - 4)}{x(x - 4)(x - 2)}$$
$$= \frac{x^2 + 2x + 4}{5x}$$

■

Division Dividing by a fraction is equivalent to multiplying by its inverse its reciprocal). That is, $\frac{a}{b} \div \frac{c}{d} = \frac{a}{b} \cdot \frac{d}{c} = \frac{ad}{bc}$.

EXAMPLE 4 *Division of algebraic fractions*

$$\frac{a^3y - ay^3}{ay - 4a} \div \frac{y^4 - a^4}{y^2 - 4y}$$
$$= \frac{a^3y - ay^3}{ay - 4a} \cdot \frac{y^2 - 4y}{y^4 - a^4}$$
$$= \frac{ay(a + y)(a - y)}{a(y - 4)} \cdot \frac{y(y - 4)}{(y^2 + a^2)(y + a)(y - a)}$$
$$= \frac{-y^2}{y^2 + a^2}$$

[To cancel, regard $(a - y)$ as $(-1)(y - a)$.] ■

COMMON DENOMINATORS

Finding a common denominator for two algebraic fractions is the first step in performing addition or subtraction. The process is illustrated below.

EXAMPLE 5 *Finding a common denominator for two algebraic fractions*

a. To find a common denominator for

$$\frac{x + 3}{x + 2} \quad \text{and} \quad \frac{x - 1}{x - 5}$$

begin by noticing that any common denominator must contain both a factor of $x + 2$ and a factor of $x - 5$. The **least common denominator**, that is, the one containing the necessary factors and no others, is $(x + 2)(x - 5)$. The fractions equivalent to the given two, and having the required denominator, are

$$\frac{(x + 3)(x - 5)}{(x + 2)(x - 5)} \quad \text{and} \quad \frac{(x + 2)(x - 1)}{(x + 2)(x - 5)}$$

b. To find a common denominator for

$$\frac{1}{x^2 + 3x} \quad \text{and} \quad \frac{1}{x^2 - 4x}$$

begin by factoring both denominators to obtain

$$\frac{1}{x(x + 3)} \quad \text{and} \quad \frac{1}{x(x - 4)}$$

A common denominator must contain factors of $x, x + 3$ and $x - 4$. It need not contain a second factor of x. Thus, the least common denominator is $x(x + 3)(x - 4)$. The fractions equivalent to the given two, and having the required denominator, are

$$\frac{x - 4}{x(x + 3)(x - 4)} \quad \text{and} \quad \frac{x + 3}{x(x + 3)(x - 4)} \qquad ■$$

ADDITION AND SUBTRACTION OF FRACTIONS

To add or subtract algebraic fractions, you must first rewrite them with a common denominator. The numerators can then be combined. That is,

$$\frac{a}{b} \pm \frac{c}{d} = \frac{ad}{bd} \pm \frac{bc}{bd} = \frac{ad \pm bc}{bd}.$$

EXAMPLE 6 *Addition and subtraction of algebraic fractions*

a. To add

$$\frac{x + 3}{x + 2} +$$

begin by finding a common denominator, as was done in the last example. The addition can then be rewritten as

$$\frac{(x + 3)(x - 5)}{(x + 2)(x - 5)} + \frac{(x + 2)(x - 1)}{(x + 2)(x - 5)}$$

The two fractions can now be added by adding their numerators.

$$\frac{(x + 3)(x - 5) + (x + 2)(x - 1)}{(x + 2)(x - 5)}$$

$$= \frac{(x^2 - 2x - 15) + (x^2 + x - 2)}{(x + 2)(x - 5)}$$

$$= \frac{2x^2 - x - 17}{(x + 2)(x - 5)}$$

b. To subtract

$$\frac{1}{x^2 + 3x} - \frac{1}{x^2 - 4x}$$

begin by finding a common denominator, as was done in the last example. The subtraction can be rewritten as

$$\frac{x - 4}{x(x + 3)(x - 4)} - \frac{x + 3}{x(x + 3)(x - 4)}$$

The two fractions can now be subtracted by subtracting their numerators.

$$\frac{(x - 4) - (x + 3)}{x(x + 3)(x - 4)} = \frac{-7}{x(x + 3)(x - 4)}$$

■

SIMPLIFICATION OF COMPLEX FRACTIONS

A **complex fraction** is one whose numerator and/or denominator are themselves fractions. A complex fraction can be written as a simple fraction by either of two commonly used techniques. The first makes use of the fact that a fraction denotes division, so that

$$\frac{\frac{a}{b}}{\frac{c}{d}} \quad \text{means} \quad \frac{a}{b} \div \frac{c}{d}$$

EXAMPLE 7 *Simplifying a complex fraction*

$$\frac{\frac{x}{x + 1}}{\frac{x^2}{x^2 - 1}} \quad \text{means} \quad \frac{x}{x + 1} \div \frac{x^2}{x^2 - 1}$$

$$= \frac{x}{x + 1} \cdot \frac{x^2 - 1}{x^2}$$

$$= \frac{x}{x + 1} \cdot \frac{(x + 1)(x - 1)}{x^2}$$

$$= \frac{x - 1}{x}$$

■

A second technique, illustrated by the following example, is often effective when the numerator or denominator of a complex fraction is itself the sum or difference of two or more fractions.

EXAMPLE 8 *Simplifying a complex fraction*

To simplify

$$\frac{\frac{2}{x} - \frac{2}{y}}{\frac{4}{x^2} - \frac{4}{y^2}}$$

begin by finding a common denominator for all the terms in the numerator and denominator. The least common denominator is x^2y^2.

Next, multiply the complex fraction by another fraction whose numerator and denominator are both $\frac{x^2y^2}{1}$.

$$\frac{\frac{2}{x} - \frac{2}{y}}{\frac{4}{x^2} - \frac{4}{y^2}} \cdot \frac{\frac{x^2y^2}{1}}{\frac{x^2y^2}{1}}$$

$$= \frac{\frac{2}{x}\frac{x^2y^2}{1} - \frac{2}{y}\frac{x^2y^2}{1}}{\frac{4}{x^2}\frac{x^2y^2}{1} - \frac{4}{y^2}\frac{x^2y^2}{1}}$$

$$= \frac{2xy^2 - 2x^2y}{4y^2 - 4x^2}$$

$$= \frac{2xy(y - x)}{4(y + x)(y - x)}$$

$$= \frac{xy}{2(y + x)}$$

■

EXERCISES *In Exercises* 1–20, *perform all operations and simplify.*

1. $\frac{4x + 20}{6x - 54} \cdot \frac{x - 9}{x + 5}$

2. $\frac{6}{x^2 - 4} \cdot \frac{3x + 6}{2x}$

3. $\frac{x - 2}{x + 3} \cdot \frac{x^2 - 9}{5x - 10} \cdot \frac{5}{x - 3}$

4. $\frac{x^2 + 4x + 4}{3x^4y^3} \cdot \frac{x^5y^2}{(2x + 4)}$

5. $\frac{x}{2x - 3} \div \frac{2x}{2x^2 - x - 3}$

6. $\frac{7}{x^2 + 2x + 4} \div \frac{14}{8 - x^3}$

7. $\frac{2x - 3}{x^2} \cdot \frac{x}{x - 4} \div (2x^2 - 5x - 12)$

8. $\frac{x^3 + 1}{1 - x^2} \div \frac{2x^2 - 2x + 2}{x + 1}$

9. $\frac{z}{2} + \frac{z}{3}$

10. $\frac{7}{x} - \frac{1}{x - 7}$

11. $\frac{a + 3}{a^2 - a} + \frac{a + 4}{a^2 - 1}$

12. $\frac{2q - 1}{8 - q} - \frac{3q + 7}{q - 8}$

13. $\frac{4x^2}{x - 5} - 4x$

14. $\frac{2x}{x - 6} + \frac{12}{6 - x}$

15. $\frac{3}{y^3 + 2y^2} + \frac{y - 1}{y^3 + 4y^2 + 4y}$

16. $\frac{x^2}{(x - 1)(x - 2)} - \frac{x}{(x - 2)(x - 3)} + \frac{1}{(x - 1)(x - 3)}$

17. $\dfrac{\dfrac{3x}{5y}}{\dfrac{12x}{15y}}$

18. $\dfrac{\dfrac{2t}{t-1}}{t^2-4t+3}$

19. $\dfrac{\dfrac{3}{x}+\dfrac{1}{2}}{\dfrac{3}{x}-\dfrac{1}{2}}$

20. $\dfrac{\dfrac{1}{x+2}-\dfrac{x}{x+5}}{\dfrac{x}{x+2}-\dfrac{1}{x+5}}$

A-17 EQUATIONS WITH ALGEBRAIC FRACTIONS

In solving an equation with algebraic fractions, the first step is often to multiply both sides by the least common denominator of all fractions in the equation. The resulting equation has no algebraic fractions, but may not be equivalent to the original one. Therefore, you need to check all solutions in the *original* equation.

EXAMPLE 1 *Solving equations with algebraic fractions*

a. To solve

$$\frac{3x+1}{6x-3}=\frac{5}{12x-6}+\frac{7}{2}$$

begin by factoring all denominators to obtain

$$\frac{3x+1}{3(2x-1)}=\frac{5}{6(2x-1)}+\frac{7}{2}$$

Next, find a common denominator for all fractional expressions in the equation. The least common denominator is $6(2x-1)$. Multiplying both sides of the equation by this quantity produces

$$\frac{3x+1}{3(2x-1)}\cdot\frac{6(2x-1)}{1}=\frac{5}{6(2x-1)}\cdot\frac{6(2x-1)}{1}+\frac{7}{2}\cdot\frac{6(2x-1)}{1}$$

$$2(3x+1)=5+21(2x-1)$$

$$6x+2=5+42x-21$$

$$18=36x$$

$$x=\frac{1}{2}$$

To obtain this solution, both sides of the equation were multiplied by the variable expression $6(2x-1)$, which has a value of 0 when $x=\frac{1}{2}$. Since multiplication of both sides of an equation by 0 does not always produce an equivalent equation, the solution may not be valid. In fact, substitution of $\frac{1}{2}$ for x in the original equation results in division by 0, so there is no solution.

b. To solve

$$\frac{3x+1}{6x-3} = \frac{-13}{12x-6} + \frac{7}{2}$$

begin by factoring all denominators:

$$\frac{3x+1}{3(2x-1)} = \frac{-13}{6(2x-1)} + \frac{7}{2}$$

As in part (a), multiply both sides by the least common denominator of $6(2x - 1)$ to obtain

$$\frac{3x+1}{3(2x-1)} \cdot \frac{6(2x-1)}{1} = \frac{-13}{6(2x-1)} \cdot \frac{6(2x-1)}{1} + \frac{7}{2} \cdot \frac{6(2x-1)}{1}$$

$$2(3x+1) = -13 + 21(2x-1)$$

$$6x + 2 = -13 + 42x - 21$$

$$36 = 36x$$

$$x = 1$$

This time, the multiplication by $6(2x - 1)$ does not invalidate the solution, since $6(2x - 1) \neq 0$ when $x = 1$. Therefore, the equation has the solution $x = 1$. ■

In general, if both sides of an equation are multiplied by a variable expression, any resulting solution is valid *unless substituting the solution into the original equation makes one or more denominators zero.*

EXERCISES *In Exercises* 1–8, *solve the equation.*

1. $\frac{2x}{3} - \frac{x}{4} = \frac{25}{12}$

2. $\frac{x}{6} + \frac{1}{3} = \frac{21}{2x}$

3. $\frac{x}{2-x} - 7 = \frac{2}{2-x}$

4. $\frac{x}{2-x} - 7 = \frac{10}{2-x}$

5. $\frac{x}{x+5} = \frac{x+5}{x}$

6. $\frac{2x-3}{x} = \frac{x+7}{x-9}$

7. $\frac{6}{x^2-9} - \frac{4}{x} = \frac{1}{x-3}$

8. $\frac{x}{x+4} + \frac{7}{x^2+x-12} = \frac{1}{x-3}$

A–18 RADICALS AND RATIONAL EXPONENTS

EVEN AND ODD ROOTS

If $n \geqslant 2$ is an integer and $x^n = a$, then x is an **nth root** of a. The number n is the **index** of the root. It is often of interest to find all real nth roots of a given number.

EXAMPLE 1 *Real nth roots of a number*

a. Both 8 and -8 are real square roots of 64, because $8^2 = 64$ and $(-8)^2 = 64$.

b. There are no real square roots of -64, because the equation $x^2 = -64$ has no real solutions.

c. The only real cube root of 64 is 4, because $4^3 = 64$, and the equation $x^3 = 64$ has no other real solutions.

d. Both 2 and -2 are real sixth roots of 64, because $2^6 = 64$ and $(-2)^6 = 64$. ■

The foregoing example can be generalized.

- If n is even, every positive real number a has two real nth roots, one positive and one negative. The positive root, called the **principal *n*th root** of a, is denoted by $\sqrt[n]{a}$. The negative root is always $-\sqrt[n]{a}$. The index n is usually omitted if $n = 2$. Thus $\sqrt{64} = 8$, $-\sqrt{64} = -8$, $\sqrt[6]{64} = 2$, and $-\sqrt[6]{64} = -2$.
- If n is even, no negative real number has a real nth root.
- If n is odd, every real number a has exactly one real nth root $\sqrt[n]{a}$. Thus $\sqrt[3]{64} = 4$ and $\sqrt[3]{-64} = -4$.
- $\sqrt[n]{0} = 0$ for every index n.

In the expression $\sqrt[n]{a}$, the symbol $\sqrt{}$ is called a **radical**, and a is the **radicand**.

SIMPLIFYING RADICALS

If n is even, then $\sqrt[n]{a^n} = |a|$ for every real number a.
If n is odd, then $\sqrt[n]{a^n} = a$ for every real number a.
For example, $\sqrt{(-5)^2} = \sqrt{25} = 5 = |-5|$, and $\sqrt[3]{(-5)^3} = \sqrt[3]{-125} = -5$.

EXAMPLE 2 *Simplifying radicals*

a. $\sqrt{25x^2} = \sqrt{(5x)^2} = |5x| = 5|x|$

b. $\sqrt{x^8} = \sqrt{(x^4)^2} = |x^4| = x^4$

c. $\sqrt[3]{8x^6} = \sqrt[3]{(2x^2)^3} = 2x^2$

d. $\sqrt[4]{x^4y^8} = \sqrt[4]{(xy^2)^4} = |xy^2| = |x|y^2$ ■

MULTIPLYING AND DIVIDING RADICALS

If $\sqrt[n]{a}$ and $\sqrt[n]{b}$ are both real, then

$$\sqrt[n]{a} \cdot \sqrt[n]{b} = \sqrt[n]{ab}$$

$$\frac{\sqrt[n]{a}}{\sqrt[n]{b}} = \sqrt[n]{\frac{a}{b}}$$

That is, to multiply or divide radicals *of the same index*, multiply or divide the radicands. In the following examples, assume that all radical expressions represent real numbers.

EXAMPLE 3 *Multiplying and dividing radicals and simplifying*

a. $\sqrt{12x^5}\sqrt{3x} = \sqrt{36x^6} = \sqrt{(6x^3)^2} = 6x^3$

(We do not need to write $6|x^3|$, because $\sqrt{12x^5}$ and $\sqrt{3x}$ are not real if $x < 0$.)

b. $\dfrac{\sqrt{12x^5}}{\sqrt{3x}} = \sqrt{\dfrac{12x^5}{3x}} = \sqrt{4x^4} = 2x^2$

c. $\sqrt[3]{6x^2}\sqrt[3]{9x^2} = \sqrt[3]{54x^4} = \sqrt[3]{2 \cdot 27 \cdot x^3 \cdot x}$

$= \sqrt[3]{3^3}\sqrt[3]{x^3}\sqrt[3]{2x} = 3\sqrt[3]{2x}$

d. $\dfrac{\sqrt{30x^6}}{\sqrt{10x}} = \sqrt{\dfrac{30x^6}{10x}} = \sqrt{3x^5} = \sqrt{3 \cdot (x^2)^2 \cdot x}$

$= x^2\sqrt{3x}$ ■

ADDING AND SUBTRACTING RADICALS

If $\sqrt[n]{a}$ is real, then

$$r\sqrt[n]{a} \pm s\sqrt[n]{a} = (r \pm s)\sqrt[n]{a}$$

That is, to add or subtract radicals *of the same index and the same radicand*, add or subtract their coefficients.

EXAMPLE 4 *Adding and subtracting radicals and simplifying*

a. $7\sqrt{2x} - 4\sqrt{2x} = 3\sqrt{2x}$

b. $\sqrt[3]{5} + \sqrt[3]{40} = \sqrt[3]{5} + \sqrt[3]{2^3 \cdot 5} = \sqrt[3]{5} + 2\sqrt[3]{5} = 3\sqrt[3]{5}$ ■

RATIONALIZING NUMERATORS AND DENOMINATORS

When working with an expression such as $\dfrac{\sqrt{3}}{\sqrt{5}}$, it is sometimes convenient to form an equivalent expression having no radicals in the denominator. To do so is to **rationalize the denominator**. At other times, we may want to **rationalize the numerator**.

EXAMPLE 5 *Rationalizing numerators and denominators*

a. To rationalize the denominator of $\dfrac{\sqrt{3}}{\sqrt{5}}$, multiply by $\dfrac{\sqrt{5}}{\sqrt{5}}$.

$$\frac{\sqrt{3}}{\sqrt{5}} \cdot \frac{\sqrt{5}}{\sqrt{5}} = \frac{\sqrt{15}}{\sqrt{25}} = \frac{\sqrt{15}}{5}$$

b. To rationalize the numerator of $\dfrac{\sqrt{3}}{\sqrt{5}}$, multiply by $\dfrac{\sqrt{3}}{\sqrt{3}}$.

$$\frac{\sqrt{3}}{\sqrt{5}} \cdot \frac{\sqrt{3}}{\sqrt{3}} = \frac{\sqrt{9}}{\sqrt{15}} = \frac{3}{\sqrt{15}}$$ ■

To rationalize a numerator (denominator) consisting of a single nth root, multiply both the numerator and denominator by another nth root so that the numerator (denominator) becomes a perfect nth power.

EXAMPLE 6 *Rationalizing numerators and denominators*

a. To rationalize the numerator of $\dfrac{\sqrt[3]{x}}{\sqrt[3]{y}}$,

$$\frac{\sqrt[3]{x}}{\sqrt[3]{y}} \cdot \frac{\sqrt[3]{x^2}}{\sqrt[3]{x^2}} = \frac{\sqrt[3]{x^3}}{\sqrt[3]{x^2y}} = \frac{x}{\sqrt[3]{x^2y}}$$

b. To rationalize the denominator of $\dfrac{1}{\sqrt[4]{3}}$,

$$\frac{1}{\sqrt[4]{3}} \cdot \frac{\sqrt[4]{27}}{\sqrt[4]{27}} = \frac{\sqrt[4]{27}}{\sqrt[4]{81}} = \frac{\sqrt[4]{27}}{3}$$

■

To rationalize a numerator or denominator having the form $\sqrt{a} + \sqrt{b}$ or $\sqrt{a} - \sqrt{b}$, use the fact that

$$\left(\sqrt{a} + \sqrt{b}\right)\left(\sqrt{a} - \sqrt{b}\right) = a - b$$

EXAMPLE 7 *Rationalizing numerators and denominators*

a. To rationalize the numerator of $\dfrac{\sqrt{7} + \sqrt{3}}{\sqrt{2}}$,

$$\frac{\sqrt{7} + \sqrt{3}}{\sqrt{2}} \cdot \frac{\sqrt{7} - \sqrt{3}}{\sqrt{7} - \sqrt{3}} = \frac{7 - 3}{\sqrt{2}\left(\sqrt{7} - \sqrt{3}\right)}$$

$$= \frac{4}{\sqrt{14} - \sqrt{6}}$$

b. To rationalize the denominator of $\dfrac{3}{\sqrt{11} - \sqrt{5}}$,

$$\frac{3}{\sqrt{11} - \sqrt{5}} \cdot \frac{\sqrt{11} + \sqrt{5}}{\sqrt{11} + \sqrt{5}} = \frac{3\left(\sqrt{11} + \sqrt{5}\right)}{11 - 5}$$

$$= \frac{3\left(\sqrt{11} + \sqrt{5}\right)}{6} = \frac{\sqrt{11} + \sqrt{5}}{2}$$

■

RATIONAL EXPONENTS

If m and n are integers, $\dfrac{m}{n}$ is in lowest terms, and $\sqrt[n]{a}$ is real, then

$$a^{m/n} = \left(\sqrt[n]{a}\right)^m = \sqrt[n]{a^m}$$

EXAMPLE 8 *Evaluating expressions involving rational exponents*

a. $8^{5/3} = \left(\sqrt[3]{8}\right)^5 = 2^5 = 32$

b. $9^{-1/2} = \dfrac{1}{9^{1/2}} = \dfrac{1}{\sqrt{9}} = \dfrac{1}{3}$

■

The laws of exponents given on page A26 apply to rational exponents.

EXAMPLE 9 *Applying the laws of exponents*

a. $x^{1/2}x^{2/3} = x^{1/2+2/3} = x^{3/6+4/6} = x^{7/6}$

b. $(x^{1/3}y^2)^{3/2} = (x^{1/3})^{3/2}(y^2)^{3/2} = x^{1/2}y^3$ ■

EXERCISES *In Exercises 1–8, simplify the radical.*

1. $\sqrt{9t^2}$ **2.** $\sqrt{100q^4}$

3. $\sqrt{u^2v^8}$ **4.** $\sqrt{49x^4y^6}$

5. $\sqrt[3]{27a^3}$ **6.** $\sqrt[3]{a^3b^6}$

7. $\sqrt[4]{16y^{12}}$ **8.** $\sqrt[5]{p^5q^{20}}$

In Exercises 9–20, perform the indicated operations and simplify.

9. $\sqrt{2}\sqrt{72}$ **10.** $\sqrt[3]{4x}\,\sqrt[3]{2x^2}$

11. $\sqrt{10x}\sqrt{15xy}$ **12.** $\sqrt[3]{x^4y}\,\sqrt[3]{x^2y^4}$

13. $\dfrac{\sqrt{189x^2}}{\sqrt{21}}$ **14.** $\dfrac{\sqrt[3]{250}}{\sqrt[3]{2z^3}}$

15. $\dfrac{\sqrt{189x^3}}{\sqrt{3}}$ **16.** $\dfrac{\sqrt[3]{500z^4}}{\sqrt[3]{2z^3}}$

17. $10\sqrt{k} + 2\sqrt{k}$ **18.** $\sqrt[4]{ab} - 5\sqrt[4]{ab}$

19. $5\sqrt{20x} - 2\sqrt{5x^3}$ **20.** $\sqrt[3]{81} - \sqrt[3]{3}$

In Exercises 21–28,

a. Rationalize the numerator.

b. Rationalize the denominator.

21. $\dfrac{\sqrt{x}}{\sqrt{6}}$ **22.** $\sqrt{\dfrac{2}{d}}$

23. $\dfrac{\sqrt[3]{4}}{\sqrt[3]{s}}$ **24.** $\sqrt[3]{\dfrac{a}{b}}$

25. $\dfrac{\sqrt{13} - \sqrt{5}}{\sqrt{5}}$ **26.** $\dfrac{\sqrt{x}}{\sqrt{x} + \sqrt{y}}$

27. $\dfrac{\sqrt{2} + 1}{\sqrt{3}}$ **28.** $\dfrac{\sqrt{2} - 1}{\sqrt{3}}$

In Exercises 29–32, evaluate the expression without using a calculator.

29. $1000^{2/3}$ **30.** $49^{-1/2}$

31. $9^{5/2}$ **32.** $\left(\dfrac{8}{27}\right)^{1/3}$

In Exercises 33–40, simplify the expression.

33. $a^6a^{2/3}$ **34.** $(a^6)^{2/3}$

35. $\dfrac{a^6}{a^{2/3}}$ **36.** $(b^5c^{10})^{3/5}$

37. $\left(\frac{4}{x^8}\right)^{3/2}$ **38.** $\left(\frac{27u}{v^3}\right)^{4/3}$

39. $p^{-1/2}(p^{2/5})^{3/2}$ **40.** $\left(\frac{r^{5/3}}{r^{1/6}}\right)^{2/3}$

A–19 EQUATIONS WITH RADICALS

In solving an equation with square roots of variable expressions, one step is often to eliminate the radicals by squaring both sides of the equation. Since this operation does not always produce an equivalent equation, you need to check all solutions in the *original* equation.

EXAMPLE 1 *Solving an equation with radicals*

To solve

$$\sqrt{2x - 5} = x - 10$$

begin by squaring both sides of the equation.

$$\left(\sqrt{2x - 5}\right)^2 = (x - 10)^2$$
$$2x - 5 = x^2 - 20x + 100$$
$$x^2 - 22x + 105 = 0$$
$$(x - 7)(x - 15) = 0$$
$$x = 7 \quad \text{or} \quad x = 15$$

Next, substitute each solution for x in the original equation.

$\sqrt{2(7) - 5} = 7 - 10$	$\sqrt{2(15) - 5} = 15 - 10$
$\sqrt{9} = -3$	$\sqrt{25} = 5$
$3 = -3$	$5 = 5$

Substituting 7 for x produces a false equation, so 7 is not a solution. (It is sometimes called an *extraneous solution.*) Substituting 15 for x produces a true equation, so the solution is $x = 15$. ■

EXAMPLE 2 *Solving an equation with radicals*

To solve

$$\sqrt{4x - 3} - 2 = \sqrt{2x - 5}$$

begin by squaring both sides.

$$\left(\sqrt{4x - 3} - 2\right)^2 = \left(\sqrt{2x - 5}\right)^2$$
$$(4x - 3) - 4\sqrt{4x - 3} + 4 = 2x - 5$$

The resulting equation still contains a radical. To eliminate it, isolate it on one side of the equation, then square both sides again.

$$2x + 6 = 4\sqrt{4x - 3}$$

$$(2x + 6)^2 = \left(4\sqrt{4x - 3}\right)^2$$

$$4x^2 + 24x + 36 = 16(4x - 3)$$

$$4x^2 - 40x + 84 = 0$$

$$4(x - 3)(x - 7) = 0$$

Now substitute each solution for x in the original equation.

$\sqrt{4(3) - 3} - 2 = \sqrt{2(3) - 5}$	$\sqrt{4(7) - 3} - 2 = \sqrt{2(7) - 5}$
$\sqrt{9} - 2 = \sqrt{1}$	$\sqrt{25} - 2 = \sqrt{9}$
$3 - 2 = 1$	$5 - 2 = 3$
$1 = 1$	$3 = 3$

Since substituting each solution for x produces a true equation, the solutions are $x = 3$ and $x = 7$. ■

EXERCISES *In Exercises* 1–16, *solve the equation.*

1. $\sqrt{3x + 1} = 7$

2. $\sqrt{1 - x} = 2$

3. $\sqrt{x - 5} = 0$

4. $\sqrt{x - 5} = -3$

5. $\sqrt{x + 5} = \sqrt{2x - 1}$

6. $\sqrt{3x} = \sqrt{x + 20}$

7. $\sqrt{x^2 - 9} = 4$

8. $\sqrt{x^2 + 12} = \sqrt{7x}$

9. $\sqrt{x} = x - 12$

10. $\sqrt{x + 5} = 2x$

11. $\sqrt{x^2 - 64} = x - 4$

12. $\sqrt{x^2 - 25} = 1 - x$

13. $\sqrt{x - 6} + \sqrt{x - 1} = 5$

14. $\sqrt{x} = 1 - \sqrt{x - 9}$

15. $\sqrt{x + 10} = \sqrt{x + 4} + 2$

16. $\sqrt{3x + 3} + 1 - \sqrt{6x + 7} = 0$

APPENDIX B

TIPS FOR GRAPHING FUNCTIONS WITH A CALCULATOR

PREREQUISITES BEFORE BEGINNING THIS APPENDIX, MAKE SURE YOU ARE FAMILIAR WITH:

The Keystrokes for Evaluating Functions, Graphing, Tracing, and Zooming on Your Calculator

To obtain accurate graphical information about a function from a calculator, you must do more than press keys.

- You must know what graphical information is provided by the equation of the function. You will develop that knowledge systematically beginning in Chapter 3.
- You must know how to represent graphical information accurately by controlling the appearance of the graph on your calculator screen. You will develop that knowledge in this appendix.

B–1 THE VIEWING WINDOW

When you graph a function on your calculator, the screen depicts a rectangular portion of the coordinate plane called the **viewing window**. You choose the viewing window by assigning minimum and maximum values to x and y. The appearance of a calculator graph is affected by both the *size* and the *shape* of the viewing window.

SIZE OF THE VIEWING WINDOW

The size of the viewing window affects both the extent of the graph and the amount of detail that you can see.

EXAMPLE 1 *Viewing a graph through windows of different sizes*

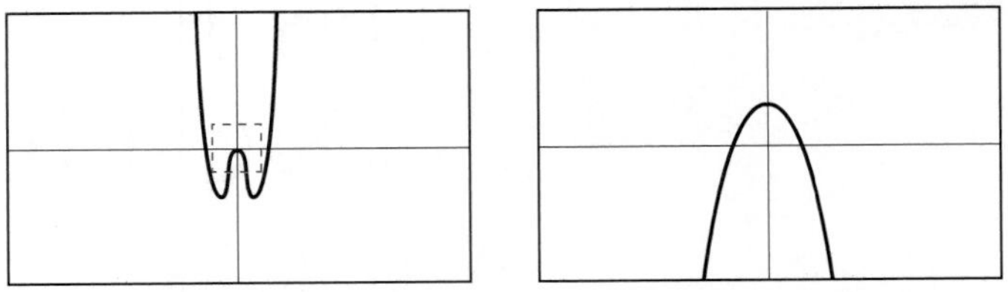

(a)
$[-100, 100]$ by $[-100, 100]$
Xscale = 10, Yscale = 10

(b)
$[-10, 10]$ by $[-10, 10]$
Xscale = 1, Yscale = 1

Figure B–1

Figure B-1 shows two graphs of the function $f(x) = 0.01x^4 - 1.21x^2 + 3$. The viewing window is indicated under each graph. Figure B-1b "blows up" the small dotted rectangle in Figure B-1a to fill the entire screen. That is, a portion of the coordinate plane 20 units wide and 20 units high (from -10 to 10 in each direction) now occupies the visual space previously occupied by a portion 200 units wide and 200 units high (from -100 to 100 in each direction). Thus we have **zoomed in** by a factor of 10 in each direction.

Which window is "better"? Neither is inherently better, but each brings out graphical features that the other does not. The window in Figure B-1a shows the overall shape of the graph, while the window in Figure B-1b reveals the number of x-intercepts. ■

EXERCISES *In Exercises* 1–2,

a. Graph the function in each viewing window.

b. Tell which window was obtained by zooming in from the other, and by what factor.

1. $y = 4x - x^2$; $[-10, 10]$ by $[-10, 10]$
$[-1, 1]$ by $[-1, 1]$

2. $y = 4x - x^2$; $[3, 5]$ by $[-1, 1]$
$[-5, 5]$ by $[-5, 5]$

In Exercises 3–6,

a. Graph the function in the given window.

b. Zoom in by a factor of 10 around the x-intercept shown in the window.

c. Zoom in a second time by a factor of 10 around the x-intercept.

d. Trace along the graph to approximate the x-intercept within 0.01.

3. $y = x^3 - 10$; $[-10, 10]$ by $[-10, 10]$

4. $y = 7 + 2x - x^2$; $[0, 10]$ by $[-10, 10]$

5. $y = 3 - \sqrt{2x}$; $[0, 10]$ by $[-5, 5]$

6. $y = 10 - 5x^3 - x^4$; $[-10, 0]$ by $[-100, 100]$

7. Match each graph with one of the viewing windows $[-10, 10]$ by $[-10, 10]$, $[-1, 1]$ by $[9, 11]$, $[9, 10]$ by $[0, 1]$. All three graphs represent the same function.

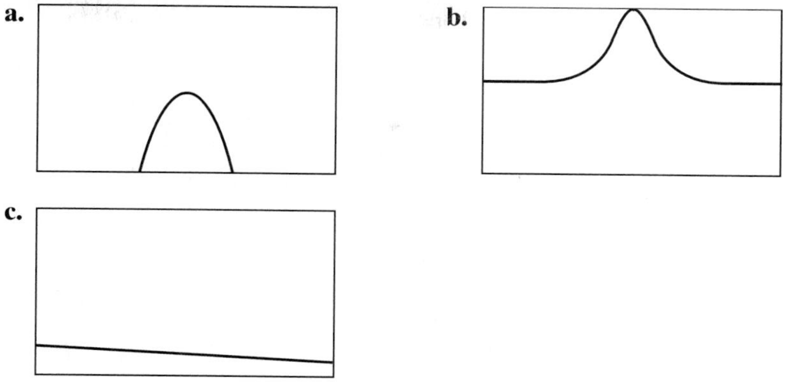

SHAPE OF THE VIEWING WINDOW

The shape of the viewing window affects the apparent steepness of a graph.

EXAMPLE 2 *Viewing a graph through windows of different shapes*

Figure B-2 shows four graphs of the function $y = x^2$. The viewing window used for each graph is indicated under it.

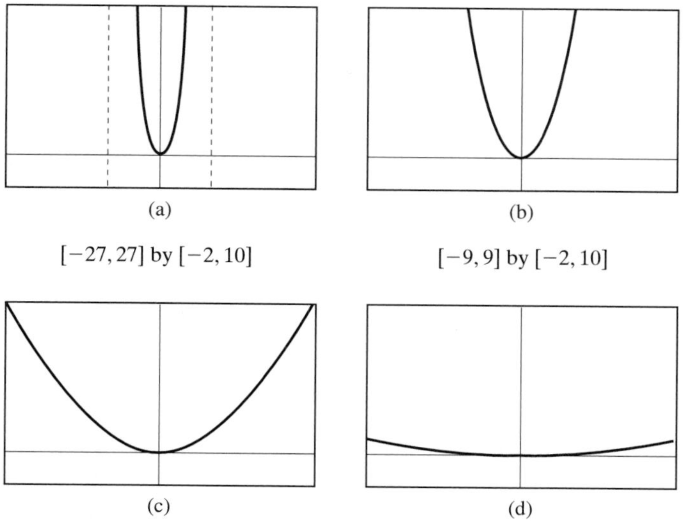

(a) $[-27, 27]$ by $[-2, 10]$

(b) $[-9, 9]$ by $[-2, 10]$

(c) $[-3, 3]$ by $[-2, 10]$

(d) $[-1, 1]$ by $[-2, 10]$

Figure B–2

All the graphs extend from -2 to 10 in the y-direction, and the window in Figure B-2a extends from -27 to 27 in the x-direction. In Figure B-2b we have zoomed in by a factor of 3 in the x-direction, thus expanding the dashed rectangular portion of the window in Figure B-2a to fill the entire screen. Zooming in repeatedly by a factor of 3 in the x-direction yields the graphs in Figures B-2c and d. Zooming in horizontally has the effect of making the graph appear horizontally stretched and, therefore, less steep. ■

EXERCISES

8. **a.** Graph the function $f(x) = x^3 - x$ in each of the windows $[-3, 3]$ by $[-50, 50]$, $[-3, 3]$ by $[-10, 10]$, $[-3, 3]$ by $[-2, 2]$, $[-3, 3]$ by $[-0.4, 0.4]$.

b. By what factor did you zoom in vertically to obtain each graph from the previous one?

c. How does zooming in vertically affect the apparent steepness of the graph?

Each of Exercises 9–10 *shows three views of a graph. Match each graph with its viewing window. The coordinate axes have not been shown in any window.*

9.

a.

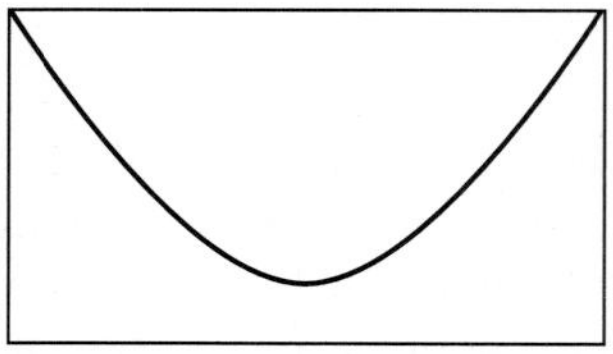

b.

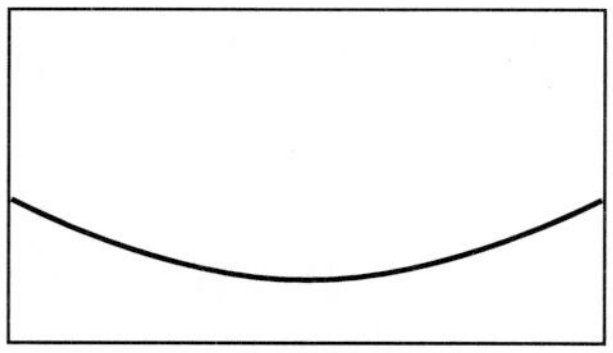

c.

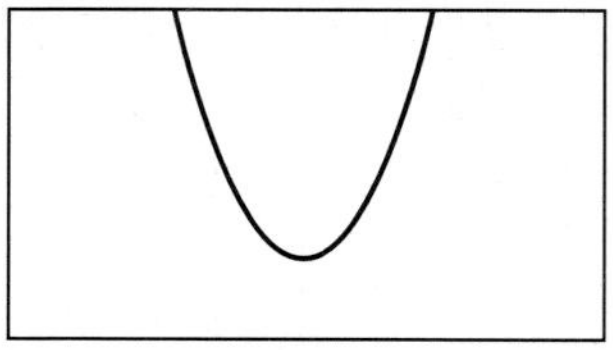

(a) $[-4, 4]$ by $[-1, 4]$
(b) $[-4, 4]$ by $[-4, 16]$
(c) $[-4, 4]$ by $[-16, 64]$

10.

a.

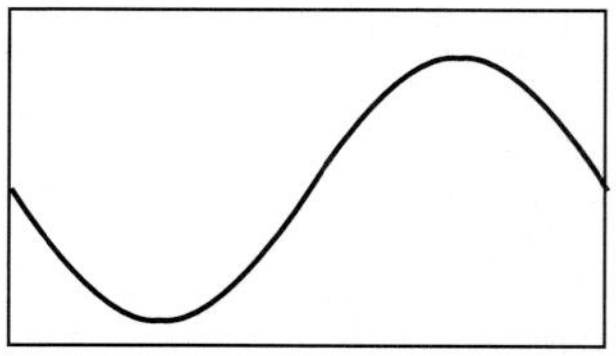

b.

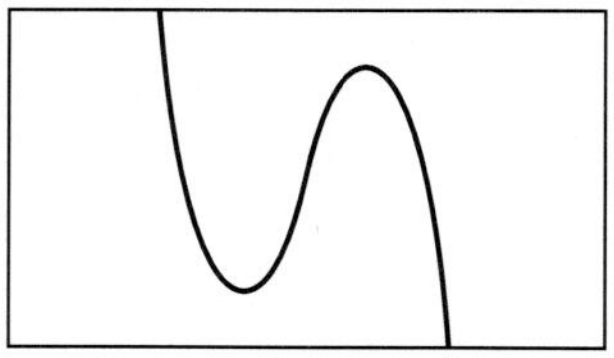

c.

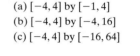

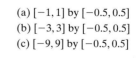

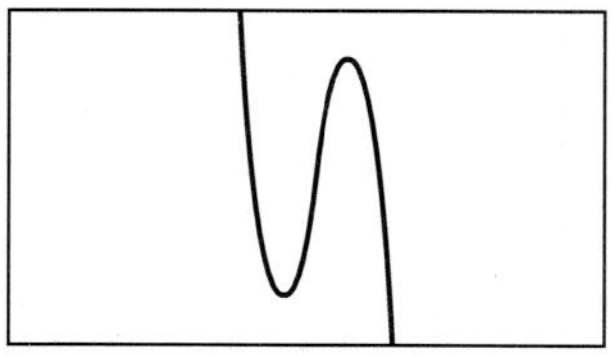

(a) $[-1, 1]$ by $[-0.5, 0.5]$
(b) $[-3, 3]$ by $[-0.5, 0.5]$
(c) $[-9, 9]$ by $[-0.5, 0.5]$

UNDISTORTED WINDOWS

You have seen that some viewing windows make graphs appear steeper or flatter than they actually are. It is often necessary to choose a viewing window that depicts steepness accurately. We will refer to such a window as an **undistorted viewing window**.

EXAMPLE 3 *Distorted and undistorted viewing windows*

Figure B-3 shows the graph of $y = x$ in two windows on a Texas Instruments TI-85 calculator.

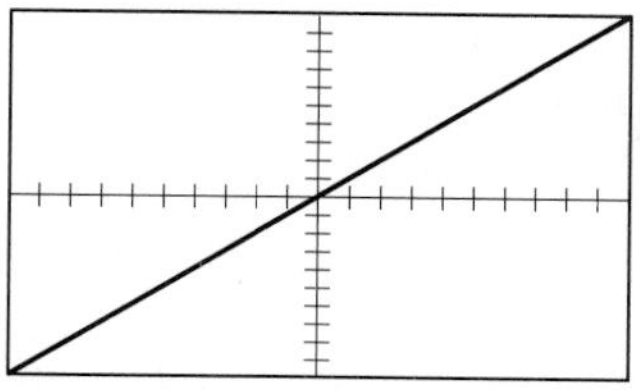

(a)

$[-10, 10]$ by $[-10, 10]$
Xscale = 1, Yscale = 1

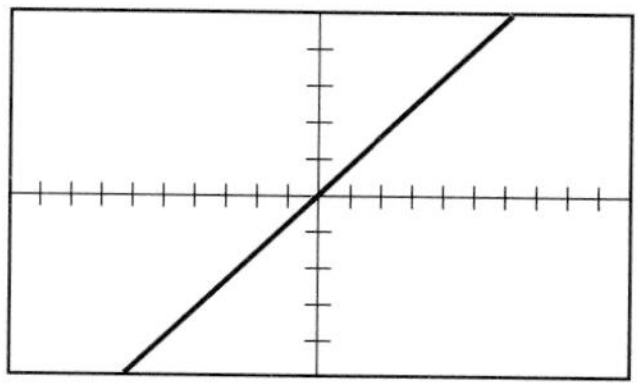

(b)

$[-10, 10]$ by $[-5, 5]$
Xscale = 1, Yscale = 1

Figure B–3

In Figure B-3a the scale marks on the x-axis are farther apart than those on the y-axis. Thus horizontal distances are magnified in comparison with vertical distances, and the slope of the graph appears to be smaller than its actual value of 1. In Figure B-3b the scale marks are equally far apart on both axes, and the slope of the graph appears to be 1.

The screen on a TI-85 calculator is twice as wide as it is high, so that its undistorted windows show regions of the plane that are twice as wide as they are high. For example, the windows $[-20, 20]$ by $[-10, 10]$ and $[0, 100]$ by $[0, 50]$ are undistorted on a TI-85. ■

Example 3 illustrates that the shape of your calculator screen determines whether a given window is undistorted.

EXERCISE **11. a.** Consult your owner's manual to discover how to obtain undistorted windows on your calculator. (The manual may refer to them as "square" windows.)

b. Find at least two undistorted windows on your calculator that are centered at $(0, 0)$, and at least one that is not.

ADJUSTING THE WINDOW

You have seen that some viewing windows make it difficult to identify important graphical features. Here are some of the most common problems, and some troubleshooting strategies that can often correct them.

Problem: The graph runs off the top or bottom of the screen and back on.
Solution: Zoom out vertically by a factor of 10.

Problem: The graph is contained in a thin horizontal strip.
Solution: Zoom in vertically by a factor of 10 around the horizontal strip. If repeated zooming yields no additional details, zoom out horizontally by a factor of 10.

Problem: The graph is contained in a thin vertical strip.
Solution: Zoom in horizontally by a factor of 10 around the vertical strip. If repeated zooming yields no additional details, zoom out vertically by a factor of 10.

Problem: The graph is contained in a small area near the center of the screen.
Solution: Zoom in by a factor of 10 around the region containing the graph.

Problem: The graph is not visible on the screen, or is visible only near one edge of the screen.
Solution: Zoom out by a factor of 10 around the center of the screen.

Problem: Any strange appearance in a graph.
Solution: Make sure you have entered the function correctly.

B-2 GRAPHING LINEAR FUNCTIONS

Although you can graph linear functions easily by hand, you may sometimes want to graph them on your calculator for several reasons. For example,

- It is easier to construct a table of values for a linear function from a calculator graph than from an equation or a hand-drawn graph.
- You must often solve problems by looking at a linear graph together with a more complicated one that would be difficult to graph manually.

Viewing windows for graphs should usually include all points of interest. For a linear function, the points of interest are the intercepts. Therefore, it is helpful to find the intercepts before graphing, especially if the coefficient of x or y is unusually large or small. A rough estimate will do.

EXAMPLE 1 *Finding a viewing window for a linear function*

To choose an appropriate viewing window for

$$y = 0.0011x - 43.5$$

begin by observing that the y-intercept is $(0, -43.5)$. The interval of y-values in the viewing window should include both -43.5 to show the y-intercept and 0 to show the x-axis. The interval $[-50, 10]$ includes both values.

Since the slope of the line is very small $\left(\text{about } \frac{1}{1000}\right)$, the line rises slowly, and the x-intercept has a large x-coordinate. You can obtain a rough estimate by recalling that the slope is the ratio $\frac{\text{change in } y}{\text{change in } x}$. From $(0, -43.5)$ to the x-intercept $(x, 0)$ the change in y is 43.5, so the change in x is about $(1000)(43.5) = 43{,}500$. Therefore, the interval of x-values in the viewing window should include both 0 and 43,500.

The window $[-10{,}000, 50{,}000]$ by $[-50, 10]$ shows both intercepts. Because the window is 60,000 units long and only 60 units high, you might want to place scale marks every 10,000 units on the x-axis and every 10 units on the y-axis. The graph is shown in Figure B-4.

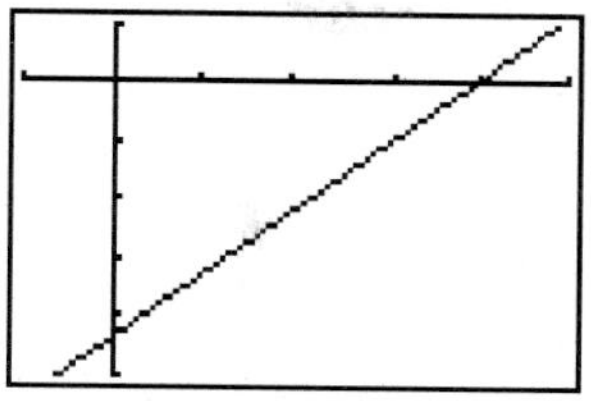

$[-10{,}000, 50{,}000]$ by $[-50, 10]$
Xscale = 10,000, Yscale = 10

Figure B–4

The window in Figure B-4 displays both intercepts clearly but distorts the steepness of the line. In general, when you choose a window to display some graphical feature, it may omit or distort some other feature. ■

Example 1 illustrates a general procedure for graphing linear functions on a calculator.

- Find the y-intercept. Choose an interval containing its y-coordinate, and also containing 0. This is the interval of y-values for your viewing window.
- Find the x-intercept. Choose an interval containing its x-coordinate, and also containing 0. This is the interval of x-values for your viewing window.
- Scale the x-axis by choosing a convenient number between $\frac{1}{20}$ and $\frac{1}{5}$ of the width of the window.
- Scale the y-axis by choosing a convenient number between $\frac{1}{20}$ and $\frac{1}{5}$ of the height of the window.

Exercises *In Exercises* 1–8,

a. Find viewing windows showing both intercepts.

b. Describe any resulting distortion in the shape of the graph.

1. $y = x + 12$ **2.** $y = 20 - 4x$

3. $y = 800 - 0.2x$

4. $y = -0.04 - 0.5x$ (Your viewing window should make it clear that the graph does not pass through the origin.)

5. $y = 0.0003x + 0.03$ **6.** $y = -7500(x + 0.4)$

7. $y = 279x + 35{,}000$ **8.** $y = 1000 - 2x$

B–3 GRAPHING QUADRATIC FUNCTIONS

For a quadratic function, the points of interest are the vertex and all intercepts. To avoid a lot of trial and error, find or estimate some of these points before graphing. On the graph of $y = ax^2 + bx + c$, the y-intercept is $(0, c)$, the x-coordinates at the x-intercepts are the solutions to $ax^2 + bx + c = 0$, and the vertex has an x-coordinate of $-\dfrac{b}{2a}$.

EXAMPLE 1 *Finding a viewing window that includes the vertex and intercepts of a quadratic graph*

To choose an appropriate viewing window for

$$y = x^2 - 92x + 180$$

begin by observing that the y-intercept is (0, 180), and the vertex has an x-coordinate of $-\dfrac{-92}{2(1)} = 46$. The x-coordinates at the x-intercepts are the solutions to

$$x^2 - 92x + 180 = 0$$
$$(x - 2)(x - 90) = 0$$
$$x = 2, 90$$

To include both axes and all points of interest, the window should include the x-values 0, 2, 46, and 90, and the y-values 0 and 180. The window $[-10, 100]$ by $[-250, 250]$, shown in Figure B-5a, shows all the intercepts, but misses the vertex.

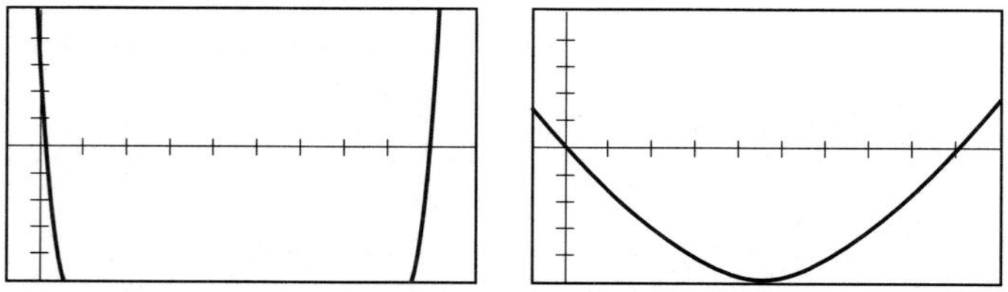

$[-10, 100]$ by $[-250, 250]$
Xscale = 10, Yscale = 50

$[-10, 100]$ by $[-2000, 2000]$
Xscale = 10, Yscale = 500

Figure B–5

Knowing the x-coordinate of the vertex, you can calculate the y-coordinate as -1936. Alternately, you can trace along the graph and observe the y-coordinates when x is near 46. Figure B-5b shows the graph in the window $[-10, 100]$ by $[-2000, 2000]$. ■

Figure B-5 illustrates that it is often impossible to find a single ideal viewing window. The window in Figure B-5b includes all the points of interest, but it distorts the steepness of the graph. The window also makes the graph appear to pass through the origin, when in fact it does not.

Example 1 illustrates a general procedure for graphing quadratic functions on a calculator.

- Find the x-coordinates of the vertex and all x-intercepts. Choose an interval containing them, and also containing 0. This is the interval of x-values for your viewing window.
- Find the y-intercept. Choose an interval containing its y-coordinate, and also containing 0. Graph the function using this interval of y-values.
- If your window does not show the vertex, trace along the graph to find its y-coordinate and expand your interval of y-values accordingly.
- Scale the axes as you would for a linear function.

EXERCISES *In Exercises* 1–6,

a. Find all intercepts and the x-coordinate of the vertex analytically.

b. Find a viewing window that shows the vertex and all intercepts.

c. Describe any graphical features that are distorted or hidden by the viewing windows you chose.

d. Zoom in until you can estimate the x-coordinate of the vertex and each x-intercept with an error of no more than 0.01. Make sure your results are consistent with those in part (a).

1. $y = x^2 + x - 12$

2. $y = 6x - x^2$

3. $y = x^2 - 6x + 20$

4. $y = -9x^2 + 37x - 38$

5. $y = x^2 + 500x - 500{,}000$

6. $y = 50 - 0.000001x^2$

B–4 GRAPHING QUADRATIC RELATIONS

In general, it is impossible to write equations of functions defined implicitly by a given relation. However, every quadratic relation can be represented by a pair of implicitly defined functions. On your calculator, it is more practical to work with the functions, rather than the relation, for two reasons:

- You can find the y-values for each x-value more easily.
- Most calculators graph only functions, and not more general relations.

Example 1 demonstrates how to express a quadratic relation

$$Ax^2 + Bxy + Cy^2 + Dx + Ey + F = 0$$

as a pair of functions.

EXAMPLE 1 *Expressing a quadratic relation as a pair of functions*

Given

$$3x^2 - 6xy + y^2 + 6x - 2y + 4 = 0$$

regroup the terms to obtain

$$y^2 + (-6x - 2)y + (3x^2 + 6x + 4) = 0$$

This equation has the form $ay^2 + by + c = 0$, with

$$a = 1, \qquad b = -6x - 2, \qquad c = 3x^2 + 6x + 4$$

Using the quadratic formula $y = \dfrac{-b \pm \sqrt{b^2 - 4ac}}{2a}$, we obtain

$$\begin{aligned} y &= \frac{-(-6x - 2) \pm \sqrt{(-6x - 2)^2 - 4(1)(3x^2 + 6x + 4)}}{2(1)} \\ &= \frac{6x + 2 \pm \sqrt{24x^2 - 12}}{2} \\ &= 3x + 1 \pm \sqrt{6x^2 - 3} \end{aligned}$$

Thus the relation can be expressed as the pair of functions

$$f_1(x) = 3x + 1 + \sqrt{6x^2 - 3}$$

and

$$f_2(x) = 3x + 1 - \sqrt{6x^2 - 3}$$

To obtain graphs of functions defined implicitly by quadratic relations, it is often most efficient to begin with the window $[-10, 10]$ by $[-10, 10]$ and adjust it as needed by using the troubleshooting strategies listed in Section B-1. Figure B-6 shows the graphs of f_1 and f_2 in the window $[-10, 10]$ by $[-10, 10]$, along with the graph of the entire relation.

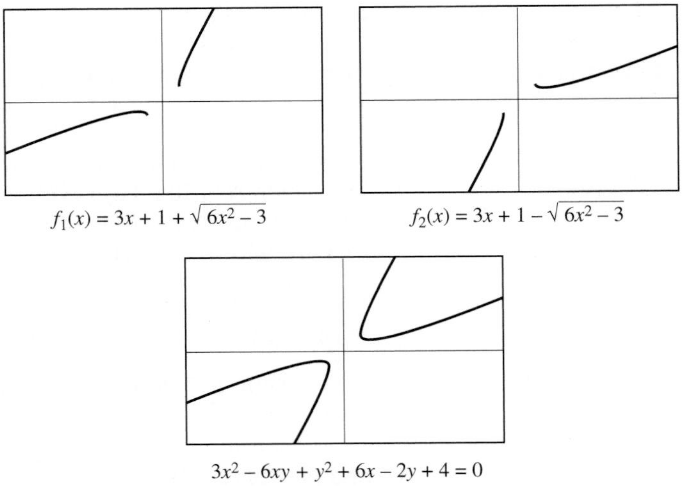

Figure B–6 ■

Example 1 illustrates a general procedure for graphing quadratic relations on a calculator.

- Express the relation as a pair of implicitly defined functions.
- Graph the functions in the window $[-10, 10]$ by $[-10, 10]$. If this window is not satisfactory, use the troubleshooting strategies in Section B-1 to adjust it.

EXERCISES *In Exercises* 1–10,

a. Rewrite the quadratic relation by expressing y as a pair of functions of x, and graph the two functions on the same screen.

b. Trace along each graph and zoom in around each intercept until you can estimate its coordinates with an error of no more than 0.01.

1. $y^2 - 4x^2 = 16$ **2.** $4x^2 + 25y^2 = 400$

3. $2y^2 - 3x + 7y - 50 = 0$ **4.** $15 = 3x^2 - 5y^2 + 12x - 20y$

5. $x^2 - 3xy + y^2 = 0$

6. $x^2 + 6xy - 2y^2 - 10x + 10y + 25 = 0$

7. $5x - xy + 3y - 2x^2 - 2y^2 = 100$

8. $xy + 4y^2 + 6x + 4y - 9 = 0$

9. $x^2 + 5xy - 3x + 2y - 10 = 0$

10. $x^2 + 12xy = y^2 - 6x - 6y$

B–5 GRAPHING POLYNOMIAL FUNCTIONS

INCLUDING POINTS OF INTEREST

For a polynomial graph, the most important points of interest are its intercepts and turning points. Since polynomial graphs of degree greater than 2 are typically more complex than linear or quadratic graphs, you cannot always know whether a given window shows all points of interest. However, your knowledge of polynomial graphs from Chapter 7 can often help you decide.

EXAMPLE 1 *Deciding whether a given viewing window depicts a polynomial graph accurately*

A graph of $g(x) = x^4 - 64x^2 - 500$ is shown in Figure B-7.

How can you tell whether other points of interest are outside the window? The window shows two x-intercepts, which is less than the maximum for a polynomial of degree 4. However, there are three turning points, which is the maximum for a polynomial of degree 4. Since there are no other turning points outside the window, there are also no other x-intercepts. Therefore, the window shows all points of interest.

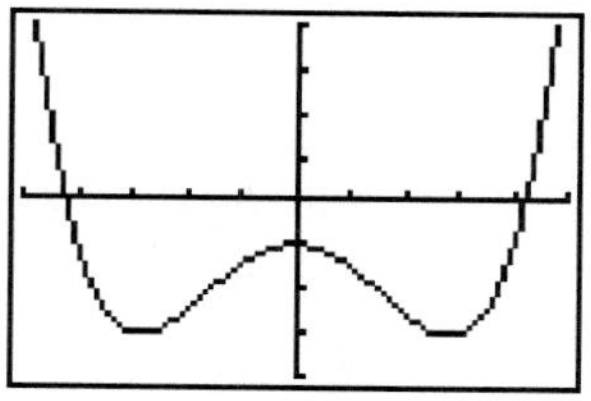

[−10, 10] by [−2000, 2000]

Figure B–7

■

EXERCISES *In Exercises* 1–6,

a. Use your knowledge of polynomial graphs and some trial and error to find a viewing window that shows all points of interest.

b. Trace along the graph and zoom in around each x-intercept and turning point until you can estimate its x-coordinate with an error of no more than 0.01.

1. $f(x) = x^4 - 14x^3 + 24x^2$

2. $f(x) = 400x - x^3$

3. $f(x) = x^3 - 12x^2 + 49x - 50$

4. $f(x) = x^6 - 7x^4 + 9x^2 - 12$

5. $f(x) = x^7 - 4x^6 - 14x^5 + 56x^4 + 49x^3 - 196x^2 - 36x + 144$

6. $f(x) = x^6 - 12x^4 + 48x^2$

The following convenient rule, popularized by textbook author John Saxon, can be used to find a window containing all intercepts and turning points of a polynomial. Calculus is needed to prove it, so we state it without proof.

> Suppose that A is the largest absolute value of any coefficient in the polynomial
>
> $$P(x) = a_n x^n + a_{n-1}x^{n-1} + a_{n-2}x^{n-2} + \cdots + a_1 x + a_0$$
>
> Then the x-values $-\dfrac{A}{|a_n|} - 1$ and $\dfrac{A}{|a_n|} + 1$ are the left and right boundaries of a viewing window containing every x-intercept and turning point of P.

EXAMPLE 2 *Applying Saxon's rule*

For $f(x) = 6x^3 - 7x^2 - 27x + 21$, $A = |-27| = 27$, and $a_n = 6$. The value of $\dfrac{A}{|a_n|} + 1$ is $\dfrac{|-27|}{|6|} + 1 = 5.5$. Therefore, some viewing window with x-values $[-5.5, 5.5]$ contains every point of interest on the graph of $y = f(x)$. To find an appropriate interval of y-values, begin with $[-10, 10]$ and adjust as needed. The window $[-5.5, 5.5]$ by $[-100, 100]$ works well for this function. ■

EXERCISES *In Exercises* 7–10, *use Saxon's rule to find a viewing window containing every point of interest on the graph of* $y = f(x)$. *Then trace and zoom to find the x-coordinate of each x-intercept and turning point with an error of no more than* 0.01.

7. $f(x) = x^4 - 14x^3 + 24x^2$

8. $f(x) = 400x - x^3$

9. $f(x) = x^6 - 7x^4 + 9x^2 - 12$

10. $f(x) = -2x^6 + 24x^4 - 96x^2$

It is often difficult to decide whether a given viewing window depicts a polynomial graph accurately. As with so many things in life, there is no substitute for experience. For the moment, however, here is a reasonable strategy for choosing windows.

- Choose an x-interval large enough to contain all points of interest. Some trial and error may be needed, but your knowledge of points of interest and end behavior, along with Saxon's rule, can make your search easier.
- View the graph in several windows, using the x-interval you have chosen and varying the size of the y-interval. Some trial and error may be needed here as well, and the best y-intervals are often very large. If your graph runs off the top or bottom of the screen and back on, try increasing the size of your y-interval by a factor of 10.

APPROXIMATING TURNING POINTS

Example 3 illustrates a method of approximating a turning point of a polynomial function with an error of less than 0.01 in both coordinates.

EXAMPLE 3 *Approximating a turning point of a polynomial*

Figure B-8 shows a graph of $P(x) = x^4 - 5x^3 + 6x^2$.

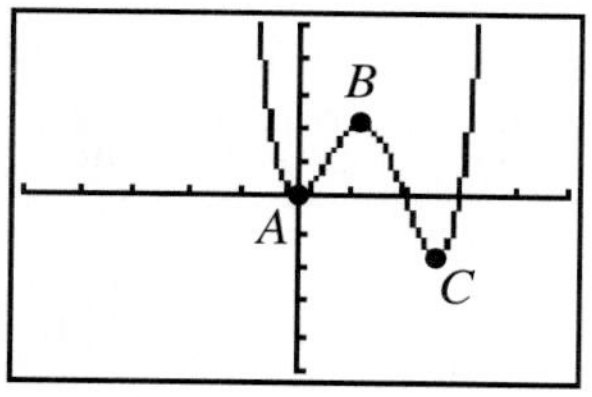

[−5, 5] by [−5, 5]
Xscale = 1, Yscale = 1

Figure B–8

To estimate the location of the maximum at B, trace along the graph near B until the y-values stop increasing and start to decrease. We obtained the following sequence of points on a Texas Instruments TI-82 calculator. Yours may yield slightly different results.

$$x = 1.0638298, \quad y = 2.0513636$$
$$x = 1.1702128, \quad y = 2.0792035$$
$$x = 1.2765957, \quad y = 2.0317722$$

Thus the local maximum value of P is at least 2.0792035 and occurs at some x-value in the interval (1.0638298, 1.2765957).

For greater accuracy, zoom in around B, as in Figure B-9a.

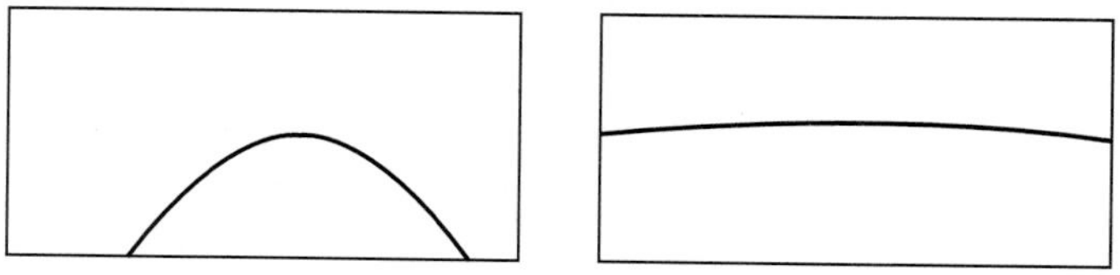

(a) [0.6, 1.6] by [1.6, 2.6]

(b) [1.15, 1.17] by [2.07, 2.09]

Figure B–9

Tracing again yields the following sequence of points:

$$x = 1.1425532, \quad y = 2.0791050$$
$$x = 1.1531915, \quad y = 2.0797443$$
$$x = 1.1638298, \quad y = 2.0796324$$

One more zoom, in Figure B-9b, confirms that the maximum occurs at $x \approxeq 1.16$, with an error of less than 0.01.

The y-value at the maximum appears to be about 2.08. To demonstrate graphically that this estimate has an error of less than 0.01, graph $y = P(x)$ in a viewing window with $x = 1.16 \pm 0.01$ and $y = 2.08 \pm 0.01$, as in Figure B-9b.

Since the maximum is clearly in this window, it is located at $(1.16, 2.08)$, with an error of less than 0.01 in each direction.

Similarly, the local minimum at A occurs at $(0, 0)$. In Exercise 11, you can find the minimum at C. ■

Example 3 illustrates a general procedure for locating turning points on a polynomial graph.

- Trace along the graph until the y-values stop increasing and start to decrease, or vice versa. Zoom in around this x-value and trace again. Repeat the process until you can locate the x-coordinate of the turning point with an error of no more than 0.01.
- Guess the y-coordinate at the maximum, and graph the function in a window whose y-values extend 0.01 units above and below your guess. If the turning point does not appear in the window, revise your guess and try again.

EXERCISES

11. Find the local minimum of $P(x) = x^4 - 5x^3 + 6x^2$ at the point C in Figure B-8.

In Exercises 12–14, *find all turning points of the polynomial function with an error of no more than* 0.01 *in each coordinate.*

12. $G(x) = 2x^4 - 3x^3 + 2x^2 - x - 3$

13. $w = 3x^3 - 10x + 6$

14. $y = x^7 - 3x^5 + 2x^2 - 1$

B-6 GRAPHING RATIONAL FUNCTIONS

To show the graph of a rational function accurately, a window should show all points of interest and discontinuities and indicate the graph's end behavior. It is often difficult to be certain that a window shows all points of interest, but the following strategy usually produces an accurate graph of a rational function $R(x) = \dfrac{N(x)}{D(x)}$.

- Find all discontinuities by setting $D(x) = 0$, and find all x-intercepts by setting $N(x) = 0$.
- Choose an interval of x-values that includes all x-intercepts and vertical asymptotes with some room to spare.
- Begin with a reasonable interval of y-values, such as $[-10, 10]$ by $[-10, 10]$, and adjust as needed to show turning points and vertical asymptotes clearly.
- Zoom out repeatedly until the graph's end behavior becomes apparent.

- If possible, choose a single viewing window that shows all graphical features accurately. If this is not possible, then choose two or more windows that depict the graph accurately together.

When graphing rational functions it is often helpful to put your calculator into "dot" mode, rather than "connected" mode. (See your owner's manual to learn how to do this.) Figure B-10 shows two graphs of $y = \dfrac{1}{x^2 - 4}$ to illustrate the difference between the modes. Figure B-10a, in connected mode, makes it appear that the vertical asymptotes are a part of the graph. Figure B-10b, in dot mode, makes it clear that they are not.

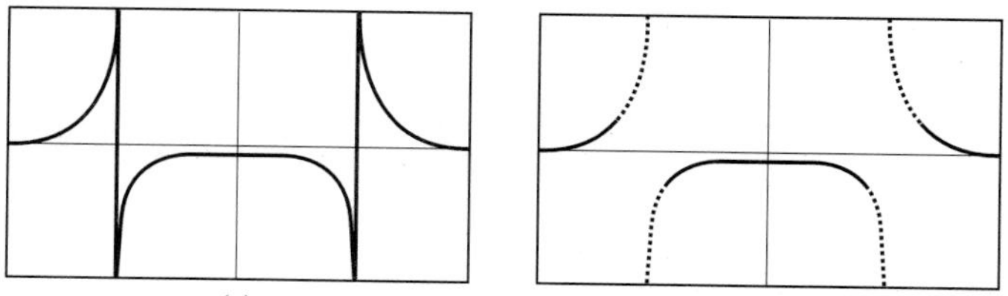

(a) $[-4, 4]$ by $[-4, 4]$

(b) $[-4, 4]$ by $[-4, 4]$

Figure B–10

EXAMPLE 1 *Choosing an appropriate window for a rational graph*

To graph

$$f(x) = \frac{x^2 - 4x + 3}{x + 1}$$

first observe that the denominator is zero when $x = -1$, and the numerator is zero when $x = 1$ or 3. Thus, $f(x)$ has a discontinuity at $x = -1$, and x-intercepts at $(1, 0)$ and $(3, 0)$, so the window should include x-values of -1, 1, and 3. The window in Figure B-11a uses a y-interval of $[-20, 20]$ because a smaller interval does not show the left side of the graph well. This window shows the intercepts, but does not clearly show the vertical asymptote or the end behavior.

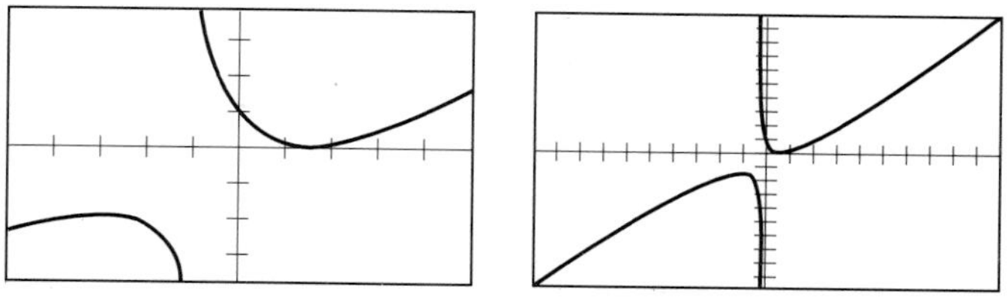

(a) $[-5, 5]$ by $[-20, 20]$ Xscale = 1, Yscale = 5

(b) $[-50, 50]$ by $[-50, 50]$ Xscale = 5, Yscale = 5

Figure B–11

The window in Figure B-11b was obtained by zooming out horizontally and adjusting the y-interval. This window shows both the discontinuity and the end behavior, but does not show the intercepts clearly. Both windows are needed to depict the graph accurately. ■

EXERCISES *In Exercises* 1–4,

a. Identify all x-intercepts and vertical asymptotes.

b. Find one or more viewing windows that show all intercepts and vertical asymptotes and the graph's end behavior.

c. Trace and zoom to locate all x-intercepts and vertical asymptotes with an error of no more than 0.01. Make sure your results agree with those in part (a).

1. $f(x) = \dfrac{x^2 - 4}{2x^2}$

2. $f(x) = \dfrac{x}{x^2 - 9}$

3. $f(x) = \dfrac{x^4}{x^3 - 25x}$

4. $f(x) = \dfrac{x^2 + 2}{x^4 + 4}$

ANSWERS TO ODD-NUMBERED EXERCISES

SECTION 1–1

1. $y = -0.020t + 0.150$ **3.** 0.091

5. a. The left portion of the graph would become less steep. It would still pass through (0.7, 0.136), since that point corresponds to the reading taken at 2:52. The y-intercept on the graph would be higher than in Figure 1–2.

b. Bill's friend's testimony favors the prosecution.

SECTION 1–2

1. Responses will vary. **3.** 10.59 inches

5. The fan speed appears to be about 3700 or 3800 rpm. This result is less precise than that obtained from the equation, but is consistent with it.

7. (Sample response) The equation provides a more precise answer than the graph, but the graph provides an estimate more quickly than the equation. The table does not provide a direct answer to either question.

9. (Sample response) I would rather have access to Table 3 because it provides reasonably precise answers to the questions that customers would ask most often.

11. Estimates will vary, but should be close to 360 parts per million (ppm) for 2000 and 420 ppm for 2050. The estimate for the year 2000 should be more trustworthy, because it assumes that the graph will show no surprising changes over a period of only 15 years beyond 1985.

13. a.

Year	CO_2 concentration (parts per million)
1960	315
1970	325
1980	335
1985	342

Answers may vary slightly.

b. 360 (Answers may vary.)

c. 420 (Answers may vary.)

15. a. $y = 10 - x$ **b.**

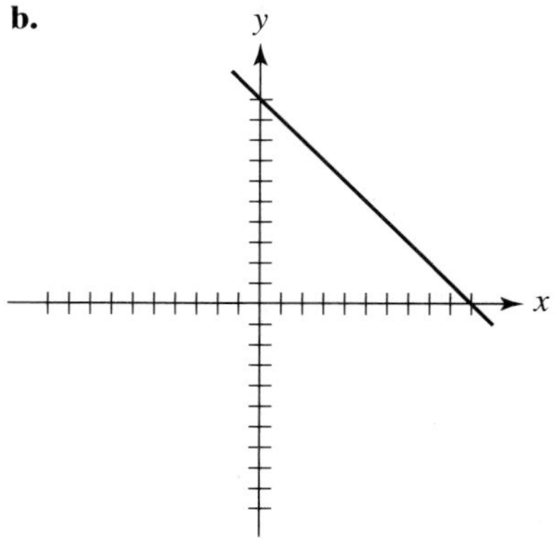

17. a. $y = \dfrac{12}{x}$ **b.**

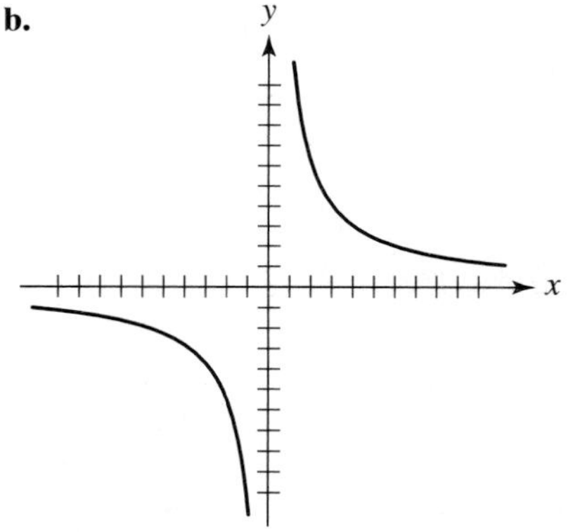

19. a.

x	y
−4	14
−2	12
0	10
2	8
4	6

b.

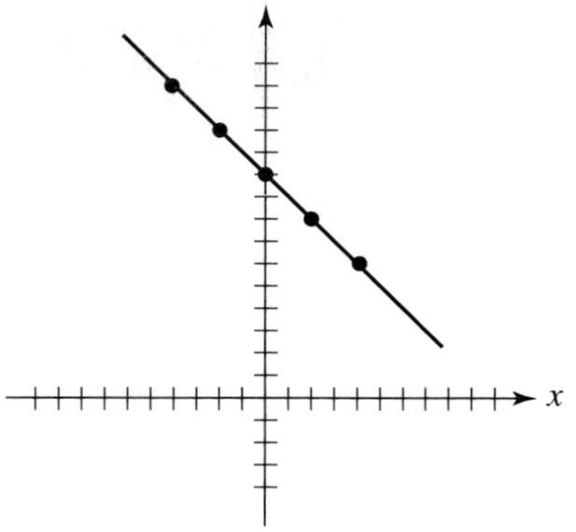

21. a.

x	y
−4	−6
−2	6
0	10
2	6
4	−6

b.

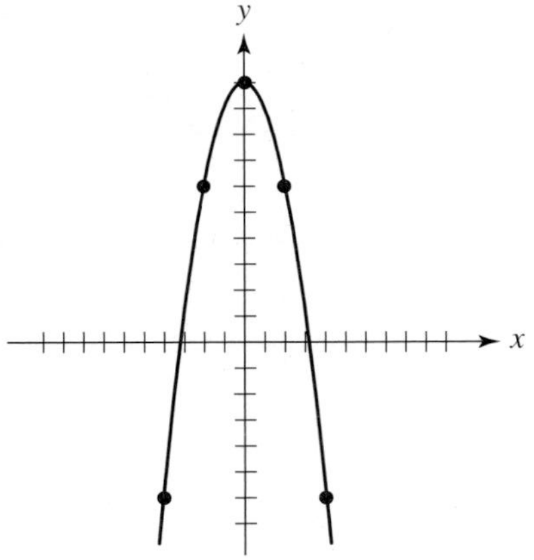

23. a.

x	y
−2	−3
−1	0
0	1
1	0
2	−3

b. $y \cong -24$

25. a.

x	y
−2	0
−1	−1
0	−2
1	−1
2	0

b. $y \cong 3$

27. a. (Sample response) The information is incomplete because the table does not show the life expectancy at every age.

b. (Sample response) The information is approximate because the values of E have almost certainly been rounded.

c. (Sample response) The information is explicit because the values of the variables can be read directly from the table.

29. a. (Sample response) The information is incomplete because it shows sea levels only for about the past 50,000 years.

b. (Sample response) The information is approximate because sea levels must be estimated from the graph.

c. (Sample response) The information is hidden because sea levels must be read from the graph.

31. a.

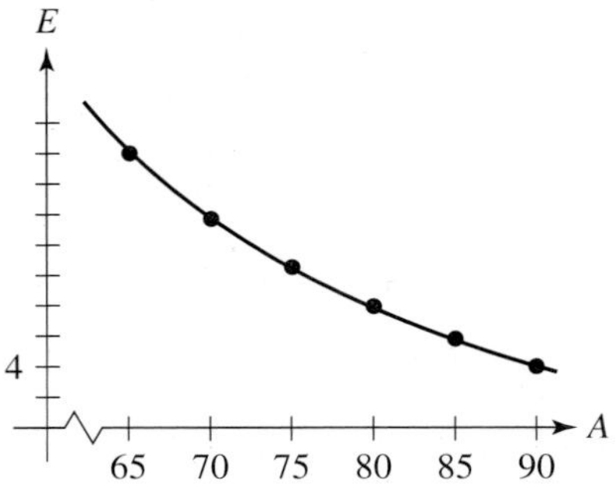

b. about 2 years (Answers will vary.)

33. a.

D	r
200	2.0
300	3.1
400	4.2
500	5.3
600	6.4

b.

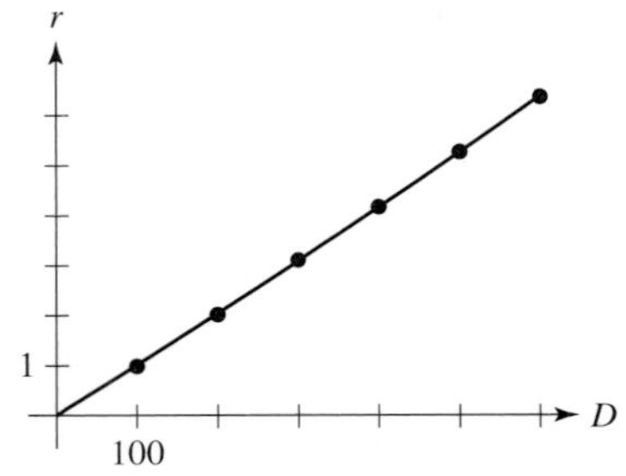

35. Responses will vary. **37.** Responses will vary.

SECTION 1-3

1. Responses will vary. **3.** $360

5. a. We cannot answer this question. We could answer if we knew the dollar amounts spent on prescription drugs in either 1990 or 1991.

b. Physician care experienced the smallest percentage increase.

c. We cannot answer this question. We could answer if we knew the dollar amounts spent on each category in either 1990 or 1991.

d. We cannot answer this question. We could answer if we knew the dollar amounts spent on each category in either 1990 or 1991.

7. 105

9. Fill the larger container, pour 3 liters from it into the smaller container, then empty the smaller container. There are now 2 liters in the larger container. Pour that into the smaller container. Then once again fill the larger container. There are now 5 liters in the larger container and 2 liters in the smaller one. Pour from the larger to the smaller until the smaller one is full. There are now 4 liters in the larger container.

11. 37 **13.** 43.01 feet **15.** 8 loaves of pumpernickel, 6 loaves of day-old bread

17. We don't know how you feel, but we would rather ride in the first elevator. The steep portion of the second graph indicates a long drop in a very short time. Our guess is that the cable on the second elevator snapped as it reached the second floor.

19. The first car has traveled farther during the hour. The two cars are going equally fast after an hour.

21. Responses will vary.

SECTION 2-1

1. a. The independent variable is F.
b. Yes, since each Celsius temperature C corresponds to a unique Fahrenheit temperature F.

3. a. y is not a function of x. **b.** y is not a function of x.

5. $y = \dfrac{9 - 4x}{2}$; y is a function of x.

7. $y = 1 - x^4$; y is a function of x.

9. a. $x = \dfrac{9 - 2y}{4}$; x is a function of y.
b. $x = 15 - 0.5y$; x is a function of y.
c. $x = \pm\sqrt[4]{1 - y}$; x is not a function of y.
d. $x = \sqrt[3]{1 - y^2}$; x is a function of y.

11. y is a function of x. **13.** y is not a function of x.

15.

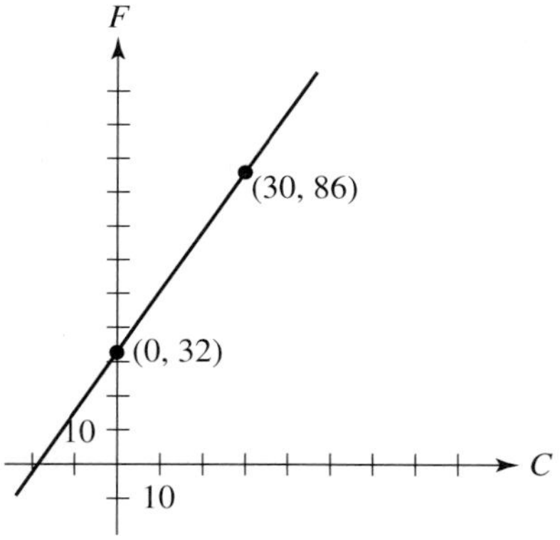

F is a function of C. This conclusion agrees with the one obtained in Exercise 10.

17. The equation $x = y^5 + y + 1$ represents y as a function of x, since no vertical line intersects its graph more than once.

19. a. y is a function of x. **b.** More information is needed.

21. a. y is a function of x. **b.** More information is needed.

23. $y = 2x^5 - 3x$; y is a function of x.

25. This equation cannot be solved for y.

27. y is not a function of x. **29.** y is a function of x.

31. y is not a function of x.

33. a. w is not a function of h; h is not a function of w.
b. The student who was 70 inches tall and weighed 155 pounds withdrew.

35. a. The equations describe processes for obtaining unique values of F and C from given values of K, so both F and C are functions of K.

b.

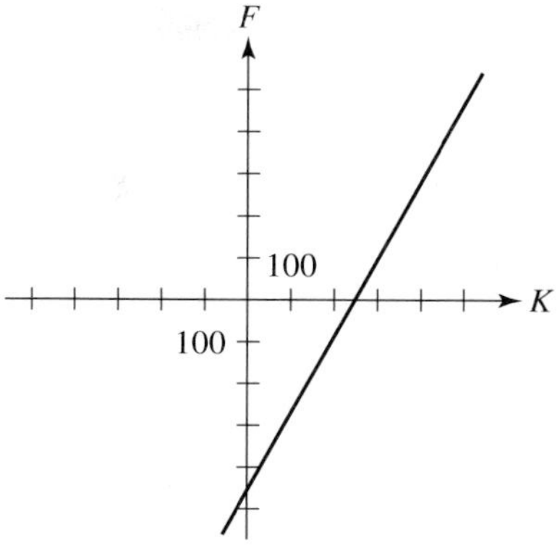

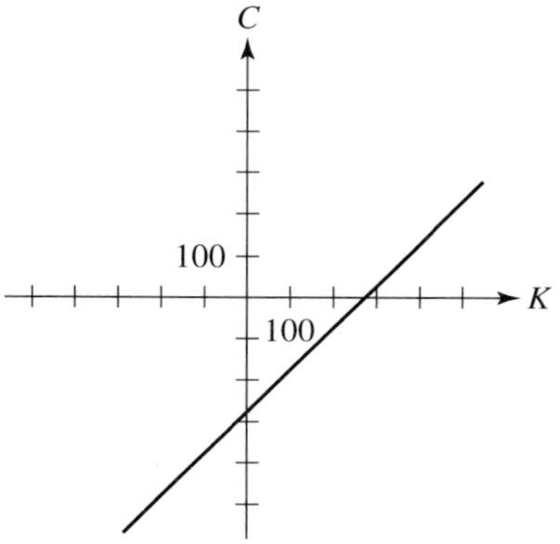

No vertical line intersects either graph more than once, so both F and C are functions of K.

c. $K = C + 273.2$; $K = \dfrac{5}{9}(F + 459.7)$; K is a function of both F and C.

37. Responses will vary.

39. (Sample response) Many tables of physical relationships, such as those found in newspapers and magazines, are not complete tables of a relationship. If such a table fits a function, you need more information before concluding that it was actually generated by a function.

41. Most responses will build on the bulleted paragraphs on page 29.

43. The correspondence is both a function from A to B and a function from B to A.

SECTION 2-2

1. $H(7) = 31; H(0) = -4; H(A) = 5A - 4;$
$H(A - 1) = 5A - 9$

3. $H(7) = \frac{7}{8}; H(0) = 0; H(A) = \frac{A}{A + 1};$
$H(A - 1) = \frac{A - 1}{A}$

5. a. $g(41) = 5$

This represents the Celsius temperature corresponding to 41°F.

b. $g(-40) = -40$

This says that −40°F and −40°C are the same temperature.

c. $g(86) = 30$

7. $(f + g)(x) = x^2 - x + 2$ $\quad (f - g)(x) = x^2 + x - 6$
$(fg)(x) = x^3 + 4x^2 + 2x - 8$ $\quad \left(\frac{f}{g}\right)(x) = \frac{x^2 - 2}{4 - x}$

9. $(f + g)(x) = \sqrt{t - 1} + \sqrt{t^2 - 1}$
$(f - g)(x) = \sqrt{t - 1} - \sqrt{t^2 - 1}$
$(fg)(x) = |t - 1|\sqrt{t + 1}$
$\left(\frac{f}{g}\right)(x) = \sqrt{\frac{1}{t + 1}}$ if $t \neq 1$

11. $p(2500) = 3.5; p(2100) = 5.5; p(1700) = 7.5;$
$p(1300) = 9.5; p(900) = 11.5$

13. $\frac{106h - 10{,}000}{h - 100}$

15. $(u \circ v)(x) = 8 - 4x^2$ $\quad (v \circ u)(x) = 64 - 64x + 16x^2$

17. $(u \circ v)(x) = \frac{1 - 3x}{x^2}$ $\quad (v \circ u)(x) = \frac{1}{x^2 - 3x}$

19. $x \rightarrow x^2 \rightarrow x^2 + 3 \rightarrow 5(x^2 + 3)$
$f(x) = 5(x^2 + 3)$

21. $x \rightarrow x + 3 \rightarrow 5(x + 3) \rightarrow [5(x + 3)]^2$
$f(x) = [5(x + 3)]^2$

23. the multiplying-by-six function, the subtracting-two function, the square-root function

25. the square-root function, the multiplying-by-six function, the subtracting-two function

27. $T(50) = 2$

29. a. h represents the process of multiplying the input by 20 and adding 13 to the result.

b. $h(4) = 93; h(0) = 13; h(-0.5) = 3$

c. $h(3Z) = 60Z + 13$

d. $(h + g)(x) = 21x + 13$ $\quad (h - g)(x) = 19x + 13$
$(hg)(x) = 20x^2 + 13x$ $\quad \left(\frac{h}{g}\right)(x) = \frac{20x + 13}{x}$

e. $(h \circ f)(x) = 20x + 93$ $\quad (f \circ h)(x) = 20x + 17$

31. a. h represents the process of subtracting 4 times the square of the input from 100 times the input.

b. $h(4) = 336; h(0) = 0; h(-0.5) = -51$

c. $h(3Z) = 300Z - 36Z^2$

d. $(h + g)(x) = 101x - 4x^2$ $\quad (h - g)(x) = 99x - 4x^2$
$(hg)(x) = 100x^2 - 4x^3$ $\quad \left(\frac{h}{g}\right)(x) = 100 - 4x$ if $x \neq 0$

e. $(h \circ f)(x) = -4x^2 + 68x + 336$
$(f \circ h)(x) = -4x^2 + 100x + 4$

33. a. h represents the process of dividing the input by the sum of the input and 1.

b. $h(4) = \frac{4}{5}; h(0) = 0; h(-0.5) = -1$

c. $h(3Z) = \frac{3Z}{3Z + 1}$

d. $(h + g)(x) = \frac{x^2 + 2x}{x + 1}$ $\quad (h - g)(x) = \frac{-x^2}{x + 1}$
$(hg)(x) = \frac{x^2}{x + 1}$ $\quad \left(\frac{h}{g}\right)(x) = \frac{1}{x + 1}$ if $x \neq 0$

e. $(h \circ f)(x) = \frac{x + 4}{x + 5}$ $\quad (f \circ h)(x) = \frac{5x + 4}{x + 1}$

35. a. h represents the process of doubling the input, adding 10 to the result, and taking the square root of that result.

b. $h(4) = 3\sqrt{2}; h(0) = \sqrt{10}; h(-0.5) = 3$

c. $h(3Z) = \sqrt{6Z + 10}$

d. $(h + g)(x) = \sqrt{2x + 10} + x$
$(h - g)(x) = \sqrt{2x + 10} - x$
$(hg)(x) = x\sqrt{2x + 10}$ $\quad \left(\frac{h}{g}\right)(x) = \frac{\sqrt{2x + 10}}{x}$

e. $(h \circ f)(x) = \sqrt{2x + 18}$ $\quad (f \circ h)(x) = \sqrt{2x + 10} + 4$

37. a. h represents the process of associating each input with itself.

b. $h(4) = 4; h(0) = 0; h(-0.5) = -0.5$

c. $h(3Z) = 3Z$

d. $(h + g)(x) = 2x$ $\quad (h - g)(x) = 0$
$(hg)(x) = x^2$ $\quad \left(\frac{h}{g}\right)(x) = 1$ if $x \neq 0$

e. $(h \circ f)(x) = x + 4$ $\quad (f \circ h)(x) = x + 4$

39. $(v \circ u)(P) = 4P^3 - 46P^2 + 172P - 210.5$
$(u \circ v)(Q) = Q^3 - 5Q^2 + 4Q - 7$

41. $x \rightarrow x^3 \rightarrow x^3 - 5 \rightarrow \frac{x^3 - 5}{2}$
$f(x) = \frac{x^3 - 5}{2}$

43. $x \rightarrow x - 5 \rightarrow \frac{x - 5}{2} \rightarrow \left(\frac{x - 5}{2}\right)^3$
$f(x) = \left(\frac{x - 5}{2}\right)^3$

45. the multiplying-by-three function, the taking-the-reciprocal function, the adding-one function

47. the multiplying-by-three function, the adding-one function, the taking-the-reciprocal function

49. a. $(q \circ p)(A)$ does not make sense in the physical context.

b. $(p \circ q)(D)$ represents the air pressure in pounds per square inch at a distance of D miles west of St. Louis.

51. a. $\left(\frac{f}{g}\right)(1) = \frac{f(1)}{g(1)} = \frac{3}{0}$, which is undefined.

b. $(f \circ g)(4) = f[g(4)] = f(8)$, which is undefined.

53. a. $\frac{9T}{5} + 41$ **b.** $82.4°$

c. $T + 278.2$ **d.** $301.2°$

55. $(u \circ c)(T)$ represents the Fahrenheit temperature of the water after T seconds, and $(u \circ c)(60)$ represents the Fahrenheit temperature of the water after 1 minute.

57. a. $f(P) = 0.85P$ **b.** $g(P) = P - 1000$

c. The composition $g \circ f$ results in the lower price, and is obtained by applying the discount first.

59. $P(x) = 8.50x - 212.50$.

SECTION 2–3

1. domain $= (-\infty, \infty)$; range $= [-5, \infty)$

3. domain $= (-\infty, \infty)$; range $= (0,1]$

5. domain $= (-\infty, \infty)$; range $= (-\infty, \infty)$

7. domain $= [-8, \infty)$; range $= [0, \infty)$

9. a.

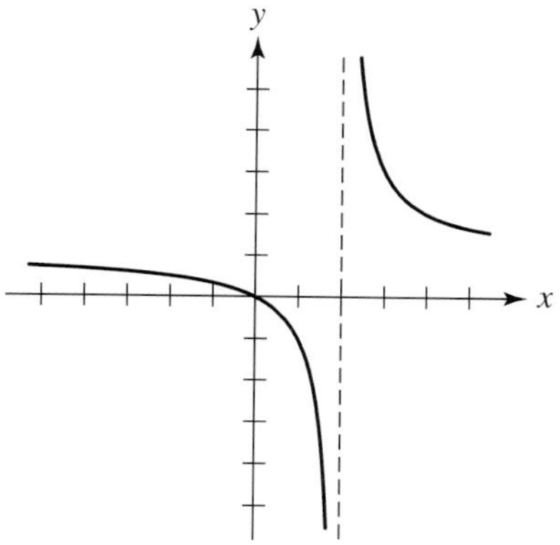

b.

x	$f(x)$
0	0
1	-1
2	undefined
3	3
4	2
5	$\frac{5}{3}$

c. The domain of f is $(-\infty, 2) \cup (2, \infty)$. The limitation $x \neq 2$ shows up in the equation as a factor of $x - 2$ in the denominator, in the graph as a vertical asymptote at $x = 2$, and in the table as an undefined y-value opposite $x = 2$.

11. $(1, 9)$ **13.** $(-\infty, 2]$

15. a. 0.000000078 **b.** 0.00078

c. 0.49 **d.** 0.4999999999

17. $[0, 0.5)$

19. 2, 4, 6, 8, 10, 12, 14, 16, 18, 20

21. 2, 0, -4, -10, -18, -28, -40, -54, -70, -88

23. 1; 2; 6; 24; 120; 720; 5040; 40,320; 362,880; 3,628,800

25. 1024, 0, 512, 256, 384, 320, 352, 336, 344, 340

27. a. 3, 7, 11, 15, 19, 23, 27, 31, 35, 39

b. $a_n = 4n - 1, a_{1000} = 3999$

29. increasing on $(-\infty, 2)$, decreasing on $(2, \infty)$

31. increasing on $(-\infty, \infty)$

33. domain $= \{0, 1, 2, 3\}$; range $= \{0, 1, 2, 3\}$

35. domain $= \{-2, 0, 2, 4, 6\}$; range $= \{-1, 2, 5, 8, 11\}$

37. domain $= (-\infty, \infty)$; range $= (-\infty, 2]$

39. domain $= [-5, 5]$; range $= [0, 3]$

41. domain $= (-\infty, \infty)$; range $= (-\infty, \infty)$

43. domain $= (-\infty, 2) \cup (2, \infty)$; range $= (-\infty, 3) \cup (3, \infty)$

45. domain $= (-\infty, \infty)$; range $= (-\infty, \infty)$

47. domain $= [2, \infty)$; range $= [2, \infty)$

49. $[-1, 3]$ **51.** $[2, \infty)$

53. domain $= (-\infty, \infty)$; range $= [0, \infty)$

55. domain $= (-\infty, 0) \cup (0, \infty)$; range $= (0, \infty)$

57. domain $= [-1, 1]$; range $= [0, 1]$

59. domain $= (-\infty, \infty)$; range $= [0, \infty)$

61. domain $= (-1, 1)$; range $= [1, \infty)$

63. domain $= [-1, 4]$; range $= [0, 65]$

65. domain $= (0, 10.58]$; range $= (0, 3600]$

67. domain $= (-273.2, \infty)$; range $= (-459.7, \infty)$

69. domain $= (0, 2000)$; range $= (0; 1{,}000{,}000)$

71. 4, 3, 2, 1, 0, -1, -2, -3, -4, -5

73. 1, 8, 27, 64, 125, 216, 343, 512, 729, 1000

75. 1, 0.1, 0.01, 0.001, 0.0001, 0.00001, 0.000001, 0.0000001, 0.00000001, 0.000000001

77. 1, -1, 1, -1, 1, -1, 1, -1, 1, -1

79. increasing on $(-\infty, \infty)$

81. increasing on $(-\infty, -1)$ and $(2, \infty)$, decreasing on $(-1, 2)$

83. $$\frac{(\sqrt{5}+1)^2 - (\sqrt{5}-1)^2}{2^2\sqrt{5}} = 1$$

$$\frac{(\sqrt{5}+1)^4 - (\sqrt{5}-1)^4}{2^4\sqrt{5}} = 3$$

$$\frac{(\sqrt{5}+1)^6 - (\sqrt{5}-1)^6}{2^6\sqrt{5}}$$

85. a. If $a_{n+1} = a_n + a_{n-1}$, then $a_4 = 3$.
If $a_{n+1} = a_1 + a_2 + \cdots + a_{n-1}$, then $a_4 = 4$.

b. If $a_{n+1} = a_n + 1$, then $a_4 = 4$.
If $a_{n+1} = a_n + a_{n-1}$, then $a_4 = 5$.

c. If $a_{n+1} = (n+1)a_n$, then $a_4 = 24$.
If $a_{n+1} = a_{n-2}$, then $a_4 = 1$.

d. If $a_{n+1} = a_{n-1} - a_n$, then $a_4 = -3$.
If $a_{n+1} = (a_n)^2 + (a_{n-1})^2$, then $a_4 = 5$.

87. Responses will vary. **89.** Many graphs are possible.

91. a. $(-\infty, \infty)$ **b.** $(-\infty, \infty)$
c–e. $(-\infty, \infty)$
f. $(-\infty, -3) \cup (-3, 2) \cup (2, \infty)$
g. $(-\infty, \infty)$ **h.** $(-\infty, \infty)$

93. a. $[0, \infty)$ **b.** $(-\infty, 4]$
c–e. $[0, 4]$ **f.** $[0, 4)$
g. $(-\infty, 4]$ **h.** $[0, 16]$

95. a. $[0, \infty)$ **b.** $(-\infty, 2) \cup (2, \infty)$
c–e. $[0, 2) \cup (2, \infty)$ **f.** $[0, 2) \cup (2, \infty)$
g. $(2, \infty)$ **h.** $[0, 16) \cup (16, \infty)$

97. The domains of $f + g, f - g$, and fg each consist of the x-values common to the domains of f and g. The domain of $\frac{f}{g}$ consists of those x-values common to the domains of f and g, for which $g(x) \neq 0$.

99. In Exercise 93, the domain of $g \circ f$ is $[0, 16]$. This is contained in the domain of f, which is $[0, \infty)$. A graph indicates that the range of $g \circ f$ is $[0, 2]$. This is contained in the range of g, which is $[0, \infty)$.

SECTION 2–4

1. $x \cong 0.73$ **3.** $x \cong -3.70, 2.70$
5. $x = 12$ **7.** $[-3.70, 2.70]$
9. $[12, \infty)$ **11. a.** $x = -1, 5$ **b.** $(-1, 5)$

13. a. $x \cong 1.59$ **b.** $[1.59, \infty)$
15. a. no solution **b.** $(-\infty, \infty)$

17. a. $x = -5, 0, 5$ **b.** $[-5, 5]$
19. a. $x = 1, 4$ **b.** $[1, 4]$

21. a.

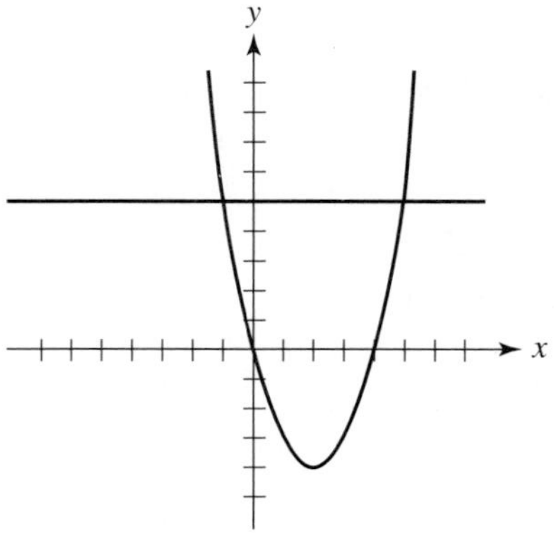

b. The two graphs intersect at a particular x-value exactly when the corresponding y-value is the same on both graphs. Since $y = x^2 - 4x$ on the first graph, and $y = 5$ on the second graph, the two graphs intersect at exactly those points for which $x^2 - 4x = 5$.

c. $x = -1, 5$

23. a. $x = 16$ **b.** $(16, \infty)$
25. a. $x = -2, 0, 2$ **b.** $(-\infty, -2] \cup [0, 2]$

27. a. $x = -1, 2$ **b.** $(-1, 2)$
29. a. 25 cakes per week **b.** more than 25 cakes per week

SECTION 3–1

1. a. 0.61 **b.** 0.61
c. Choices will vary. **d.** 0.61
e. Calculations will vary.

3. not linear **5.** linear

7. The rate of change in S with respect to T is 0.61. This means that each increase of 1° in temperature results in an increase of 0.61 meters per second in the speed of sound.

9. The rate of change in the fan speed with respect to the motor pulley diameter is 340. This represents the increase in rpm corresponding to each 1-inch increase in the diameter of the pulley.

11. The rate of change in cost with respect to distance driven is 0.15. This represents the cost in dollars of each additional mile driven.

13. a. The points are collinear. **b.** 0.61

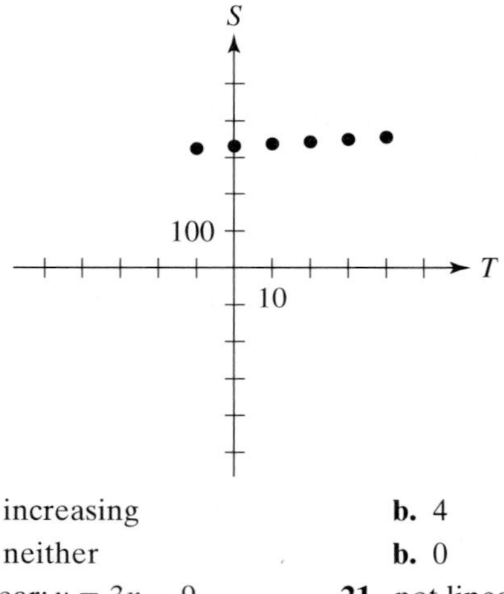

15. a. increasing **b.** 4

17. a. neither **b.** 0

19. linear; $y = 3x - 9$ **21.** not linear

23. not linear

25. The graphs in Exercises 19, 20, 22, and 24 should appear to be lines.

27. a. $y = 4x + 4$ **b.** $y = -x + 2$
c. $y = -2$ **d.** $y = -\frac{2}{3}x - 2$

29. a. -1 **b.** decreasing

31. a. 0 **b.** neither

33. a. $\frac{1}{4}$ **b.** increasing

c. The rate of change represents the rise in temperature corresponding to each additional chirp in a 1-minute period.

35. a. not linear

37. a–b. linear; $f(x) = 0.5x - \sqrt{2}$ **c–d.** increasing; 0.5

39. a–b. linear; $\Sigma(x) = \pi x + 0$ **c–d.** increasing; π

41. a–b. linear; $a(x) = \frac{1}{3}x + 0$ **c–d.** increasing; $\frac{1}{3}$

43. a–b. linear; $z(x) = 6x + 9$ **c–d.** increasing; 6

45. a. linear **b.** 4
c. increasing **d.** $y = 4x + 9$

47. a. linear **b.** -1
c. decreasing **d.** $y = 100 - x$

49. a. not a function

51. a. If v = the number of valentines purchased, and c = the cost of the valentines in dollars, then $c = 2.50v$.
b. 2.50

53. a. $s = 100 - 2.5m$ **b.** -2.5

55. a. $A = -1800t + 36{,}000$
b. The rate of change in A with respect to t is -1800. This says that the airliner descends 1800 feet each minute.
c. $A = 360d$
d. The rate of change in A with respect to d is 360. This says that as the airliner gets 1 mile closer to the airport, it descends 360 feet.

57. When $x = 3, y = 84$. When $x = 5, y = 68$.

59. When $x = 4, y = 34$. When $y = 5, x = 16.5$.

61. No, because the calculation of the missing entries was made possible by the constancy of the ratio $\frac{\Delta y}{\Delta x}$.

63. a. $\frac{\Delta y}{\Delta x} = 1$ **b.** Choices will vary.

65. The points from Exercises 2, 4, and 5 are collinear.

SECTION 3-2

1. If $C = 10$, then $M = \frac{10 - 0.48}{0.67} \cong 14.21$. However, if Carrie talks for 14.21 minutes, she must pay for 15 minutes. Therefore, no call will cost exactly \$10.

3. a. If $C = 10$, then $M \cong 9.28$. No call will cost exactly \$10.
b. $M \leq \frac{C - 0.63}{1.01}$; $M \leq 9.28$. In the context of the problem, $M \leq 9$.

5. 65 **7.** 300

9. Yes; $y = 4x$. **11.** Yes; $y = 0.2x$.

13. a. The rate of change in M with respect to W is constant. Specifically, each time W increases by 1 week, M increases by 0.25 miles.
b. The line relating W and M has a slope of 0.25, since that is the rate of change in M with respect to W. It also passes through (1, 1), since Tim runs 1 mile per day during the first week. The equation in point-slope form is therefore $M - 1 = 0.25(W - 1)$. In slope-intercept form, it is $M = 0.25W + 0.75$.

15. Yes; $a_n = 4 + 3(n - 1)$, $a_{425} = 1276$.

17. Yes; $a_n = 7.5 - 1.5(n - 1)$, $a_{425} = -628.5$.

19. \$3240

21. 1,002,000 **23.** 10,950

25. If the terms in the series are written down twice, as in Figure 3-3, there are N pairs, each with a sum of $(a_1 + a_N)$. The sum of all the pairs is $N(a_1 + a_N)$, and this is twice the required sum, so $S_N = \frac{N(a_1 + a_N)}{2}$.

27. 5610

29. a. If d = horizontal distance from the low end in feet, and h = elevation in feet, then $h = 10 + 0.04d$.
b–c. 250 feet **d.** Responses will vary.

31. 72 people **33.** 76

35. 43.4

37. a. Yes, because doubling the burning time doubles the cost.
b. Yes, because doubling the number of people doubles the water usage.
c. No, because doubling the number of people cuts each person's closet space in half.
d. Yes, because doubling the number of pages doubles the weight of the book.

39. a. Arithmetic **b.** $a_n = 7 + 3(n - 1)$
c. 304; 1279 **d.** 15,550; 273,275

41. a. Arithmetic **b.** $a_n = 3487.6 - 0.1(n - 1)$
c. 3477.7; 3445.2 **d.** 348,265; 1,473,220

43. a. Not arithmetic

45. a. Arithmetic **b.** $a_n = \pi - 2 + (\pi - 1)(n - 1)$
c. $100\pi - 101$; $425\pi - 426$
d. $5050\pi - 5150$; $90{,}525\pi - 90{,}950$

47. a. The difference between consecutive terms is 32.
b. 1904 feet **c.** 57,600 feet

SECTION 3-3

1. The ratio $\frac{\Delta y}{\Delta x}$ is not the same for all pairs of points.

3.

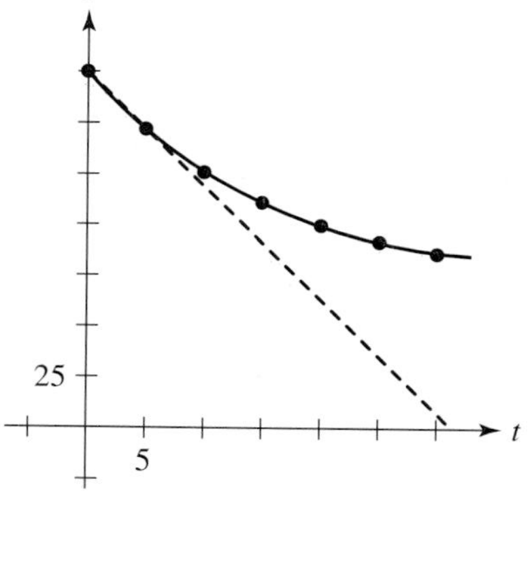

5. Entries in the second column will vary. The missing y-values from Exercise 4 are 169.4, 163.8, 158.2, and 152.6.

7. about 4.46 minutes

9.

Year	Value
1980	$52,000
1981	55,000
1982	58,000
1983	61,000
1984	64,000
1985	67,000
1986	70,000
1987	73,000
1988	76,000
1989	79,000

11.

n	$\sqrt{n}$	n	$\sqrt{n}$
16	4.0	26	5.0
17	4.1	27	5.1
18	4.2	28	5.2
19	4.3	29	5.3
20	4.4	30	5.4
21	4.5	31	5.5
22	4.6	32	5.6
23	4.7	33	5.7
24	4.8	34	5.8
25	4.9	35	5.9
		36	6.0

13. The average rate of change is -1.4, indicating that the coffee cooled at an average rate of 1.4° per minute during the time interval [20, 30].

15. The average rate of change in hours of daylight with respect to the number of days past June 21 is about -0.096, indicating that on the average, each day between June 21 and September 21 Yellowknife receives about 0.096 hours less daylight than on the previous day.

17. a. 7.45 **b.** $7.45
c. From part (a), as your sales level increases from 500 to 510 shirts, each additional shirt sold produces about $7.45 in additional profit. At a sales level of 505 shirts, your marginal profit is $7.45.

19. Increasing your sales level from 505 to 506 shirts increases your profit more.

21. a. $L(x) = \frac{1}{3}x + \frac{1}{3}$ **b.** $L(2) = 1 = f(2)$; $L(5) = 2 = f(5)$
c. $L(3) - f(3) \cong -0.081$; $L(4) - f(4) \cong -0.065$

23. a. $L(x) = x - 10$ **b.** $L(2) = -8 = f(2)$; $L(5) = -5 = f(5)$
c. $L(3) - f(3) = 2$; $L(4) - f(4) = 2$

25. 100 **27.** 0

29. a. -0.095 (Answers will vary slightly.) The number of farms in the United States decreased by an average of about 0.095 million each year.
b. 0.965 million.

31. a. 0.91. The percentage of persons between the ages of 18 and 24 enrolled in college increased by an average of 0.91 each year.
b. 29.84 **c.** 35.31
d. 98.4
e. The estimate in part (d) is probably not very accurate because it is a long-term prediction based on a short-term trend.

33. (Sample response) It is assumed that the airplane descends at a constant angle, and that it travels at a constant speed. The first assumption is probably reasonable, but the second is probably not, since the airplane must decrease its speed to land.

35. a. milepost 286
b. The assumption of linearity means that the rate of change in milepost numbers with respect to time remains constant. In other words, it means that I am driving at a constant speed.

37. a. $x \cong 15.18$ minutes **b.** $x = 26.25$ minutes
c. (Sample response) The interval (92, 100) is much closer to 90 than is (147, 175). We should therefore expect that the cooling of the coffee through the range of temperatures between 92° and 100° is a more accurate predictor of the time when the temperature will reach 90°.

SECTION 3-4

1. The first entry in the second column is (12)($73.30) = $879.60, representing the total cost of 12 monthly premiums. Under this plan Sarah must pay all of her medical expenses up to $1000. Therefore, the other entries in the second column represent $879.60 in premiums plus the indicated medical expenses in the first column.

3. a. If Sarah's medical expenses are $1000 or less, the cost of the plan is the amount of her medical expenses plus the amount of her premiums. Symbolically, this is $x + 879.60$.
b. If Sarah's medical expenses are more than $1000, the cost of the plan is $1879.60 plus 20% of her medical expenses above $1000. Symbolically, this is $1879.60 + 0.20(x - 1000) = 0.20x + 1679.60$.

5. a. **b.**

x	$f(x)$	x	$f(x)$
2.5	4	2.95	4.9
2.6	4.2	2.96	4.92
2.7	4.4	2.97	4.94
2.8	4.6	2.98	4.96
2.9	4.8	2.99	4.98
3.0	5	3.00	5
3.1	4.9	3.01	4.99
3.2	4.8	3.02	4.98
3.3	4.7	3.03	4.97
3.4	4.6	3.04	4.96
3.5	4.5	3.05	4.95

c. Yes, because $f(3) = 5$ and $f(x)$ is near 5 when x is near 3.

7.

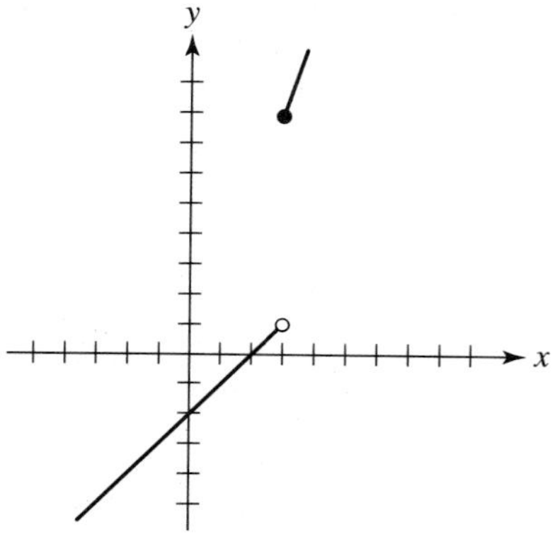

9.

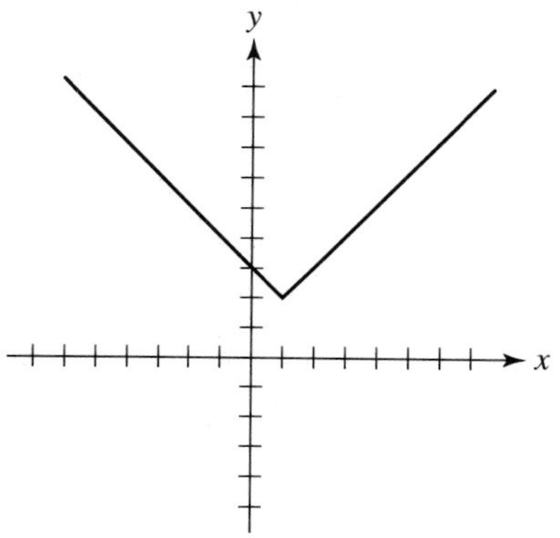

11.

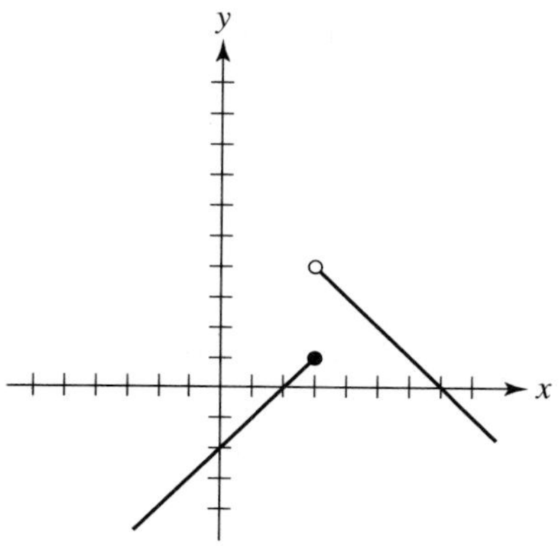

13. (Sample response) At an x-value where the rule defining the function changes, evaluate the two linear expressions that define the function on each side of that x-value. If the results are the same, the pieces join to form a continuous graph. Otherwise they don't.

15. a. 0.1 cm; 0.05 cm; 0.05 cm; 0.1 cm

b. The deviation is the *size* of the difference $h - 13.90$ between actual height and designed height, without regard to the *sign* of the difference. Thus it is $|h - 13.90|$.

c. The difference between the actual capacity and the designed capacity is $\pi(4.280)^2 h - \pi(4.280)^2 (13.90) = \pi(4.280)^2(h - 13.90)$. The deviation is the size of this difference, which is its absolute value $|\pi(4.280)^2(h - 13.90)| = \pi(4.280)^2|h - 13.90| \cong 57.55|h - 13.90|$.

17. $g(x) = \begin{cases} -3x + 4 & \text{if } x < 0 \\ 3x + 4 & \text{if } x \geqslant 0 \end{cases}$

19. $H(z) = \begin{cases} 5 + z & \text{if } z < 0 \\ 5 - z & \text{if } z \geqslant 0 \end{cases}$

21. a. $y = \begin{cases} -2x - 4 & \text{if } x < 0 \\ 2x - 4 & \text{if } x \geqslant 0 \end{cases}$

b. The slope of the left piece is -2.
The slope of the right piece is 2.
The vertex is $(0, -4)$.

c.

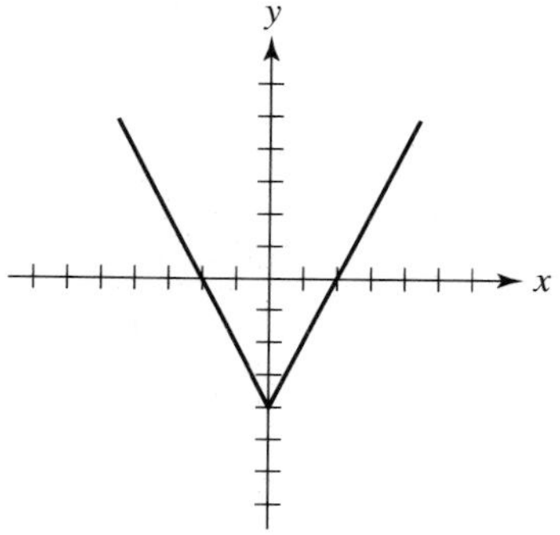

d. The axis of symmetry is the line $x = 0$ (the y-axis).
The range is $[-4, \infty)$.

23. a. $y = \begin{cases} (x - 6) + 3 & \text{if } x < 6 \\ -(x - 6) + 3 & \text{if } x \geqslant 6 \end{cases}$

b. The slope of the left piece is 1.
The slope of the right piece is -1.
The vertex is $(6, 3)$.

c.

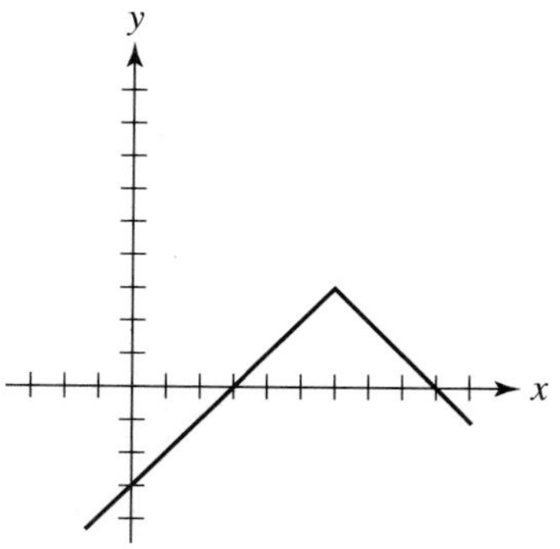

d. The axis of symmetry is the line $x = 6$. The range is $(-\infty, 3]$.

25. a.

x	y
2	4
3	2
4	0
5	2
6	4

b.

x	y
−2	0
−1	−2
0	−4
1	−2
2	0

c.

x	y
−7	$\frac{5}{3}$
−6	$\frac{4}{3}$
−5	1
−4	$\frac{4}{3}$
−3	$\frac{5}{3}$

d.

x	y
4	1
5	2
6	3
7	2
8	1

e. The tables indicate the slope of each piece of the graph and the location of the vertex.

27. a–d.

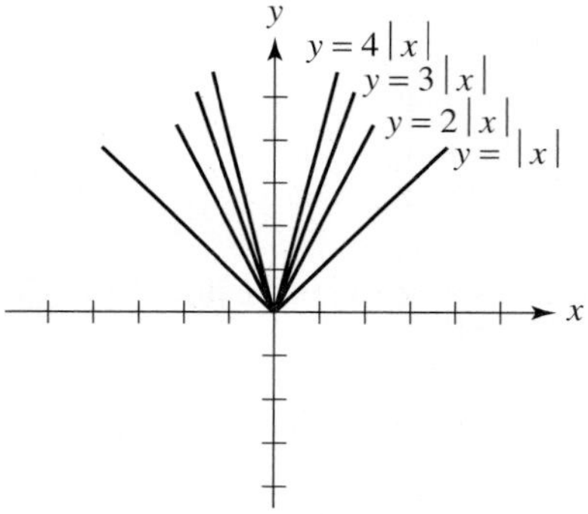

e. The graph must be stretched vertically by a factor of a.

29. a–d.

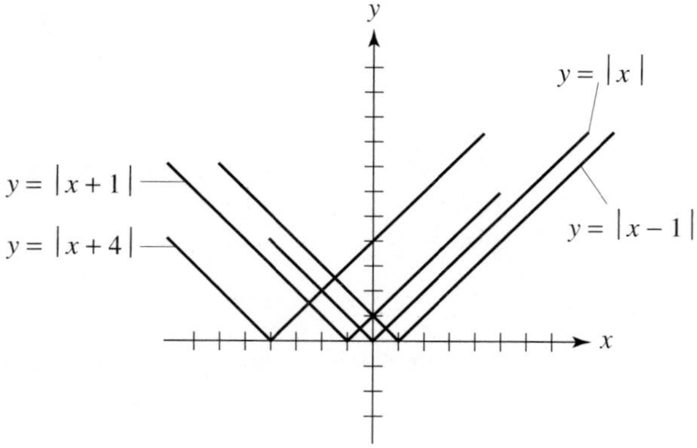

e. The graph must be shifted h units to the right. If $h < 0$, the actual shift is to the left.

31. a.

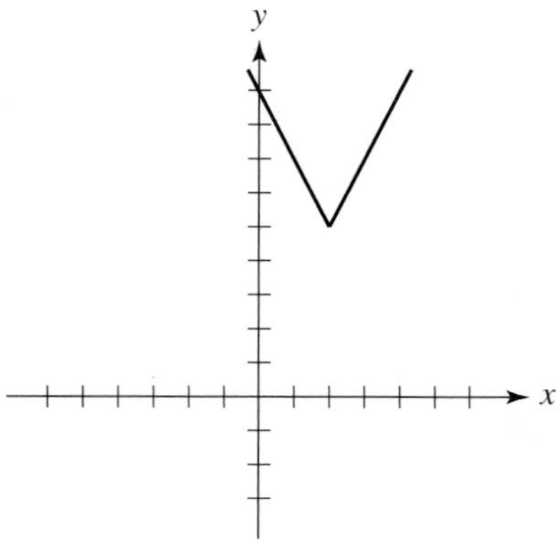

Stretch the graph of $y = |x|$ vertically by a factor of 2, then shift it 2 units to the right and 5 units up.

b.

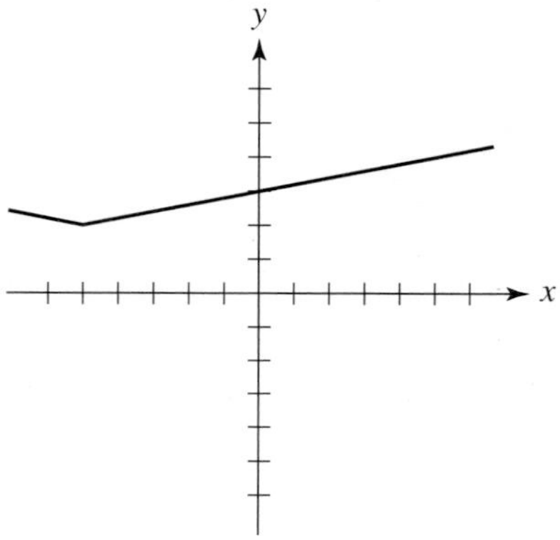

Compress the graph of $y = |x|$ vertically by a factor of 5, then shift it 5 units to the left and 2 units up.

c.

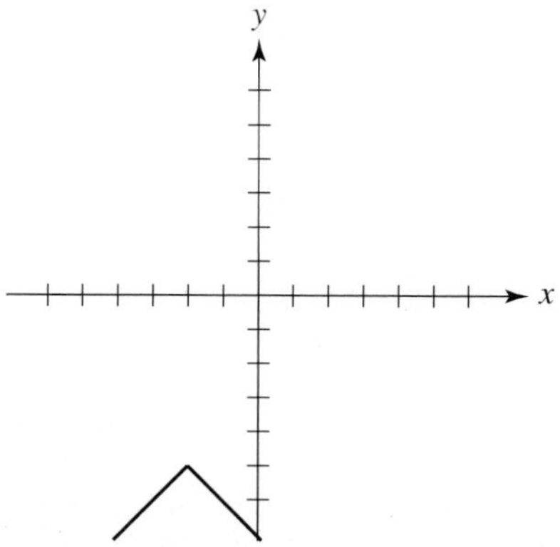

Reflect the graph of $y = |x|$ in the x-axis, then shift it 2 units to the left and 5 units down.

d.

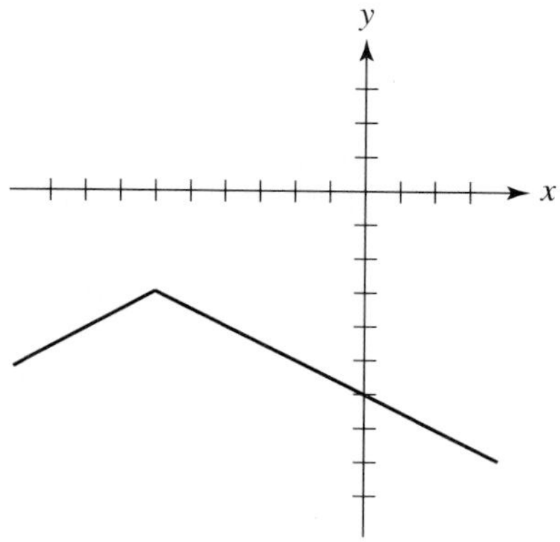

Compress the graph of $y = |x|$ vertically by a factor of 2, reflect it in the x-axis, then shift it 6 units to the left and 3 units down.

33. $t = -2, 3$

35. $(-\infty, -22.4] \cup [-17.6, \infty)$

37. a.

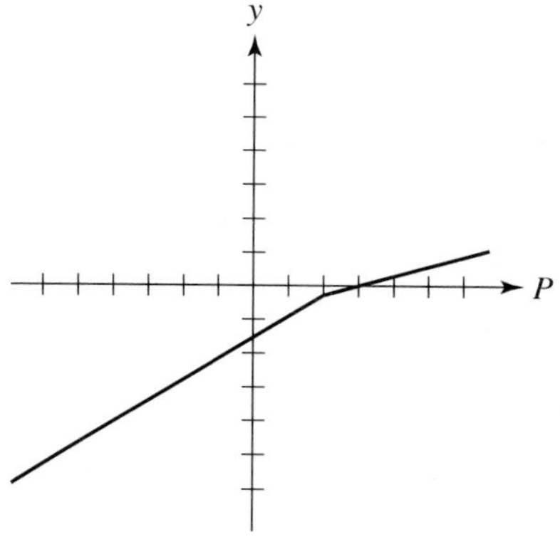

b. domain $= (-\infty, \infty)$; range $= (-\infty, \infty)$

c. increasing on $(-\infty, \infty)$

39. a.

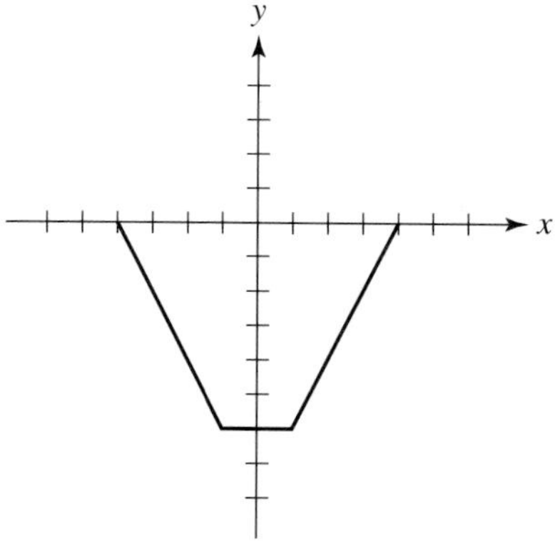

b. domain $= [-4, 4]$; range $= [-6, 0]$

c. decreasing on $[-4, -1)$; increasing on $(1, 4]$

41. a. The slope of the left piece is -6.
The slope of the right piece is 6.
The vertex is $(0, 0)$.

b.

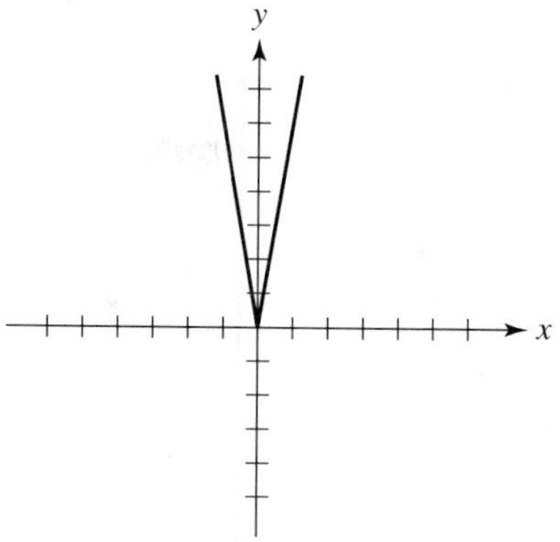

c. The axis of symmetry is the line $x = 0$ (the y-axis). The range is $[0, \infty)$.

d. The function is decreasing on $(-\infty, 0)$ and increasing on $(0, \infty)$.

e. The graph of $y = |x|$ must be stretched vertically by a factor of 6.

43. a. The slope of the left piece is -1.
The slope of the right piece is 1.
The vertex is $(6, 0)$.

b.

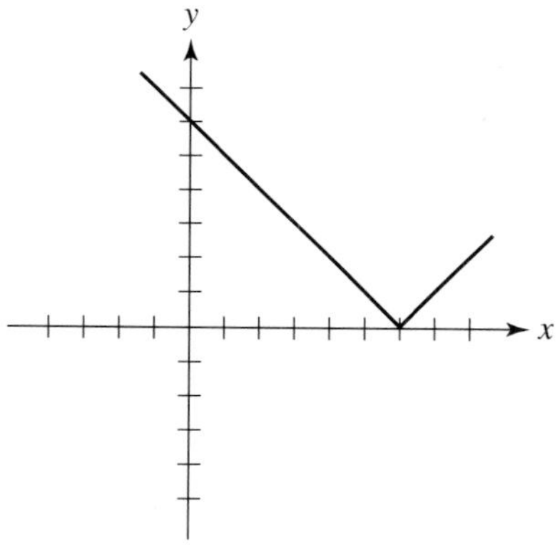

c. The axis of symmetry is the line $x = 6$. The range is $[0, \infty)$.

d. The function is decreasing on $(-\infty, 6)$ and increasing on $(6, \infty)$.

e. The graph of $y = |x|$ must be shifted 6 units to the right.

45. a. The slope of the left piece is -3.
The slope of the right piece is 3.
The vertex is $(-2, 0)$.

b.

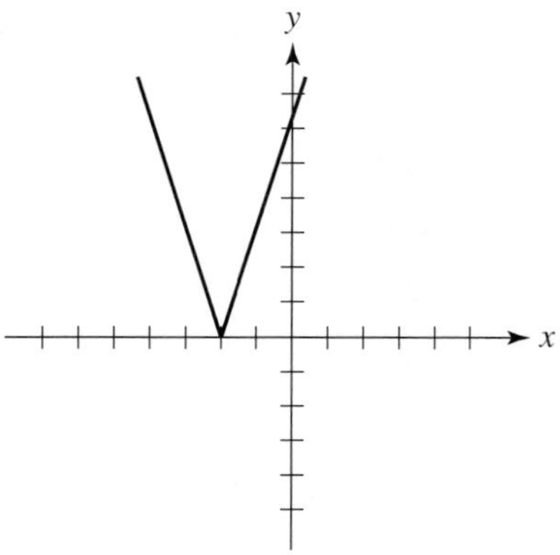

c. The axis of symmetry is the line $x = -2$. The range is $[0, \infty)$.

d. The function is decreasing on $(-\infty, -2)$ and increasing on $(-2, \infty)$.

e. The graph of $y = |x|$ must be stretched vertically by a factor of 3, then shifted 2 units to the left.

47. a. The slope of the left piece is -5.
The slope of the right piece is 5.
The vertex is $(-1, 7)$.

b.

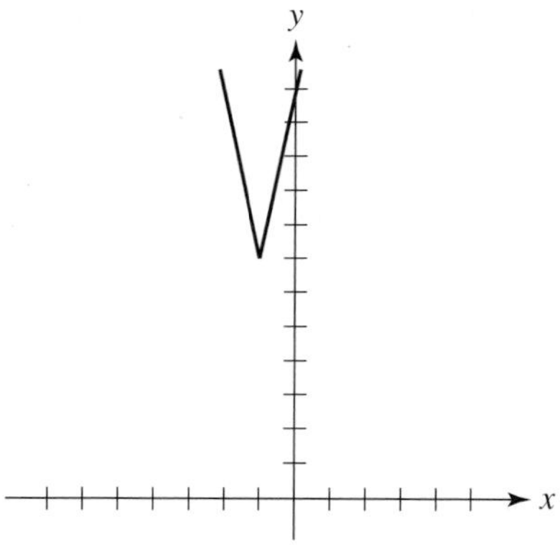

c. The axis of symmetry is the line $x = -1$. The range is $[7, \infty)$.

d. The function is decreasing on $(-\infty, -1)$ and increasing on $(-1, \infty)$.

e. The graph of $y = |x|$ must be stretched vertically by a factor of 5, then shifted 1 unit to the left and 7 units up.

49. $x = -\frac{5}{3}, \frac{5}{3}$

51. $x = 15$

53. $\left(-\frac{26}{3}, 4\right)$

55. $(-\infty, 0) \cup (4, \infty)$

57. $(-\infty, \infty)$

59. For Plan 2, $g(x) = \begin{cases} x + 979.20 & \text{if } 0 \le x \le 500 \\ 0.20x + 1379.20 & \text{if } x > 500 \end{cases}$

For Plan 3, $h(x) = \begin{cases} x + 1065.60 & \text{if } 0 \le x \le 250 \\ 0.20x + 1265.60 & \text{if } x > 250 \end{cases}$

61. Responses will vary.

63.

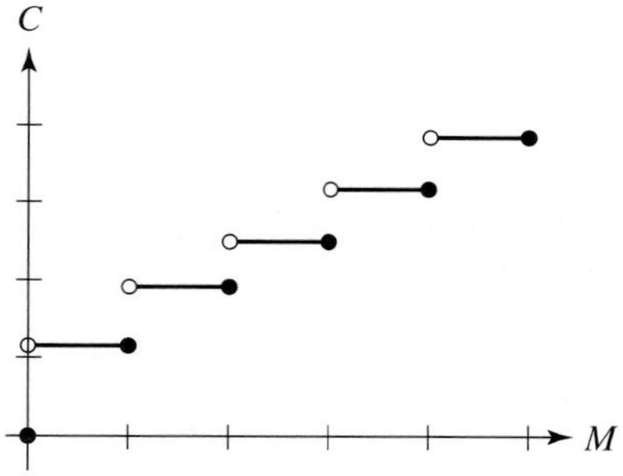

65. $E(t) = |t - 8832|$

67. a. $E(L) = 4|L - 129|$ **b.** $128.5 \le L \le 129.5$

69. a. $E(r) = 13.90\pi|r^2 - 4.280^2|$ **b.** $4.25 \le r \le 4.31$

71. a–b.

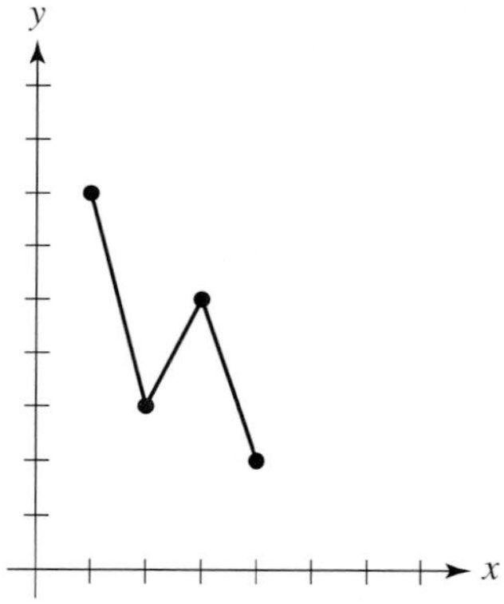

c. $y = -4x + 11$
$y = 2x - 1$
$y = -3x + 14$

d. $y = \begin{cases} -4x + 11 & \text{if } 1 \le x < 2 \\ 2x - 1 & \text{if } 2 \le x < 3 \\ -3x + 14 & \text{if } 3 \le x \le 4 \end{cases}$

SECTION 4–1

1.

Price per shirt	Demand	Supply
\$ 3.50	2500	400
5.50	2100	600
7.50	1700	800
9.50	1300	1000
11.50	900	1200
13.50	500	1400

3. If x represents the price per shirt, in dollars, and y represents the demand for shirts, then $y = 3200 - 200x$.

5. $x = 10.50, y = 1100$

7. $(1, 5)$ **9.** $(3, 2)$

11. For the system in Exercise 8 the solution is $(-3.30, -5.15)$. For the system in Exercise 9, the solution is $(3, 2)$.

13. The terms $0.65L$, $0.80M$, and $0.65C$ represent the number of days per month Paul and his helpers will spend on landscaping, masonry, and carpentry, respectively. The equation $0.65L + 0.80M + 0.65C = 72$ expresses the condition that the total amount of time Paul and his helpers spend working should equal 72 days per month.

15. Everyone can be utilized to full capacity if the company schedules 60 days of landscaping work, 12 days of masonry, and 36 days of carpentry each month.

17.

x	y (1st equation)	y (2nd equation)
1	0	0
2	1	1
3	2	2
4	3	3

The system has infintely many solutions. The table indicates that for every value of x, the corresponding values of y in the two equations are equal.

19. dependent **21.** dependent

23. The system in Exercise 19 can be rewritten

$$3S - 5T = -4$$
$$9S - 15T = -12$$

Multiplying the first equation by 3 yields

$$9S - 15T = -12$$
$$9S - 15T = -12$$

Subtracting yields $0 = 0$, so the system is dependent.

25. $(4.25, 5)$ **27.** $(T, V) = (30, 4)$

29. no solution **31.** $(x, y, z) = (10, 1, 5)$

33. $(p, q, r) = (-0.1, 0.3, 0.2)$

35. consistent and independent

37. dependent

39. a.

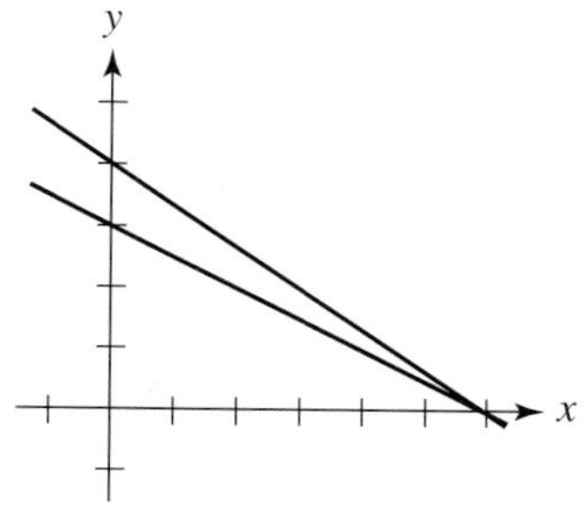

b.

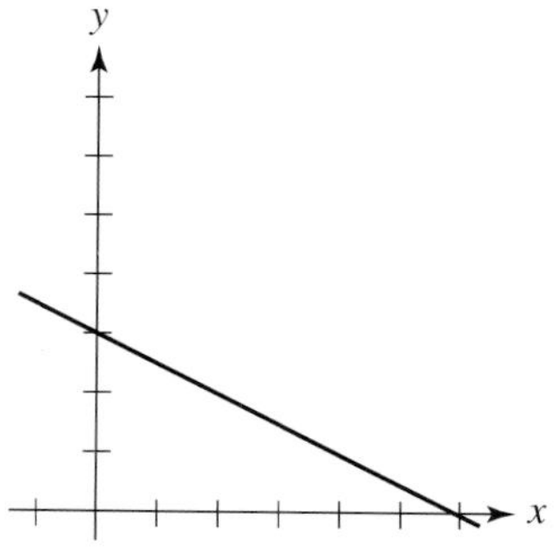

c.

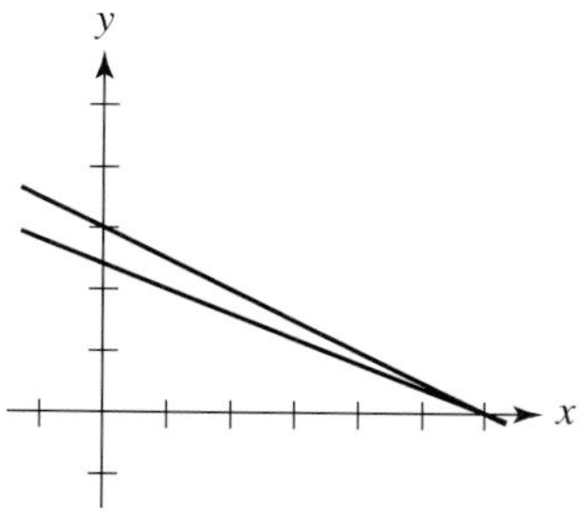

The system in part (b) is dependent

41. a.

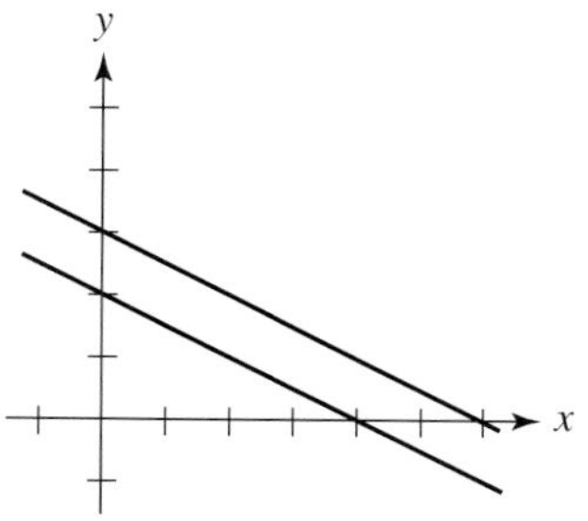

b.

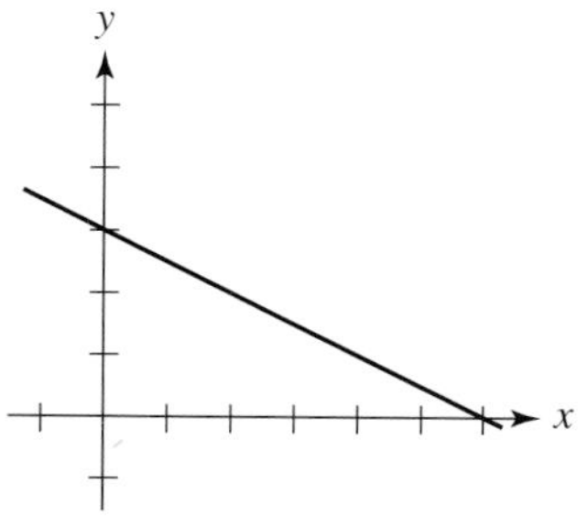

c.

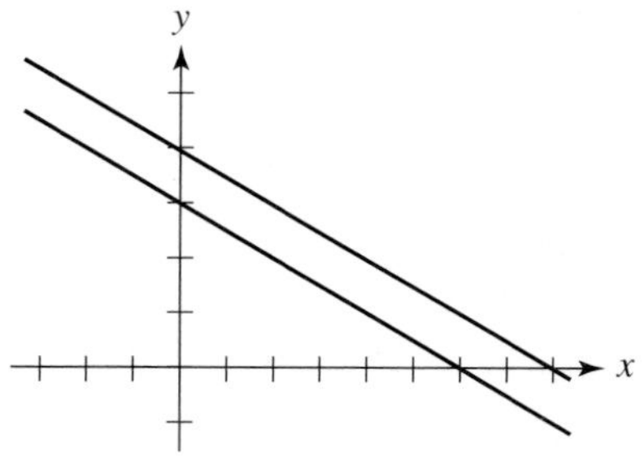

The system in part (b) is dependent.

43. \$3333.33

45. 18%

47. 30,000 pounds of the first alloy and 20,000 pounds of the second alloy

49. Responses will vary.

51. A system of three linear equations in two variables usually has no solutions.

53. A system of one linear equation in two variables has infinitely many solutions.

55. none

57. infinitely many

59. none; infinitely many; one

SECTION 4–2

1. $\begin{bmatrix} 3 & 4 & 8 \\ 1 & 7 & 5 \end{bmatrix}$

3. $\begin{bmatrix} 2 & 3 & 5 \\ -1 & -4 & 0 \\ 3 & 0 & 12 \\ 0 & 4 & -4 \end{bmatrix}$

5. If the variables are s and t,
$$\begin{aligned} 23s + 6t &= -2 \\ 7s + 31t &= 12 \end{aligned}$$

7. If the variables are x, y, and z,
$$\begin{aligned} x + 3y + 9z &= 5 \\ \tfrac{1}{3}x + 4y - 13z &= 7 \\ -2y + 3z &= 16 \end{aligned}$$

9. The third row was multiplied by $\frac{1}{60}$.

11. The matrix is not in reduced row-echelon form because the third column, which contains the leading nonzero entry in the third row, contains other nonzero entries.

13. The matrix is in reduced row-echelon form.

15. $\begin{bmatrix} 1 & 0 & 2 \\ 0 & 1 & -1 \end{bmatrix}$

17. $\begin{bmatrix} 1 & 0 & 0 & 1.5 \\ 0 & 1 & 0 & 11 \\ 0 & 0 & 1 & 5 \end{bmatrix}$

19. $(x, y) = (2, -1)$

21. $(x, y, z) = (1.5, 11, 5)$

23.
$$\begin{aligned} 0.25L + 0.15M + 0.20C &= 24 \\ 0.60L + 0.80M + 0.70C &= 56 \\ 0.15L + 0.05M + 0.10C &= 8 \end{aligned}$$

25. a–b. no solution

c. The graph shows that the system has no solution because the graphs of the two equations in the system have no point in common. The matrix shows the same thing because the system is equivalent to one containing the self-contradictory equation $0 = -5$.

27. Here is one possible sequence of operations. Multiply row 1 by 4, add -0.60 times row 1 to row 2, add -0.15 times row 1 to row 3, interchange rows 2 and 3, multiply row 2 by -25, add -0.60 times row 2 to row 1, add -0.44 times row 2 to row 3.

29. $x = 8 - 2z_0$
$y = 7 - 4z_0$
$z = z_0$ for any real value of Z.

31. $w = 0.5 - 10y_0$
$x = 2.5 - 20y_0$
$y = y_0$ for any real value of Z.
$z = 4$

33. $A = -1$
$B = -1$
$C = 3 - d$
$D = d$ for any real value of d

35. $p = 0$
$q = s_0$
$r = 0$
$s = s_0$ for any real value of s_0

37. $\begin{bmatrix} 1 & 1 & 1 & 1 & 0 \\ 3 & -1 & 19 & 6 & \frac{1}{7} \\ 2 & 0 & 0 & 0 & 8 \end{bmatrix}$

39. $\begin{bmatrix} 2 & -3 & -4 & -5 & 0 & 0.001 \\ 0 & 2 & -3 & -4 & -5 & -0.001 \end{bmatrix}$

41. No.

43. Yes.

45. $\begin{bmatrix} 1 & -\frac{7}{2} & 3 \\ 0 & 1 & -\frac{2}{3} \end{bmatrix}$

47. $\begin{bmatrix} 1 & 0 & 0 & 0 \\ 0 & 1 & 0 & 0 \\ 0 & 0 & 1 & 0 \end{bmatrix}$

49. no solution

51. $x = -\frac{2}{3}z_0 + \frac{3}{2}$
$y = \frac{1}{3}z_0 + 2$
$z = z_0$ for any real value of z_0

53. $R = 2s + 7$
$S = s$ for any real value of s
$T = 2$

55. no solution

57. $x = -2$
$y = 4 - 0.1z_0$
$z = z_0$
for any real value of z_0

59. $v = 25 - 3x_0$
$w = -x_0$
$x = x_0$
$y = 15 - 5z_0$
$z = z_0$
for any real values of x_0 and z_0

61. a. $x_1 + x_2 = 250$; $y_1 + y_2 = 300$; $x_1 + y_1 = 200$; $x_2 + y_2 = 350$

b. $x_1 = -100 + y_2$
$x_2 = 350 - y_2$
$y_1 = 300 - y_2$
$y_2 = y_2$ for any real value of y_2

c. Some choices for (x_1, x_2, y_1, y_2) are $(100, 150, 100, 200)$, $(150, 100, 50, 250)$, $(50, 200, 150, 150)$, and $(0, 250, 200, 100)$.

d. One possibility would be to set up a counter on the straightaway between C and D, positioned between the exit to B and the entrance from A as shown.

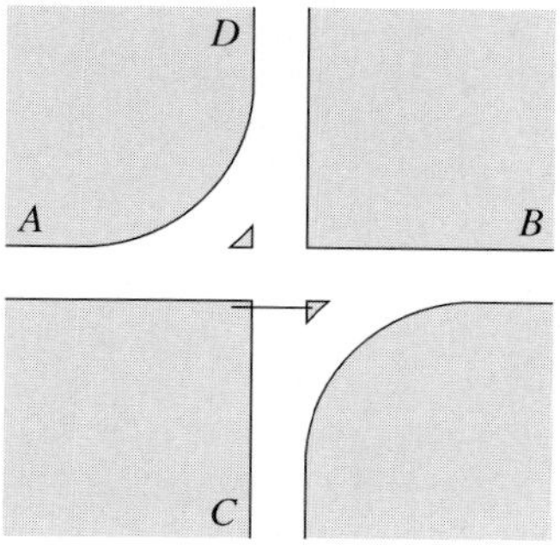

e. One possibility would be to move the counter at C to the position shown in part (d).

63. a. $0.20C + 0.25L + 0.15M = 16$
$0.70C + 0.60L + 0.80M = 56$
$0.10C + 0.15L + 0.05M = 8$

b. A system in which all the variables are leading cannot have infinitely many solutions. Therefore, the system in ***Sunsilk 3*** must have at least one nonleading variable.

65. No.

SECTION 4–3

1. The cost of x bales of timothy at \$1.50 per bale is $1.50x$ dollars, and the cost of y bales of alfalfa at \$2.50 per bale is $2.50y$ dollars. The combined cost is $1.50x + 2.50y$ dollars, which must be equal to her expenditure of \$165.

3. (Sample response) Every linear inequality can be written in one of the forms $y > f(x)$, $y < f(x)$, $y \geq f(x)$ or $y \leq f(x)$. The inequality $y > f(x)$ is true at all points above the graph of $y = f(x)$. The inequality $y < f(x)$ is true at all points below the graph of $y = f(x)$. The inequality $y \geq f(x)$ is true at all points on or above the graph of $y = f(x)$. The inequality $y \leq f(x)$ is true at all points on or below the graph of $y = f(x)$.

5.

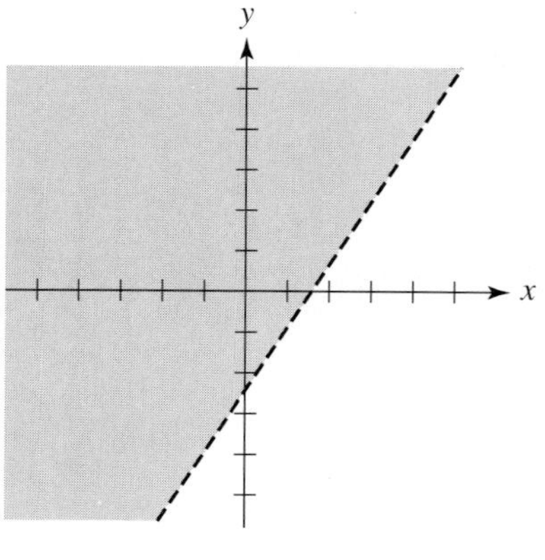

7.

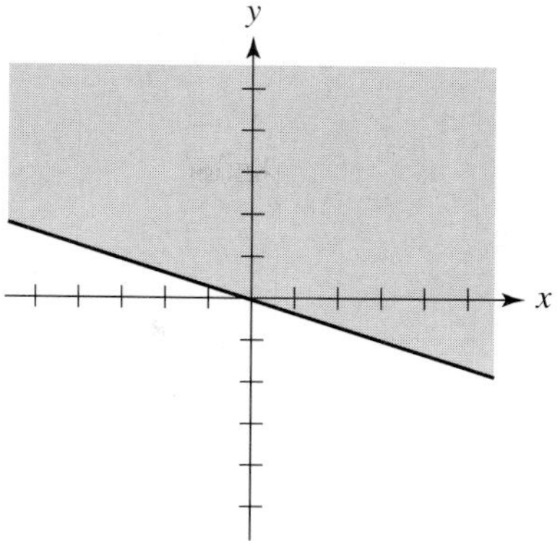

9.

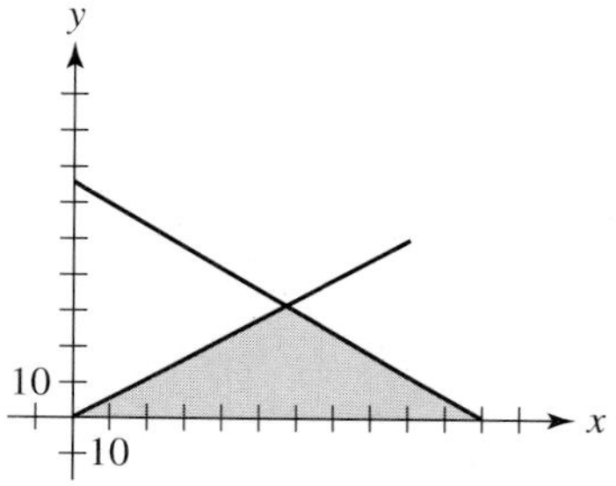

11.

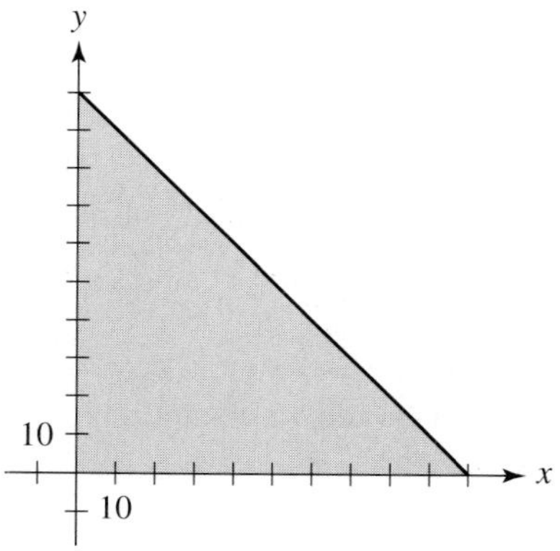

13. The feasible region represents the combinations of at most 100 bales of timothy and alfalfa Carrie can buy for no more than \$165, containing at least twice as much timothy as alfalfa.

15.

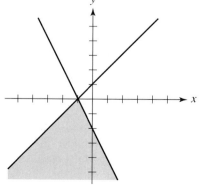

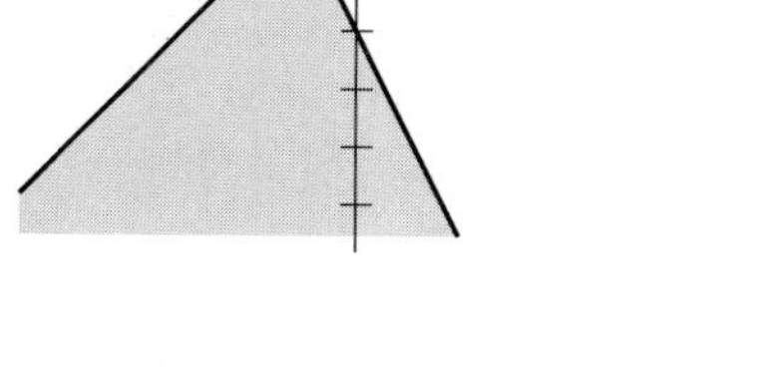

17.

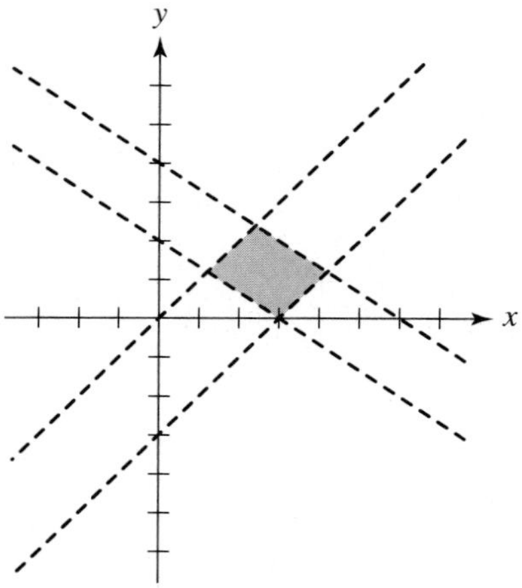

19.

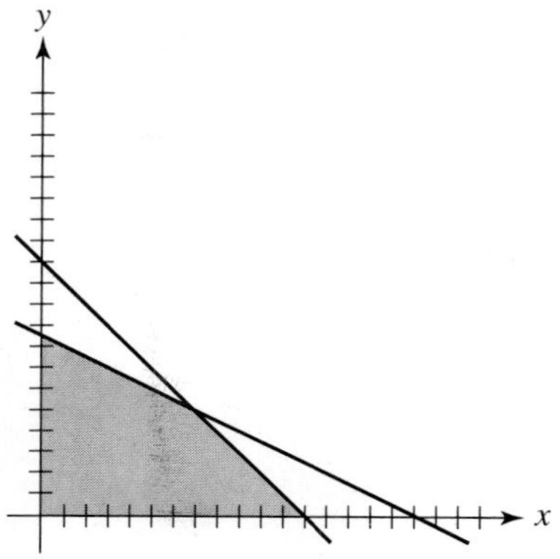

21. At $(0, 0)$, $50 - 2x - 3y = 50$.
At $(17, 0)$, $50 - 2x - 3y = 16$.
At $(0, 12)$, $50 - 2x - 3y = 14$.
At $(7, 5)$, $50 - 2x - 3y = 21$.
The minimum value is 14, and occurs at $(0, 12)$.

23. -27

25.

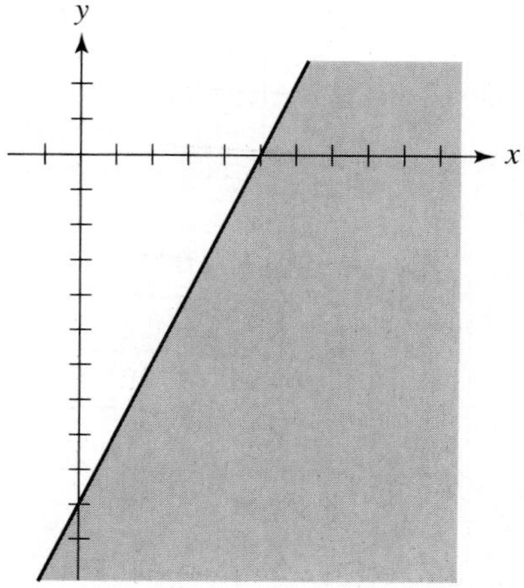

27.

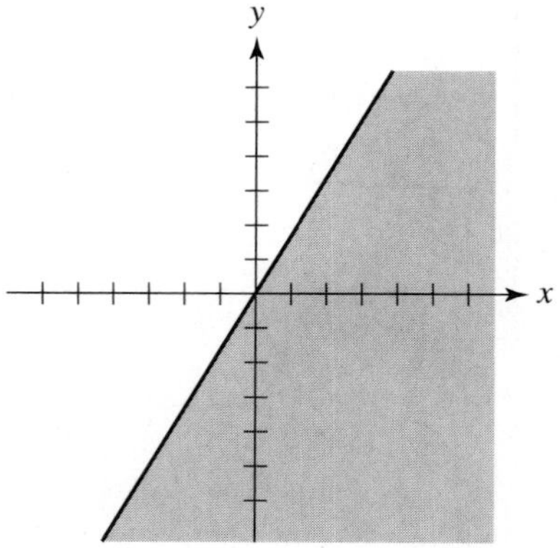

29.

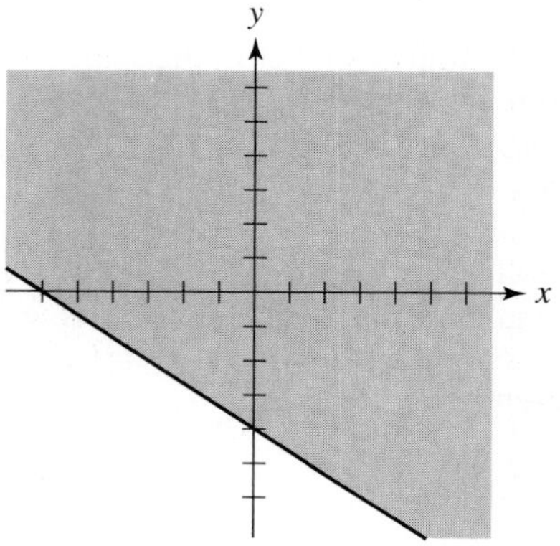

31.

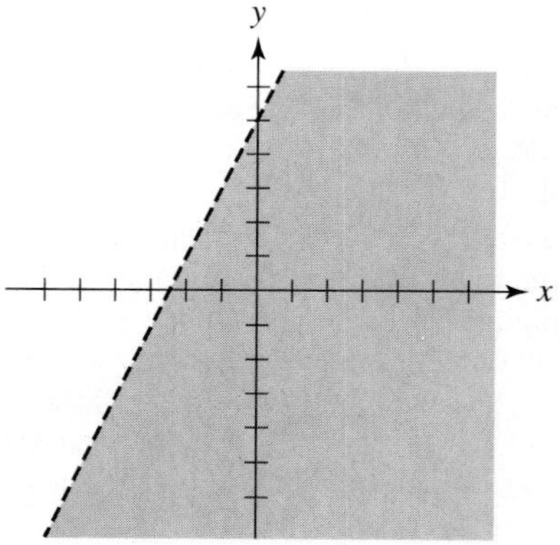

33.

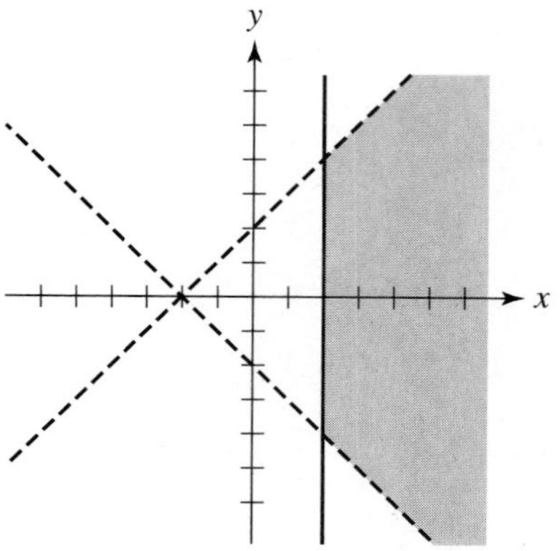

35.

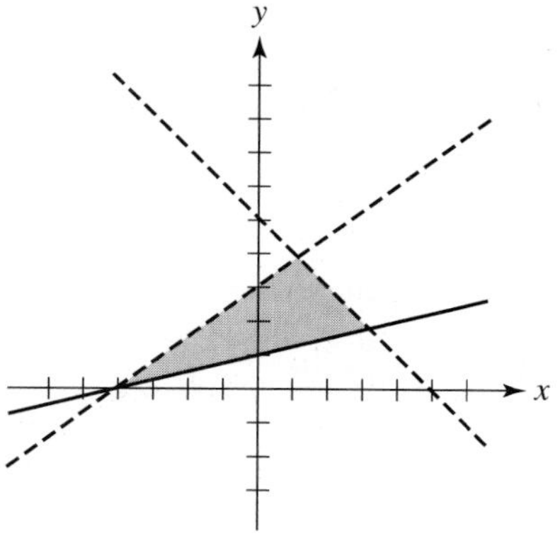

37.

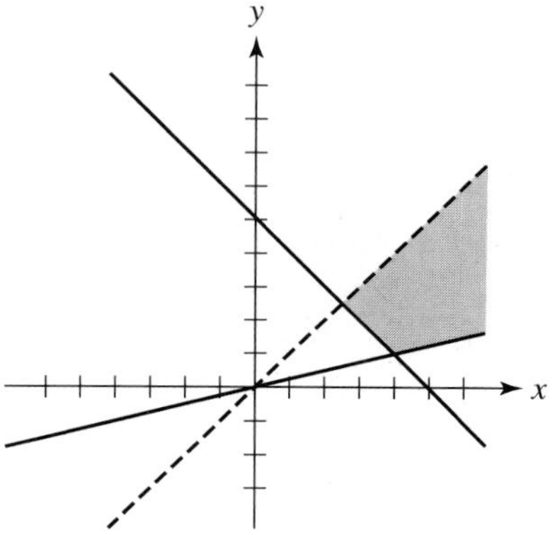

39. -5.75

41. 9

43. 250 newspaper ads, 10 TV spots

SECTION 5–1

1. **a.** At each sales level x, the profit is the y-coordinate on the graph of $y = P(x)$. The point V has the largest y-coordinate on the graph and, therefore, represents the sales level with the maximum profit.
   b. Estimates will vary slightly, but should be around 1250 shirts.
3. The maximum profit appears to occur halfway between the x-intercepts.
5. **a.** $(0, 0)$ **b.** steeper if $|a| > 1$ and flatter if $|a| < 1$
   c. up if $a > 0$ and down if $a < 0$
7. **a–d.**

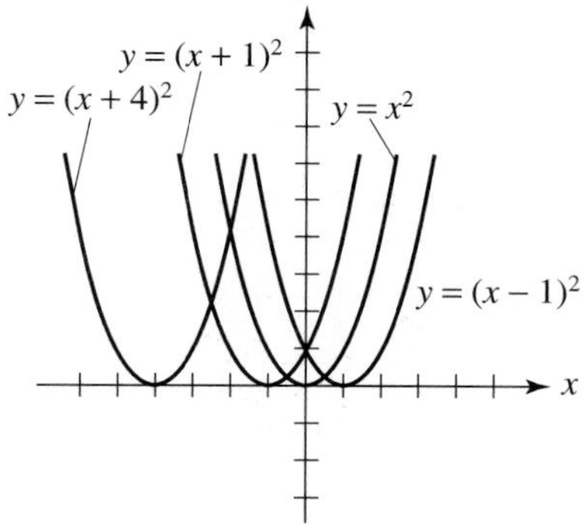

9. **a–d.**

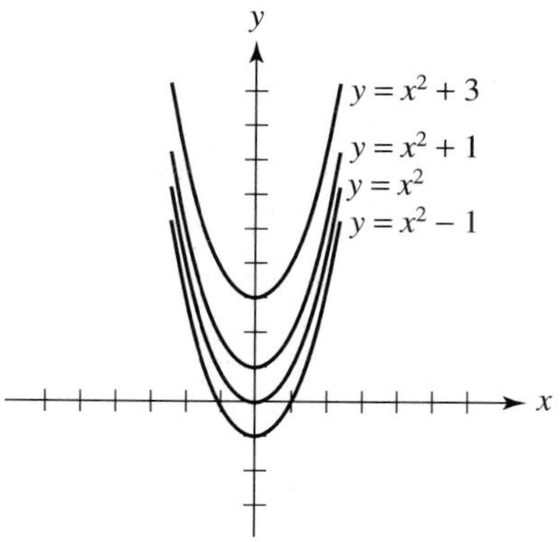

11. **a.**

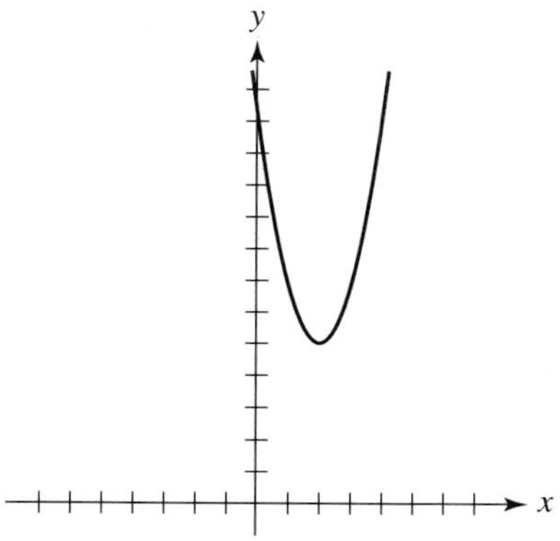

b.

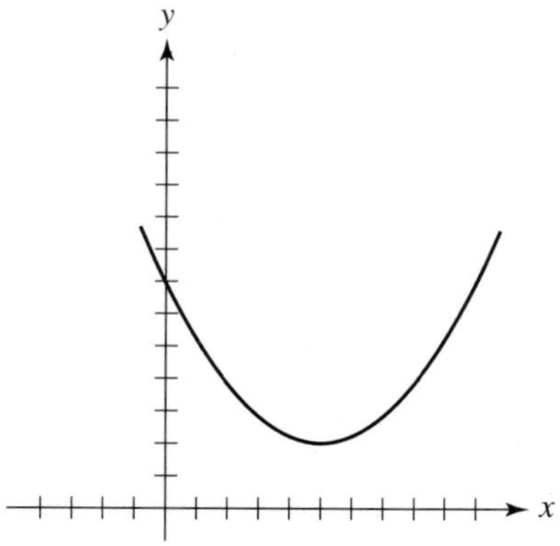

c.

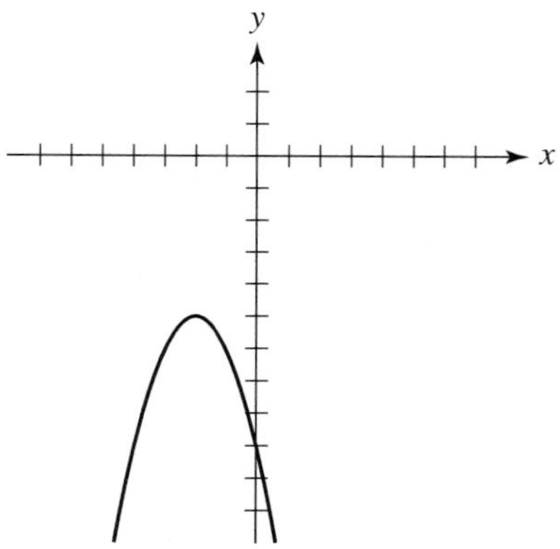

d.

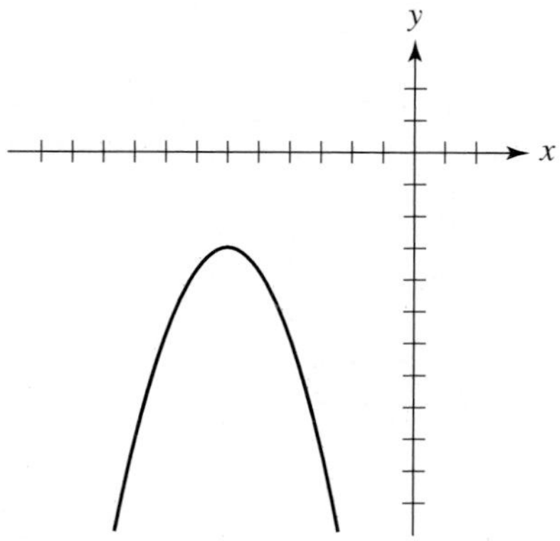

13. a. The graph of $y = x^2$ must be stretched vertically by a factor of 2, then shifted 4 units to the right and 1 unit up.

b.

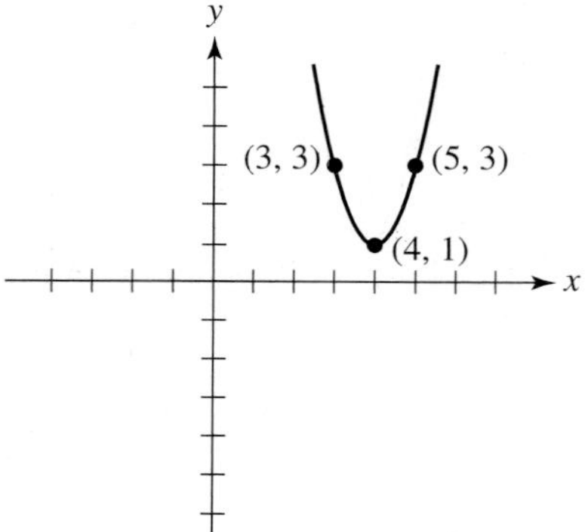

15. a. The graph of $y = x^2$ must be compressed vertically by a factor of $\frac{4}{3}$, reflected in the x-axis, then shifted 3 units down.

b.

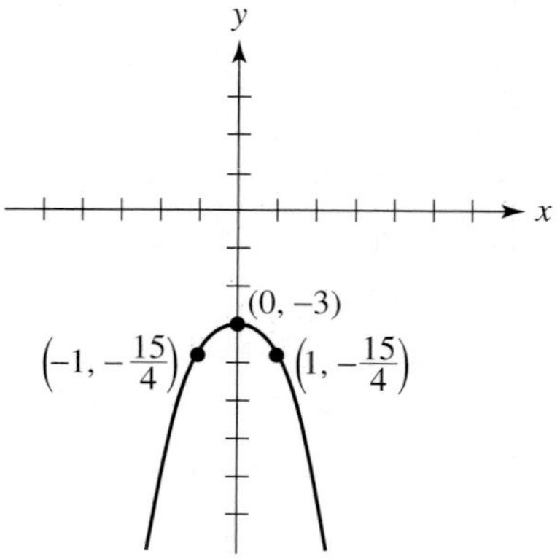

17. vertex $= (-4, -7)$; axis of symmetry $x = -4$; range $= [-7, \infty)$

19. vertex $= (0, 9)$; axis of symmetry $x = 0$; range $= (-\infty, 9]$

21. a. $x = \frac{1}{2}, \frac{9}{2}$ **b.** $x = \frac{5}{2}$

c. $x = \frac{5}{2} \pm 2i$

23. A quadratic equation can have 0, 1, or 2 real solutions. The number of real solutions to a quadratic equation $f(x) = 0$ is equal to the number of x-intercepts on the graph of $y = f(x)$.

25. discriminant $= -16$; no x-intercepts

27. discriminant $= 0$; one x-intercept

29. $f(x) = 3x\left(x + \frac{8}{3}\right)$ **31.** $P(t) = 6\left(t + \frac{5}{3}\right)\left(t - \frac{1}{2}\right)$

33. $f(x) = 15\left(x + \frac{12}{5}\right)\left(x + \frac{1}{3}\right)$

35. $Q(x) = \left[x - \left(1 + \frac{\sqrt{6}}{2}\right)\right]\left[x - \left(1 - \frac{\sqrt{6}}{2}\right)\right]$

37. $Q(x) = [x - (-4 + \sqrt{6})][x - (-4 - \sqrt{6})]$

39. $P(x) = -0.005(x - 50)(x - 2450)$; $(50, 0)$ and $(2450, 0)$

41. 3

43. vertex $= (0, 7)$; axis of symmetry $x = 0$; range $= (-\infty, 7]$

45. vertex $= (1, 18)$; axis of symmetry $x = 1$; range $= [18, \infty)$

47. a. $y = 0.5(x + 4)^2 + 1$

b. vertex $= (-4, 1)$
axis of symmetry $x = -4$
range $= [1, \infty)$

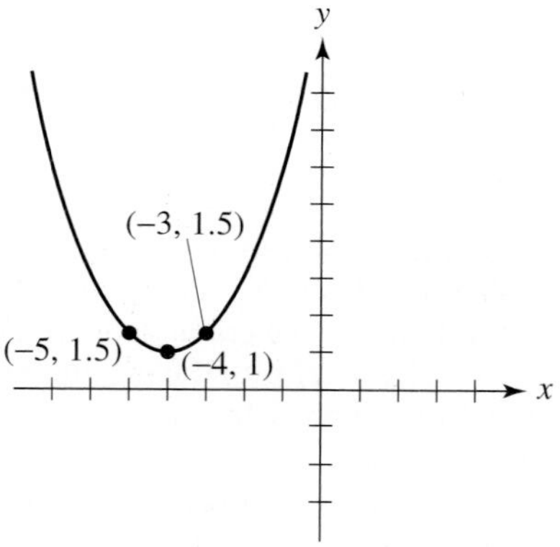

49. a. $w = -3(z - 1)^2 + 3$

b. vertex $= (1, 3)$
axis of symmetry $z = 1$
range $= (-\infty, 3]$

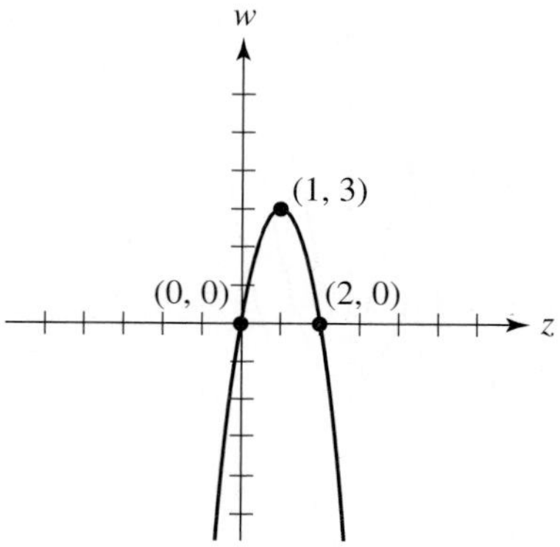

51. $(-4, 0), (-8, 0)$ **53.** $\left(-3 \pm \frac{1}{\sqrt{2}}, 0\right)$

55. If a and k are both positive, then $a(x-h)^2 \geqslant 0$ for all x, so $a(x-h)^2 + k > 0$ for all x. Similarly, if a and k are both negative, then $a(x-h)^2 + k < 0$ for all x. In each case, $a(x-h)^2 + k$ is never 0, so the graph has no x-intercepts.

57. a.

x	$y = 3x^2$	First differences	Second differences
1.0	3.00		
		3.75	
1.5	6.75		1.5
		5.25	
2.0	12.00		1.5
		6.75	
2.5	18.75		1.5
		8.25	
3.0	27.00		

b.

x	$y = ax^2$	First differences	Second differences
-1	a		
		$-a$	
0	0		$2a$
		a	
1	a		$2a$
		$3a$	
2	$4a$		$2a$
		$5a$	
3	$9a$		

c.

x	$y = ax^2$	First differences	Second differences
x_0	ax_0^2		
		$2ahx_0 + ah^2$	
$x_0 + h$	$ax_0^2 + 2ahx_0 + ah^2$		$2ah^2$
		$2ahx_0 + 3ah^2$	
$x_0 + 2h$	$ax_0^2 + 4ahx_0 + 4ah^2$		$2ah^2$
		$2ahx_0 + 5ah^2$	
$x_0 + 3h$	$ax_0^2 + 6ahx_0 + 9ah^2$		$2ah^2$
		$2ahx_0 + 7ah^2$	
$x_0 + 4h$	$ax_0^2 + 8ahx_0 + 16ah^2$		

59. quadratic **61.** quadratic

63. a. 5.1. For speeds between 30 and 40 mph, every increase of 1 mph in speed results in an increase of about 5.1 feet in the required stopping distance.

b. The average rate of change is 6.2. As your speed increases, increases of 1 mph in speed produces increasingly larger increases in the required stopping distance.

65. a. The graph of $y = x^2$ must be shifted 2.5 units to the right and 6.8 units down.

b–c.

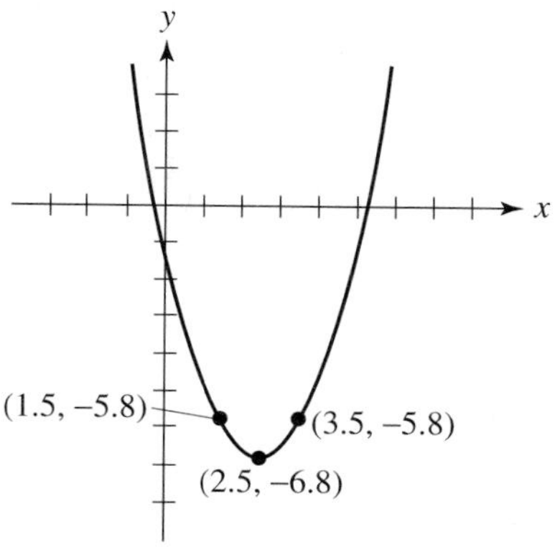

d. axis of symmetry $x = 2.5$
range $= [-6.8, \infty)$

e. decreases on $(-\infty, 2.5)$, increases on $(2.5, \infty)$

67. a. The graph of $y = r^2$ must be stretched vertically by a factor of 2π, then shifted 1 unit to the left and 2π units down.

b–c.

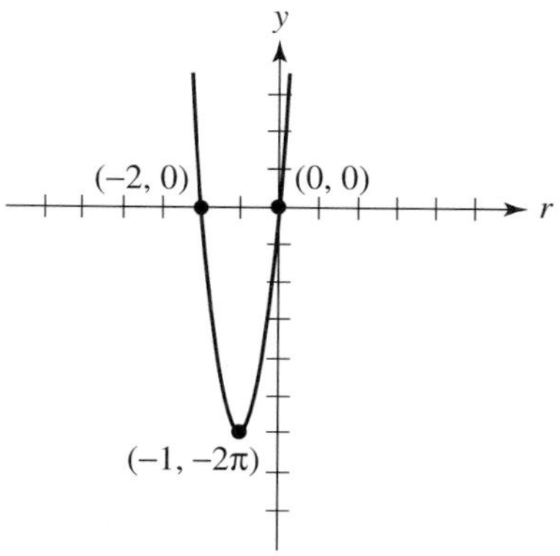

d. axis of symmetry $r = -1$
range $= [-2\pi, \infty)$

e. decreases on $(-\infty, -1)$, increases on $(-1, \infty)$

69. a. The graph of $q = p^2$ must be stretched vertically by a factor of 100, then shifted 3 units to the left and 5 units down.

b–c.

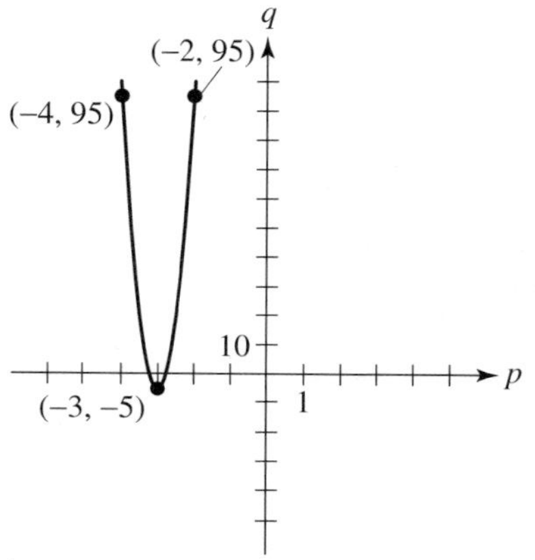

d. axis of symmetry $p = -3$
range $= [-5, \infty)$
e. decreases on $(-\infty, -3)$, increases on $(-3, \infty)$

71. a. x-intercepts $(-2, 0), (3, 0)$
vertex $= \left(\frac{1}{2}, \frac{25}{4}\right)$

b.

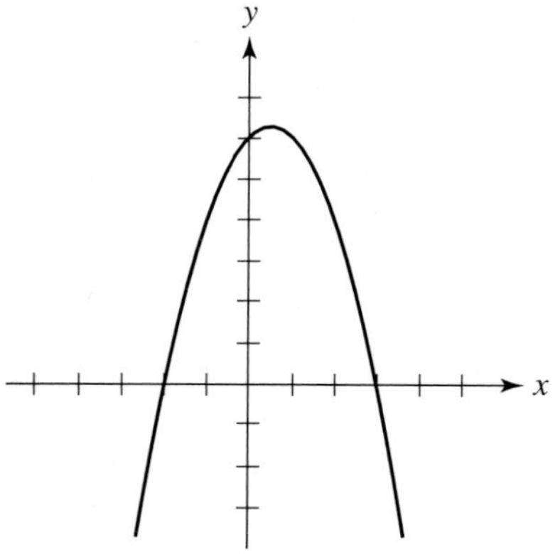

c. axis of symmetry $x = \frac{1}{2}$
range $= \left(-\infty, \frac{25}{4}\right]$
d. increases on $\left(-\infty, \frac{1}{2}\right)$, decreases on $\left(\frac{1}{2}, \infty\right)$

73. a. t-intercepts $(\pm 5, 0)$
vertex $(0, 400)$

b.

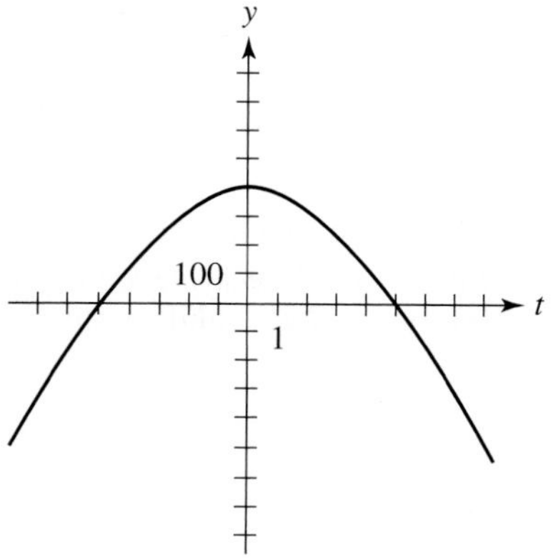

c. axis of symmetry $t = 0$
range $= (-\infty, 400]$
d. increases on $(-\infty, 0)$,
decreases on $(0, \infty)$

75. a. x-intercepts $\left(\frac{3 \pm \sqrt{5}}{2}, 0\right)$
vertex $\left(\frac{3}{2}, -\frac{5}{4}\right)$

b.

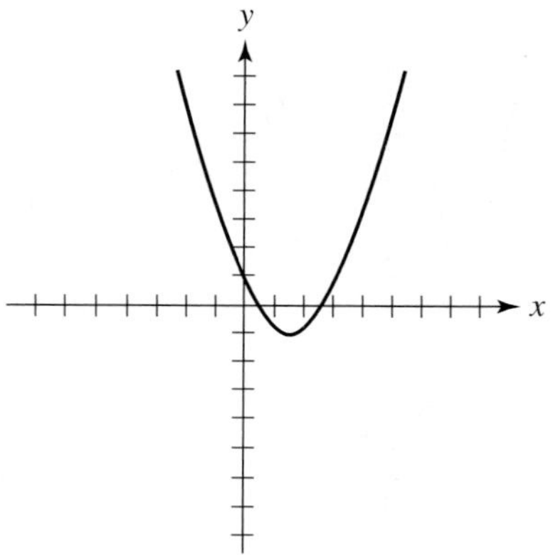

c. axis of symmetry $x = \frac{3}{2}$
range $= \left[-\frac{5}{4}, \infty\right)$
d. decreases on $\left(-\infty, \frac{3}{2}\right)$, increases on $\left(\frac{3}{2}, \infty\right)$

77. a. u-intercepts $(-6, 0), (0, 0)$
vertex $(-3, -18)$

b.

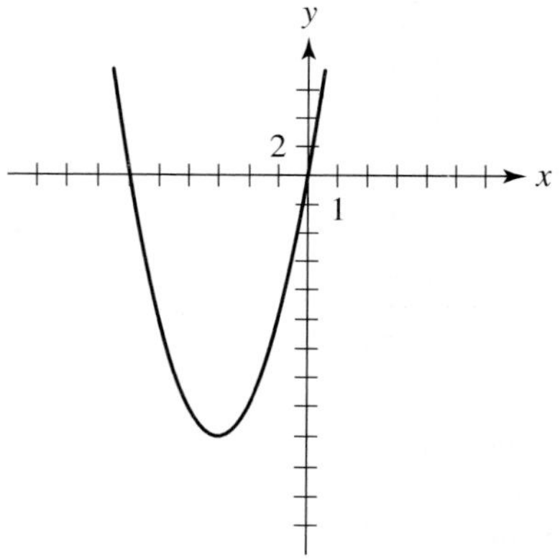

c. axis of symmetry $u = -3$
range $= [-18, \infty)$
d. decreases on $(-\infty, -3)$, increases on $(-3, \infty)$

79. a. x-intercepts $(\pm \sqrt{5}, 0)$
vertex $(0, -5)$

b.

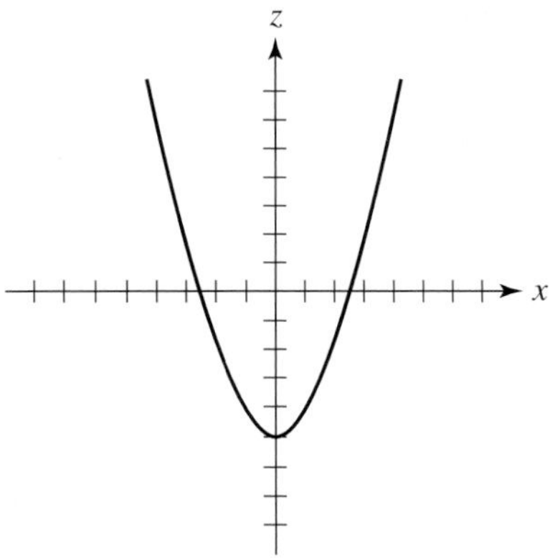

c. axis of symmetry $x = 0$
range $= [-5, \infty)$
d. decreases on $(-\infty, 0)$, increases on $(0, \infty)$

81. a. no α-intercepts
vertex $= (2, 80)$, other points $(1, 85)$ and $(3, 85)$
b.

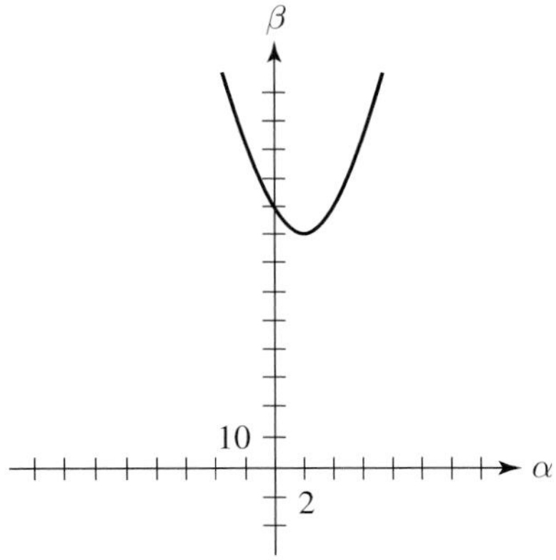

c. axis of symmetry $\alpha = 2$
range $= [80, \infty)$
d. decreases on $(-\infty, 2)$, increases on $(2, \infty)$

83. $y = -\frac{2}{3}(x + 3)(x - 2)$ **85.** $y = -\frac{5}{16}x(x - 100)$

87. $y = 0.12(x + 4)^2$ **89.** $y = -(x + 10)^2 - 5$

91. It is the vertex.

93. a.
$$\begin{aligned} f(x) &= ax^2 + bx + c \\ &= a\left(x^2 + \frac{b}{a}x\right) + c \\ &= a\left(x^2 + \frac{b}{a}x + \frac{b^2}{4a^2} - \frac{b^2}{4a^2}\right) + c \\ &= a\left(x^2 + \frac{b}{a}x + \frac{b^2}{4a^2}\right) - \frac{b^2}{4a} + c \\ &= a\left(x + \frac{b}{2a}\right)^2 - \frac{b^2 - 4ac}{4a} \end{aligned}$$

b. From the standard graphing form in part (a) the vertex is $\left(-\frac{b}{2a}, \frac{b^2 - 4ac}{4a}\right)$.

95. a. $y = g(x)$ is steepest, $y = h(x)$ is flattest.
b. The graphs do not appear to support the conclusions in part (a). This is because the windows distort the steepness of each graph.
c. Yes.

97. $F(x) = [x - (2 + 3i)][x - (2 - 3i)]$

99. $F(x) = \left[x - \left(-2 + \frac{\sqrt{2}}{2}i\right)\right]\left[x - \left(-2 - \frac{\sqrt{2}}{2}i\right)\right]$

SECTION 5–2

1. a. $s(t) = 0$ when the height of the rocket is 0, that is, when the rocket is at ground level. Since the flight lasts 10 seconds, $s(10) = 0$.
b. $s(t) = -16t^2 + 160t$ **c.** $(5, 400)$; 400 feet

3. $(-1, 3)$; minimum **5.** $(-6, 1176)$; maximum

7. The entire graph is above the x-axis.

9. $(-\infty, -1) \cup \left(\frac{1}{2}, \infty\right)$ **11.** $[-5, 20]$

13. $(a, b, c) = (0.055, 1.25, -10)$
$0.055(30)^2 + 1.25(30) - 10 = 77$
$0.055(40)^2 + 1.25(40) - 10 = 128$
$0.055(50)^2 + 1.25(50) - 10 = 190$
$0.055(60)^2 + 1.25(60) - 10 = 263$

15. $y = -0.5x^2 + 9x - 22$ **17.** $y = 4x^2 - 12x + 9$

19. $h = -2.66t^2 + 90t$

21. $(1, 0)$; minimum **23.** $(-2, +1)$; minimum

25. $(-6, 38)$; maximum **27.** $(3, 108)$; maximum

29. $[-5, -13]$ **31.** $(-\infty, -4.5) \cup (1, \infty)$

33. $(-\infty, \infty)$ **35.** no solutions

37. quadratic; $y = x^2 + 2$ **37.** not quadratic

41. quadratic; $y = \frac{1}{8}x^2 - \frac{1}{2}x + \frac{185}{2}$

43. 52.5 feet on each side **45.** $852 \leqslant x \leqslant 2348$

47. between 67 and 90 people **49.** 223.61 feet

51. $S(5) = -2.38$, indicating that a car traveling at an initial speed of 5 mph has a stopping distance of -2.38 feet. Because this result makes no physical sense, we can conclude that the function S does not describe the relationship between initial speed and stopping distance for speeds near 5 mph.

53. a. $Q(30) = 78$; $Q(40) = 128$; $Q(50) = 190$; $Q(60) = 264$
b. The graphs are close over the entire interval, so the stopping distances predicted by S and Q are also close.

- **c.** The graphs are not very close when x is near 0. Since the graph of Q has no negative y-values, it predicts more accurate stopping distances for low initial speeds.
- **d.** Responses will vary. Most students will choose Q because it is a better model for low initial speeds, but some may argue that S is a better model for initial speeds between 30 and 60 mph, since it fits Table 19 exactly.

SECTION 6–1

1. Any vertical line between $x = -1$ and $x = 1$ intersects the graph twice.

3. The legs of the right triangle in the diagram have lengths x and y, and the hypotenuse has length 1. By the Pythagorean theorem, $x^2 + y^2 = 1$.

5. a. Any horizontal line between $y = -1$ and $y = 1$ intersects the graph twice.

b. The first and last rows have the same second entries and different first entries.

c. Solving for x yields $x = \pm\sqrt{1 - y^2}$. Some values of the independent variable y generate more than one value of the dependent variable x.

7. Yes. **9.** No.

11. No. **13.** Yes.

15.

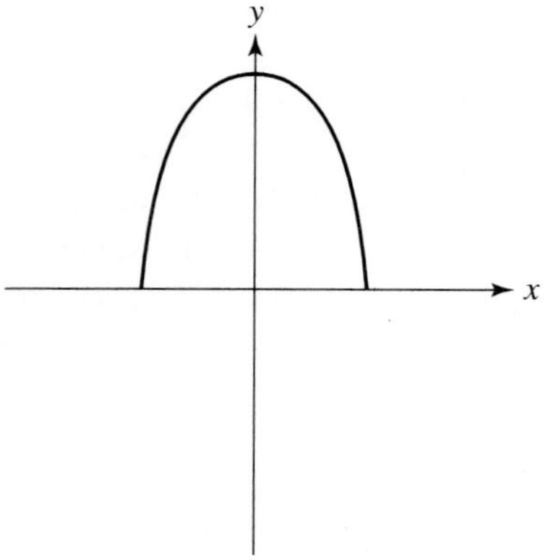

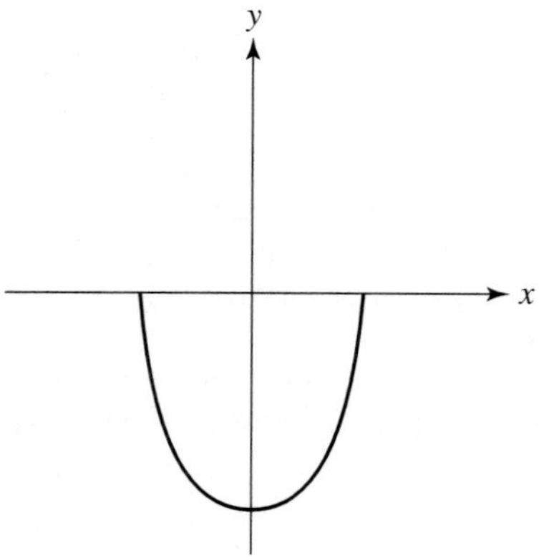

17.

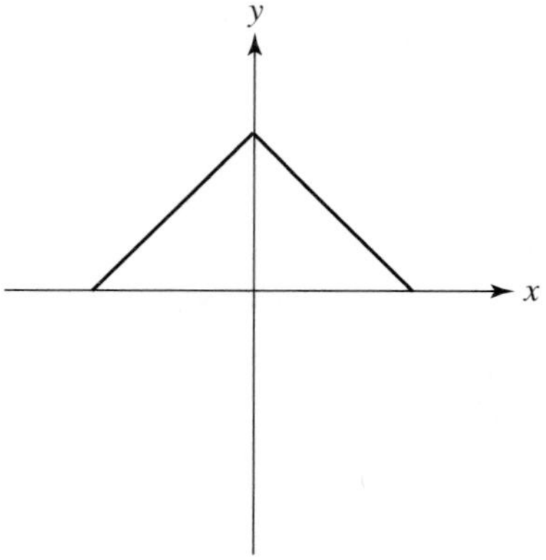

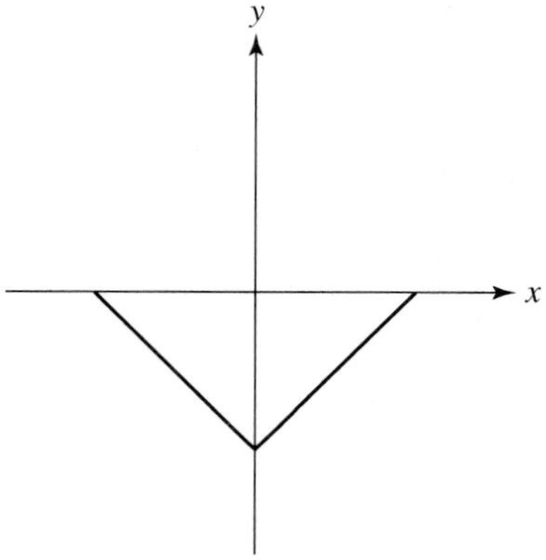

19. $y = 4 - x^2$; yes

21. $y = 1 \pm \sqrt{x^2 + 1}$; no

23. $y = \pm\frac{1}{2}\sqrt{4 - x^2}$

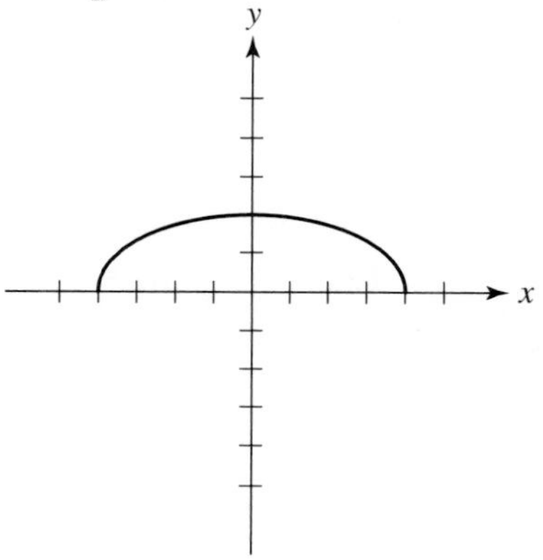

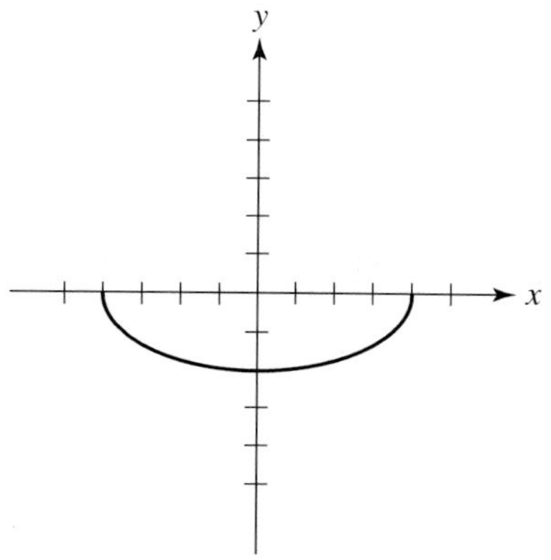

25. $y = \pm\frac{1}{2}\sqrt{4 + x^2}$

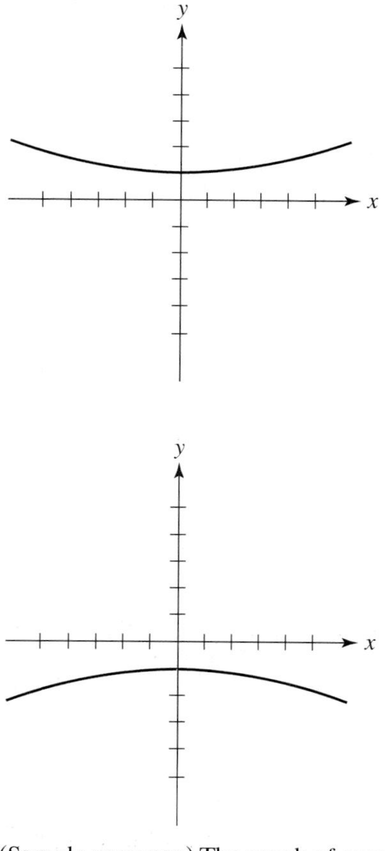

27. (Sample response) The graph of every quadratic relation consists of at most two pieces, each of which is the graph of a function. No vertical line can intersect the graph of either function more than once, so no vertical line can intersect the graph of the relation more than twice.

29. $x = [f(x)]^2$
$x = (\sqrt{x})^2$
$x = x$

31. $9x^2 + [f(x) - 5]^2 = 36$
$9x^2 + [(5 - 3\sqrt{4 - x^2}) - 5]^2 = 36$
$9x^2 + (-3\sqrt{4 - x^2})^2 = 36$
$9x^2 + 9(4 - x^2) = 36$
$9x^2 + 36 - 9x^2 = 36$
$36 = 36$

33. a. $y = \pm\frac{1}{3}\sqrt{4x^2 - 36}$

b.

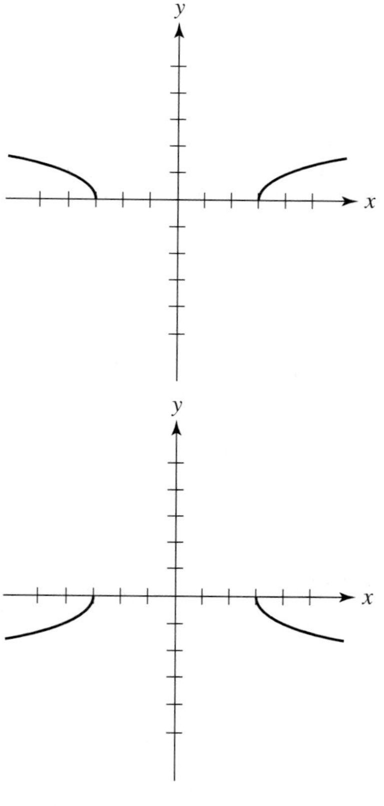

35. a. $y = \pm\sqrt{\frac{x + 6}{2}}$

b.

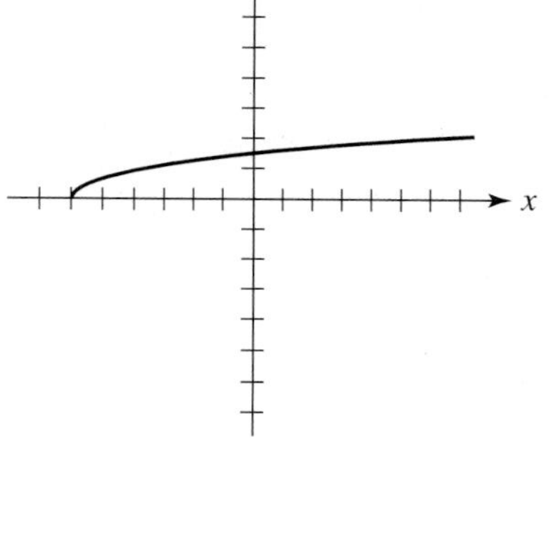

b.

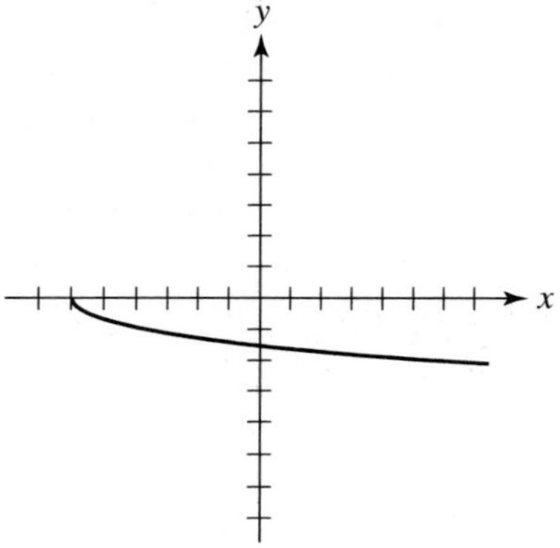

37.

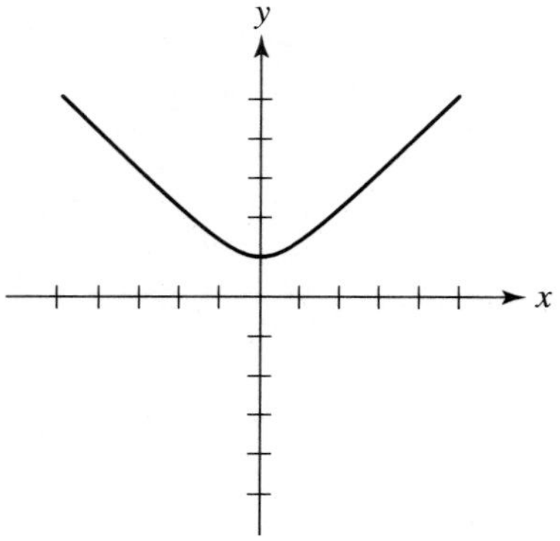

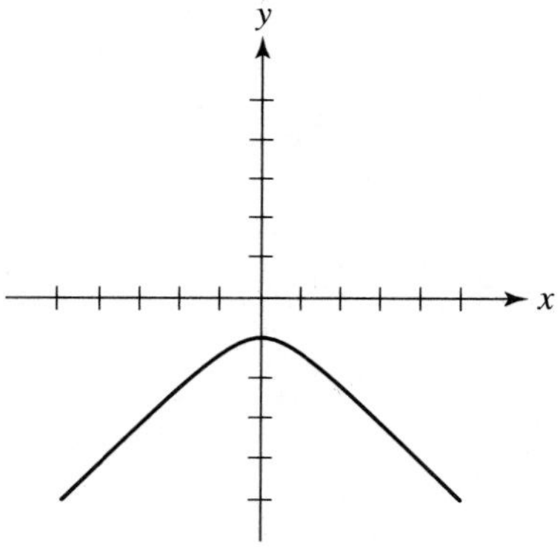

39.

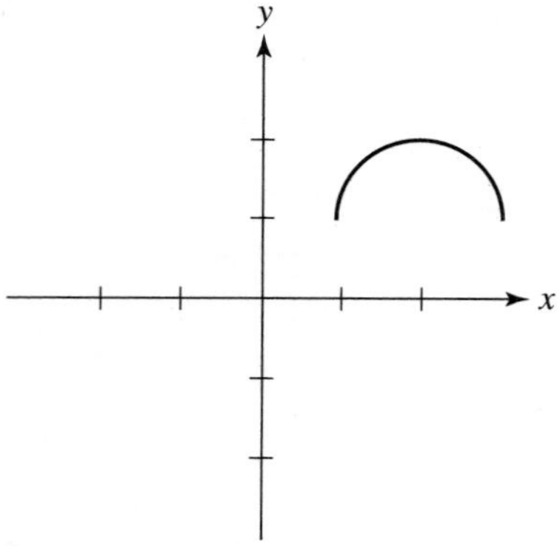

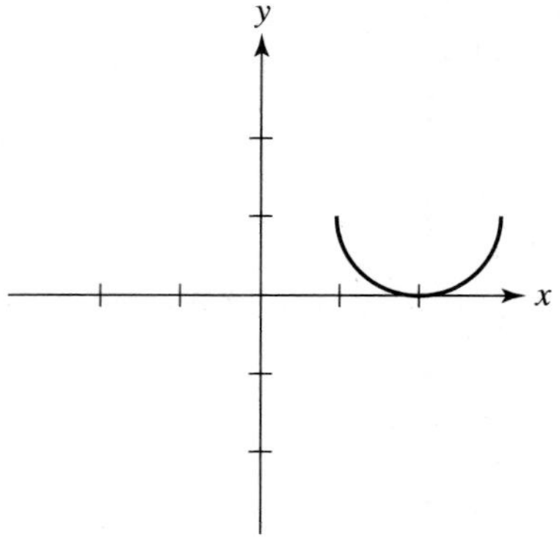

41. a. $y = \pm\sqrt{100 - x^2}$

b. $f_1(x) = \sqrt{100 - x^2}$

$f_2(x) = -\sqrt{100 - x^2}$

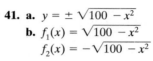

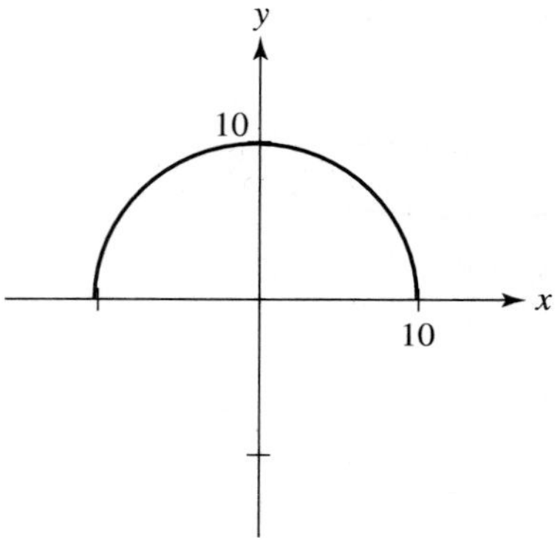

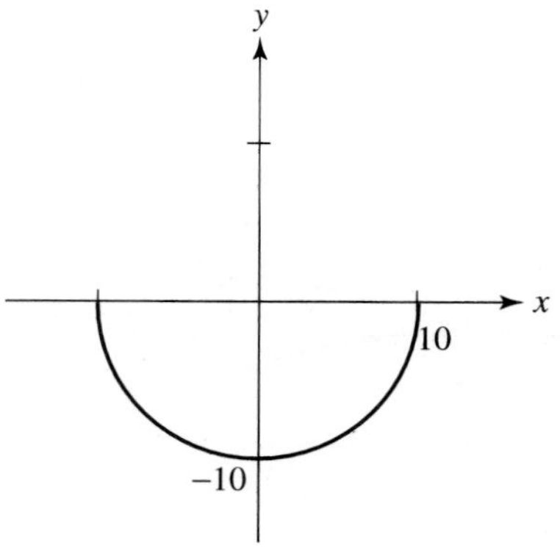

43. a. $4x^2 + y = 4x + 15$
$y = -4x^2 + 4x + 15$

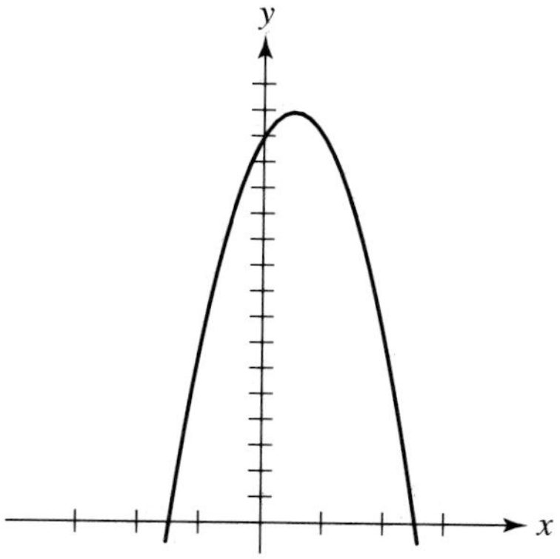

45. a. $y = 3 \pm \sqrt{-x^2 + 6x + 9}$

b. $f_1(x) = 3 + \sqrt{-x^2 + 6x + 9}$
$f_2(x) = 3 - \sqrt{-x^2 + 6x + 9}$

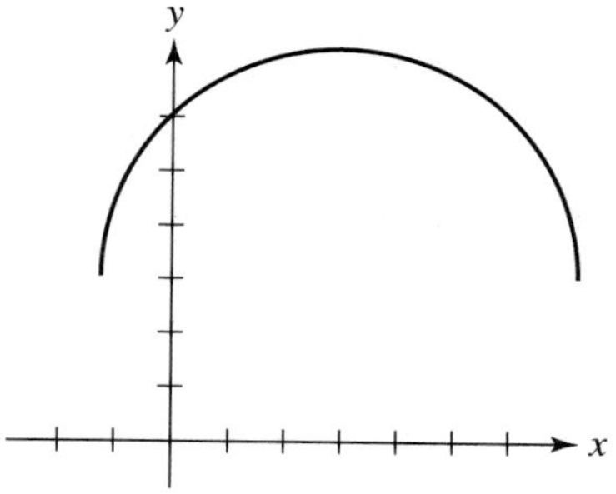

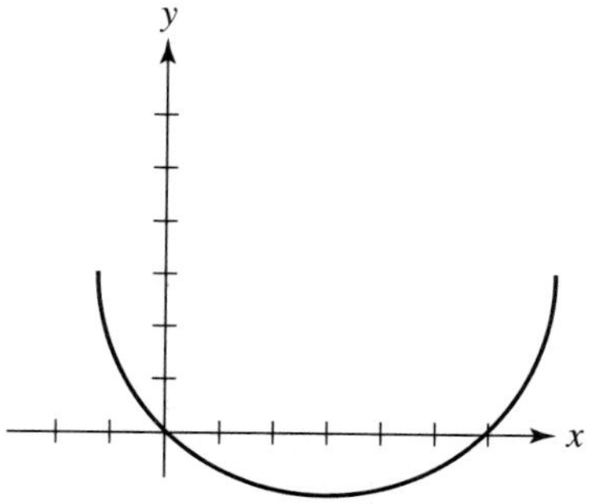

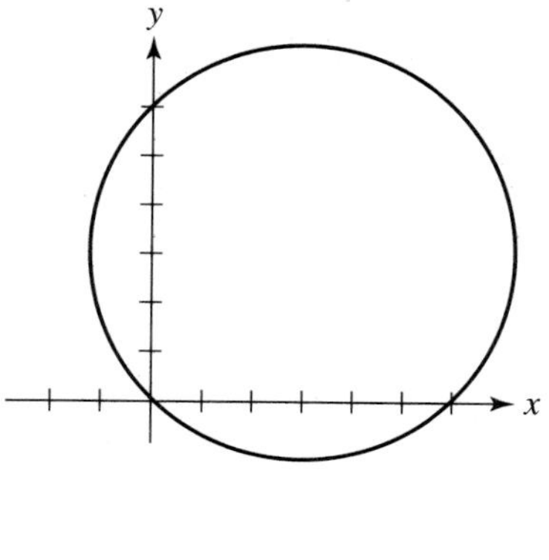

47. a. $y = \sqrt[3]{x^2 - 1}$

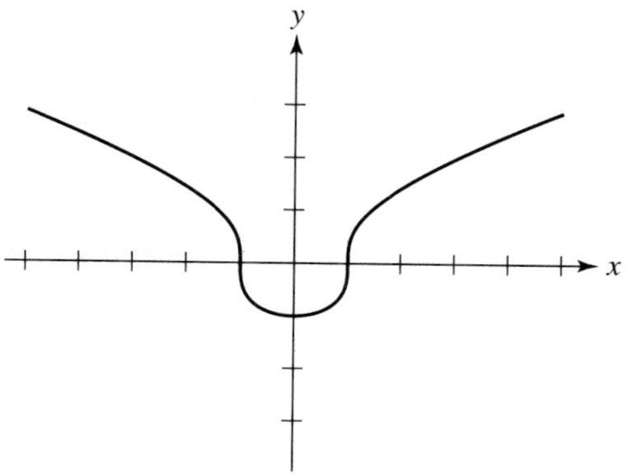

49. $3x - [f(x)]^2 = 6$

$3x - (\sqrt{3x - 6})^2 = 6$

$3x - (3x - 6) = 6$
$6 = 6$

51. $x^2 - 4[f(x) - 1] = 4x$

$x^2 - 4\left[\frac{1}{4}(x - 2)^2 - 1\right] = 4x$

$x^2 - [(x - 2)^2 - 4] = 4x$
$x^2 - [(x^2 - 4x + 4) - 4] = 4x$
$x^2 - (x^2 - 4x) = 4x$
$4x = 4x$

53. (Sample response) If a function $y = f(x)$ is defined implicitly by a relation, then the equation of the relation is true whenever the equation $y = f(x)$ is true. Substituting $f(x)$ for y in the relation produces an equation in x that is always true.

55. a. $y = \frac{1}{2}(-x - 5 \pm \sqrt{-3x^2 - 6x - 15})$

b. The graph is empty.

57. a. $y = \frac{1}{11}(-5\sqrt{3}\,x \pm 8\sqrt{x^2 + 11})$

b.

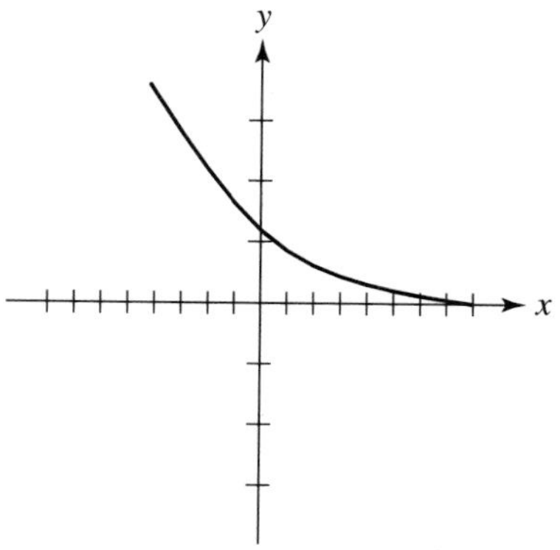

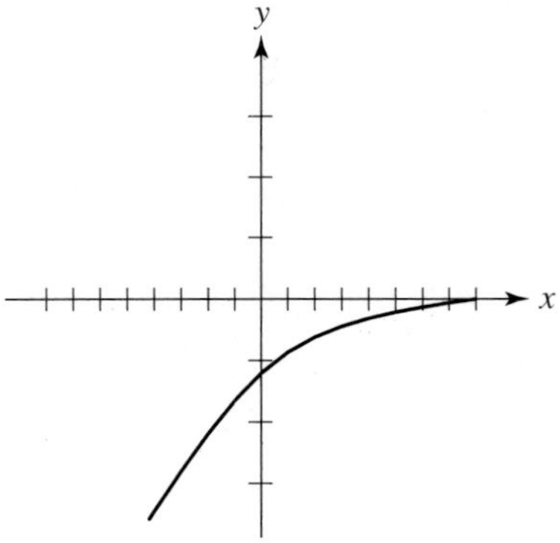

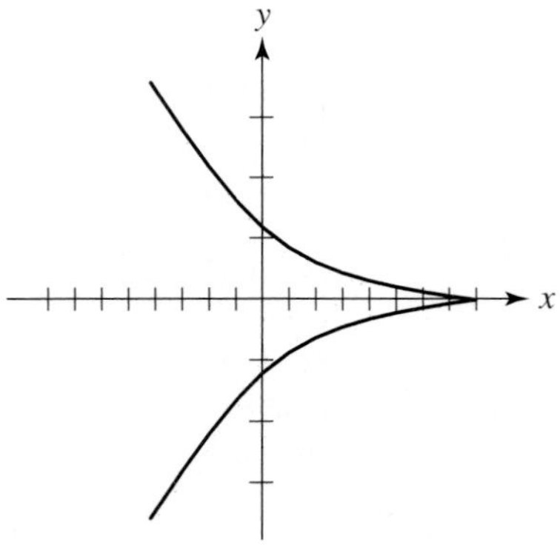

SECTION 6–2

1.

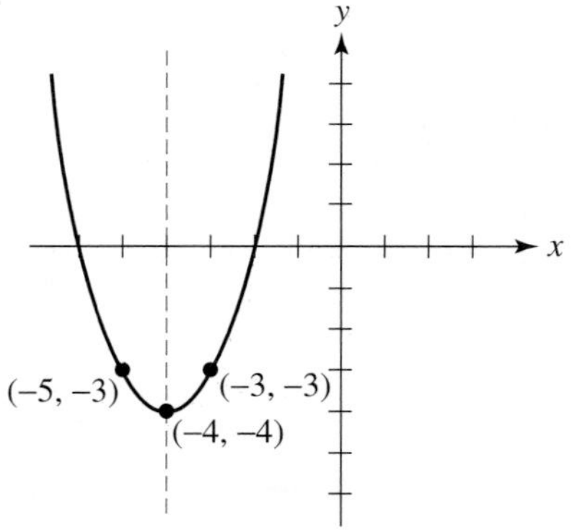

3.

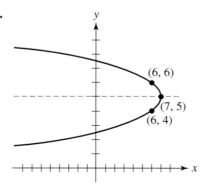

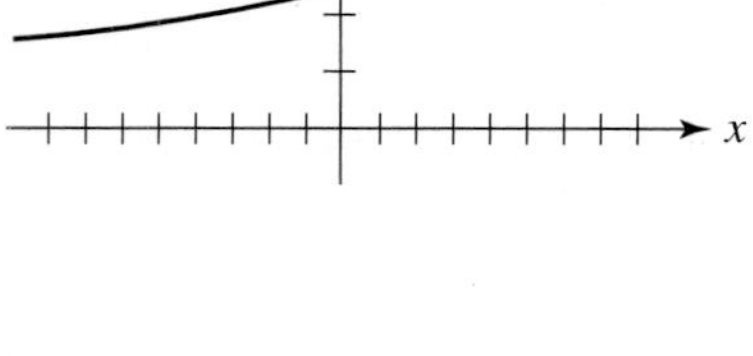

5. **a.** If $|x| > 3$, then $\left(\frac{x}{3}\right)^2 > 1$, so $\left(\frac{y}{2}\right)^2 < 0$ and there are no real solutions for y. This means that there are no points on the graph for which $|x| > 3$, so the entire graph lies on or between the lines $x = -3$ and $x = 3$.

b. If $|y| > 2$, then $\left(\frac{y}{2}\right)^2 > 1$, so $\left(\frac{x}{3}\right)^2 < 0$ and there are no real solutions for x. This means that there are no points on the graph for which $|y| > 2$, so the entire graph lies on or between the lines $y = -2$ and $y = 2$.

c. If $y = 0$, then $\left(\frac{x}{3}\right)^2 = 1$.

$$\frac{x}{3} = \pm 1$$
$$x = \pm 3$$

If $x = 0$, then $\left(\frac{y}{2}\right)^2 = 1$.

$$\frac{y}{2} = \pm 1$$
$$y = \pm 2$$

7. **a.** If $\left|\frac{x-10}{2}\right| > 1$, then $\left(\frac{x-10}{2}\right)^2 > 1$, so $\left(\frac{y-7}{3}\right)^2 < 0$, and there is no real solution for y. This means that there are no points on the graph for which $\left|\frac{x-10}{2}\right| > 1$. Thus all points on the graph have $\left|\frac{x-10}{2}\right| \leqslant 1$

$$-1 \leqslant \frac{x-10}{2} \leqslant 1$$
$$-2 \leqslant x - 10 \leqslant 2$$
$$8 \leqslant x \leqslant 12$$

b. If $\left|\frac{y-7}{3}\right| > 1$, then $\left(\frac{y-7}{3}\right)^2 > 1$, so $\left(\frac{x-10}{2}\right)^2 < 0$, and there is no real solution for x. This means that there are no points on the graph for which $\left|\frac{y-7}{3}\right| > 1$. Thus all points on the graph have

$$\left|\frac{y-7}{3}\right| \leqslant 1$$
$$-1 \leqslant \frac{y-7}{3} \leqslant 1$$
$$-3 \leqslant y - 7 \leqslant 3$$
$$4 \leqslant y \leqslant 10$$

c. If $y = 7$, then $\left(\frac{x-10}{2}\right)^2 = 1.$

$$\frac{x-10}{2} = \pm 1$$
$$x - 10 = \pm 2$$
$$x = 8, 12$$

If $x = 10$, then $\left(\frac{y-7}{3}\right)^2 = 1.$

$$\frac{y-7}{3} = \pm 1$$
$$y - 7 = \pm 3$$
$$y = 4, 10$$

9. $\left(\frac{x}{4}\right)^2 + \left(\frac{y}{8}\right)^2 = 1$

center = $(0, 0)$
vertices = $(0, \pm 8)$
covertices = $(\pm 4, 0)$
major axis $x = 0$ (y-axis)
minor axis $y = 0$ (x-axis)

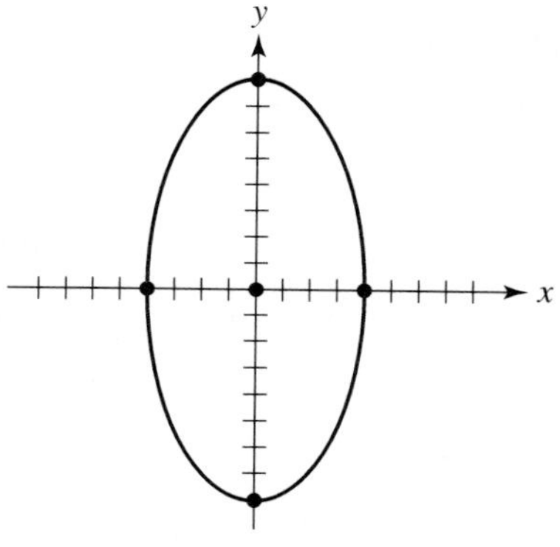

11. $\left(\frac{x+4}{4}\right)^2 + \left(\frac{y-2}{2}\right)^2 = 1$

center = $(-4, 2)$
vertices = $(-8, 2), (0, 2)$
covertices = $(-4, 0), (-4, 4)$
major axis $y = 2$
minor axis $x = -4$

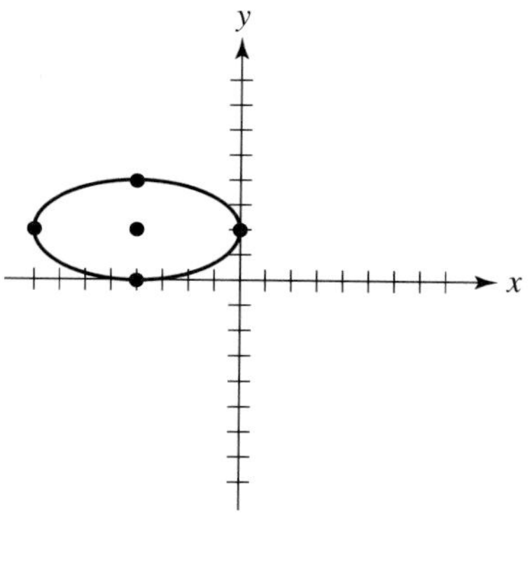

13. $\left(\frac{x-30}{\sqrt{6}}\right)^2 + (y+25)^2 = 1$

center = $(30, -25)$
vertices = $(30 \pm \sqrt{6}, -25)$
covertices = $(30, -26), (30, -24)$
major axis $y = -25$
minor axis $x = 30$

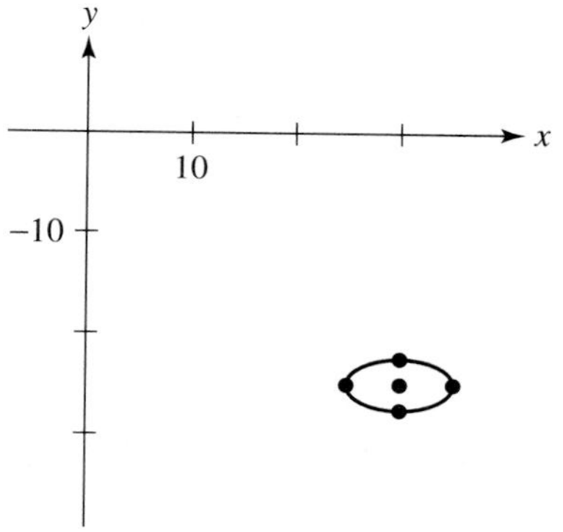

15. center = $(0, 0)$
radius = 5

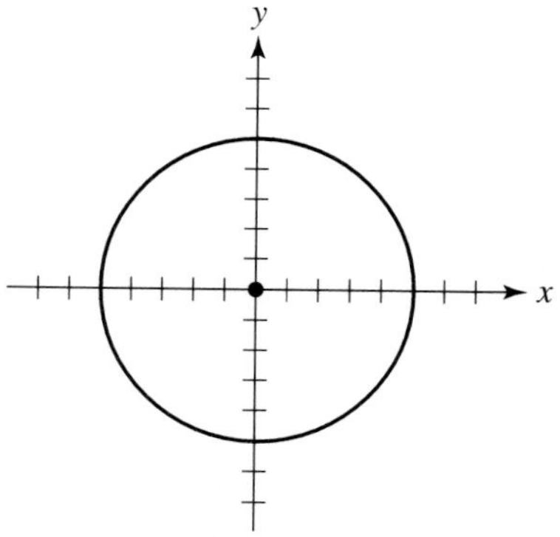

17. center = $(-6, 2)$
radius = 2

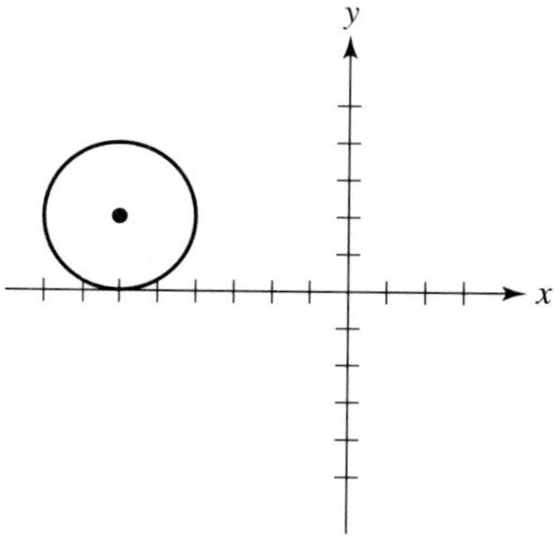

19. center = $(4, 3)$
radius = $\sqrt{2}$

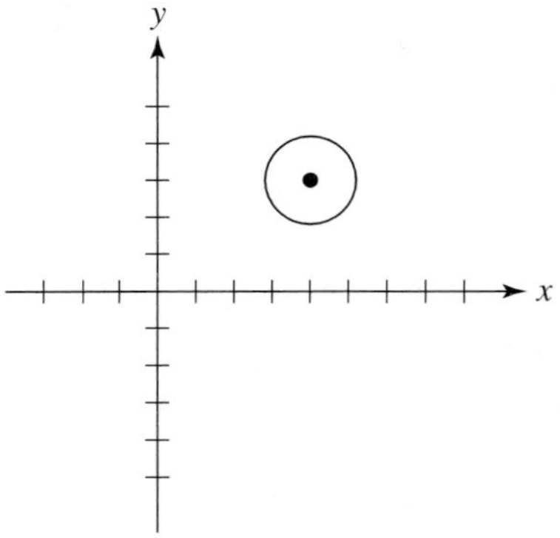

21. $x^2 - y^2 = 1$
$y^2 = x^2 - 1$
$y = \pm\sqrt{x^2 - 1}$
In the first quadrant the last equation is equivalent to $y = \sqrt{x^2 - 1}$ because $y \geqslant 0$. In general, however, y can be negative, so the equations are not equivalent.

23.

x	y
1	0
2	$\sqrt{3} \cong 1.73$
10	$\sqrt{99} \cong 9.95$
100	$\sqrt{9999} \cong 99.99$

25. If (a, b) is on the graph of $x^2 - y^2 = 1$, then $a^2 - b^2 = 1$. In that case, it is also true that $(-a)^2 + b^2 = 1$ and $a^2 + (-b)^2 = 1$, so the points $(-a, b)$ and $(a, -b)$ are also on the graph.

27. center = $(-5, -6)$
vertices = $(-8, -6), (-2, -6)$
transverse axis $y = -6$
conjugate axis $x = -5$
asymptotes $y + 6 = \pm\frac{4}{3}(x + 5)$

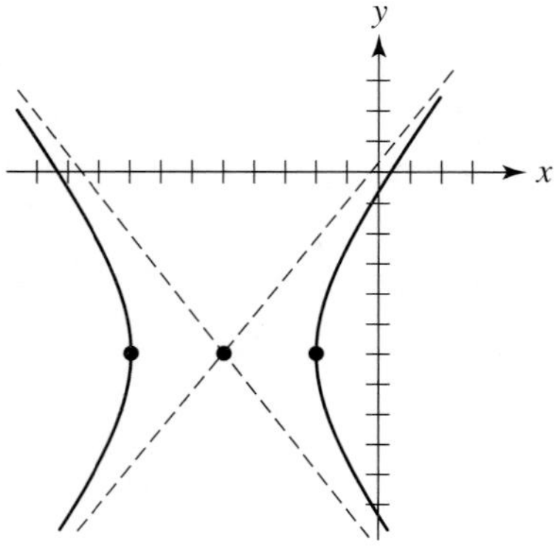

29. center = $(-129, 85)$
vertices = $(-129, 50), (-129, 120)$
transverse axis $x = -129$
conjugate axis $y = 85$
asymptotes $y - 85 = \pm\frac{7}{10}(x + 129)$

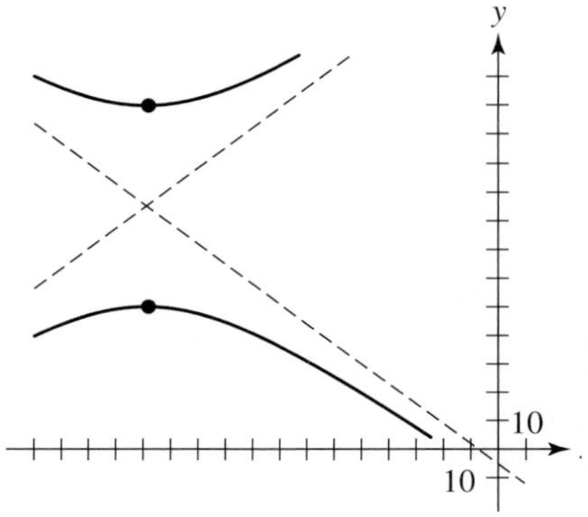

31. The graph is a single point at the origin.

33. The graph consists of the two lines $y = x$ and $y = -x$.

35. The graph is empty.

37. The graph consists of the lines $y = \pm\frac{1}{2}(x - 3)$.

39. $x = 3y^2$
vertex = $(0, 0)$
axis of symmetry $y = 0$ (x-axis)

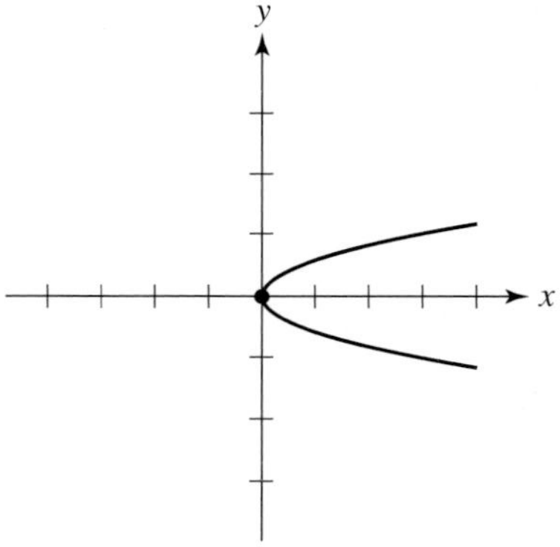

41. $\left(\frac{x}{\sqrt{14}}\right)^2 + \left(\frac{y}{2}\right)^2 = 1$

center = $(0, 0)$
vertices = $(\pm\sqrt{14}, 0)$
covertices = $(0, \pm 2)$
major axis $y = 0$ (x-axis)
minor axis $x = 0$ (y-axis)

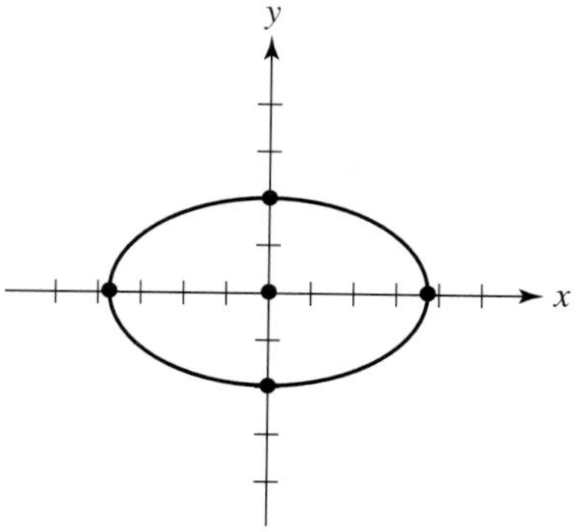

43. $(x + 1)^2 + (y - 2)^2 = \frac{9}{4}$

center = $(-1, 2)$
radius = $\frac{3}{2}$

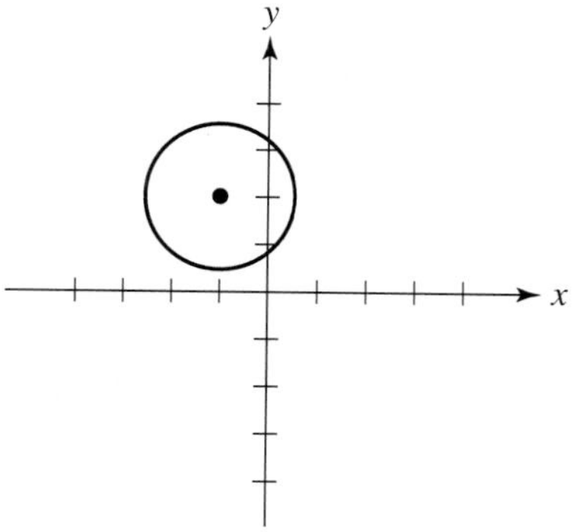

45. $y = (x + 1)^2 - \frac{1}{4}$

vertex $\left(-1, -\frac{1}{4}\right)$
axis of symmetry $x = -1$

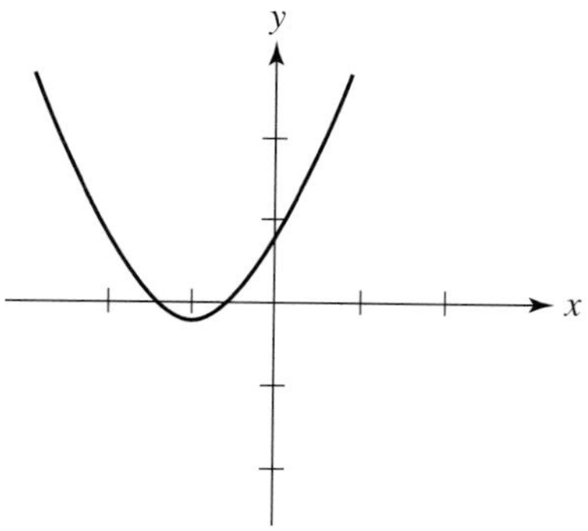

47. $\left(\frac{x - 7}{2}\right)^2 + \left(\frac{y}{\sqrt{20}}\right)^2 = 1$

center = $(7, 0)$
vertices = $(7, \pm 2\sqrt{5})$
covertices = $(5, 0), (9, 0)$
major axis $x = 7$
minor axis $y = 0$ (x-axis)

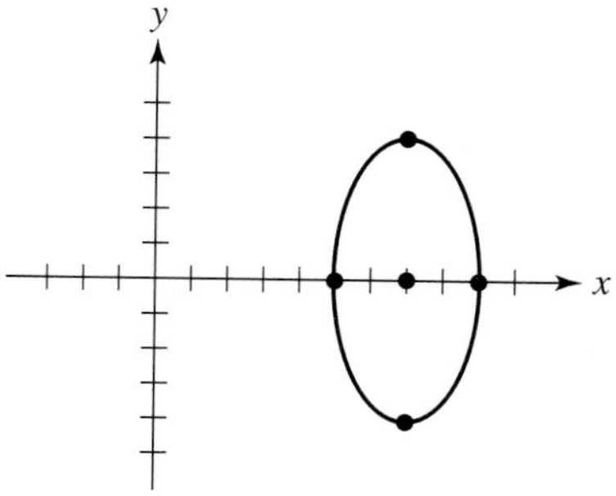

49. $y = 2(x - 3)^2 - 8$

vertex = $(3, -8)$
axis of symmetry $x = 3$

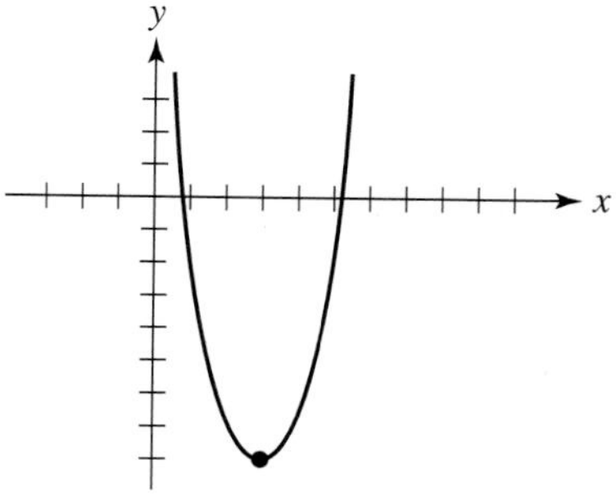

51. The graph is a pair of intersecting lines, with equations $y + 4 = \pm 2(x + 3)$.

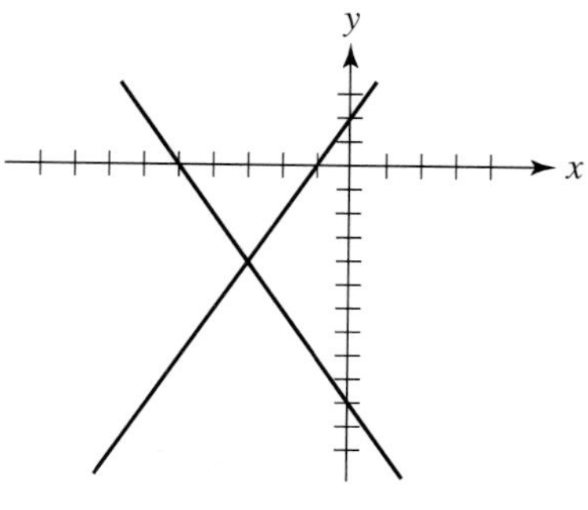

53. $\left(\dfrac{x + 0.1}{10}\right)^2 + \left(\dfrac{y - 0.6}{5}\right)^2 = 1$

center $= (-0.1, 0.6)$
vertices $= (-10.1, 0.6), (9.9, 0.6)$
covertices $= (-0.1, -4.4), (-0.1, 5.6)$
major axis $y = 0.6$
minor axis $x = -0.1$

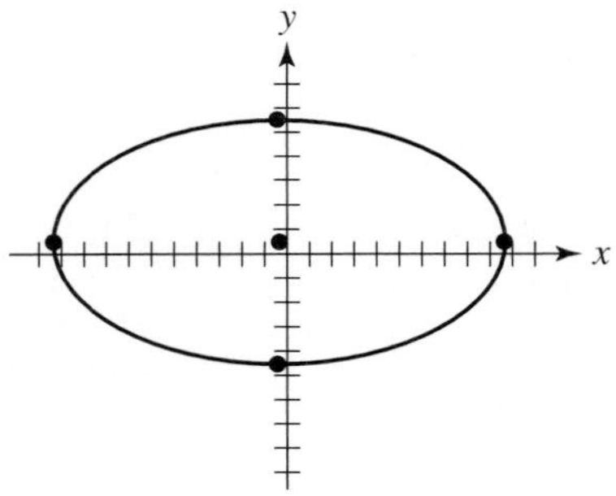

55. $\left(\dfrac{x + 3}{5}\right)^2 + \left(\dfrac{y - 1}{2}\right)^2 = 1$

center $= (-3, 1)$
vertices $= (-8, 1), (2, 1)$
covertices $= (-3, -1), (-3, 3)$
major axis $y = 1$
minor axis $x = -3$

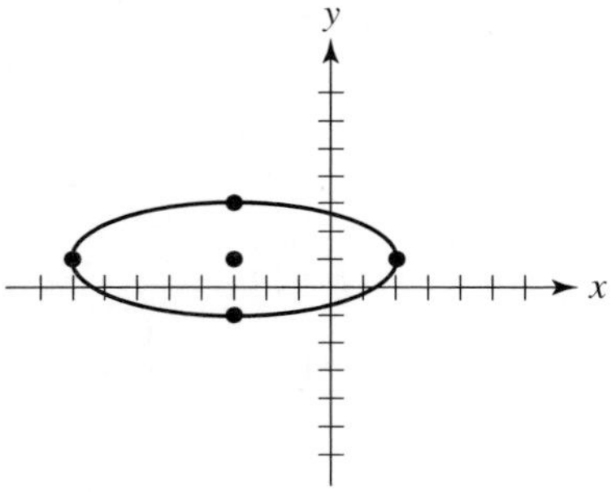

57. The graph is empty.

59. $\left(\dfrac{x + 3}{5}\right)^2 - \left(\dfrac{y - 1}{2}\right)^2 = 1$

center $= (-3, 1)$
vertices $= (-8, 1), (2, 1)$
transverse axis $y = 1$
conjugate axis $x = -3$
asymptotes $\dfrac{y - 1}{2} = \pm\dfrac{x + 3}{5}$

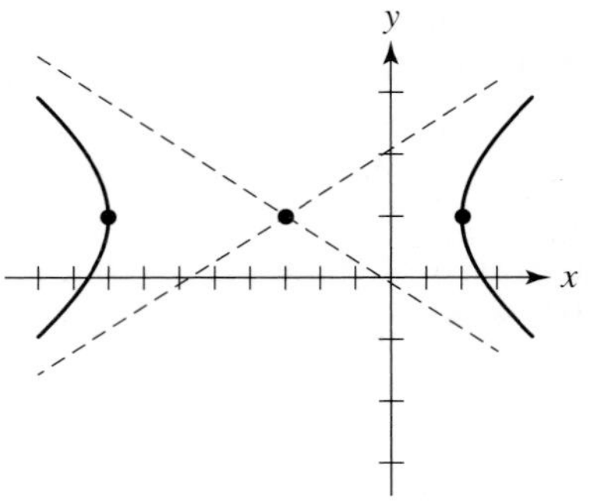

61. $(y + 2)^2 - \left(\dfrac{x + 3}{\sqrt{5}}\right)^2 = 1$

center $= (-3, -2)$
vertices $= (-3, -3), (-3, -1)$
transverse axis $x = -3$
conjugate axis $y = -2$
asymptotes $y + 2 = \pm\dfrac{x + 3}{\sqrt{5}}$

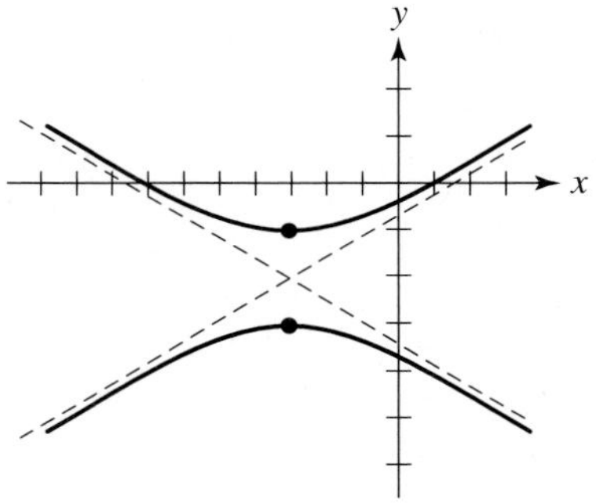

63. $x = \dfrac{3.88}{(3.26)^2}y^2$

SECTION 6-3

1. a. $\left(\dfrac{-2}{2}\right)^2 + (0)^2 = 1; \left(\dfrac{0}{2}\right)^2 + (1)^2 = 1;$
$\left(\dfrac{0}{2}\right)^2 + (-1)^2 = 1; \left(\dfrac{2}{2}\right)^2 + (0)^2 = 1$

b. (Sample response) The expression $\dfrac{x}{2}$ in the new equation plays the same role as x in the unit circle. Therefore, if an ordered pair (a, b) satisfies the equation of the unit circle and (c, b) satisfies the new equation, then $\dfrac{c}{2} = a$ or, equivalently, $c = 2a$.

c. Each x-coordinate on the new graph is twice the corresponding x-coordinate on the unit circle.

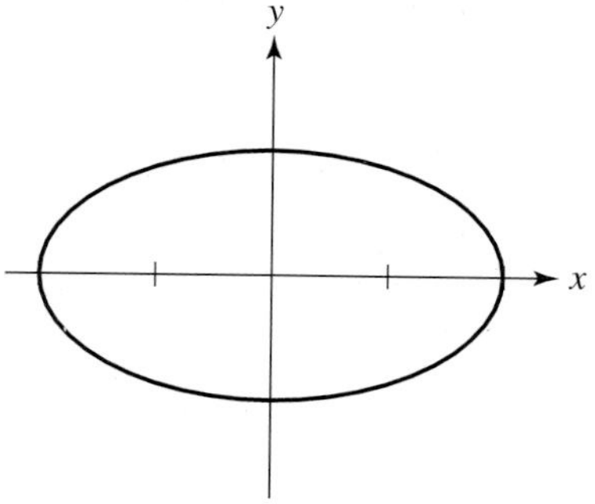

d. The circle was stretched horizontally by a factor of 2.

e. $y = \pm\sqrt{1 - \left(\frac{x}{2}\right)^2}$

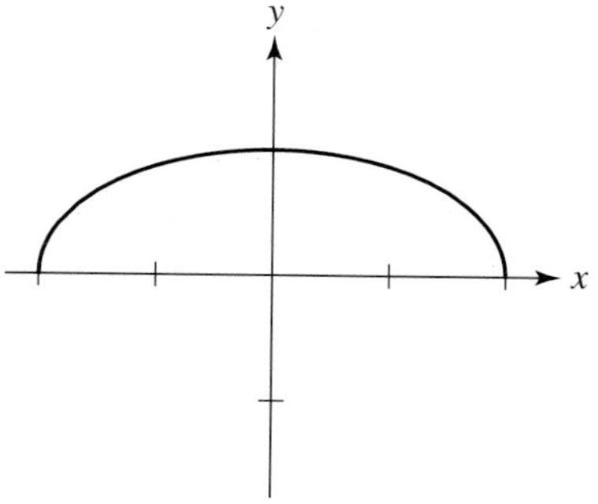

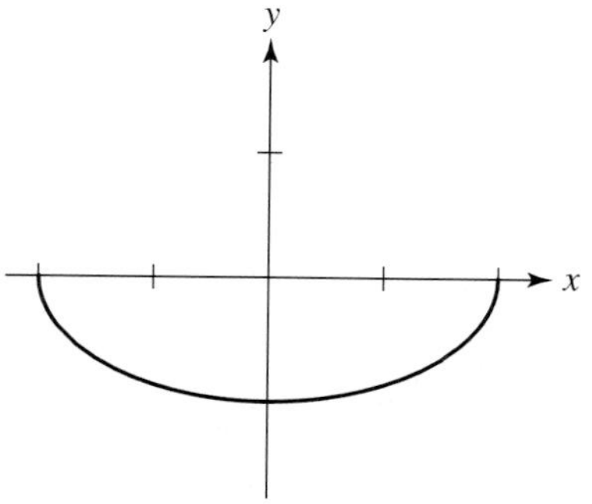

3. a.

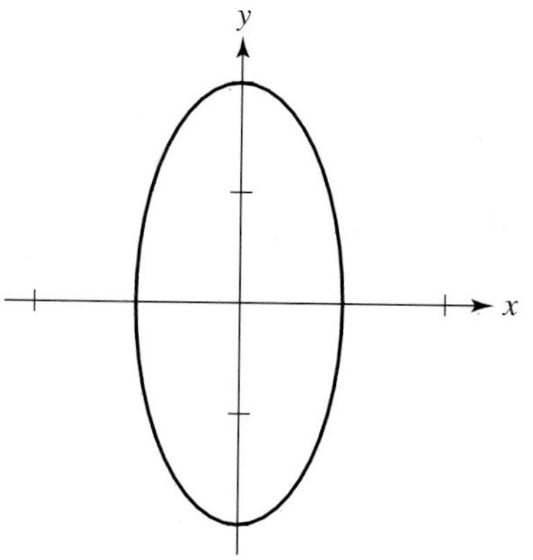

b. The circle was stretched vertically by a factor of 2.

c. $y = \pm 2\sqrt{1 - x^2}$

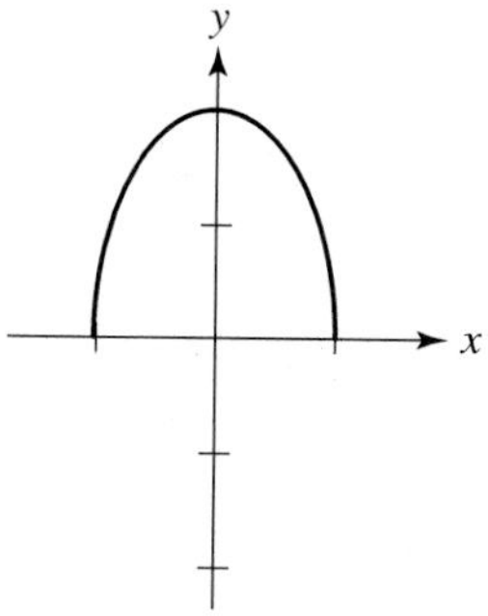

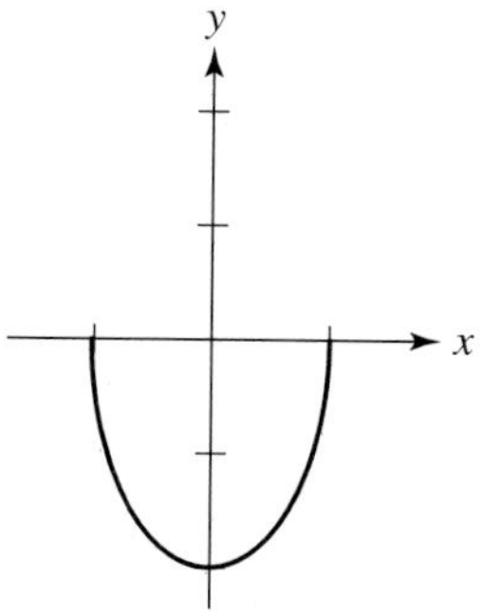

5. The graph is compressed horizontally by a factor of 3.

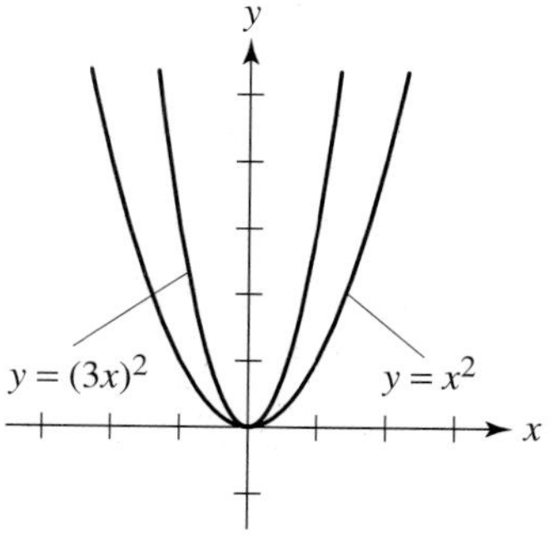

7. The graph is stretched horizontally by a factor of 2.

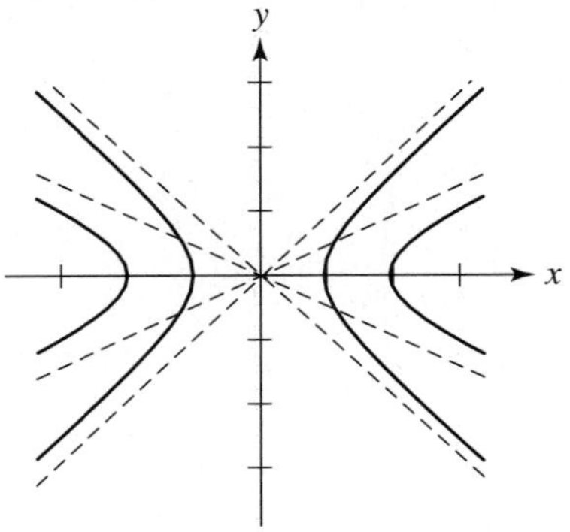

9. The graph is stretched horizontally by a factor of 3 and vertically by a factor of 4.

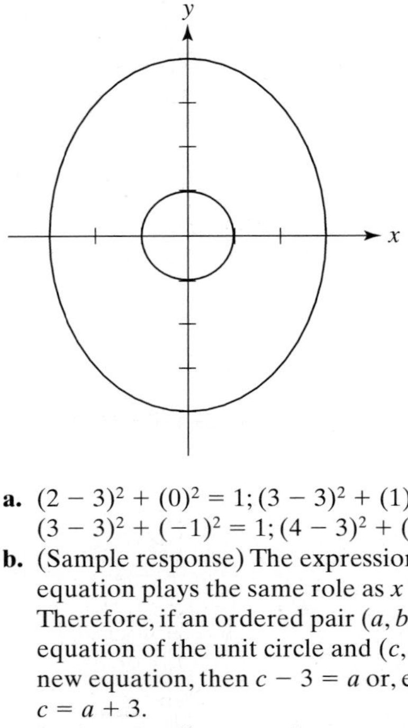

11. a. $(2-3)^2+(0)^2=1; (3-3)^2+(1)^2=1;$
$(3-3)^2+(-1)^2=1; (4-3)^2+(0)^2=1$

b. (Sample response) The expression $x-3$ in the new equation plays the same role as x in the unit circle. Therefore, if an ordered pair (a, b) satisfies the equation of the unit circle and (c, b) satisfies the new equation, then $c-3=a$ or, equivalently, $c=a+3$.

c. Each x-coordinate on the new graph is three more than the corresponding x-coordinate on the unit circle.

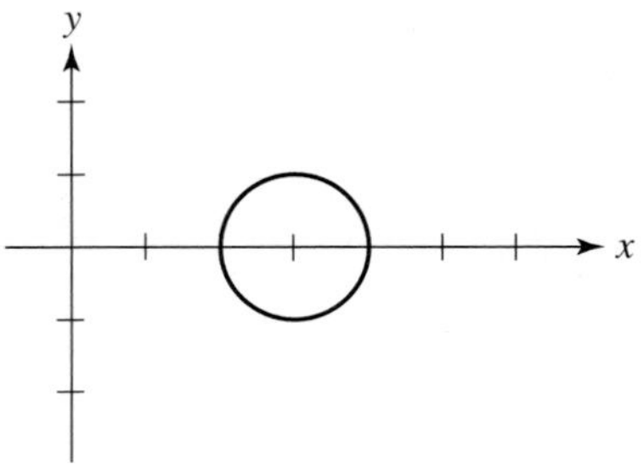

d. The circle was shifted 3 units to the right.

e. $y=\pm\sqrt{1-(x-3)^2}$

13. a.

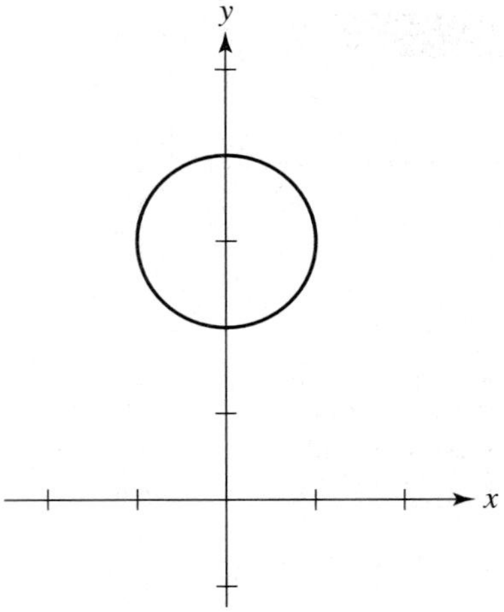

b. The circle was shifted 3 units up.

c. $y=3\pm\sqrt{1-x^2}$

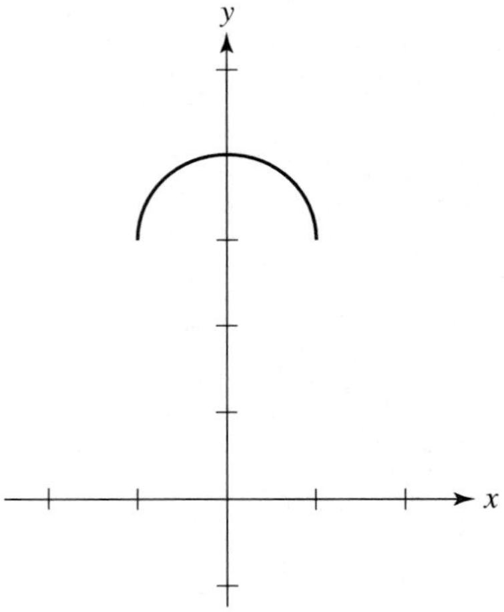

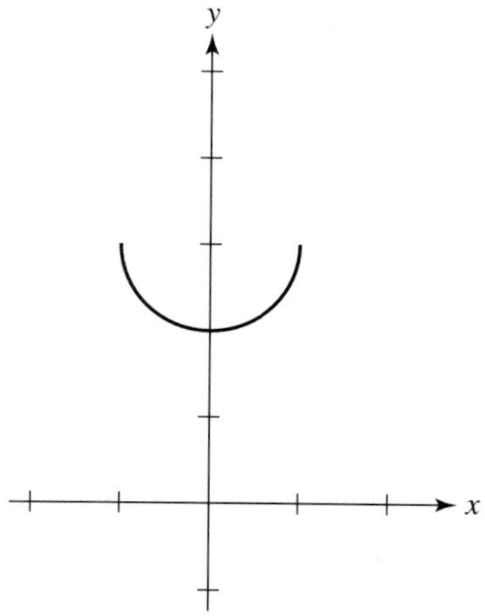

15. The graph is shifted 3 units to the right.

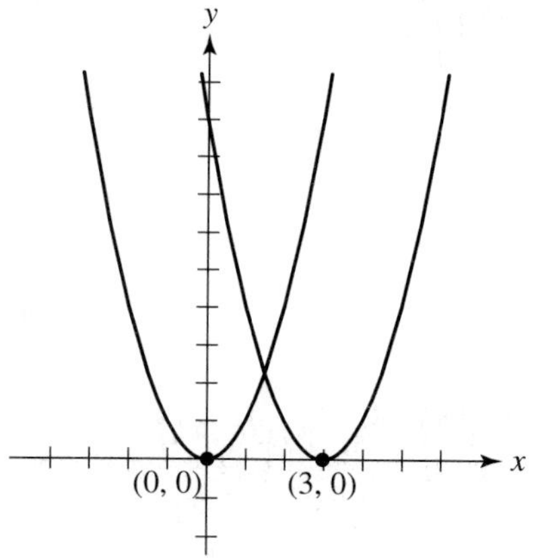

17. The graph is shifted 2 units to the left.

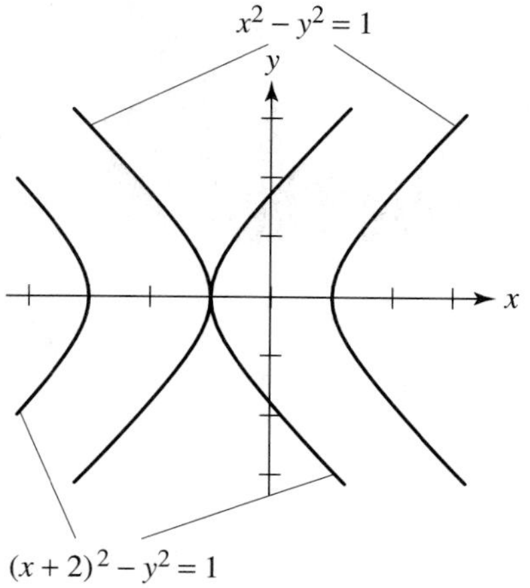

19. The graph is shifted 3 units to the left and 4 units down.

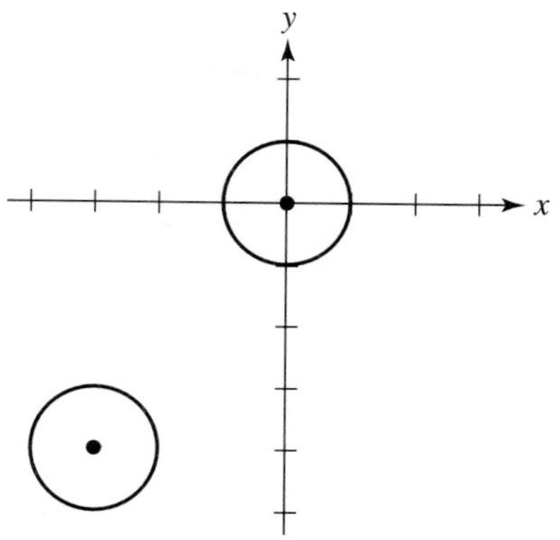

21. $x^2 + y^2 = 1; x^2 + y^2 = 1$; the graph of each equation is the unit circle.

23. **a.** $x^2 + (-y - 2)^2 = 1$
$x^2 + [-(y + 2)]^2 = 1$
$x^2 + (y + 2)^2 = 1$

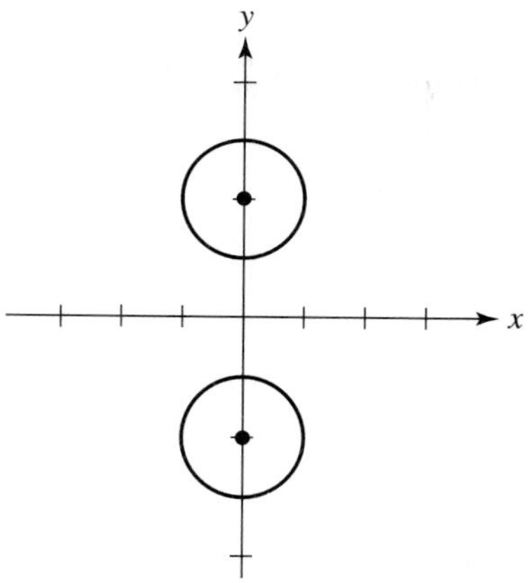

b. The second graph is the reflection of the first graph in the x-axis.

25. symmetric about the x-axis

27. symmetric about both axes

29. **a.**

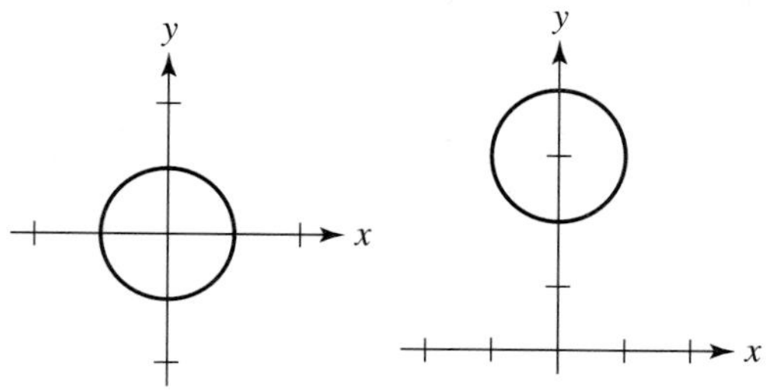

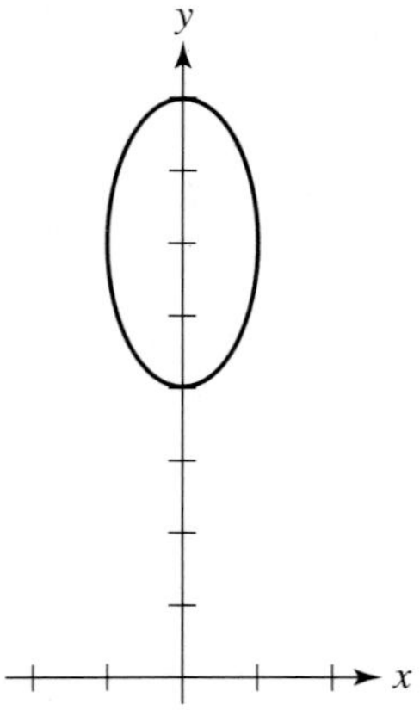

b. $x^2 + \left(\frac{y}{2} - 3\right)^2 = 1$

31. center $(-129, 85)$
vertices $= (-179, 85), (-79, 85)$
covertices $= (-129, 50), (-129, 120)$
major axis $y = 85$
minor axis $x = -129$

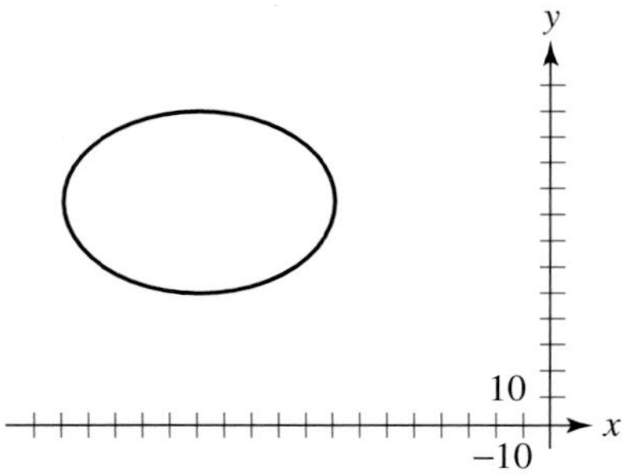

33. center $= (0.5, 1.5)$
vertices $= (0.5, -0.5), (0.5, 3.5)$
covertices $= (-0.5, 1.5), (1.5, 1.5)$
major axis $x = 0.5$
minor axis $y = 1.5$

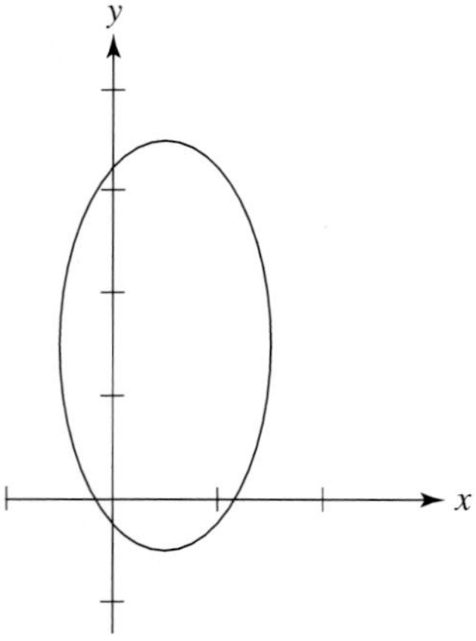

35. center $= (-129, 85)$
vertices $= (-179, 85), (-79, 85)$
transverse axis $y = 85$
conjugate axis $x = -129$
asymptotes $\frac{y - 85}{35} = \pm \frac{x + 129}{50}$

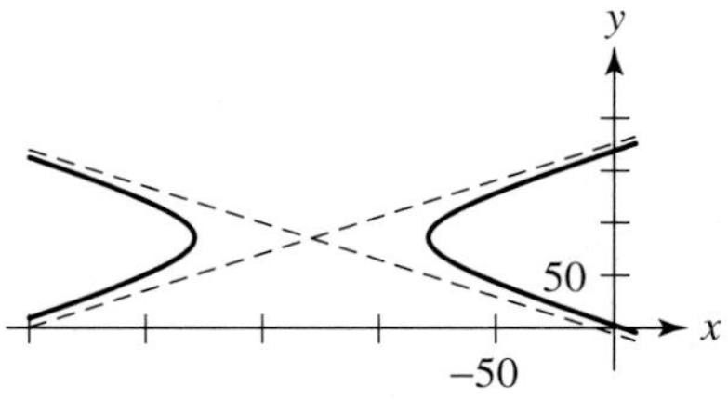

37. center $= (0.5, 1.5)$
vertices $= (0.5, -0.5), (0.5, 3.5)$
transverse axis $x = 0.5$
conjugate axis $y = 1.5$
asymptotes $\frac{y - 1.5}{2} = \pm (x - 0.5)$

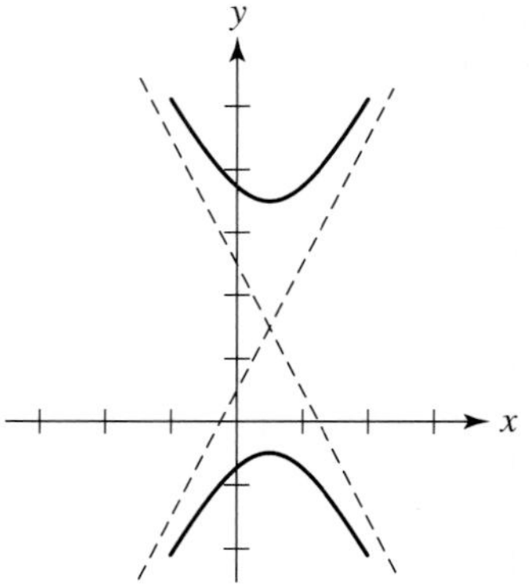

39. The graph of $x = y^2$ is reflected in the y-axis and stretched horizontally by a factor of 5.
vertex $= (0, 0)$
axis of symmetry $y = 0$ (x-axis)

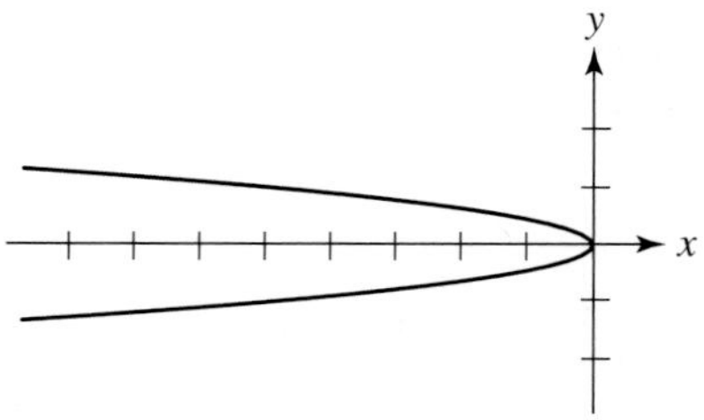

41. The graph of $x^2 + y^2 = 1$ is stretched horizontally by a factor of $\sqrt{6}$ and vertically by factor of $\sqrt{5}$.
center $= (0, 0)$
vertices $= (\pm\sqrt{6}, 0)$
covertices $= (0, \pm\sqrt{5})$
major axis $y = 0$ (x-axis)
minor axis $x = 0$ (y-axis)

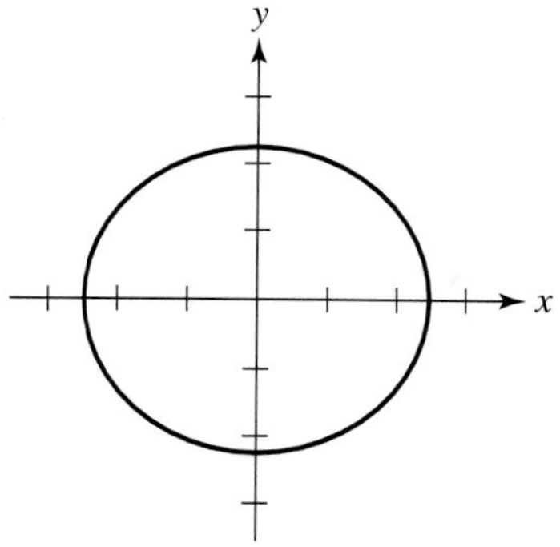

43. The graph of $x^2 + y^2 = 1$ is stretched both horizontally and vertically by a factor of 5, then shifted 10 units to the left and 20 units up.
center $= (-10, 20)$
radius $= 5$

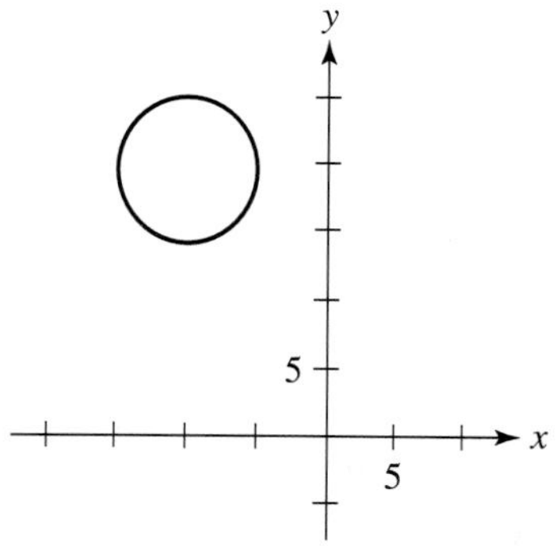

45. The graph of $y = x^2$ is shifted 10 units to the left and 5 units down.
vertex $= (-10, -5)$
axis of symmetry $x = -10$

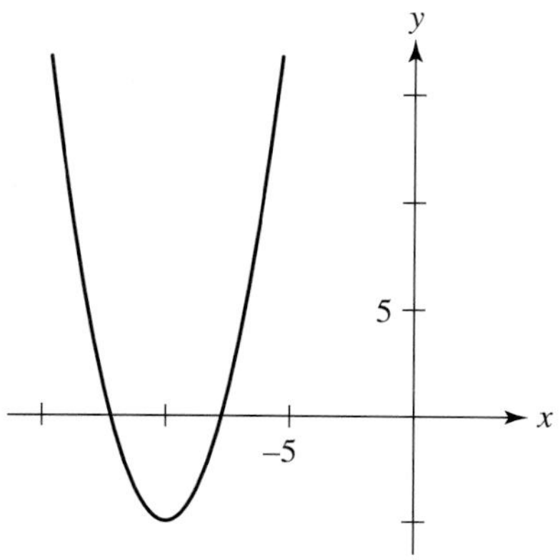

47. The graph of $x^2 + y^2 = 1$ is stretched horizontally by a factor of 4 and vertically by a factor of 6, then shifted 2 units to the left and 3 units down.
center $= (-2, -3)$
vertices $= (-2, -9), (-2, 3)$
covertices $= (-6, -3), (2, -3)$
major axis $x = -2$
minor axis $y = -3$

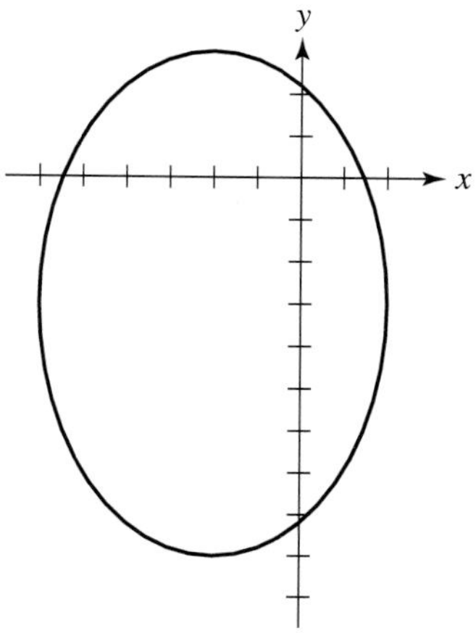

49. The graph of $x^2 - y^2 = 1$ is stretched horizontally by a factor of 4 and vertically by a factor of 6, then shifted 2 units to the left and 3 units down.
center $= (-2, -3)$
vertices $= (-6, -3), (2, -3)$
transverse axis $y = -3$
conjugate axis $x = -2$
asymptotes $\dfrac{y+3}{6} = \pm\dfrac{x+2}{4}$

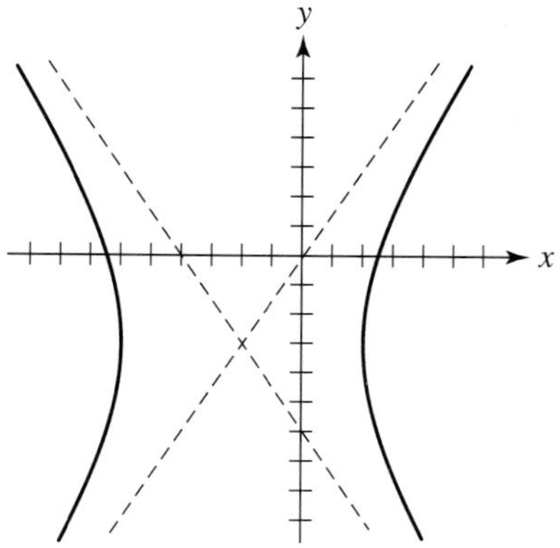

51. symmetric about the y-axis
53. not symmetric about either axis
55. symmetric about the x-axis

57. **a.** $(x - 5)^2 + (y + 2)^2 = 1$ **b.** $\left(\frac{x}{2}\right)^2 + (2y)^2 = 1$

c. $\left(\frac{x + 1}{3}\right)^2 + y^2 = 1$ **d.** $\left(\frac{x}{3} + 1\right)^2 + y^2 = 1$

59.

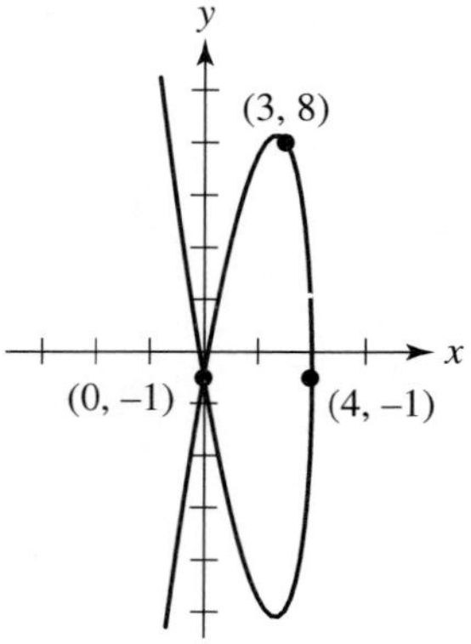

61.

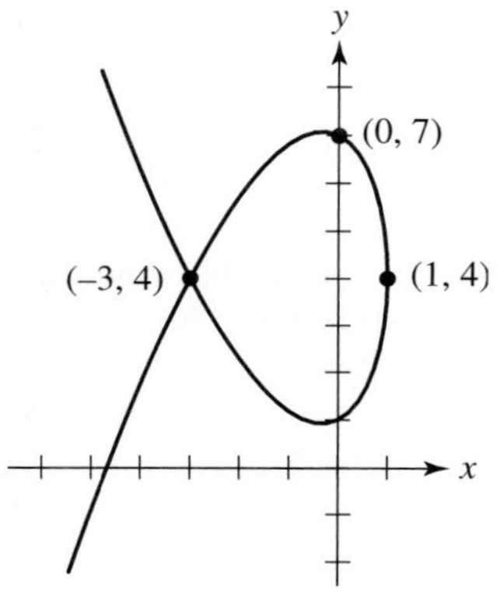

63. $y = (-2.4 \times 10^{-5})x^2 - (3.2 \times 10^{-9})x^3$

65. **a–b.**

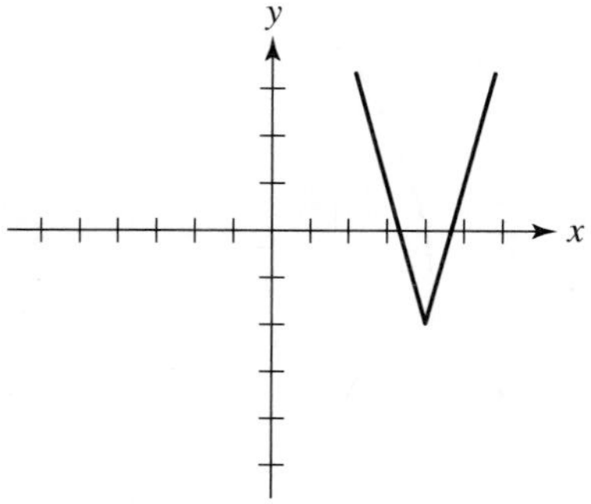

c. $\frac{y + 2}{3} = |x - 4|$

$y + 2 = 3|x - 4|$

$y = 3|x - 4| - 2$

67.

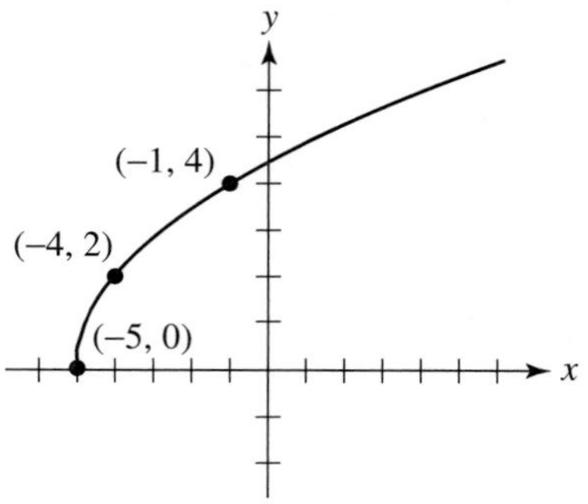

69.

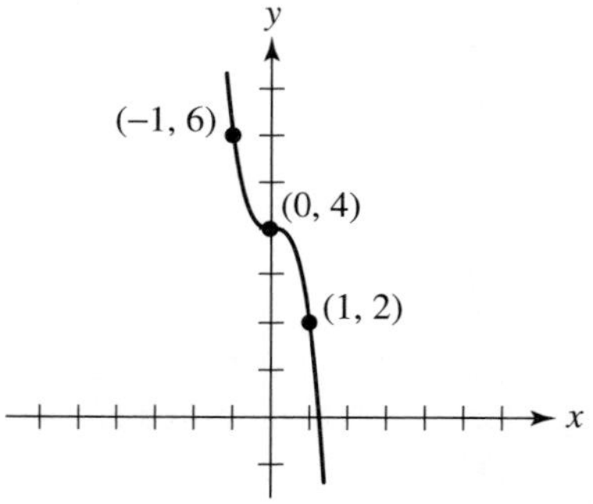

71. **a.**

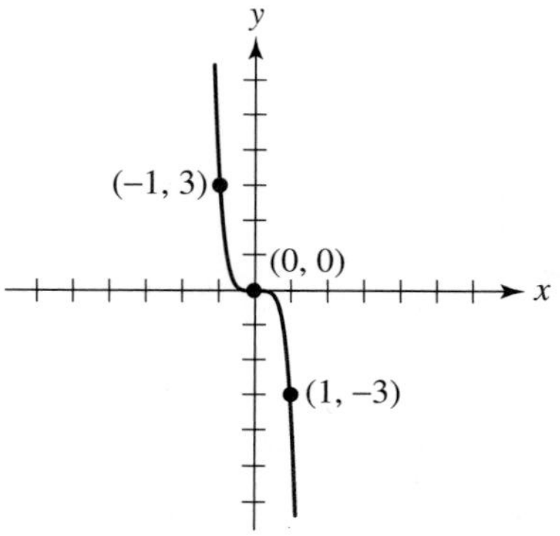

b.

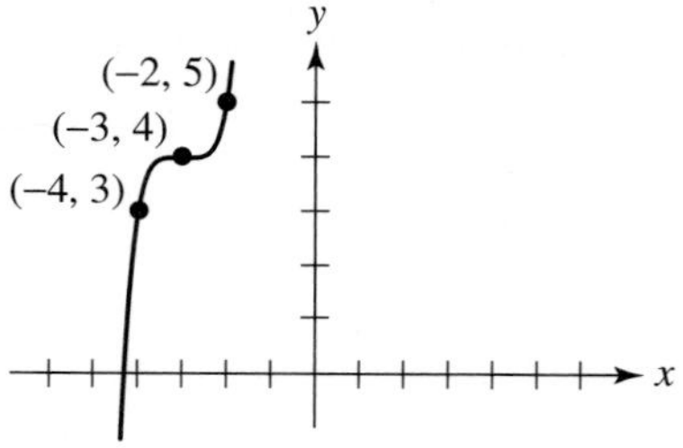

c.

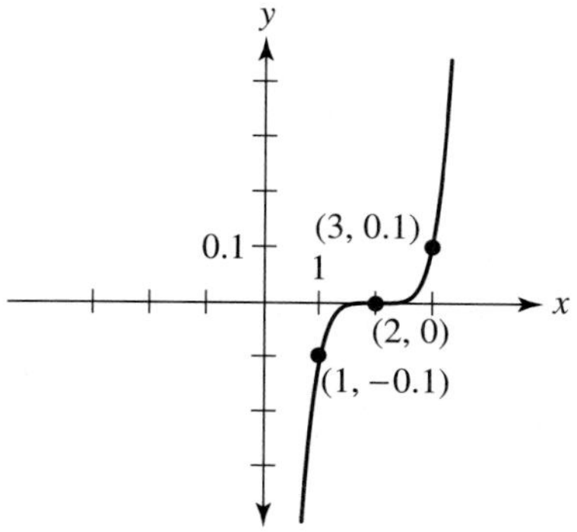

d.

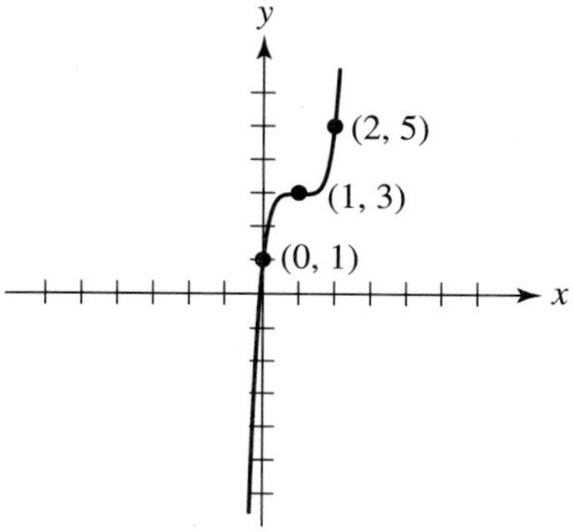

73. a.

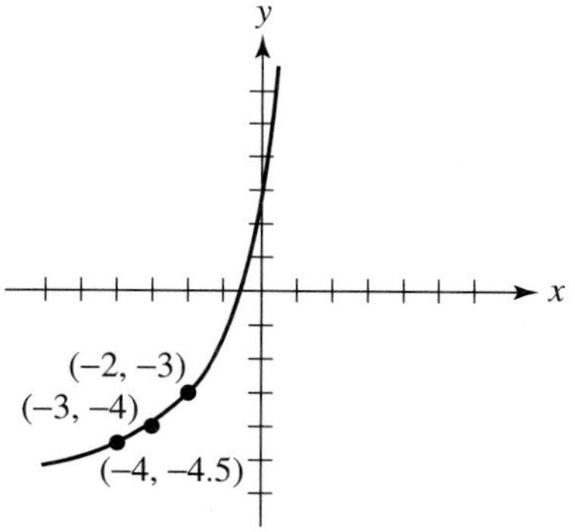

b.

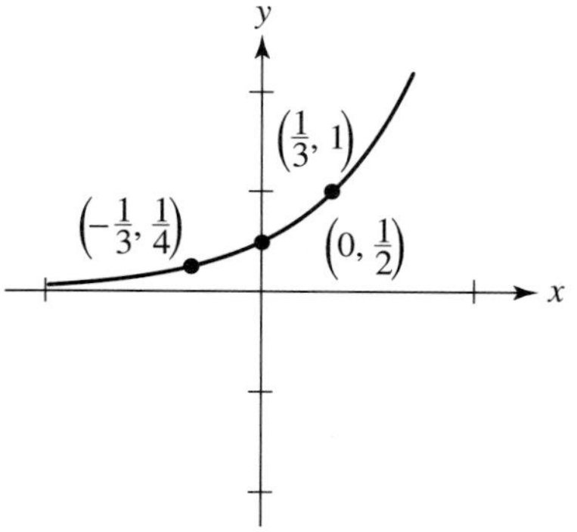

c.

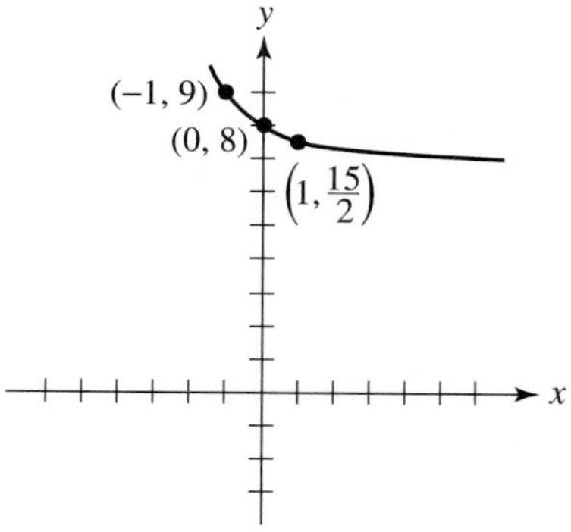

d.

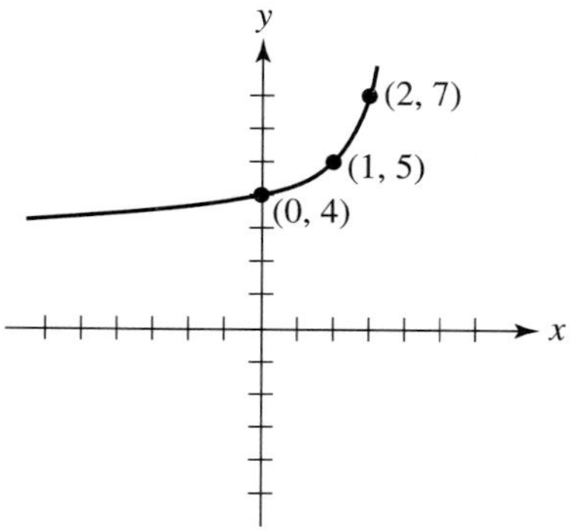

SECTION 6-4

1. Folding the paper as shown in Figure 6-25 superimposes point D on point F, while leaving point P stationary. Thus the line segment PD is superimposed on PF, so the two segments must have the same length.

3. Since both sides of the equation in Exercise 2 are positive, squaring both sides produces an equivalent equation:

$$\begin{aligned} x^2 + (y - p)^2 &= (y + p)^2 \\ x^2 + y^2 - 2py + p^2 &= y^2 + 2py + p^2 \\ x^2 &= 4py \\ y &= \frac{1}{4p}x^2 \end{aligned}$$

5. focus = $(0, 1)$
directrix $x = 4$

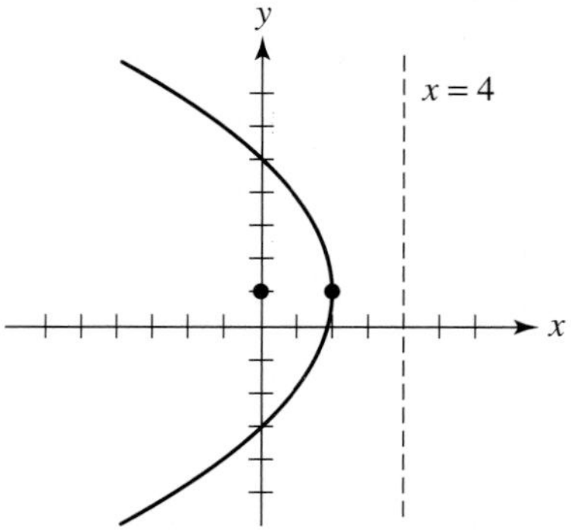

7. focus = $(0.4, 0.08)$
directrix $y = -0.02$

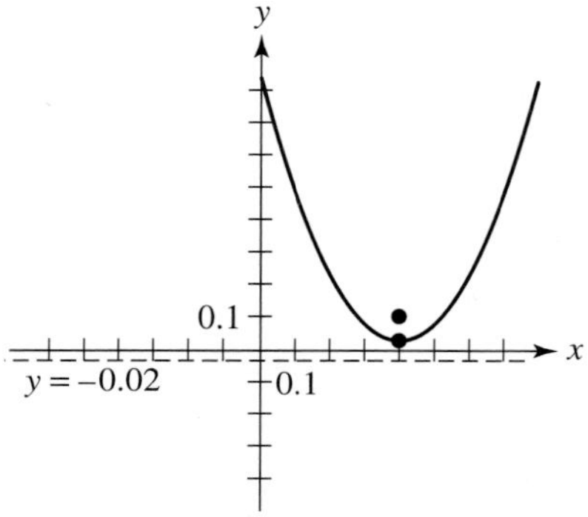

9. about 4.17 feet

11. The points P, F_1, and F_2 have coordinates (x, y), $(-c, 0)$, and $(c, 0)$, respectively. By the distance formula, $PF_1 = 2\sqrt{[x - (-c)]^2 + (y - 0)^2}$
and $PF_2 = \sqrt{(x - c)^2 + (y - 0)^2}$
Since $\overline{PF_1} + \overline{PF_2} = 2a$, we have
$\sqrt{(x + c)^2 + y^2} + \sqrt{(x - c)^2 + y^2} = 2a$

13. a. It is more nearly circular.

b. It would be a circle.

15. foci = $(\pm 3, 0)$
eccentricity = $\frac{3}{5}$

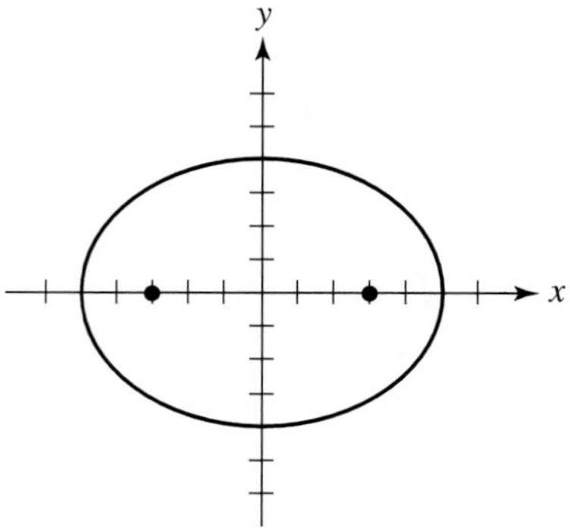

17. foci = $\left(-1, -3 \pm 2\sqrt{5}\right)$
eccentricity = $\frac{2\sqrt{5}}{6}$

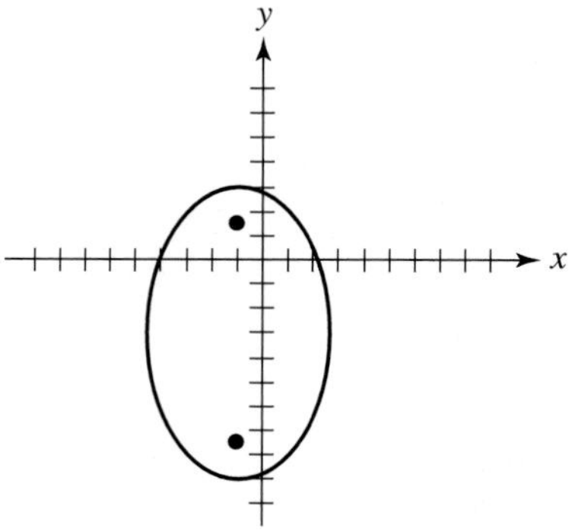

19. at the foci, about 7.07 feet from the center

21. The points P, F_1, and F_2 have coordinates (x, y), $(-c, 0)$, and $(c, 0)$, respectively. By the distance formula, $PF_1 = \sqrt{[x - (-c)]^2 + (y - 0)^2}$
and
$PF_2 = \sqrt{(x - c)^2 + (y - 0)^2}$
Since $PF_1 - PF_2 = 2a$, we have
$\sqrt{(x + c)^2 + y^2} - \sqrt{(x - c)^2 + y^2} = 2a$

23. foci = $(\pm\sqrt{41}, 0)$

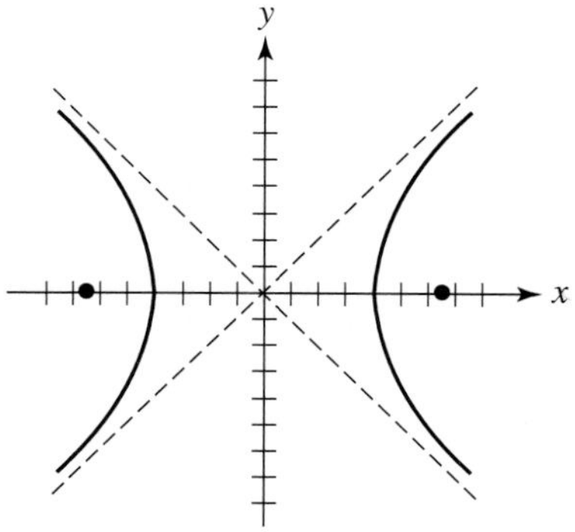

25. foci = $(-1 \pm 2\sqrt{13}, -3)$

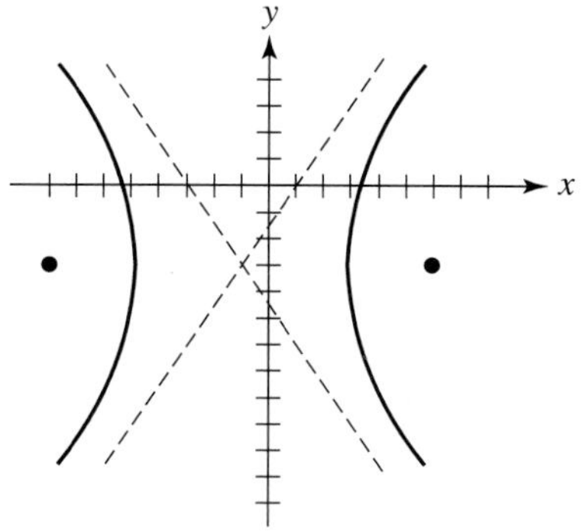

27. The primary shock wave took 0.5 seconds longer to reach Petaluma than to reach Sacramento. Since the shock wave travels at 25 km/sec, this means that the epicenter was 12.5 km closer to Sacramento than to Petaluma. The points satisfying that condition lie on one branch of a hyperbola with foci at Sacramento and Petaluma. The value of c is half the distance between the two cities, or 55. The value of a is half the difference in the distances from the epicenter, or 6.25. This value could also be calculated as 12.5 times the difference of 0.5 seconds in arrival times.

29. center = $(5, 0)$
vertices = $\left(5 \pm \sqrt{14}, 0\right)$
foci = $(1, 0), (9, 0)$
asymptotes $\dfrac{x - 5}{\sqrt{14}} = \pm \dfrac{y}{\sqrt{2}}$

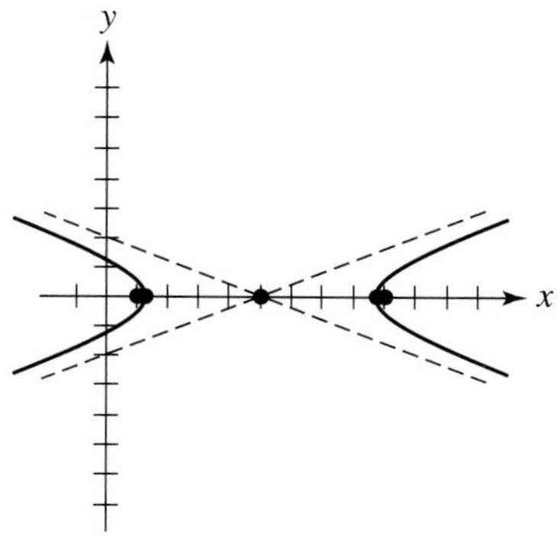

31. center = $(5, 0)$
vertices = $\left(5 \pm \sqrt{14}, 0\right)$
covertices = $\left(5, \pm \sqrt{2}\right)$
foci = $\left(5 \pm 2\sqrt{3}, 0\right)$

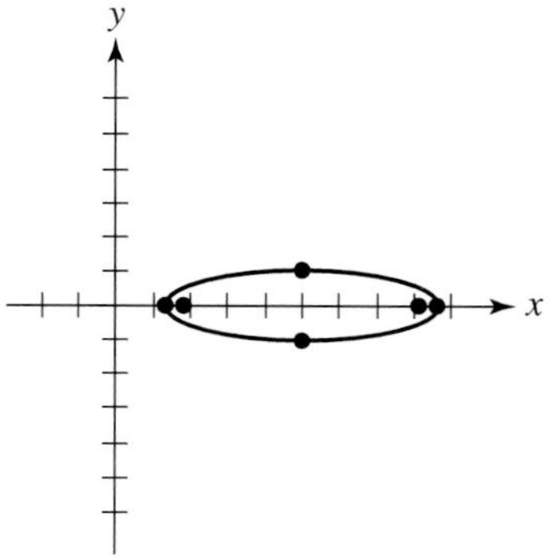

33.

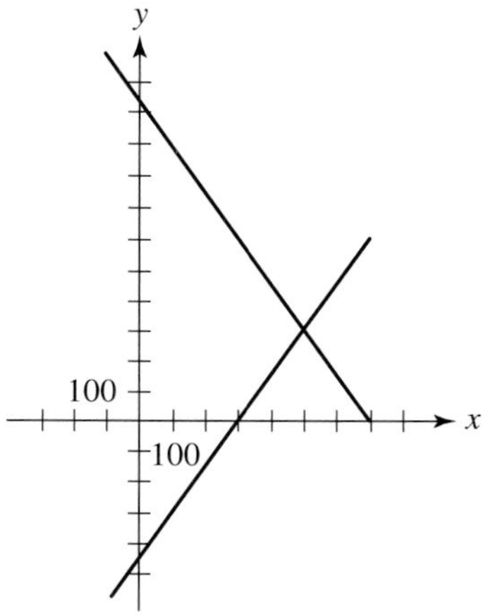

35. center $= \left(\frac{1}{2}, -\frac{1}{2}\right)$

vertices $= \left(\frac{1}{2} \pm \frac{9}{\sqrt{2}}, -\frac{1}{2}\right)$

covertices $= \left(\frac{1}{2}, -\frac{1}{2} \pm \frac{3}{\sqrt{2}}\right)$

foci $= (-5.5, -0.5), (6.5, -0.5)$

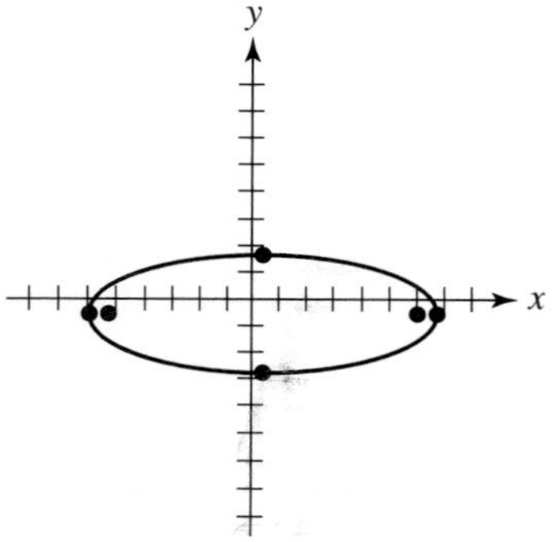

37. center $= (69.5, 46.5)$
radius $= \sqrt{6992.5}$

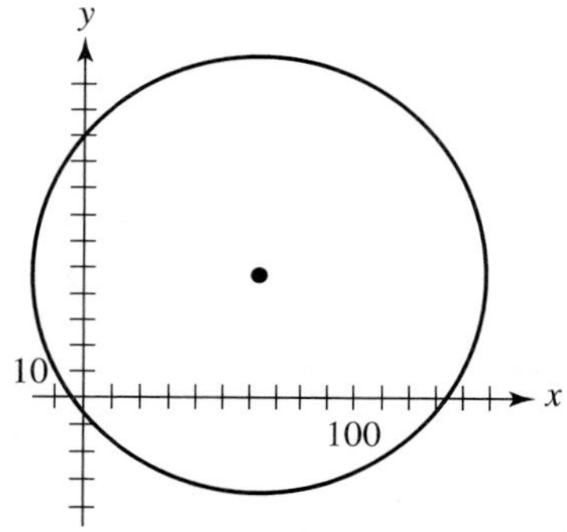

39. a. If the bulb is at the focus of the parabola and the axis of symmetry points straight ahead of the car, then all light rays coming from the bulb will strike the reflective surface and will be reflected straight ahead of the car, resulting in maximum illumination.

b. about 0.68 inches in front of the vertex

41. a. 0.0090

b. $a = 30.06, c = 0.27$

c. $(x - 0.27)^2 + y^2 = 30.06^2$

43.

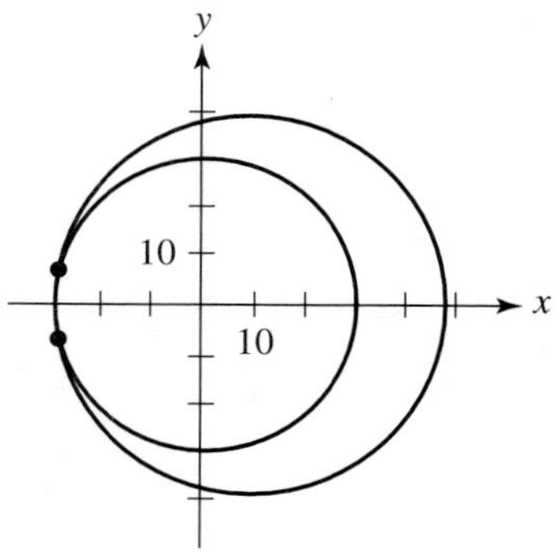

The graphs intersect at approximately $(-28.80, \pm 7.64)$. Estimates of the fraction of Pluto's orbit inside Neptune's orbit will vary.

45. Since PD is vertical, its length is $y - (k - p)$. By the distance formula, $\overline{PF} = \sqrt{(x - h)^2 + [y - (k + p)]^2}$. Therefore,

$$\sqrt{(x - h)^2 + [y - (k + p)]^2} = y - (k - p)$$

Since both sides of the equation represent positive distances, squaring both sides produces an equivalent equation:

$$(x - h)^2 + [y - (k + p)]^2 = y^2 - 2(k - p)y + (k - p)^2$$

$$(x - h)^2 + y^2 - 2(k + p)y + (k + p)^2 = y^2 - 2(k - p)y + (k - p)^2$$

$$(x - h)^2 + y^2 - 2ky - 2py + k^2 + 2pk + p^2 = y^2 - 2ky + 2py + k^2 - 2pk + p^2$$

$$(x - h)^2 = 4py - 4pk$$

$$4py = (x - h)^2 + 4pk$$

$$y = \frac{1}{4p}(x - h)^2 + k$$

47. By the distance formula,

$$\overline{PF_1} = \sqrt{[x - (h - c)]^2 + (y - k)^2} = \sqrt{[(x - h) + c]^2 + (y - k)^2}$$

and

$$\overline{PF_2} = \sqrt{[x - (h + c)]^2 + (y - k)^2} = \sqrt{[(x - h) - c]^2 + (y - k)^2}.$$

Thus, $\sqrt{[(x - h) + c]^2 + (y - k)^2} - \sqrt{[(x - h) - c]^2 + (y - k)^2} = 2a$

$$\sqrt{[(x - h) + c]^2 + (y - k)^2} = 2a + \sqrt{[(x - h) - c]^2 + (y - k)^2}$$

$$\left\{\sqrt{[(x - h) + c]^2 + (y - k)^2}\right\}^2 = \left\{2a + \sqrt{[(x - h) - c]^2 + (y - k)^2}\right\}^2$$

$$[(x - h) + c]^2 + (y - k)^2 = 4a^2 + 4a\sqrt{[(x - h) - c]^2 + (y - k)^2} + [(x - h) - c]^2 + (y - k)^2$$

$$4a\sqrt{[(x-h)-c]^2+(y-k)^2} = -4a^2-[(x-h)-c]^2+[(x-h)+c]^2$$

$$4a\sqrt{[(x-h)-c]^2+(y-k)^2} = -4a^2-(x-h)^2+2c(x-h)-c^2+(x-h)^2 + 2c(x-h)+c^2$$

$$4a\sqrt{[(x-h)-c]^2+(y-k)^2} = -4a^2+4c(x-h)$$

$$a\sqrt{[(x-h)-c]^2+(y-k)^2} = -a^2+c(x-h)$$

$$\{a\sqrt{[(x-h)-c]^2+(y-k)^2}\}^2 = [-a^2+c(x-h)]^2$$

$$a^2\{[(x-h)-c]^2+(y-k)^2\} = a^4-2a^2c(x-h)+c^2(x-h)^2$$

$$a^2[(x-h)^2-2c(x-h)+c^2+(y-k)^2] = a^4-2a^2c(x-h)+c^2(x-h)^2$$

$$a^2(x-h)^2-2a^2c(x-h)+a^2c^2+a^2(y-k)^2 = a^4-2a^2c(x-h)+c^2(x-h)^2$$

$$a^2c^2-a^4 = c^2(x-h)^2-a^2(x-h)^2-a^2(y-k)^2$$

$$(c^2-a^2)(x-h)^2-a^2(y-k)^2 = a^2(c^2-a^2)$$

$$\frac{(x-h)^2}{a^2}-\frac{(y-k)^2}{c^2-a^2} = 1$$

SECTION 6–5

1. The ratio $\frac{d}{s^2} = \frac{1}{15}$ for any data point.

3. The first differences of the s-values are not constant.

5. As the water becomes shallower, the value of d decreases, so the value of $s = \sqrt{15d}$ also decreases.

7.

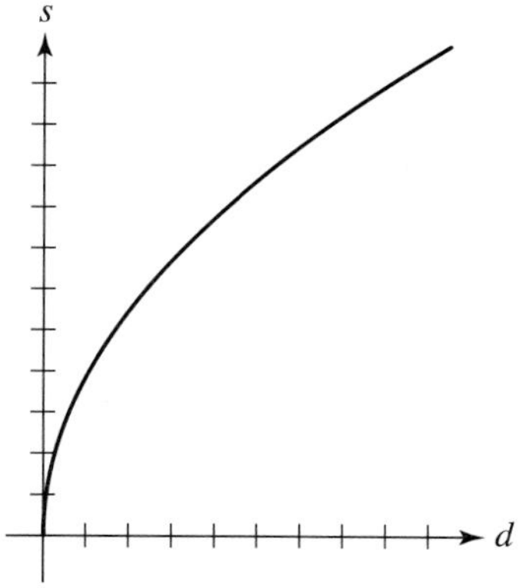

The graph is steeper when d is small.

9. a. The domain is $[0, \infty)$ because the vertical line $x = x_0$ contains a point of the graph if and only if $x_0 \geq 0$. Similarly, the range is $[0, \infty)$ because the horizontal line $y = y_0$ contains a point of the graph if and only if $y_0 \geq 0$.

b. The expression $y = \sqrt{x}$ is defined if and only if $x \geq 0$, so the domain is $[0, \infty)$. The value of $\sqrt{x}$ can be any nonnegative real number, so the range is also $[0, \infty)$.

11. a.

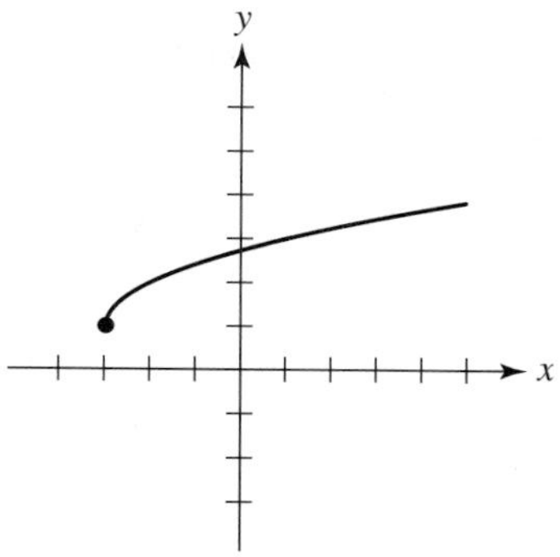

b. domain $= [-3, \infty)$
range $= [1, \infty)$

13. a.

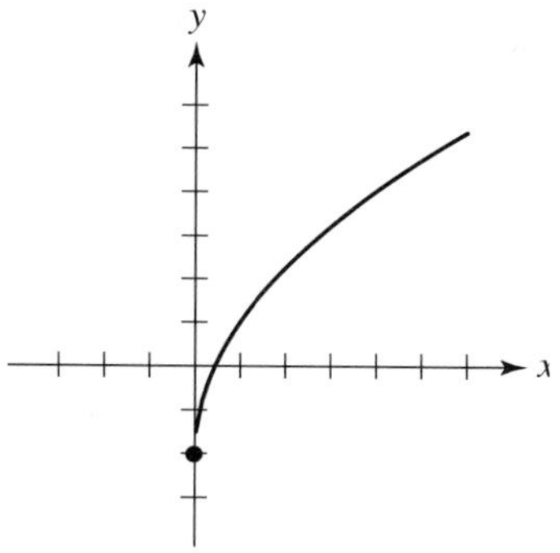

b. domain $= [0, \infty)$
range $= [-2, \infty)$

15. $y = \sqrt{x^2 + 2053x}$

17. $\left(\frac{x-1026.5}{1026.5}\right)^2 - \left(\frac{y}{1026.5}\right)^2 = 1$

a.

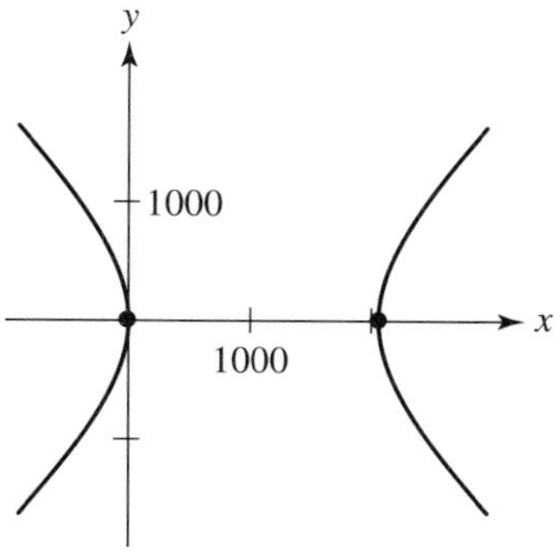

b.

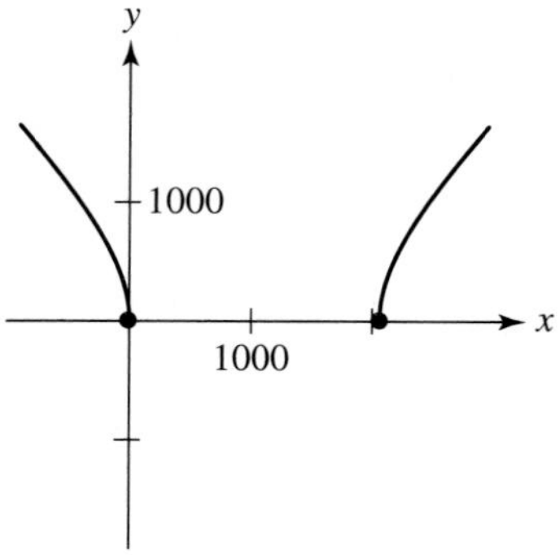

c.

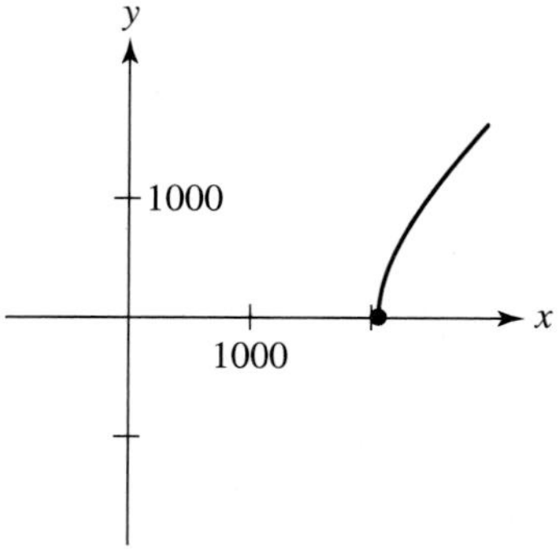

19. a. $\left(\frac{x}{2}\right)^2 + \left(\frac{y}{4}\right)^2 = 1$

b.

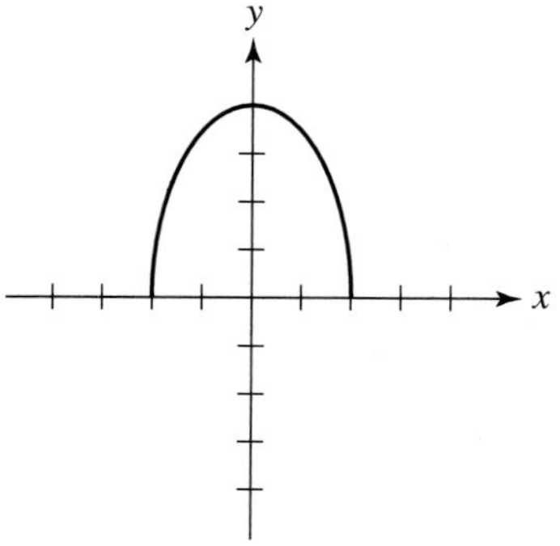

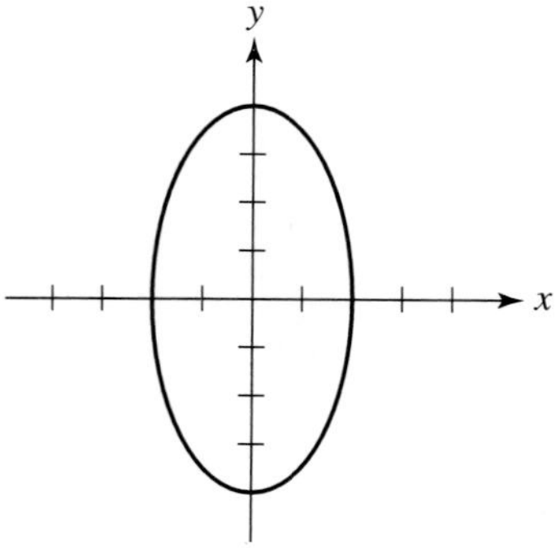

c. domain = $[-2, 2]$
range = $[0, 4]$

21. a. $\left(\frac{y}{10}\right)^2 - \left(\frac{x}{5}\right)^2 = 1$

b.

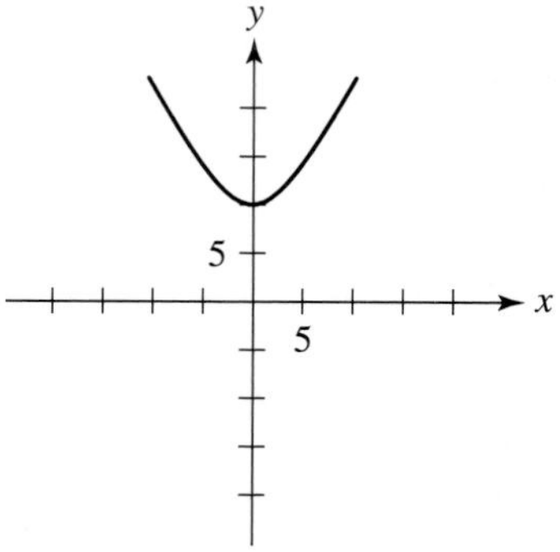

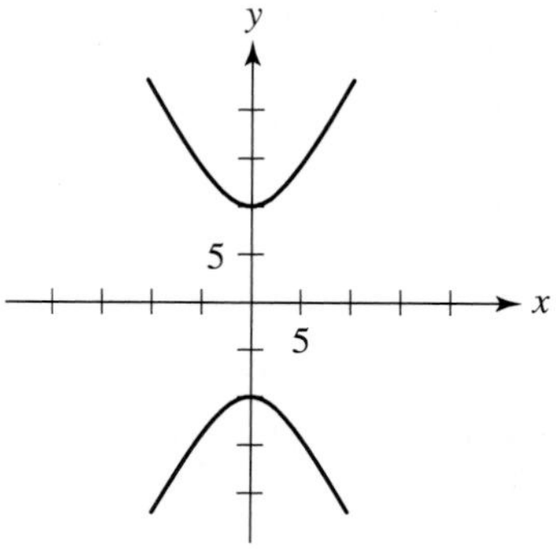

c. domain = $(-\infty, \infty)$
range = $[10, \infty)$

23. a.

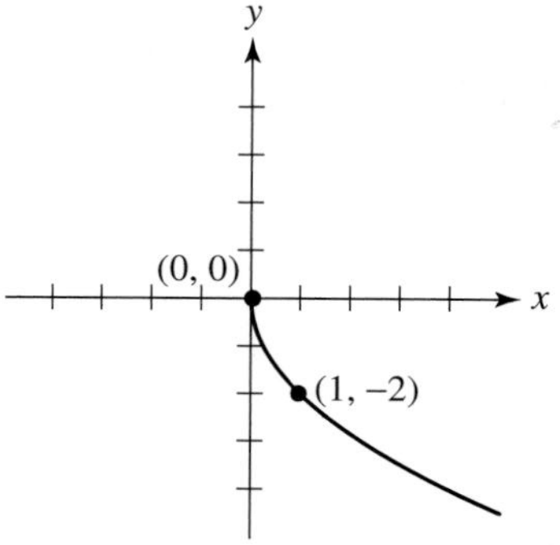

b. domain = $[0, \infty)$
range = $(-\infty, 0]$

25. a.

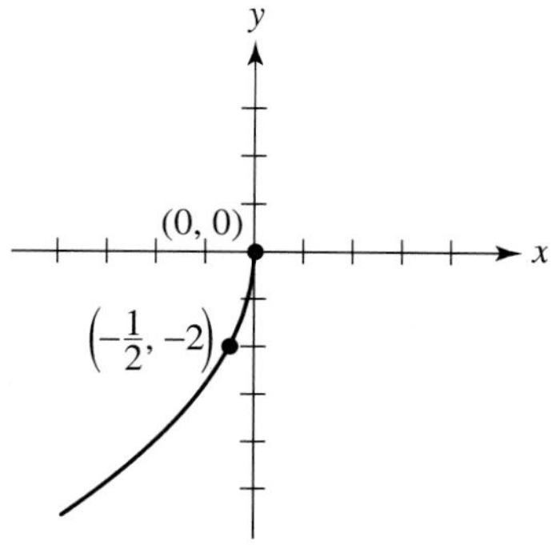

b. domain = $(-\infty, 0]$
range = $(-\infty, 0]$

27. a.

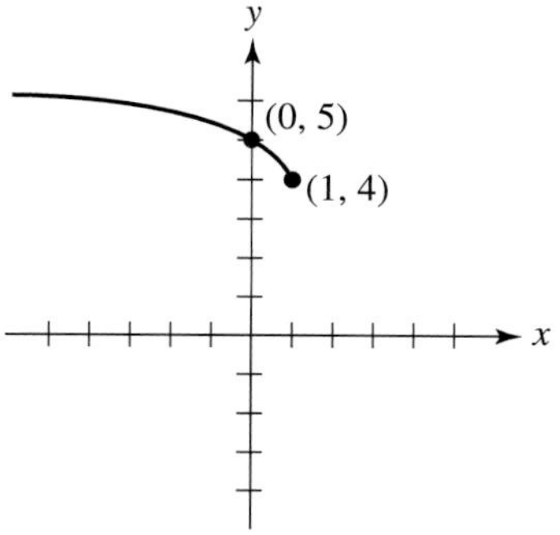

b. domain = $(-\infty, 1]$
range = $[4, \infty)$

29. a.

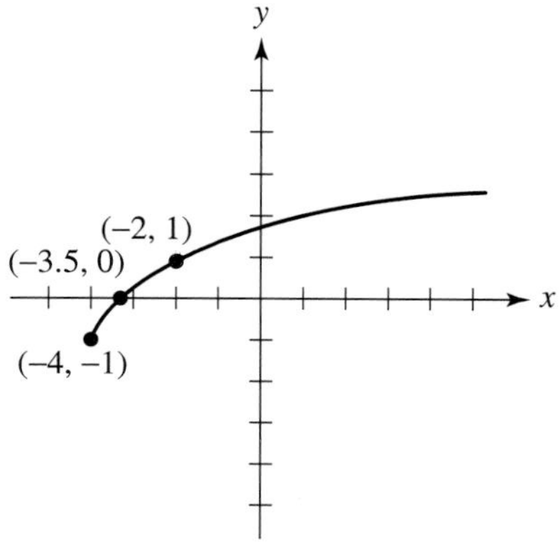

b. domain = $[-4, \infty)$
range = $[-1, \infty)$

31. a. $x^2 + \left(\frac{y}{3}\right)^2 = 1$

b.

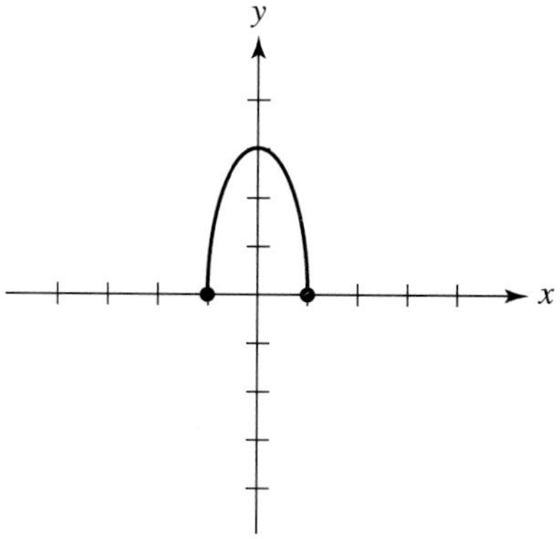

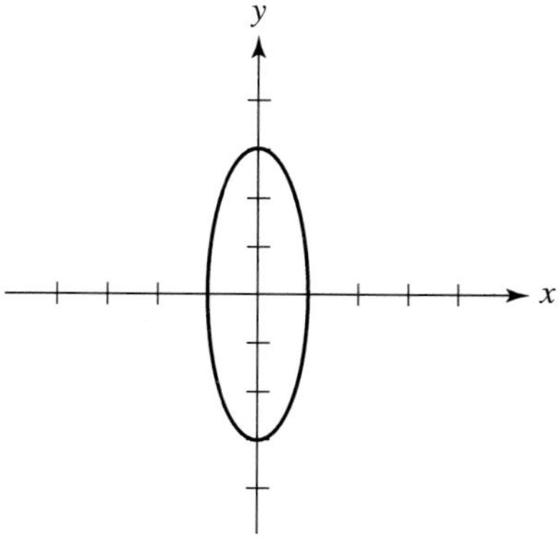

c. domain = $[-1, 1]$
range = $[0, 3]$

33. a. $\left(\frac{y}{3}\right)^2 - x^2 = 1$

b.

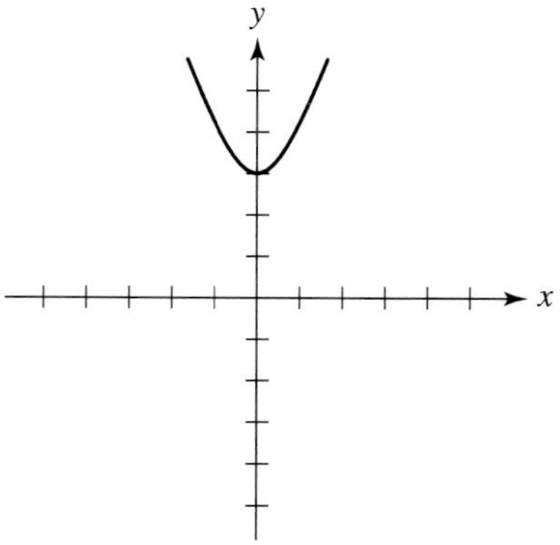

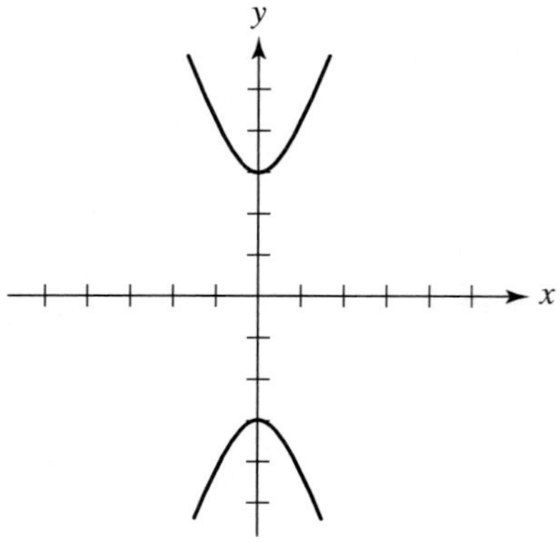

c. domain = $(-\infty, \infty)$
range = $[3, \infty)$

35. a. $\left(\frac{x}{2}\right)^2 - \left(\frac{y-5}{6}\right)^2 = 1$

b.

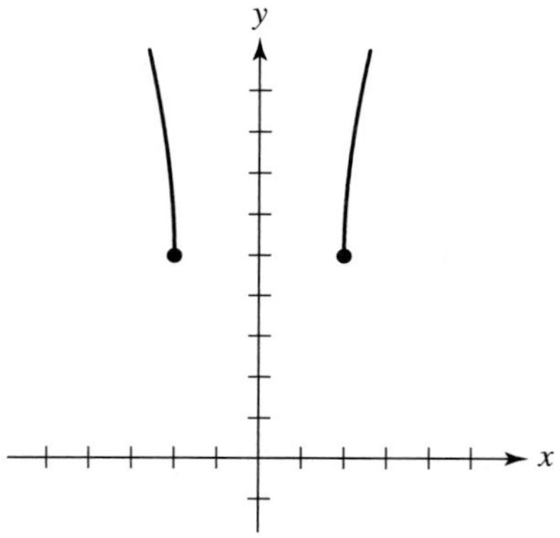

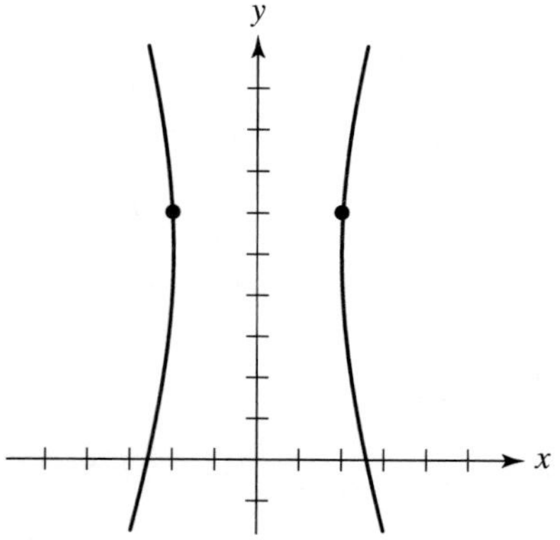

c. domain = $(-\infty, -2] \cup [2, \infty)$
range = $[5, \infty)$

37. a. $(x-5)^2 + (y+6)^2 = 4$

b.

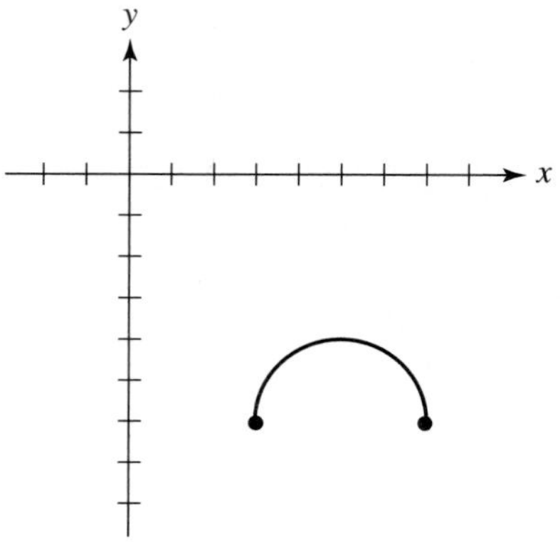

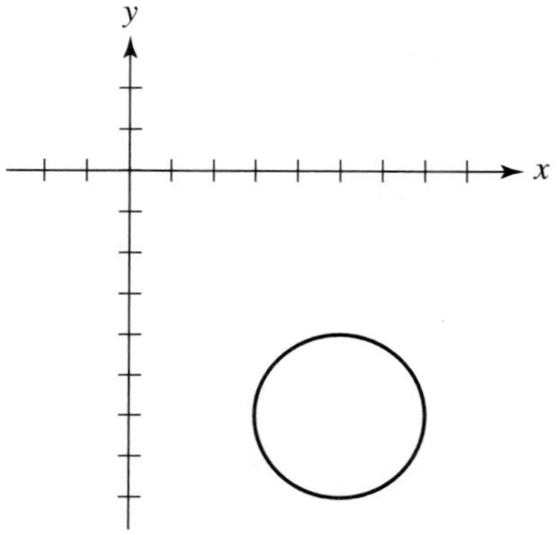

c. domain = $[3, 7]$
range = $[-6, -4]$

39. a. domain = $[-9, \infty)$, range = $[-0.2, \infty)$
b. domain = $[0, \infty)$, range = $[0, \infty)$

41. $y = \sqrt{x^2 + 900}$

43. $\left(\frac{y}{30}\right)^2 - \left(\frac{t-5}{1/3}\right)^2 = 1$

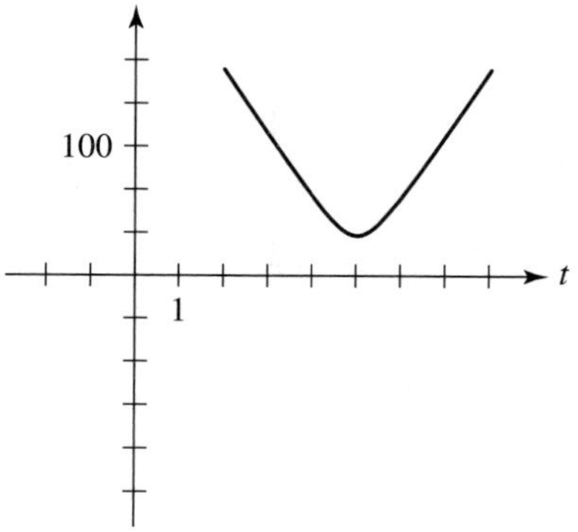

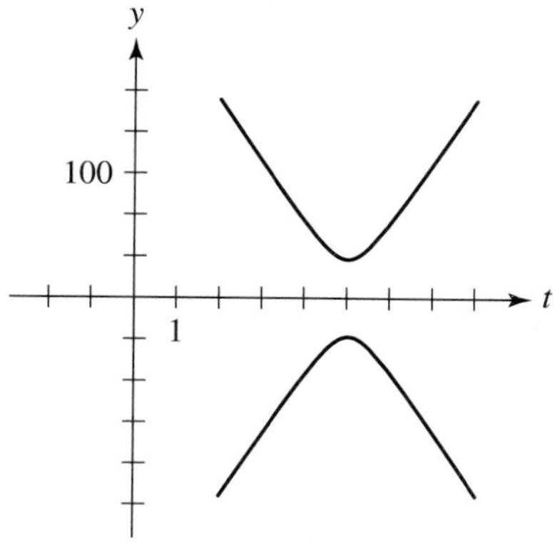

SECTION 6-6

1. $\dfrac{(y+55)^2}{50^2} - \dfrac{x^2}{22.91^2} = 1$; lower branch

3. $(0, 1)$ **5.** $(-4, -6), (4, 6)$

7. $(-\sqrt{8}, -\sqrt{11}), (-\sqrt{8}, \sqrt{11}), (\sqrt{8}, -\sqrt{11}),$ $(\sqrt{8}, \sqrt{11})$

9. $(-\sqrt{5}, -\sqrt{3}), (-\sqrt{5}, \sqrt{3}), (\sqrt{5}, -\sqrt{3}), (\sqrt{5}, \sqrt{3})$

11. $(-\sqrt{13}, -12\sqrt{13}), (-2\sqrt{3}, -26\sqrt{3}), (2\sqrt{3}, 26\sqrt{3}),$ $(\sqrt{13}, 12\sqrt{13})$

13. $(-6, 1), (-2, 3), (2, -3), (6, -1)$

15. a.

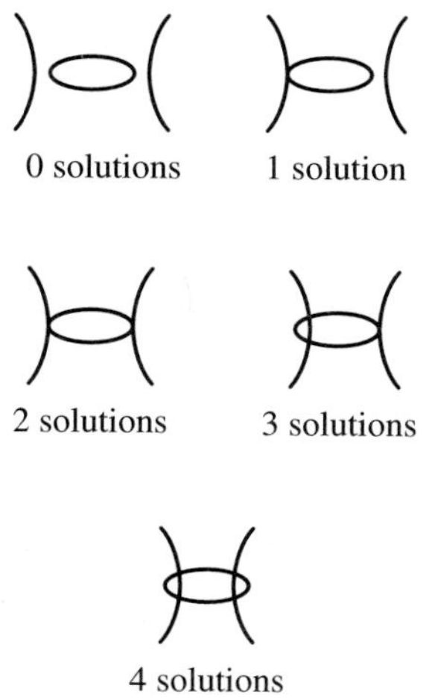

b.

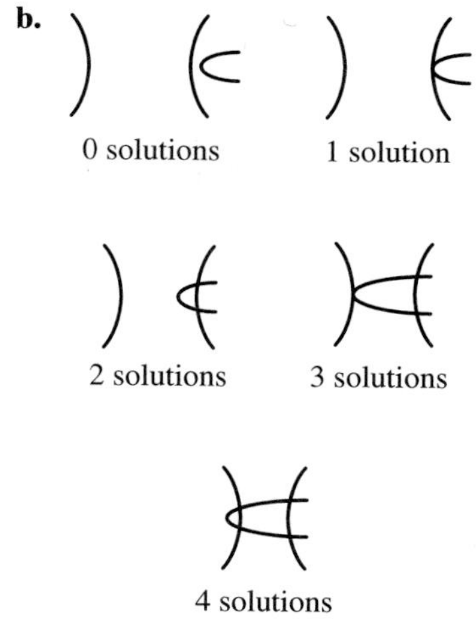

17. $(-4.47, 0.53), (0.73, 5.73)$ **19.** no real solutions

21. a–b. The ship lies on or inside the circle $x^2 + (y+40)^2 = 400$.

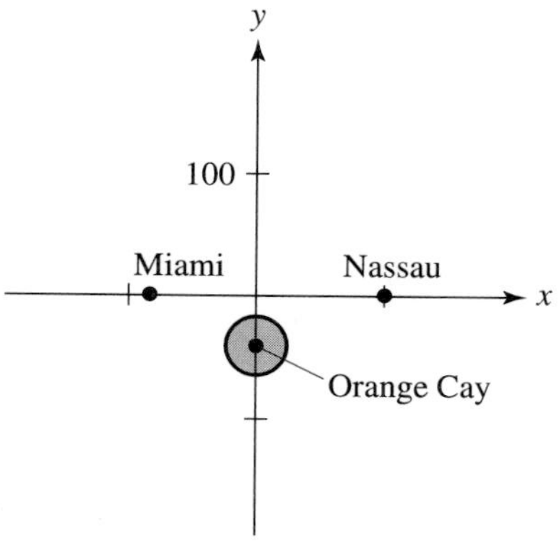

23. No. The region in Exercise 22 does not include the entire region in Exercise 21.

25.

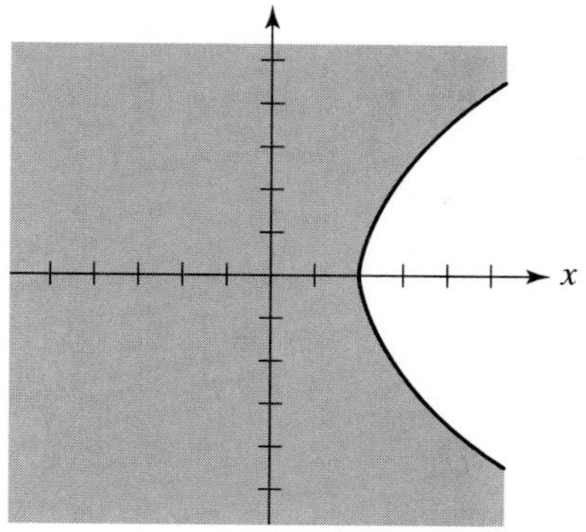

27.

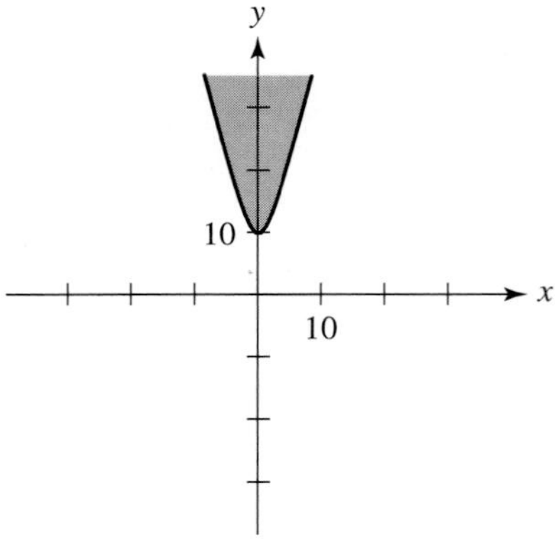

29.

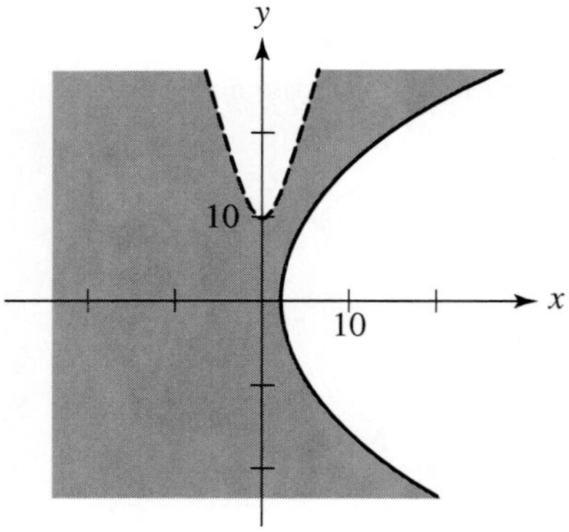

31.

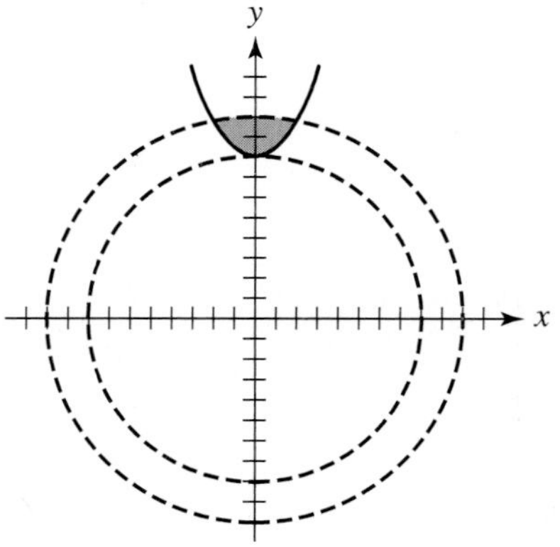

33. $x^2 + (y + 40)^2 \leq 400$
$(x - 100)^2 + y^2 \leq 10{,}000$

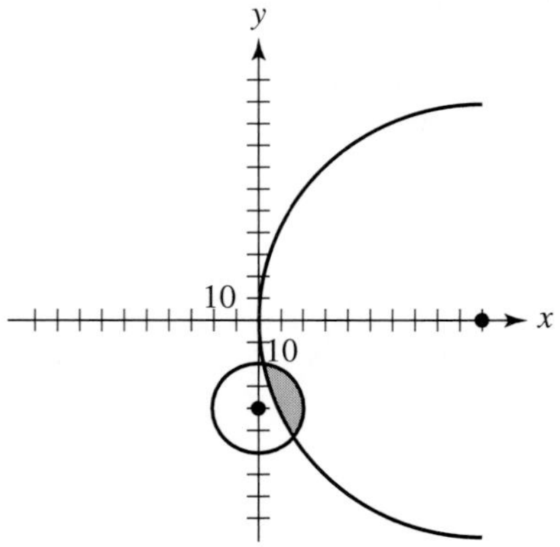

35. Responses will vary.

37. $(-2\sqrt{5}, 0.4\sqrt{5}), (-2, 2), (2, -2), (2\sqrt{5}, -0.4\sqrt{5})$

39. no real solutions

41. $(-3.32, 5), (0, -6), (3.32, 5)$ **43.** $(0, -2), (4, 0)$

45.

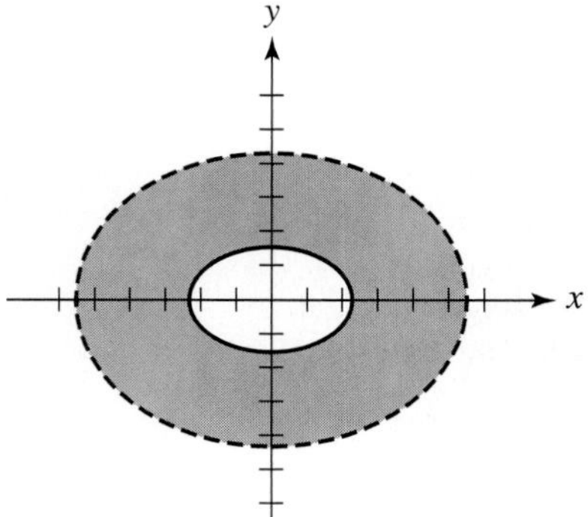

47.

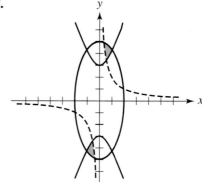

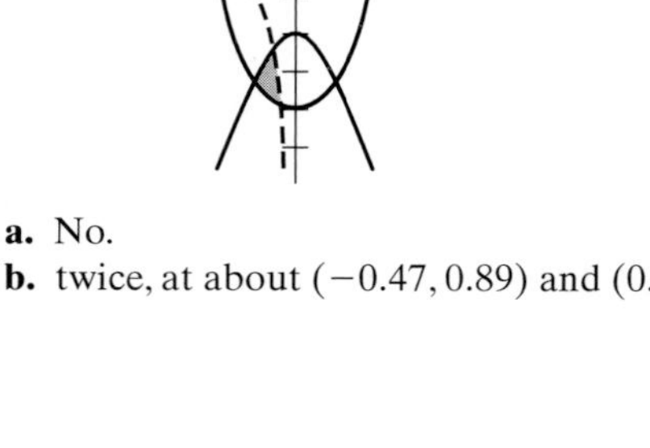

49. **a.** No.
b. twice, at about $(-0.47, 0.89)$ and $(0.91, -0.48)$

51. **a.** The time for the batted ball to reach the fielder plus the time for the thrown ball to reach first base must be no more than 3 seconds. Since the ball always travels at 120 feet per second, the total distance traveled by the ball must be 360 feet. Thus $\overline{AB} + \overline{BC} = 360$. This means that the fielder is on an ellipse with foci at A and C and with $a = 180$.

b. $\left(\frac{x - 45}{180}\right)^2 + \left(\frac{y}{174.28}\right)^2 \leq 1$

$x > 0$

$y > 0$

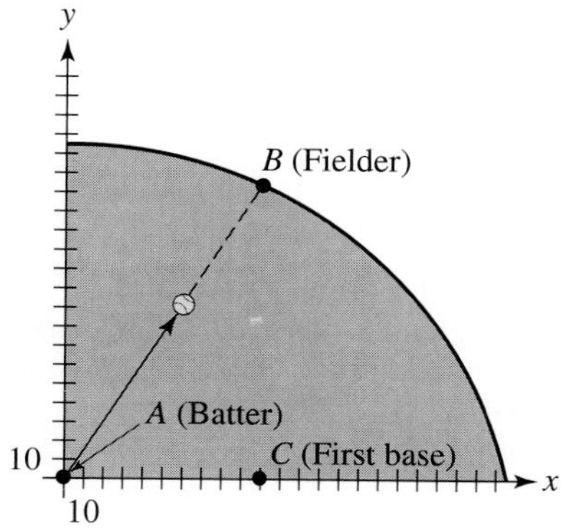

SECTION 7–1

1. The graphs all pass through $(-1, -1)$, $(0, 0)$, and $(1, 1)$. The left tail of each graph points down, and the right tail points up. For larger powers, the graph is flatter near the origin, and steeper when $|x| > 1$.

3.

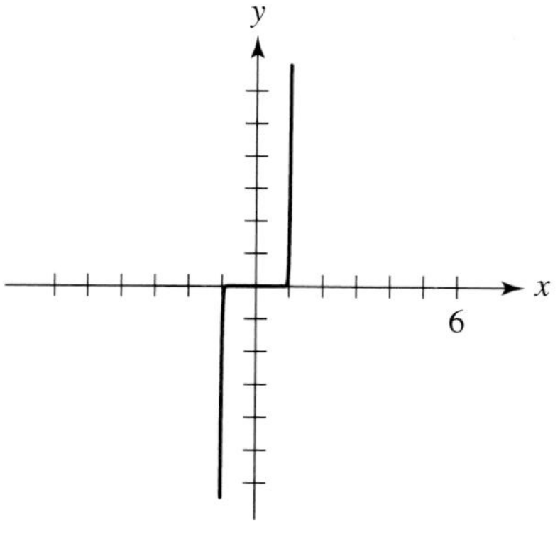

$y = x^{49}$

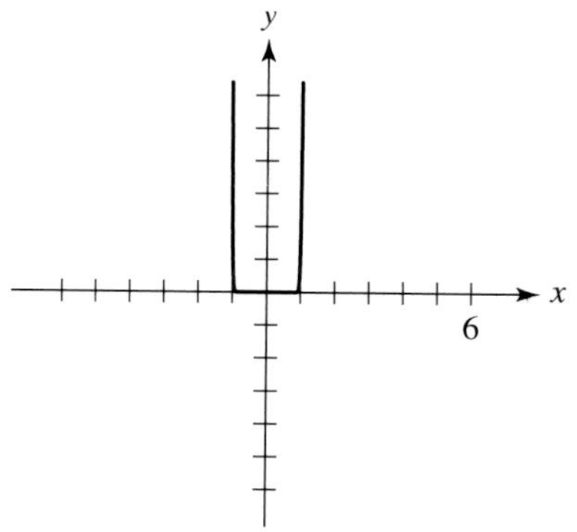

$y = x^{50}$

5.

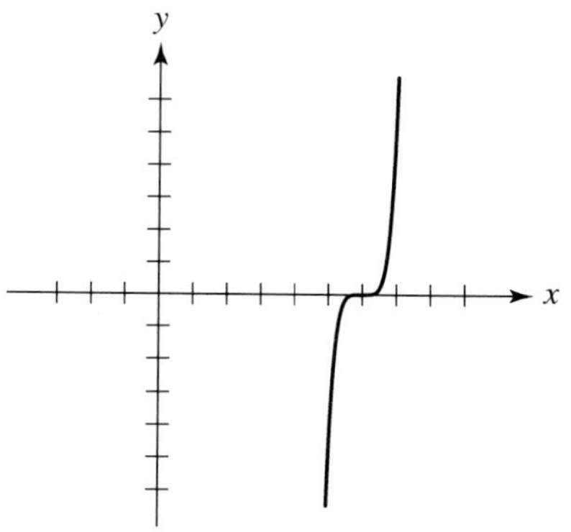

7.

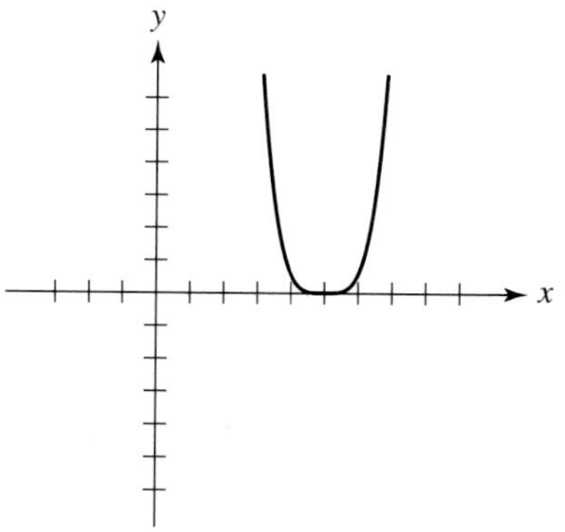

9. a.

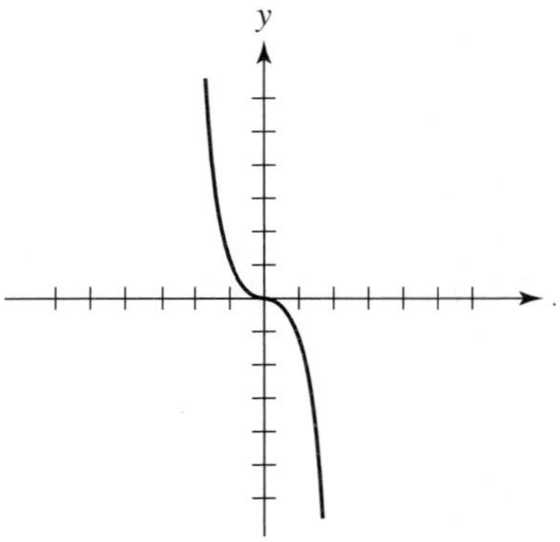

b. $y = (-x)^3$

11. a.

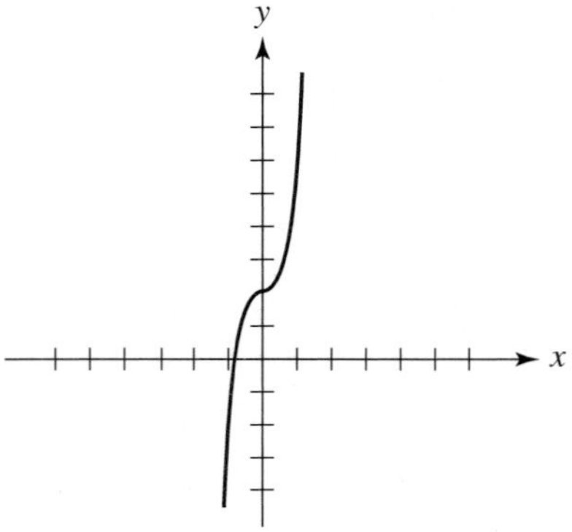

b. $y = 4x^3 + 2$

13. In Section 3-2 it is stated:

> If $f(x) = mx$ for some real m, then whenever the input x is doubled, the output $f(x)$ is also doubled.

In Section 7-1 it is stated:

> For nth power polynomial variations, doubling x multiplies y by 2^n.

Replacing n by 1 in the second statement yields the first statement.

15. Yes; $y = 2x^3$.

17. Yes; $y = 0.5x^4$.

19. a. $W \cong 0.14C^2$

b. 68,600 pounds

21. a.

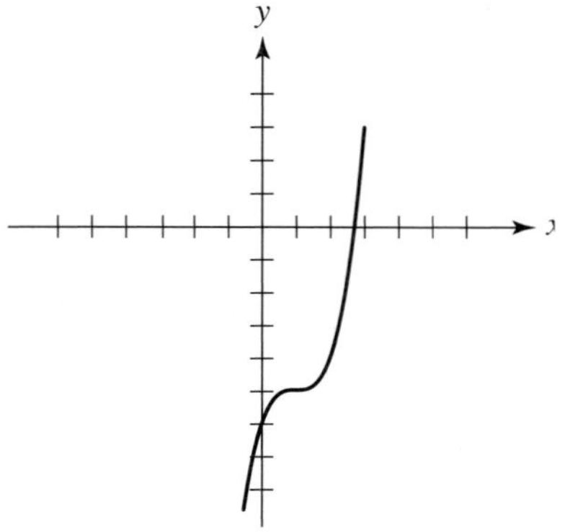

b. domain $= (-\infty, \infty)$
range $= (-\infty, \infty)$

23. a.

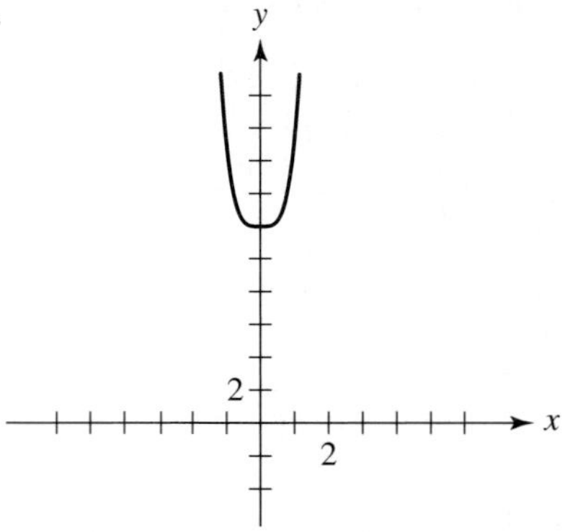

b. domain $= (-\infty, \infty)$
range $= [12, \infty)$

25. a. The domain and range are unchanged.

b. The domain is unchanged. If n is odd, the range is unchanged. If n is even, the range changes from $[0, \infty)$ to $(-\infty, 0]$.

c. The domain and range are unchanged.

d. The domain is unchanged. If n is odd, the range is unchanged. If n is even and the new equation is $y = x^n + k$, the range changes from $[0, \infty)$ to $[k, \infty)$.

27. a.

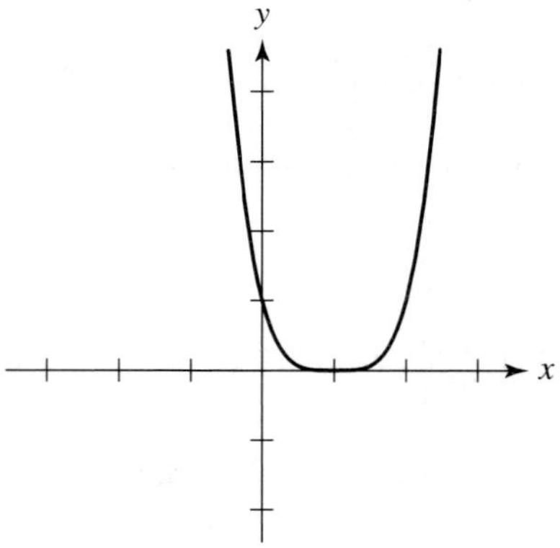

b. $y = (x - 1)^4$

29. a.

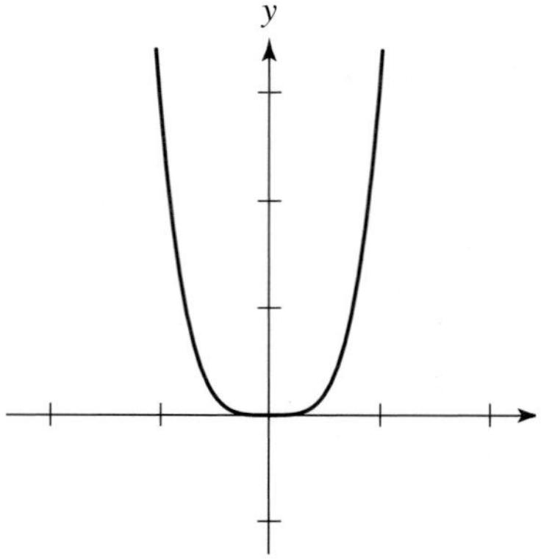

b. $y = 3x^4$

31. a.

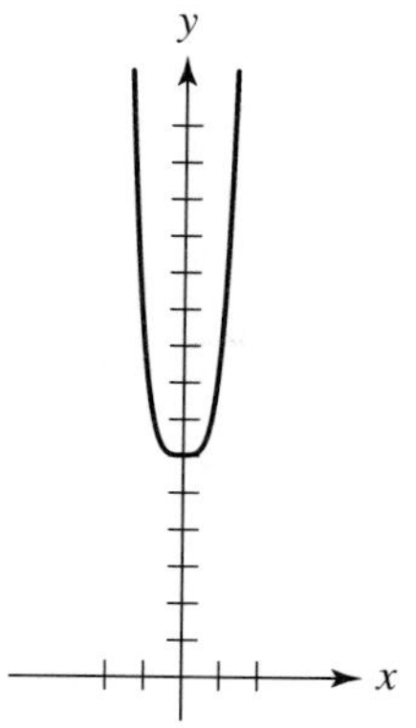

b. $y = 3(x^4 + 2)$

33. Yes; $y = 15x$.

35. No.

37. The satellite would need to be about 236 kilometers under the surface of Mars. Tell your colleagues that their plans cannot be carried out without some alterations.

SECTION 7-2

1. a. $x = -3, 2$
b. $x = 0, 2, 3$
c. $x = \pm 1, -2$
d. $x = \pm 1, \pm 3$

3. a. $x = 4$
b. $x = 0, \pm 3$
c. $x = 1, \pm i$
d. $x = \pm\sqrt{2}$

5. $-3, -1, 2$

7. $-2, \dfrac{1 \pm \sqrt{5}}{2}$

9. a.

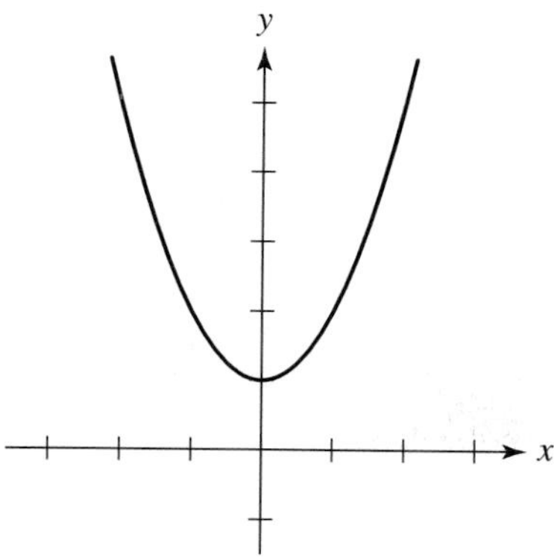

b. Each real zero must correspond to an x-intercept on the graph. Since the graph has no x-intercepts, the function has no real zeros.

c. $x = \pm i$

11. In Exercise 3a, degree = 2, number of zeros = 2.
In Exercise 3b, degree = 5, number of zeros = 5.
In Exercise 3c, degree = 3, number of zeros = 3.
In Exercise 3d, degree = 4, number of zeros = 4.

13. $0, -7, 6$

15. $3, \pm\sqrt{2}$

17. a. $g(-1) = 0$; -1 is a zero of g.
b. $g(x) = x^3(x + 1)(x - 1)(x^2 + 2)$; $x = 0, \pm 1, \pm\sqrt{2}\,i$

19. a. $g(-1) = -1$; -1 is not a zero of g.
b. $g(x) = (2x^2 + 1)(x^2 + 2)$; $x = \pm\sqrt{\dfrac{1}{2}}\,i, \pm\sqrt{2}\,i$

21. $b = 4$, multiplicity 3
$b = -\pi$, multiplicity 5

23. 4, multiplicity 3 each;
$r = -5, 3$, multiplicity 2 each

25. $z = \pm\sqrt{2}$, multiplicity 2 each
$z = \pm\sqrt{2}\,i$, multiplicity 1 each

27. $x = -1 \pm 2i$; degree = 2, number of zeros = 2

29. $r = 0, \pm\sqrt{3}\,i$; degree = 5, number of zeros = 5

31. a. The degree of the polynomial is 3, and we have found only one solution. The number of solutions counted according to multiplicity must be 3.
b. $x = 11 \pm \sqrt{23}\,i$

33. a.
$$P(x) = \left(x - \frac{1}{2}\right)^2 (x^2 - 3)$$
$$= \left(x^2 - x + \frac{1}{4}\right)(x^2 - 3)$$
$$= x^4 - x^3 - \frac{11}{4}x^2 + 3x - \frac{3}{4}$$

b.

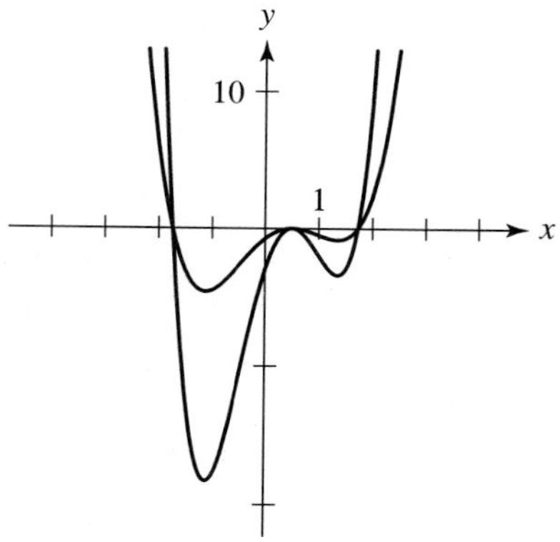

The two graphs have the same x-intercepts.

c. $P(x) = 0$ if and only if $4P(x) = 0$.

d. If the zeros of Q are the same as those of P with the same multiplicities, then Q also has factors of $\left(x - \frac{1}{2}\right)^2$, $(x + \sqrt{3})$, and $(x - \sqrt{3})$. Since Q has no other zeros, it can have no other linear factors. Thus Q must be a constant multiple of
$$\left(x - \frac{1}{2}\right)^2 (x + \sqrt{3})(x - \sqrt{3}).$$

35. $P(x) = (x - 3)^2(2x - 1)$

37. $P(x) = (x + 5)(2x + 3)^2$

39. $P(x) = (x^2 - 2)^2$

41. $4, \dfrac{-3 \pm \sqrt{29}}{2}$

43. $-\dfrac{7}{3}, \dfrac{1}{2}, 3$

45. $4, -1 \pm \sqrt{5}$

47. $-3, 1.80$

SECTION 7-3

1. a. $x = -4$, multiplicity 1; $x = -1$, multiplicity 1; $x = 3$, multiplicity 3
$(-4, 0), (-1, 0), (3, 0)$
b. $x = 0$, multiplicity 1
$(0, 0)$
c. $x = 0$, multiplicity 2; $x = \pm 2\sqrt{7}$, multiplicity 1 each
$(0, 0), (-2\sqrt{7}, 0), (2\sqrt{7}, 0)$
d. $x = \pm 2$, multiplicity 1 each
$(-2, 0), (2, 0)$

3. a. $P(x) = (x + 2)^2(x - 1)^2(x - 3)^2$; $(-2, 0), (1, 0), (3, 0)$
b. $P(x) = x^3 - 6x$; $(0, 0), (\pm\sqrt{6}, 0)$
c. $P(x) = (2x - 3)^2(3x + 1)$; $\left(\frac{3}{2}, 0\right)$
d. $P(x) = (x^2 - 2)^2(x^2 + 2)$; $(\pm\sqrt{2}, 0), (\pm\sqrt{3}, 0)$

5. $f(x) = (x + 4)^3(x - 1)^2(x - 3)$

7. $f(x) = x^2(x + 4)^2(x - 4)^2$

9. local minimum at $(-2, -16)$, local maximum at $(2, 16)$ decreasing on $(-\infty, -2)$ and $(2, \infty)$; increasing on $(-2, 2)$

11. local maximum at about $(-1.05, 13.03)$, local minimum at about $(1.05, -1.03)$ increasing on $(-\infty, -1.05)$ and $(1.05, \infty)$, decreasing on $(-1.05, 1.05)$

13. a. radius $\cong$ 11.46 inches, length $\cong$ 36.00 inches
b. about 11.17 ounces

15. about 8.76 ounces; less

17. When $|x|$ is large, the absolute value of $-100x^3$ is small in comparison with that of x^4, so the sign of the sum agrees with that of x^4.

19. left tail down, right tail up

21. left tail up, right tail down

23. both tails down **25.** both tails up

27. symmetric about the y-axis

29. symmetric about the origin

31. symmetric about the y-axis

33. (Sample response) According to the test described in Section 6-3, a graph is symmetric about the y-axis if replacing x by $-x$ produces an equivalent equation. If y is a function of x, this amounts to saying that $f(-x) = f(x)$. The test in Section 6-3 is more general, because it can be applied even if y is not a function of x.

35. $y = (x - 1)(x - 4)$

37. $y = (x - 0.1)(x - 1)(x - 10)(x - 100)$

39. $y = (x + 5)(x + 3)(x + 1)(x - 2)(x - 4)(x - 6)$

41. $y = x^3(x + 5)(x + 2)(x - 4)$

43. a. $\pm\sqrt{3 \pm \sqrt{5}}$
b. local maximum at $(0, 4)$
absolute minima at $(\pm 1.73, -5.00)$
c. both tails up
d. Choices will vary.
e. decreasing on $(-\infty, -1.73)$ and $(0, 1.73)$, increasing on $(-1.73, 0)$ and $(1.73, \infty)$

45. a. 0.58, 4.75, 17.67
b. local minimum at $(2.52, -65.38)$
local maximum at $(12.81, 478.57)$
c. left tail up, right tail down
d. Choices will vary.
e. decreasing on $(-\infty, 2.52)$ and $(12.81, \infty)$, increasing on $(2.52, 12.81)$

47. a. $-3.19, 2.19$
b. absolute maximum at $(-1.85, 40.29)$
c. both tails
d. Choices will vary.
e. increasing on $(-\infty, -1.85)$, decreasing on $(-1.85, \infty)$

49. even **51.** neither

53. a. 2.08, 7.58, [3, 3.5] **b.** [3, 3.5]
c. $-1.92, -2.92, [-0.5, 0.5]$ **d.** $[-0.5, 0.5]$

55. domain $(-\infty, \infty)$, range $= (-\infty, \infty)$

57. domain $(-\infty, \infty)$, range $= (-\infty, 0]$

59. domain $(-\infty, \infty)$, range $= [64, \infty)$

61. (Sample response) In this extremely large viewing window (that is, *from a distance*), details such as turning points are too small to see, and the two graphs appear identical. In particular, their end behavior is the same.

SECTION 7-4

1. a.

x	$f(x)$	First differences	Second differences	Third differences
1	3			
		7.125		
1.5	10.125		6.75	
		13.875		2.25
2.	24		9	
		22.875		2.25
2.5	46.875		11.25	
		34.125		2.25
3.	81		13.5	
		47.625		
3.5	128.625			

b.

x	$f(x)$	First differences	Second differences	Third differences
-1	$-a$			
		a		
0	0		0	
		a		$6a$
1	a		$6a$	
		$7a$		$6a$
2	$8a$		$12a$	
		$19a$		$6a$
3	$27a$		$18a$	
		$37a$		
4	$64a$			

c.

x	$f(x)$	First differences	Second differences	Third differences
x_0	ax_0^3			
		$3ax_0^2h + 3ax_0h^2 + ah^3$		
$x_0 + h$	$ax_0^3 + 3ax_0^2h + 3ax_0h^2 + ah^3$		$6ax_0h^2 + 6ah^3$	
		$3ax_0^2h + 9ax_0h^2 + 7ah^3$		$6ah^3$
$x_0 + 2h$	$ax_0^3 + 6ax_0^2h + 12ax_0h^2 + 8ah^3$		$6ax_0h^2 + 12ah^3$	
		$3ax_0^2h + 15ax_0h^2 + 19ah^3$		$6ah^3$
$x_0 + 3h$	$ax_0^3 + 9ax_0^2h + 27ax_0h^2 + 27ah^3$		$6ax_0h^2 + 18ah^3$	
		$3ax_0^2h + 21ax_0h^2 + 37ah^3$		$6ah^3$
$x_0 + 4h$	$ax_0^3 + 12ax_0^2h + 48ax_0h^2 + 64ah^3$		$6ax_0h^2 + 24ah^3$	
		$3ax_0^2h + 27ax_0h^2 + 61ah^3$		
$x_0 + 5h$	$ax_0^3 + 15ax_0^2h + 75ax_0h^2 + 125ah^3$			

3. 3 **5.** 2

7. (Sample response) A table containing four data points with equally spaced x-values has only one third difference. Thus the third differences are constant by default, so the table must fit a cubic polynomial function.

9. Some possibilities are (4, 26), (5, 61), (6, 120), (7, 209), (8, 334), (9, 501).

11. $f(x) = \frac{4}{3}x^3 - 6x^2 + \frac{20}{3}x + 1$

13. $f(x) = x^4 - 2x^3 - 3x + 1$

15. 1.37 **17.** 1.62

19. 1, 3 **21.** 1.84, 1.20, 1.52

23. 6 **25.** 2

27. $f(x) = \frac{25}{6}x^3 - \frac{41}{2}x^2 + \frac{79}{3}x$

29. $f(x) = \frac{1}{2}x^2 - \frac{13}{2}x + 18$

31. 2.28 **33.** 4.38

35. 1.34 **37.** −1.27, 2

SECTION 7-5

1. $(1.20, 2.55) \cup (4.25, \infty)$ **3.** $\{-1, 3\}$

5. (Sample response) According to the new cost function, $C(x)$ represents the cost of producing $100x$ shirts. According to the former cost function, the cost of producing $100x$ shirts is $C(100x) = 3.50(100x) + 612.50 = 350x + 612.50$. Thus the new cost function should be $C(x) = 350x + 612.50$. Similarly, according to the former revenue function, the revenue produced by the sale of $100x$ shirts is $R(100x) = -0.005(100x)^2 + 16(100x) = -50x^2 + 1600x$.

7. a. $P(x) = -0.08x^3 - 44x^2 + 1250x - 612.50$
b. $0.50 \leq x \leq 26.59$; between 50 and 2659 shirts.
c. The third x-intercept is about $(-577.10, 0)$, and the graph is above the x-axis to the left of this intercept. The solution is $(-\infty, -577.10] \cup [0.50, 26.59]$. This differs from the solution in part (b) because the negative values of x do not make sense in the physical context.

9. $(-\infty, -2.83] \cup \{0\} \cup [2.83, \infty)$ **11.** $[0, \infty)$

13. $(-\infty, -7] \cup \left[-\frac{4}{3}, \infty\right)$ **15.** $\left(1, \frac{7}{2}\right) \cup \left(\frac{7}{2}, 5\right) \cup (5, \infty)$

17. $9.16 \leq r \leq 13.49$ **19.** $(-\infty, -5) \cup (0, 3)$

21. $[-2, 2]$ **23.** $(-\infty, -5) \cup (0, 3)$

25. $[-2, -1] \cup [1, 2]$ **27.** $\left(-\infty, -\frac{5}{2}\right) \cup \left(\frac{1}{3}, 7\right)$

29. $(-\infty, -4) \cup (12, \infty)$

31.

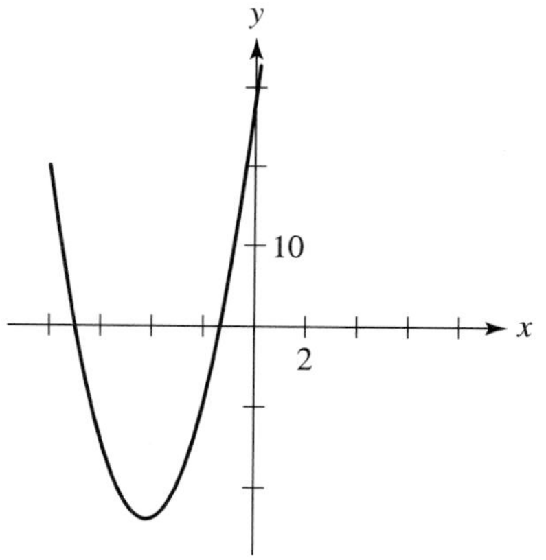

33.

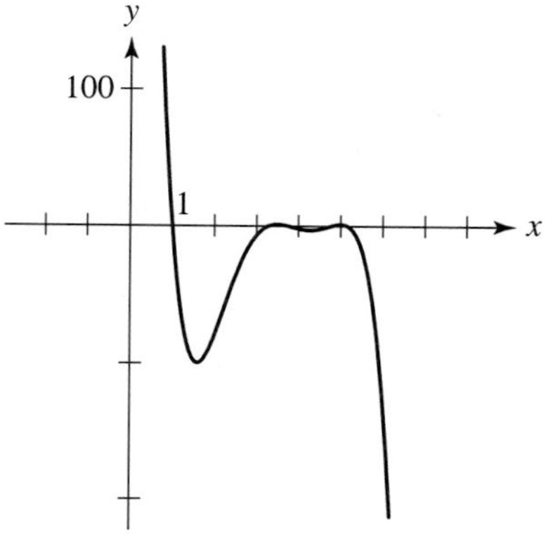

35. (Sample response) The sign pattern in Exercise 15 indicates that the graph is above the x-axis when $x < 1$ and below it when $x > 1$ except at the x-intercepts $\left(\frac{7}{2}, 0\right)$ and $(5, 0)$. Conversely, the graph in Exercise 33 indicates that $y < 0$ in $\left(1, \frac{7}{2}\right) \cup \left(\frac{7}{2}, 5\right) \cup (5, \infty)$.

37.

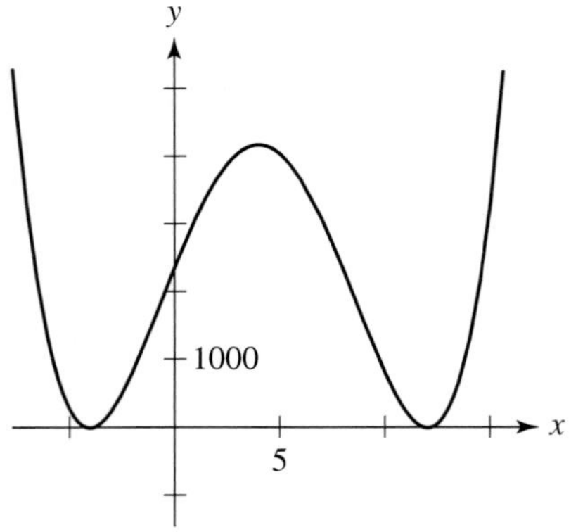

39.

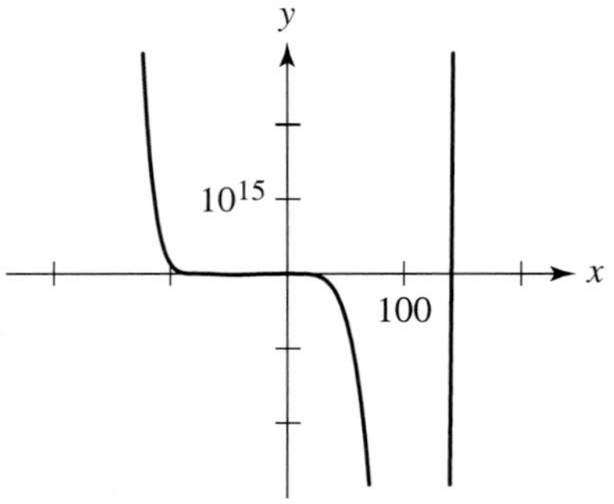

SECTION 8-1

1. $g(s) = \dfrac{450}{s}$

3. a. about 1 hour and 18 minutes
b. You save about 58 minutes at a cost of about \$2.59 per minute

5. a.

x	$f(x)$
1	1
10	0.1
100	0.01
1000	0.001

b.

x	$f(x)$
−1	−1
−10	−0.1
−100	−0.01
−1000	−0.001

c.

x	$f(x)$
1	1
0.1	10
0.01	100
0.001	1000

d.

x	$f(x)$
−1	−1
−0.1	−10
−0.01	−100
−0.001	−1000

7. The tables illustrate that when $|x|$ is large, both $|f(x)|$ and $|g(x)|$ are small. They also illustrate that when $|x|$ is small, both $|f(x)|$ and $|g(x)|$ are large.

9. a. 45 minutes **b.** 2.7 minutes
c. 0.0024 seconds

11. a. 7.5 days **b.** 5.14 years
c. 4500 years

13. Yes; $y = \dfrac{84}{x}$.

15. Yes; $y = \dfrac{900}{x^2}$.

17. 12

19. 2

21. The graph of $y = \dfrac{1}{x}$ has a vertical asymptote at $x = 0$ (on the y-axis). When the graph is shifted 1 unit to the left, the asymptote shifts 1 unit to the left, to the line $x = -1$. Similarly, when the graph is shifted 2 units up, the horizontal asymptote shifts from $y = 0$ to $y = 2$.

23.

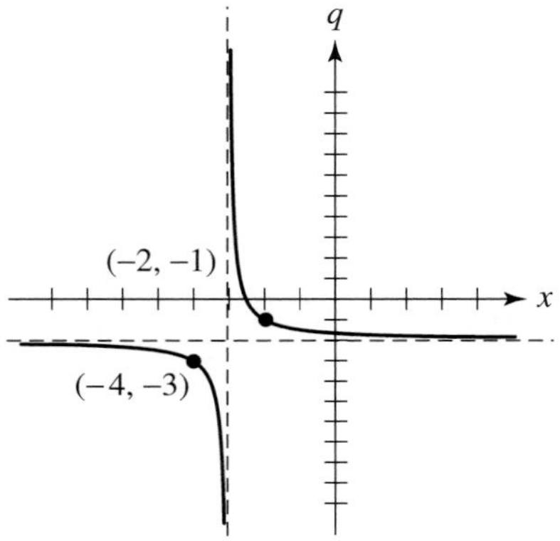

vertical asymptote $x = -3$
horizontal asymptote $y = -2$
domain $= (-\infty, -3) \cup (-3, \infty)$
range $= (-\infty, -2) \cup (-2, \infty)$

25.

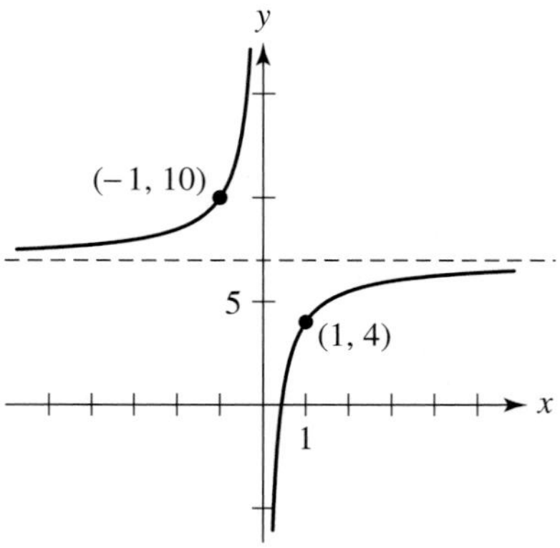

vertical asymptote $x = 0$ (y-axis)
horizontal asymptote $y = 7$
domain $= (-\infty, 0) \cup (0, \infty)$
range $= (-\infty, 7) \cup (7, \infty)$

27.

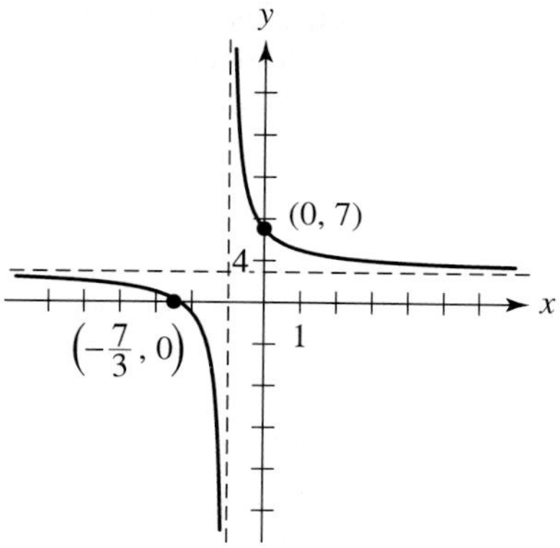

vertical asymptote $x = -1$
horizontal asymptote $y = 3$

29.

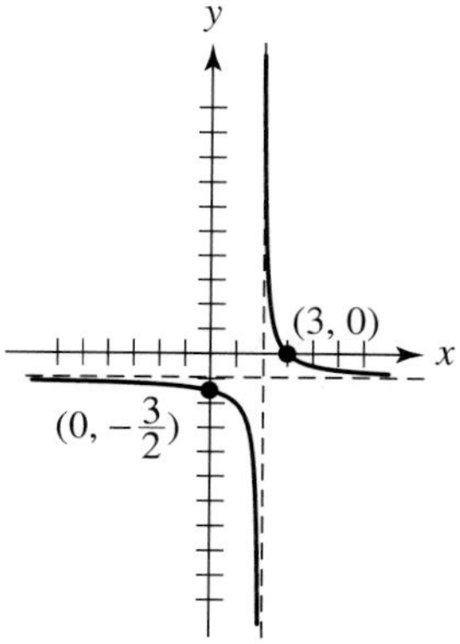

vertical asymptote $x = 2$
horizontal asymptote $y = -1$

31. Yes; $y = \dfrac{0.1}{x^2}$.

33. Yes; $y = \dfrac{4096}{x^2}$.

35. 18

37. $\dfrac{1}{2}$

39. a. 60 lb/in^2 **b.** $\dfrac{1}{5}$

41. 0.45 newtons

43.

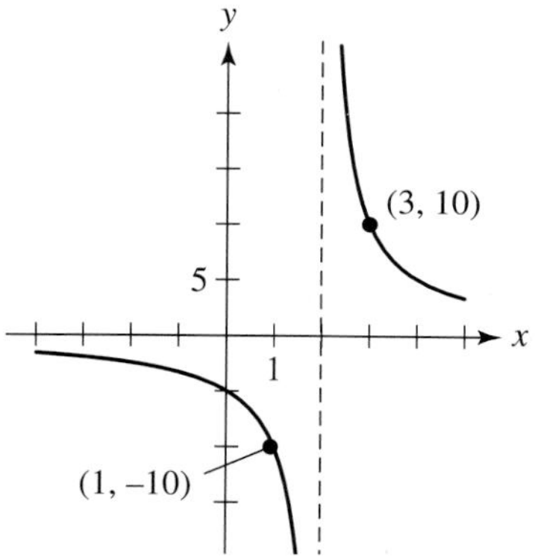

vertical asymptote $x = 2$
horizontal asymptote $y = 0$ (x-axis)
domain $= (-\infty, 2) \cup (2, \infty)$
range $= (-\infty, 0) \cup (0, \infty)$

45.

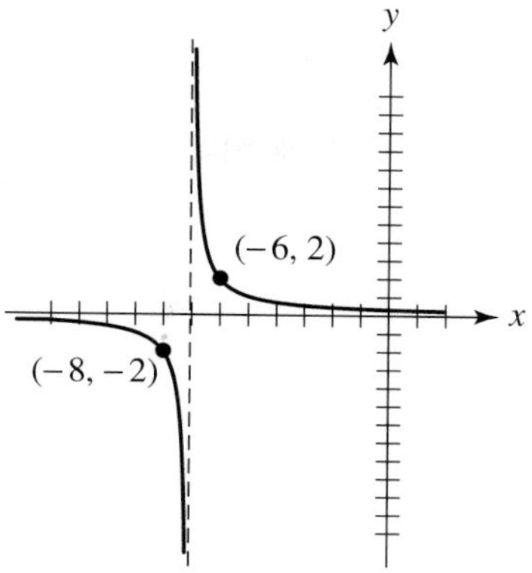

vertical asymptote $x = -7$
horizontal asymptote $y = 0$ (x-axis)
domain $= (-\infty, -7) \cup (-7, \infty)$
range $= (-\infty, 0) \cup (0, \infty)$

47.

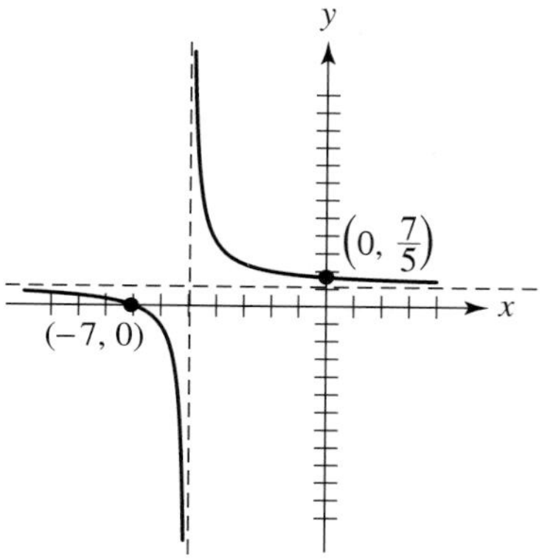

vertical asymptote $x = -5$
horizontal asymptote $y = 1$

49.

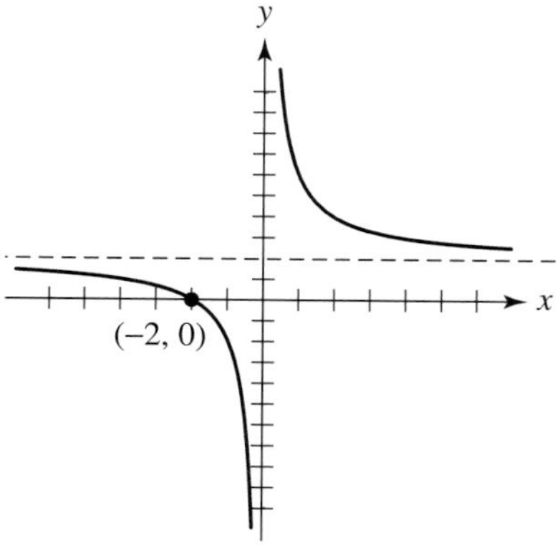

vertical asymptote $x = 0$ (y-axis)
horizontal asymptote $y = 2$

51.

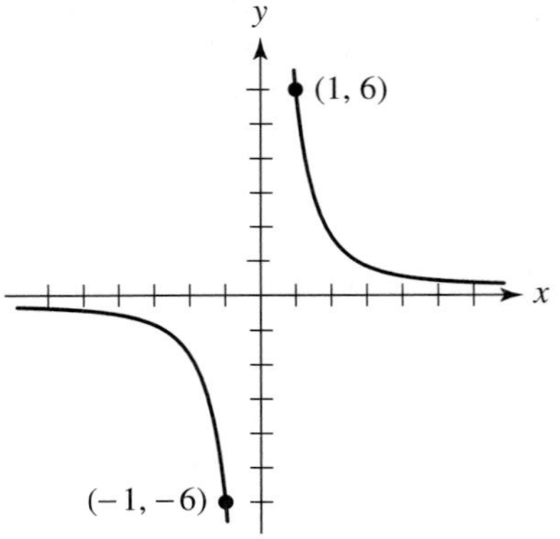

vertical asymptote $x = 0$ (y-axis)
horizontal asymptote $y = 0$ (x-axis)
domain $= (-\infty, 0) \cup (0, \infty)$
range $= (-\infty, 0) \cup (0, \infty)$

53.

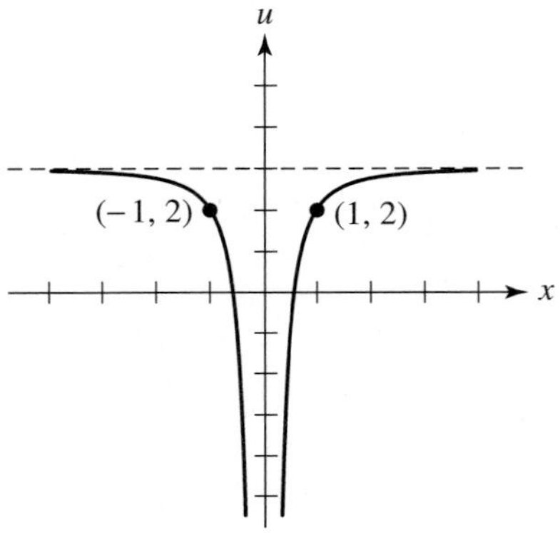

vertical asymptote $x = 0$ (u-axis)
horizontal asymptote $u = 3$
domain $= (-\infty, 0) \cup (0, \infty)$
range $= (-\infty, 3) \cup (3, \infty)$

SECTION 8-2

1. a. The equation $x^2 + 1 = 0$ has only the solutions $x = \pm i$, so $x^2 + 1$ is not 0 for any real value of x.
b. The graph has no x-intercepts.

3. a.

x	$h(x)$
0	−1
0.9	−10
0.99	−100
0.999	−1000
1	undefined
1.001	1000
1.01	100
1.1	10
2	1

b. As the values of x become close to 1, $|h(x)|$ becomes arbitrarily large.

5. (Sample response) If x is near 1, then the value of $x - 1$ is near 0. The reciprocal of a number near 0 has a large absolute value, so $|h(x)| = \left|\frac{1}{x-1}\right|$ is large.

7. vertical asymptote at $x = -4$

9. missing point at $x = 5$

11. (Sample response) As the level of production increases, the given description indicates that the cost of advertising increases slowly at first, then more rapidly. Thus the cost function does not exhibit a constant rate of change. As x increases, the graph of the cost function becomes steeper, suggesting a parabolic shape.

13. a.

x	y
100	11.50
10	73.60
1	703.51
0.1	7003.50

Each row shows the average cost per shirt at a given level of production.

b. When the production level is low, the average cost per shirt is very high. On the graph of $y = A(x)$, x-values near 0 correspond to large y-values. This suggests that the graph has a vertical asymptote at $x = 0$.

15. a. 9.50
b. If you increase production, each additional shirt will cost about \$9.50 to produce. Since you are selling shirts for only \$9 each, you will be losing money on the additional shirts. Therefore, you should not increase production.

17. a. $\frac{x-6}{x-1} = 1 - \frac{5}{x-1}$
b.

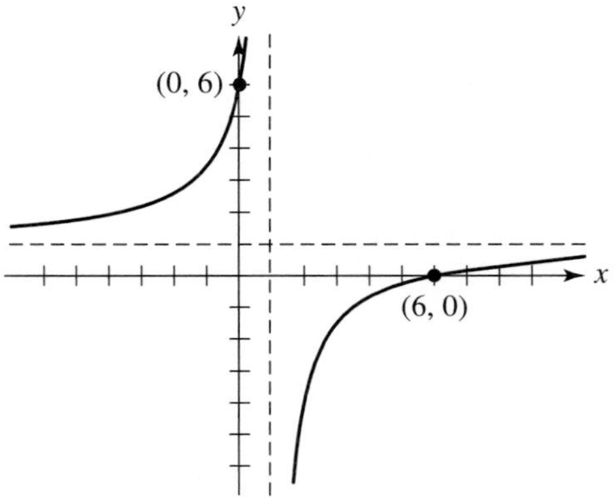

horizontal asymptote $y = 1$

19. a. (Sample response) When $|x|$ is large, then $|x - 1|$ is also large. Since the reciprocal of a large number is near 0, any expression $\left|\frac{\text{constant}}{x-1}\right|$ is small when $|x|$ is large.

b. (Sample response) Since $h(x)$ is the sum of the expressions $x + 2$ and $-\frac{4}{x-1}$, and since the second expression is near 0 when $|x|$ is large, $h(x)$ must be near $x + 2$ when $|x|$ is large.

21.

x	$g(x)$ (approximate)
10	0.44
100	0.95
1,000	0.995
10,000	0.9995

The table suggests that if $|x|$ is large, $g(x)$ is near 1.

23. Yes; $y = \frac{1}{3}$.

25. No; both tails down.

27. a. $P(t) = 2000t$ **b.** $Q(t) = 500 + 50t$

c. $f(t) = \frac{2000t}{500 + 50t}$; this represents the ratio of open-edition pieces to original and limited-edition pieces in Robert Arnold's work.

29. a–b. vertical asymptote at $x = 4$ **c.** Yes; $y = 1$.

31. a–b. missing point at $x = 0$

c. No; left tail down, right tail up.

33. a–b. vertical asymptote at $x = -3$ **c.** Yes; $y = -1$.

35. a–b. vertical asymptote at $x = \pm\sqrt{3}$

c. Yes; $y = 0$.

37. a–b. missing point at $x = 1$ **c.** Yes; $y = 0$.
vertical asymptote at $x = 7$

39. a–b. missing point at $x = -2$

c. No; left tail down, right tail up.

41. a–b. vertical asymptotes at $x = \pm\sqrt{2}$

c. Yes; $y = -1$.

43. a–b. no discontinuities **c.** Yes; $y = 0$.

45. The discontinuity occurs when $50t + 500 = 0$, that is, when $t = -10$. The discontinuity may be interpreted as follows. Robert Arnold had produced 500 original and limited-edition pieces prior to 1990. If he had produced them at his current rate of 50 per year, he would have begun in 1980, at $t = -10$.

47. a. $\frac{21.7}{\pi r^2}$ **b.** $S(r) = 2\pi r^2 + 2\pi r\left(\frac{21.7}{\pi r^2}\right)$

c. S is discontinuous when $r = 0$, and its graph has a vertical asymptote there. This says that if the radius of the can is near 0, the surface area is extremely large (because the height must be large to produce a volume of 21.7 in^3).

d. The end behavior of S agrees with that of $S = r^2$, so the right tail of the graph points up. This says that if the radius of the can is large, the surface area is also large.

49. a. The function $f(x)$ is the sum of the expressions $x^2 + 2$ and $\frac{3x - 2}{2x^2 + x - 4}$. Exercise 48 shows that the second expression is near 0 if $|x|$ is large, so that $f(x)$ is near $x^2 + 2$.

b. The function $f(x)$ is the sum of $P(x)$ and the remainder from the division. The remainder is a rational function in which the denominator has a larger degree than the numerator. By the Highest Degree Theorem, the remainder is near 0 when $|x|$ is large, so $f(x)$ is near $P(x)$.

51. $y = x$ **53.** $y = 3x + 4.5$

55. $y = 2x + 2$ **57.** $y = 0.5x$

SECTION 8-3

1. 4 **3.** no solution

5. a. 200 and 350

b. If your production level is either 200 or 350 shirts per month, the average cost per shirt is $9.

7. $[-3, 2) \cup [7, \infty)$ **9.** $(1, \infty)$

11. a. $(-\infty, -3) \cup [-1, 3)$ **b.** $[-3, 2) \cup [7, \infty)$

c. $(-\infty, -5) \cup (-3, \infty)$ **d.** $(1, \infty)$

13. a. $(-\infty, -3) \cup [-1, 3)$ **b.** $[-3, 2) \cup [7, \infty)$

c. $(-\infty, -5) \cup (-3, \infty)$ **d.** $(1, \infty)$

15. $(-3, 0) \cup (4, \infty)$ **17.** $[-3, -2) \cup [1, \infty)$
The inequality is true in $[-3, -2) \cup [1, \infty)$.

19. $(0, 1)$ **21.** $[-\sqrt{6}, 0) \cup [\sqrt{6}, \infty)$

23. $(-\infty, -20) \cup (-20, -10) \cup (10, 30) \cup (30, \infty)$

25. $[0, \infty)$ **27.** $(-\infty, 0) \cup \left[\frac{1}{2}, \infty\right)$

29. $(-\infty, 1) \cup (7, \infty)$ **31.** $(-\infty, -2) \cup \{0\}$

33. a. $1.46 < r < 4.77$ **b.** 2.79

c. (Sample response) A can of the optimal shape is difficult to hold and drink from. Furthermore, cans of a nonstandard shape would be hard to recognize in a supermarket, and would not fit most vending machines. Finally, the cost of retooling a canning plant would probably outweigh any savings in materials.

SECTION 9-1

1. a. (Sample response) Each entry in the second column is obtained by multiplying the previous entry by 1.02. Thus the entry opposite $t = 4$ can be obtained by starting with the entry of 6 opposite $t = 0$ and multiplying by 1.02 four times. The final result is $6(1.02)^4$.

b. (Sample response) If Table 29 is continued, the entry opposite any value of t can be obtained by starting with the entry of 6 opposite $t = 0$ and multiplying by 1.02t times. The final result is $6(1.02)^t$.

3. Yes; 3. **5.** No.

7. a.

x	y
5	243
6	729
7	2,187
8	6,561
9	19,683
10	59,049

When x is large, y is extremely large.

b.

x	y (approx.)
−1	0.33
−2	0.11
−3	0.037
−4	0.012
−5	0.0041
−6	0.0014
−7	0.00046
−8	0.00015
−9	0.000051
−10	0.000017

When $x < 0$ and $|x|$ is large, y is extremely close to 0.

9. a. $(0, \infty)$

b. The left tail points up, and the right tail approaches the x-axis.

c. The left tail approaches the x-axis, and the right tail points up.

11. In Exercise 3, $y = \dfrac{2}{243}3^x$. In Exercise 4, $y = 64\left(\dfrac{1}{2}\right)^x$.

In Exercise 6, $y = \left(-\dfrac{16}{3}\right)\left(-\dfrac{3}{2}\right)^x$.

13. (Sample response) Whenever x is increased by 1, the power of b is increased by 1, so the value of the function is multiplied by b. The ratio of the new value of f to the previous value is b.

15. a. 14.75 **b.** 36.46

17. a. (Sample response) Raising $\dfrac{1}{2}$ to a large positive power is equivalent to taking the reciprocal of 2 to a large positive power. Since the reciprocal of a large positive number is a small positive number, the value of $g(x)$ is positive and extremely near 0 when x is large.

b. (Sample response) Raising $\dfrac{1}{2}$ to a large negative power is equivalent to raising 2 to a large positive power. Thus when $x < 0$ and $|x|$ is large, $g(x)$ is extremely large.

c. (Sample response) The results taken together suggest that as the exponent x increases, the value of $\left(\dfrac{1}{2}\right)^x$ decreases. This, in turn, suggests that if $0 < b < 1$, $g(x) = b^x$ is a decreasing function, becoming arbitrarily large when $x < 0$ and $|x|$ is large and becoming arbitrarily close to 0 when x is large.

The results also show that $\left(\dfrac{1}{2}\right)^x > 0$ for all values of x, and thus suggest that the range of g is $(0, \infty)$.

19. c. domain $= (-\infty, \infty)$, range $= (0, \infty)$

d. y-intercept $= (0, 1)$, no x-intercepts

e. $f(1) = b$, $f(-1) = \dfrac{1}{b}$

f. horizontal asymptote on the right side of the x-axis

g. decreasing on $(-\infty, \infty)$

21.

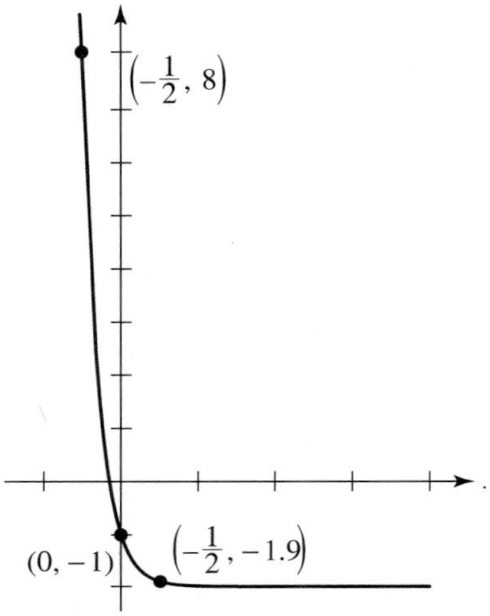

domain $= (-\infty, \infty)$
range $= (-2, \infty)$
decreasing

23.

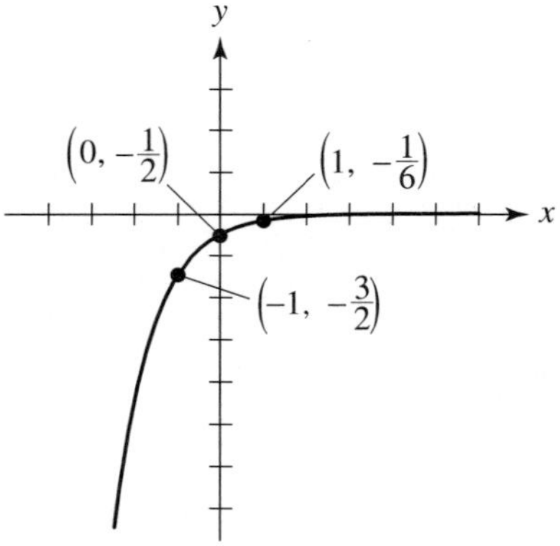

domain $= (-\infty, \infty)$
range $= (-\infty, 0)$
increasing

25. $f(0) = 1, f(1.71) \cong 2, f(3.42) \cong 4, f(5.13) \cong 8$; 1.71

27. $f(-2.32) \cong 1, f(-1.32) \cong 2, f(-0.32) \cong 4, f(0.68) \cong 8$; 1.00

29. $s = 2^n$

31. (Sample response) One reason is that by asking 20 yes/no questions, a child can distinguish among more than a million objects. Another reason is that most children choose familiar objects for their opponents to identify, so that the pool of likely objects is usually much smaller than a million.

33. $a = \left\{2, -\frac{2}{3}, \frac{2}{9}, -\frac{2}{27}, \frac{2}{81}, \cdots\right\}$; yes.

35. $a = \left\{\frac{1}{5}, \frac{2}{5}, \frac{4}{5}, \frac{8}{5}, \frac{16}{5}, \cdots\right\}$; yes.

37. No. **39.** Yes; 2.59.

41. a. $Ar + Ar^2 + Ar^3 + \cdots + Ar^n + Ar^{n+1}$

b. $A(1 - r^{n+1})$ **c.** $S_n = \dfrac{A(1 - r^{n+1})}{1 - r}$

43. 12.80, 12.80, 12.80 **45.** 12.21, 31.87, 114.55

47. In Exercise 42, $\dfrac{A}{1 - r} = 9$. In Exercise 43, $\dfrac{A}{1 - r} = 12.8$.

49. Yes; $y = (0.021)(9^x)$. **51.** No.

53. $y = 7(\sqrt{7})^x$ **55.** $(47.54)(0.94)^x$

57.

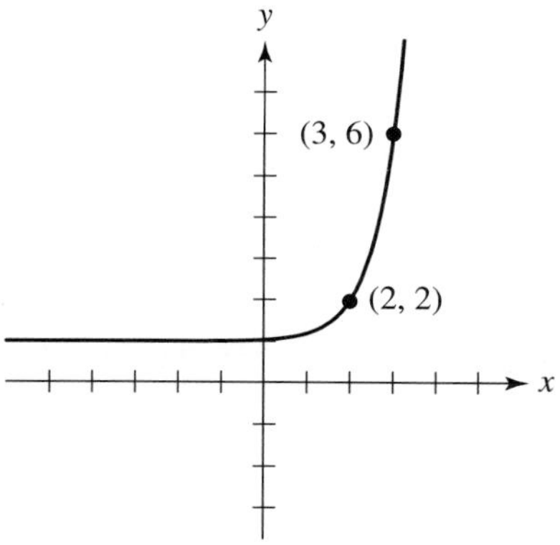

domain = $(-\infty, \infty)$
range = $(1, \infty)$
increasing

59.

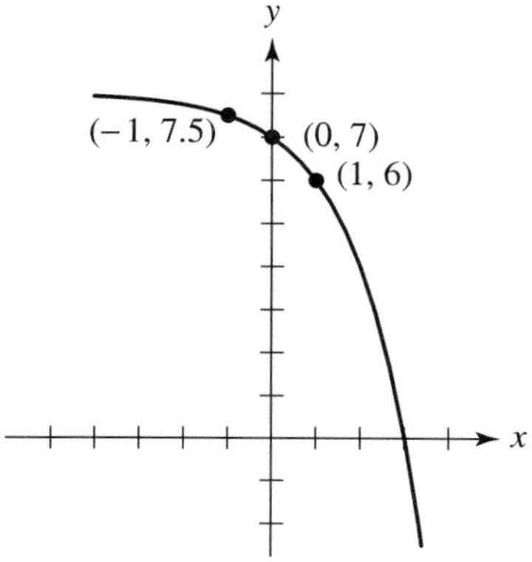

domain = $(-\infty, \infty)$
range = $(-\infty, 8)$
decreasing

61. a.

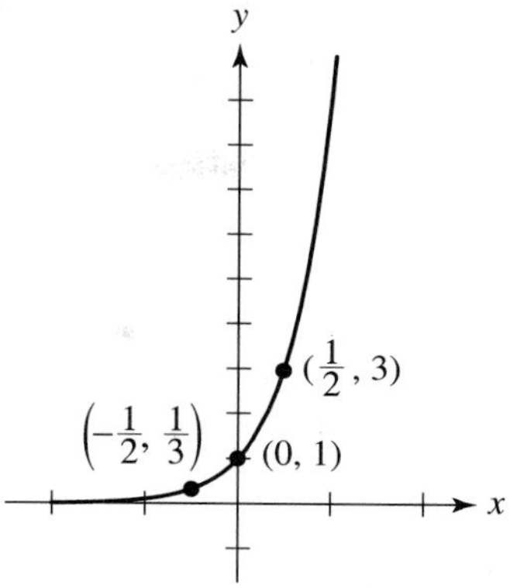

b. 64

63. a. Yes. **b.** $-(0.9^n)$; -0.43
c. -10

65. a. Yes. **b.** $10 \cdot 5^n$; 3,906,250
c. No finite sum exists.

67. a. Yes. **b.** $\left(-\frac{3}{4}\right)^n$; 0.10
c. $\frac{4}{7}$

69. a. No. **71.** $f(n) = 220\left(\sqrt[12]{2}\right)^n$

73. 29 days

SECTION 9-2

1. a. After 1 year you owe the original \$10,000, plus interest amounting to 5.9% of \$10,000. The total amount you owe, in dollars, is $10{,}000 + 0.059(10{,}000) = 10{,}000(1 + 0.059)$.

b.

Interest periods	Amount owed
0	10,000
1	10,590
2	11,214.81
3	11,876.48
4	12,577.20
5	13,319.25

c. $10{,}000(1.059)^t$

3. a. After one-third of a year you owe the original \$10,000, plus interest amounting to one-third of 5.9% of \$10,000. The total amount you owe, in dollars, is $10{,}000\left(1 + \dfrac{0.059}{3}\right)$.

b.

Interest periods	Amount owed
0	10,000
1	10,196.67
2	10,397.20
3	10,601.68
4	10,810.18
5	11,022.78
6	11,239.56
7	11,460.61
8	11,686.00
9	11,915.82
10	12,150.17
11	12,389.12
12	12,632.77
13	12,881.22
14	13,134.55
15	13,392.86

c. $10{,}000\left(1 + \frac{0.059}{3}\right)^{3t}$

5. \$13,431

7.

n	$\left(1 + \frac{1}{n}\right)^n$
1,000	2.717
1,000,000	2.718280
1,000,000,000	2.718281827

9. **a.** annually, \$6067.64; monthly, \$6237.27; continuously, \$6254.46
b. annually, \$594,107.13; monthly, \$653,813.37; continuously, \$659,856.91
c. annually, \$866.03; monthly, \$1260.81; continuously, \$1359.14

11.

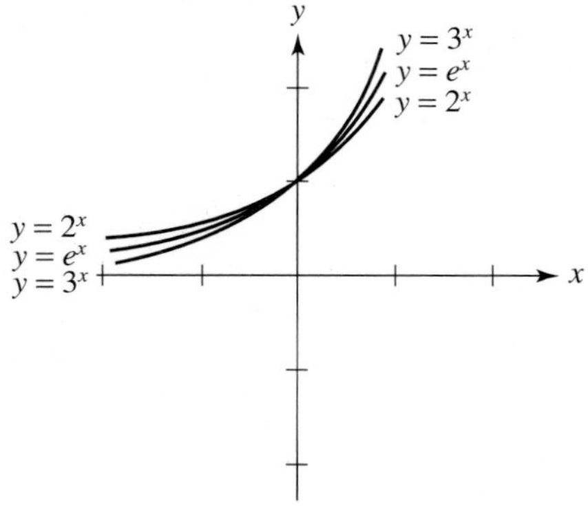

13. (Sample response) If the independent variable is time, then the average rate of change in f over any time interval is the average rate at which f grows over that interval. If the average rate of change is always equal to the value of f at some point within the interval, then f grows at a rate equal to its size.

15. 44%

17. $y \cong e^{1.10t}$

19. $y \cong 3e^{-0.69t}$

21.

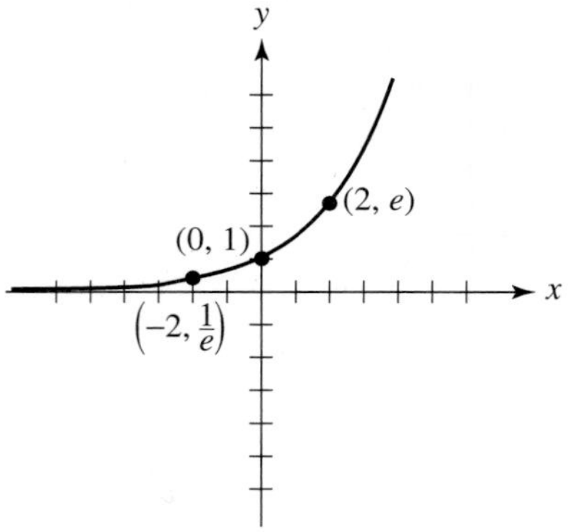

23.

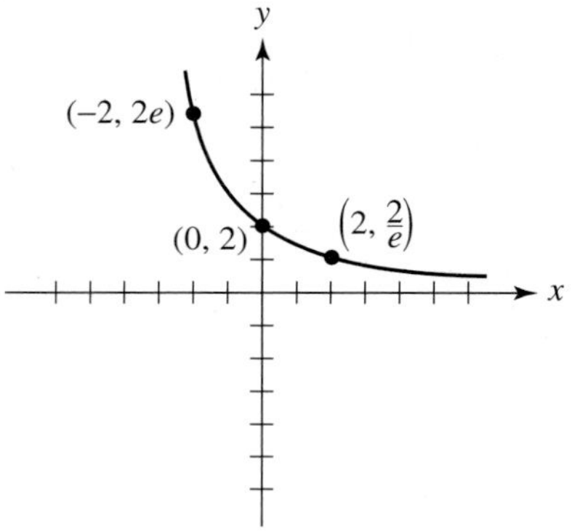

25. The graph in Exercise 21 is reflected in the y-axis to generate the graph in Exercise 22. That graph is stretched vertically by a factor of 2 to generate the graph in Exercise 23. That graph is shifted 3 units down to generate the graph in Exercise 24.

27. **a.** $f(t) = (10{,}114.32)e^{0.0485t/12}$
b. \$10,270.86

29. **a.** \$10.31 **b.** \$21,124.14

31. **a.** $f(t) = 2^{(-2.22 \times 10^{-10})t}$ **b.** 97%
c. about 2 billion years

33. $6! = 720$; $7! = 5040$; $10! = 3{,}628{,}800$; $25! \cong 1.55 \times 10^{25}$

SECTION 9-3

1. $(F \circ G)(x) = F[G(x)] = F(x - 2) = (x - 2) + 2 = x$
$(G \circ F)(x) = G[F(x)] = G(x + 2) = (x + 2) - 2 = x$

3. No. **5.** No.

7. Yes. **9.** No.

11. Yes. **13.** Yes.

15. $(f \circ g)(x) = x$; $(g \circ f)(x) = x$; both results can be simplified to x for all real values of x.

17. a.

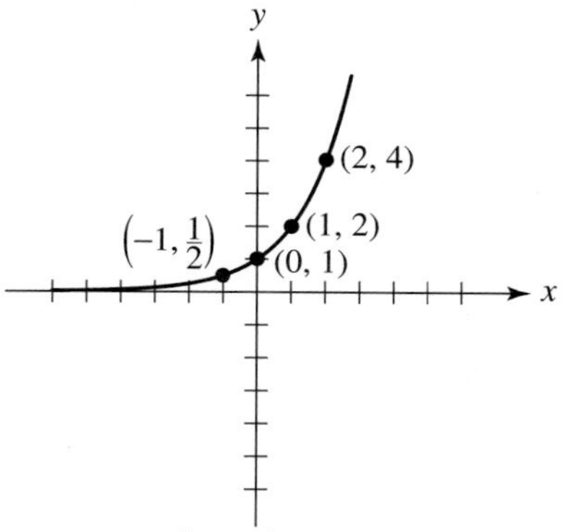

b. The points $\left(-1, \frac{1}{2}\right)$, (0, 1), (1, 2), and (2, 4) make up a table generated by $f(x) = 2^x$. The corresponding table generated by $y = f^{-1}(x)$ contains those points with their coordinates reversed.

c. The plotted points suggest that the points (a, b) and (b, a) are always reflections of each other in the line $y = x$.

d.

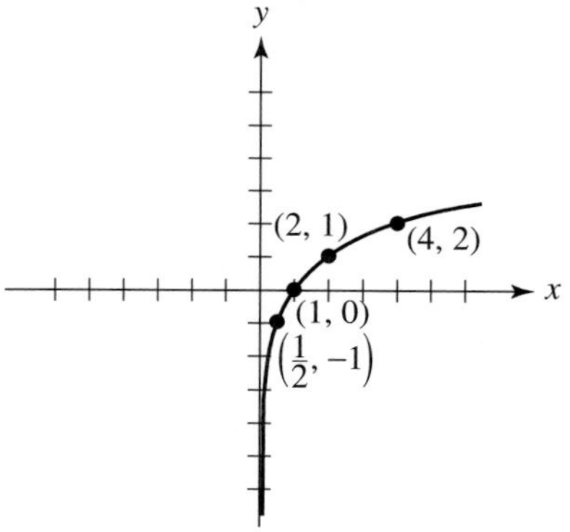

19. a. $(-\infty, -1) \cup (-1, \infty)$

b. $f[g(x)] = f\left(-\dfrac{2x}{x-1}\right)$

$$= \frac{-\dfrac{2x}{x-1}}{-\dfrac{2x}{x-1} + 2} = \frac{-\dfrac{2x}{x-1}}{-\dfrac{2x}{x-1} + 2} \cdot \frac{x-1}{x-1}$$

$$= \frac{-2x}{-2x + 2(x-1)} \text{ (if } x \neq 1) = \frac{-2x}{-2} = x$$

for all x in the domain of g

21. a.

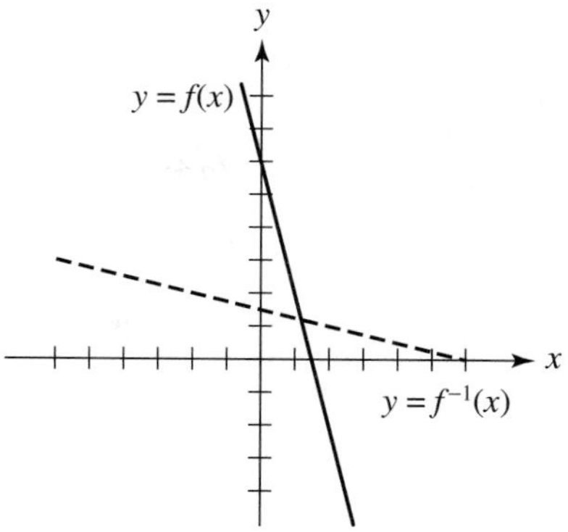

b. $f^{-1}(x) = \dfrac{6 - x}{4}$

23. a.

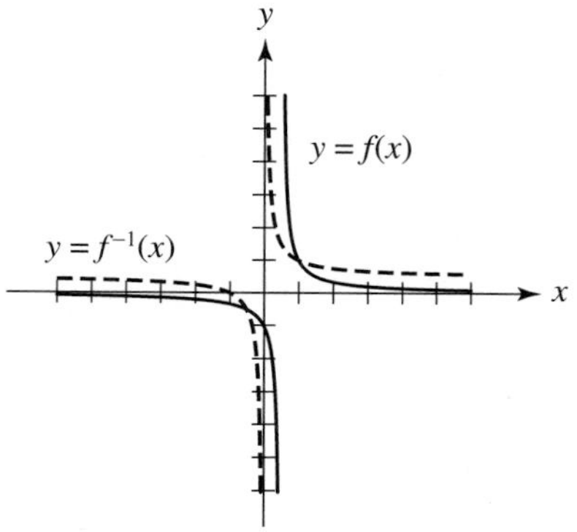

b. $f^{-1}(x) = \dfrac{x + 1}{2x}$

25. a.

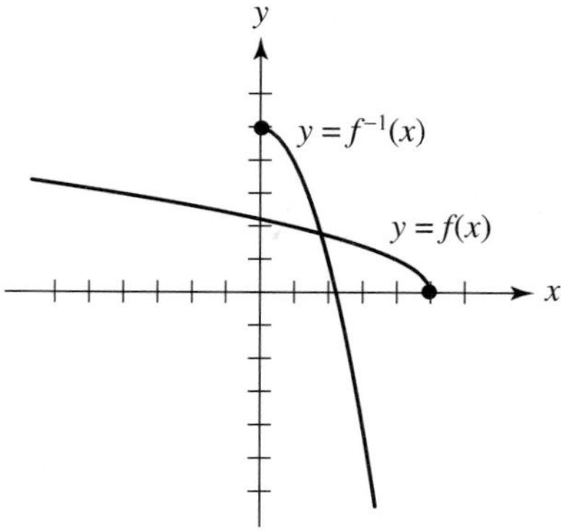

b. $f^{-1}(x) = 5 - x^2$ for $x \geqslant 0$

27. a.

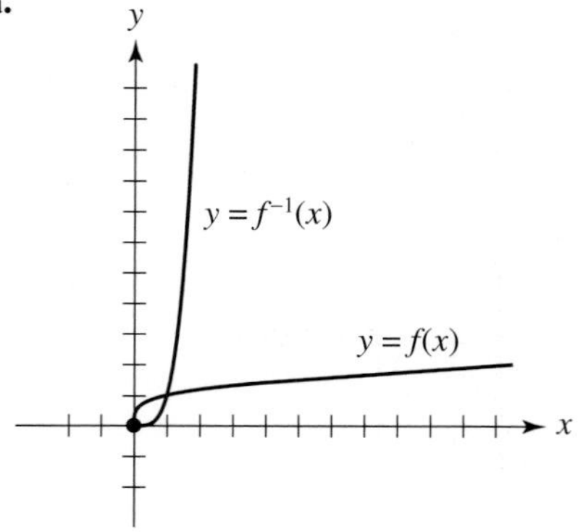

b. $f^{-1}(x) = x^4$ for $x \geqslant 0$

29. a. $v = \dfrac{-20 \pm 5\sqrt{16 + 6d}}{3}$

b. The function $d = 0.06v^2 + 0.8v$ is one-to-one on $[0, \infty)$. Its inverse is found by reversing the roles of the independent and dependent variables, as was done in part (a).

31. $f_1(x) = 10 - 2x^2$ for $x \leqslant 0$; $f_2(x) = 10 - 2x^2$ for $x \geqslant 0$

$f_1^{-1}(x) = -\sqrt{\dfrac{10 - x}{2}}$; $f_2^{-1}(x) = \sqrt{\dfrac{10 - x}{2}}$

33. $f_1(x) = 3(x - 1)^2 + 2$ for $x \leqslant 1$; $f_2(x) = 3(x - 1)^2 + 2$ for $x \geqslant 1$

$f_1^{-1}(x) = 1 - \sqrt{\dfrac{x - 2}{3}}$; $f_2^{-1}(x) = 1 + \sqrt{\dfrac{x - 2}{3}}$

35. Answers will vary.

37. Yes.

39. Yes.

41.

x	$f^{-1}(x)$
$\sqrt{3}$	1
12,946	2.8
0	7π
e	$\frac{2}{3}$

$f^{-1}[f(1)] = 1$ $\quad f[f^{-1}(\sqrt{3})] = \sqrt{3}$

$f^{-1}[f(2.8)] = 2.8$ $\quad f[f^{-1}(12{,}946)] = 12{,}946$

$f^{-1}[f(7\pi)] = 7\pi$ $\quad f[f^{-1}(0)] = 0$

$f^{-1}\left[f\left(\frac{2}{3}\right)\right] = \frac{2}{3}$ $\quad f[f^{-1}(e)] = e$

43. No.

45. Yes.

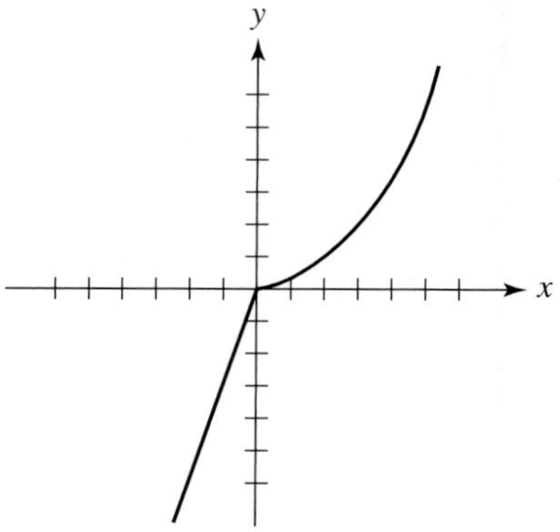

47. Yes.

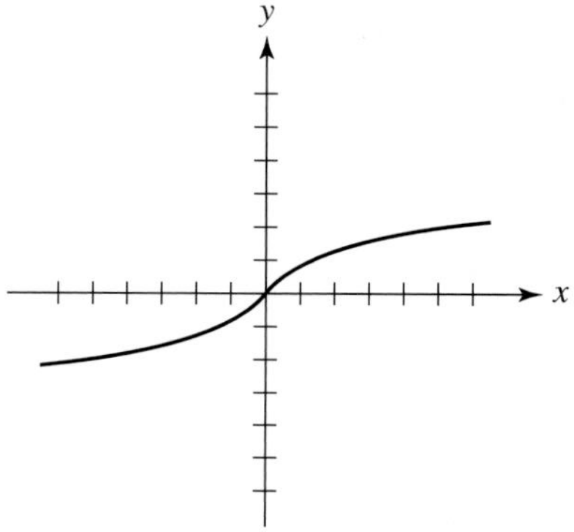

49. Yes.

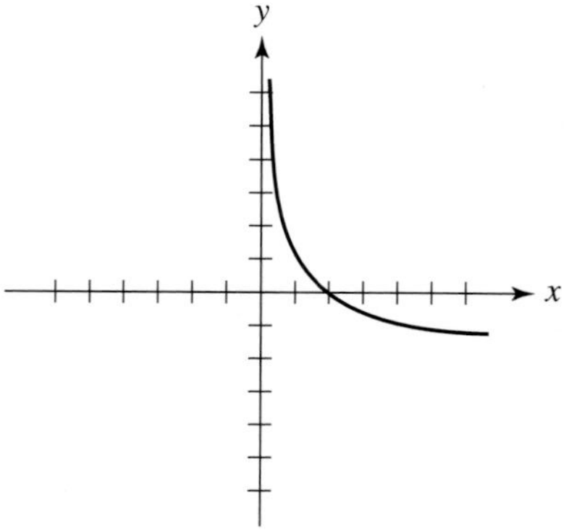

51. No.

53. Yes.

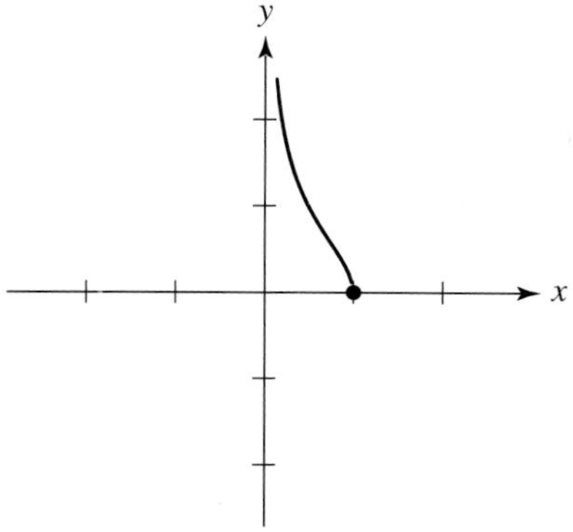

55. Yes.

57. Yes.

59. No.

61. No.

63. $f^{-1}(x) = 5x + 4$

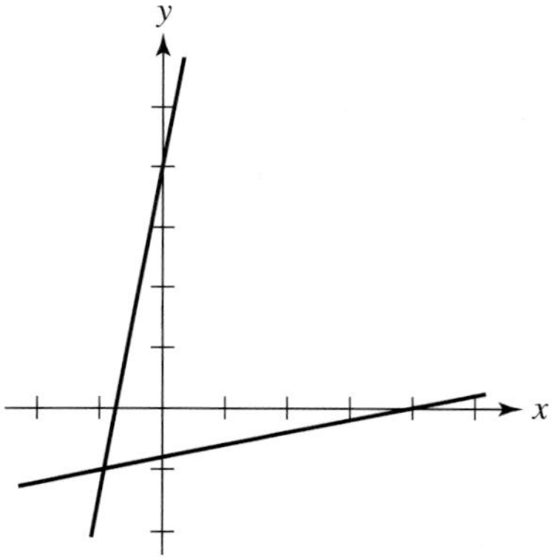

65. $f^{-1}(x) = \sqrt[3]{3x - 8}$

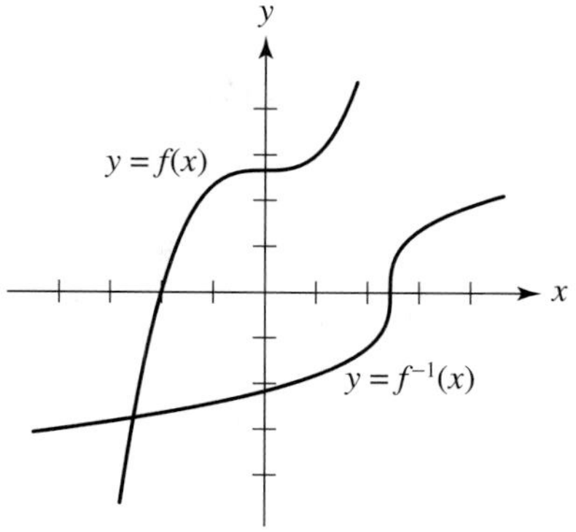

67. $f^{-1}(x) = \dfrac{5x + 5}{x}$

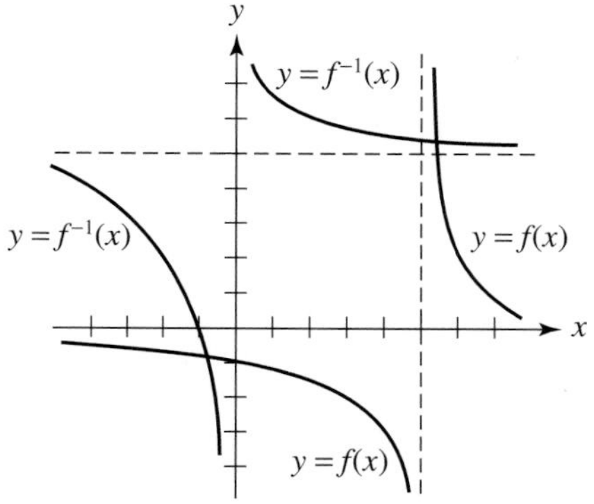

69. $f^{-1}(x) = 10 - x^2$ for $x \geqslant 0$

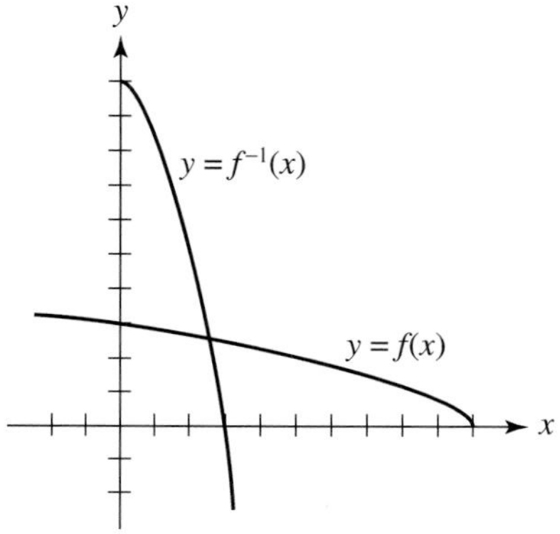

71. $f_1(x) = 7 - 3x^2$ for $x \leqslant 0$; $f_2(x) = 7 - 3x^2$ for $x \geqslant 0$
$f_1^{-1}(x) = -\sqrt{\dfrac{7 - x}{3}}$; $f_2^{-1}(x) = \sqrt{\dfrac{7 - x}{3}}$

73. $f_1(x) = (x + 2)^2 + 1$ for $x \leqslant -2$; $f_2(x) = (x + 2)^2 + 1$ for $x \geqslant -2$
$f_1^{-1}(x) = -\sqrt{x - 1} - 2$; $f_2^{-1}(x) = \sqrt{x - 1} - 2$

75. $(-\infty, -1), (-1, 0), (0, 1), (1, \infty)$

77. $(-\infty, -3), (-3, 0), (0, 3), (3, \infty)$

SECTION 9–4

1. $y = 20 + 17e^{-0.051t}$

3. a. No horizontal line intersects the graph of $y = 2^x$ more than once.

b. (Sample response) Solving $2^x = 7$ is equivalent to finding the point $(x_0, 7)$ on the graph of $y = g(x)$. This, in turn, is equivalent to finding the point $(7, x_0)$ on the graph of $y = g^{-1}(x)$, so that $x_0 = g^{-1}(7)$.

5.

x	$y = \log_5 x$
5	1
25	2
125	3
625	4

7. It must be true that $\log_b 1 = 0$ for any base $b > 0$ because $b^0 = 1$ for any $b > 0$. Similarly, $\log_b b = 1$ because $b^1 = b$ for any $b > 0$.

9. Yes.

11. Yes.

13. a–b.

x	$h(x)$
−2	1
−1	3
0	9
1	27
2	81

x	$h^{-1}(x)$
1	−2
3	−1
9	0
27	1
81	2

c. The table for the inverse of $h(x) = 9 \cdot 3^x$ in Exercise 13b and the table for $y = \log_3 x - 2$ in Exercise 12c are identical.

15.

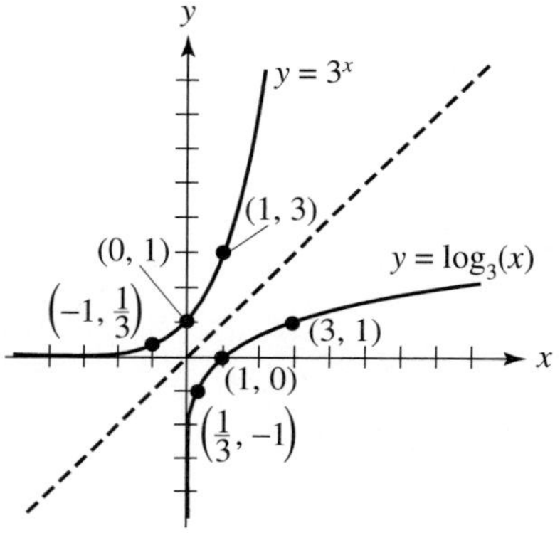

17. a. If $c = \log_{1/b} x$, then $x = \left(\frac{1}{b}\right)^c$, so it is also true that $x = b^{-c}$, which means that $-c = \log_b x$. Thus $\log_{1/b} x = -\log_b x$.

b.

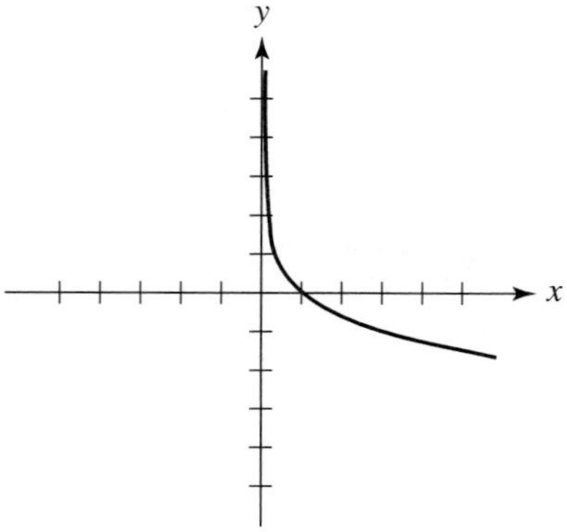

19.

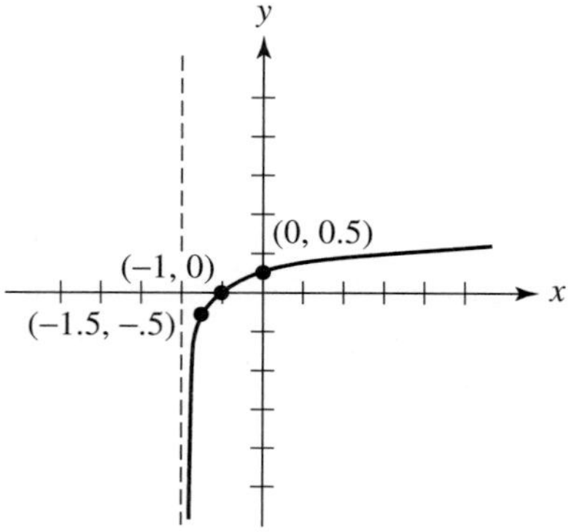

domain $= (-2, \infty)$
range $= (-\infty, \infty)$
increasing

21.

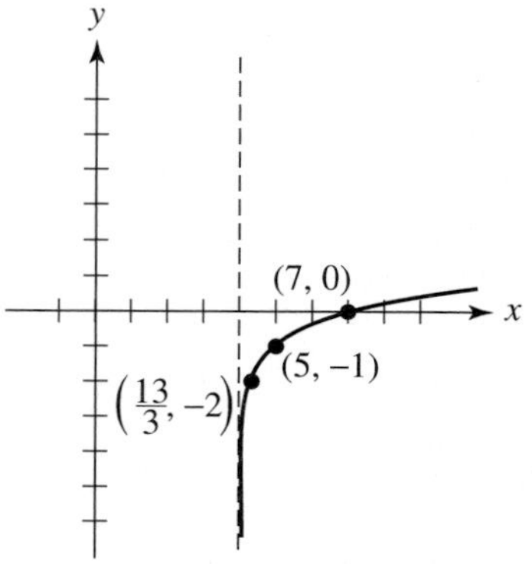

domain $= (4, \infty)$
range $= (-\infty, \infty)$
increasing

23.

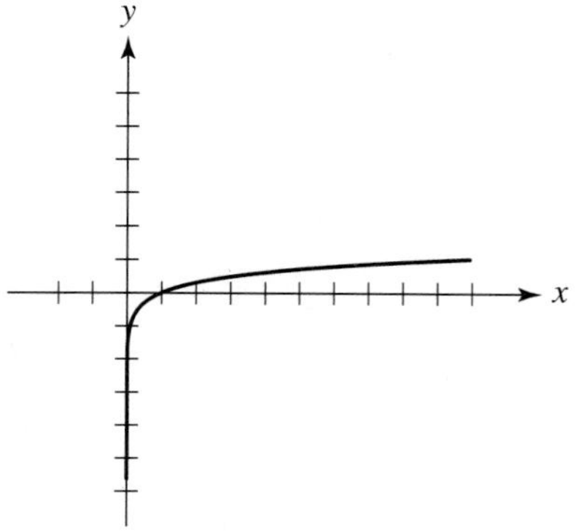

25.

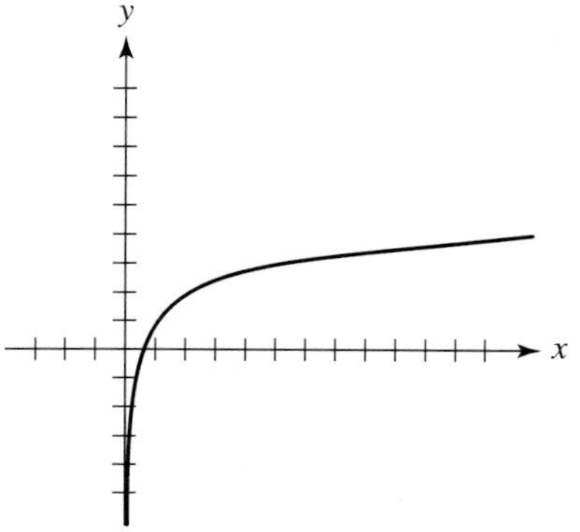

27. $\dfrac{\ln 25.7}{5}$

29. $\log\left(\dfrac{2.87}{2}\right)$

31. $\dfrac{e^{0.57} + 100}{4}$

33. $\dfrac{-\ln 0.45}{0.00012} \cong 6654$

35. **a.** To see why $\log_b\left(\dfrac{m}{n}\right) = \log_b m - \log_b n$, let's look at the expressions $b^{\log_b (m/n)}$ and $b^{(\log_b m - \log_b n)}$.
If these two powers of b are equal, then the exponents must be equal. (This is another way of saying that the function $y = b^x$ is one-to-one.)
The first expression reduces to $\dfrac{m}{n}$, and the second can be rewritten as $\dfrac{b^{\log_b m}}{b^{\log_b n}}$, which also reduces to $\dfrac{m}{n}$. This establishes that $\log_b\left(\dfrac{m}{n}\right) = \log_b m - \log_b n$.

b. Look at the expressions $b^{\log_b(m^a)}$ and $b^{a \log_b m}$.
The first expression reduces to m^a, and the second can be rewritten as $(b^{\log_b m})^a$, which also reduces to m^a. This establishes that $\log_b(m^a) = a \log_b m$.

37. $\log a - (\log b + \log c)$

39. $2 \log a + \log b$

41. $x = Ab^y$
$\log_b x = \log_b(Ab^y)$
$\log_b x = \log_b A + \log_b b^y$
$\log_b x = \log_b A + y$
$y = \log_b x - \log_b A$
This equation has the form $y = \log_b x + k$, with $k = -\log_b A$.

43. 1

45. 16

47. **a.** 1.40 **b.** 0.71
c. −6.33 **d.** 1
e. 0.74 **f.** −3.03
g. 2.61 **h.** 0.23

49.

x	y
1	0
10	1
100	2
1000	3

51. **a.** $\dfrac{1}{2}$ **b.** 2
c. $\dfrac{1}{2}$ **d.** −4
e. −7 **f.** −4
g. 2 **h.** 4

53. Yes.

55. Yes.

57.

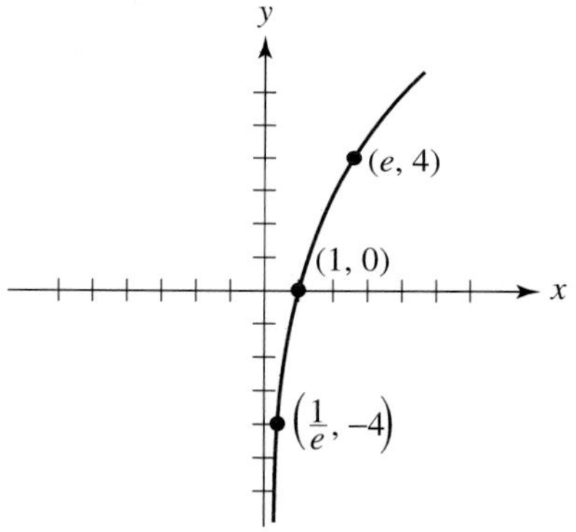

domain = $(0, \infty)$
range = $(-\infty, \infty)$
increasing

59.

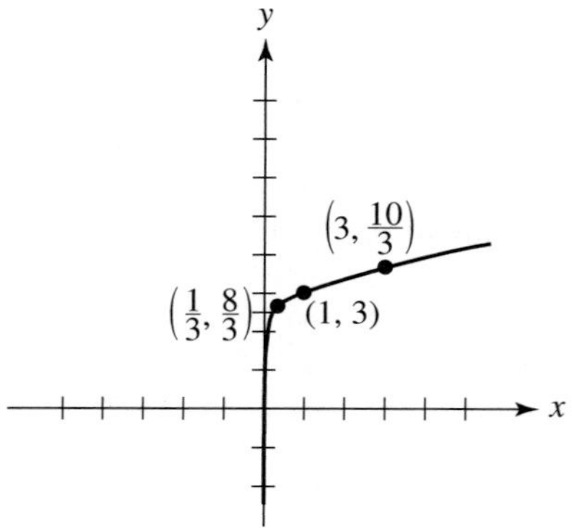

domain = $(0, \infty)$
range = $(-\infty, \infty)$
increasing

61.

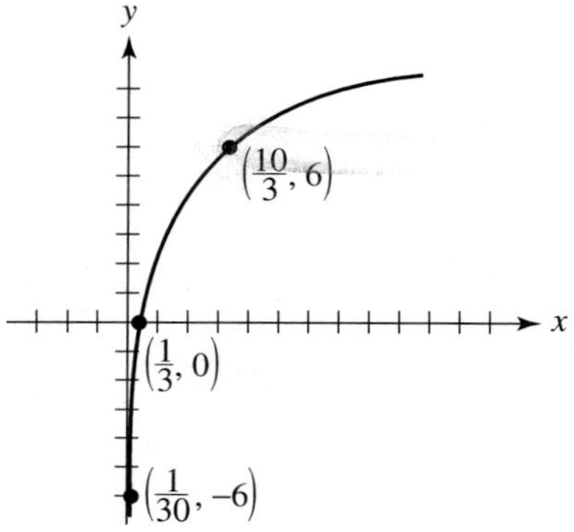

domain = $(0, \infty)$
range = $(-\infty, \infty)$
increasing

63.

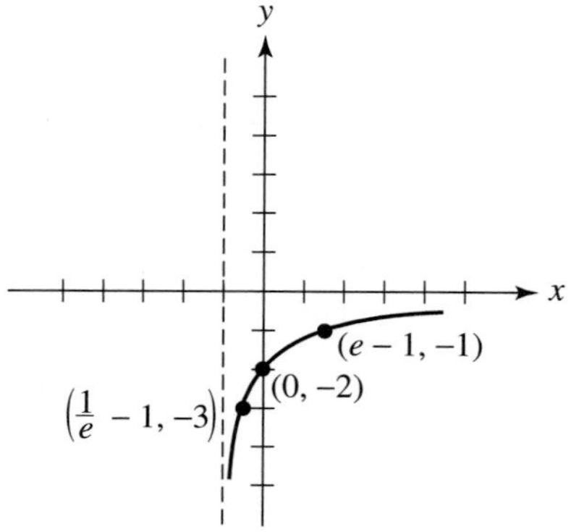

domain = $(-1, \infty)$
range = $(-\infty, \infty)$
increasing

65. $\dfrac{\log_5 9 + 4}{3}$

67. $\log_4\left(\dfrac{2.21}{3}\right) - 5$

69. $\ln 2 - 6$

71. 3

73. e^{10}

75. 2

77. 4

79. 5

81. $\dfrac{1}{2}\log_4 x$

83. $\log_4 x - 2$

85. $-3\log_4 x$

87. $\log_4 x$

89. **a.** true **b.** false
c. false **d.** true
e. true **f.** true

91. **a.** $\dfrac{I_1}{I_2} = (2.512)^{M_2 - M_1}$

$$\log\left(\frac{I_1}{I_2}\right) = \log(2.512)^{M_2 - M_1}$$

$$\log I_1 - \log I_2 = (M_2 - M_1)\log 2.512$$

$$M_2 - M_1 = \frac{1}{\log 2.512}(\log I_1 - \log I_2)$$

$$M_2 - M_1 \cong 2.50(\log I_1 - \log I_2)$$

b. about 1.59 billion **c.** about 2.50

93. **a.** about 31.62 **b.** 1,000,000
c. about 1.51×10^{24}, about 2.51×10^{24}

95. The 2 on the lower ruler and the 3 on the upper ruler represent lengths of log 2 and log 3, respectively. The position of the slide rule demonstrates that the sum of these two lengths is log 6, since the 3 on the upper ruler is opposite the 6 on the lower ruler. Thus $\log 6 = \log 2 + \log 3 = \log(2 \cdot 3)$, so $6 = 2 \cdot 3$.

SECTION 9-5

1. **a.**

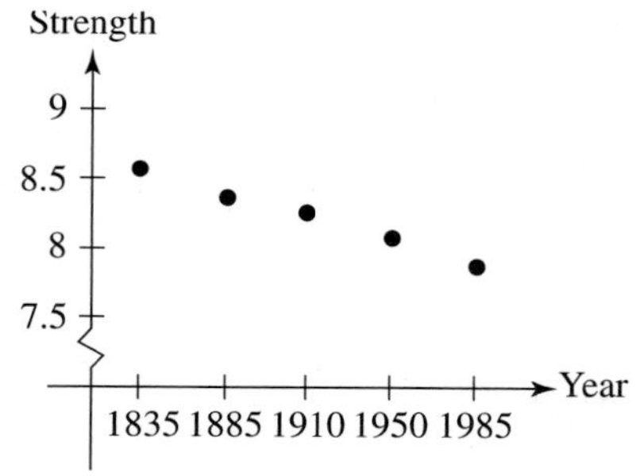

b. Lines will vary.

3. Answers will vary.

5. a.

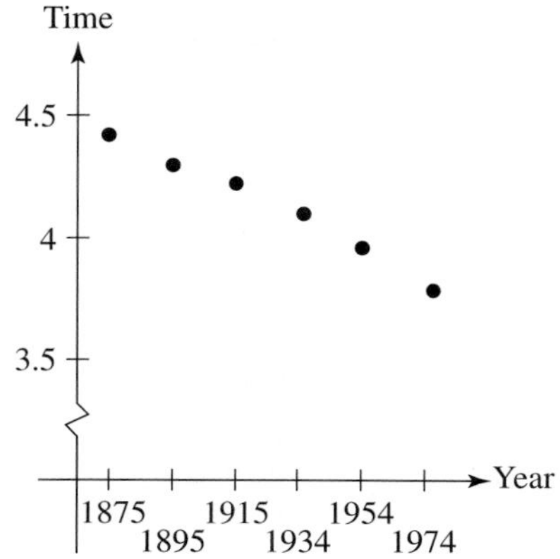

Lines and equations will vary.
b. Estimates will vary. **c.** Answers will vary.
d. Answers will vary.

7. a. $y = -2.07x + 7.98$ **b.** $y = 43.90x + 382.87$
c. $y = 0.01x + 0.5$

9. Answers will vary. **11.** Estimates will vary.

13. a.

t	y	$Y = \ln y$
0	335	5.81
1	1,285	7.16
2	3,933	8.28
3	9,015	9.11
4	18,195	9.81
5	31,452	10.36
6	46,053	10.74
7	55,388	10.92
8	78,215	11.27

b. Equations will vary. **c.** Calculations will vary.
d. Graphs will vary.

15. a. $Y = 0.65t + 6.67$ **b.** $y = 788.40e^{0.65t}$
c.

t	y
0	788
1	1,510
2	2,893
3	5,541
4	10,615
5	20,333
6	38,949
7	74,608
8	142,915

17. The exponential model constructed in Exercise 15 was not an accurate predictor of the number of deaths from AIDS in 1991. This indicates that the actual relationship between t and y may not be exponential.

19. a. If $y = Ax^m$, then $\ln y = \ln(Ax^m) = \ln A + \ln x^m$.
b. Because $\ln x^m = m \ln x$, the equation can be written $y = \ln A + m \ln x$.
c. The equation is $\ln y = m \ln x + \ln A$. This has the form $\ln y = (\text{constant})\ln x + (\text{constant})$, with m and $\ln A$ as the two constants.

21. Tables will vary. **23.** $y \cong 0.00055x^{3/2}$

25. a. $y \cong 0.0034x - 1.94, r \cong 0.99986$
b. $y \cong 39.78rx - 306.17, r \cong 0.92$
c. $y \cong 6.59(1.000077)^x, r \cong 0.92$
d. $y \cong 0.0016x^{1.071}, r \cong 0.99994$; this gives the best fit.

27. Answers will vary. **29.** Equations will vary.

31.

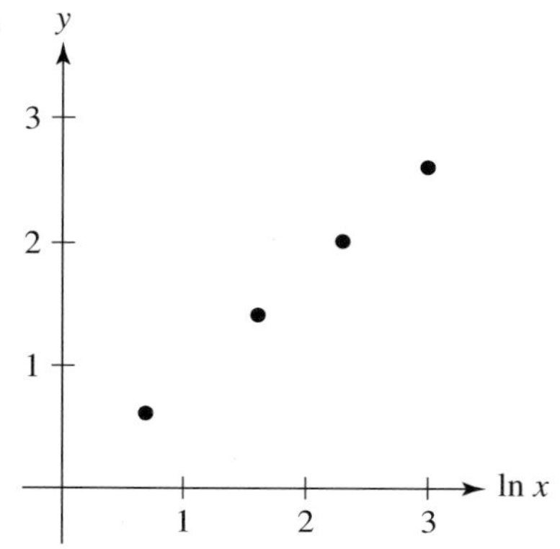

Yes.

33.

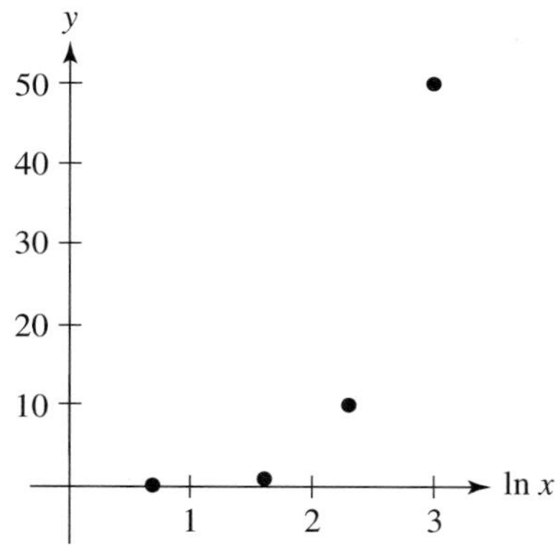

No.

35. a.

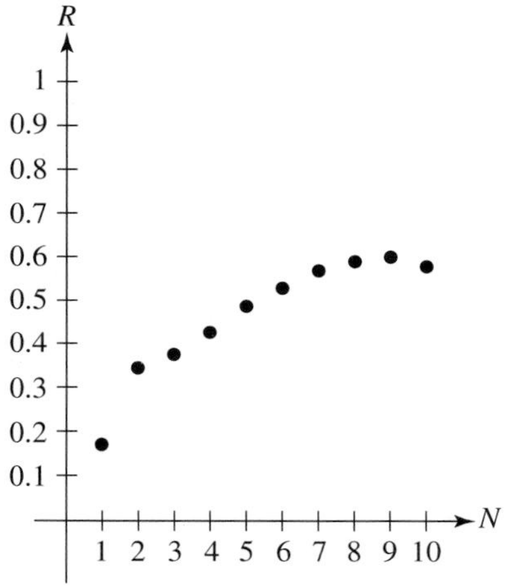

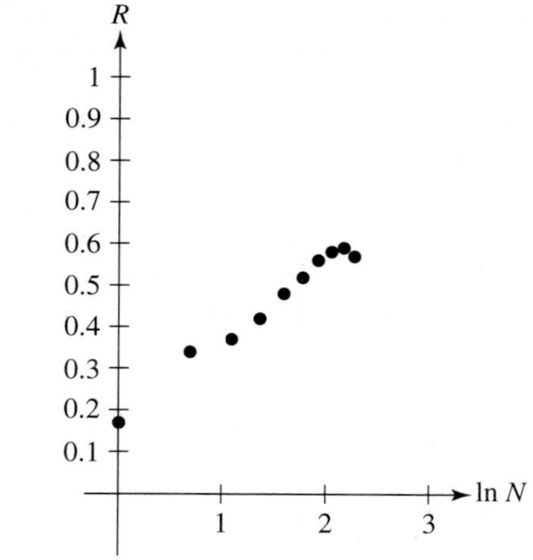

The data looks more nearly logarithmic.

b. Equations will vary.

37.

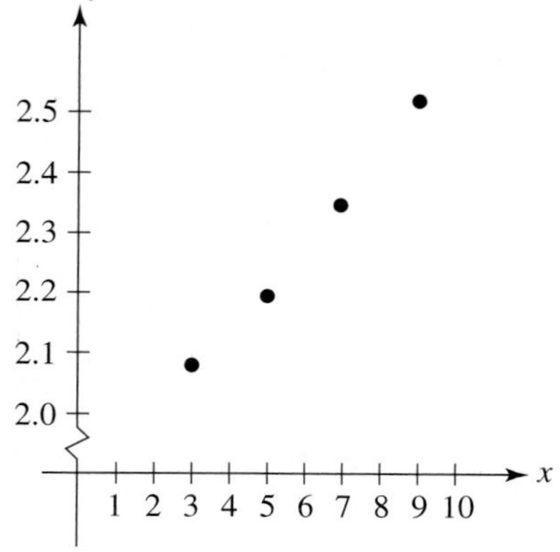

Yes.

39.

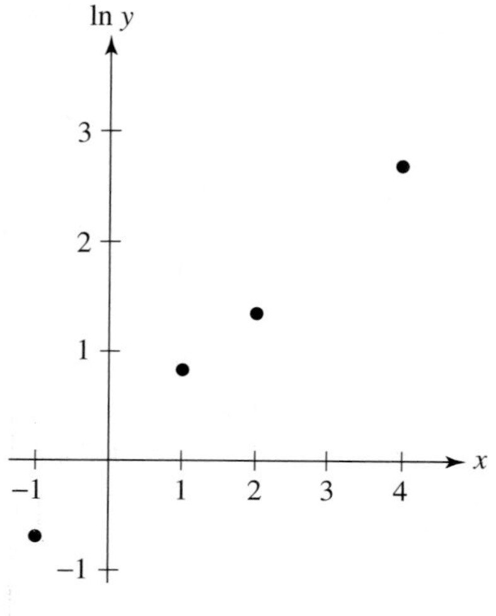

Yes.

41. Equations will vary.

43.

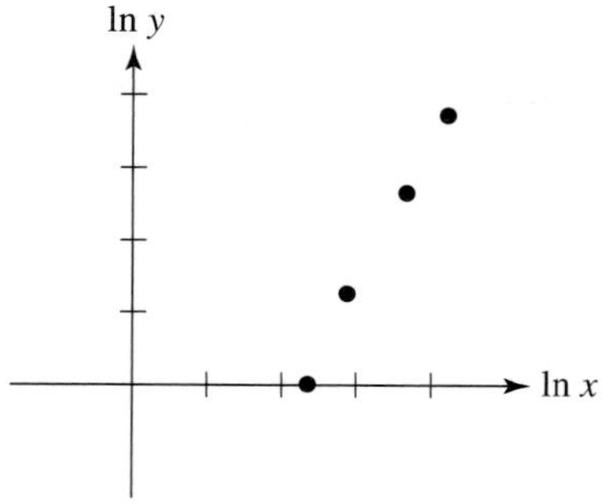

Yes.

45.

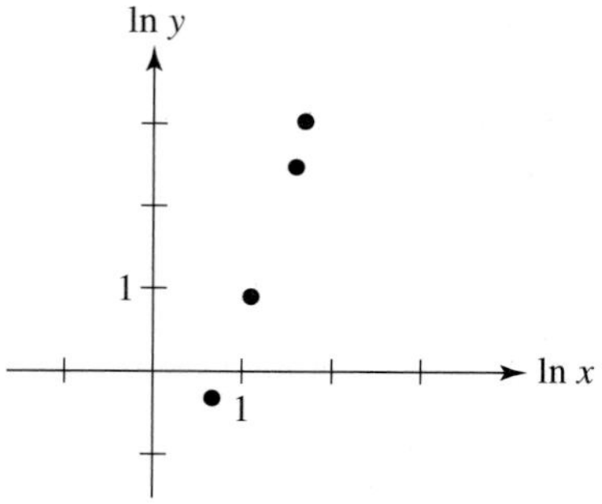

Yes.

47. linear ($y \cong 11.51 - 1.05x$)

49. power ($y \cong 1.04x^{0.49}$)

51. a. $y \cong 3.77(10^{-17})(1.03)^t$

b. The equation predicts an 1860 population of about 31,392,000 people and an 1870 population of about 42,211,000 people.

c. The Civil War would account for some of the shortfall.

53. a. The horizontal and vertical distances on semilog paper are proportional to x and $\log y$, respectively. The plot is linear if the relationship between x and $\log y$ is linear, that is, if x and y are related by an exponential function.

b. The horizontal and vertical distances on log-log paper are proportional to $\log x$ and $\log y$, respectively. The plot is linear if the relationship between $\log x$ and $\log y$ is linear, that is, if x and y are related by a power function.

SECTION A–1

1. a. 8230 **b.** between 8225 and 8235

c. 8.23×10^3

3. a. 0.0032 **b.** between 0.00315 and 0.00325

c. 3.2×10^{-3}

5. a. 61,400.009

b. between 61,400.0085 and 61,400.0095

c. 6.1400009×10^4

7. a. $0.0\underline{10203}$

b. between 0.0102025 and 0.0102035

c. 1.0203×10^{-2}

9. 0.13 **11.** 8.8

13. 90 **15.** 8.80

17. 86.141 **19.** 8.801

SECTION A-2

1. -6 **3.** $\frac{13}{3}$

5. -24 **7.** 0

9. Both sides of the equation were divided by 2. Then 1 was added to both sides of the equation and x was subtracted from both sides of the equation. Both sides of the equation were divided by 3.

SECTION A-3

1. $A = (5, 2)$ $B = (-5, 2)$ $C = (-3, -1)$
$D = (4, -3)$ $E = (2, 0)$ $F = (0, -4)$

3. quadrant III **5.** quadrant IV

7. quadrant II **9.** x-axis

SECTION A-4

1. 8 **3.** 13

5. $\sqrt{300}$ **7.** $\sqrt{10}$

9. 5 **11.** 2

13. $\sqrt{193}$ **15.** 41

SECTION A-5

1. $(-1, 4), (5, 10)$ **3.** $(3, 4), (5, 0)$

5.

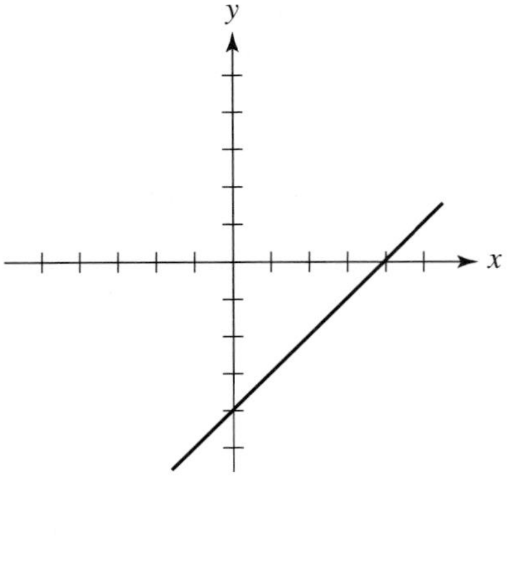

7.

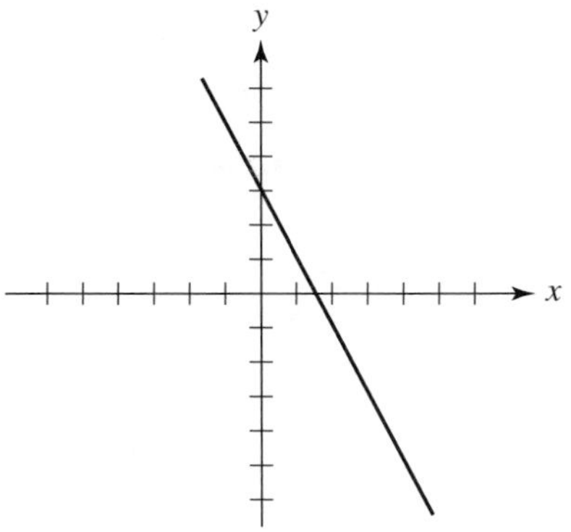

9.

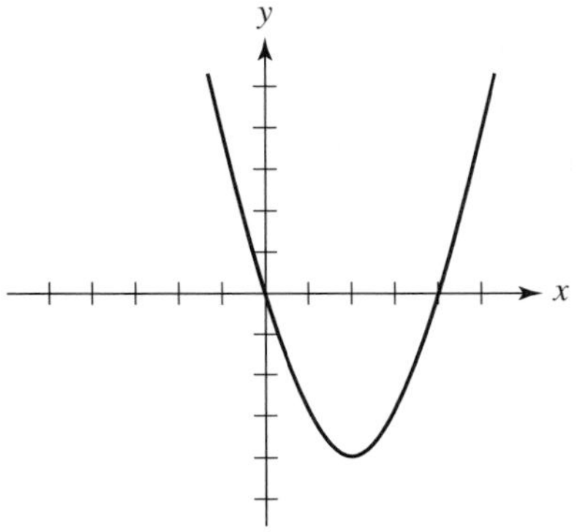

11.

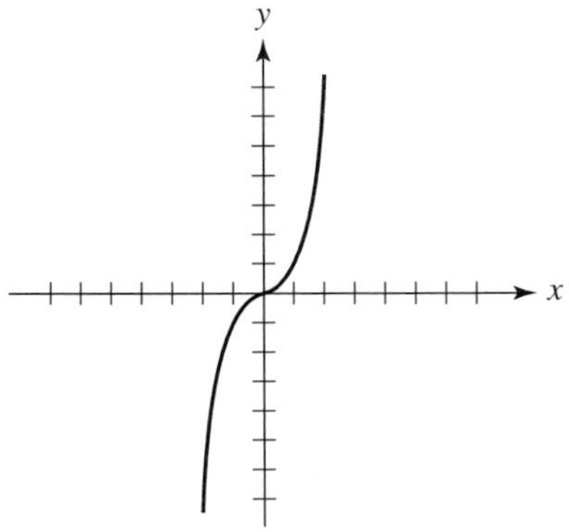

13. a.

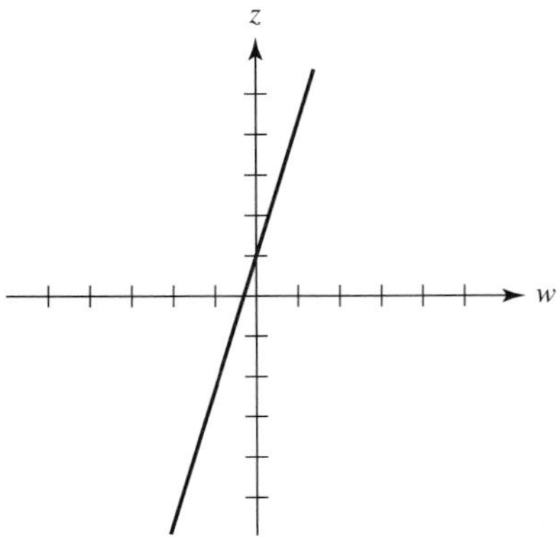

13. b.

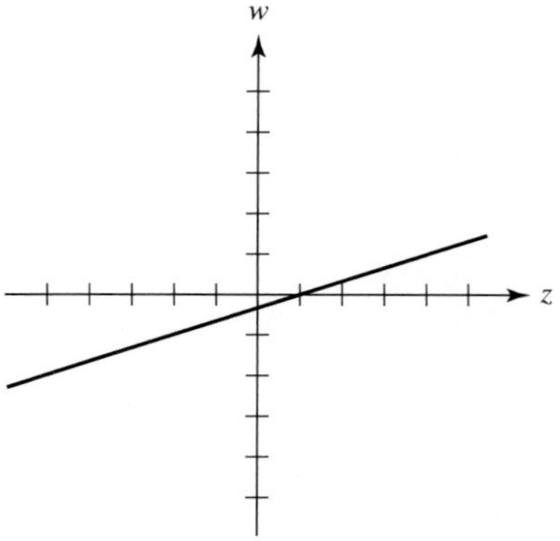

SECTION A-6

1.

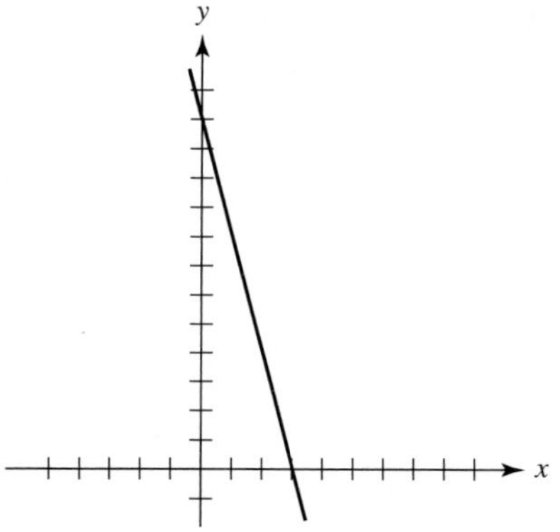

3.

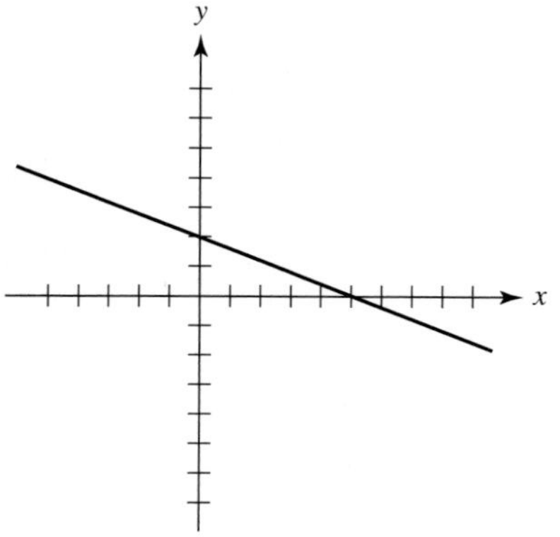

5.

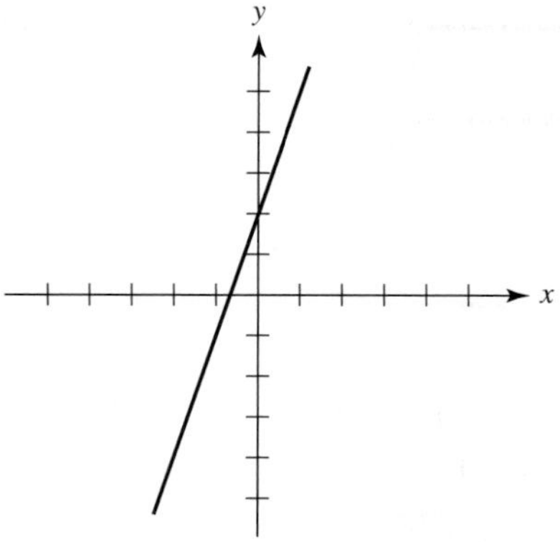

7.

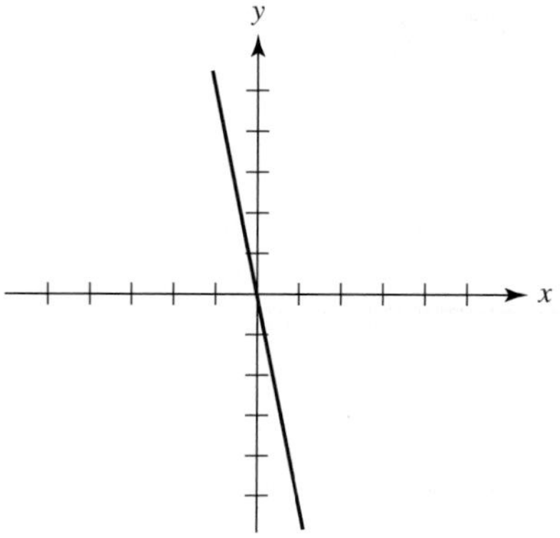

9.

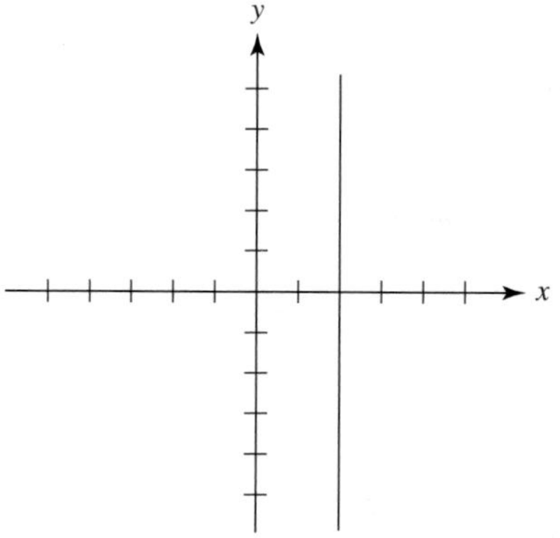

11. $y = x - 4$ **13.** $y = \frac{2}{3}x - 3$

15. $y = 4x - 25$ **17.** $y = -x$

SECTION A-7

1. 

$(1, 5]$

3. 

$(-40, 40)$

5. 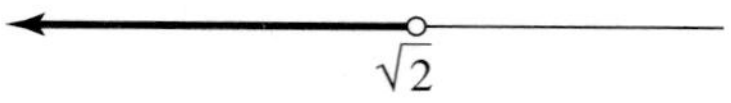

$(-\infty, \sqrt{2})$

7. 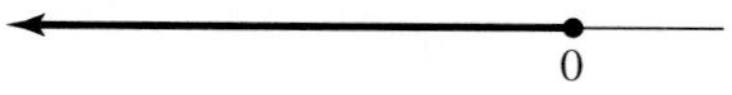

$(-\infty, 0]$

9. 

$-2 < x < 2$

11. 

$0 \leq x < 1.5$

13. 

$x < 1$

15. 

$-\frac{1}{2} \leq x < \frac{2}{3}$

17. $-2 < x \leqslant 3$ $(-2, 3]$ **19.** $-2 \leqslant x \leqslant 3$ $[-2, 3]$

21. $x > 3$ $(3, \infty)$ **23.** $x \geqslant -2$ $[-2, \infty)$

25. 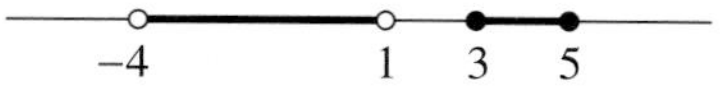

27. 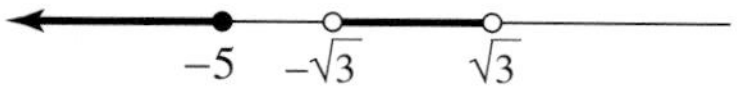

29. $(-\infty, -2] \cup [3, \infty)$ **31.** $(-\infty, -2] \cup (3, \infty)$

SECTION A–8

1. $(-6, \infty)$

3. $\left[\frac{13}{3}, \infty\right)$

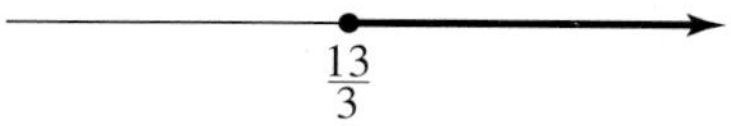

5. $(-\infty, -24]$

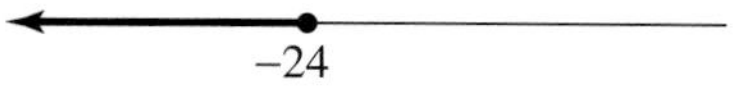

7. $(0, \infty)$

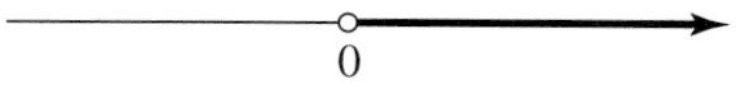

SECTION A–9

1. a. ± 7 **b.** $[-7, 7]$
c. $(\infty, -7) \cup (7, \infty)$

3. a. $5, 8$ **b.** $[5, 8]$
c. $(-\infty, 5) \cup (8, \infty)$

5. a. $-6, \frac{8}{3}$ **b.** $\left[-6, \frac{8}{3}\right]$
c. $(-\infty, -6) \cup \left(\frac{8}{3}, \infty\right)$

7. a. no solution **b.** no solution
c. $(-\infty, \infty)$

SECTION A–10

1. $(x, y) = (2, 1)$ **2.** $(x, y) = (18, -22)$

5. $(x, y) = (3, -2)$ **7.** $(x, y) = (2, 2)$

9. $(x, y, z) = (-2, 1, 1)$

11. $(x, y, z) = \left(\frac{23}{36}, \frac{1}{18}, -\frac{1}{9}\right)$

13. $(a, b, c) = (0, 0, 0)$ **15.** $(x, y, z) = (5, 11, -7)$

SECTION A–11

1. 32 **3.** 1

5. 1 **7.** $\frac{1}{8}$

9. k **11.** $8b^7$

13. $\frac{z^4}{4}$ **15.** $4x^2y^8$

17. $c + 3 - \frac{5}{c^2}$ **19.** $v^{5/6}$

SECTION A–12

1. $xyz(x + y + z)$ **3.** $(c + 5d)(c - 5d)$

5. $(r + 7)(r + 3)$ **7.** $(x + 5)(x - 6)$

9. $(3x + 11)(x - 1)$ **11.** $(2p - 1)(p - 2)$

13. $xy(x + 3y)(x^2 - 3xy + 9y^2)$

15. $ax(x^2 - 2ax + 5a^2)$

17. $(x^2 + 4)(x + 2)(x - 2)$

19. $(q + 1)(q^2 - q + 1)(q - 1)(q^2 + q + 1)$

SECTION A–13

1. $-7, 4$ **3.** $-\frac{7}{4}, 2$

5. $-3, 9$ **7.** $-\frac{5}{2} \pm \frac{\sqrt{61}}{2}$

9. $-4 \pm \sqrt{30}$ **11.** $-1, 2$

13. $-3 \pm 2\sqrt{6}$ **15.** $\pm \frac{\sqrt{35}}{7} i$

17. 0.1 **19.** $-6, -1$

21. $-\frac{10}{3}, 1$ **23.** $-12, 5$

25. no real solutions; $\frac{5 \pm \sqrt{23}\, i}{6}$

27. 2 real solutions; $7 \pm \sqrt{79}$

29. one real solution; $\frac{5}{2}$

SECTION A–14

1. $2 + 7i$ **3.** $2\sqrt{2}\,i$

5. $-34 + 22i$ **7.** 27

9. $-\frac{14}{41} - \frac{38}{41}i$ **11.** $-\frac{8}{7} + \frac{3}{7}i$

13. $-2 + 2i$ **15.** 1

SECTION A–15

1. $\frac{1}{2}x^3 - 2x^2 - \frac{7}{2}$ **3.** $2x - \frac{7}{x} + \frac{8}{x^2}$

5. $x^4 + 6x^2 - x + 12 + \dfrac{-7x^2 + 22x - 8}{x^3 - 2x + 1}$

7. $x^2 + 3 + \dfrac{7}{x - 4}$

9. $2x - 4$ **11.** $1 + \dfrac{-12x + 6}{x^3 + 3x^2 + 4x - 5}$

SECTION A–16

1. $\frac{2}{3}$ **3.** 1

5. $\dfrac{x + 1}{2}$ **7.** $\dfrac{2x - 3}{x(2x + 3)(x - 4)^2}$

9. $\dfrac{5z}{6}$ **11.** $\dfrac{2a^2 + 8a + 3}{a(a + 1)(a - 1)}$

13. $\dfrac{20x}{x - 5}$ **15.** $\dfrac{y^2 + 2y + 6}{y^2(y + 2)^2}$

17. $\frac{3}{4}$ **19.** $\dfrac{6 + x}{6 - x}$

SECTION A–17

1. 5 **3.** no solution

5. $-\frac{5}{2}$ **7.** $\dfrac{-12}{5}$

SECTION A–18

1. $3|t|$ **3.** $|u|v^4$

5. $3a$ **7.** $2|y|^3$

9. 12 **11.** $5x\sqrt{6y}$

13. $3|x|$ **15.** $3x\sqrt{7x}$

17. $12\sqrt{k}$ **19.** $(10 - 2x)\sqrt{5x}$

21. a. $\dfrac{x}{\sqrt{6x}}$ **b.** $\dfrac{\sqrt{6x}}{6}$

23. a. $\dfrac{2}{\sqrt[3]{2s}}$ **b.** $\dfrac{\sqrt[3]{4s^2}}{s}$

25. a. $\dfrac{8}{\sqrt{65} + 5}$ **b.** $\dfrac{\sqrt{65} - 5}{5}$

27. a. $\dfrac{1}{\sqrt{6} - \sqrt{3}}$ **b.** $\dfrac{\sqrt{6} + \sqrt{3}}{3}$

29. 100 **31.** 243

33. $a^{20/3}$ **35.** $a^{16/3}$

37. $\dfrac{8}{x^{12}}$ **39.** $p^{1/10}$

SECTION A–19

1. 16 **3.** 5

5. 6 **7.** ± 5

9. 16 **11.** 10

13. 10 **15.** $-\frac{15}{4}$

SECTION B–1

1. a.

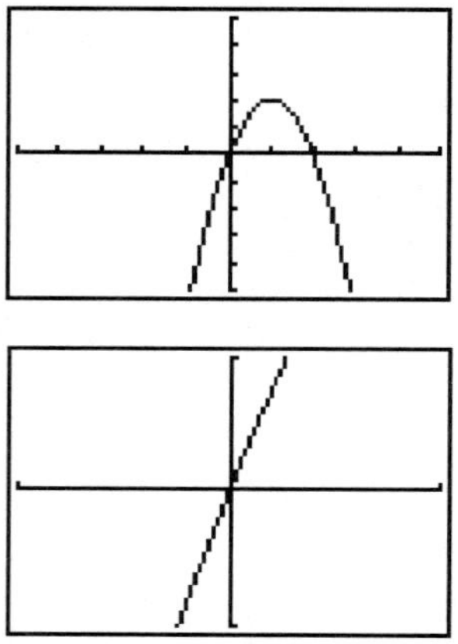

b. The second window was obtained by zooming in from the first window by a factor of 10.

3. a.

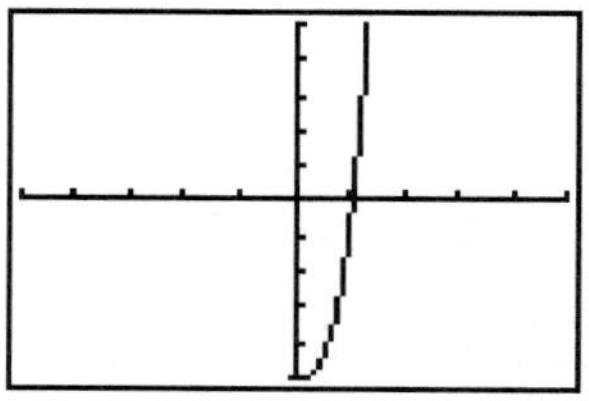

b–d. $(2.15, 0)$

5. a.

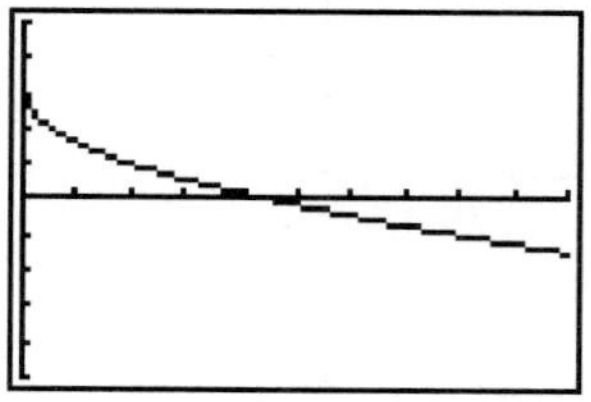

b–d. $(4.50, 0)$

7. a. [−1, 1] by [9, 11] **b.** [−10, 10] by [−10, 10]
c. [9, 10] by [0, 1]

9. a. [−4, 4] by [−4, 16] **b.** [−4, 4] by [−16, 64]
c. [−4, 4] by [−1, 4]

11. Responses will vary.

SECTION B–2

1. a. Choices will vary. **b.** Responses will vary.

3. a. Choices will vary. **b.** Responses will vary.

5. a. Choices will vary.
b. The graph appears flatter than it really is.

7. a. Choices will vary.
b. The graph appears flatter than it really is.

SECTION B–3

1. a. y-intercept $= (0, -12)$; vertex at $x = -\frac{1}{2}$; x-intercepts at $x = -4, 3$
b–c. Responses will vary. **d.** See part (a).

3. a. y-intercept $= (0, 20)$; vertex at $x = 3$; no x-intercepts
b–c. Responses will vary. **d.** See part (a).

5. a. y-intercept $= (0, -500{,}000)$; vertex at $x = -250$; x-intercepts at $x = -1000, 500$
b–c. Responses will vary. **d.** See part (a).

SECTION B–4

1. a. $y = \pm\sqrt{4x^2 + 16}$
b. y-intercepts $\cong (0, \pm 4.00)$
no x-intercepts

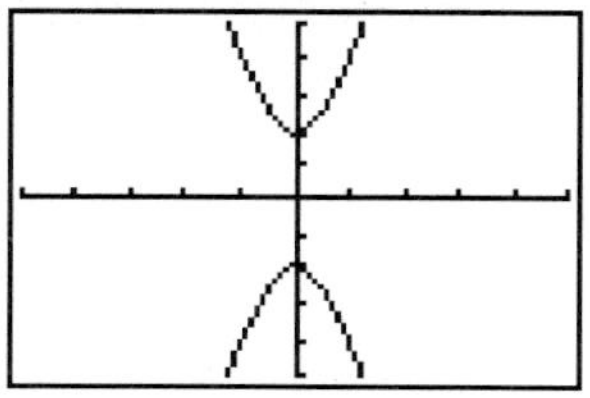

3. a. $y = -\frac{7}{4} \pm \frac{\sqrt{24x + 449}}{4}$
b. y-intercepts $\cong (0, -7.05), (0, 3.55)$
x-intercepts $\cong (-16.67, 0)$

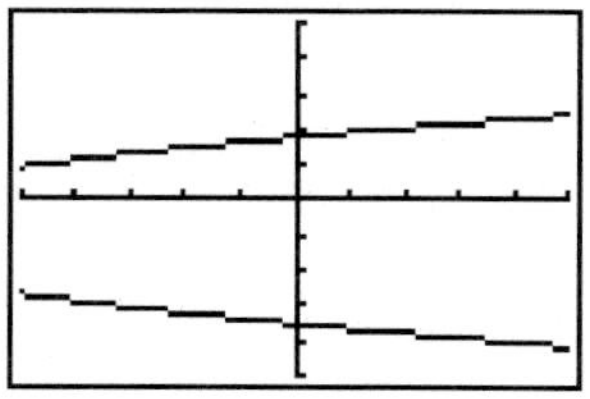

5. a. $y = \frac{3}{2}x \pm \frac{\sqrt{5}}{2}|x|$
b. intercept $= (0, 0)$

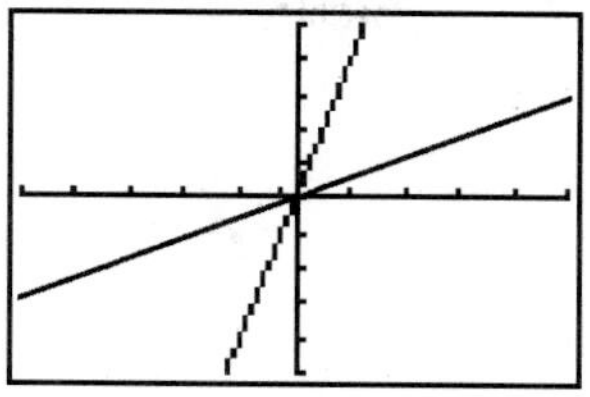

7. a. $y = \frac{3}{4} - \frac{1}{4}x \pm \frac{\sqrt{-15x^2 + 34x - 791}}{4}$
b. no intercepts

9. a. $y = \frac{-x^2 + 3x + 10}{5x + 2}$
b. y-intercept $\cong (0, 5.00)$
x-intercepts $\cong (-2.00, 0), (5.00, 0)$

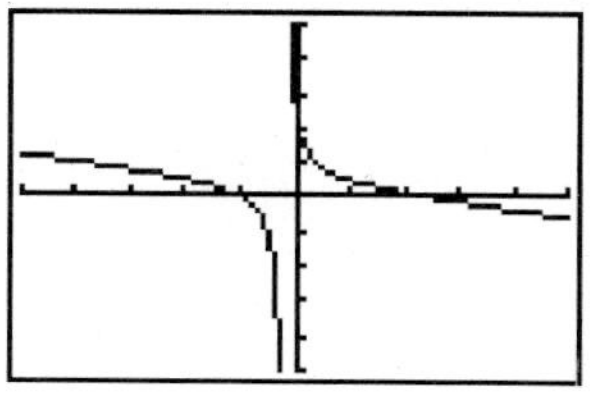

SECTION B–5

1. a. Choices will vary.
b. (0, 0), (2, 0), (12, 0)

3. a. Choices will vary.
b. (1.51, 0)

5. a. Choices will vary.
b. (−3, 0), (−2, 0), (−1, 0), (1, 0), (2, 0), (3, 0), (4, 0)

7. [−25, 25]; x-intercepts (0, 0), (2, 0), (12, 0);
turning points (0, 0), (1.31, 12.66), (9.19, −1706.35)

9. [−13, 13]; x-intercepts (±2.41, 0);
turning points (0, −12), (±0.88, −8.76), (±1.97, −24.05)

11. (2.59, −1.62)

13. local maximum $\cong (-1.05, 13.03)$
local minimum $\cong (1.05, -1.03)$

SECTION B–6

1. a. x-intercepts $(\pm 2, 0)$
vertical asymptote $x = 0$
b. Choices will vary.

3. a. no x-intercepts
vertical asymptotes
$x = \pm 5$
b. Choices will vary.

INDEX

Entries appear in index, as follows:
bold entry, **bold pages**: term is **defined**
Bold Italic Entry*: *Mathematical Looking Glass
Italic entry: *Pólya strategy* or *reference title*